Progress in Inorganic Chemistry

Volume 59

Advisory Board

PROGRESS IN INORGANIC CHEMISTRY

Edited by

Kenneth D. Karlin

DEPARTMENT OF CHEMISTRY
JOHNS HOPKINS UNIVERSITY
BALTIMORE, MARYLAND

VOLUME 59

Cover Design: Wiley

Published by John Wiley & Sons, Inc., Hoboken, New Jersey.
Published simultaneously in Canada.

Library of Congress Cataloging-in-Publication Data is available.

Library of Congress Catalog Number: 59013035

ISBN: 978-1-118-87016-7

Printed in the United States of America.

10 9 8 7 6 5 4 3 2 1

Contents

Iron Catalysis in Synthetic Chemistry

SUJOY RANA, ATANU MODAK, SOHAM MAITY, TUHIN PATRA, AND DEBABRATA MAITI

Department of Chemistry, Indian Institute of Technology Bombay, Powai, Mumbai, India

CONTENTS

Progress in Inorganic Chemistry, Volume 59, First Edition. Edited by Kenneth D. Karlin.

I. INTRODUCTION

During the last few decades, transition metal catalysts, especially those on precious metals [e.g., palladium (Pd), rhodium (Rh), iridium (Ir), and ruthenium (Ru)] have proven to be efficient for a large number of applications. The success of transition metal based organometallic catalysts lies in the easy modification of their environment by ligand exchange. A very large number of different types of ligands can coordinate to transition metal ions. Once the ligands are coordinated, the reactivity of the metals may change dramatically. However, the limited availability of these metals, in order of decreasing risk (depletion): Au > Ir, Rh, Ru > Pt, Re, Pd), as well as their high price (Fig. 1) and significant toxicity, makes it desirable to search for more economical and environmental friendly alternatives. A possible solution to this problem could be the increased use of catalysts based on first-row transition metals, especially iron (Fe) (1). In contrast to synthetic precious metal catalysts, iron takes part in various biological systems as an essential key element and electron-transfer reactions.

Due to its abundance, inexpensiveness, and environmentally benign nature, use of iron has increased significantly in the last two decades for synthetic transformation both in asymmetric synthesis and reaction methodology. This development encouraged us to summarize the use of iron catalysis in organic synthesis, which includes cycloadditions, C–C, C–N bond formation, redox, and other reactions. This chapter has been divided into different sections based on the reaction type.

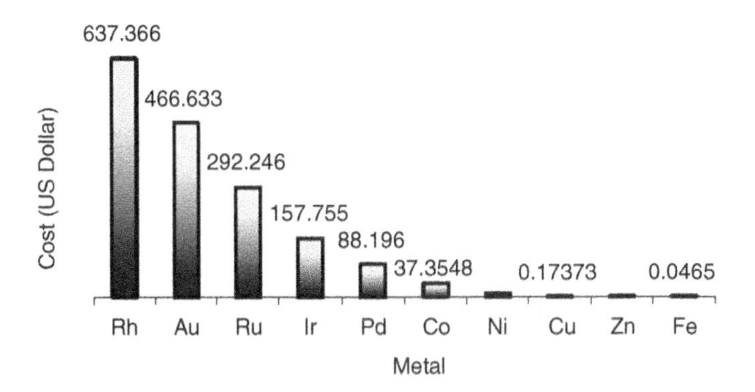

Figure 1. Comparison of prices for different transition metals (Sigma Aldrich).

II. ADDITION REACTIONS

A. Cycloadditions

1. *The [2 + 2] Cycloaddition*

In 2001, Itoh and co-workers (2) demonstrated the [2 + 2] cyclodimerization of *trans*-anethol catalyzed by alumina supported iron(III) perchlorate. A C_2 symmetric cyclobutane derivative was obtained in excellent yield (92%) at room temperature (rt), though longer reaction time was required. They applied the same catalytic system for the cycloaddition of styrenes and quinones. However, 2,3-dihydrobenzofuran derivatives were obtained in excellent yields in place of the desired [2 + 2] cycloadduct (Scheme 1) (3). Earlier, in 1982, Rosenblum and Scheck (4) showed that the $CpFe(CO)_2$ cation, where Cp = cyclopentadienyl, could afford the unsaturated bicycle through the reaction of alkenes and methyl tetrolate, though the yields obtained were inferior.

Significant improvement in iron-catalyzed [2 + 2] cycloaddition was achieved in 2006 by Chirik and co-workers (5). They reported an intramolecular [2 + 2] cycloaddition of the dienes resulting in the formation of [0.2.3] heptane derivatives catalyzed by a bis(imino)-pyridine iron(II) bis(dinitrogen) complex and only cis product was obtained. Further, labeling experiments confirmed the reaction to be stereospecific. A number of dienes containing different amine and ester functional groups reacted efficiently, but the presence of secondary amine and an $SiMe_2$ group inhibited the reaction. This reaction can also be performed in the dark, clearly indicating the process to be thermally driven, rather than a photochemical one. A mechanism of this catalytic process was proposed where iron is assumed to maintain its ferrous oxidation state throughout the reaction with the help of redox active *i*PrPDI ligand (Scheme 2).

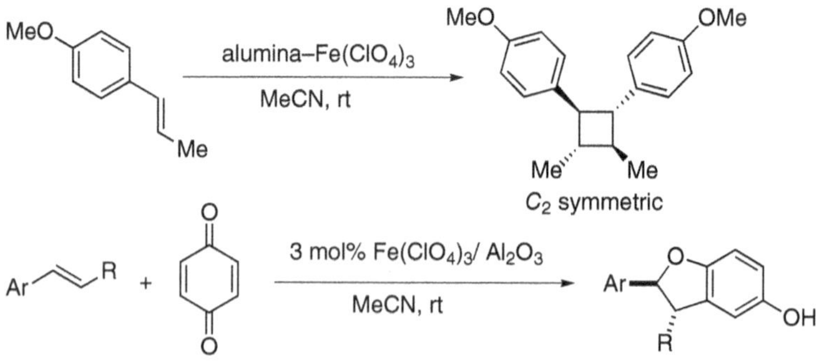

MeO Me alumina–$Fe(ClO_4)_3$ MeCN, rt MeO OMe Me Me C_2 symmetric

Ar R + O O 3 mol% $Fe(ClO_4)_3$/ Al_2O_3 MeCN, rt Ar O OH R

Scheme 1. Early examples of iron-catalyzed [2 + 2] cycloaddition.

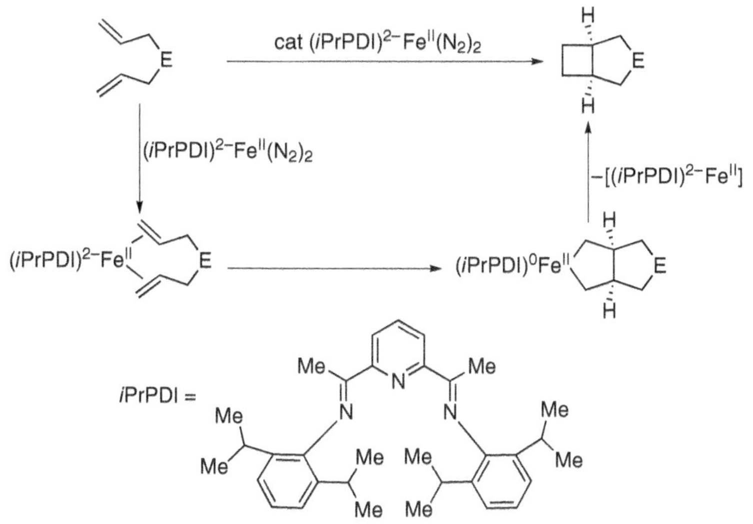

Scheme 2. Plausible mechanism involving the iron(II) oxidation state [PDI = (*N*,*N*′,E,*N*,*N*′,E)-*N*,*N*′-(1,1′)-(pyridine-2,6-diyl)bis(ethan-1-yl-1-ylidine))bis(2,6-diisopropylaniline)].

A combination of ethylene and butadiene resembles a thermally allowed [4 + 2] cycloaddition reaction, namely, the Diels–Alder reaction. Using their redox-active bis(imino)-pyridine supported iron catalysts, Chirik and co-workers (6) reported the more challenging [2 + 2] cycloaddition from the same set of starting materials that furnished vinylcyclobutane in an excellent 95% yield. The protocol turned out to be substrate specific, as with insertion of a methyl group in the 2- position of diene, no cycloadduct was observed; rather it resulted in a 1,4-addition product. To shed light on their plausible mechanism, several labeling experiments were carried out with different substrates. They were successful in intercepting one iron metallocyclic intermediate, which resulted from ethylene insertion into the coordinated diene. The same species was also prepared by reacting vinylcyclobutane, the product of the [2 + 2] cycloaddition, with the iron catalyst. Thus the reaction proved to be reversible with iron metallocycle as an intermediate, and the backward reaction demonstrated a rare example of sp^3–sp^3 C−C bond activation with an iron catalyst under mild conditions. Isolation of the metallocycle intermediate and labeling experiments led to a proposed mechanism for [2 + 2] cycloaddition and 1,4-addition. The reaction initiated by displacement of dinitrogen ligands by diene an η^4 complex, and ethylene insertion, which furnished the isolable metallocycle intermediate. In the next step, butadiene-induced reductive-elimination resulted in vinylcyclobutane along with regeneration of an iron butadiene intermediate. However, with isoprene, β-hydrogen elimination followed by C−H reductive elimination resulted in the 1,4-addition product (Scheme 3).

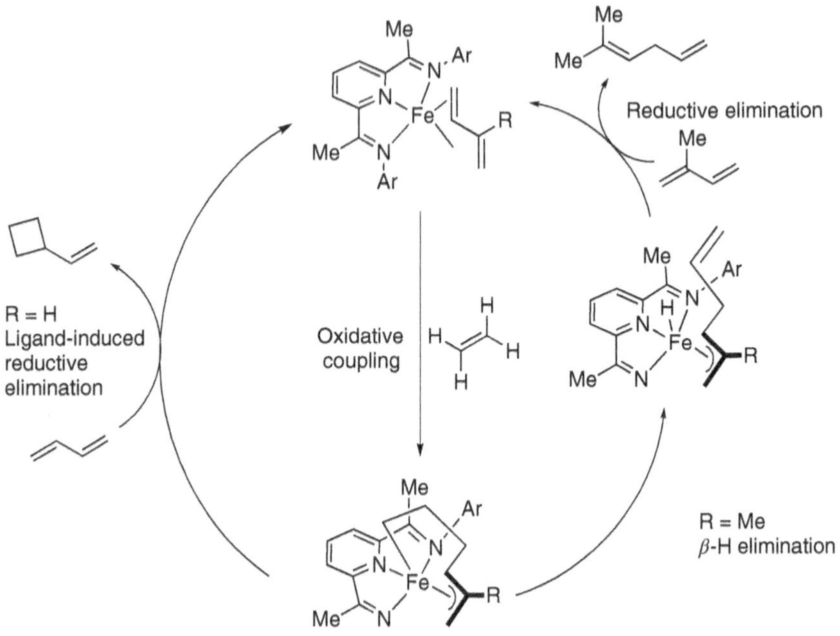

Scheme 3. Proposed mechanism for [2 + 2] cycloaddition of ethylene and butadiene.

2. *The [3 + 2] Cycloaddition*

In 2012, Plietker and co-workers (7) reported an iron-catalyzed [3 + 2] cycloaddition of vinylcyclopropanes (VCP) and activated olefins or *N*-tosyl imines to generate functionalized vinylcyclopentanes and cyclopyrrolidines in high yields and regioselectivities. The activation of VCP by the electron-rich ferrate, Bu_4N-$[Fe(CO)_3(NO)]$ (TBAFe) (TBA = tetrabutylammonium), resulted in the formation of an intermediate allyl–Fe complex, which can be regarded as an a1, a3, d5-synthon (Scheme 4). Subsequent Michael addition onto activated olefins generated another carbanion, which readily attacked the carbocationic part of the intermediate to generate VCP derivatives. The scope of VCPs was tested with 1,1-bis(phenylsulfonyl)ethylene as the Michael acceptor where different functional groups like esters, nitriles, and amides were tested. Likewise, a variety of Michael acceptors containing esters, sulfones, nitriles, amides, and ketones were successfully employed in this reaction. Further, they tried to incorporate imines as the Michael acceptor to extend their methodology. However, only *N*-tosylarylimines reacted successfully while *N*-Ph and *N*-Boc protected imines gave no or undesired products. Notably, activation of a carbon–carbon bond of VCP by an inexpensive iron catalyst would encourage further investigation on other strained systems (e.g., cyclobutanes, aziridines, and oxiranes).

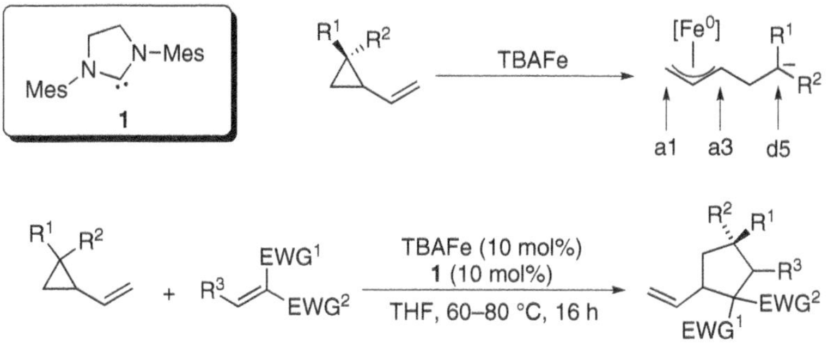

Scheme 4. Iron-catalyzed [3 + 2] cycloaddition of VCPs and activated olefins [Mes = mesylate, EWG = electron-withdrawing group, THF = tetrahydrofuran (solvent)].

Simple $FeCl_3$ acts as a Lewis acid catalyst to assist the ring opening of another strained system, *N*-tosylaziridines (NTs), which in the presence of base reacts efficiently with terminal aryl alkynes to generate substituted 2-pyrrolines (Scheme 5) (8).

Further, a one-pot synthesis of γ-amino ketones from 2-pyrrolines was achieved by treatment with H_2O at rt for 12 h. However, the scope of the reaction was limited to Cl, F, and OMe containing arylalkynes and only NTs reacted successfully. Internal alkyne resulted in a lower yield (48%), while alkylalkyne, as well as electron-deficient aziridines, gave no product. Recently, Wang and co-workers (9) reported an Fe(II)/N, O ligand-catalyzed asymmetric [3 + 2] cycloaddition reaction of *in situ* generated azomethineylides and electron-deficient alkenes (Scheme 6). Only 10 mol% $FeCl_2$ in the presence of diarylprolinol and Et_3N efficiently catalyzed the cycloaddition to afford a five-membered heterocyclic endo adduct stereoselectively in good-to-moderate yield.

In 2002, Kundig et al. (10) reported the first asymmetric [3 + 2] cycloaddition of nitrones and enals to generate isooxazolidines catalyzed by the Lewis acidic iron complex (*R,R*)-**3** (Scheme 7). The role of a Lewis acid was crucial as an *α,β*-unsaturated aldehyde had to be activated in preference to stronger Lewis basic nitrones having two coordination sites against one point coordinating enals. However, they were successful in discovering such a reactive yet selective

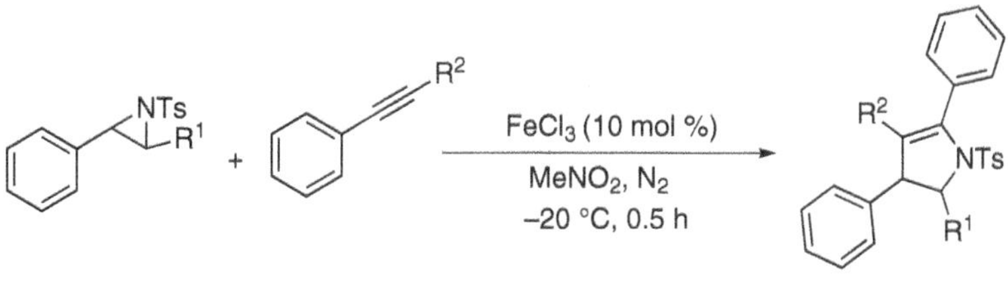

Scheme 5. Iron chloride acts as a Lewis acid catalyst in [3 + 2] cycloaddition.

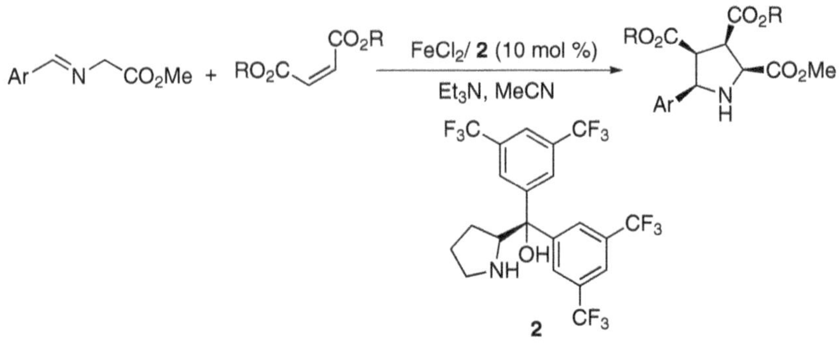

Scheme 6. Iron(II)/N, O ligand-catalyzed asymmetric [3 + 2] cycloaddition.

iron- and ruthenium-based Lewis acidic complex. The iron complex turned out to be the more beneficial choice.

In the presence of 2, 6-lutidine, which acts as a scavenger of acidic impurities, C, N diarylnitrones and heterocyclic *N*-oxides reacted efficiently with methacrolein to generate an endo adduct selectively. Notably, this transformation was also achieved by an elegant organocatalytic pathway with a high degree of enantioselectivity by the MacMillan group in 2000 (11).

3. *The [2 + 2 + 2] Cycloaddition*

Inter- and intramolecular [2 + 2 + 2] cycloaddition reactions of alkynes and nitriles catalyzed by transition metals have been considered as the most straightforward and convenient approach to synthesize six-membered arenes and highly

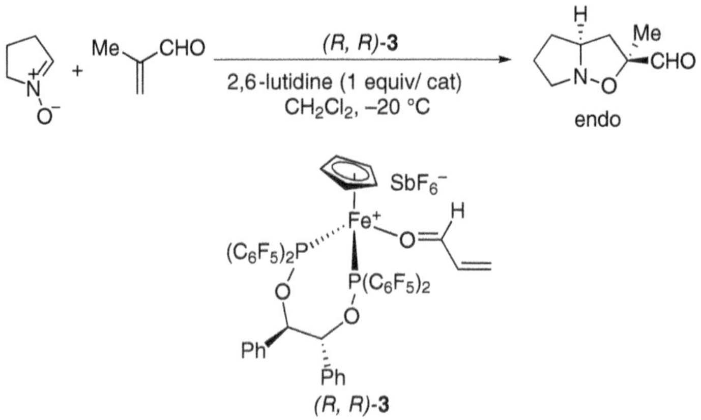

Scheme 7. Iron-catalyzed [3 + 2] cycloaddition of nitrones and enals.

substituted pyridines. Importantly, a number of functional groups (e.g., alcohols, amines, ethers, esters, and halogens) can be tolerated while several C–C bonds are formed in a single step. For these transformations, several transition metals ranging from Co, Ru, Rh, Ni, Ti to bimetallic systems (e.g., Zr/Ni and Zr/Cu) have been used in recent decades. Iron catalysis has also played a crucial role in this reaction, though until very recently, methods were limited by poor chemo- and regioselectivity, as well as difficulty in preparation and handling of the catalysts.

In 2000, Pertici and co-workers (12) reported a cyclotrimerization reaction of terminal alkynes catalyzed by Fe(η^6-CHT)(η^4-COD), where CHT = cyclohepta-1,3,5-triene and COD = 1,5-cyclooctadiene, respectively, to generate various multisubstituted benzene derivatives. The method lacked regioselectivity as a mixture of two regioisomers was formed for most of the terminal alkynes in an ~1:1 ratio. Meanwhile, Zenneck and co-workers (13) developed a [2 + 2 + 2] cycloaddition reaction of two molecules of alkynes and nitriles catalyzed by an Fe (0) complex to generate pyridines. This reaction was also limited by poor chemoselectivity, as well as a complex procedure of catalyst preparation.

However, better chemoselectivity was achieved by Guerchais and co-workers in 2002 (14) as they employed iron bis(acetonitrile) and tris(acetonitrile) complexes to catalyze the cycloaddition reactions of carbon–carbon and carbon–nitrogen triple bonds (Scheme 8). Three equivalents of alkynes cyclotrimerized to

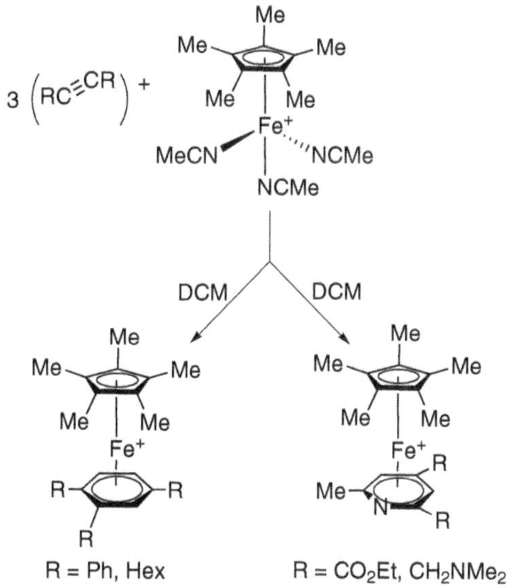

3 (RC≡CR) + [$Cp^*Fe^+(NCMe)_3$]

DCM / DCM

R = Ph, Hex

R = CO_2Et, CH_2NMe_2

Scheme 8. Iron-catalyzed cycloaddition reaction of C–C and C–N triple bonds (DCM = dichloromethane, CH_2Cl_2).

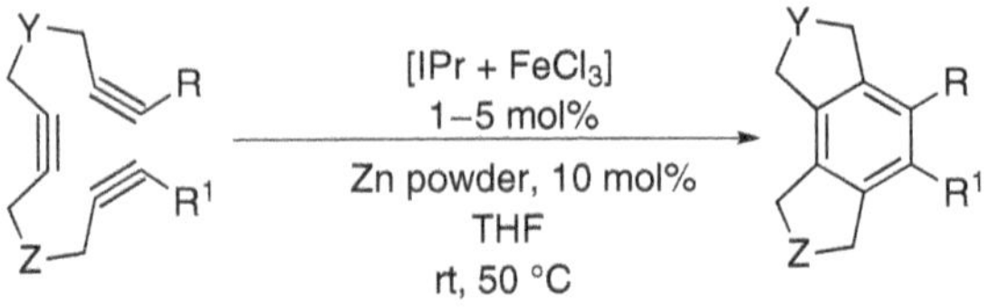

$$\text{triyne (Y, Z; R, } R^1) \xrightarrow[\substack{\text{Zn powder, 10 mol\%} \\ \text{THF} \\ \text{rt, 50 °C}}]{\substack{[\text{IPr} + FeCl_3] \\ \text{1–5 mol\%}}} \text{fused arene (Y, Z; R, } R^1)$$

Scheme 9. Intramolecular cyclotrimerization of triynes catalyzed by bench-stable iron salt [IPr = 1,3-bis(2,6-diisopropylimidazolium)-2,3-dehydro-1*H*-imidazole].

produce arene complexes in the presence of an iron tris(acetonitrile) complex in CH_2Cl_2 solvent at rt. Under the same condition, alkynes having heteroatoms bonded to the propargylic position afforded pyridine complexes instead of previously observed arene complexes, by the heterocyclotrimerization of two alkynes and one metal-bound acetonitrile ligand. When MeCN was used as solvent, in place of CH_2Cl_2, only ethyl propiolate reacted among the alkynes as the carbonyl group successfully coordinated with the metal center competing with inhibiting acetonitriles to provide a free pyridine derivative in 73% yield, rather than generating the metal–pyridine complex. It was evident that nature of the solvent had dramatically altered the outcome of the reaction as no organometallic product was detected in this case.

On the other hand, in 2005 iron-catalyzed intramolecular cyclotrimerization of triynes was reported by Okamoto and co-workers (15), which was less problematic in terms of regioselectivity (Scheme 9). So far, the iron catalysts that have been discussed are based on iron arene or iron 1,5-cylooctadiene and cycloheptatriene complexes. An alternate approach with simple iron salts is advantageous, as preparation and storage of expensive organometallic iron complexes can be avoided. Further, this approach rendered the related processes much more economical, as a fewer stabilizing ligands were required while reactions were performed under milder condition with high efficiency. Inspired by such an approach, Okamoto and co-workers (15) preferred commercially available iron and cobalt salts, which in the presence of suitable ligands and reducing agent can act as efficient catalysts for such transformations. They tested a number of commercially available iron, cobalt, and nickel salts in the presence of an imidazolium carbene ligand, and observed that cyclotrimerization occurred efficiently only at rt or at 50 °C under a reducing condition. Zinc powder was the reducing agent of choice, which supposedly converted the *in situ* generated metal complexes to their corresponding low-valent complexes so as to initiate the process by formation of a metallacycle intermediate. Further, they showed the advantages of their method by efficient formation of carbocyclic, *O*-heterocyclic and biaryl compounds. In another report from the same group, N-based bidentate ligands (e.g., 1,2-diimines or 2-iminomethylpyridines) were utilized in iron-catalyzed chemo- and regioselective cyclotrimerization of triynes (16).

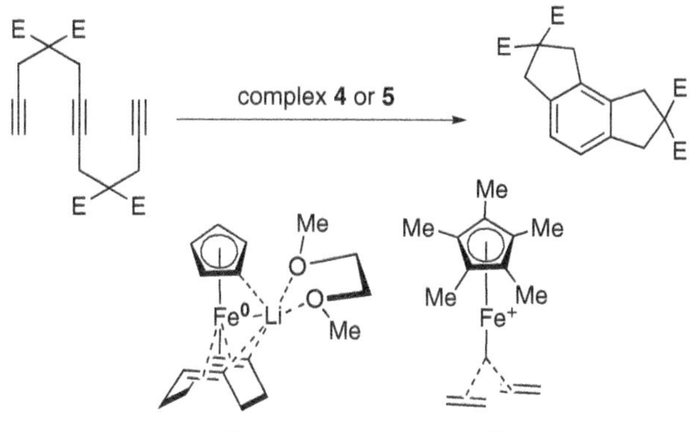

Scheme 10. Cyclotrimerization catalyzed by low-valent iron–olefin complexes.

Recently, Furstner et al. (17) synthesized a fine blend of iron complexes of formal oxidation states −2, 0 and +1 from readily available ferrocene. Among these low-valent iron olefin complexes, complex **4** turned out to be a very efficient catalyst in the cyclotrimerization reaction. This outcome is not surprising as Fe(0) complexes are isoelectronic with Co(I) and Rh(I) species, which are arguably the most widely used catalysts in transition metal catalyzed [2 + 2 + 2] cycloaddition reactions. Interestingly, complex **5** with a formal oxidation state of (+1) is also found to be effective, though a higher catalyst loading and longer reaction time is required (Scheme 10). To gain mechanistic insights, 1,2-diphenylacetylene (tolane) was reacted with a series of iron complexes.

Although significant advances have been made in recent years regarding transition metal catalyzed [2 + 2 + 2] cycloaddition, an efficient iron-catalyzed protocol for chemoselective synthesis of pyridines eluded the researchers for a long time. The crucial role of low-valent iron complexes in realizing efficient [2 + 2 + 2] cycloaddition lies in the fact that it facilitates the formation of a metallocyclic intermediate by oxidative cyclization, subsequent reductive elimination that generate arenes or pyridines. In 2006, Holland and co-workers (18) revealed that alkyne binding to a low-valent iron metal center is particularly stronger than that of phosphine. Inspired by this report, Wan and co-workers (19) developed an iron catalyst comprising of readily available FeI_2 and dppp [1,3-bis(diphenylphosphino)propane] as the phosphine ligand in the presence of Zn dust, which served as the reducing agent (Scheme 11). Efficient synthesis of pyridines was observed only at rt starting from diynes and a slight excess of nitriles in THF solvent. They initially postulated that both ferracyclopentadiene, as well as the azaferracyclopentadiene intermediate, might be operating in the catalytic system and two plausible pathways were proposed. A competitive experiment using

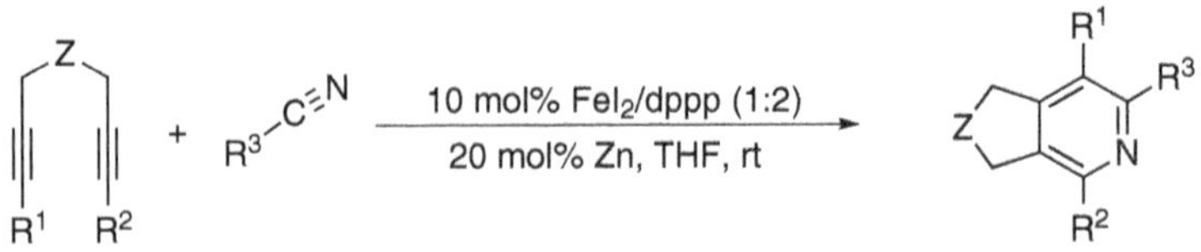

Scheme 11. An efficient iron-catalyzed [2+2+2] cycloaddition for pyridine synthesis.

an unsymmetrical diyne and acetonitrile indicated a ferracyclopentadiene intermediate that might not be involved in the overall catalytic system. Further, another competitive experiment with acetonitrile and 3 equiv acetylenes confirmed that formation of such a ferracyclopentadiene intermediate is inhibited in the presence of nitriles.

At the same time, Louie and co-workers (20) reported another efficient method of iron-catalyzed pyridine synthesis. The $Fe(OAc)_2$ in the presence of a sterically hindered bis(imino)pyridine ligand catalyzes the cycloaddition of a different substrate class, alkyne nitriles and alkynes, to form a number of pyridine derivatives (Scheme 12).

4. *The [4+2] Cycloaddition*

The [4+2] cycloaddition reaction serves as an efficient and powerful tool for synthesizing six-membered ring compounds by forming carbon–carbon and carbon–heteroatom bonds. According to the Woodward–Hoffmann rule, the concerted suprafacial $[\pi 4_s + \pi 2_s]$ addition of diene with a dienophile is thermally allowed and the reaction rate or feasibility of the reaction is strongly dependent on

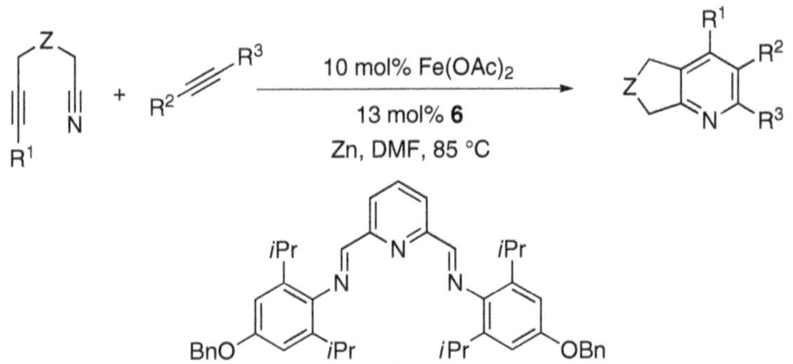

Scheme 12. Pyridine synthesis by iron-catalyzed [2+2+2] cycloaddition of alkyne nitriles and alkynes (DMF = *N,N*-dimethylformamide).

the energy gap of the frontier orbitals of the reacting species. Generally, it is classified into two distinct categories: a normal Diels–Alder reaction that involves interaction between highest occupied molecular orbital (HOMO) of the diene and lowest unoccupied molecular orbital (LUMO) of the dienophile and Diels–Alder reaction with inverse electron demand involving the HOMO of the dienophile and the LUMO of the diene. In a normal Diels–Alder reaction, if the LUMO of the dienophile can be further lowered in energy, the reaction would be much faster and can proceed at a significantly lower temperature. One way to lower the energy is to coordinate the heteroatom present in the EWG of the dienophile by Brønsted or Lewis acids. In this regard, transition metal complexes (e.g., iron complexes) can facilitate the reaction by applying the same concept and can also induce chirality into the reaction by using stabilizing chiral ligands. However, only a few reports are in the literature regarding an iron-catalyzed Diels–Alder reaction.

In 1991, Corey et al. (21) reported the first iron-catalyzed asymmetric Diels–Alder reaction between cyclopentadiene and 3-acryloyl-1,3-oxazolidin-2-one. For this asymmetric catalytic system, FeX_3 was chosen as the Lewis acidic metal component, along with a C_2 symmetric bis(oxazoline) ligand, which imposed the chiral environment. This metal–ligand (FeI_3) complex, was further activated by insertion of molecular I_2 into the reaction mixture, which significantly accelerated the rate of the reaction even at –50 °C. The endo adduct was preferentially obtained in preparatively useful yield (85%). Further, the chiral ligand was found to be readily recoverable and recyclable, which emphasized the synthetic utility of this protocol. Use of a fluxional additive with a similar catalyst system comprising of $Fe(ClO_4)_2$ and the ligand improved the enantioselectivity further (up to 91% ee) (22). Here ee = enantiomeric excess. Khiar (23) in 1993 and Imamoto and co-workers (24) in 2000 devised other bidentate ligands, such as C_2 symmetric bis (sulfoxides) and diphosphine oxides, respectively, for an asymmetric Diels–Alder reaction that resulted in lower diastereo- and enantioselectivity for the reaction (Scheme 13).

Practical utility of the asymmetric Diels–Alder reaction was further enhanced when Kanemasa et al. (25, 26) unveiled a series of cationic aqua. complexes comprising of transition metal perchlorates and C_2 symmetric tridentate ligand DBFOX/Ph (**10**) (Scheme 14). The use of a tridentate ligand was particularly beneficial as it remained strongly bound to the metal by competitive coordination with the substrate and created an attractive chiral environment in which the metal was embodied. This in turn disfavored the aggregation or oligomerization of the complex, yet it induced a high degree of asymmetry in the reaction outcome. Further, the stability of the complexes in water made this catalytic system advantageous.

In 2004, Shibasaki and co-workers (27) devised an efficient iron-catalyzed Diels–Alder reaction that resulted in the formation of highly substituted acyl cyclohexene derivatives in high enantiomeric purity (up to 92% ee) (Scheme 15).

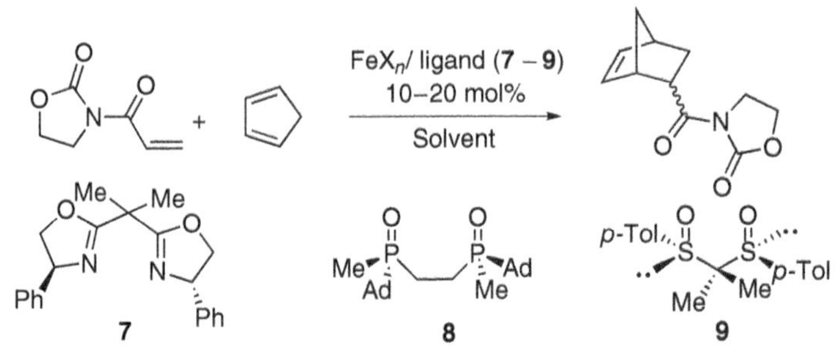

	7	8	9
Yield	85%	85%	78%
de	94%	16%	92%
ee	80%	endo 72% exo 74%	56%
	Corey, (1991) (21)	Imamoto co-workers (2000) (24)	Khiar et al. (1993) (23)

Scheme 13. The N, P, and S based ligand system for iron-catalyzed [4 + 2] cycloaddition (Diasteriometric excess = de, Tol = tolyl, Ad = adamantyl).

A 1.2:1 combination of tridenetate aryl-pybox ligands (**11**) and $FeBr_3$ in conjunction with $AgSbF_6$ provided an efficient catalyst that reacted with trisubstituted and tetrasubstituted diene with equal ease. Further, this protocol was successfully applied in the synthesis of biologically relevant natural product *ent*-hyperforin by the same group in 2010 (28).

In search of an efficient asymmetric Diels–Alder reaction, Kundig and co-worker (29) prepared a series of chiral phosphine ligands from an iron–

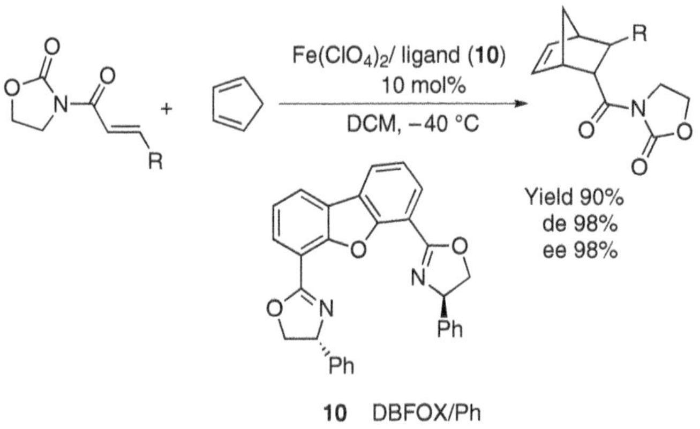

Scheme 14. Asymmetric Diels–Alder reaction-catalyzed cationic iron aquo complexes [DBFOX = 4,6-dibenzofurandilyl-2,2′-bis(4-phenyloxazoline)].

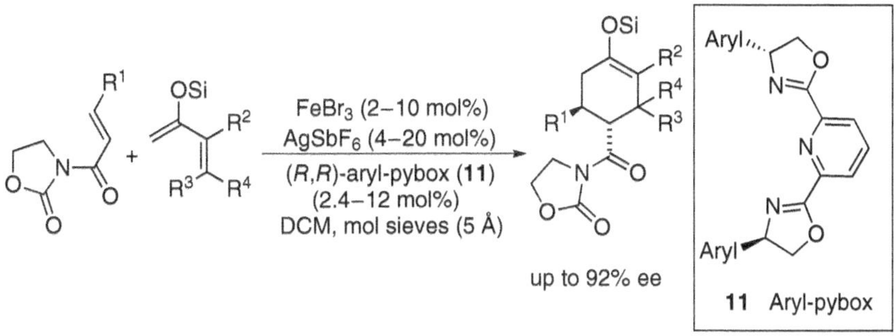

Scheme 15. An efficient iron-catalyzed Diels–Alder reaction [pybox = bis(oxazolinyl)pyridine].

cyclopentadienyl complex with a cyclopentane diol and a hydrobenzoin backbone. These C_2 symmetric ligand systems were compatible with iron, as well as with ruthenium, and cycloaddition between cyclopentadiene and enals were realized in high diastereo– and enatioselectivity.

Alkynes were used as the dienophile as well. In 1992, Jacobsen and co-workers (30) reported an iron-catalyzed [4 + 2] cycloaddition of 1,3-butadiene and alkynes involving a "bare" Fe^+ cation. Experiments were performed in a Fourier transform mass spectrometer (Nicolet FTMS-1000), where Fe^+ was generated by laser desorption–ionization from a high-purity iron foil. The *in situ* generated Fe(1,3-butadiene)$^+$ reacted rapidly with ethyne (and propyne) via a proposed η^3-complex to form Fe(1,4-hexadiene)$^+$, which upon subsequent dehydrogenation yielded the Fe(benzene)$^+$ complex. However, with alkenes or nitriles, no cycloaddition was observed in this case. Alkynes were also used in a stoichiometric reaction with vinylketeneiron (0) to generate catechol derivatives in moderate yields and regioselectivity.

The Hetero-Diels–Alder reaction, which is regarded as a convenient route to access six-membered heterocyclic compounds between aldehydes and dienes, are limited by the usage of either activated aldehydes (e.g., glyoxylates) or electron-rich dienes e.g., Danishefsky's diene and Rawal's diene. Further, strong Brønsted or Lewis acid had to be employed to overcome the poor reactivity of unactivated dienes. These drawbacks were successfully addressed by Matsubara and co-workers (31) in 2012, as they reported an unprecedented [4 + 2] cycloaddition of unactivated aldehydes and simple dienes catalyzed by iron(III)–porphyrin complex under mild and neutral conditions. A wide array of aldehydes and dienes containing various functional groups were reacted efficiently in the presence of 5 mol% of $[Fe(TPP)]BF_4$. In addition, highly substituted pyran scaffolds were generated in excellent yields and diastereoselctivities (Table I). High chemoselectivity, tolerance of water in the reaction medium, and mild reaction conditions made this method advantageous.

TABLE I
Scope of Hetero-Diels–Alder Reaction Catalyzed by $[Fe(TPP)]BF_4$[a]

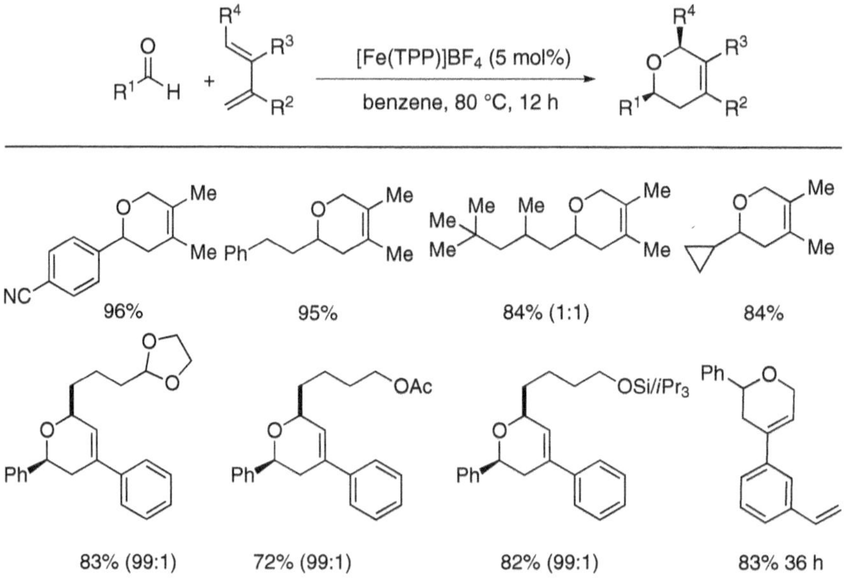

$[Fe(TPP)]BF_4$ (5 mol%)
benzene, 80 °C, 12 h

96% 95% 84% (1:1) 84%

83% (99:1) 72% (99:1) 82% (99:1) 83% 36 h

[a] Here TPP = tetraphenylporphyrin.

B. Cyclopropanation

Small ring molecules are potentially important to influence the pharmaceutical properties of many bioactive drugs (32). In this respect, cyclopropyl moieties achieved more attention due to its ubiquitous presence in many natural products (33), insecticides, modern pharmaceuticals, and in critical synthetic intermediates (34). So far, the traditional process of cyclopropanation is the [2 + 1] addition of different carbenes with olefins via radical pathways (35). In this respect, transition metal (35) (Ru, Rh, Co, Cr, Mo, W, the Fischer–Tropsch carbene-transfer process) mediated transfer of carbene to olefin from the stoichiometric carbene source is one of the efficient pathways.

In 1966, Jolly and Pettit (36) first reported cyclopropanation (Scheme 16) by an iron complex to an olefin. Importantly, this was the first example of a metal–carbene complex acting as a carbene-transfer agent. Treatment of cyclohexene in the presence of $CpFe(CO)_2CH_2OMe$ (**12**) and acid gave norcarane in 46% yield. It was proposed that the reaction was accomplished by the intermediacy of $CpFe(CO)_2CH_2^+$.

Recently, the cyclopropanation reaction was further developed. Most of the time, the process of carbene transfer was hampered by low selectivity with the different types of catalysts used. In 1993, Hossain and co-workers (37) developed the first iron-based cyclopropanation reaction in a catalytic manner (Scheme 17). The Lewis

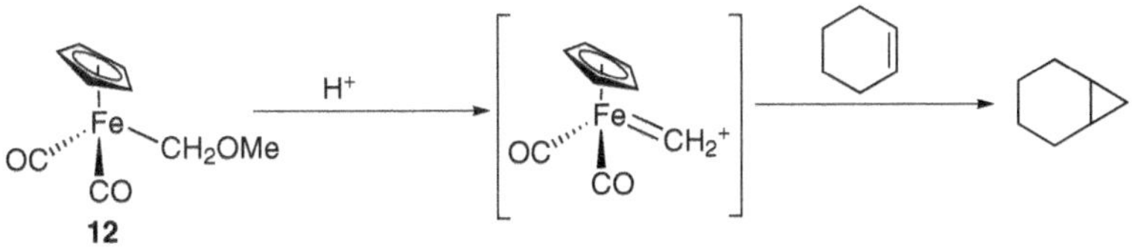

Scheme 16.

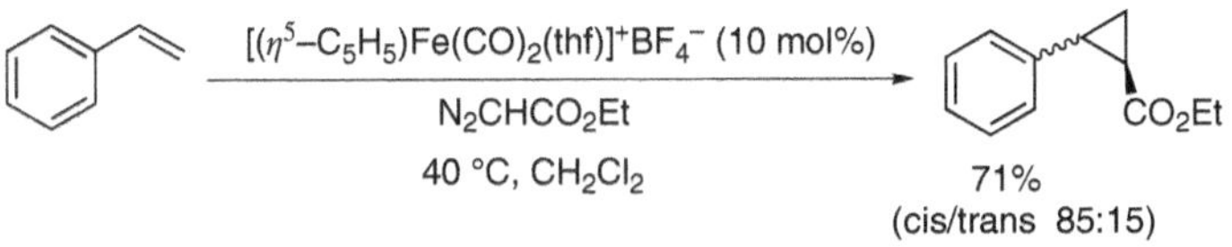

Scheme 17.

acidic iron center in $[(\eta\text{-}C_5H_5)Fe(CO)_2(thf)]^+BF_4^-$ can act as an efficient catalyst to cyclopropanate styrene analogues in the presence of ethyldiazoacetate (EDA) as the carbene source. After several rounds of optimization, it was found that 10 mol% of catalyst at 40 °C with 5 equiv of styrene were the optimal requirement.

In the proposed mechanism (Scheme 18), THF was dissociated first from the iron Lewis acid to generate cationic intermediate (**13**), which reacted with EDA to form

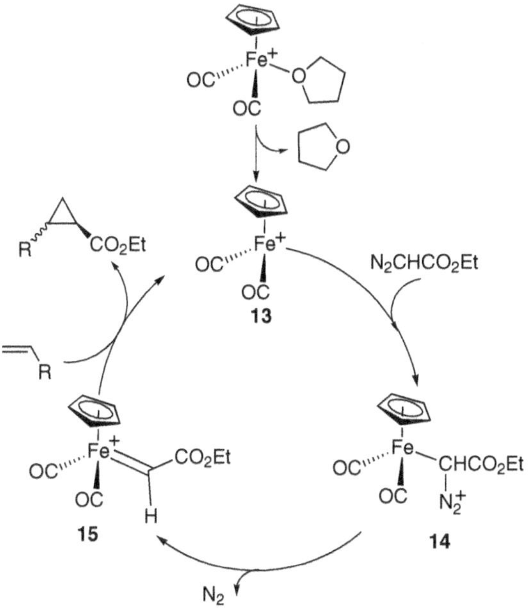

Scheme 18.

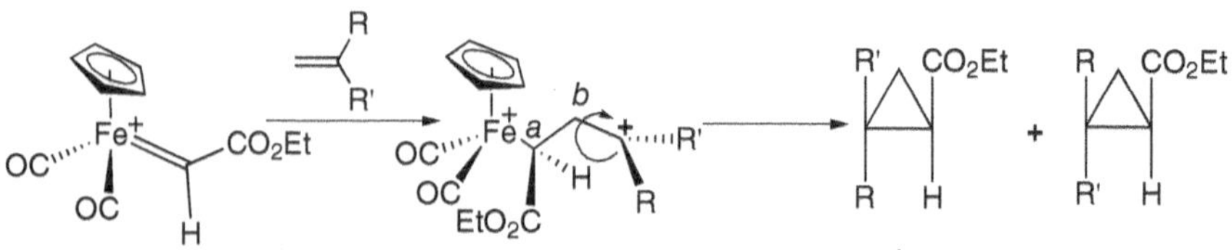

R	R'	cis	trans
H	Ph	85	15
H	*p*-MePh	60	40
Me	OMe	55	45

Scheme 19.

an intermediate complex (**14**) followed by extrusion of nitrogen to give an extremely reactive iron–carbene complex (**15**). The new complex readily transferred the carbene moiety to styrene. Several controlled experiments further supported this plausible mechanism.

The iron–carbene complex reacted with styrene to form a 5.6:1 mixture of *cis/trans*-1-phenyl-2-carboxycyclopropane (Scheme 19). This reaction indicated the presence of the short-lived γ-carbocation, which was rearranged to give the expected product. The ratio of cis and trans products was mainly dependent on the electronic property of the substituents attached to the intermediate. When electron-donating groups were present, the rotation of the $C_\beta{-}C_\gamma$ bond was greater. Consequently, the cis/trans selectivity was less for *p*-methylstyrene and 2-methoxypropene.

In 2002, Nguyen and co-workers (38) reported the olefin cyclopropanation using μ-oxo-bis[(salen)iron(III)] complexes [salen = *N,N'*-bis(3,5-di-*tert*-butylsalicylidene)-1,2-cyclohexanediamine] (**16–20**) (Scheme 20). Thus, this (salen)iron complex (**16**) can be used as an efficient, selective, and inexpensive metal alternative to a widely used ruthenium(II) salen complex.

The ethyldiazoacetate can act as an efficient reducing agent, and can break the μ-oxo bridge to produce the active (salen)iron(II) complex (**16**) for cyclopropanation. An optimized condition for cyclopropanation referred 5 mol% of catalyst with dry benzene or toluene as the solvent under refluxing temperature (Scheme 21) (38). By varying the diamine backbone of the complex, different yields of the product were obtained with subsequent increased or decreased reactivities. The reaction was fastest with the least sterically hindered backbone (e.g., 1,2-ethanediamine). But a bulkier 1,2-dimethyl-1,2-ethanediamine backbone gave the slowest reaction.

In 2002, Morise et al. (39) reported the cyclopropanation reaction using *trans*-$[(CO)_3Fe(\mu\text{-}L^{P,N})_2Cu]BF_4$ (**21**), which was the first metal–metal bonded six-membered ring system with P,N donors (Scheme 22). In this complex, formally

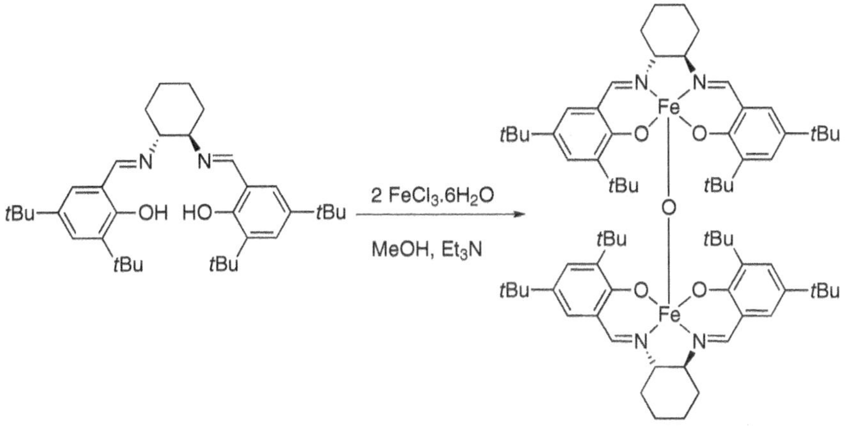

Scheme 20.

a zero-valent Fe center became attached to the Cu(I) center via the nitrogen of the flanked oxazoline moiety from the phosphine group. This complex can be used as an efficient catalyst for the cyclopropanation of styrene with ethyl diazoacetate. The reaction was carried out using 1 mol% of the catalyst with DCM as solvent at rt. The *trans*- and *cis*- ethyl-2-phenyl-1-cyclopropanecarboxylates were obtained in 91% isolated yield in a 70:30 ratio. The complex **21** was the first metal–metal bonded heterometallic catalyst for cyclopropanation.

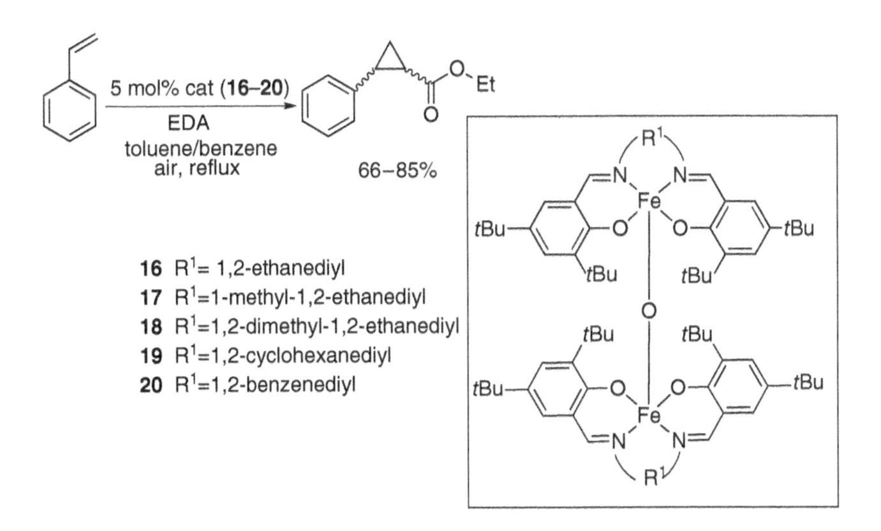

Scheme 21.

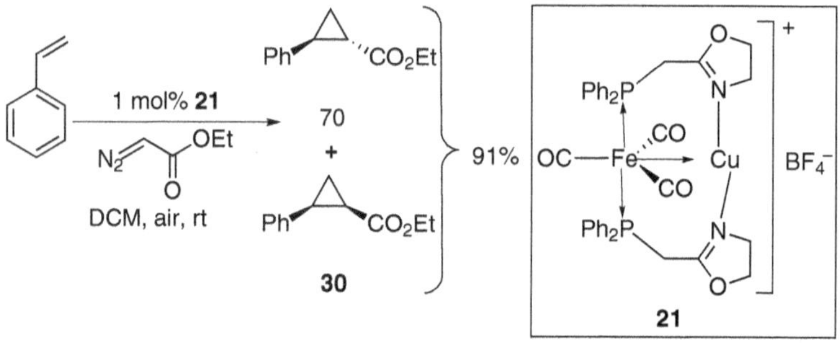

Scheme 22.

In 1995, Woo and co-workers (40) reported the asymmetric cyclopropanation reaction of styrenes using different iron(II) complexes with chiral macrocyclic (porphyrin-based) ligands (Scheme 23). These ligands provided the auxiliary stereogenic centers in close proximity to the active metal sites and also made those complexes as an efficient catalyst. The reaction was useful for the production of industrially important trans cyclopropyl ester derivatives.

In order to get the mechanistic insight into the reaction, labeling experiments were performed for styrene and styrene-d_8. The reaction was presumed to go via iron(II), which was originally *in situ* generated by the reductant ethyl diazo acetate. In contrast to other cyclopropanating catalysts, iron(II)(TTP) where TTP = *meso*-tetra-*p*-tolylporphyrin, was less electrophilic. In the transition state, the alkene possesses some carbocataionic character, which was in accordance with the reverse secondary kinetic isotopic effect (KIE).

According to the proposed transition state (Scheme 24), the selectivity was mainly dependent on the orientation of the alkene with the porphyrin plane. The shape selectivity of the alkene was mainly dependent on the presence of the

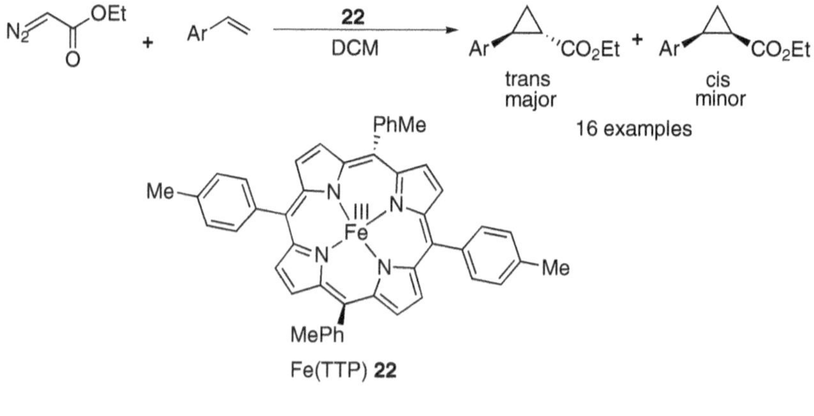

Scheme 23.

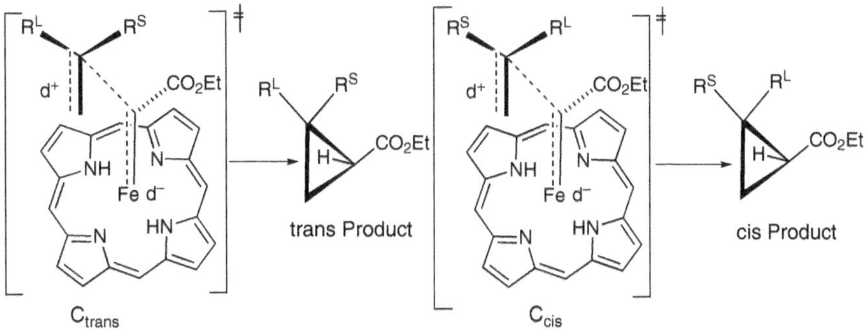

Scheme 24.

substituents on the nearest carbon to the macrocyclic plane. The trans product was dominated due to the interaction between macrocycle and the large group (R_L). The proposed model also depicted the increasing trans/cis ratio in some donor solvents, which could coordinate axially to iron. Such coordination can reduce the electrophilicity of the complex and therefore trans selectivity can be increased.

In 2002, the same group (41) developed some iron(II) complexes with different macrocyclic ligands and iron(II) porphyrin complexes, like iron(II) (D_4-TpAP) and Fe($\alpha_2\beta_2$-BNP) for asymmetric cyclopropantion (Scheme 25, BNP = bis

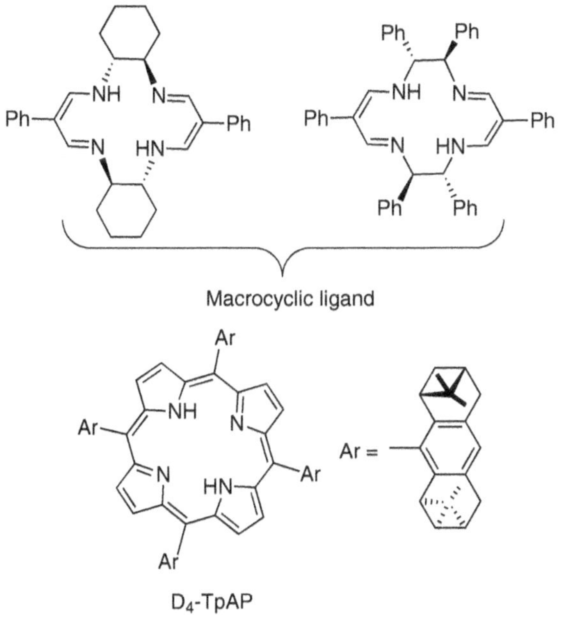

Scheme 25.

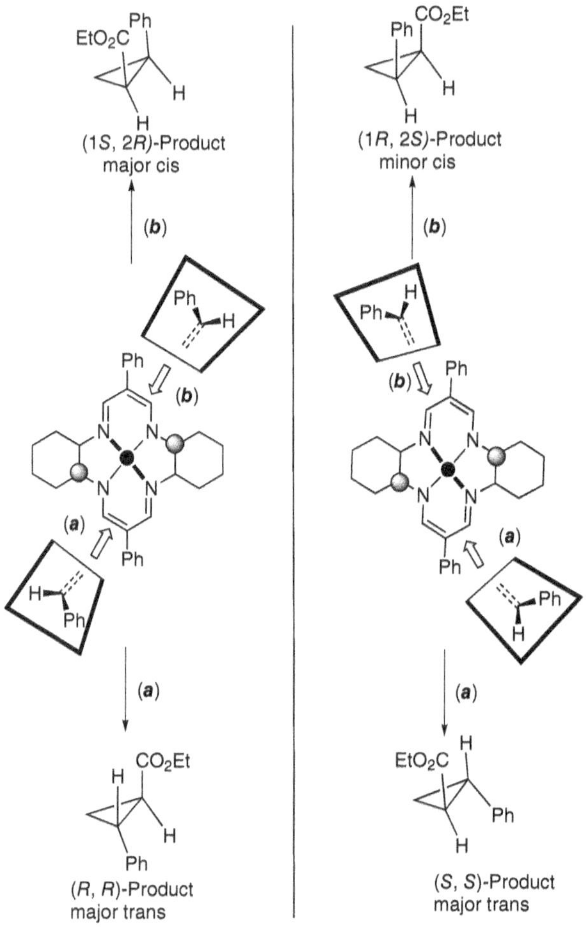

Scheme 26.

(binaphthylporphyrin)) with EDA. The reactions were carried out using 0.1–0.4 mol% of an Fe–porphyrin catalyst and 1–2 mol% of an Fe–macrocycle catalysts. Predominantly trans products were obtained compared to cis.

The enantioselectivity for this reaction was solely attributed to the orientation of carbene, as well as the olefins. For the catalysts containing macrocycles, selectivity appeared due to the parallel orientation of the C=C axis with the M=C bond (Scheme 26). Minimized steric interactions between the ester group and the axial proton of the chiral cyclohexyl group were achieved, when the olefin approached from path *a*. Thus, the observed product had an (*R,R*) configuration and was

TABLE II
Cyclopropanation Using the Halterman Catalyst

23
Fe–Halterman catalyst

Substrate	4-Me-styrene	4-MeO-styrene	4-Cl-styrene	4-Br-styrene
	53%	58%	58%	58%
trans/cis	96:4	93:7	94:6	92:8

obtained as a major trans isomer. A similar sense of chirality was introduced when Fe–porphyrin catalysts were used. The olefin approached from the right side of the carbene plane and as a result both (*S,S*) and (*R,R*) products were favored.

In 2009, Simonneaux and co-workers (42) reported the asymmetric intermolecular cyclopropanation of styrene analogues using an aryl diazoketone as the carbene source and a chiral Halterman iron–porphyrin complex (**23**) as catalyst. The initial attempt was made using the iron chloride Fe(TPP)Cl as a catalyst at rt. But low yield and major side products made those attempts unsuccessful. When the Fe–Halterman catalyst was applied the yields, as well as the selectivity (76% for trans), were increased. Different electronically and sterically demanding substrates were successful in that process, with moderate-to-good yield (Table II).

Safe and environmentally benign methodologies are always in demand for synthetic chemistry. Especially when the reactive intermediates are toxic and explosive. Considering these facts, Morandi and Carreira recently developed (43) a new procedure for cyclopropanation that minimized the risk as well as the time and

TABLE III
Cyclopropanation Using an Iron–Porphyrin Complex in Water

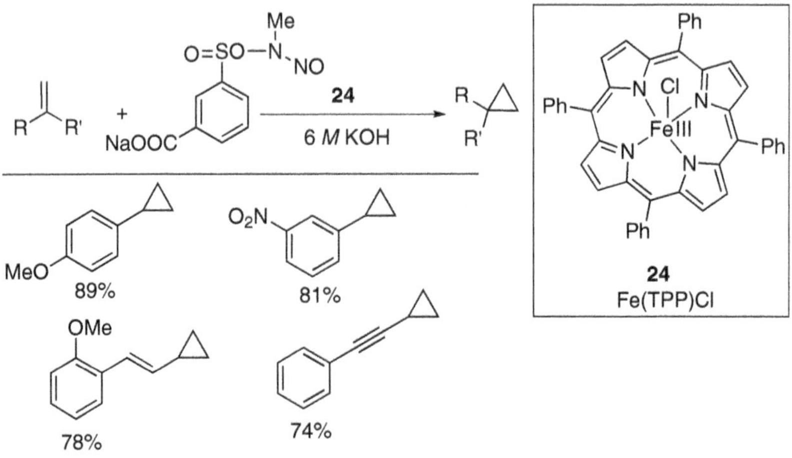

24
6 *M* KOH
NaOOC
Me
NO
R
R'
89%
81%
78%
74%
MeO
O_2N
OMe
Ph
Fe(III)
Cl
24
Fe(TPP)Cl

effort. A water-soluble diazald derivative (i.e., *N*-methyl-*N*-nitroso-*p*-toluenesulfonamide), which showed low toxicity compared to other diazomethane precursors, released diazomethane *in situ* on treatment with a 6 molar KOH solution. Tandem cyclopropanation occurred in the presence of the Fe(TPP)Cl catalyst (**24**) when the ejected diazomethane was transferred to the organic layer. Except for being hydrophilic, both the electron-rich and the electron-poor subtrates were well tolerated under optimal reaction condition (Table III).

Carreira and co-workers (44) developed a cyclopropanation reaction (Scheme 27, dr = diastereomeric ratio) using the same catalyst Fe(TPP)Cl (**24**) and glycine ethyl ester hydrochloride as the inexpensive and safe carbene source to yield the *trans*-cyclopropyl ester selectively.

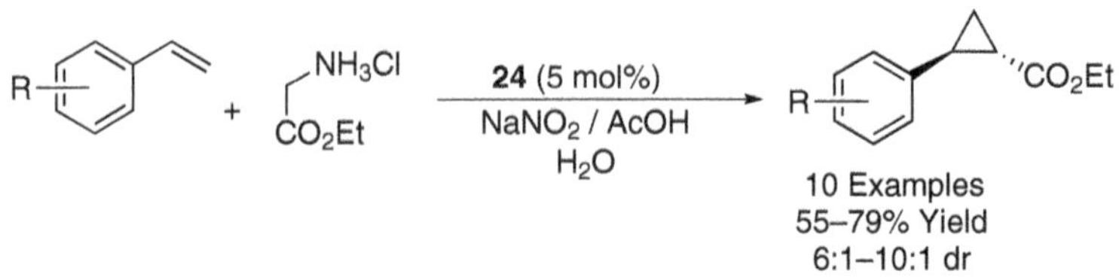

NH_3Cl
CO_2Et
24 (5 mol%)
$NaNO_2$ / AcOH
H_2O
R
CO_2Et
10 Examples
55–79% Yield
6:1–10:1 dr

Scheme 27.

TABLE IV
Iron-Catalyzed Cyclotrifluoromethylation[a]

24 (3 mol%), DMAP (10 mol%)
$CF_3CH_2NH_3Cl$ (1.5 equiv)
$NaNO_2$ (20 mol%), H_2SO_4 (10 mol%)
H_2O, rt, 14 h

Me	MeO	F_3C	Br	
99%	89%	77%	95%	10%

[a] 4-Dimethylaminopyridine = DMAP.

Trifluoromethylated cyclopropanes are important compounds in drug delivery (45), though very few synthetic methods were reported for their preparation. Carreira and co-workers (46, 47) recently reported potentially applicable methods for the synthesis of trifluoromethylated cyclopropanes by using trifluoroethylamine hydrochloride as the carbene source. Tandem cyclopropanation occurred in the presence of 3 mol% Fe(TPP)Cl (**24**) and saturated $NaNO_2$ solution to generate carbene. Both electron-rich and electron-deficient dienes were good substrates (Table IV) for this transformation, but it was unable to cyclopropanate 1,2-trans-substituted double bonds.

C. Aziridination and Aziridine Ring-Opening Reactions

Synthesis of various nitrogen-based compounds, particularly α-amido ketones, can be achieved by ring-opening aziridination (48). Therefore, development of sustainable and effective methods for aziridination is highly desirable. Bolm and co-workers (49) developed iron-catalyzed aziridination. They synthesized α-*N*-arylamido ketones by using 2.5 mol% $Fe(OTf)_2$ as catalyst and PhINTs as a nitrene source (Table V), where OTf = trifluoromethanesulfonate.

Reaction conditions for a 0.25-mmol scale: $Fe(OTf)_2$ (2.5 mol%), enol silyl ether (2 equiv), MeCN (1 mL), rt, 1 h.

Sulfonamide and iodosylbenzene or iodobenzne diacetate in the presence of magnesium oxide gave styrene aziridine derivatives in good yields. Use of MgO could be avoided by using less acetonitrile. The reaction gave moderate-to-good yields for styrene derivatives as substrates and moderate yields for internal olefins (Table VI).

TABLE V
Synthesis of α-*N*-Arylamido Ketones[a]

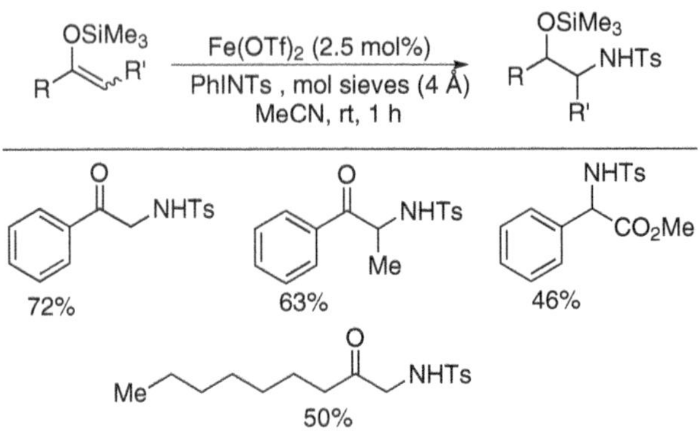

OSiMe$_3$, R, R'
$Fe(OTf)_2$ (2.5 mol%)
PhINTs , mol sieves (4 Å)
MeCN, rt, 1 h
OSiMe$_3$, R, NHTs, R'

O, NHTs — 72%
O, NHTs, Me — 63%
NHTs, CO_2Me — 46%
Me, O, NHTs — 50%

[a] *N*-Tosyliminobenzyliodinane = PhINTs.

TABLE VI
Iron-Catalyzed Aziridination[a,b,c,d]

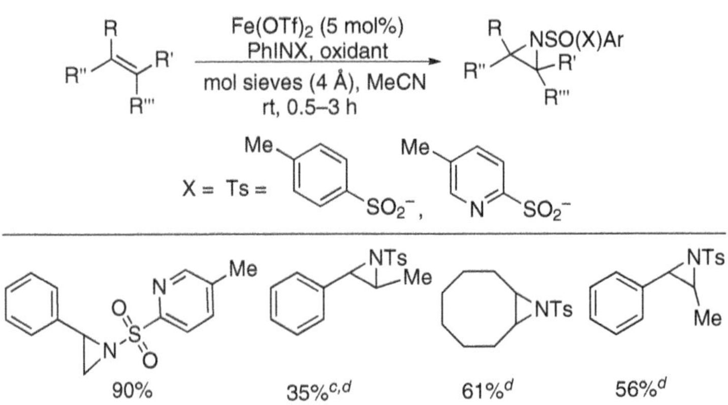

R, R', R'', R'''
$Fe(OTf)_2$ (5 mol%)
PhINX, oxidant
mol sieves (4 Å), MeCN
rt, 0.5–3 h
R, NSO(X)Ar, R', R'', R'''

X = Ts = Me, SO_2^- , Me, N, SO_2^-

N, Me, O, S, N, O — 90%
NTs, Me — 35%[c,d]
NTs — 61%[d]
NTs, Me — 56%[d]

[a] Reaction conditions for a 0.25-mmol scale: $Fe(OTf)_2$ (2.5 mol%), enol silyl ether 1 (2 equiv), MeCN (1 mL) rt, 1 h. Ts = tosyl.
[b] The known products were identified by comparison of their analytical data with those of previous reports.
[c] Only trans-product was obtained selectively.
[d] Use of 10 mol% of $Fe(OTf)_2$.

TABLE VII
Asymmetric Aziridination of Styrenes[a]

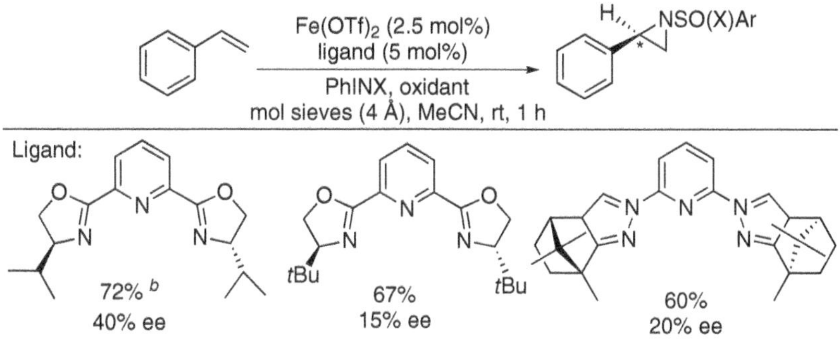

$Fe(OTf)_2$ (2.5 mol%)
ligand (5 mol%)
PhINX, oxidant
mol sieves (4 Å), MeCN, rt, 1 h

H,,, NSO(X)Ar

Ligand:

Ligand	Yield	ee
iPr-pybox	72% [b]	40% ee
tBu-pybox	67%	15% ee
bis(pyrazolyl)pyridine	60%	20% ee

[a] Reaction conditions for a 0.25-mmol scale: $Fe(OTf)_2$ (2.5 mol%), ligand (5 mol%), styrene (20 equiv), MeCN (1 mL), rt, 1 h.
[b] Both $Fe(OTf)_2$ (5 mol%) and a chiral ligand (30 mol%) were used.

Asymmetric synthesis of aziridine was achieved in the presence of chiral nitrogen ligands based on 2,6-bis(*N*-pyrazolyl)pyridines (Table VII). A radical mechanism was proposed from observed isomerization of *cis*-stilbene to the cis and trans isomer under the reaction condition.

Further improvement in yield was obtained by using quinaldic acid in the presence of ionic liquids, such as ethyl methyl imidazolium bis[(trifluoromethyl) sulfonyl]-amide (emim BTA) or LiBTA (Scheme 28) (50).

Ring opening of aziridines by a nucleophile can generate stereospecific *β*-functionalized amines. In this context, Schneider and co-workers (51) developed an iron-catalyzed method to synthesize *β*-functionalized amines (Scheme 29). Different N-substituted aziridines and aniline derivatives were tolerated under this reaction condition.

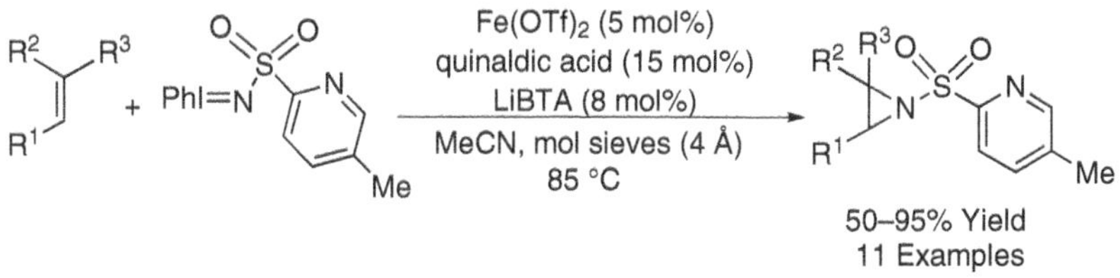

R^2, R^3, R^1 + PhI=N–SO_2–(5-methylpyridin-2-yl)

$Fe(OTf)_2$ (5 mol%)
quinaldic acid (15 mol%)
LiBTA (8 mol%)
MeCN, mol sieves (4 Å)
85 °C

50–95% Yield
11 Examples

Scheme 28. Effect of ionic liquid in aziridination reaction.

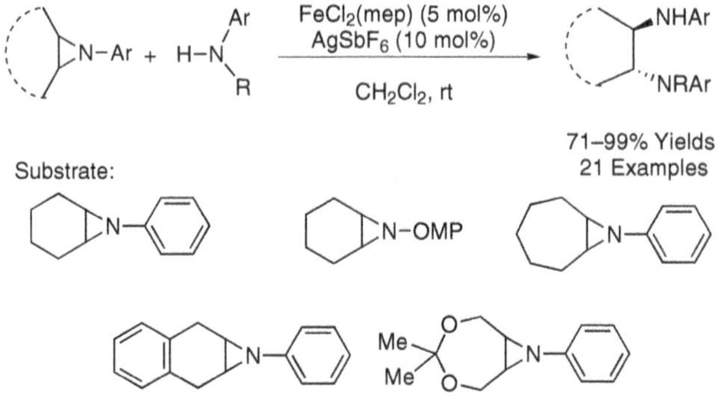

Scheme 29. Reaction scheme of ring-opening aziridination [mep = *N*,*N*′-dimethyl-*N*,*N*′-bis(2-pyridylmethyl)-ethane, OMP = *o*-methoxyphenyl].

D. Carbometalation of C–C Unsaturated Bond

A carbometalation reaction is an addition reaction of an organometallic compound to an unsaturated carbon–carbon bond resulting in a new carbon–carbon and carbon–metal bond formation. Generally, catalytic iron salt in the presence of an organometalic compound forms an organo-iron species, which accelerates the addition reaction to the unsaturated C–C bond. In 1977 Lardicci and co-workers (52) first used $FeCl_3$ as a catalyst in the alkylation of hex-1-yne by organoaluminium compounds. This protocol showed regiospecificity toward 2-alkyl-alk-1-ene (**26**) and trialkylbuta-1,3-dienes (**29**) with a small amount of oligomer and other cycilc trimers. With an optically active alkyl substituent at the α-position of the triple bond, high stereospecificity was noticed upon carbometalation. With [1-D]hex-1-yne, no deuterium transfer was detected after hydrolysis (Scheme 30).

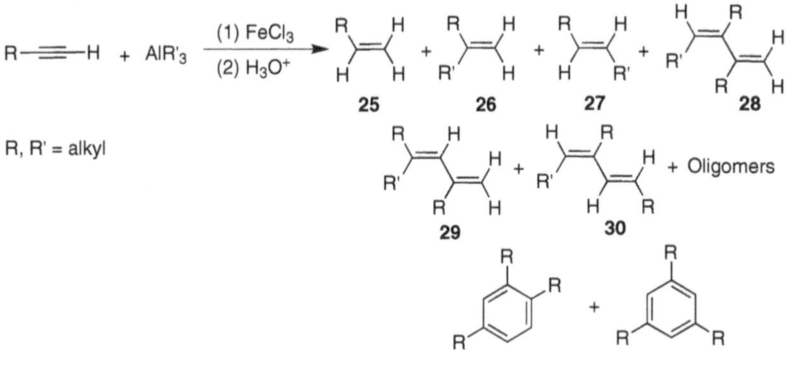

Scheme 30. Iron-catalyzed organoalumination of alphatic alk-1-ynes.

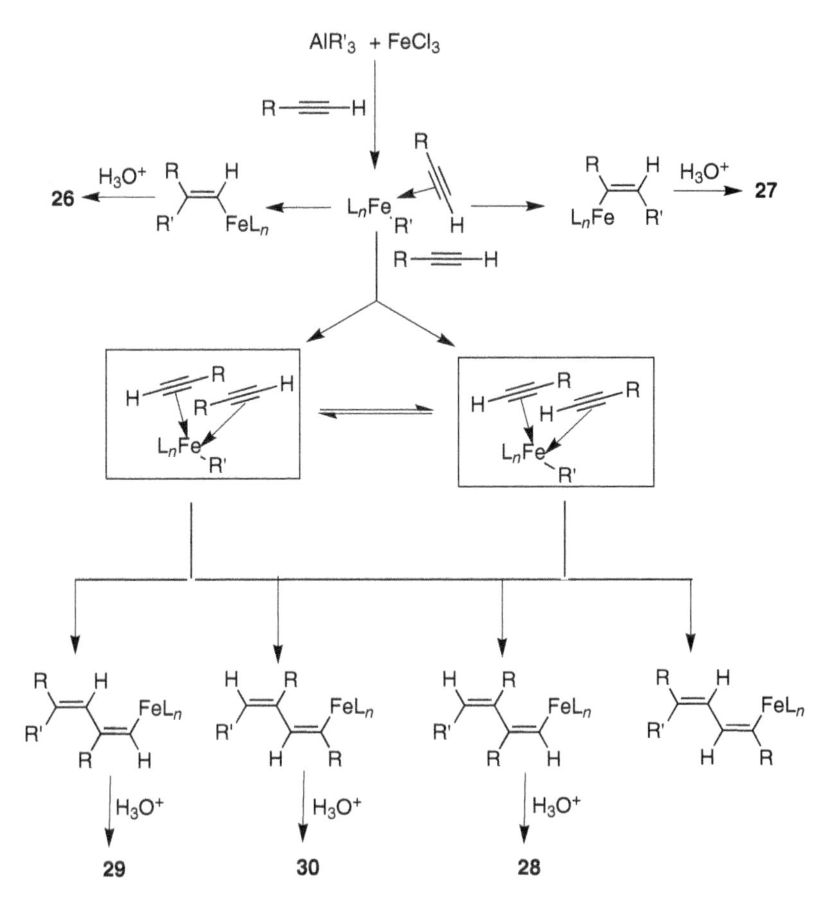

Scheme 31. Proposed pathways for iron-catalyzed organoalumination of alphatic alk-1-ynes.

To begin with, the iron center performed ligand exchange with an excess of organoaluminium compound, which underwent a cis-addition to the triple bond forming an iron–carbon single bond. Compound **26** was preferentially formed over **27** due to steric and electronic reasons. According to the proposed mechanism, the dienyl species were formed from two probable π-alkyne–iron species, which resulted in four organoiron compounds. Among them, compound **29** was more favorable due to stereoelectronic factors. Alkylation to the alkyne played a competitive role with cyclic trimer formation. Finally, increased steric hindrance on the alkyne moiety led to cyclization (Scheme 31) (52b).

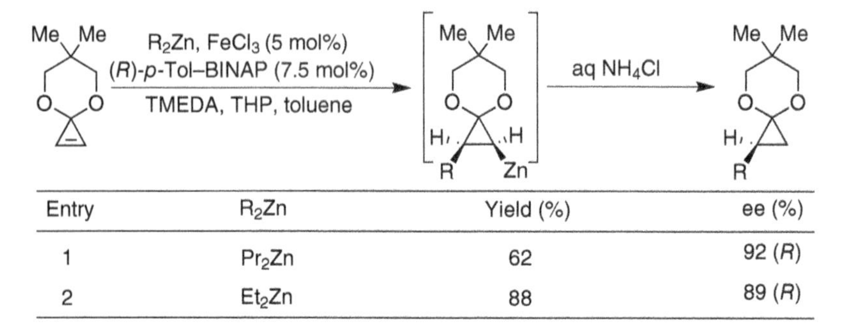

Entry	R_2Zn	Yield (%)	ee (%)
1	Pr_2Zn	62	92 (*R*)
2	Et_2Zn	88	89 (*R*)

Scheme 32. Iron-catalyzed olefin carbometalation (THP = tetrahydropyran).

In 2000, Nakamura et al. (53) reported $FeCl_3$ catalyzed olefin carbometalation using a Grignard reagent or organozinc complexes (Scheme 32). A cyclopropene moiety was easily carbometalated by this method. The carbometalated intermediate was also trapped with different carbon electrophiles. Enantioselective carbozincation was achieved by applying a number of bidentate phosphine ligands. The optimized condition with (*R*)-*p*-Tol-BINAP (2,2′-bis(diphenylphosphino)-1,1′-binaphthyl) and TMEDA [*N,N,N′N′*-tetramethylethane-1,2-diamine (solvent)] produced carbometalation with high enantioselectivity.

Hosomi's and co-workers (54) reported iron-catalyzed stereo- and regioselective carbolithiation of alkynes using a catalytic amount of $Fe(acac)_3$ (Scheme 33). They proposed that iron catalysis was going through an iron-ate complex. Under this reaction condition, alkynyl ether and alkynyl amines were well tolerated, but reaction with a simple alkyne (e.g., 6-dodecyne) totally failed.

Through iron-catalyzed alkyne carbometalation of propargylic and homopropergylic alcohol with Grignard reagent, a class of substituted allylic and

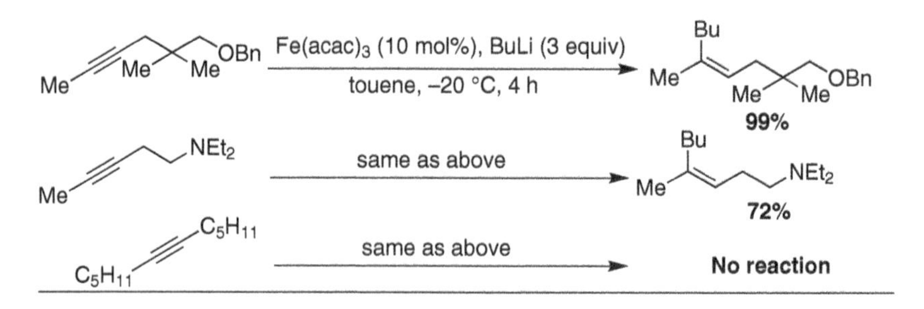

Scheme 33. Iron-catalyzed region- and stereoselective carbolithiation of alkynes (acac = acetylacetonate).

TABLE VIII
Stereo- and Regioselective Carbometalation of Propargylic and Homoproporgylic Alcohol[a]

Entry	Condition	Product	Yield (%)
1	$Fe(ehx)_3$ (0.2 equiv) dppe (0.2 equiv)		75
2	$Fe(acac)_3$ (0.2 equiv)		75
3	$Fe(acac)_3$ (0.4 equiv)		63

[a] 1,2-Bis(diphenylphosphino) ethane = dppe; ehx = 2-ethylhexanoate.

homoallylic substrates were synthesized stereoselectively (Table VIII). In 2005, Zhang and Ready (55) demonstrated regio- and stereoselective carbometalation by use of a catalytic amount of iron(III) salt. A small amount of dialkylated alkene and, in some cases, a hydrometalated species, was detected as a side product. Further, a vinyl Grignard intermediate was trapped with a different electrophile to produce tetrasubstituted allylic or homoallylic alcohols (Table IX).

The iron(III) center is reduced through a fast ligand-exchange process with the Grignard reagent. Then alkoxide directed carbometalation occurred forming a cyclic vinyl-iron intermediate. Subsequent metathesis with the Grignard reagent formed a vinyl-magnessium species, which was responsible for electrophilic substitution. After β-hydride elimination from the FeR_n species, an iron–hydride complex was generated that performed the hydrometalation of the alkyne moiety (Scheme 34) (55).

With the prospect of alkyne carbometalation, in 2005, Hayashi and co-workers (56) reported an iron–copper cooperative catalytic system, which successively

TABLE IX
Intermediate Trapping of an Electrophile of a Carbometalation Reaction[a]

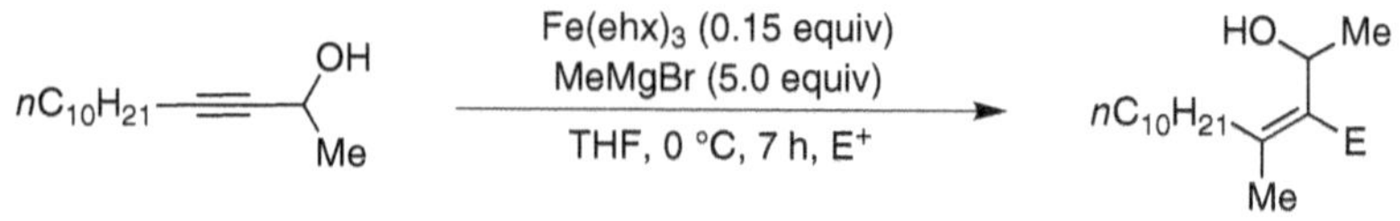

$Fe(ehx)_3$ (0.15 equiv), MeMgBr (5.0 equiv), THF, 0 °C, 7 h, E^+

Entry	E^{+a}	Product	Yield (%)
1		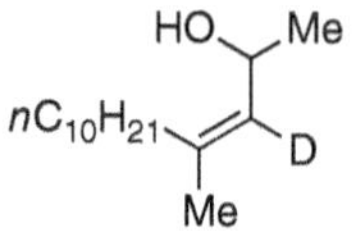	75 (92 D)
2	$ZnCl_2$.NBS	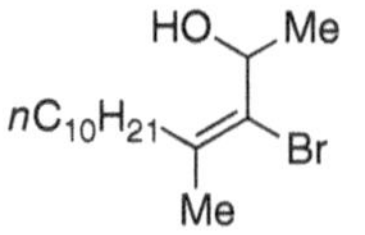	65
3	DMF	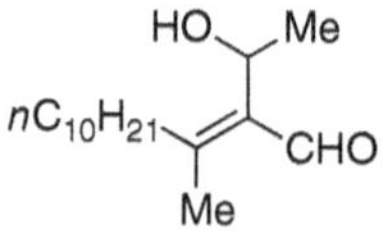	50
4	CuCN.LiCl. allyl bromide	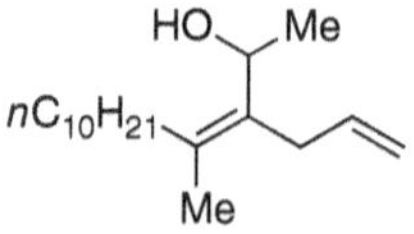	61

[a] N-Bromosuccinimide = NBS.

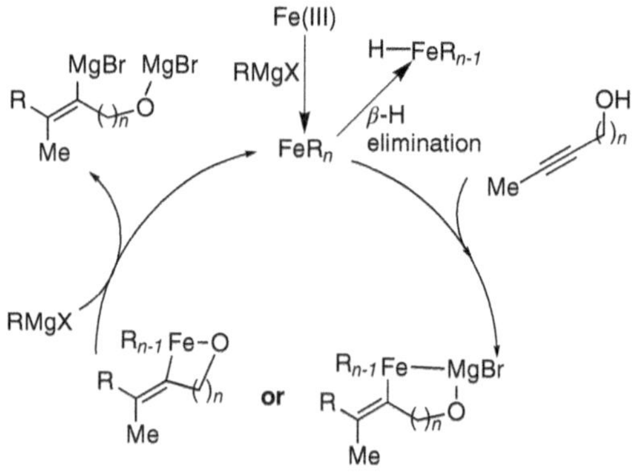

Scheme 34. Proposed mechanism for the carbometalation of propargylic and homopropargylic alcohol.

TABLE X
Alkyne Carbometalation Using an Iron–Copper Cooperative Catalytic System

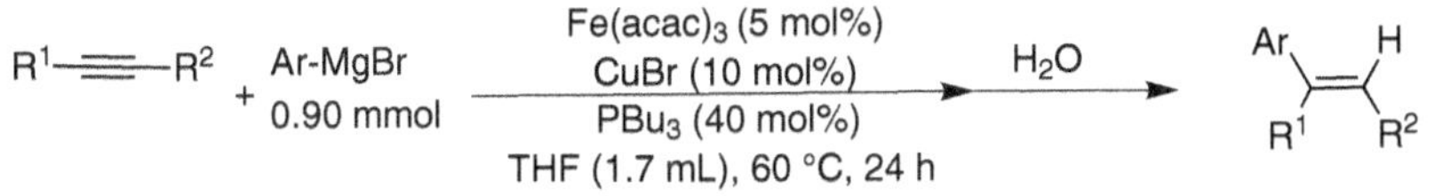
R1—≡—R2
+ Ar-MgBr
0.90 mmol
Fe(acac)3 (5 mol%)
CuBr (10 mol%)
PBu3 (40 mol%)
THF (1.7 mL), 60 °C, 24 h
H2O
Ar
H
R1
R2

Entry	Ar	R^1	R^2	Yield (%)	(*E*):(*Z*)
1	Ph	Pr	Pr	62	97:3
2	3-MeC_6H_4	Pr	Pr	70	95:5
3	3-OMeC_6H_4	Pr	Pr	56	97:3
4	4-FC_6H_4	Pr	Pr	40	95:5
5	$3,5\text{-Me}_2C_6H_3$	H	Ph	34	91:5
6	$3,5\text{-Me}_2C_6H_3$	Me	$SiMe_3$	56	72:26

performed aryl magnesiation with a Grignard reagent (Table X). An aryl magnesium bromide with an electron-donating as well as electron-withdrawing aryl group was successfully employed.

An aryl–iron species was proposed through the ligand exchange between the iron salt and the arylmagnesium bromide. The aryl–iron complex accomplished a cis addition with the alkyne forming vinyl–iron complex. Upon transmetalation this comlex gave vinyl-cuprate. A subsequent transmetalation with a Grignard reagent formed the alkenylmagnesium bromide (Scheme 35) (56, 57).

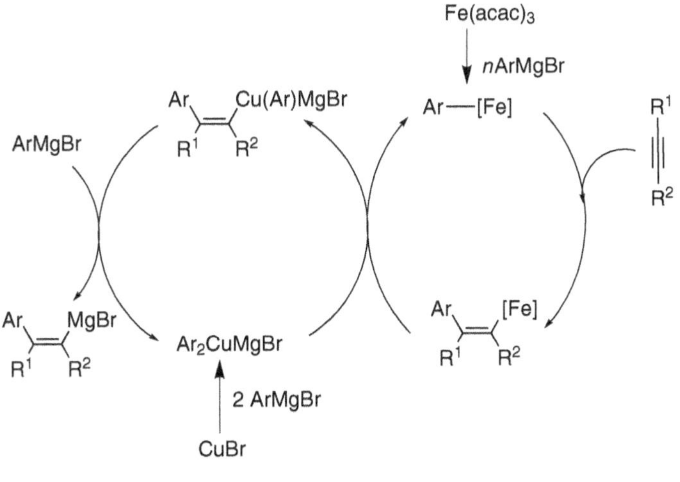
Fe(acac)3
nArMgBr
Ar—[Fe]
Ar
Cu(Ar)MgBr
R1
R2
ArMgBr
R1
R2
Ar
MgBr
R1
R2
Ar2CuMgBr
2 ArMgBr
CuBr
Ar
[Fe]
R1
R2

Scheme 35. Mechanism of arylmagnesiation of alkynes with an Fe/Cu cooperative catalytic system.

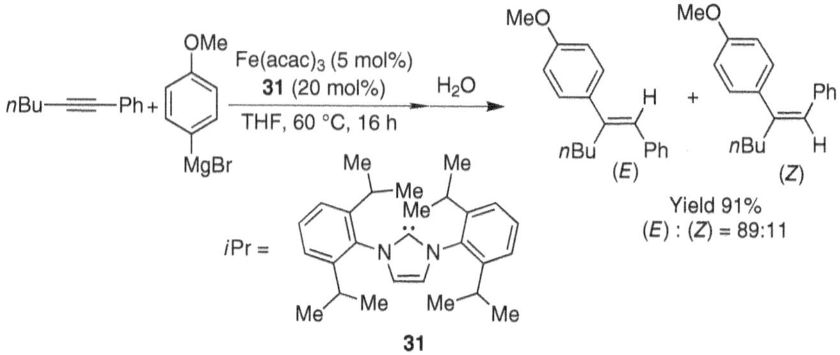

Scheme 36. Iron-catalyzed arylmagnesiation of alkynes in the presence of the NHC ligand.

In 2007, the same group reported arylmagnesiation of aryl(alkyl)acetylenes in the presence of a catalytic amount of $Fe(acac)_3$ and an N-heterocyclic carbene (NHC) ligand (Scheme 36) (58). The aryl–iron species preferentially promoted cis addition to the alkyne forming an alkenyliron intermediate. The NHC ligated intermediate gave the desired product upon transmetalation with Grignard reagent.

Again in 2009, Hiyashi and co-workers (59) reported iron-catalyzed carbolithiation of alkynes in the presence of a catalytic amount of TMEDA (Table XI). In this system, they demonstrated alkyllithiation of aliphatic and aromatic substituted alkyne with good yield and stereoselectivity. The alkene-lithiated intermediate was

TABLE XI
Carbolithiation of Alkynes in the Presence of a Catalytic Amount of TMEDA

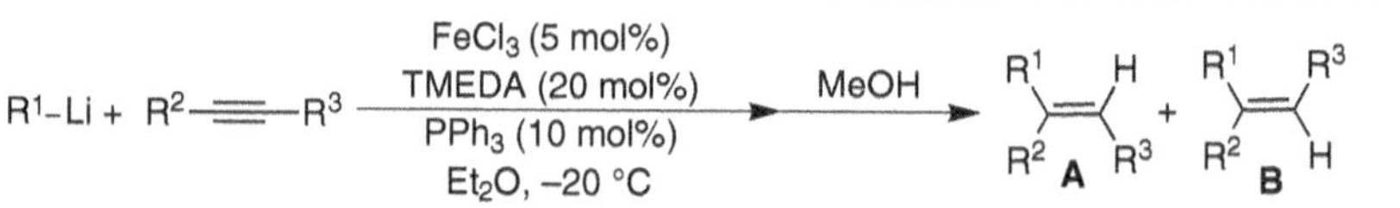

Entry	R^1	R^1	R^3	Time (h)	Yield (%)	**A**: **B**
1	Bu	Bu	Ph	1.5	81	
2	Bu	Bu	3-$CF_3C_6H_4$	1.5	82	
3	Bu	Bu	2-$MeOC_6H_4$	1.5	82	
4	Bu	Et	Ph	1.0	81	98:2
5	*i*Bu	Me	Ph	0.25	72	>99:1

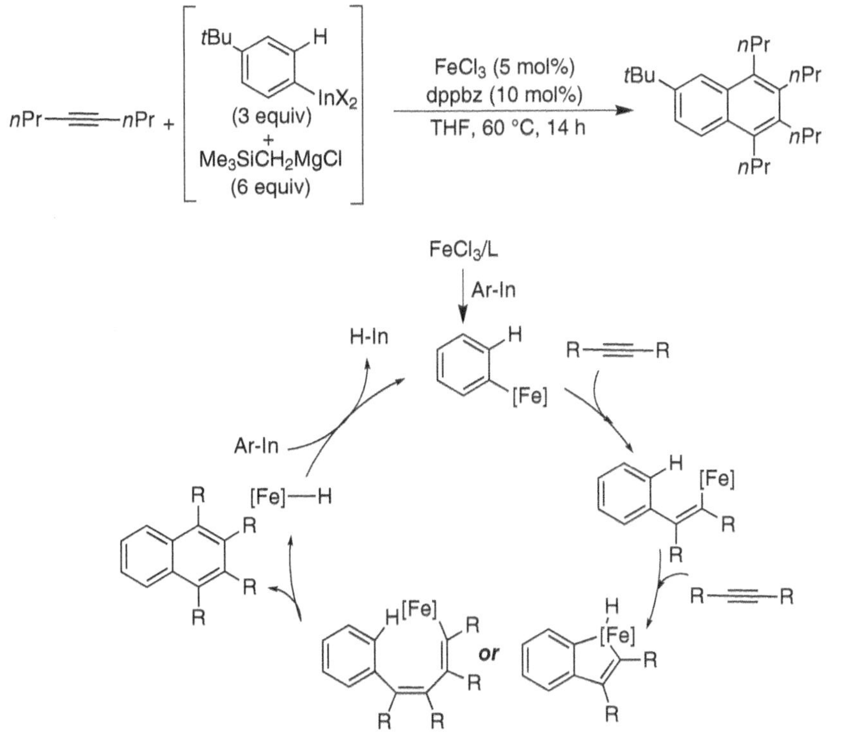

Scheme 37. Iron-catalyzed annulation reaction of aryllindium reagents and alkynes (dppbz = 1,2-bis (diphenylphosphino)benzene

also used for further electrophilic substitution with aldehyde, alkyl bromide, and so on.

An aromatic ring can be constructed through carbometalation of two alkynes in the presence of a catalytic amount of iron chloride. Upon formation of an aryl–iron complex from an arylindium reagent, reaction with alkyne was generated from an alkenyl–iron species. An intramolecular C–H activation involving another alkyne resulted in ring annulations (Scheme 37) (60).

Carbometalation in oxa- and azabicyclic alkene moieties are often problematic due to the ring-opening reaction through β-heteroatom elimination (Scheme 38). Ito and Nakamura (61) reported an iron-catalyzed diastereoselective organozincation of oxa- and azabicyclic alkenes in the presence of dppbz based ligands with much less conversion to the ring-opening product (Table XII).

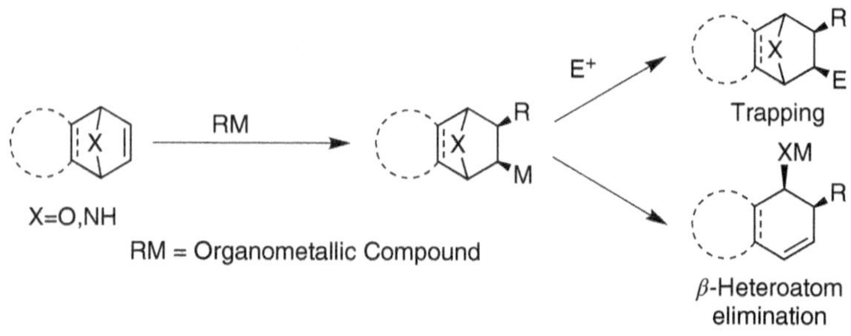

Scheme 38. Schematic diagram for carbometalation of oxa- and azabicyclic alkene.

TABLE XII

Carbometalation in Oxa- and Azabicyclic Alkene Moieties[a,b]

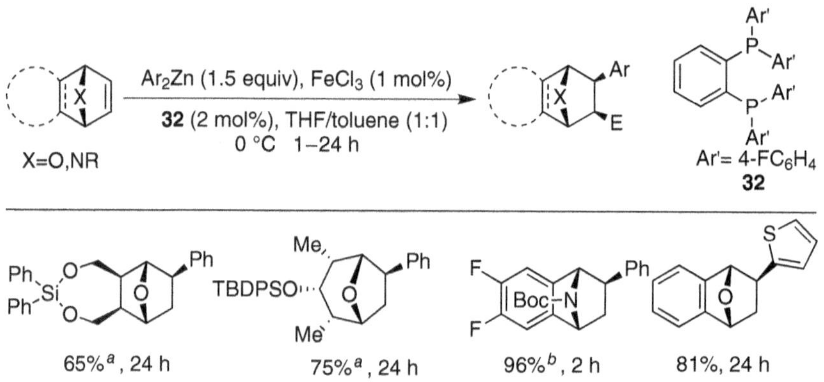

X=O,NR; Ar_2Zn (1.5 equiv), $FeCl_3$ (1 mol%), **32** (2 mol%), THF/toluene (1:1), 0 °C 1–24 h; Ar' = 4-FC_6H_4 (**32**)

65%[a], 24 h	75%[a], 24 h	96%[b], 2 h	81%, 24 h

[a] At 40 °C, $FeCl_3$ (3 mol%), ligand (6 mol%).

[b] At 25 °C (Boc = *tert*-Butyloxycarbonyl; TBDPS = *tert*-butyldiphenylsilane).

E. Michael Addition

The Michael addition is a useful pathway for C–C bond formation in the synthesis of an organic molecule. The convenient base-catalyzed Michael addition affords a number of side-product formations. To avoid the use of base, several methods using a transition metal have been developed. A number of homogeneous, as well as heterogeneous, iron-catalyzed Michael addition reactions have enriched this field with prospects for the future.

TABLE XIII
Michael Addition of Amines onto Acrylate Acceptors

R_2NH + CH$_2$=CHC(O)X —[$FeCl_3$ (10 mol%), 25 °C; 42 h, DCM (solvent)]→ NR_2CH$_2$CH$_2$C(O)X

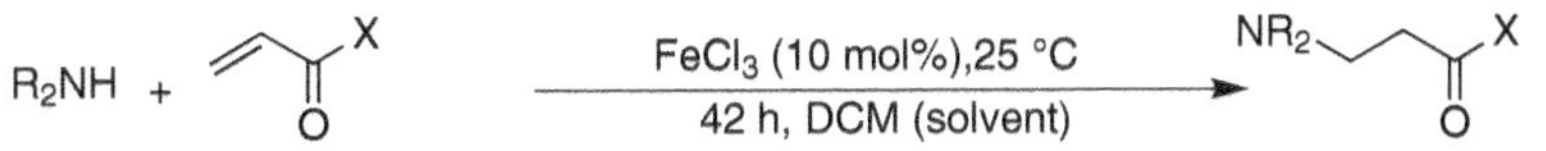

Entry	Amines	X	Yield (%)
1	Diethylamine	OEt	96
2	Diethylamine	H	Polymerization
3	Piperidine	OEt	97
4	Morpholine	OEt	90
5	Pyrrolidine	OEt	95
6	*n*-Butylamine	OEt	79

In 1989, Laszlo et al. (62) reported an $FeCl_3$ catalyzed Michael addition of amines onto acrylate acceptors (Table XIII). Herein, $FeCl_3$ acted as a Lewis acid, which coordinated with the carbonyl oxygen of the acrylate acceptor and catalyzed the reaction toward a thermodynamically favored 1,4- addition.

Iron(III) salts have proven to be an effective catalyst for Michael addition between 1,3-dicarbonyl compounds and vinyl ketones. In 1997, Christoffers (63) reported the $FeCl_3$ catalyzed Michael addition reaction at rt (Scheme 39).

First, the enone substrate interacted with the iron center through a vacant coordination site of a 1,3-dicarbonyl ligated iron complex. Then the center carbon of the dionato ligand performed the nucleophilic attack to the enone in 1,4-fashion. Since an olefin moiety should be in close contact with the dioneto ligand for the alkylation of enone, the (*S*)-trans enone strongly disfavored the reaction (Scheme 40) (64).

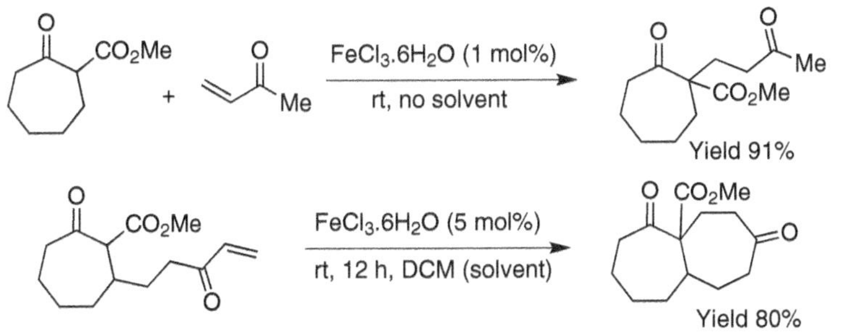

CO_2Me + Me — $FeCl_3.6H_2O$ (1 mol%), rt, no solvent → Me, CO_2Me, Yield 91%

CO_2Me — $FeCl_3.6H_2O$ (5 mol%), rt, 12 h, DCM (solvent) → CO_2Me, Yield 80%

Scheme 39. Iron-catalyzed Michael reaction of 1,3-dicarbonyl compounds and enones.

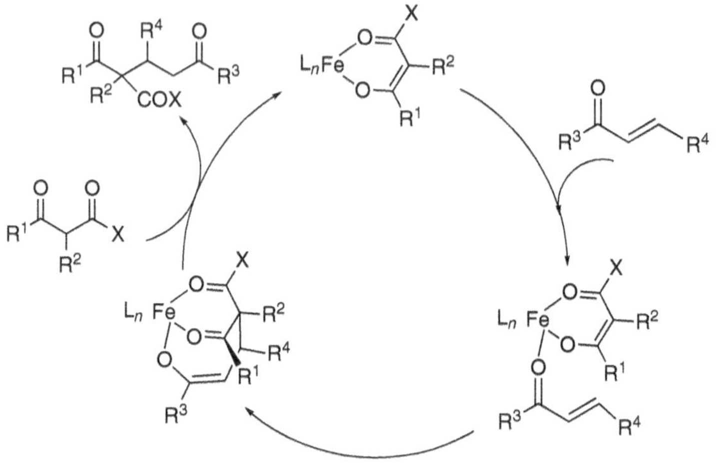

Scheme 40. Proposed mechanism for an iron-catalyzed Michael reaction of 1,3-dicarbonyl compounds and enones (COX = acyl halide).

A new class of Michael addition product was generated. The 2-acceptor substituted cycloalkenones with an iron salt formed a stable enolate, which acted as a Michael vinylogous donor toward the acceptor methyl vinyl ketone. Some amount of aldol product of the desired Michael addition was also formed as a side product (Scheme 41) (65).

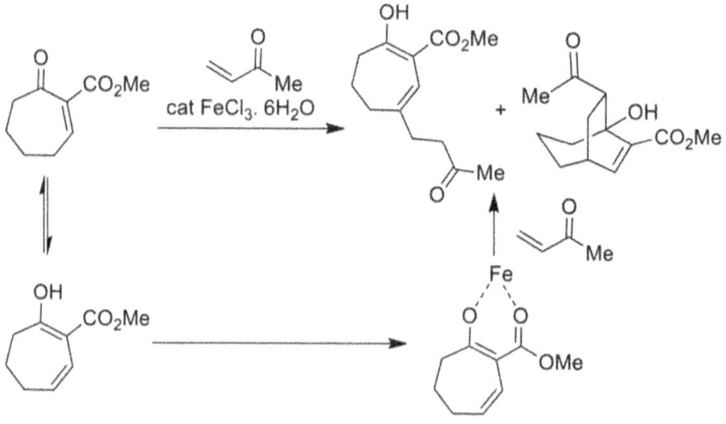

Scheme 41. Iron-catalyzed Michael reaction with a vinylogous donor molecule.

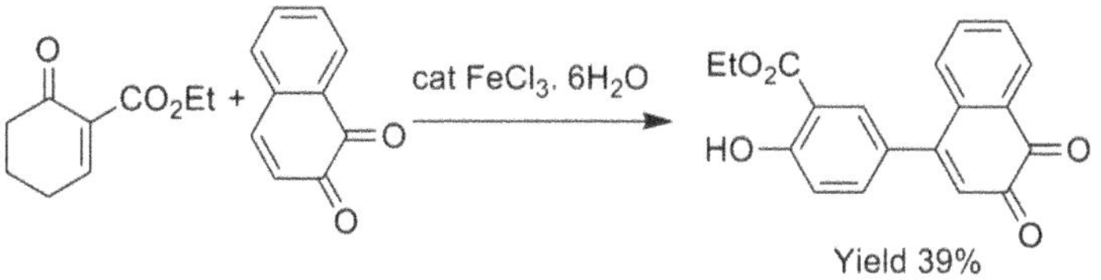

Scheme 42. Synthesis of biaryl compounds by iron-catalyzed Michael reaction.

By using a quinine derivative as the acceptor of a vinylogous Michael addition, biaryl cross-coupled products were formed after overoxidation of the donor and acceptor moieties of the addition product (Scheme 42) (66).

Asymetric Michael reaction with a chiral ligand was reported by Christoffers and co-workers (67). In 2003, an iron-catalyzed Michael reaction on a solid support was reported by Kitayama and co-worker (68).

F. Barbier-Type Reaction

Another popular way to construct a C–C bond is to attack the carbonyl center with a nucleophilic carbon center. In this regard, the Barbier reaction has drawn much attention for the preparation of alcohols from carbonyl compounds with simultaneous C–C bond formation. Generally, the nucleophilic carbon center is generated *in situ* from an alkyl or aryl halide, using various reducing agents (e.g., alkali and alkaline earth metals or lanthanides and their salts). Among various lanthanides, Molander and Harris (69) found SmI_2 and YbI_2, to be very effective in promoting various types of intramolecular Barbier-type reactions in the presence of an iron catalyst. Several 2-(*n*-iodoalkyl) afforded good-to-excellent yield of the corresponding bicyclic alcohols in the presence of SmI_2 and catalytic iron tris-(dibenzoylmethane) [Fe(dbm)$_3$] (Scheme 43, DBM = dibenzoylmethane) (70).

In particular, entries 1 and 4 in Table XIV show almost exclusive formation of one stereoisomer. Lack of stereoselectivity for entry 2 is due to the ease of attack from both the equatorial and axial side for equatorial side chains, whereas an axial side has only one option for equatorial attack, leading to a single diastereomer (Scheme 44) (70a).

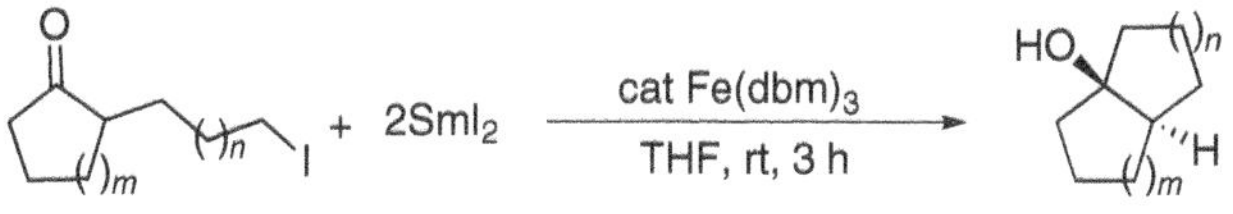

Scheme 43.

TABLE XIV
Iron-Catalyzed Intramolecular Barbier Reaction

Entry	*m*	*n*	% GC Yield (Isolated Yield)[a]	cis: trans
1	1	1	90 (60)	>99.5:<0.5
2	2	1	100 (75)	1.3: 1
3	3	1	85 (77)	2.0: 1
4	1	2	67	18: 1
5	2	2	95 (75)	1: 1.5
6	3	2	(83)	1: 2.3

[a]Gas chromatography = GC.

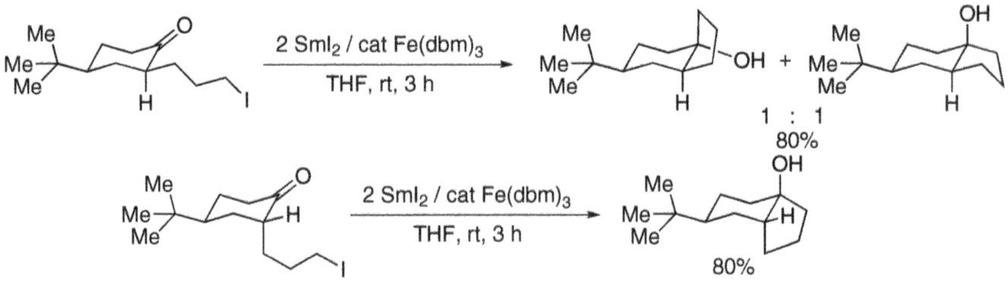

Scheme 44.

Even for the 3-(alkyl)cycloalkanonenes, the desired intramolecular rearranged bicyclic product was obtained in moderate-to-good yield (71). In all these cases, Fe $(dbm)_3$ was used as an effective catalyst due to its nonhygroscopic, air-stable nature and easy solubility in THF. In addition, other iron complexes including $FeCl_2$, $FeCl_3$, and $Fe(acac)_3$ were found to catalyze this reaction quite effectively (Scheme 45).

Acyclic acyl derivatives (e.g., esters, amides, and thioesters) were successfully converted into cyclic ketones (Scheme 46) (72).

Cyclization by C–C bond formation followed by lactone ring opening via C–O bond cleavage can be obtained via intramolecular nucleophilic acyl substitution (INAS) with specific stereoselectivity (Scheme 47) (73).

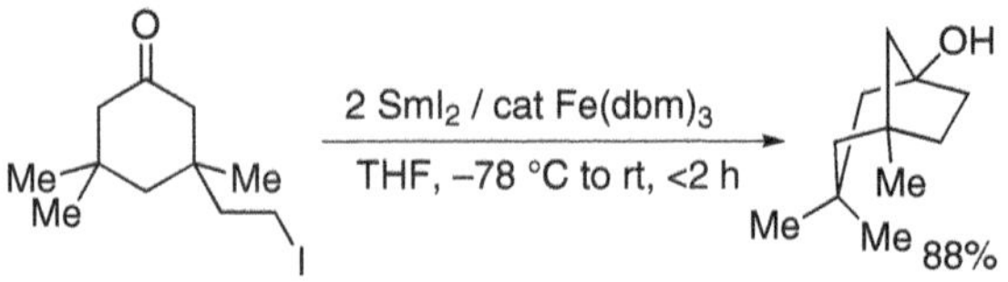

Scheme 45.

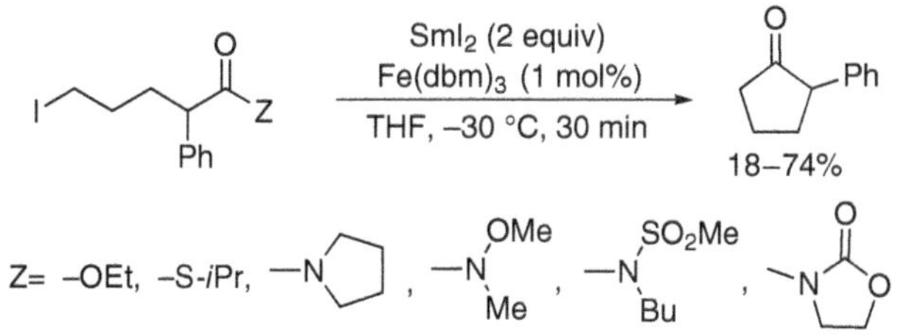

SmI_2 (2 equiv)
Fe(dbm)$_3$ (1 mol%)
THF, –30 °C, 30 min
18–74%

Z= –OEt, –S-*i*Pr, ...

Scheme 46.

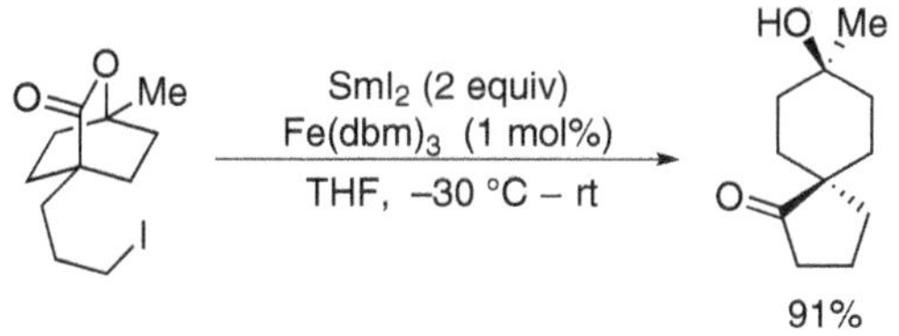

SmI_2 (2 equiv)
Fe(dbm)$_3$ (1 mol%)
THF, –30 °C – rt
91%

Scheme 47.

G. Kharasch Reaction

Construction of a new C–C bond can also be done by simple addition of polyhalogenated compounds with an alkenyl double bond by cleaving a C–X (X = Cl, Br, I) bond to form new geminal C–X and C–C single bonds (74). This type of transformation, popularly known as the Kharasch reaction, mainly goes through a radical pathway and starts by homolytic cleavage of a C–X bond in the presence of a radical initiator. Then newly generated alkyl radical attacks the alkene followed by final abstraction of a halide radical from a new polyhalide molecule leading to the final coupled product. Due to this radical pathway, several iron catalysts were found to be particularly efficient to initiate this transformation.

The first breakthrough in an iron-catalyzed Kharasch-type reaction came from the Hogeveen group (75). In their study, $Me_3NFe(CO)_4$ was preferred to $Fe_2(CO)_9$ for addition of CCl_4 in various alkene (Scheme 48). Interestingly, a strained tricyclic compound was formed in excellent yield from norbornadiene (Scheme 48).

Dimeric $[Fe_2(CO)_4(Cp)_2]$ was also found to catalyze this reaction at an elevated temperature (76). A detailed kinetic study showed first-order dependence on halocarbon and a dimeric iron complex, but a more complex dependence over the concentration of alkene. Interestingly, when simple halocarbon was replaced

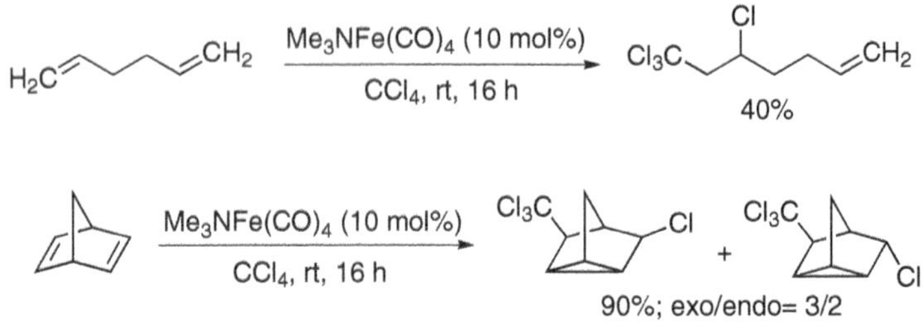

Me3NFe(CO)4 (10 mol%)
CCl4, rt, 16 h
40%
Me3NFe(CO)4 (10 mol%)
CCl4, rt, 16 h
90%; exo/endo= 3/2

Scheme 48.

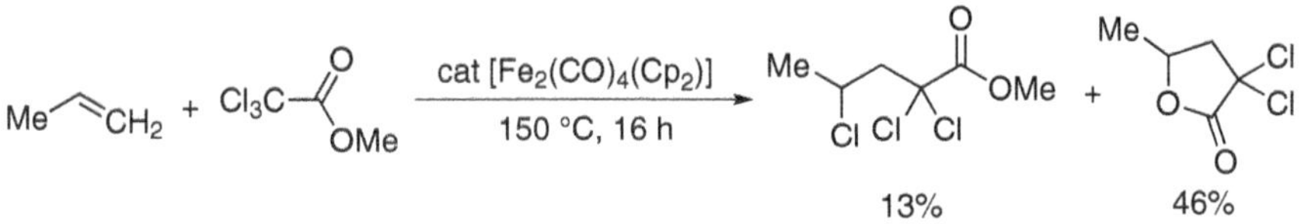

cat [Fe2(CO)4(Cp2)]
150 °C, 16 h
13%
46%

Scheme 49.

by methyltrichloroacetate, a mixture of products were obtained with lactone as the major product (Scheme 49) (76b).

Elemental iron also has been found to catalyze this type of transformation. A bromo radical is preferred over chloro in this halogen-atom transfer radical addition to alkene process (Scheme 50 DCE = 1,2-dichloroethane) (77, 78).

At elevated temperature, α-dichloroester also can give the same type of compound by breakage of the C$-$Cl bond. The proposed mechanism involves single-electron transfer (SET) from Fe(0) to the LUMO of the haloester with concomitant expulsion of halide anion (Scheme 51) (78).

An intramolecular Kharasch-type reaction also can be performed by using a catalytic amount of $FeCl_2[(P(OEt)_3]_3$ from α,α-dichloroester in both diastereoisomers (79). Relative yields of the isomers depend on the catalyst loading and time. When haloacid was used in place of haloester, lactone was observed as the major product (Scheme 52).

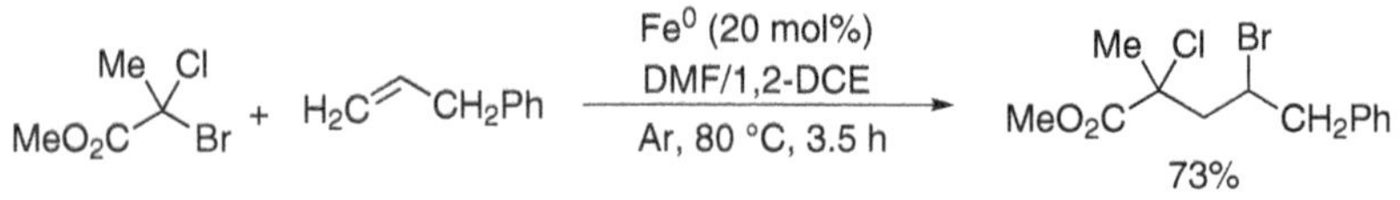

Fe0 (20 mol%)
DMF/1,2-DCE
Ar, 80 °C, 3.5 h
73%

Scheme 50.

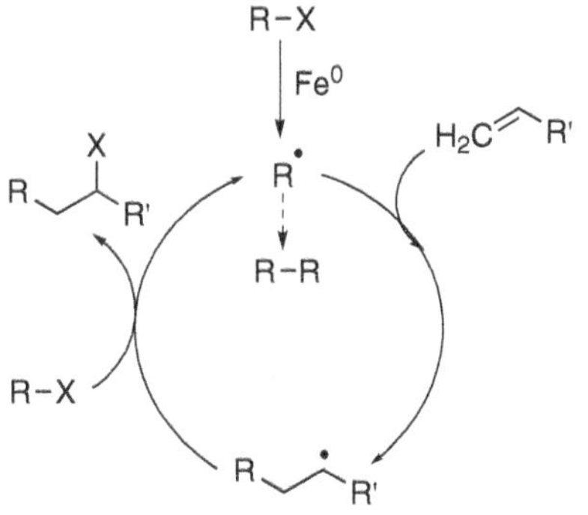

Scheme 51.

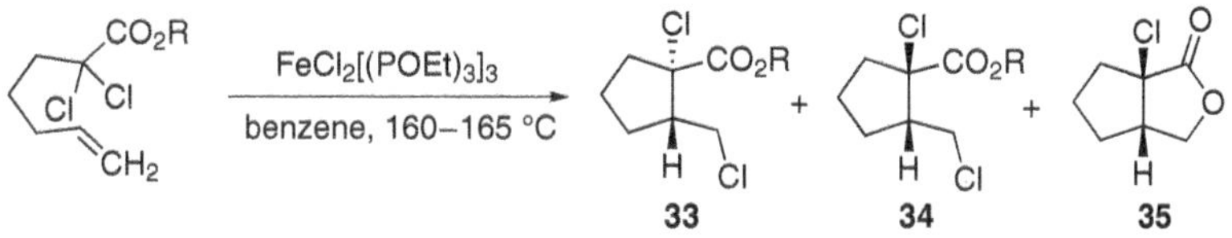

R	Catalyst (mol%)	Time (h)	**33** Yield (%)	**34** Yield (%)	**35** Yield (%)
Et	5.7	8	51	24	0
Et	8.7	20	32	51	0
H	4.0	24	0	0	66

Scheme 52.

III. THE C–C BOND FORMATIONS VIA C–H FUNCTIONALIZATION

A. The C–H Arylation

Biaryl structural motifs are ubiquitous in nature as well as in synthetic chemistry. From the days of the Ullmann reaction to traditional cross-coupling reactions, transition metals have played the key role in forming biaryls through C–C coupling. With advancement in coupling reaction chemistry, chemists have devised a method where one organometallic reagent has been replaced by unactivated, simple arenes thereby minimizing the related preactivation, extra synthetic step, and waste. This approach is termed a "direct arylation" strategy. In recent decades, progress in this field has been achieved primarily with second- and

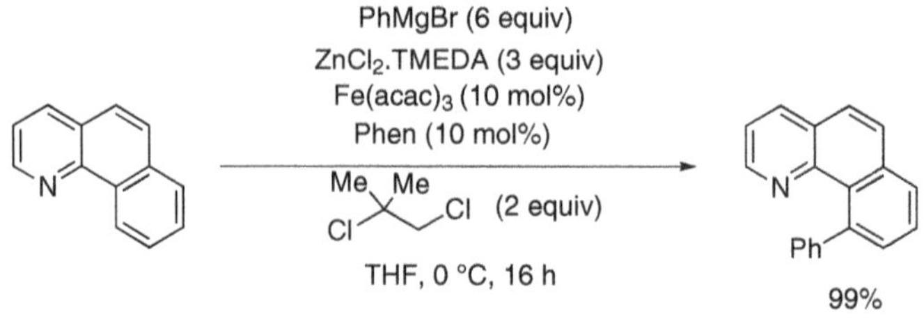

PhMgBr (6 equiv)
$ZnCl_2$.TMEDA (3 equiv)
$Fe(acac)_3$ (10 mol%)
Phen (10 mol%)
Me Me Cl Cl (2 equiv)
THF, 0 °C, 16 h
Ph
99%

Scheme 53. Iron-catalyzed direct arylation, where Phen = 1,10-phenanthroline.

third-row transition metal catalysts (e.g., Pd, Rh, Ru). The utilization of first-row transition metals, which are less expensive and environmentally benign, is challenging, though in recent years significant improvements have been observed with Cu, Ni, or Fe catalysts. In Section III.A.1, recent advances in iron-catalyzed direct arylation of unactivated arenes will be discussed briefly.

1. *Direct Arylation With Organometallic Reagents*

In 2008, Nakamura and co-workers (80) reported the first iron-catalyzed direct arylation with an *in situ* generated organozinc reagent in the presence of a suitable oxidant. Under optimized conditions, only 10 mol% $Fe(acac)_3$–phen resulted in an excellent yield of the desired product at a remarkably low temperature (0 °C) (Scheme 53).

The compounds $FeCl_2$ and $FeCl_3$ also were found to be equally reactive, but Fe $(acac)_3$ was chosen due to its ease of handling. Choice of a proper ligand system was extremely crucial, as $Fe(acac)_3$ alone was inefficient to catalyze the reaction. Bidentate nitrogen-based ligand phen was found to be the most efficient after extensive experimentations with various nitrogen-based bi- and tridentate ligands. Similarly, several dihalide oxidants were tested and 1,2-dichloroisobutane (DCIB) was found to promote this transformation in quantitative yield. Also, the combination of PhMgBr and $ZnCl_2$.TMEDA (in 2:1 ratio), was essential for the success of the reaction, as Ph_2Zn or PhZnBr were unable to promote the reaction.

Regioselectivity of arylation was determined by the presence of a nitrogen-atom in the substrates, which serves as the anchoring group for the metal. With benzoquinoline as the substrate, arylation was observed exclusively at the C10 position, as it was the only available position for arylation. With 2-phenylpyridine, an aryl group was inserted at the ortho position(s) only to give primarily the monoarylated product, though diarylation was also observed to some extent with symmetrical substrates (Scheme 54).

The variation of electronically different substituents on 2-phenylpyridine and in an aryl zinc reagent did not affect the yield of the desired product. However, the

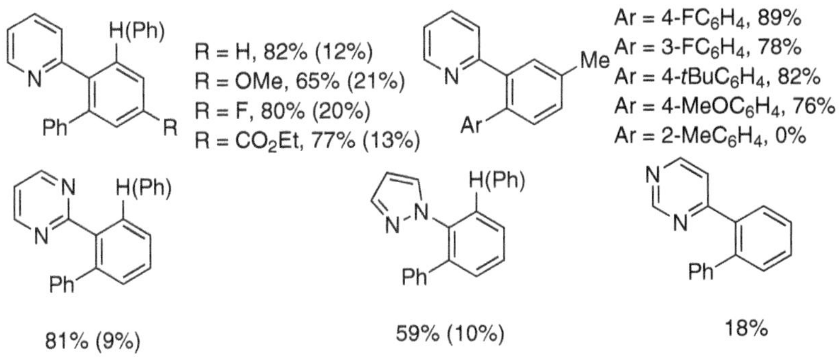

R = H, 82% (12%)
R = OMe, 65% (21%)
R = F, 80% (20%)
R = CO_2Et, 77% (13%)

Ar = 4-FC_6H_4, 89%
Ar = 3-FC_6H_4, 78%
Ar = 4-tBuC_6H_4, 82%
Ar = 4-MeOC_6H_4, 76%
Ar = 2-MeC_6H_4, 0%

81% (9%)

59% (10%)

18%

Scheme 54. Iron-catalyzed direct arylation of 2-aryl pyridine derivatives.

method was found to be very sensitive to the steric requirement of the substrate, as well as of the reagent. With a methyl group in the 3 position on the phenyl ring of 2-phenylpyridine, arylation occurred exclusively at the less hindered position and with 2-tolylzinc no product was observed.

In 2009, Nakamura and co-workers (81) also reported a related method, where the aryl group was selectively inserted into the ortho C–H bond of aryl imines (Scheme 55). In this report, dtbpy (4,4′-di-*tert*-butyl-2,2′-bipyridine) was used as the ligand in place of phen with $Fe(acac)_3$. The 1,2-dichloroisobutane (DCIB) was used as the oxidant to generate arylated ketimines or the corresponding ketones. Mono-arylated products were obtained for a number of aryl and heteroaryl imines in excellent yields. Interestingly, under the applied condition traditional leaving groups of cross-coupling chemistry (e.g., Cl, Br, OTs, and even OTf) were tolerated, thus exhibiting the possibility of orthogonal arylation or sequential arylation–functionalization. This protocol was also found to be sensitive to the steric requirements of the substrates like the previous method. Now, the mechanism of this reaction remained unclear and a detailed investigation is still required. Further, they enhanced the applicability of their methods by replacing DCIB by

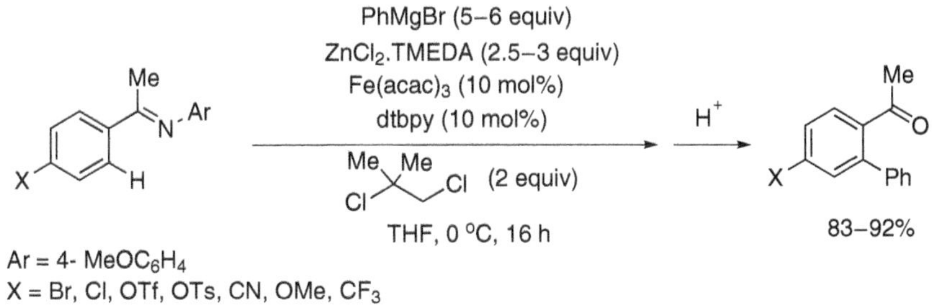

PhMgBr (5–6 equiv)
$ZnCl_2$.TMEDA (2.5–3 equiv)
$Fe(acac)_3$ (10 mol%)
dtbpy (10 mol%)
(2 equiv)
THF, 0 °C, 16 h
H^+

83–92%

Ar = 4- MeOC_6H_4
X = Br, Cl, OTf, OTs, CN, OMe, CF_3

Scheme 55. Iron-catalyzed directed ortho arylation of aryl imines.

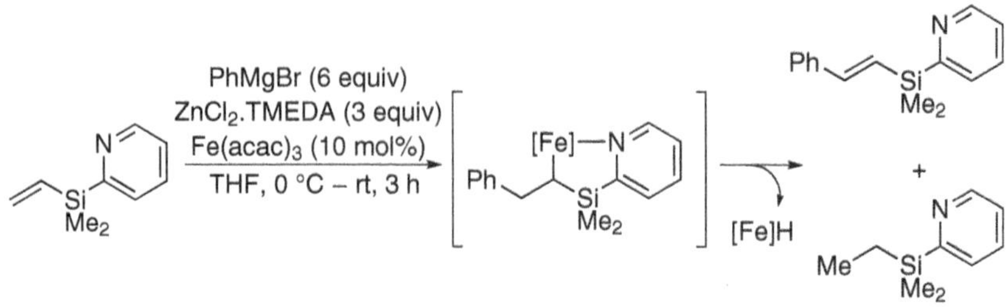

Scheme 56. Iron-catalyzed oxidative Heck reaction.

molecular oxygen as the stoichiometric oxidant, where sequential slow diffusion of oxygen into the reaction medium at 12-h intervals turned out to be advantageous.

In 2010, by applying similar conditions, Ilies et al. devised an oxidative Heck reaction in which the iron-catalyzed arylation of olefins was affected by organozinc and Grignard reagents through a proposed ferracycle intermediate (Scheme 56) (82). Here, 1-bromo-2-chloroethane was used as the optimal oxidant in place of DCIB to significantly suppress the reduced byproduct. Substrates without a directing group like 1-octene and styrene, as well as vinyl acetates or 2-vinyl pyridines, having a directing group, remained unreactive.

Encouraged by the success of iron-catalyzed sp^2 C–H functionalization, Nakamura and co-workers (83) further investigated the sp^3 C–H bond activation with an iron catalyst. In 2010, they reported an intermolecular coupling of aliphatic amines containing an *N*-(2-iodophenyl) methyl group (*N*-IBn) (acting as an internal trigger for C–H bond cleavage) and an organometallic reagent to generate an α-substituted product with an iron-based catalyst (Scheme 57). The reaction

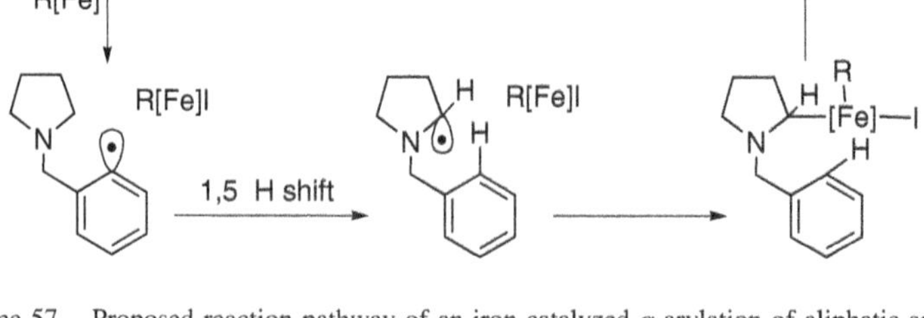

Scheme 57. Proposed reaction pathway of an iron-catalyzed α-arylation of aliphatic amines.

n = 1, 84%
n = 2, 81%
n = 3, 83%
n = 4, 82%

X = H,F; 91%
X = Cl; 88%
X = OMe; 90%

88%

79%

54% 70% 69% 47%

Scheme 58. Scope of an iron-catalyzed α-arylation of aliphatic amines.

proceeded by metal-mediated activation of a C–I bond, followed by intramolecular 1,5-hydrogen transfer and a subsequent C–C bond formation through a putative organoiron intermediate. Under optimized conditions, various cyclic and acyclic aliphatic amines reacted smoothly, though with a bromo analogue, lower yield of the desired product was observed (Scheme 58, Bn = benzyl). In the case of unsymmetrical aliphatic amines, arylation occurred preferentially at a more substituted position, thereby further enhancing the probability of a radical pathway.

Control experiments suggested iron as the sole catalyst, as reaction occurred successfully with other iron sources (e.g., $FeCl_3$ or $FeCl_2$), but in the absence of the catalyst, no conversion was observed. Labeling experiments clearly indicated the 1,5-H shift to be strictly intramolecular and no H/D crossover was observed. Also, a competition experiment ruled out the possibility of C–H bond cleavage as the rate-determining step of the reaction (Scheme 59).

Tran and Daugulis (84) also reported an iron-catalyzed deprotonative alkylation of five- and six-membered heteroarenes (e.g., thiophenes, furans, pyridines, and electron-deficient arenes) in the presence of a Grignard reagent with alkyl halides (Scheme 60). Grignard reagent primarily serves as a proton abstractor and a number of alkyl bromides and iodides undergo the transformation successfully.

Without using Grignard–organozinc reagents, iron mediated direct arylation of unactivated arenes with boronic acids was reported by Yu and co-workers (85) for the first time in 2008 (Scheme 61). A number of substituted boronic acids and different unactivated arenes were coupled efficiently at a moderately high temperature. The mechanism for this reaction still remained unclear, though the intermediacy of radical species was ruled out by this investigation. However, this method needs further modification, as it requires a stoichiometric amount of iron salt and excess unactivated arenes.

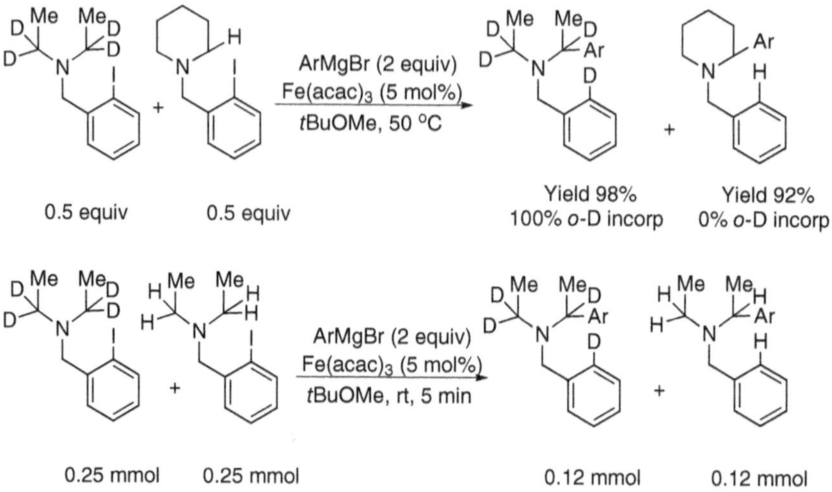

0.5 equiv 0.5 equiv

ArMgBr (2 equiv)
$Fe(acac)_3$ (5 mol%)
*t*BuOMe, 50 °C

Yield 98%
100% *o*-D incorp

Yield 92%
0% *o*-D incorp

0.25 mmol 0.25 mmol

ArMgBr (2 equiv)
$Fe(acac)_3$ (5 mol%)
*t*BuOMe, rt, 5 min

0.12 mmol 0.12 mmol

Scheme 59. Labeling experiments for mechanistic study.

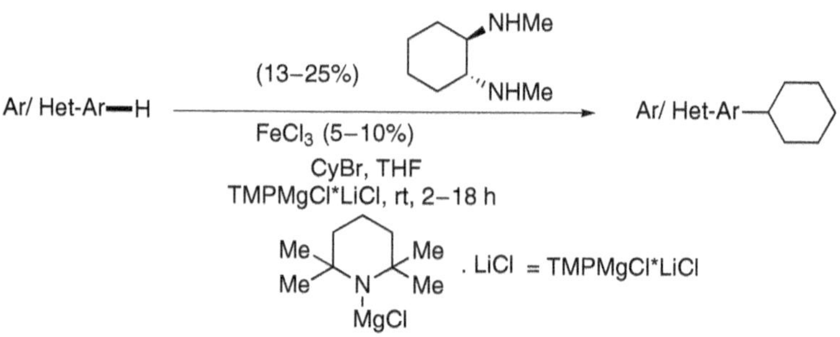

Ar/ Het-Ar—H

(13–25%)

NHMe

NHMe

$FeCl_3$ (5–10%)
CyBr, THF
TMPMgCl*LiCl, rt, 2–18 h

Ar/ Het-Ar—

Me Me Me Me N MgCl . LiCl = TMPMgCl*LiCl

Scheme 60. Iron-catalyzed deprotonative alkylation of heteroarenes (TMP = 2,2,6,6-tetramethylpiperidinyl).

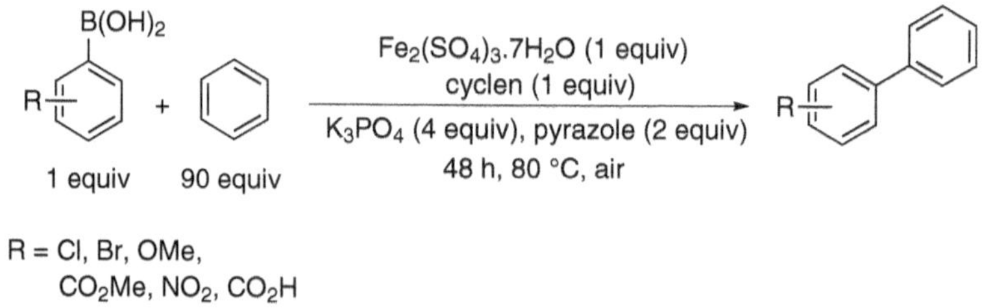

$B(OH)_2$

R

1 equiv 90 equiv

$Fe_2(SO_4)_3.7H_2O$ (1 equiv)
cyclen (1 equiv)
K_3PO_4 (4 equiv), pyrazole (2 equiv)
48 h, 80 °C, air

R

R = Cl, Br, OMe, CO_2Me, NO_2, CO_2H

Scheme 61. Iron-mediated direct arylation of arenes by arylboronic acids (cyclen = 1,4,7,10-tetraazacyclododecane).

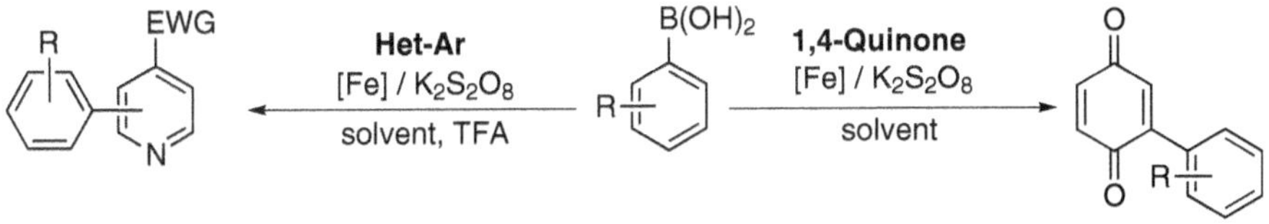

Scheme 62. Iron-catalyzed arylation of quinones and heteroarenes (TFA = trifluoroacetic acid).

Recently, Yu and co-workers (85) reported an iron mediated C–C coupling reaction in 2012, that started from aryl boronic acids and electron-deficient pyridines and quinones (Scheme 62). In the presence of a suitable oxidizing agent, $K_2S_2O_8$, a number of quinone derivatives were efficiently arylated by FeS. Further, no exogenous ligand is required for the success of this reaction. Notably, all these transformations significantly improve the applicability of iron catalysis to achieve green and sustainable reaction protocols in coupling chemistry.

2. *Direct Arylation With Aryl Halides*

In 2010, Lei and Charette (86) independently reported iron-catalyzed direct arylation of unactivated arenes with aryl halides. In Charette's report, arylation with aryl iodides and simple arenes was affected efficiently without the need of any directing group or stoichiometric organometallic reagents (Scheme 63) (86).

The optimized condition required 5 mol% $Fe(OAc)_2$ as the catalyst and bathophenanthroline (10 mol%) as the ligand in the presence of KO*t*Bu (2 equiv) at 80 °C. Even a 0.5 mol% catalyst loading was effective at an elevated temperature. Various aryl and heteroaryl iodides were found to give the desired product in moderate-to-excellent yields. Electron-rich iodides were found to be more reactive, as even at rt moderate yields were obtained with a prolonged reaction time. Notably, the protocol was efficient only with aryl iodides, as aryl bromide proved to be less reactive while aryl chloride remained unreactive.

In a preliminary mechanistic investigation, a KIE of 1.04 was found thus ruling out the possibility of C–H bond cleavage as the rate-determining step. A radical pathway was proposed and an iron(II)–iron(III) redox cycle was thought to be operative (Scheme 64).

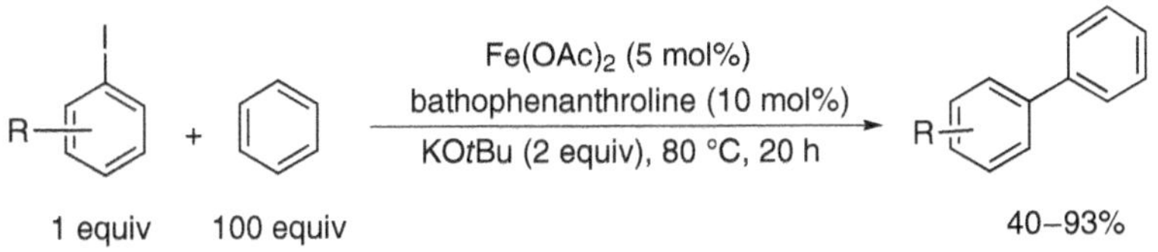

Scheme 63. Iron-catalyzed direct arylation of arenes with aryl iodides.

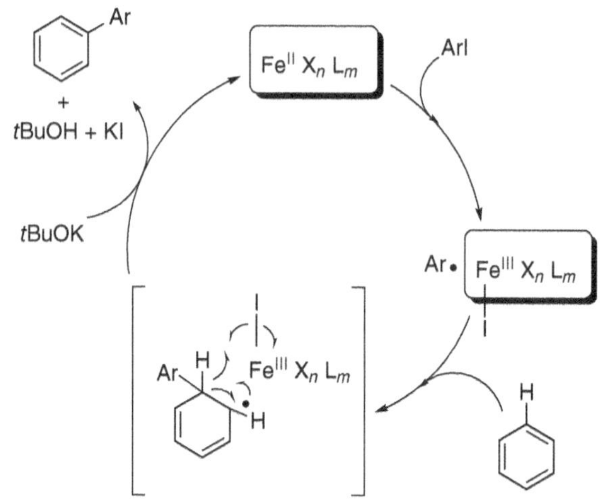

Scheme 64. Proposed catalytic cycle.

In this proposed mechanism, one-electron oxidation of the iron center activated the aryl–halogen bond and generated a radical species. In the next step, a radical addition on arene was followed by abstraction of a halogen atom from the metal center and KO*t*Bu quenched the resulting HI to form *t*BuOH, which had been detected in the reaction medium. This radical pathway was further supported by reactions with radical scavengers (e.g., TEMPO, 2,2,6,6-tetramethyl-piperidin-1-yl)oxyl, and galvinoxyl), which practically inhibited the reaction. Control experiments clearly indicated the prominent role of iron as the sole catalyst regardless of its purity and no product was obtained in the absence of $Fe(OAc)_2$.

A similar method was demonstrated by the Lei and co-workers (87) group in the same year, where $FeCl_3$ was the catalyst of choice along with *N,N'*-dimethylethylinediomine (DMEDA) as the ligand (Scheme 13). At moderate temperature, aryl halides were efficiently reacted depending on the nature of the base used. With a strong base, LiHMDS (2 equiv), the less reactive aryl bromides exhibited good reactivity, while aryl chlorides were found to be moderately reactive (Scheme 65). In the presence of excess KO*t*Bu (3 equiv), more reactive aryl iodides reacted efficiently to form the desired product in satisfactory yields. Electron-rich aryl bromides gave a better result compared to electron-deficient ones. Also, substituent in the ortho position resulted in a lower yield, indicating that steric hindrance had a detrimental effect on reactivity. Notably, the same trend regarding the stereoelectronic effect of the substituents was observed with aryl iodides in Charette's report (86).

Detailed mechanistic studies were not conducted in this case, however, preliminary studies ruled out any intermediacy of a benzyne analogue. A labeling experiment with [d_6]-benzene yielded a KIE value of 1.7 and thus a radical

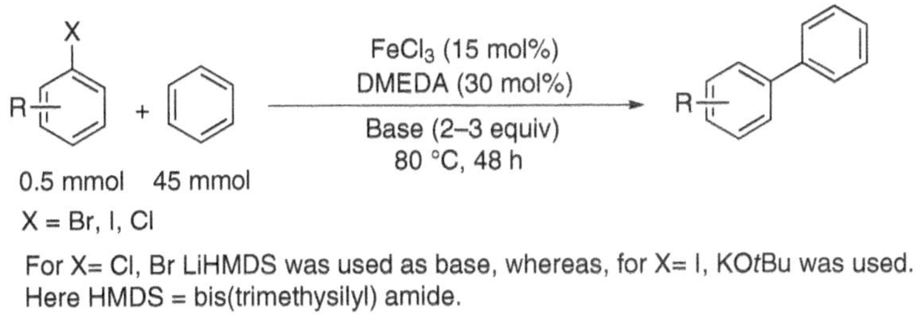

FeCl3 (15 mol%)
DMEDA (30 mol%)
Base (2–3 equiv)
80 °C, 48 h
0.5 mmol 45 mmol
X = Br, I, Cl
For X= Cl, Br LiHMDS was used as base, whereas, for X= I, KO*t*Bu was used.
Here HMDS = bis(trimethysilyl) amide.

Scheme 65. Iron-catalyzed direct arylation of benzene with aryl halides.

pathway was considered. Also, in accordance with the report of Buchwald and Bolm (88), reaction was performed with ultrapure $FeCl_3$ under standard conditions and the same results were obtained. The arene coupling partner was required to be in excess for the reaction. In conclusion, in direct arylation chemistry these methods clearly indicate the significant development where less expensive iron salts exhibited unprecedented reactivity.

B. The C–C Bond Formation Via Cross-Dehydrogenative Coupling

A cross-coupling reaction is one of the most powerful tools in organic synthesis. Several approaches toward the desired cross-coupled products are known. The straightforward approach will be direct coupling of two different C–H bonds instead of using either organic halides–pseudohalides or organometallic partners or both (89). This technique is popularly known as cross-dehydrogenative coupling (CDC). Li (90) made a pioneering contribution in CDC by using a first-row transition metal catalyst. After their initial work with copper, first iron-catalyzed CDC was reported in 2007 and an iron catalyst was found to be more efficient than a copper catalyst (91).

1. The CDC Between Two sp^3 C–H Bonds

Intially, diphenylmethane was used as one of the coupling partners for the double activation at the benzylic position. Similarly, another coupling partner was 1,3-dicarbonyl with an activated α-C–H bond. It was found that $FeCl_2$ was best among various iron salts and di(*tert*-butyl)peroxide (DTBP) replaced *tert*-butyl hydrogen peroxide (TBHP) with a further increase in yield (Scheme 66) (91).

The proposed mechanism involved homolysis of DTBP and a single–electron transfer from iron(II) with generation of a *tert*-butoxyl radical and a *tert*-butoxide anion, which was then followed by generation of an active benzyl radical and iron(III) enolate by proton abstraction. Then electrophilic radical attack led to the desired cross-coupled product with the regeneration of an iron(II) species to facilitate the catalytic cycle (Scheme 67) (91).

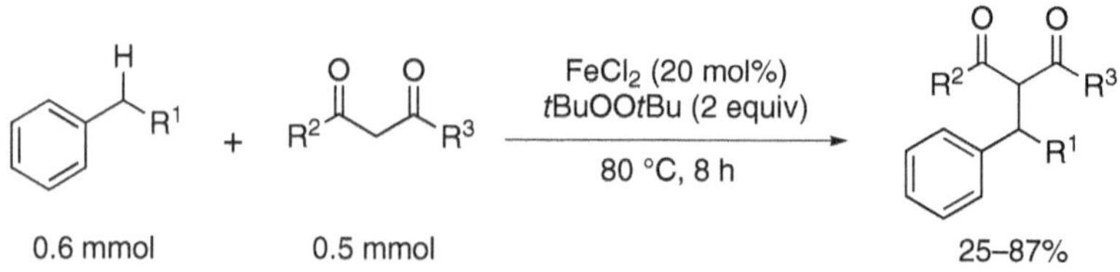

Scheme 66.

Instead of C_{sp^3}–H at the activated benzylic position, the α-C_{sp^3}–H bond of nitrogen in amine or oxygen in ether can also be activated (90, 92) by using an iron catalyst. In this case, $FeCl_2$, $FeBr_2$, $Fe(OAc)_2$, and $Fe_2(CO)_9$ showed a catalytic property, but $Fe_2(CO)_9$ was found to be the best in combination with DTBP as the oxidizing agent. Both acyclic and cyclic ether along with thioether and tertiary amine gave a moderate-to-excellent yield of the desired CDC product with 1,3-dicarbonyl compounds (Scheme 68) (93).

The presence of nitrogen or oxygen at the α-position was crucial due to the formation of iminium or oxonium ions as the key intermediate via SET. Further attack from a carbon nucleophile led to the product (Scheme 69) (90).

Extension of this work by Li et al. (94) using *N,N*-dimethylaniline as the source of the methylene group produced dialkylation with 1,3-dicarbonyl and a bridged bis-1,3-dicarbonyl compound was generated (Scheme 70).

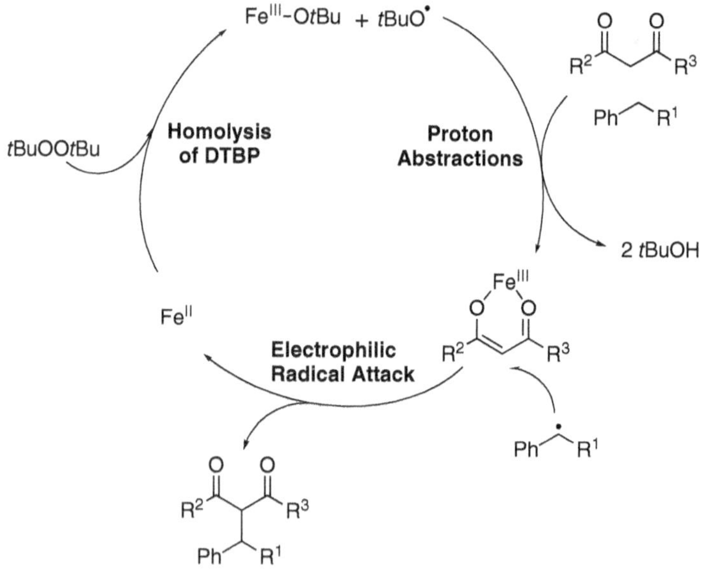

Scheme 67.

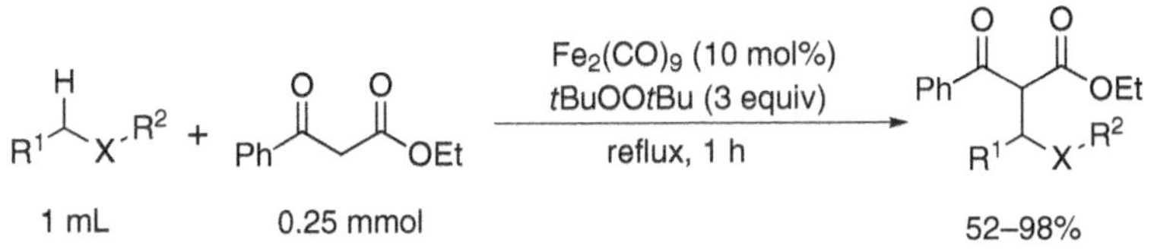

Scheme 68.

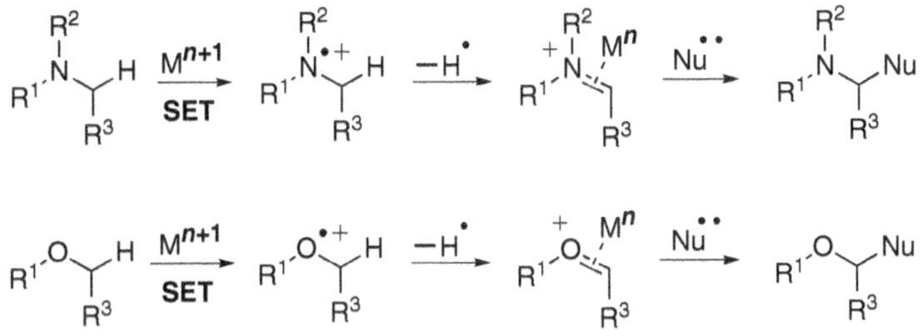

Scheme 69.

Several possible pathways were proposed for this transformation. In Li's proposal, first *N,N*-dimethylaniline underwent CDC with one molecule of the 1,3-dicarbonyl compound. Then, in one route direct S_2N attack might occur where *N*-methylaniline acted as a leaving group. In other route, cope elimination led to an *α,β*-unsaturated carbonyl moiety, where further Michael addition of a second 1,3-dicarbonyl compound led to the final product (94). Also *in situ* generation of formaldehyde from *N,N*-dimethylaniline and further coupling with two molecules of 1,3-dicarbonyl compound cannot be ruled out (Scheme 71).

Activation of the alkylic $C_{sp^3}-H$ bond is most challenging due to low reactivity, selectivity problems, and the lack of a coordination site for the transition metal catalyst for $C_{sp^3}-H$ bonds. In all previous reports, both coupling partners had specific activated $C_{sp^3}-H$ bonds to participate in cross-coupling reactions. Till now, only one successful example with iron had been reported, where unactivated alkane was used as a substrate (Scheme 72) (95).

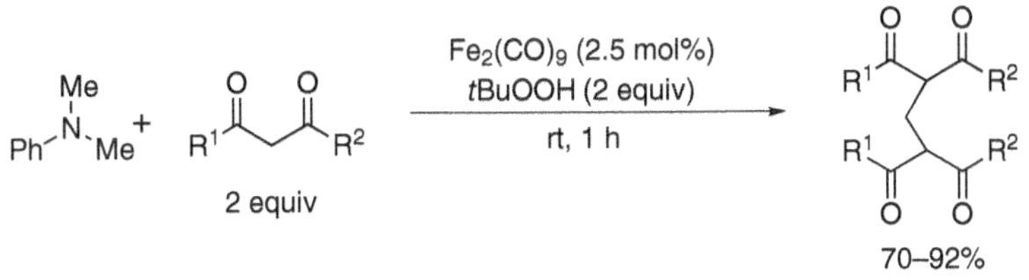

Scheme 70.

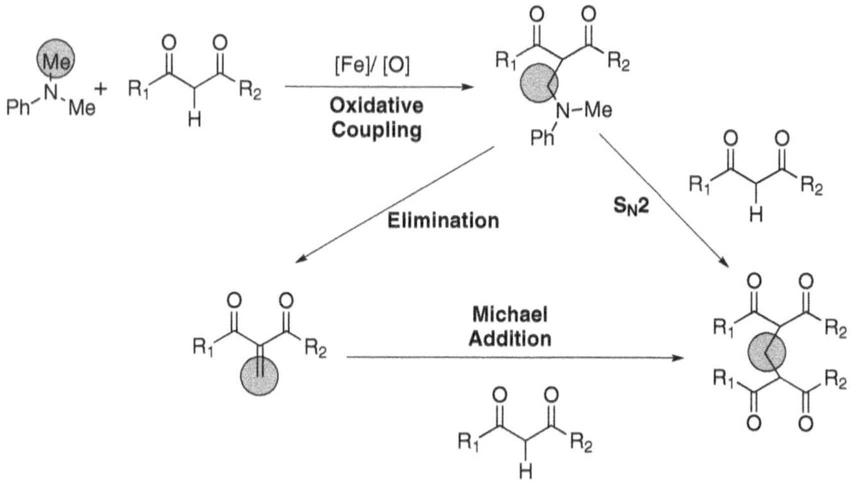

Scheme 71.

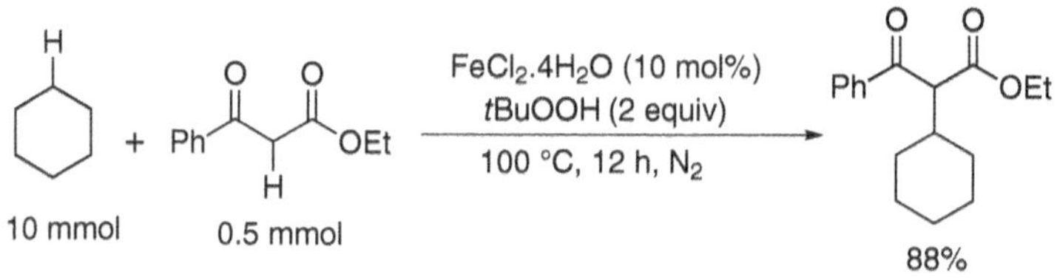

Scheme 72.

2. *The CDC Between sp^3 and sp^2 C–H Bonds*

In accordance with the mechanism of CDC between the 1,3-dicarbonyl compound and dibenzylmethane, as in Li's report, other nucleophiles can be considered to quench the intermediate formed from dibenzylmethane through the iron-catalyzed SET process. Shi and co-workers (96) reported the cross-dehydrogenative arylation (CDA) of dibenzylmethane with electron-rich arenes. Use of DDQ (dichlorodicyanobenzoquinone) in place of DTBP as oxidant gave a significant increase in yield, and DCE as solvent reduced the amount of dibenzylmethane used for the reaction. Excellent regioselectivity was observed for a class of electron-rich arenes, but double CDA was also observed for more electron-rich arenes (Scheme 73 DDQ = 2,3-dichloro-5,6-dicyano-1,4-benzoquinone).

Based on preliminary mechanistic evidences by deuterium labeling, KIE, and the side products obtained, first iron(II) assisted SET oxidation was proposed to produce a dibenzyl radical and reduced quinone. Then again, dibenzyl cation was generated

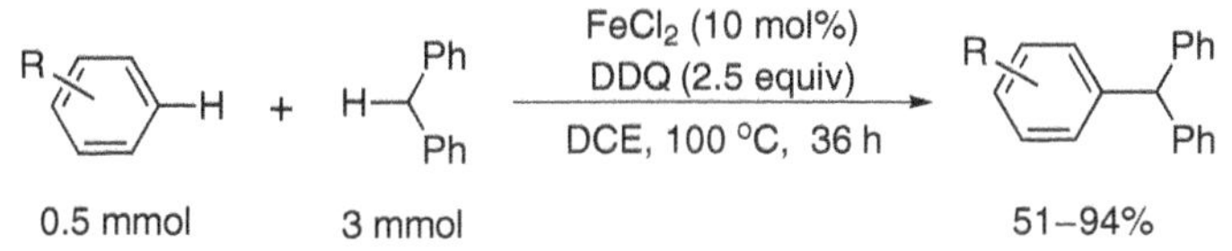

Scheme 73.

from dibenzyl radical through SET with regeneration of an iron(II) species. Finally, attack of this dibenzyl cation to an electron-rich arene followed by proton abstraction by reduced hydroquinone afforded the coupling product (Scheme 74) (96).

Similarly, electron-rich alkenes were considered as the active nucleophile in Shi's other report to quench the benzyl radical or cation. Surprisingly, only styrene was found to give the coupled product, but another styrene derivative did not give the desired product.

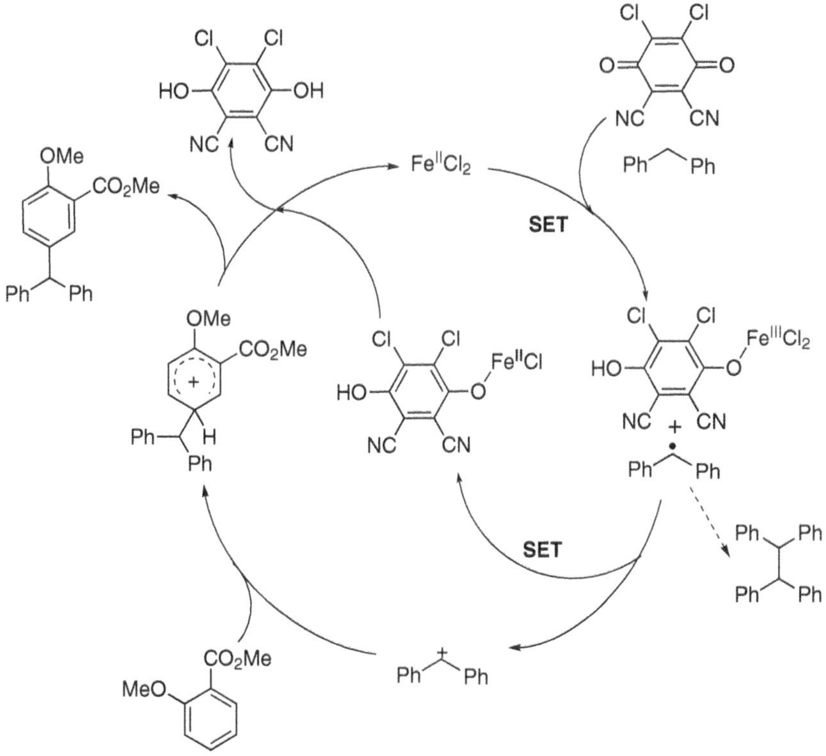

Scheme 74.

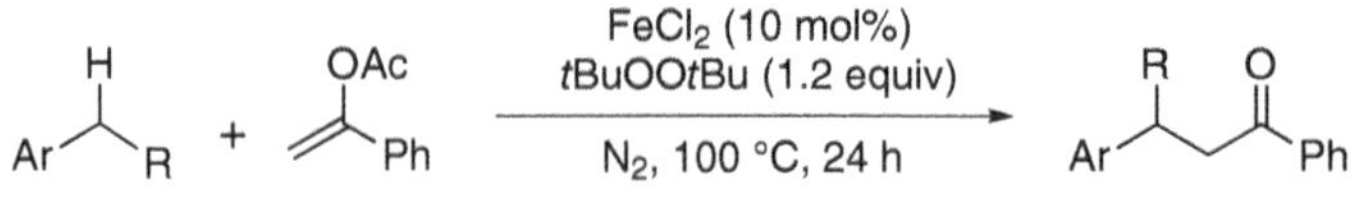

Scheme 75.

Later, 1-aryl-vinylacetate was found to give the coupled keto product with removal of the acetate group. The $FeCl_2$ and DTBP (*t*BuOO*t*Bu) combination was found to be excellent for this type of transformation (Scheme 75) (97).

Two possible catalytic pathways were proposed with SET from iron as the key step. In pathway A, a *tert*-butoxyl radical and a benzyl radical attacked vinyl acetate simultaneously to give an iron-coordinated radical intermediate, which upon β-cleavage gave the desired coupled product (Scheme 76). In pathway B, a benzyl radical was further oxidized into a benzyl cation through SET and was

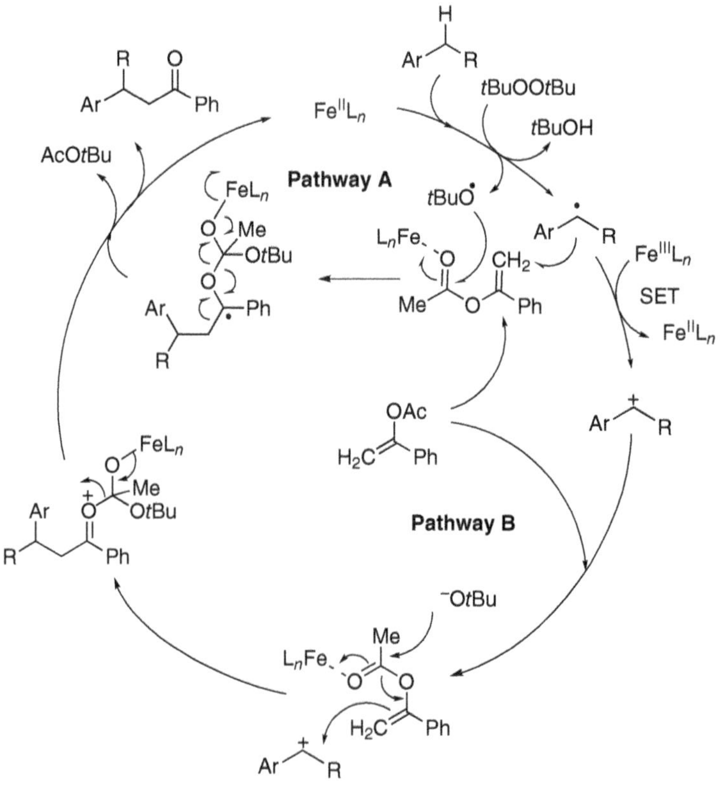

Scheme 76.

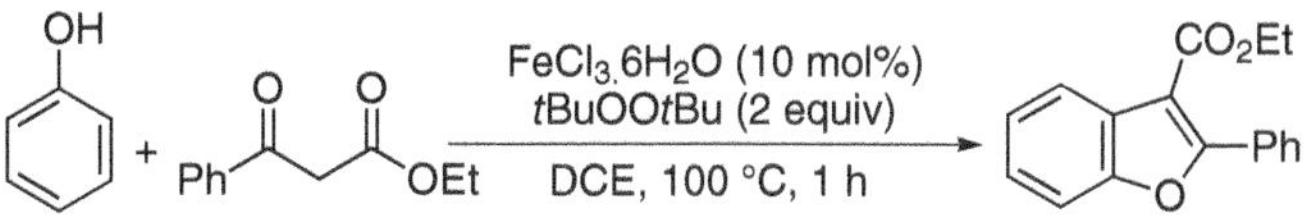

Scheme 77.

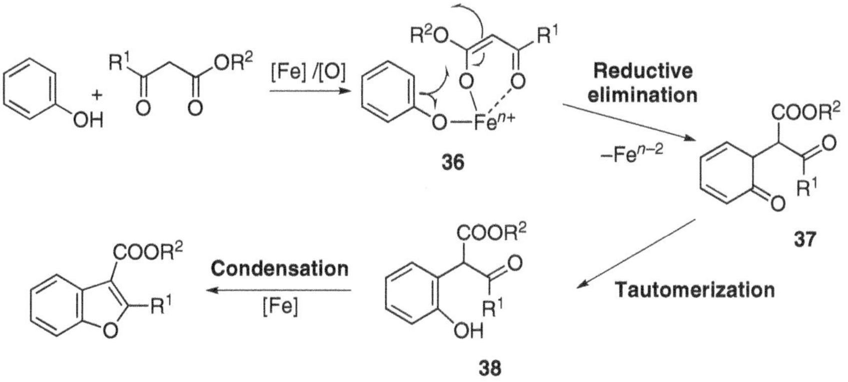

Scheme 78.

quenched by vinyl acetate. A proton-abstraction process may be involved in the rate-determining step in either pathway as indicted by an intermolecular isotopic competitive study ($K_{H/D} = 2.4$) (Scheme 76) (97).

Yet another breakthrough came from Li and co-workers (98), wherein an iron-catalyzed CDC reaction and subsequent annulation of phenol and β-keto ester led to the polysubstituted benzofuran. Notably, various iron salts did not alter the yield too much, but the presence of water in an iron catalyst was crucial for a good yield (Scheme 77).

The hypothesized mechanism showed formation of an Fe^{n+} chelated complex, followed by reductive elimination, tautomerization, and condensation leading to the final product (98). Formation of the desired product from prepared compound **36** supported this hypothesis (Scheme 78).

3. The CDC Between sp^3 and sp C–H Bonds

Besides aryl- and alkene-type substrates, terminal alkynes were also investigated. Instead of 1,3-dicarbonyl systems, an α-sp^3 C–H bond from nitrogen was found effective in CDC with terminal alkynes (Scheme 79) (99).

Both aromatic and aliphatic tertiary amines led to the desired product with $FeCl_2$ and DTBP combination. Iron-catalyzed SET led to generation of an iminium

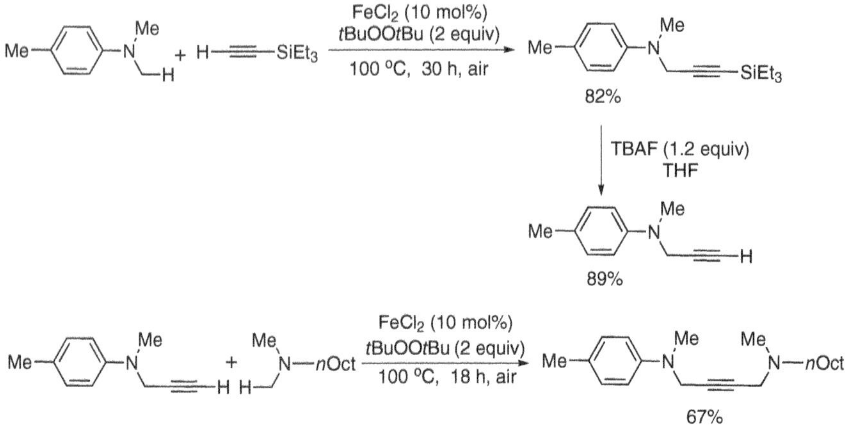

Scheme 79.

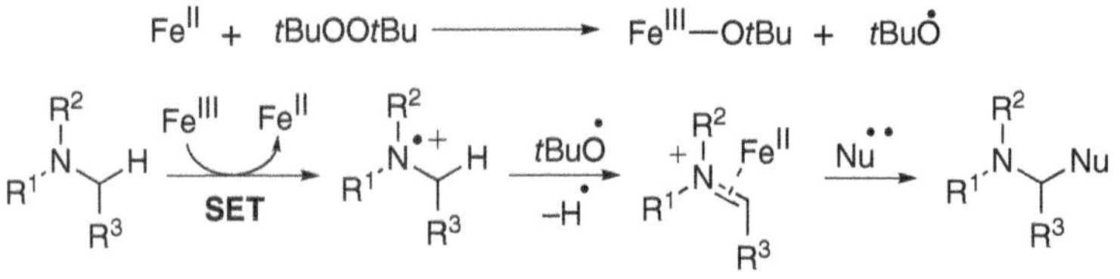

Scheme 80.

ion intermediate and quenching by an alkynyl carbanion proposed to be involved in this transformation (Scheme 80).

C. The C−C Bond Formation via Cross-Decarboxylative Coupling

Another potential approach for C−C bond formation is through C−H transformation and intermolecular–intramolecular decarboxylative coupling. The ready availability, low toxicity, low cost of carboxylic acids made it attractive as one of the coupling partners for selective cross-coupling (96, 100). A proline derivative was found to couple decarboxylatively with substituted β-naphthols in the presence of $FeSO_4$ and DTBP (101). Besides β-naphthols, electron-rich α-naphthols and indoles were also suitable nucleophiles for this decarboxylative coupling (Scheme 81).

The proposed mechanism involved homolysis of peroxide in the presence of iron(II) salts, followed by hydrogen-atom abstraction by a *tert*-butyl radical from the carboxylic acid to initiate decarboxylation with generation of 2-pyrrolidinyl

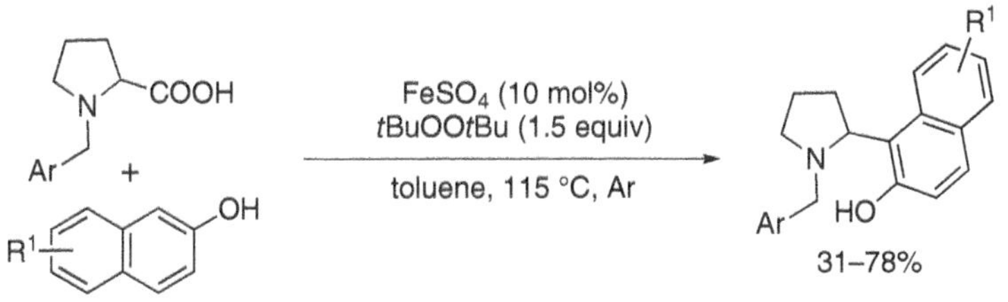

Scheme 81.

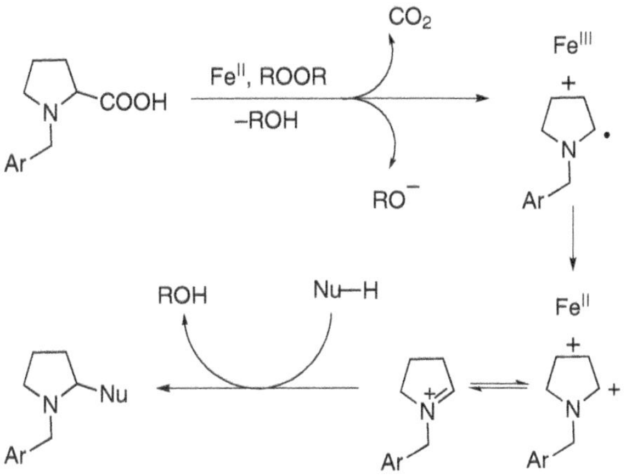

Scheme 82.

radical. In the presence of Iron(III), this radical was further oxidized to the 2-pyrrolidinium cation and was attacked by nucleophile generated by proton abstraction from electron-rich naphthols or indoles (Scheme 82) (101).

D. The C–C Bond Formation via Alkene Insertion

Another design for C–C bond formation relating to this well-known technique of radical or cation generation by iron-catalyzed SET processes followed by quenching by an electron-rich compound could be using unsaturated alkenes or alkynes instead of electron-rich arenes in the absence of any radical quencher. With the use of the relatively high activity of the α-C–H bond of a heteroatom, Tu and co-workers (102) reported this type of radical quenching using terminal alkenes (Scheme 83).

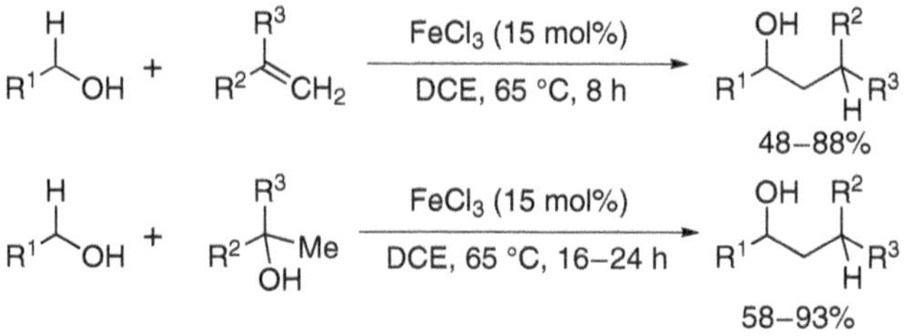

Scheme 83.

Controlled experiments, deuterium labeling, and cross-over experiments were conducted to give an idea about the possible mechanism. Failure of aldehyde to give the desired product indicated that the oxidation–hydroacylation–reduction pathway was not followed (Scheme 84, Eq. 1). Quantitative deuterium incorporation in alkene (Scheme 84, Eq. 2) and deuterium scrambling (Scheme 84, Eq. 3)

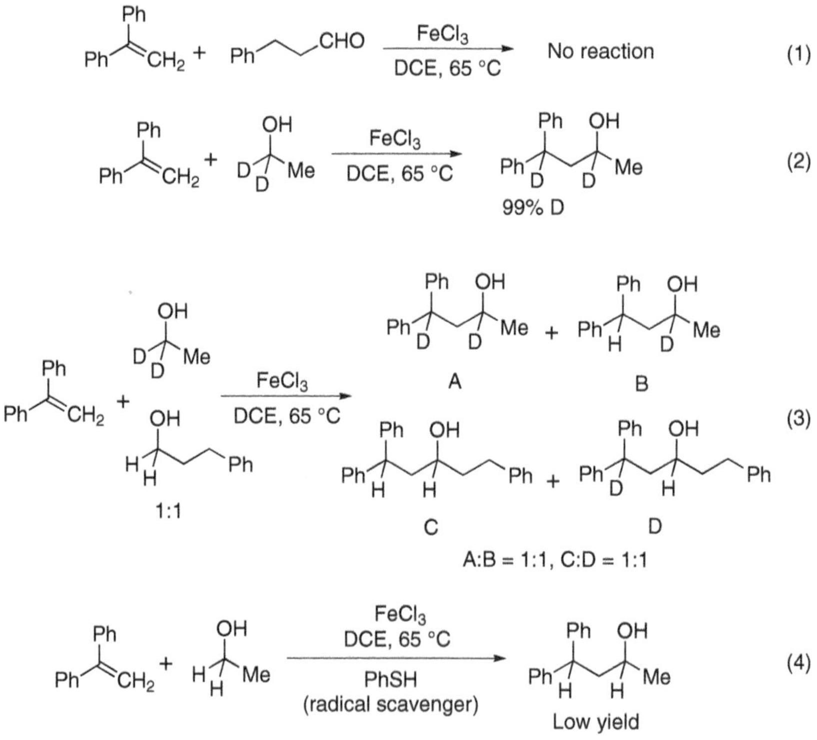

Scheme 84.

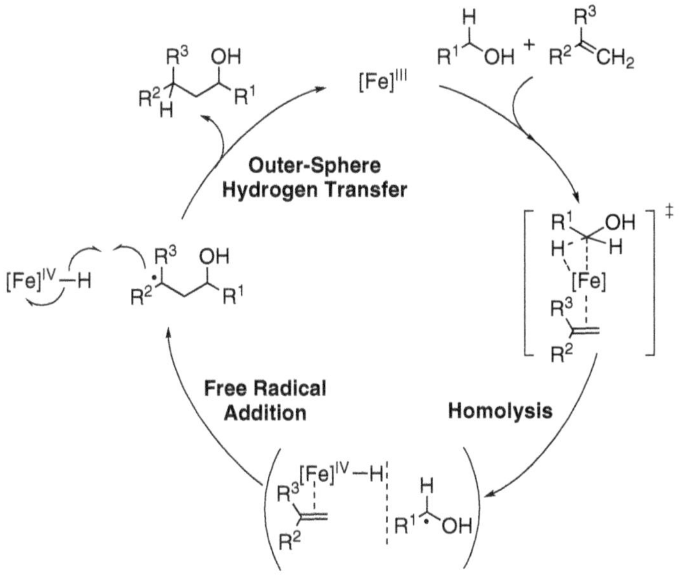

Scheme 85.

indicated that hydrogen transfer from alcohol to alkene occurred in discretely intermolecular fashion. Again low yield of the desired product in the presence of a radical scavenger (Scheme 84, Eq. 4) implied that radical pathway was followed.

In accordance with all of these results, the proposed catalytic cycle showed formation of a Fe^{IV}–H active species, which was involved in an outer-sphere-type hydrogen radical transfer leading to the desired coupling product (Scheme 85) (102).

E. Oxidative Coupling of Two C–H Bonds

Irrespective of the different methods available for iron-catalyzed C–C coupling via C–H bond activation, direct application in total synthesis was restricted due to lack of control in selectivity and efficiency. Baran et al. (103) reported examples of iron mediated C–C bond formation by C–H transformation, which was directly applied in total synthesis (Scheme 86, LDA = lithium diisopropylamide).

In the total synthesis of anti-cancer natural products (e.g., stephacidins), Baran et al. (103) reported an unprecedented intermolecular oxidative C–C bond formation using $Fe(acac)_3$ (103). In order to construct the bicyclic core of stepacidin alkaloids, this type of C–C coupling was critical especially with proper chemo- and stereoselectivity. A unique oxidation potential of the iron-based

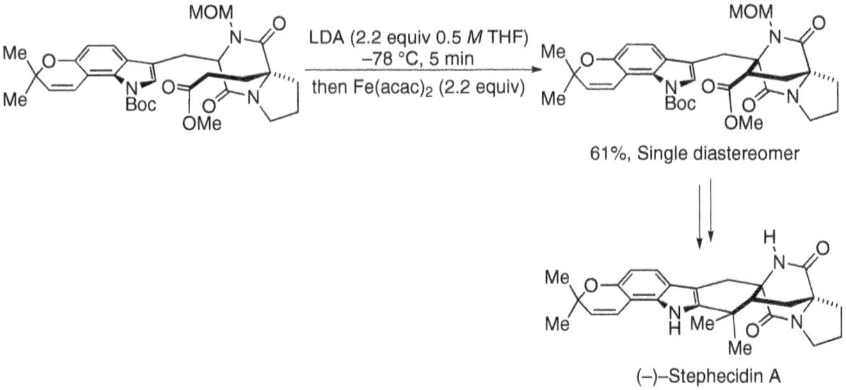

LDA (2.2 equiv 0.5 *M* THF)
−78 °C, 5 min
then $Fe(acac)_2$ (2.2 equiv)

61%, Single diastereomer

(−)–Stephecidin A

Scheme 86.

oxidant was proposed to control chemoselectivity. An easy formation of an iron-chelated transition state was proposed to control high stereoselectivity.

In another example, they reported iron-mediated intermolecular oxidative heterocoupling of two different carbonyl species to give 1,4-dicarbonyl species with a specific stereoselectivity due to the different coordinating nature of iron compared to copper (Scheme 87) (104). Mechanistic studies revealed nontemplated iron(III) mediated heterodimerization in this process with suppression of other possible side products from homodimerization, cross-homo Claisen condensation, cross-homo-aldol condensation, overoxidation, dehydrogenation, and α-oxidataion.

The same group reported another example of oxidative coupling between indinones and carvones, during the search for an alternate protocol to synthesize the core unit of hapalindoles, fischerindoles, and welwitindolinones, a class of indole-based heterocycles with a monoterpene unit attached at the C3 position (Scheme 88, MOM = methoxymethyl ether). It was found that with the indole type of moiety copper was best suited, where as for indinone the Fe based Fe(*t*BuCOCHCOMe)$_3$ was found to give the highest yield (105).

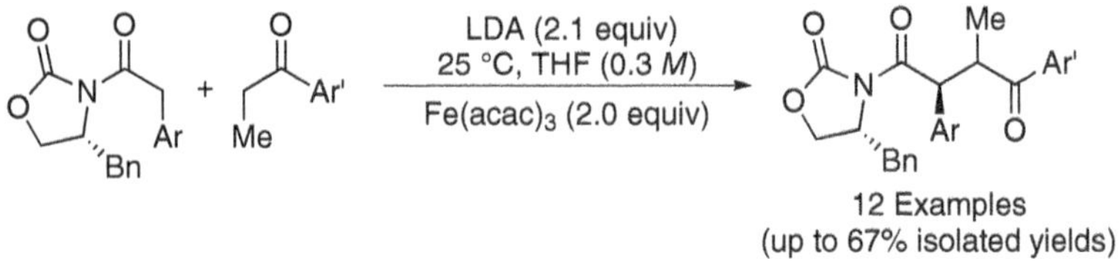

LDA (2.1 equiv)
25 °C, THF (0.3 *M*)
$Fe(acac)_3$ (2.0 equiv)

12 Examples
(up to 67% isolated yields)

Scheme 87.

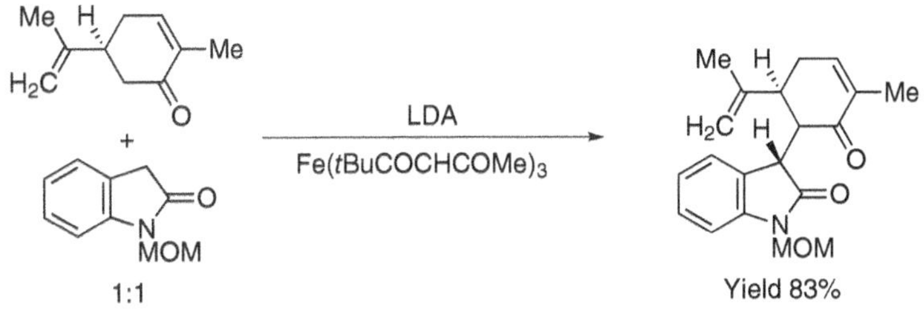

Scheme 88.

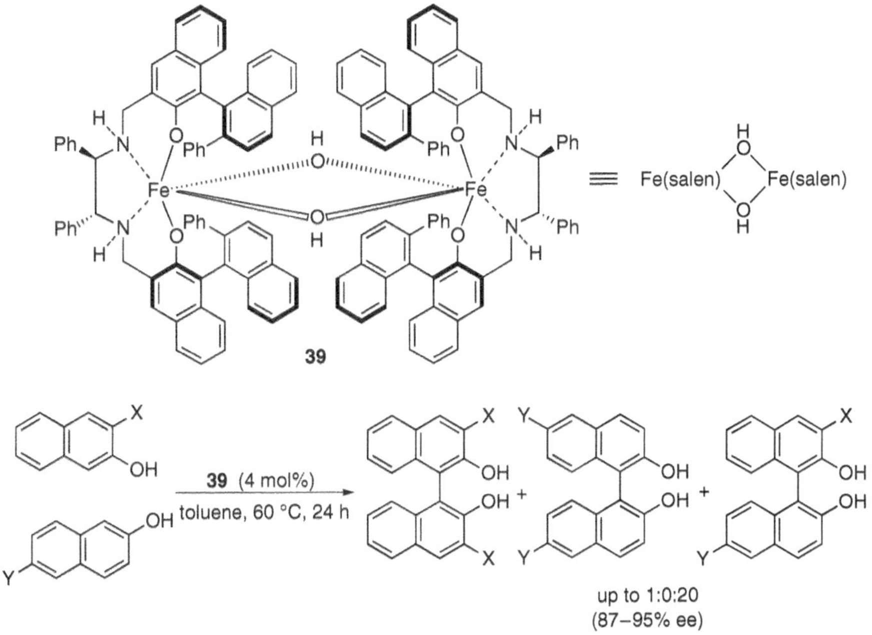

Scheme 89.

Another case of oxidative cross-coupling of β-naphthols to give C1-symmetric 1,1′–bi-2-naphthols (BINOLs) was reported by Katsuki and co-workers (106) using the [Fe(salen)] complex (**39**) as an active catalyst (Scheme 89). High yield of cross-coupled product compared to competing homocoupled product, as well as high stereoselectivity proved this method to be advantageous compared to existing stereoselective homocoupling methodologies.

IV. THE C–H BOND OXIDATION

A. Hydroxylation

Various transformations in nature occur via C–H oxidation C–H activation. Different transition metal based enzymes accomplish C–H oxidation via the SET process (107). Most of these enzymes contain "Cu" and "Fe" as a metal cofactor. In organic transformation iron-catalyzed C–H oxidation was reported one century ago, which was summarized as Gif chemistry, Fenton chemistry, and other non-heme mimic systems (108). Recent iron-catalyzed C–H oxidations were mostly focused on sulfide oxidation, epoxidation, and olefin dihydroxylation.

Recent breakthroughs in iron-catalyzed C–H oxidation made this field more fascinating. In 2007, Nakanishi and Bolm (109), developed an efficient protocol for the C–H oxidation of benzylic compounds to their corresponding ketone under mild reaction conditions (109). The protocol was clean and efficient as it used an aqueous of TBHP as an oxidant. This reaction was compatible with various functionalized benzylic compounds (Table XV).

TABLE XV
Iron-Catalyzed Benzylic Oxidation[a]

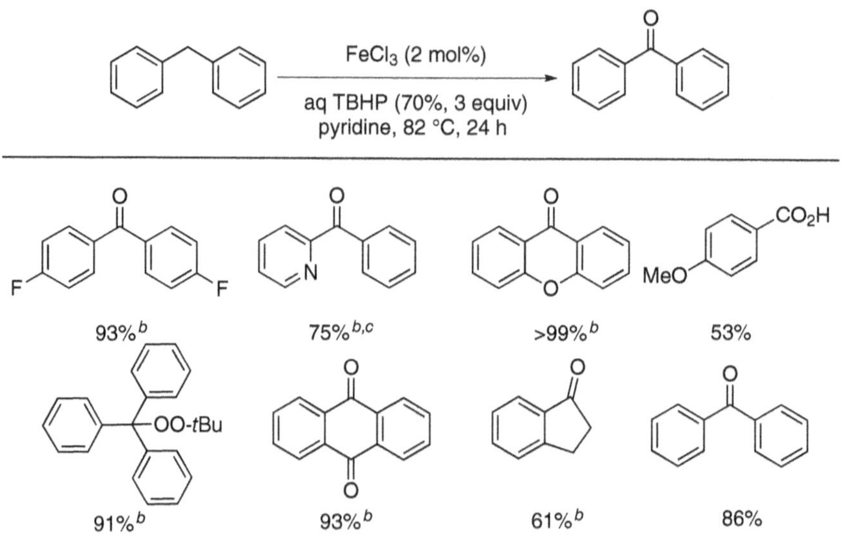

$FeCl_3$ (2 mol%)
aq TBHP (70%, 3 equiv)
pyridine, 82 °C, 24 h

F F
93%[b]

N
75%[b,c]

O
>99%[b]

MeO CO_2H
53%

OO-tBu
91%[b]

93%[b]

61%[b]

86%

[a]The compounds $FeCl_3.6H_2O$ (2 mol%), TBHP (70%) in H_2O (3 equiv), and pyridine, 82 °C, 24 h.
[b]All products were identified by comparison of their analytical data with those of previous reports or commercial materials.
[c]At 110 °C, TBHP (70%) in H_2O (6 equiv).

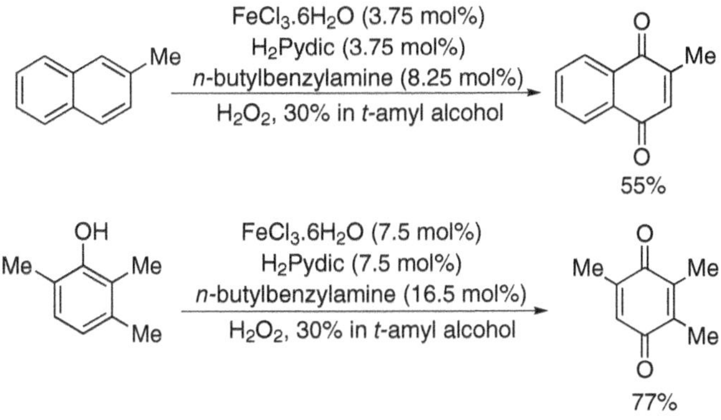

$FeCl_3.6H_2O$ (3.75 mol%)
H_2Pydic (3.75 mol%)
n-butylbenzylamine (8.25 mol%)
H_2O_2, 30% in *t*-amyl alcohol
55%

$FeCl_3.6H_2O$ (7.5 mol%)
H_2Pydic (7.5 mol%)
n-butylbenzylamine (16.5 mol%)
H_2O_2, 30% in *t*-amyl alcohol
77%

Scheme 90. The C–H hydroxylation by a non-heme iron catalyst.

Interestingly, when 4-methylanisole and diphenylmethanol were tested, corresponding overoxidation products, 4-methoxy benzoic acid and benzophenone, were obtained in high yields. Surprisingly the oxidation of triphenyl afforded *tert*-butyl triphenyl-peroxide in 91% yield instead of the corresponding alcohol product (Table XV).

Later in 2010, Beller and co-workers (110) reported an oxidation protocol for the sp^2 C–H bonds of phenols and arenes with a three-component catalyst system consisting of $FeCl_3.6H_2O$, pyridine-2, 6-dicarboxylic acid (H_2pydic), and *n*-butylbenzylamine. A green and nontoxic oxidant H_2O_2 was used (110). Under optimum conditions the oxidation of 2,3,6-trimethylphenol (TMP) and 2-methylnaphthalene gave 77% and 55% isolated yields, respectively (Scheme 90).

Notably, the 2,3,6-trimethylbenzoquinone was the key intermediate for vitamin E synthesis, whereas the oxidation of 2-methyl naphthalene provided menadione (viz, vitamin K_3, menanaphthone), which serves as a precursor for the synthesis of various vitamin K derivatives (111).

It was a long-standing goal to develop a highly chemoselective and efficient alkane C–H oxidation by non-heme mimic systems. Recent improvements were more focused on developing synthetic models using a "green" oxidant (e.g., oxygen, air, or some hyperoxides). Que and co-workers (112) significantly contributed to non-heme systems in terms of building a mechanistic frame to understand metal-based C–H hydroxylation in non-heme systems. They reported the selective C–H alkane hydroxylation by the non-heme iron catalyst/H_2O_2 in 1997. Que and co-workers (113) used multidentate nitrogen-containing ligand TPA [tris(2-pyridylmethyl)amine]. Later on, they developed different types of multidentate nitrogen-containing ligands with a modification in steric and electronic properties. These catalysts showed significantly improved activity toward C–H hydroxylation. Later on, Britovsek and co-workers (114) reported a valuable

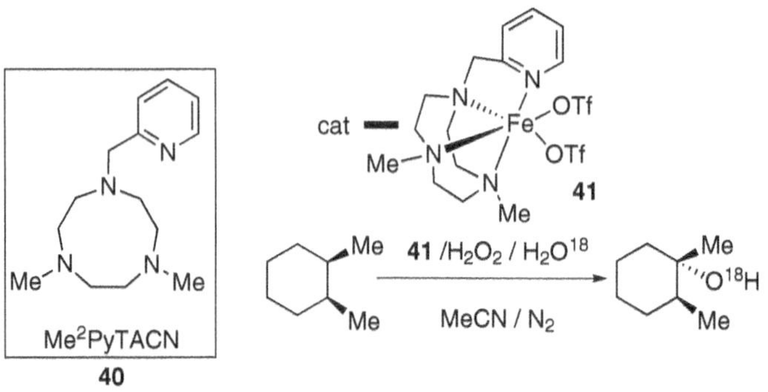

Scheme 91. Mechanism of iron-catalyzed C–O bond formation.

investigation on the synthesis and characterization of tetradentate nitrogen ligands based iron complexes's, and their catalytic activity toward C–H oxidation. High-valent iron–oxo species could be accessed in non-heme ligand systems. A mechanistic framework also had been well established for non-heme iron-catalyzed C–H hydroxylation. Other non-heme enzymes (e.g., methane mono-oxygenase and Rieske dioxygenases) could successfully catalyze the C–H oxidation in nature, which inspired us to develop synthetic models for alkane C–H oxidation (108, 115). Que and Tolman (116) also discussed the critical fundamental principles, which were important for efficacy of their catalyst. Based on these principles great progress had been made toward the development of new ligands–catalysts and their catalytic application for alkane C–H bond oxidation.

Later, Que and co-workers (117) reported another N_4-tetradentate ligated iron complex [Fe^{II} $(CF_3SO_3)_2(^{Me_2}PyTACN)$], PyTACN = bis(2-pyridylmethyl)-1,4,7-triazacyclononane, with a distorted octahedral center. These complexes showed high catalytic activity for the oxidation of cyclohexane, as well as other tertiary C–H bonds of *cis*-1,2-dimethylcyclohexane, adamentane, and 2,3-dimethylbutane (Scheme 91).

The high levels of H_2O^{18} incorporations were obtained from the H_2O^{18} labeling experiment during the oxidation of alkanes with tertiary C–H bonds [e.g., adamantane (74%), *cis*-dimethylcyclohexane (*cis*-DMCH, 79%), and 2, 3-dimethylbutane (76%)] the % of O^{18} incorporations were denoted in the first brackets. The results were independent of the substrate concentration, which suggested that the $OH{-}Fe^{V}{=}O$ species was the only key oxidant toward alkane oxidation with iron complex [Fe^{II} $(CF_3SO_3)_2(^{Me_2}PyTACN)$] (**41**). Obviously this oxidation process was significantly different from the Fe(TPA) complex [TPA = tris(2-pyridylmethyl) amine]. The new results were explained by an unusual rebound-like mechanism (Scheme 92). Notably, cis configuration of the two hydroxy groups in this catalytic system were also different from the stereo requirement of the trans configuration in

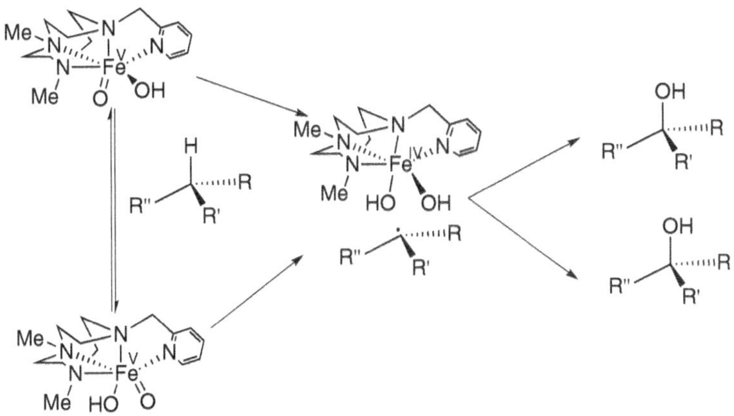

Scheme 92. Sets of iron complexes with TACN backbone ligands.

heme complexes and it was a unique feature in these non-heme iron catalytic systems.

Costas and co-workers (118) explored the iron complexes of the tetradentate ligand ^{Me_2}TACN to a "TACN" family, which had a common methylpyridine derived triazacyclononane (TACN) backbone (Scheme 92). These new families of iron complexes show unprecedented efficiency in the stereospecific oxidation of alkane C–H bonds and also dihydroxylation or epoxydation of alkenes.

In 2008, Reedijk and co-workers (119) described a simpler and more efficient method for oxidation of cyclohexane (CyH) by using a catalytic amount of iron salts $Fe(ClO_4)_2$ in acetonitrile and hydrogen peroxide as oxidant under mild conditions. The oxidation protocol afforded both cyclohexanol and cyclohexanone as a product (Scheme 93). Interestingly when the simple tridentate 2,6-bis[1-(benzylamino)ethyl]pyridine (dapb) was used in the reaction, the selectivity toward major products increased slightly.

Later in 2007, Chen and White (120) reported a selective aliphatic C–H oxidation by using the Fe(PDP) complex (**42**) '[PDP=2-({(S)-2-[(S)-1-(pyridin-2-ylmethyl) pyrrolidin-2-yl]pyrrolidin-1-yl}methyl)pyridine]. The rigid backbone of the PDP ligand in complex **42** (Scheme 94) played a vital role for selectivity in C–H hydroxylation (120).

The process was simple and clean. The electron-rich C–H bonds were selectively oxidized in the presence of electron-deficient C–H bonds, for example, the tertiary sp^3 C–H bond was preferentially oxidized over primary and secondary C–H bonds (Scheme 94). The presence of electron-withdrawing group on α or β to the carbon of the C–H bonds decreases the reactivity. Hence, the presence of electron-withdrawing groups (e.g., carbonyl, ester, acetate, and

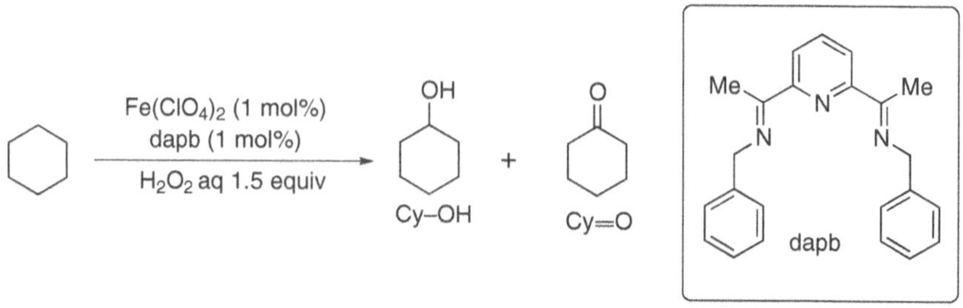

Entry	Catalyst	Major Products	Desired Product (%)	Byproducts (%)
1	With ligand	Cy–OH(37) : Cy=O(54)	91	9
2	Without ligand	Cy–OH(27): Cy=O(64)	87	13

Scheme 93. Iron-catalyzed C–H oxidation of cyclohexane with and without ligand (Cy = cyclohexyl).

halogen functional groups) provided a better selectivity. The tertiary C–H bonds remote to these groups showed a much better reactivity (Scheme 95).

Based on the stereoselectivity outcome of these iron-catalyzed C–H oxidations, a concerted mechanism mediated by an electrophillic oxidant can be visualized. Further they applied their method to complex molecules like a natural product (+)-artemisinin in which the oxidation occur at the most electron-rich and least sterically hindered site. In some cases, these iron-catalyzed methods were high yielding and the reaction completed quickly compared to the enzymatic reaction (Scheme 96*a*).

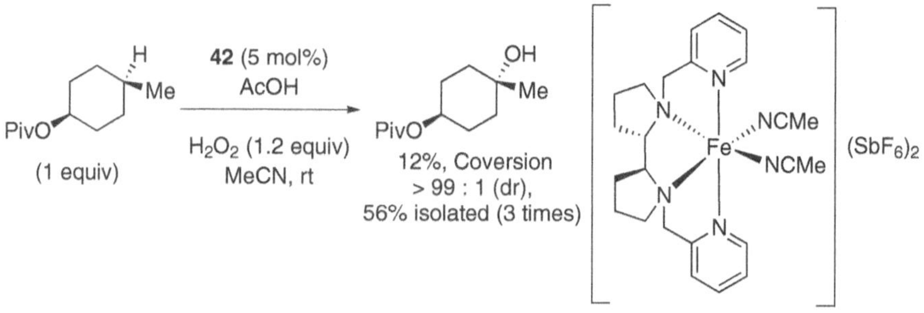

Scheme 94. Iron-catalyzed hydroxylation of unactivated sp^3 C–H bonds (Piv = pivaloyl)

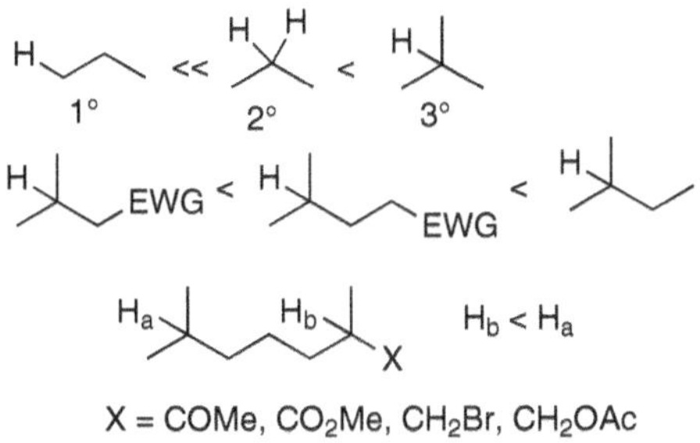

1° << 2° < 3°

EWG < EWG <

$H_b < H_a$

X = COMe, CO_2Me, CH_2Br, CH_2OAc

Scheme 95. Reactivity pattern of different types of C–H bonds.

The presence of a carboxylic group acted as a directing group where the above discussed selectivity principle did not work. It generally formed lactone as the major product. For example, the tetrahydrogibberillic acid analogue under their reaction protocol produced lactone in 52% yield. Subsequently, a slow addition protocol was reported (121). Under a slow addition protocol, the reaction provided high conversion without decrease in site selectivity (Scheme 96*b*). This method

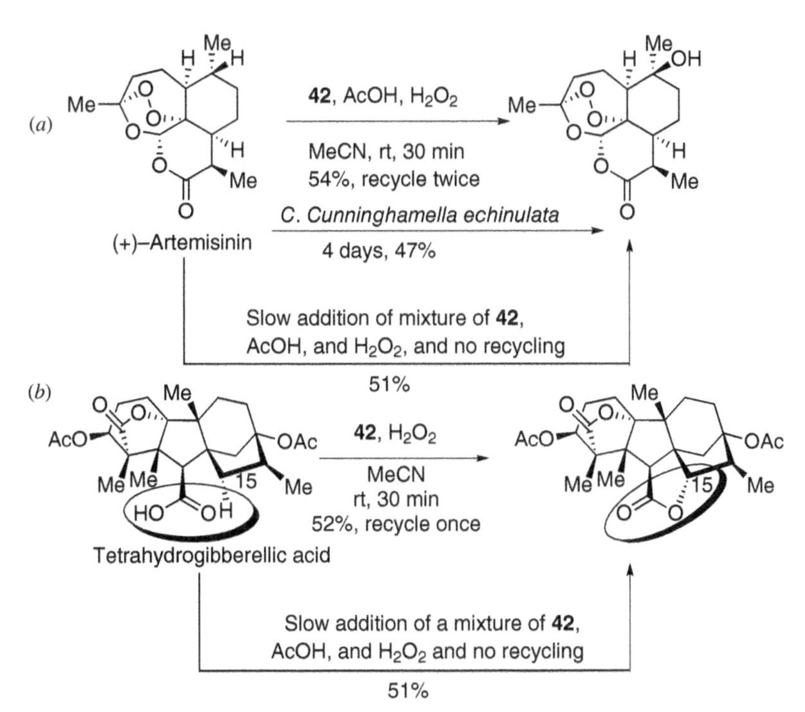

Scheme 96. Application of iron-catalyzed aliphatic C–H hydroxylation.

was applied in various diversified natural products, which showed a promising role in drug discovery.

Inspired by the initial success of developing a versatile method of iron-catalyzed sp^3 C–H hydroxylation, Chen and White (122) further explored their method with a new class of substrates, where a combined effect plays an important role for oxidation of secondary sp^3 C–H bonds (122). Clearly, a secondary C–H bond has less electron density compared to a tertiary C–H bond, but a secondary C–H bond has a higher steric accessibility than a tertiary C–H bond. Furthermore, secondary C–H bonds were present in the ring system (e.g., cyclohexane and decahydronaphthalene derivatives), where the combined effects were well pronounced in terms of their reactivity and selectivity. Contextually, the selective oxidation by a combined effect might provide a very useful method for methylene functionalization. Electronic, steric, stereoelectronic, and functional groups could determine the selectivity of C–H oxidation.

The diterpenoid-derived dione with 14 secondary C–H bonds and two tertiary C–H bonds were tested under optimum reaction conditions (Scheme 97). Based on selectivity principles, two tertiary C–H bonds were electronically and sterically deactivated due to the electron-withdrawing carbonyl group. The side chain and the B ring are electronically deactivated due to the presence of an electron-withdrawing carbonyl group. The C2 position in the A ring was the least bulky site, which was expected to be the most activated C–H bond in this case and the results from the reaction were also consistent with the prediction. The C2 oxidation product was obtained in 53% isolated yield and the C3 oxidation product in 28% yield, whereas 12% starting material was recovered.

Their method functioned well with a complex pleuromutilin derivative (Scheme 98). Dihydropleuromutilin derivative gave a C7 equatorial C–H bond hydroxylation product along with other oxidized products.

In 2009, Costas and co-workers (123) reported iron-catalyzed C–H oxidation with a different ligand backbone that afforded good regioselectivity with better

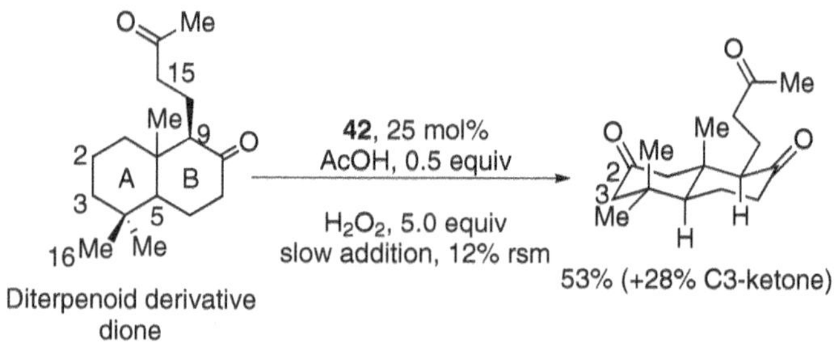

Scheme 97. Role of the combined effect toward high selectivity for secondary C–H bond hydroxylation (RSM = recovered starting material).

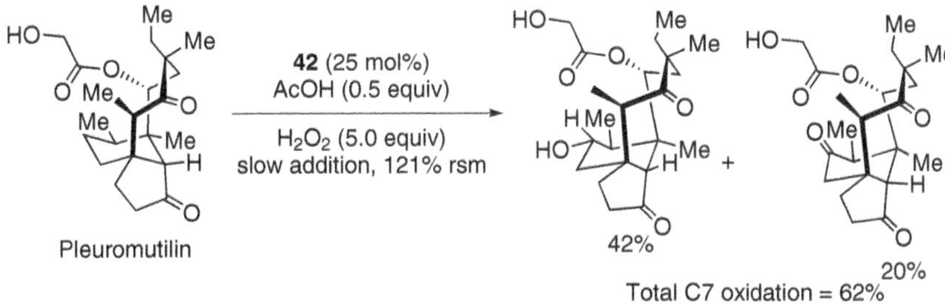

Scheme 98. Role of a combined effect toward the high diastereoselectivity of methylene oxidation.

yield (Scheme 99). Bulky hydrocarbon groups were introduced at the remote position of the pyridine ring, which helped to form a robust cavity. The iron complex [Fe(CF_3SO_3)$_2$(*S,S,R*)-mcpp)] (**43**, mcpp = N1,N2-dimethyl-N1,N2-bis (pyridin-2-ylmethyl)cyclohexane-1,2-diamine) provided better selectivity and efficiency compared to White's catalyst. Costas's catalyst (**43**) with a much lower catalyst loading (1 mol%) accomplished the hydroxylation of (−)-acetoxy-*p*-menthane preferentially at a more accessible (C1)−H bond. In this case, both electronic and steric factors played an important role in distinguishing different C−H bonds.

In 2012, Kodera and co-workers (124) reported another alternative protocol for an alkane C−H hydroxylation with a different type of iron complex (**45**) [iron(III)–monoamidate complex] as a catalyst and H_2O_2 as the oxidant. In the regioselective hydroxylation of *cis*-4-methylcyclohexyl-1-pivalate, catalyst **45** (Scheme 100*a*)

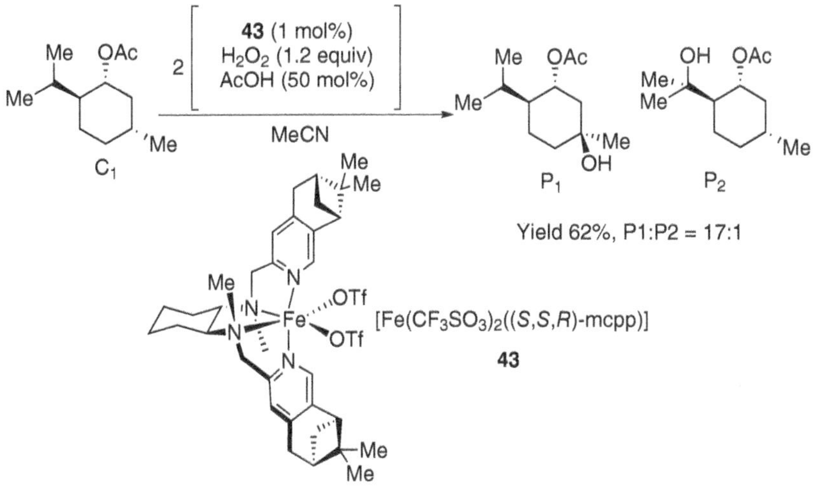

Scheme 99. Regiospecific hydroxylation by a robust non-heme iron catalyst.

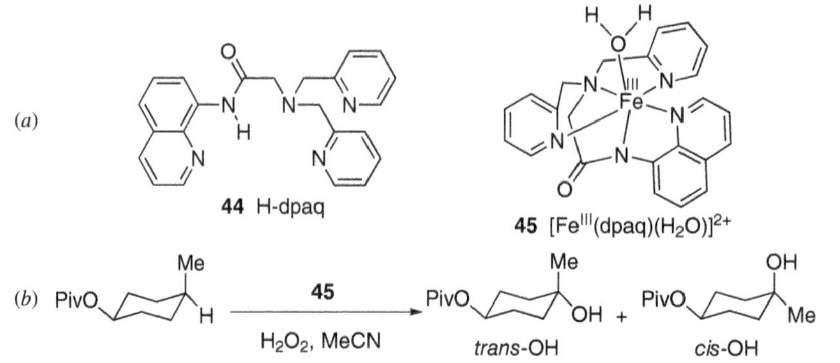

(*a*) **44** H-dpaq **45** $[Fe^{III}(dpaq)(H_2O)]^{2+}$

(*b*) **45**, H_2O_2, MeCN → *trans*-OH + *cis*-OH

Scheme 100. Regiospecific hydroxylation by an iron(III)–monoamidate complex [dpaq = 2-(bispyridin-2-ylmethyl)amino)-*N*-(quinolin-8-yl)acetamide] (**45**).

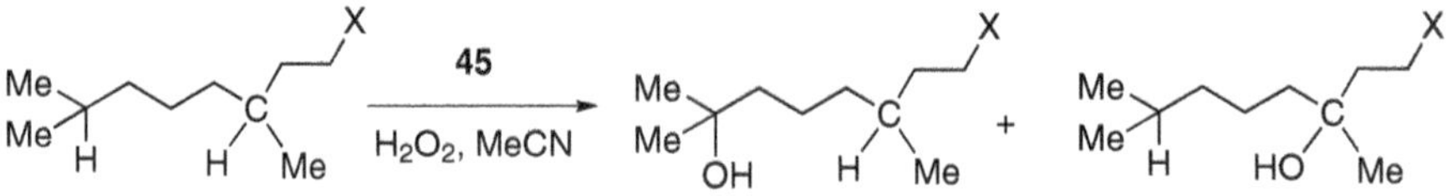

45, H_2O_2, MeCN → 7-OH + 3-OH

X		7-OH 3-OH
X = H	39	3 : 2
OAc	53	9 : 1
Br	38	15 : 1

Scheme 101. Regiospecific hydroxylation by an iron(III)–monoamidate complex (**45**).

gave better selectivity (94% retention of configuration in 38% yield) in the formation of the *trans*-OH product compared to White and Costas's catalysts, but the yield was comparable to the White catalyst and lower than Costas catalyst (Scheme 100*b*).

The hydroxylation of 1-substituted 3,7-dimethyloctane with Kodera's catalyst provided better yields and selectivity compared to others. Preferential formation of 7-OH product over 3-OH (Scheme 101) is consistent with the selectivity rule discussed above. Unlike White and Costas catalyst, no cis labile position is available in Kodera's catalyst and acetic acid is not required for catalytic activity.

B. Epoxidation

Epoxides are an important class of compounds often used as the key precursors for synthesis of various drugs and natural products (125). Sharpless epoxidation of

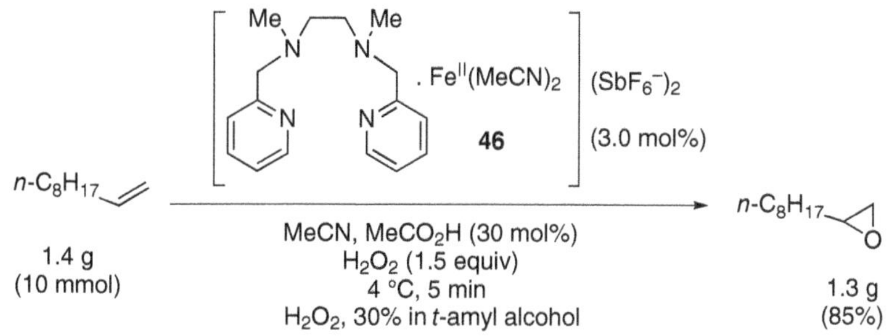

Scheme 102. Iron-catalyzed epoxidation

allylic alcohol and Katsuki–Jacobsen asymmetric epoxidation are historically significant in the development of epoxidation chemistry (126). First, iron-catalyzed epoxidation was reported in 2001 by Jacobsen and co-workers (127) with Fe–mep (**46**) in the presence of a catalytic amount of acetic acid (Scheme 102). Acetic acid plays a key role in order to form an iron-μ-oxo carboxylate bridged iron complex, which was most likely to be the active species as proposed by Jacobsen and co-workers (127).

Later on, Stack and co-workers (128) reported epoxidation by an μ-oxo-iron(III) dimer, $[((Phen)_2(H_2O)Fe^{III})_2]$ $(\mu\text{-}O)]ClO_4)_4$ in the presence of peracetic acid as the terminal oxidant. Their method can tolerate a wide range of alkenes including terminal alkenes (Scheme 103).

In 2002, Que and co-workers (129) developed a non-heme iron catalysts, $[Fe^{II}$ $(5\text{-}Me_3\text{-}TPA)(MeCN)_2](ClO_4)_2$ and $[Fe^{II}(bpmen)(MeCN)_2](SbF_6)_2$, where bpmen = *N,N'*-dimethyl-*N,N'*-bis(pyridin-2-ylmethyl)ethane-1,2-diamine, for epoxidation of alkenes under neutral conditions. But in these cases, cis-diols were obtained as the major product and epoxides as minor products. Interestingly, the introduction of acetic acid as an additive suppressed the formation of cis-diols and provided epoxides as the major product. A high-valent iron oxo was proposed to be the active species for this reactions (130).

In this field, the major breakthrough came from Beller and co-workers (131) in 2007. They developed epoxidation using $FeCl_3.6H_2O$ in combination with pyridine-2,6-dicarboxylic acid and pyrrolidine in the presence of H_2O_2 as the terminal

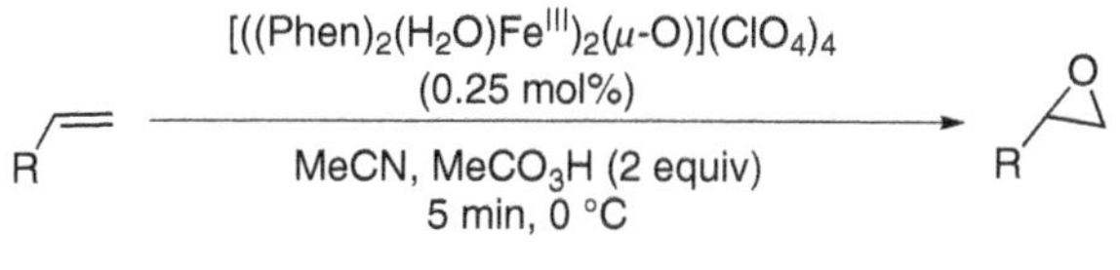

Scheme 103. Iron-μ-oxo carboxylate-bridged complex-catalyzed epoxidation.

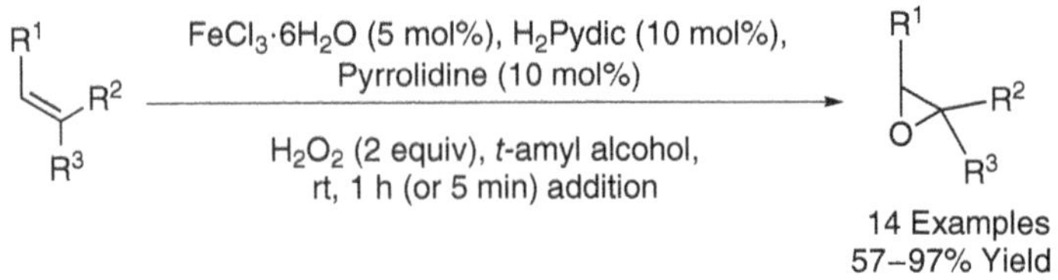

$FeCl_3 \cdot 6H_2O$ (5 mol%), H_2Pydic (10 mol%), Pyrrolidine (10 mol%)

H_2O_2 (2 equiv), *t*-amyl alcohol, rt, 1 h (or 5 min) addition

14 Examples
57–97% Yield

Scheme 104. The $FeCl_3.6H_2O$ catalyzed epoxidation.

oxidant (Scheme 104). Aromatic olefins and 1,3-cyclooctadiene gave moderate-to-excellent yields under their standard reaction protocol.

Later on, Beller and co-workers (132) developed a method, that could generate epoxides from both aliphatic and aromatic alkenes. Iron complexes in combination with various nitrogenous ligands (**47–50**) (Scheme 105) provided epoxides with good selectivity.

The formation of a red precipitate was observed under reaction conditions. From mass spectrometry analysis of the filtrate, it was found that the pybox ligand (**47**) decomposes into H_2Pydic and phenyl glycerol. After filtration, both the residue and filtrate showed similar reactivity, which implied that the filtrate contained decomposed ligand as well as iron (10). Based on these observations, they further optimized the iron source, H_2Pydic derivatives, and amine ligands in order to get better conditions.

Later they reported a few modified methods using different nitrogenous ligands (133). With formamidines (**51**) as ligands, they developed epoxidation for aromatic olefins and 1,3-dienes. Notably, both aromatic and aliphatic alkenes were tolerated with good selectivity (Scheme 106).

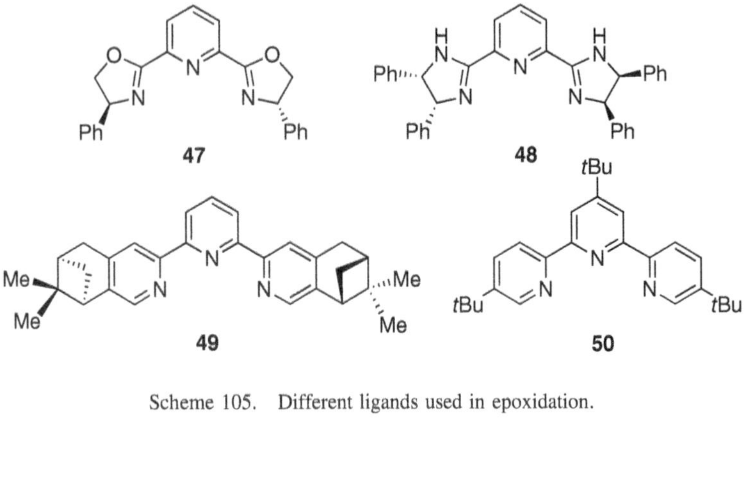

47 **48** **49** **50**

Scheme 105. Different ligands used in epoxidation.

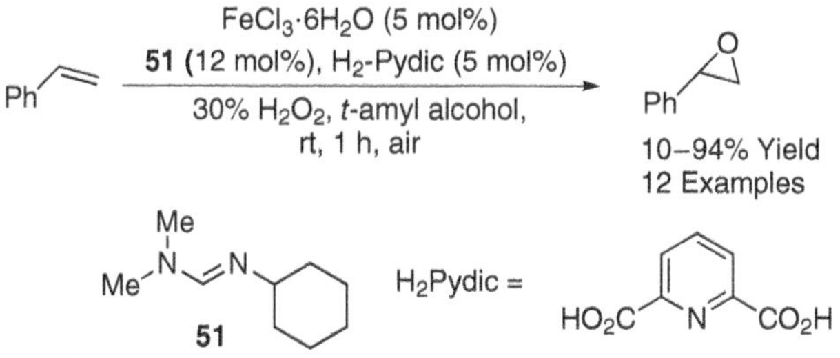
FeCl3·6H2O (5 mol%)
51 (12 mol%), H2-Pydic (5 mol%)
30% H2O2, t-amyl alcohol,
rt, 1 h, air
Ph
Ph
O
10–94% Yield
12 Examples
Me
N
N
Me
51
H2Pydic =
HO2C
N
CO2H

Scheme 106.

Control experiments with a radical inhibitor like TEMPO, butylphenylnitrone (BPN), and duroquinone suppressed the product yield greatly. Notably, the presence of 1 equiv of TEMPO in their standard reaction stopped the reaction completely. A further control experiment with *β*-pinene as substrate gave the rearrangement products (**56** and **55**), which were in accord with the radical mechanism (Scheme 107). In 2012, Kozak and co-workers (134) developed

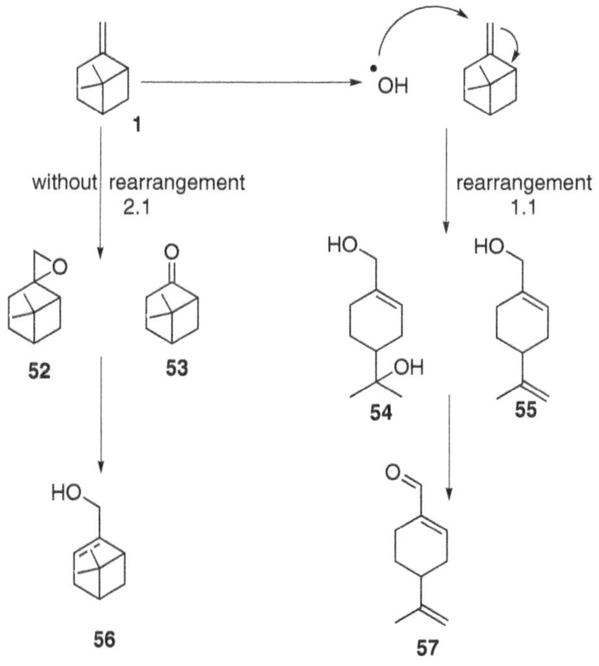
OH
1
without rearrangement
2.1
rearrangement
1.1
O
O
52
53
HO
HO
OH
54
55
HO
O
56
57

Scheme 107.

TABLE XVI
Iron-Catalyzed Asymmetric Epoxidation

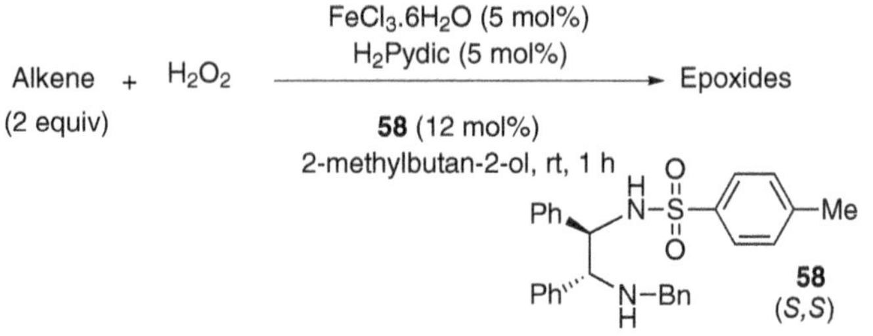

Alkene (2 equiv) + H_2O_2 → Epoxides

$FeCl_3.6H_2O$ (5 mol%), H_2Pydic (5 mol%), **58** (12 mol%), 2-methylbutan-2-ol, rt, 1 h

Ph, N–H, O=S=O, Me, Ph, N–Bn, H, **58** (*S*,*S*)

*t*Bu, O, *t*Bu	O, OSiPh$_3$	Me, O, Me
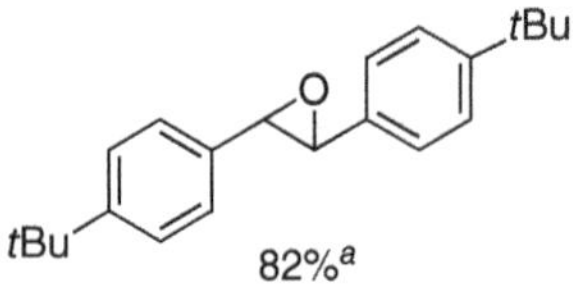 82%[a]	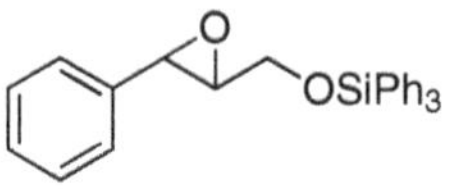 67%[a]	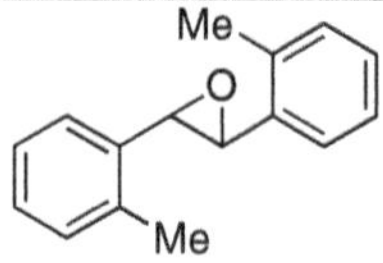 57%[a]
ee 81%[b]	ee 35%[b]	ee 28%[b]
(+)-(2*R*, 3*R*)[c]	(+)-(2*R*, 3*R*)[c]	(+)-(2*R*, 3*R*)[c]

[a] Yield of isolated product.
[b] Assigned by comparing the retention times of enantiomer.
[c] Determined by high-performance liquid chromatography (HPLC) on chiral columns by comparing the sign of the optical rotation.

epoxidation methods using *N*-methylimidazole as the organic base in acetone. Although their method showed poor selectivity for styrene, better selectivity was observed for larger alkenes.

First, the iron-catalyzed method for asymmetric epoxidation was reported by Que and co-workers (129) using [Fe(bpmcn)(CF_3SO_3)$_2$] (bpmcn = *N,N'*-bis(2-pyridyl-methyl)-*N,N'*-dimethyl-1,2-cyclohexanediamine ligand). While studying the dihydroxylation of *trans*-2-heptene with H_2O_2 as oxidant, 12% ee for the epoxide byproduct was observed. In 2007, Beller and co-workers (135) developed an efficient method for iron-catalyzed asymmetric epoxidation using H_2O_2 as oxidant in the presence of a chiral sulfoxamide ligand (**58**). Various internal alkenes were tolerated under this protocol with good-to-excellent enantioselectivity (Table XVI).

In 2011, Beller and co-workers (136) reported epoxidation using molecular O_2 as oxidant, which showed high reactivity as well as selectivity in the presence of β-keto ester and imidazole as additives. Nishikawa and Yamamoto (137) developed asymmetric epoxidation for β,β-disubstituted enones by using Fe(OTf)$_2$ with a chiral phenanthroline ligand (**59**) in the presence of peracetic acid as the terminal oxidant (Scheme 108). The reaction protocol provided epoxides of enones with high enantioselectivity (up to 92% ee).

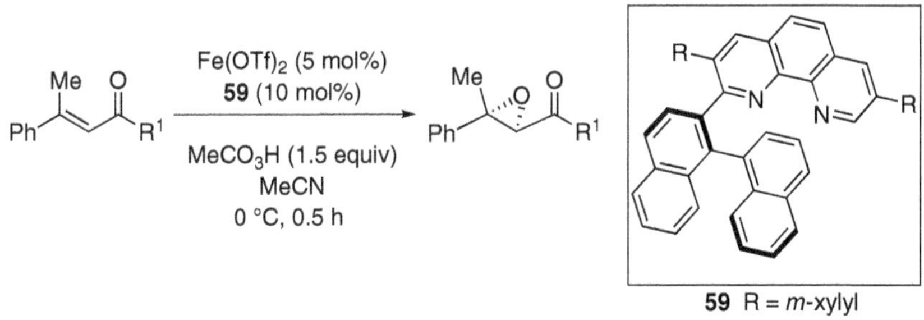

Fe(OTf)$_2$ (5 mol%)
59 (10 mol%)
MeCO$_3$H (1.5 equiv)
MeCN
0 °C, 0.5 h

59 R = *m*-xylyl

Scheme 108.

C. cis-Dihydroxylation

In nature, Riesky dioxygenases carry out dihydroxylation where iron is a metal cofactor (138). Inspired from nature, Que and co-workers (129) in 2002 developed the first iron-catalyzed synthetic non-heme systems for alkene dihydroxylation. They synthesized iron(II) complexes of the TPA family of ligands **60–65** (Scheme 109) and performed cis-dihydroxylation of 1-octene in MeCN using H_2O_2 as the terminal oxidant. Among these, TPA (**60**), TPA-5-Me$_3$ (**61**), and TPA-3-Me$_3$ (**62**) iron complexes gave 69–87% yield of the cis-dihydroxylation product. Minor amounts of epoxides were obtained as the byproduct.

Later on, they reported an asymmetric version of this cis-dihydroxylation reaction by using a chiral ligand backbone (Scheme 110) [1,1′-Bis(pyridin-2-

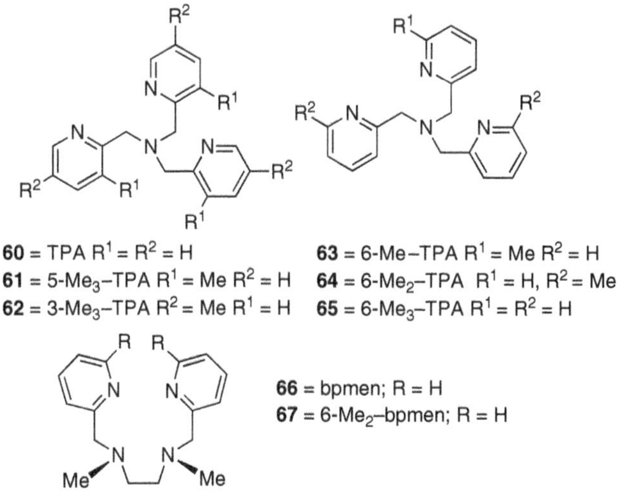

60 = TPA R^1 = R^2 = H
61 = 5-Me$_3$–TPA R^1 = Me R^2 = H
62 = 3-Me$_3$–TPA R^2 = Me R^1 = H
63 = 6-Me–TPA R^1 = Me R^2 = H
64 = 6-Me$_2$–TPA R^1 = H, R^2 = Me
65 = 6-Me$_3$–TPA R^1 = R^2 = H

66 = bpmen; R = H
67 = 6-Me$_2$–bpmen; R = H

Scheme 109.

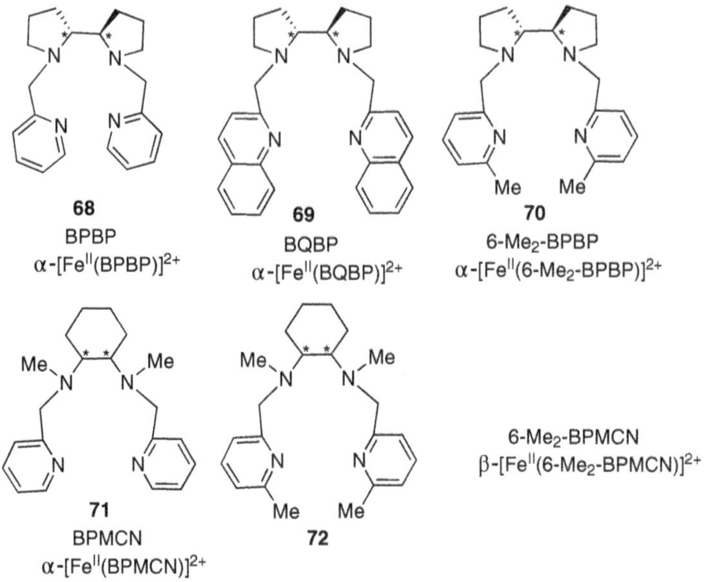

68
BPBP
α-[Fe^II^(BPBP)]^2+^

69
BQBP
α-[Fe^II^(BQBP)]^2+^

70
6-Me2-BPBP
α-[Fe^II^(6-Me2-BPBP)]^2+^

71
BPMCN
α-[Fe^II^(BPMCN)]^2+^

72
6-Me2-BPMCN
β-[Fe^II^(6-Me2-BPMCN)]^2+^

Scheme 110.

ylmethyl)2,2′-bipyrrolidine (BPBP); 1,1′-bis(quinolin-2-yemethyl)-2,2′-bipyrolidine (BQBP); *N,N′*-bis(2-pyridyl-methyl)-*N,N′*-dimethyl-1,2-cyclohexanediamine (BPMCN)] (139).

Among these sets of iron complexes (**68**–**72**), catalyst **70** was the most effective (97% ee) (Table XVII). Note that in all these cases there was a competition between epoxidation and cis-diol product formation.

TABLE XVII
Asymmetric cis-Dihydroxylation of Olefins[a]

Catalyst	Substrate	cis-Diol (%ee)[b]	Diol/Epoxide
68	*trans*-2-Heptene	38 (3)	1:4.6
69	*trans*-2-Heptene	78(3)	4:1
70	*trans*-2-Heptene	97(1)	26:1
70	*trans*-4-Octene	96(1)	13:1
70	1-Octene	76(1)	64:1
71	*trans*-2-Heptene	29	1:18
72[c]	*trans*-2-Heptene	79	3.2:1

[a] Reaction conditions: A 70-mm solution of H_2O_2 (10 equiv) in MeCN was delivered by syringe pump over a period of 20 min to a degassed and stirred solution of catalyst (0.7-mm) and substrate (0.35 m) at ambient temperature in air for **68** and **69** and under an Ar atmosphere for **70**.
[b] Percent of ee of the predominant diol isomer.
[c] Results were normalized to 10 equiv H_2O_2.

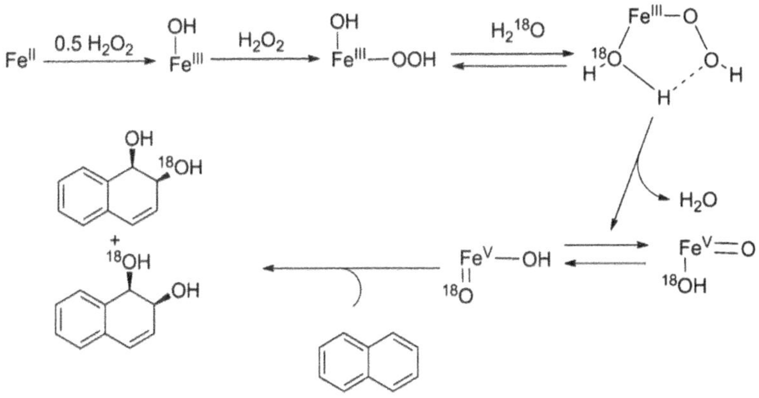

Scheme 111.

Later, in 2009 Que and co-workers (140) reported cis-dihydroxylation of the aromatic double bond of naphthalene using H_2O_2 as oxidant. Such a reaction mimics the activities of naphthalene dioxygenases. In order to establish the mechanism, they carried out an ^{18}O labeling experiment. In the presence of either 10 equiv of 2% $H_2{}^{18}O_2$ (100 equiv H_2O per $H_2{}^{18}O_2$) or 10 equiv H_2O_2 and 1000 equiv $H_2{}^{18}O$ as oxidant > 90% singly labeled diol product was observed. This control experiment indicated that one O atom in cis-dihydroxylated product came from H_2O and another came from H_2O_2 (Scheme 111).

V. CROSS-COUPLING REACTIONS

Metal-assisted cross-coupling reactions had gained immense importance toward the formation of C–C, C–X bonds over the last three to four decades due to their application in the synthesis of key precursors of various natural products and pharmaceuticals (141). In this context, iron-catalyzed coupling reactions are important, as most of the iron salts are environmentally benign and less expensive compared to the late transition metals that were well known for cross-coupling reactions (142). Nickel and palladium catalyzed cross-coupling reactions involving a wide range of organometallic reagents and organic electrophiles were widely studied (143). In 1972, Corriu and Masse (144a) Kumada and co-workers (144b) independently reported nickel-catalyzed coupling of Grignard reagents with alkenyl and aryl halides. However, iron-catalyzed coupling reaction of alkyl magnesium reagents with vinyl bromide was reported prior to 1972 (145). Following this work, iron sources were applied as catalyst for the cross-coupling reactions involving various coupling partners that were discussed in this subsection.

A. Alkenyl Derivatives as Coupling Partners

After Kochi's pioneering work, $Fe(dbm)_3$ catalyzed cross-coupling between phenyl magnesium bromide and β-styrene bromide in bromobenzene was reported. The reaction showed selectivity toward the vinyl derivative yielding *trans*-stilbene and a trace amount of homocoupled biphenyl [Scheme 112, DME = dimethoxyethane, dbm = dibenzoylmethane (ligand)] (146). Bromothiophenes were also used as coupling partner with *i*PrMgCl, affording stereospecific trans- product in contrast to nickel- or palladium-catalyzed reactions where a mixture of isomers was observed (Scheme 113) (147).

Vinyl sulfones could be tolerated as electrophiles under these reaction conditions. However, they gave a mixture of (*E,Z*) isomers. The 1,4-addition product and reduction compounds were obtained as a byproduct (148). This methodology was applied in the synthesis of natural product pheromones (Scheme 114). Notably this reaction provided an olefin compound in the presence of *n*BuLi (149).

High selectivity (92% retention of configuration) was obtained when MeLi was used as the transmetalating agent (Scheme 115) (150).

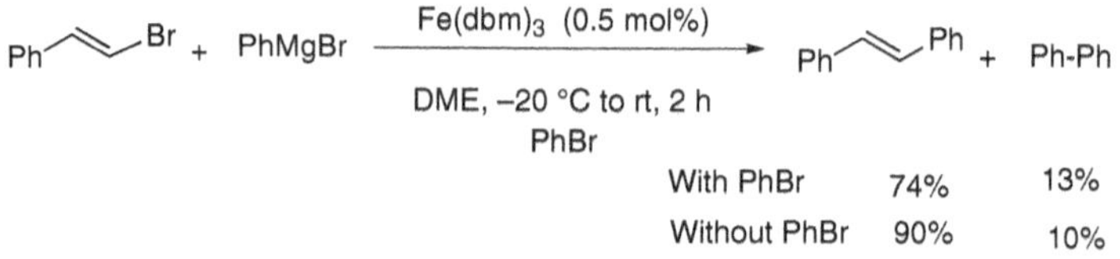

Ph–CH=CH–Br + PhMgBr → ($Fe(dbm)_3$ (0.5 mol%), DME, –20 °C to rt, 2 h, PhBr) → Ph–CH=CH–Ph + Ph-Ph

With PhBr	74%	13%
Without PhBr	90%	10%

Scheme 112.

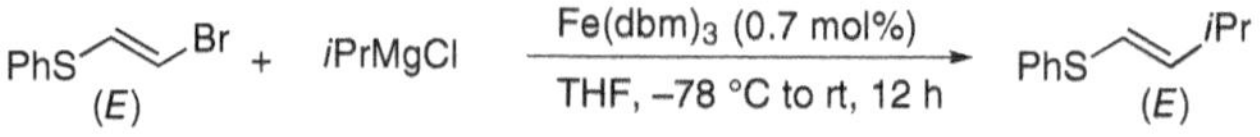

PhS–CH=CH–Br (*E*) + *i*PrMgCl → ($Fe(dbm)_3$ (0.7 mol%), THF, –78 °C to rt, 12 h) → PhS–CH=CH–*i*Pr (*E*)

Scheme 113.

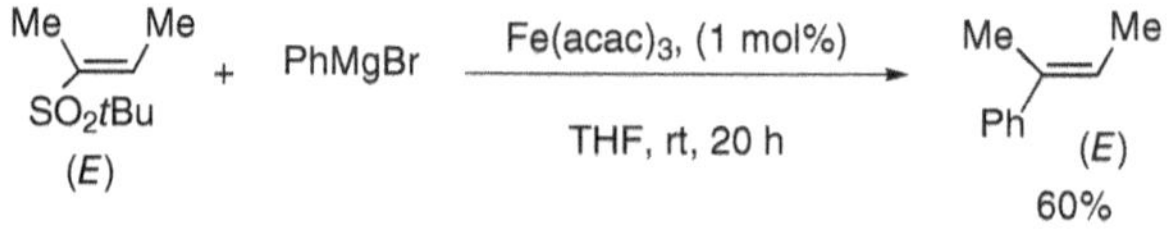

Me, Me, SO_2tBu (*E*) + PhMgBr → ($Fe(acac)_3$, (1 mol%), THF, rt, 20 h) → Me, Me, Ph (*E*) 60%

Scheme 114.

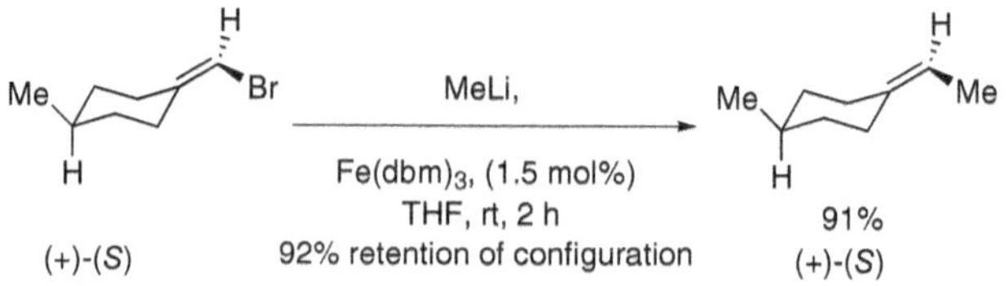

Me, H, H, Br (+)-(*S*) → (MeLi, $Fe(dbm)_3$, (1.5 mol%), THF, rt, 2 h, 92% retention of configuration) → Me, H, H, Me 91% (+)-(*S*)

Scheme 115.

TABLE XVIII
Iron-Catalyzed Cross-Coupling Reactions of Grignard[a] or Organomanganese[b] Reagents (RMX) With Alkenyl Halides (R′X)

$$RMX + R'X \xrightarrow{Fe(acac)_3 cat} R{-}R'$$

Entry	RMX	R′X	R–R′ Yield (%)
1	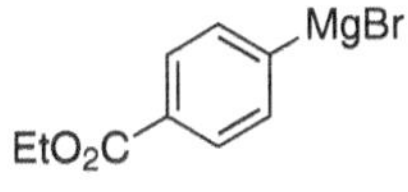 $4\text{-}EtO_2C\text{-}C_6H_4MgBr$	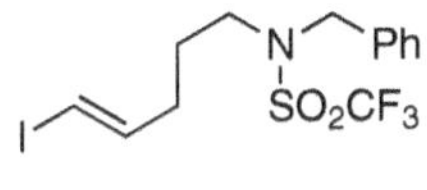 $I\text{-}CH{=}CH\text{-}(CH_2)_3\text{-}N(SO_2CF_3)CH_2Ph$	69
2	*i*PrMgBr	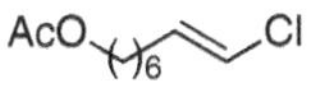 $AcO(CH_2)_6CH{=}CHCl$	72
3	*n*-BuMgBr	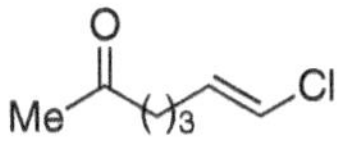 $MeC(O)(CH_2)_3CH{=}CHCl$	80
4[c]	$c\text{-}C_6H_{11}MgBr$	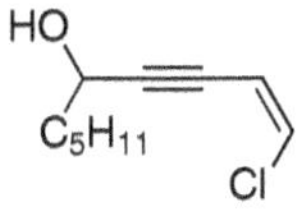 $C_5H_{11}CH(OH)C{\equiv}C\text{-}CH{=}CHCl$	82
5	*n*-BuMnCl	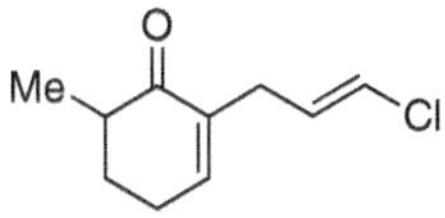 2-(3-chloroallyl)-6-methylcyclohex-2-en-1-one (Me, O, Cl)	72
6	MeMgBr	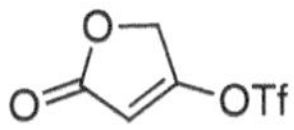 4-(OTf)-furan-2(5H)-one	70
7	*n*-BuMgBr	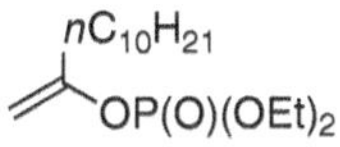 $CH_2{=}C(nC_{10}H_{21})OP(O)(OEt)_2$	78
8	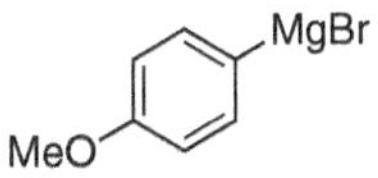 $4\text{-}MeO\text{-}C_6H_4MgBr$	 $CH_2{=}CH\text{-}SPh$	68

[a] Reactions performed with RMgBr (1.4 equiv) and $Fe(acac)_3$ (5 mol%) at −20 °C for 15 min in THF/NMP or at −5 °C for 15 min.
[b] Reaction performed with *n*-BuMnCl (1.4 equiv) and $Fe(acac)_3$ (3 mol%) in THF at rt 1 h (entry 6).
[c] Used 3 equiv of RMgBr.

Cahiez and co-workers (151) contributed significantly to improve the scope of an iron-catalyzed cross-coupling reaction by utilizing functionalized aryl Grignard reagents and vinyl halides (Table XVIII). Use of *N*-methylpyridine (NMP) as the solvent additive to THF dramatically increased the yield of the

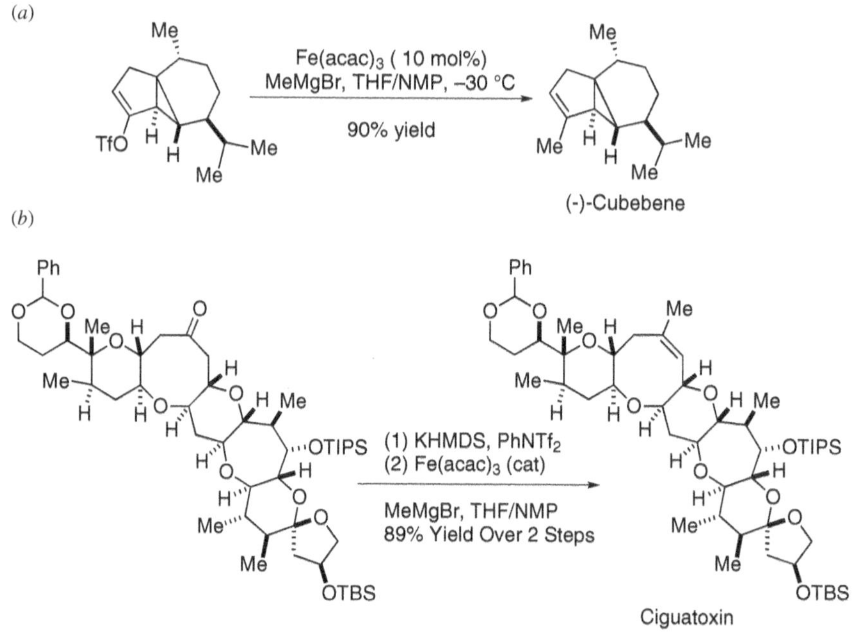

Scheme 116. Alkenyl triflate cross-coupling reactions in total synthesis [$PhNTf_2$ = *N*-phenyl-bis (trifluoromethansulfonamide; TBS = *tert*-butyldimethylsilyl; TIPS = triisopropylsilyl; KHMDS = potassium bis(trimethylsilyl)amide)].

cross-coupled product from 5 to 80%. Organomanganese reagents were tolerated as a coupling partner when $Fe(acac)_3$ was used as the catalyst (152). Different alkenyl derivatives (e.g., alkenyl sulfones, sulfides, phosphates, triflates, esters, ketones, enones, carbamates, and acetals) were well tolerated (151b, 153). These protocols were applied for the total synthesis of ciguatoxin(−)-cubebene (Scheme 116) and enantiopure bicyclic diene ligands (154). Notably, in these cases better yields were observed with iron compared to the Ni/Pd system (154b).

B. Aryl Derivatives as Coupling Partners

Application of iron-based organometallic compounds as a precatalyst in a cross-coupling reaction was explored by Fürstner and Leitner (155) (Table XIX). They used aryl halides in the presence of different iron sources (5 mol%, as a precatalyst), such as $Fe(acac)_3$, Fe(salen)Cl under mild reaction conditions (rt to −30 °C and < 1-h reaction time). Organomanganese, zinc, and magnesium derivatives

TABLE XIX
Cross-Coupling Reactions of Organometallic Reagents (RMX) With Ar–X[a]

Entry	Ar–X	RMX	Yield (%) of Product Ar–R
1	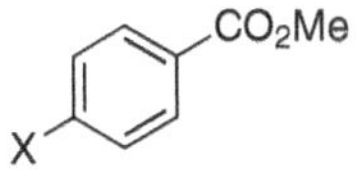 CO_2Me X	n-$C_6H_{13}MgBr$	83 (X = OTS)
2		$Et_3ZnMgBr$	93
3		$C_{14}H_{29}MnCl$	96
4		$(C_{14}H_{29})_2MnCl$	98
5		$(C_{14}H_{29})_2MnMgCl$	98
6		n-C_4H_9Li	0
7	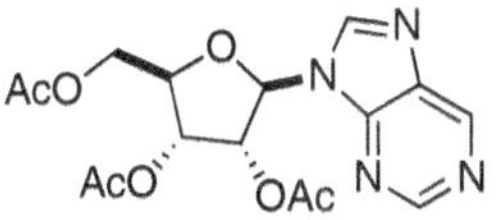 AcO O N N AcO OAc N N	n-$C_{14}H_{29}MgBr$	72
8	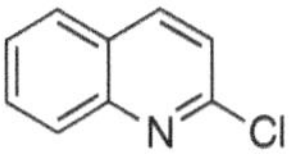 N Cl	PhMgBr	71
9		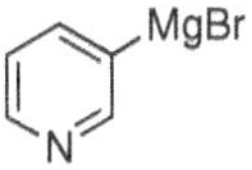 MgBr N	63

[a] Reactions performed with RM (1.2–2.3 equiv) and $Fe(acac)_3$ (5 mol%) at 0 °C or at rt in THF/NMP for 10 min (Entries 1–6) or in THF at −30 °C for 1 h (Entries 7–9).

(both alkyl and aryl) were well tolerated, but no product was obtained in the case of organolithium derivatives. Aryl triflates, tosylates, and unactivated aryl derivatives (e.g., pyridines) were also compatible as electrophiles under these reaction conditions. Reaction in a THF/NMP mixture as solvent was completed within 15 min. The reaction was susceptible to steric effects as ortho substituted arenes gave a lower yield compared to para substituted arenes. Interestingly, selective monosubstition was observed in the presence of more than one halide in the electrophile (153b). Unfortunately, the reaction suffered from the homocoupling of an aryl Grignard reagent, and is limited to only electron-deficient haloarenes.

The reaction protocol was applied in the synthesize of the intermediate of the immunosuppressive agent FTY720 (156). This method was applied in the total synthesis of spermidine alkaloid (−)-isooncinotine (Scheme 117*a*) and the olfactory macrocycle (+)-muscopyridine (Scheme 117*b*) (157, 158).

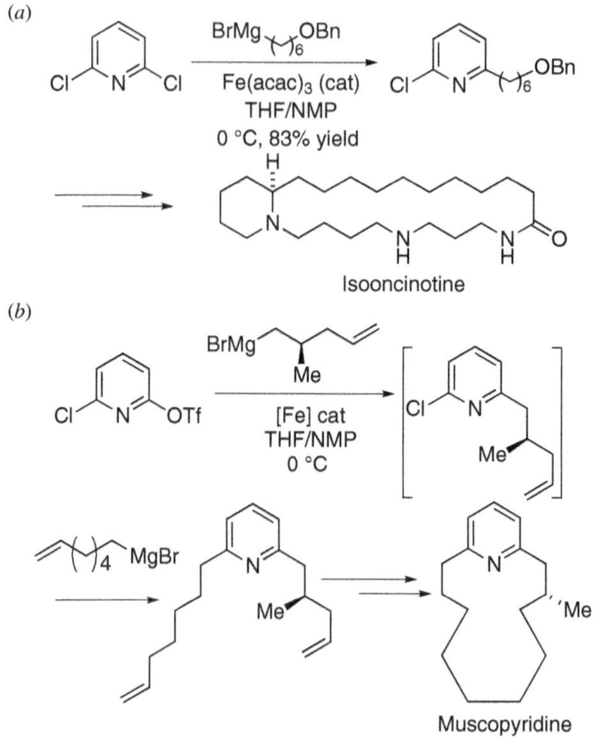

Scheme 117. Aryl electrophiles's cross-coupling reactions in total synthesis.

In 2007, Hatakeyama and Nakamura (159) reported that the presence of FeF_3 in combination with an *N*-heterocyclic carbene as ligand greatly reduced homocoupling of Grignard reagent (Scheme 118*a*). Nakamura's method was used by Hocek and Dvorakova (160) in regioselective monomethylation at the 6-position of 2,6-dichloropurines, with MeMgCl, without any modification. In 2009, Lamaty and co-workers (161) reported both the alkylation and arylation of 4-chloro-pyrrolo[3,2-*c*]quinoline in the presence of a catalytic amount of iron salts (Scheme 118*b*).

In 2012, Garg and co-workers (162) reported alkylation of aryl sulfamates and carbamates using iron catalysts and described the C_{sp^2}–C_{sp^3} bond formation. Sulfamate and carbamate functional groups could be used as directing groups for various synthetic transformations in iron-catalyzed cross-couplings due to their high stability, easy availability, and low reactivity toward conventional Pd(0) catalyzed methods (Scheme 119*a*). Recently, Knochel and co-workers (163) developed iron-catalyzed sp^2–sp^2 cross-coupling between *N*-heterocyclic chlorides/bromides and arylmagnesium reagents (Scheme 119*b*).

(*a*)

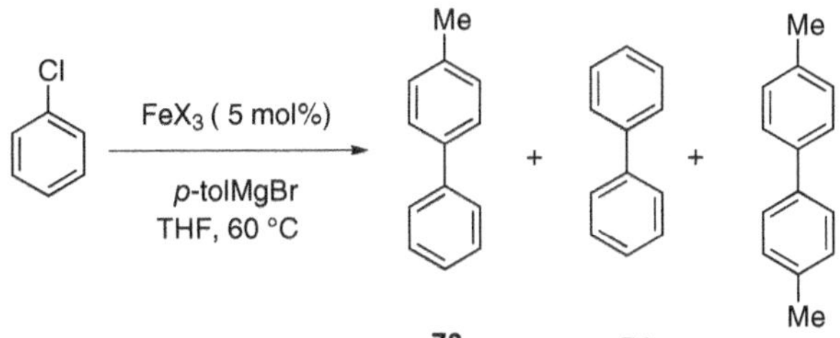

Cl

FeX_3 (5 mol%)

p-tolMgBr
THF, 60 °C

Me + + Me Me

73 74 75

Entry	FeX_3	L (mol %)	**73:74:75**
1	$FeF_3.3H_2O$	SIPr.HCl (15)	98:<1:4
2	$FeF_3.3H_2O$	–	6:trace:4
3	$FeCl_3$	SIPr.HCl (15)	32:2:32
4	$FeCl_3$	KF (20) SIPr.HCl (15)	92:1:8

(*b*)

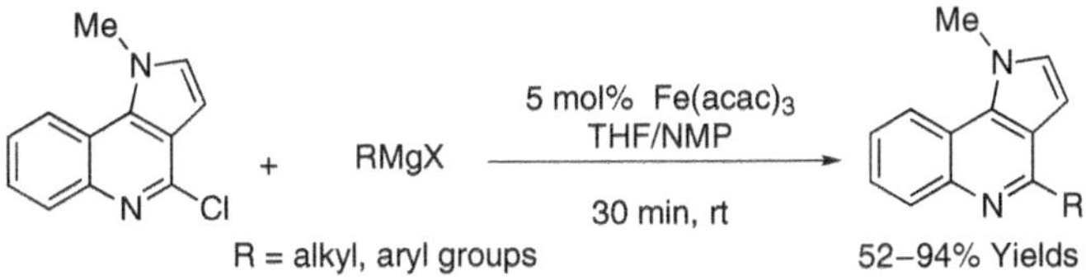

Me N N Cl + RMgX

5 mol% Fe(acac)$_3$
THF/NMP

30 min, rt

Me N N R

R = alkyl, aryl groups

52–94% Yields

Scheme 118. In this scheme SIPr = 1,3-bis(2,6-dispropylphenyl)-imidazolidinium.

(*a*)

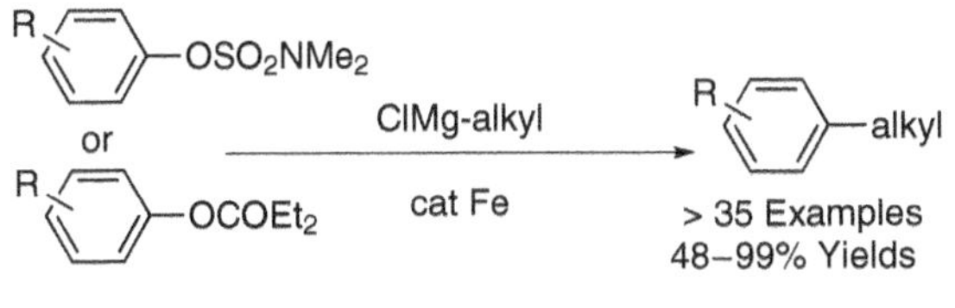

R $-OSO_2NMe_2$

or

R $-OCOEt_2$

ClMg-alkyl

cat Fe

R —alkyl

> 35 Examples
48–99% Yields

(*b*)

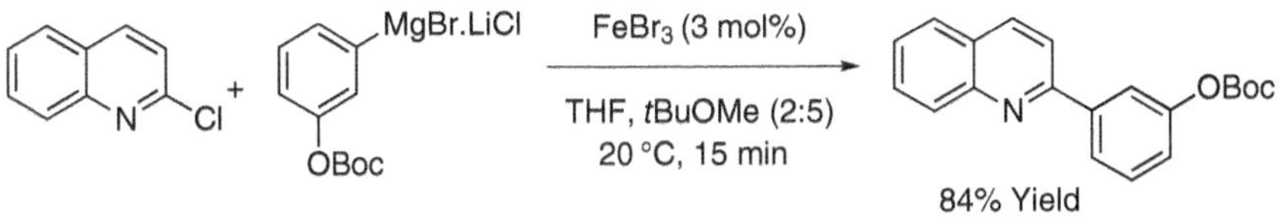

N Cl + MgBr.LiCl OBoc

$FeBr_3$ (3 mol%)

THF, *t*BuOMe (2:5)
20 °C, 15 min

N OBoc

84% Yield

Scheme 119.

C. Alkyl Derivatives as Coupling Partners

Alkyl halides are traditionally challenging substrates for cross-coupling reactions due to their high-energy barrier in oxidative addition and favorable β-hydride elimination (164). Some early examples of iron-catalyzed cross-coupling were reported with alkyl halides, but these reactions suffered from homocoupling and conversion of the alkyl halide to the corresponding alkenes and alkanes (165). Efficient iron-catalyzed cross-coupling by using Grignard reagent and alkyl halide was reported by Nakamura et al. (166) (method A, Table XX). Iron(III) chloride (5 mol%) was used as catalyst in the presence of TMEDA (Scheme 120*a*, entry 3).

Nagano and Hayashi (167) used $Fe(acac)_3$ for the same reaction using Et_2O under reflux conditions (method B). Use of solvent combination THF/NMP played a key role in the observed selectivity. In 2007, Cosy and co-workers (168a) (method C) and Cahiez and co-workers (168b) (method D) explored the scope of this reaction by using different alkyl halides as well as different alkenyl Grignard

TABLE XX
Cross-Coupling Reactions of Alkyl Halides (RX) With ArMgBr

Entry	R–X	ArMgBr	Method[a]	Yield (%)
1	cyclohexyl-X	Ph	A	99 (X = I, Br, Cl)
2	*n*-oct-X	Ph	A	91 (X = Br) 45 (X = Cl)
3	*n*-Hex–CH(Me)–Br	*p*-toyl	B	73 97 (X = 1)
4	Br–CH₂CH₂–C≡C–TIPS	Me–C(=CH₂)–MgBr	C	80
5	Me–(CH₂)₁₀–Br	Me₃Si–C(=CH₂)–MgBr	C	6:2:2 (86)
6	Me–(CH₂)₅–CH(Br)–Me	Me₂C=CH–MgBr	D	72

[a] Method A: $FeCl_3$(5 mol%), RX (1 equiv), ArMgBr (1.2 equiv, slow addition), and TMEDA (1.2 equiv, slow addition), in THF (−78 to 0 °C), 0.5 h. Method B: $Fe(acac)_3$ (5 mol%), RX (1 equiv), and ArMgBr (1.04 equiv) in refluxing Et_2O, 0.5 h.

(*a*)

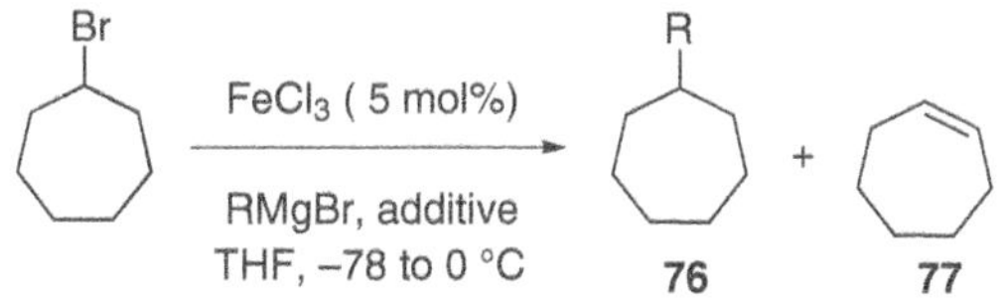

Br; $FeCl_3$ (5 mol%); RMgBr, additive; THF, −78 to 0 °C; R; **76**; + **77**

Entry	Additive	Yield (%) of **76** (R = Ph)	Yield (%) of **76** (R = H)	Yield (%) of **77**
1	None	5	0	79
2	NMP	15	Trace	3
3	TMEDA	71	3	19

(*b*)

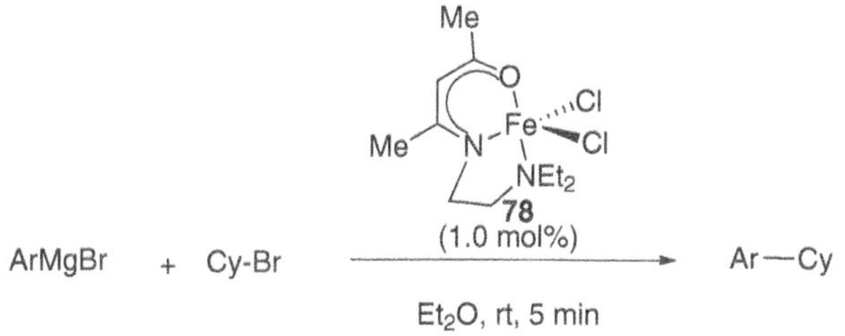

ArMgBr + Cy-Br → Ar—Cy

78 (1.0 mol%); Et_2O, rt, 5 min

Me; O; Fe; Cl; Cl; Me; N; NEt_2

(*c*)

F + MgBr, Me → **79** → Me + Me; 24%; 29%

RCH_2F + X–MgBr → **79** → X–R; Yield 64–98%

Me; *i*Pr; N; Me; N; *i*Pr; Dipp; N; Fe; Fe; N; Dipp; *i*Pr; N; Me; N; Me; *i*Pr; **79**

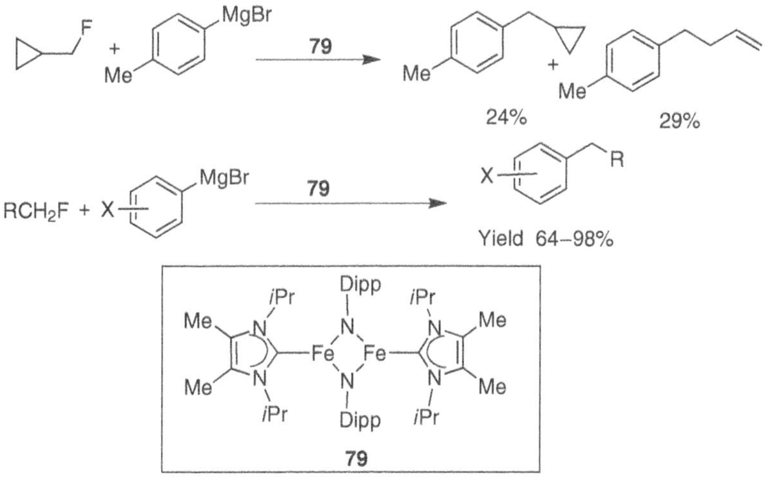

Scheme 120.

reagents as coupling partner in the presence of TMEDA as the additive (Table XX).

Later on this year (2007) Chai and co-workers (169) developed cross-coupling for unactivated alkyl halides and alkyl Grignard reagents in the presence of phosphine-based ligands, Xanthphos. In 2011, Asami and co-workers (170) reported the tridentate β-aminoketonato iron complex (**78**) as an efficient catalyst for a cross-coupling reaction between aryl magnesium bromides and alkyl halides (Scheme 120*b*).

Another approach for cross-coupling reactions involving nonactivated primary alkyl fluoride with aryl Grignard reagents was developed by Deng and co-workers (171) (in 2012) using a low-coordinate dinuclear iron complex $[(IPr_2Me_2)Fe(\mu_2\text{-}NDipp)_2Fe(IPr_2Me_2)]$ (**79**, where Dipp = 2,6-diisopropylphenyl) as catalyst. This method for C_{sp^3}–F bond arylation worked well for functionalized Grignard reagents and primary alkyl fluorides (Scheme 120*c*).

1. Low-Valent Iron Complex in Cross-Coupling Reactions

Iron-catalyzed alkylation of alkyl, allyl, and propargyl halides with an aryl Grignard reagent has been developed by using the low-valent ferrate complex $[Fe(C_2H_4)][Li(tmeda)]_2$ as a precatalyst (172). This reaction tolerated various functional groups (e.g., esters, ketone, nitriles, isocyanides, or *tert*-amines) (Table XXI). A low-valent iron-catalyst can be synthesized on a large scale as

TABLE XXI
Alkylation Catalyzed by a Low-Valent Ferrate Complex

$$RX + PhMgBr \xrightarrow[\text{THF, } -20°C]{[Fe(C_2H_4)_4][Li(tmeda)]_2 (5\text{ mol\%})} R\text{–}Ph$$

Entry	R–X	Product	Yield (%)
1	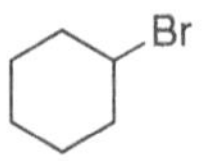 Br	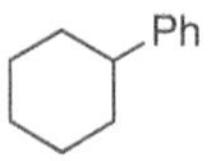 Ph	94
2	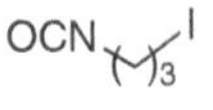 $OCN(CH_2)_3$–I	 $OCN(CH_2)_3$–Ph	90
3	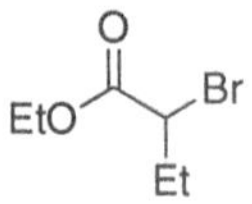 EtO, O, Br, Et	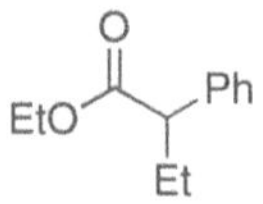 EtO, O, Ph, Et	87
4	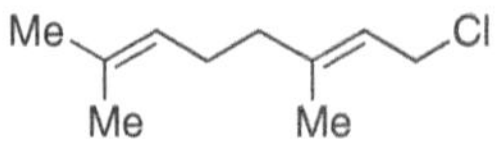 Me, Me, Me, Cl	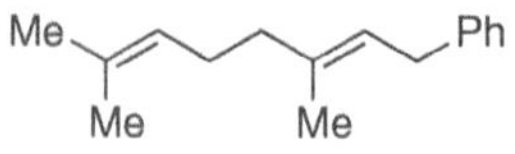 Me, Me, Me, Ph	87

an air-sensitive crystalline material on treatment of ferrocene with lithium under an ethene atmosphere (173). Following the initial report from Fürstner and co-workers (172), a variety of catalyst systems were synthesized by only varying the ligand sets. Subsequently, the catalyst system was modified further to efficiently couple the alkyl electrophile and PhMgBr (174). The reaction protocol was found to be a good alternative of Jin and Nakamura's (175), condition, where TMEDA was used in more than stoichiometric amounts (176).

D. Acyl Derivatives as Coupling Partners

Selective monoaddition of a nucleophile to an activated acid derivative is a challenging task. In 1953, Cook and co-workers (177) successfully discovered the selective monoaddition of a nucleophile to an acid chloride catalyzed by $FeCl_3$. Such a reaction was used to synthesize 1,*n*-dicarbonyl compounds using di-Grignard reagents (178). Various functionalized acid chlorides, cyanides, and thioesters with primary, secondary, and tertiary alkyl and aryl Grignard reagents–diorganozinc derivatives afforded the corresponding products in good-to-excellent yields under mild reaction condition (Table XXII) (153b, 179). Total synthesis of (*Z*)-jasmone and dihydrojasmone were done using this reaction (180).

TABLE XXII
Cross-Coupling of Acid Chloride Derivatives With RMgX

Entry	R–X	RM	Product	Yield (%)
1	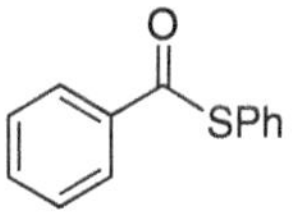 O, SPh	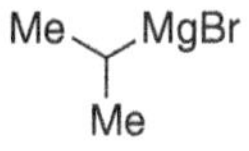 Me, MgBr, Me	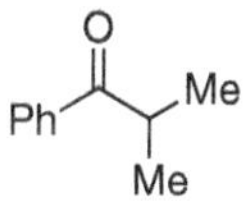 O, Ph, Me, Me	89
2	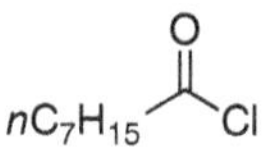 O, nC_7H_{15}, Cl	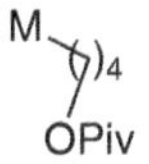 M,)4, OPiv (M = ZnR^2)	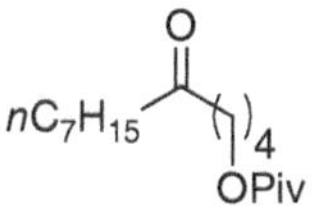 O, nC_7H_{15},)4, OPiv	89
3	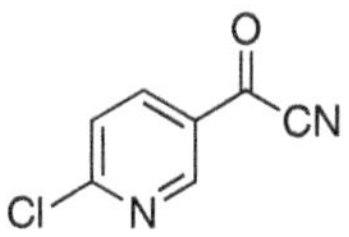 O, CN, Cl, N	PhMgBr	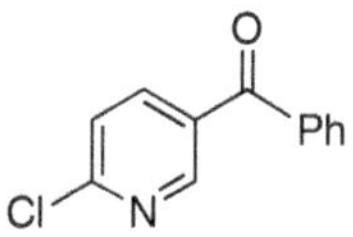 O, Ph, Cl, N	79
5	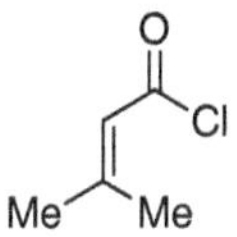 O, Cl, Me, Me	$nC_6H_{13}MgBr$	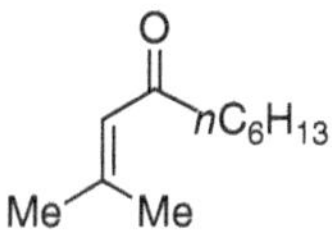 O, nC_6H_{13}, Me, Me	90

(a)

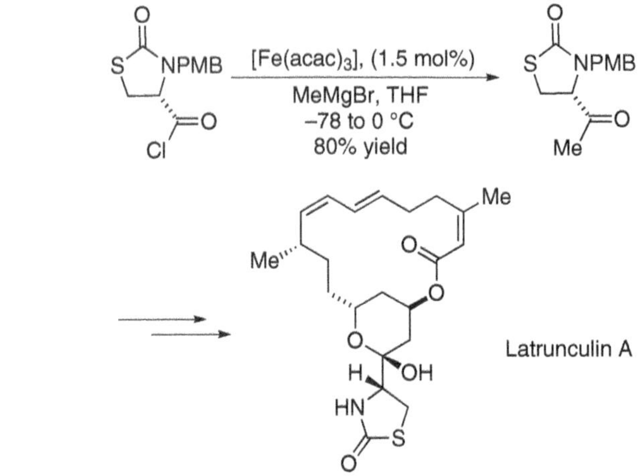

[Fe(acac)3], (1.5 mol%)
MeMgBr, THF
–78 to 0 °C
80% yield

Latrunculin A

(b)

[FeCl2(dppbz)2]
80

81 (R = tBu)
82 (R = TMS)

80–82 (1.0 mol%)
THF, –78 °C
1 h

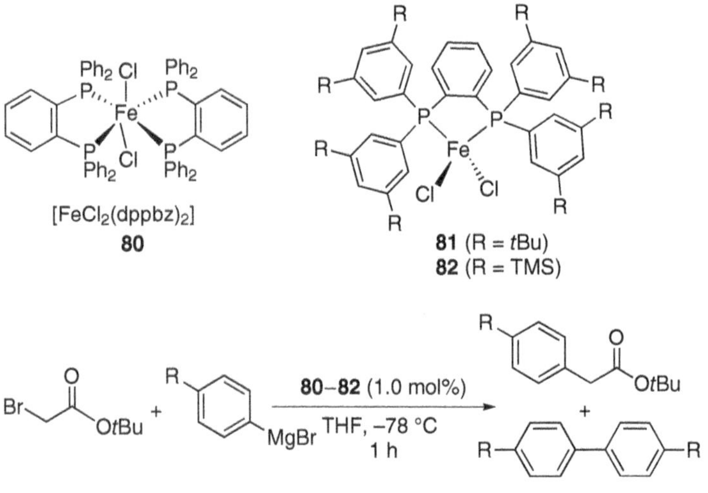

Scheme 121. Iron-catalyzed acylation toward the synthesis of natural products (TMS = trimethylsilyl).

Fürstner et al. (181) synthesized the key building blocks for a concise total synthesis of the actin-binding macrolides of the latrunculin family, like latrunculin B (Scheme 121*a*) and the musk odorant (*R,Z*)-5-muscenone. In 2011, Jin and Nakamura (175) reported a simple and effective cross-coupling reaction of α-bromocarboxylic acid derivatives with aryl Grignard reagents. Better yields were obtained by using bisphosphine iron(II) complexes (**80–82,** Scheme 121*b*). The reaction proceeds at low temperature in a chemoselective manner to produce coupling products in good-to-excellent yields.

E. Iron-Catalyzed C–O, C–S, and C–N Cross-Coupling Reaction

Iron-catalyzed methods were also applicable for a carbon–heteroatom bond formation including C–N, C–O, and C–S. A cooperative catalyst system composed of $Fe(acac)_3/CuO$ allowed *N*-arylation of aromatic heterocycles and lactams (182). Later, this type of C–N bond formation was done using $FeCl_3/Fe_2O_3$ in the presence of *N,N'*-dimethylethylenediamine (dmeda, ligand) (183). Although iron can catalyze an array of C-heteroatom coupling reactions, effect of metal contaminants in these cross-coupling reactions needs to be verified in light of recent developments reported by Buchwald and Bolm (88).

Amine-based ligand dmeda was also used for C–S bond formation (Table XXIII) (184). Recently, the phosphine-based ligand xantphos has also

TABLE XXIII
Iron-Catalyzed Carbon–Heteroatom Bond Formation

Entry	R–X	Ar–X	Product	Yield (%)
1	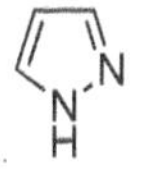	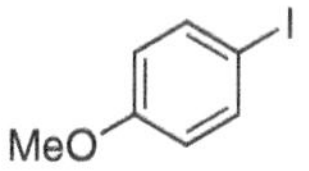	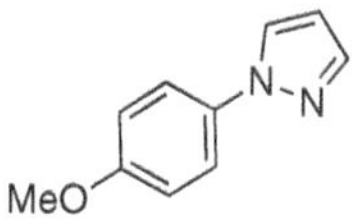	98[a]
2	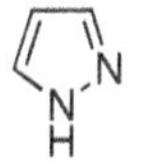	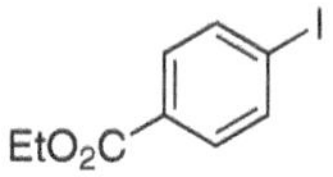	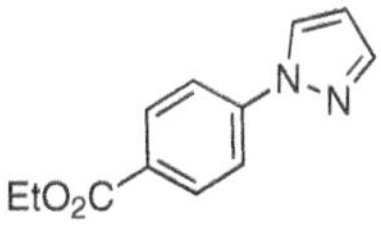	74[b]
3	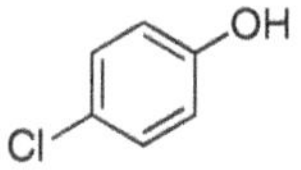	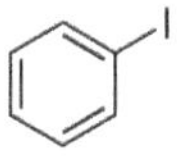	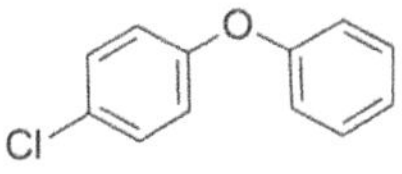	87[c]
4	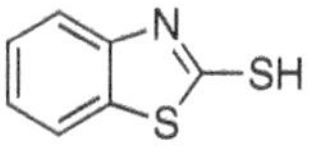	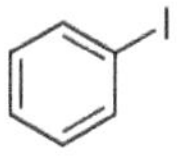	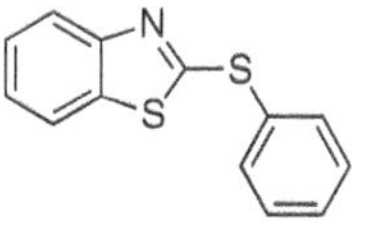	91[d]
5	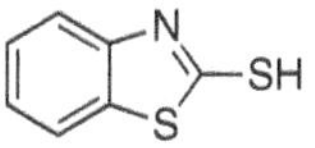	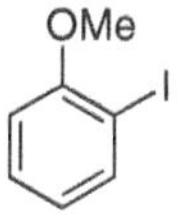	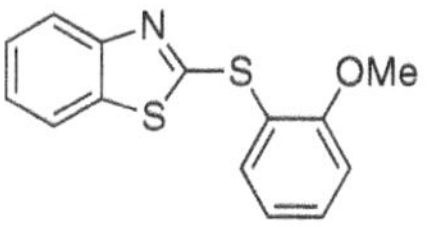	91[d]

[a] Conditions: CuO (10 mol%), $Fe(acac)_3$ (30 mol%), Cs_2CO_3 (2 equiv), DMF, 90 °C, 30 h.
[b] Conditions: $FeCl_3$ (10 mol%), dmeda (20 mol%), K_3PO_4, PhMe, 135 °C, 24 h.
[c] Conditions: $FeCl_3$ (10 mol%), $(t\text{BuCO})_2CH_2$ (20 mol%), Cs_2CO_3, DMF, 135 °C.
[d] Conditions: $FeCl_3$ (10 mol%), 20 mol% dmeda, NaO*t*Bu, toluene, 135 °C.

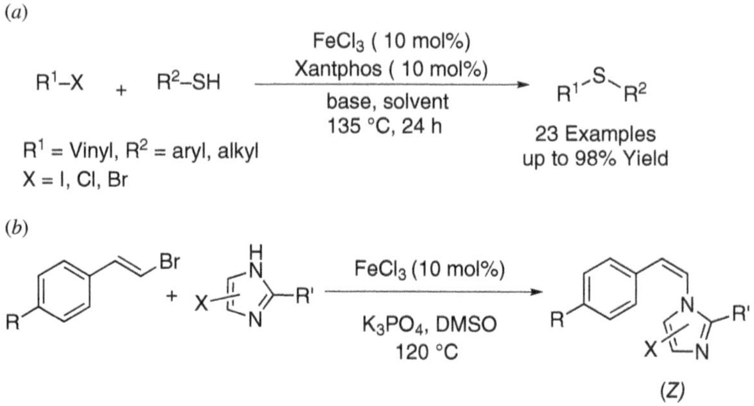

(*a*)

R^1–X + R^2–SH → $FeCl_3$ (10 mol%), Xantphos (10 mol%), base, solvent, 135 °C, 24 h → R^1–S–R^2

R^1 = Vinyl, R^2 = aryl, alkyl
X = I, Cl, Br

23 Examples
up to 98% Yield

(*b*)

$FeCl_3$ (10 mol%)
K_3PO_4, DMSO
120 °C

(*Z*)

Scheme 122.

been used for C–S bond formation with broad substrates studied by Lee and co-workers (185) (Scheme 122*a*). *S*-Vinylation is limited to 1-(2-bromovinyl)benzene and 1-(2-chlorovinyl) benzene. Related C–O bond formation was developed by using $FeCl_3$/2,2,6,6-tetramethyl-3,5-heptadione system (Table XXIII) (186).

In 2009, Mao et al. (187) reported an iron-catalyzed cross-coupling reaction of vinyl bromides/chlorides with imidazoles in the absence of ligands and additives (Scheme 122*b*, dimethyl sulfoxide = DMSO, solvent). Notably (*E*)-vinyl bromides predominantly led to (*Z*)-products, while (*E*)-vinyl chlorides predominantly afforded (*E*)-isomers (Scheme 122*b*).

F. Iron-Catalyzed Mizoraki–Heck Reaction

One of the important methods for C–C bond formation is the Mizoraki–Heck reaction (188). Traditionally, palladium catalysts were used for this reaction. However, iron-catalyzed methods could provide a greener and sustainable approach. Consequently, the iron-catalyzed Mizoraki–Heck reaction between aryl–heteroaryl iodide and styrene was developed by Vogel and co-workers (189) in 2008. A proline or picolinic acid ligated iron(II) species (20 mol%) generated (*E*)-alkene in a stereoselective manner (Scheme 123).

G. Iron-Catalyzed Negishi Coupling Reaction

An iron-catalyzed Negishi reaction was reported by Nakamura M. and co-workers (190) in 2009. The polyfluorinated arylzinc reagents (Scheme 124*b*) and

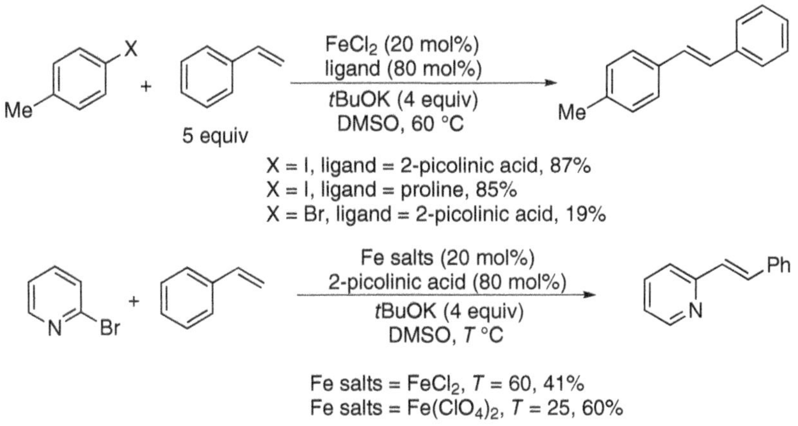

Scheme 123.

alkyl halides (Table XXIV) were coupled by using either iron-1,2-bis(diphenylphosphino)benzenedichloride (**80**) or a combination of $FeCl_3$ and dppbz[1,2-bis (diphenylphosphino)benzenedichloride].

In 2009, Bedford et al. (191) explored the scope of an iron-catalyzed Negishi reaction with 2-halopyridine and pyrimidine substrates in a THF–toluene mixture (Table XXV). They proposed formation of an Fe–Zn bimetallic intermediate (**83**) for the success of **80** in the Negishi reaction. The arylzinc reagent helped to stabilize the putative active catalyst **84** (Scheme 124*a*) by preventing catalyst decomposition (191).

Nakamura and co-workers (192) reported a cross-coupling protocol between primary–secondary alkyl sulfonates and arylzinc reagents in the presence of $FeCl_3$

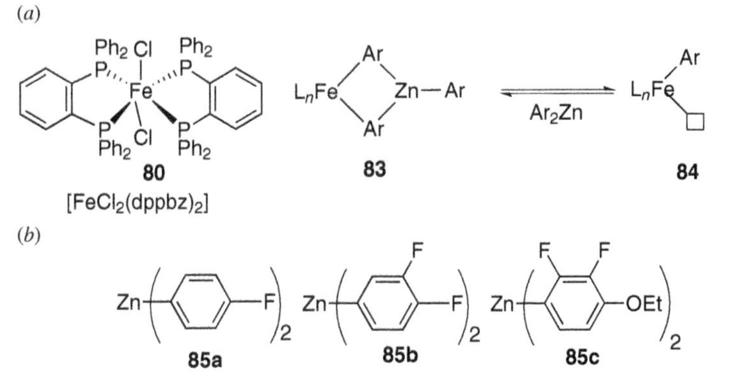

Scheme 124.

TABLE XXIV
Cross-Coupling of Arylzinc Reagents and Alkyl Halides[a]

$$R{-}X \xrightarrow[\text{THF, 60°C, 3 h}]{ZnAr_2(1.2\ \text{equiv})FeCl_3\ (3\ \text{mol\%})\ \text{with additive or}\ FeCl_2(dppbz)_2\ (3\ \text{mol\%})} R{-}Ar$$

Entry	R–X	Ar_2Zn	Product	Yield (%)[b]	Condition
1	*c*Hep-Br	**85a**	cHep–C6H4–F	92	(60 °C, 3 h)
2	Br–C6H4–CH2CH2–Br	**85b**	Br–C6H4–CH2CH2–C6H3F2	77[c]	(60 °C, 15 h)
3[d,e]	$C_{10}H_{21}I$	**85c**	$C_{10}H_{21}$–C6H2F2–OEt	91	(60 °C, 24 h)
4	BuO–CH(CH2I)–O–CH2CH=CH2	**85b**	BuO–(tetrahydrofuran, Me, Me)–CH2–C6H3F2	85	(60 °C, 15 h)

[a] Reactions were carried out on a 1.0–2.0-mmol scale.
[b] Isolated yield.
[c] The ^1H NMR yield.
[d] Used 3 mol% of $FeCl_3$ and 9 mol% of dppbz.
[e] Used 1.5 equiv of Ar_2Zn.

TABLE XXV
Iron-Catalyzed Negishi Coupling of Halopyridines[a]

$$\text{Het-X} + Zn(4\text{-tolyl})_2 \xrightarrow{\mathbf{80}\ (5\ \text{mol\%})} \text{Het-Ar}$$

Entry	Heteroaryl Halide	Product	Yield (%)
1	2-X-pyridine	2-(4-MeC6H4)pyridine	65[b]
2	2-Br-pyrimidine	2-(4-MeC6H4)pyrimidine	58(58)
3	2,5-dibromopyridine	5-Br-2-(4-MeC6H4)pyridine	56(50)

[a] Conditions: Heteroarylhalide (1.0 mmol), $Zn(4\text{-tolyl})_2$ (0.25 *M* in THF, 4.8 mL), 1 (0.05 mmol), toluene (6 mL), 100 °C, 4 h.
[b] Used 1 equiv Zn(4-tolyl).

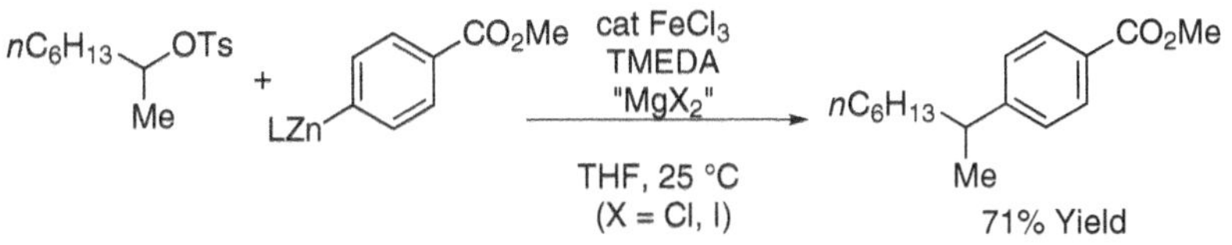

Scheme 125.

and TMEDA. The arylzinc reagent was prepared from aryllithium or magnesium reagents with ZnI_2. *In situ* formation of alkyl halides from sulfonates avoided discrete preparation of secondary alkyl halides and gave high product selectivity (Scheme 125).

H. Suzuki–Miyaura Coupling Reaction

Nakamura and co-workers (193) developed iron-catalyzed Suzuki–Miyaura coupling using iron(II) chloride–diphosphine complexes (**81**, **82**) (Scheme 126) as catalyst with magnesium bromide, lithium aryl borates. Primary and secondary alkyl halides were successfully employed as the coupling partners for this reaction.

The reaction protocol was applicable for arylboronic acid and pinacol esters possessing methoxy, dimethylamino, halides, and alkoxy carbonyl functional groups (Table XXVI). The reaction occurred via formation of *in situ* borate with butyl- or *tert*-butyllithium (192).

A stereospecific cross-coupling between alkenylboronates and alkyl halides catalyzed by iron–bisphosphine complexes (**81**) was reported in 2012 by Nakamura and co-workers (194) (Table XXVII). An *in situ* generated lithium alkenylborate was the active alkenylation agent for this reaction.

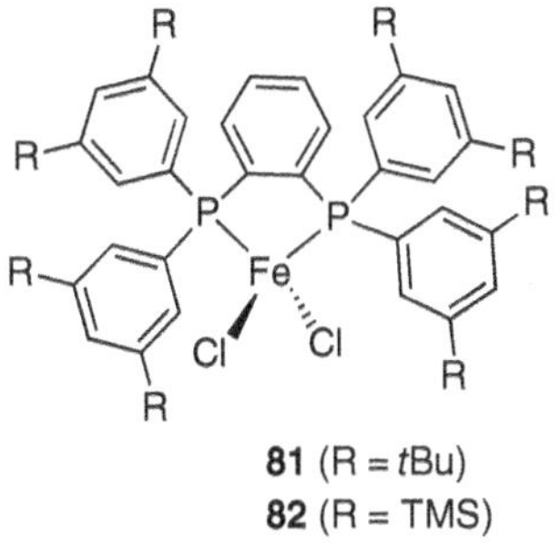

81 (R = *t*Bu)
82 (R = TMS)

Scheme 126. Iron-bisphosphine complexes (192).

TABLE XXVI

Substrate Scope of Suzuki–Miyaura Coupling of Alkyl Halides

R'—X + [Ar(R)B(pinacolate)]$^-$ M^+ → R'—Ar (81 or 82 (1–5 mol%), $MgBr_2$ (0–20 mol%), THF, 0–40 °C)

R' = 1°, 2°, alkyl, R = Et, Bu, *t*Bu, Ar = aromatic group, M = Li or MgBr

Entry[a]	Alkyl–X	Coupling Product	Yield (%)[b]
1[c]			99 (R″ = H)
2	*c*Hept–Br	*c*Hept–C_6H_4–R″	98 (R″ = OMe)
3[d–f]			90 (R″ = CO_2Me)
4[d–f]			81 (R″ = CO_2Et)
5[d–f]			74 (R″ = $CO_2i Pr$)
6	Indol-1-yl–$(CH_2)_2$–CH_2I	Indol-1-yl–$(CH_2)_2$–CH_2–(2-F-C_6H_4)	79
7[f,g]	EtO–C(O)–$(CH_2)_5$–CH_2Br	EtO–C(O)–$(CH_2)_5$–CH_2Ph	90
8[e]	AcO–C_6H_{10}–Br	AcO–C_6H_{10}–C_6H_4–OPiv	83[h]

[a]Reaction was carried out at 40 °C for 3–4 h on a 0.44–1.0-mmol scale using 3 mol% **82**. Arylborates were prepared from arylboronic acid pinacol ester and *t*BuLi.

[b]Isolated yield.

[c]Using 1 mol% **82**.

[d]Using 1 mol% **82**.

[e]Used 5 mol% **82**.

[f]At 25 °C for 3–8 h.

[g]Used BuLi instead of *t*BuLi.

[h]The cis–trans ratio: 42:58.

The reaction protocol was also applied to alkyl–alkyl Suzuki–Miyaura coupling in the presence of [Fe(acac)$_3$] (3 mol%)–xanthphos(6 mol%) and a stoichiometric amount of *i*PrMgCl (Scheme 127*a*). Here *i*PrMgCl acted as an efficient activator for trialkylboranes by forming magnesium tetraalkylborate, which was the key

TABLE XXVII
Substrate Scope of Alkyl-Alkenyl Suzuki–Miyaura Coupling

R–X + [Me, Me, Me, Me, O, O, B, tBu, R^2, R^1, R^3]$^-$ Li^+ → (**81** or **82** (5 mol%), $MgBr_2$ (20 mol%), THF, –20 to 40 °C, 24 h) → R, R^2, R^3, R^1

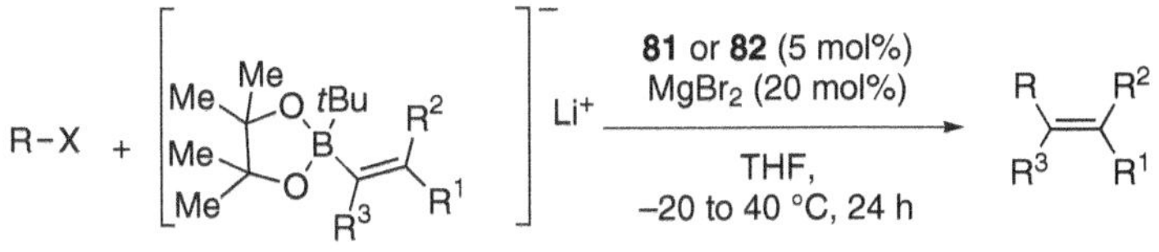

Entry	R–X	Coupling Product	Yield (%) (*E/Z*)
1	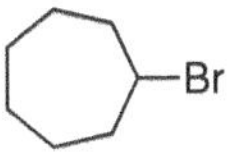 Br	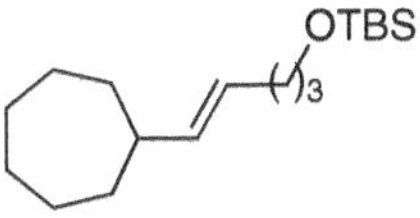 OTBS, ( )3	98 (>99% *E*)
2	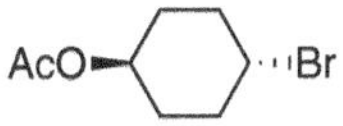 AcO, Br	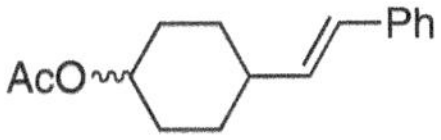 AcO, Ph	98 (>99% *E*)
3	 NC, ( )5, Br	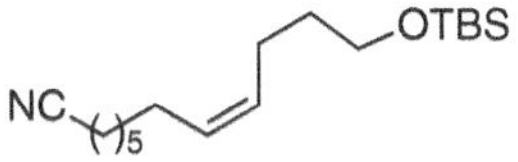 NC, ( )5, OTBS	98 (>99% *E*)
4	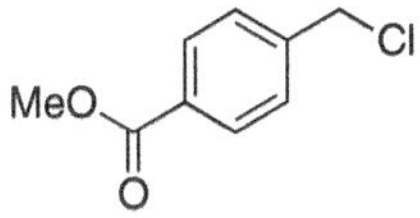 MeO, O, Cl	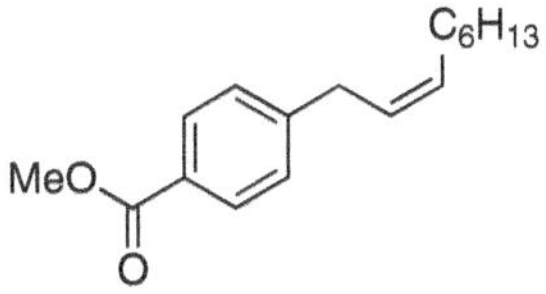 MeO, O, C_6H_{13}	98 (>99% *E*)
5	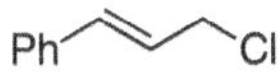 Ph, Cl	 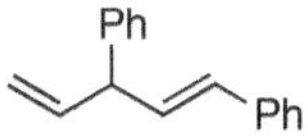 Ph, Ph Ph, Ph	98 (>99% *E*) 98 (>99% *E*)

transmetalating agent for this reaction (195). A radical clock experiment (Scheme 127*b*) gave a ring-opening cross-coupling product, which indicated the formation of an alkyl radical intermediate.This radical intermediate was suggested to trigger the Suzuki–Miyaura coupling. Recently, a modified protocol with iron–bisphosphine complexes (**81, 82**) was developed with unprotected nonactivated halohydrins and aryl aluminium reagents as the coupling partners. In this report, a free alcohol group formed an alkoxide, which in turn accelerated the reaction and enhanced the diastereoselectivity (Scheme 128) (196).

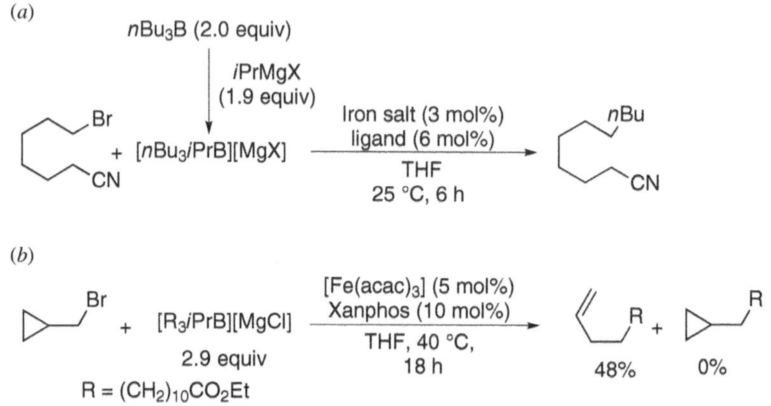

(a)

nBu_3B (2.0 equiv); iPrMgX (1.9 equiv); + $[nBu_3iPrB][MgX]$; Iron salt (3 mol%), ligand (6 mol%), THF, 25 °C, 6 h; Br; CN; nBu; CN

(b)

Br; + $[R_3iPrB][MgCl]$ 2.9 equiv; R = $(CH_2)_{10}CO_2Et$; $[Fe(acac)_3]$ (5 mol%), Xanphos (10 mol%), THF, 40 °C, 18 h; R + R; 48%; 0%

Scheme 127.

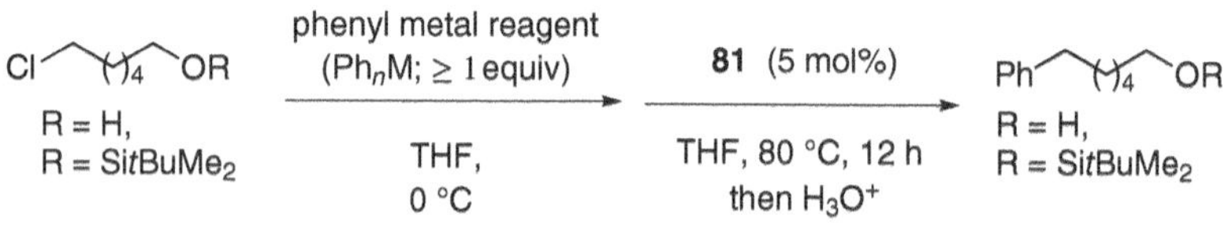

Cl $(\,)_4$ OR; R = H, R = SitBuMe$_2$; phenyl metal reagent (Ph_nM; ≥ 1equiv), THF, 0 °C; **81** (5 mol%), THF, 80 °C, 12 h then H_3O^+; Ph $(\,)_4$ OR; R = H, R = SitBuMe$_2$

Scheme 128.

I. Sonogashira Reaction

The iron-catalyzed Sonogashira reaction was developed in 2008 by Bolm and co-workers (197), with the use of $FeCl_3$, an amine-based ligand dmeda, and Cs_2CO_3 as a base in toluene (Scheme 129*a*). Use of an iron-based catalyst for this reaction instead of Pd/Cu made this method economical. Various functional groups were well tolerated on both the coupling partners. Furthermore, iron-catalyzed domino Sonogashira/hydroxylation of alkynes also was reported (Scheme 129*b*). Notably, in 2009, Buchwald and Bolm (88) reported the effect of metal contaminants in these cross-coupling reactions and concluded that these reactions were likely catalyzed by trace copper impurities rather than iron. Their correspondence raised the question about the role of iron in the Sonogashira reaction and related cross-coupling reactions.

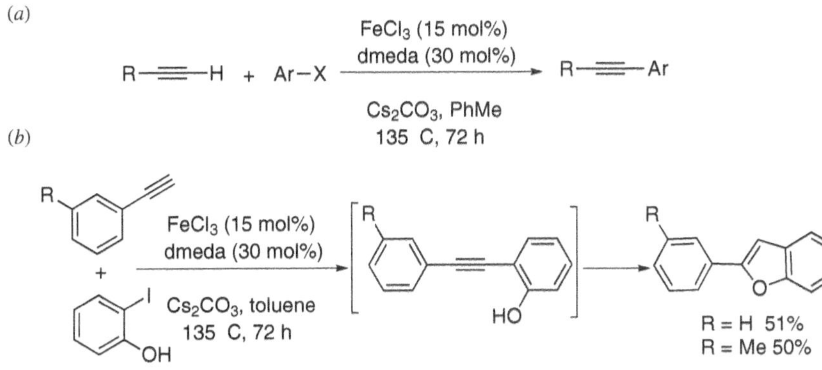

(*a*)

R—≡—H + Ar—X → R—≡—Ar

$FeCl_3$ (15 mol%)
dmeda (30 mol%)
Cs_2CO_3, PhMe
135 C, 72 h

(*b*)

R
+
I
OH
$FeCl_3$ (15 mol%)
dmeda (30 mol%)
Cs_2CO_3, toluene
135 C, 72 h
R
HO
R
O
R = H 51%
R = Me 50%

Scheme 129. Iron-catalyzed Sonogashira reaction.

J. Mechanism of Cross-Coupling Reactions

Initial mechanistic studies by Kochi (145) suggested that the iron(II) and iron(III) species rapidly oxidize Grignard reagent to give a reduced soluble form of iron that was most likely Fe(I). They also proposed that cross-coupling was independent of the concentration of alkyl magnesium halide and first order with respect to alkenyl halide and iron catalyst. Based on these observations, Kochi proposed a catalytic pathway involving oxidative (ox) addition, transmetalation, and then reductive elimination (Scheme 130*a* and *b*) similar to Pd catalyzed cross-coupling reactions.

Fürstner et al. (199) proposed the formation of a low-valent ferrate complex as the active species for their cross-coupling reactions based on their experimental data and literature reports (173, 198). At first, there was an *in situ* formation of $[Fe^{(-II)}(MgX)_2]$ (ferrate complex) and subsequently it was oxidatively added between the R−X bond. The formation of this type of ferrate complex was reported in the literature with the X-ray crystal structure of [Cp(dppe)Fe (MgBr).3THF]. It has a covalent-bond character between the Fe and Mg centers. Such an observation supported the idea that iron can remain covalently bonded to magnesium in $[Fe(MgX)_2]$, the "Inorganic Grignard Reagent" (173, 198). They found that finely dispersed Fe(0)* particles in THF dissolved slowly on treatment with an excess of nC$_{14}$H$_{29}$MgBr and the resulting solution catalyzed the cross-coupling reaction (Scheme 131). Further, iron complexes of different oxidation states −2, 0, +1, +2 and +3 were prepared, which were devoid of stabilizing ligands, and were tested for their activity toward the cross-coupling reaction. It was observed that nucleophiles (e.g., MeLi, PhLi, or PhMgBr) were unable to undergo β-hydride elimination. Rather, they rapidly reduce Fe^{3+} to Fe^{2+} and then alkylated

(*a*)

Initiation

Fe^{III} + 2 R-MgX ⟶ Fe^{I} + R_{OX}

Propagation–Oxidative Addition–Reduction Elimination

Fe^{I} + R-MgX ⟶ R-Fe^{1-} + MgX

R-Fe^{1-} + R'-Br ⟶ R-Fe^{III}(R') + Br^{-}

Propagation (II) Substitution

Fe^{I} + R'-Br ⟶ R-Fe(I)Br

R-Fe^{I}Br + R-MgX ⟶ R-Fe^{III}(R') + Br^{-}

Termination

R-Fe^{III}(R') ⟶ R-R' + Fe^{I}

(*b*)

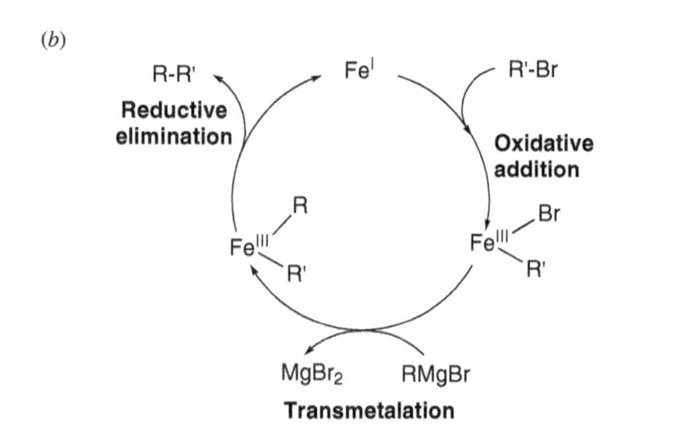

Scheme 130.

the metal center. The resulting homoleptic organoferrate complexes, like [(Me_4Fe)(MeLi)]-[Li(OEt_2)$]_2$, which was characterized by X-ray crystallography, transfered their organic ligand to activate electrophilic partner (173). However, the nucleophile, which underwent β-hydride elimination, was likely to follow a low-valent ferrate mechanism. Consequently, the reaction mechanism was dependent on substrates and the redox couple of iron in the solution. Based on these detailed studies, they had proposed both an organoferrate manifold and a low-valent redox manifold for these coupling reactions (Scheme 132) (199).

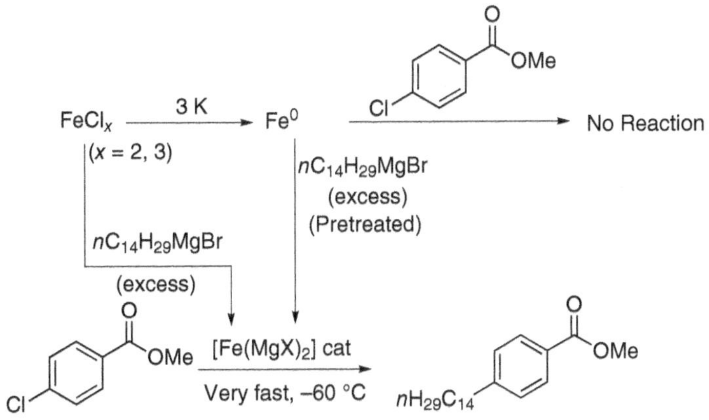

$FeCl_x$ (x = 2, 3) —3 K→ Fe^0 —(4-Cl-C_6H_4-CO_2Me)→ No Reaction

$FeCl_x$ —$nC_{14}H_{29}MgBr$ (excess)→ $[Fe(MgX)_2]$ cat

Fe^0 —$nC_{14}H_{29}MgBr$ (excess) (Pretreated)→ $[Fe(MgX)_2]$ cat

4-Cl-C_6H_4-CO_2Me —$[Fe(MgX)_2]$ cat, Very fast, –60 °C→ 4-($nH_{29}C_{14}$)-C_6H_4-CO_2Me

Scheme 131.

Later, Holzer and Hoffman (200) found a sufficient amount of racemization for chiral substrates under iron-catalyzed cross-coupling in contrast to Ni/Pd catalyzed reactions, which gave products without any loss in optical purity. On the basis of these observations, an SET mechanism (radical mechanism) was proposed rather than transmetalation.

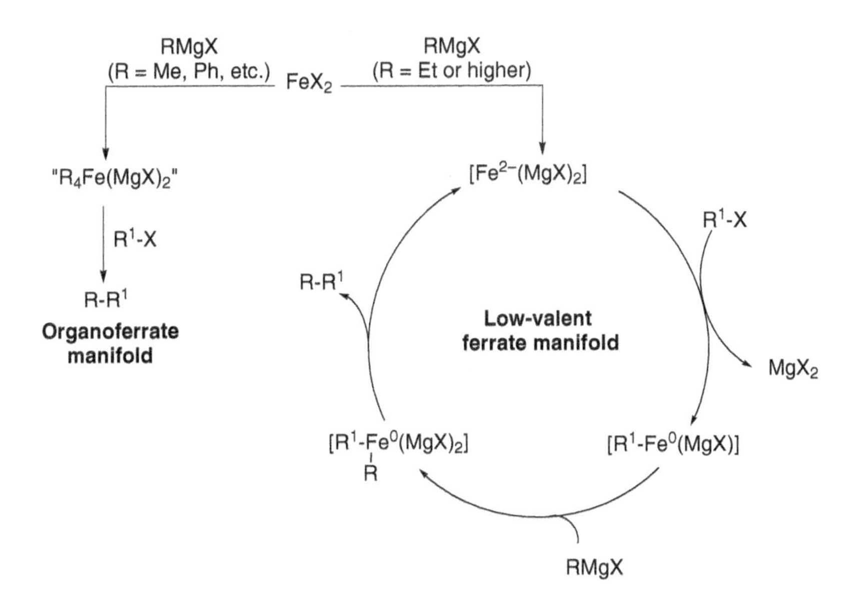

Scheme 132.

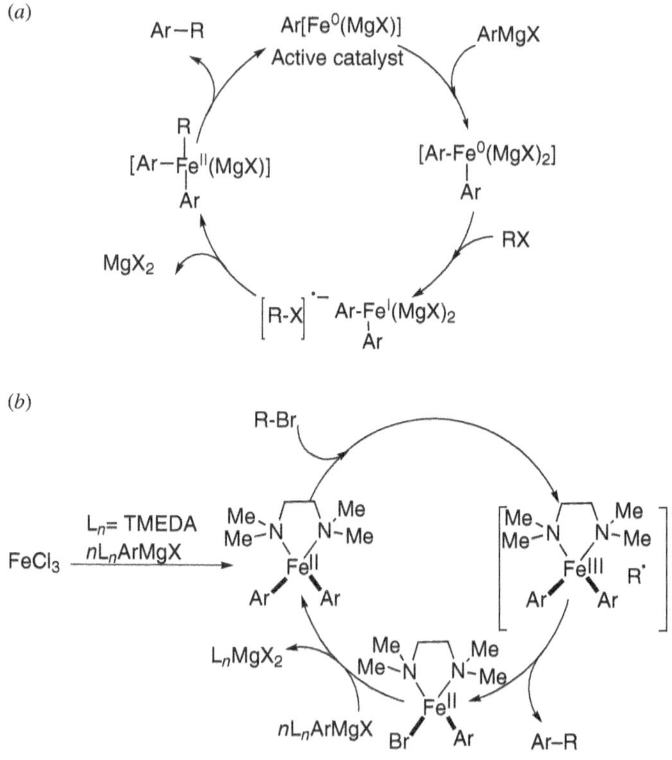

Scheme 133.

Cahiez et al. (176) proposed the Fe(0)/iron(II) cycle via formation of an alkyl radical anion (Scheme 133*a*). Cossy and co-workers (168a) also proposed a radical-based oxidative addition and a "radical clock" experiment was successfully carried out that further supported their hypothesis.

In 2009, Nagashima and co-workers (201) proposed an iron(II)–iron(III) catalytic cycle on the basis of the isolation of catalytically competent intermediates of both iron(II) and iron(III) complexes. Once again, their "radical clock" experiment confirmed a radical intermediate during these reactions (Scheme 133*b*).

In 2008 and 2009, Norrby and co-workers (202) supported the iron(I)/iron(III) cycle, which was proposed 40 years ago by Kochi on the basis of their mechanistic and computational studies. More recently in 2012, Bedford and co-workers (203) reported that iron(I) was the lowest kinetically reasonable oxidation state in a representative Negishi cross-coupling reaction with aryl zinc reagents and benzyl bromide on the basis of the isolation of the catalytically competent Fe(I) species.

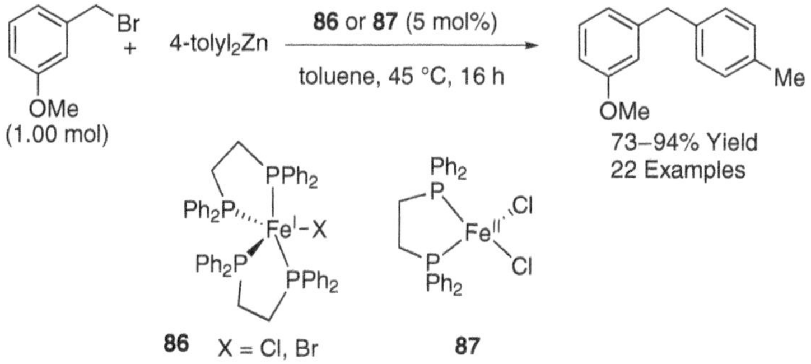

Scheme 134.

Interestingly, better results were obtained in the presence of iron(I) phosphine compared to the iron(II) phosphine complexes for reaction between aryl/alkyl halide and an arylzinc reagent (204) (Scheme 134).

As discussed above, the iron-catalyzed cross-coupling reaction mechanism is versatile and varies with the nature of the coupling partners the and on the reaction condition. Formation of a radical intermediate is likely, and the iron(I)–iron(III) catalytic cycle is more accepted compared to others.

K. Hydrocarboxylation

The iron-catalyzed hydrocarboxylation reaction was first reported by Greenhalgh and Thomas (205). Bench-stable $FeCl_2$ was used as an iron precatalyst in the presence of different amine ligands (Scheme 135; **88–91**). It was observed that $FeCl_2$–bis(imino)pyridine ligands (**88–89**) and hydride source (EtMgBr, 1.2

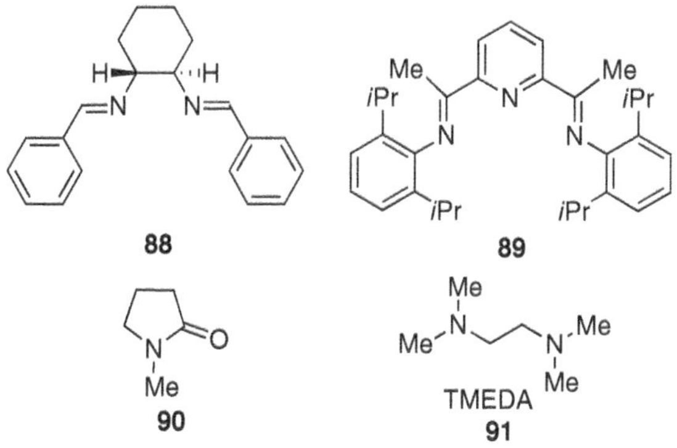

Scheme 135.

TABLE XXVIII
Substrates Scope of Hydrocarboxylation[a]

Ar–CH=CH$_2$ → (1) $FeCl_2$ (1 mol%), **89** (1 mol%), EtMgBr (120 mol%), THF, 2 h, rt (2) CO_2 → Ar–CH(Me)–CO_2H

Yield[b]
(Reaction yield α:β)[c]

Substrate	Yield
4-Me-C$_6$H$_4$–CH(Me)–CO_2H	90% (93% >, 30:1)
4-*t*Bu-C$_6$H$_4$–CH(Me)–CO_2H	83% (85% >, 40:1)
3-OBn-C$_6$H$_4$–CH(Me)–CO_2H	72% (72% >, 100:1)
Me-substituted aryl (Me, Me)–CH(Me)–CO_2H	67%[d] (78%, 1:6)

[a] Condition: 0.7 mmol of 1, 1 mol% $FeCl_2$, 1 mol% 7, THF (0.15 *M*), rt; (1) 120 mol% EtMgBr (3 *M* in Et_2O), 2 h. (2) CO_2, 30 min.
[b] Isolated yield α-product.
[c] Reaction yield and regioselectivity determined by proton nuclear magnetic resonance (^{1}H NMR) using an internal standard.
[d] Used 120 mol% cyclopentylmagnesium bromide (2 *M* in Et_2O); isolated yield of β-product.

equiv) gave the best yield of α-aryl carboxylic acid in the presence of CO_2 at atmospheric pressure (Table XXVIII).

A mechanistic investigation was carried out with methanol-d_4 instead of CO_2. In this experiment, 1-deuteroethylbenzene was observed as the main product, which indicated the possibility of α-aryl organometallic species formation. Such α-aryl organometallic species (**93**) could be formed from **92**. Subsequently, **93** underwent β-hydride elimination, hydrometalation, and transmetalation to form α-aryl organometallic species (**96**). The nucleophile **96** attacked CO_2 to give the desired product (Scheme 136).

L. Enyne Cross-Coupling Reaction

Conjugated enynes were the key structural motif of various bioactive molecules, drug intermediates, and organic electronic materials (206). Generally, Pd was known to catalyze the synthesis of these types of molecules via the $C_{sp}-C_{sp^2}$ coupling reaction. Nakamura and co-workers (207) first reported the $FeCl_3$ catalyzed enyne cross-coupling in 2008 (Table XXIX, TBDMS = *tert*-butyldimethylsilyl). Lithium bromide was used as the crucial additive and $FeCl_3$ efficiently catalyzed the cross-coupling of alkynyl Grignard reagent with alkenyl bromides–triflates.

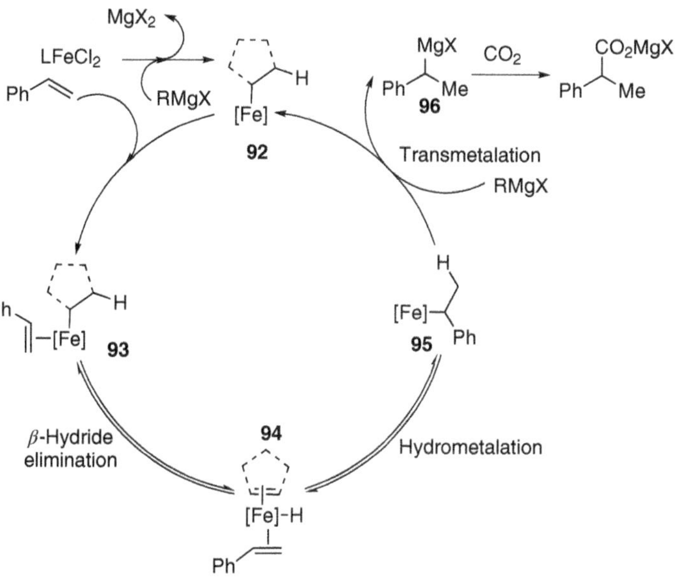

Scheme 136. Mechanism of hydrcarboxylation.

TABLE XXIX
Iron-Catalyzed Enyne Cross-Coupling[a]

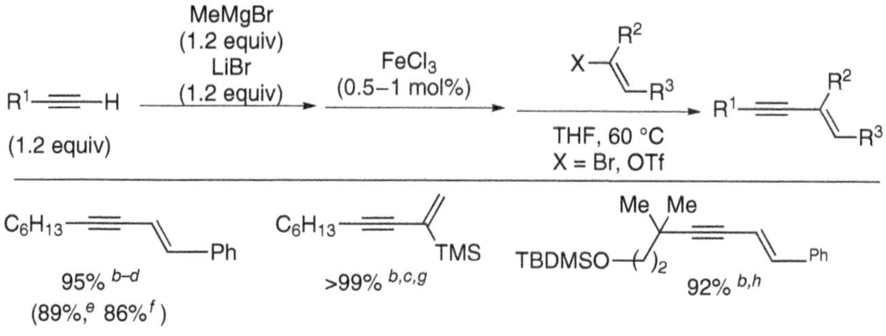

R^1—≡—H (1.2 equiv) → MeMgBr (1.2 equiv), LiBr (1.2 equiv) → $FeCl_3$ (0.5–1 mol%) → X–C(R^2)=CH–R^3, THF, 60 °C, X = Br, OTf → R^1—≡—C(R^2)=CH–R^3

Product	Product	Product
C_6H_{13}—≡—CH=CH—Ph 95% [b–d] (89%,[e] 86%[f])	C_6H_{13}—≡—C(=CH$_2$)TMS >99% [b,c,g]	TBDMSO—(CH_2)$_2$—C(Me)(Me)—≡—CH=CH—Ph 92% [b,h]

[a] Reactions were carried out at 60 °C for 24 h on a 1.0 mmol scale in the presence of 1 mol% of $FeCl_3$ unless otherwise noted (TBDMS = *tert*-butyldimethylsilyl).
[b] Isolated yield.
[c] Used 0.5 mol% of $FeCl_3$.
[d] Ratio was (*E/Z*) 88:12.
[e] Used 0.5 mol% of $FeCl_2$.
[f] Used 0.5 mol% of Fe(acac)$_3$.
[g] Reaction time was 12 h.
[h] Ratio was (*E*)/(*Z*) 85:15.

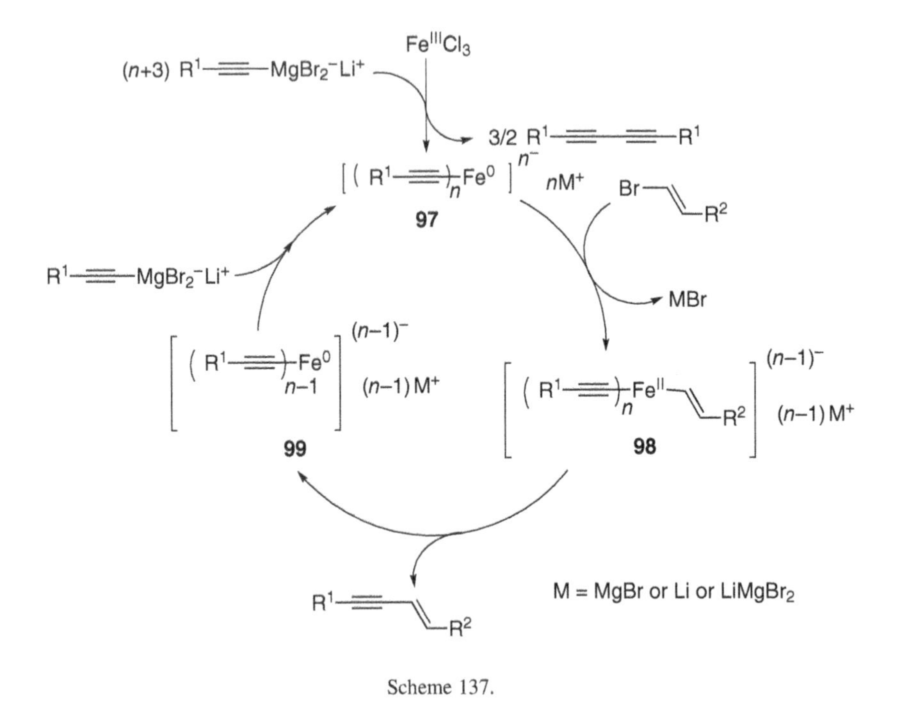

Scheme 137.

They proposed the mechanism of enyne cross-coupling via formation of the alkenyl iron complex **97**. Subsequently, oxidative addition of alkenyl bromide to **97** resulted in the formation of high-valent ferrate (**98**), which underwent reductive elimination to give the desired product. The presence of a lithium salt was important to reduce Fe^{III} to the low-valent ferrate complex **97** (Scheme 137).

VI. DIRECT C–N BOND FORMATION VIA C–H OXIDATION

Transition metal mediated C–N bond formation methods via C–H activation are important for the synthesis of nitrogen-containing organic compounds (208). Nitrene intermediates or derivatives were used as the nitrogen source in most cases (1d, 209). Often, hazardous compound PhI=NTs is used for C–N bond formation. Consequently, use of amines and amides as the nitrogen source is challenging and more sustainable (1d).

In 2008, Fu and co-workers (210) reported an efficient, inexpensive, and air-stable catalyst–oxidant ($FeCl_2$–NBS) system for amidation of benzylic sp^3 C–H in ethyl acetate under mild conditions.The reaction tolerated a variety of substrates with variations in benzyl sp^3 C–H bonds (**100a–100c**) and carboxamide–sulfonamide (Table XXX). The NBS played a vital role as an oxidant and radical initiator.

TABLE XXX
Substrate Scope for C–H Amidation[a]

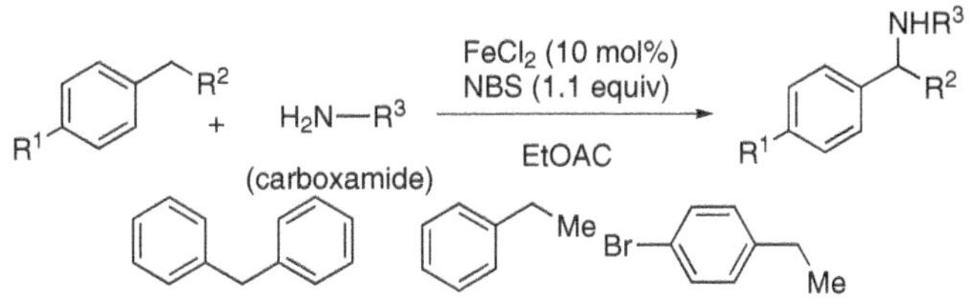

R¹ R² + H_2N—R³ (carboxamide) $FeCl_2$ (10 mol%) NBS (1.1 equiv) EtOAC → NHR³ R² R¹

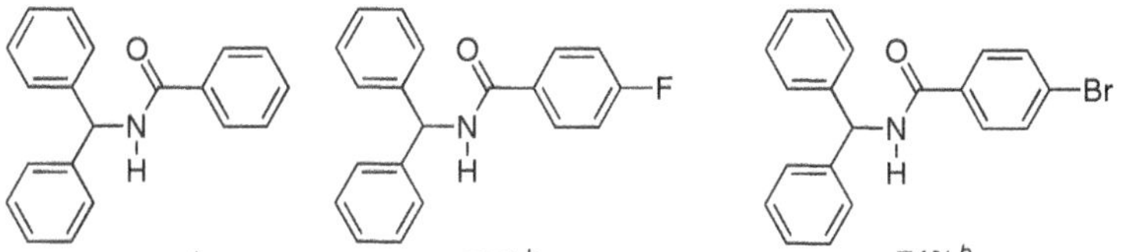

100a	100b	100c
	Me	Br, Me

68%[b]	77%[b]	74%[b]
O, N–H	O, N–H, F	O, N–H, Br

[a] Reaction conditions: benzylic reagent (1.2 mmol), amide–sulfonamide (1.0 mmol), NBS (1.1 mmol), and $FeCl_2$ (0.1 mmol).
[b] Isolated yield.

The reaction occurred via formation of the *N*-bromocarboxamide and *N*-bromosulfonamide intermediate **101**. Subsequently, it formed intermediate **102** as an active species through proton exchange. Then it formed the iron–nitrene complex **103**, which upon reacting with benzylic sp^3 C–H bonds, formed intermediate **104** to provide the desired amidation product (Scheme 138).

Later in 2010, Li and co-workers (211) developed an iron-catalyzed C–N bond-formation method without using a nitrene source (Scheme 139). Their method included oxidative C–N bond formation of azoles and ether with good-to-excellent yields. Iron(III) chloride was used as the catalyst and TBHP as the oxidant (and radical initiator too). This combination activated the α-C–H bonds of ethers. Subsequently, nucleophillic attack from azoles generated an oxidative C–N bond with ethers. Under standard reaction protocol, imidazole and its derivatives gave a high yield for such oxidative coupling with THF (Scheme 139*a*). It was also observed that benzimidazoles underwent oxidative C–N bond formation with ethers in moderate-to-good yield (Scheme 139*b*).

The reaction occurred via formation of a radical and hydroxyl anion from TBHP in the presence of iron(II). The hydroxyl anion abstracted a proton from azole and the *t*BuO• generated a radical in ether. This ether was oxidized to an oxonium ion in the presence of iron(III). Subsequently, it was attacked by an azole anion in order to from a C–N bond (Scheme 140).

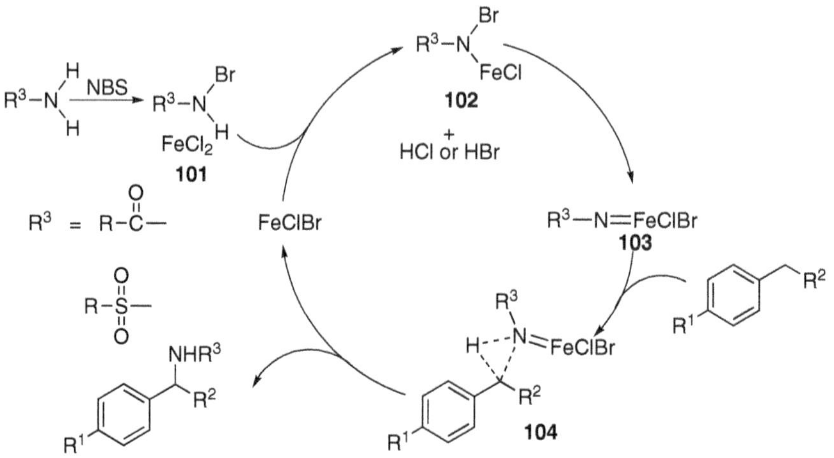

Scheme 138. Proposed mechanism of benzylic sp^3 C−H amidation.

Later in 2012, Xia and Chen (212) reported oxidative C−N bond formation between azole derivatives and amides–sulfonamides via activation of sp^3 C−H bond adjacent to the nitrogen atom, with good-to-excellent yields (Scheme 141).

Formation of a radical adjacent to nitrogen (**105**) was proposed, which was followed by oxonium ion (**106**) formation. Subsequently, nucleophillic attack occurred from azole, which gave the desired product (Scheme 142).

Iron-catalyzed direct amination was developed in 2011 by Yu and co-workers (213). They used $FeCl_3$–benzoxazoles with formamide and different secondary amines as nitrogen sources via decarbonylation (Table XXXI).

They proposed a Lewis acid catalyzed mechanism where $FeCl_3$ coordinated to a nitrogen atom and facilitated nucleophillic attack (Scheme 143).

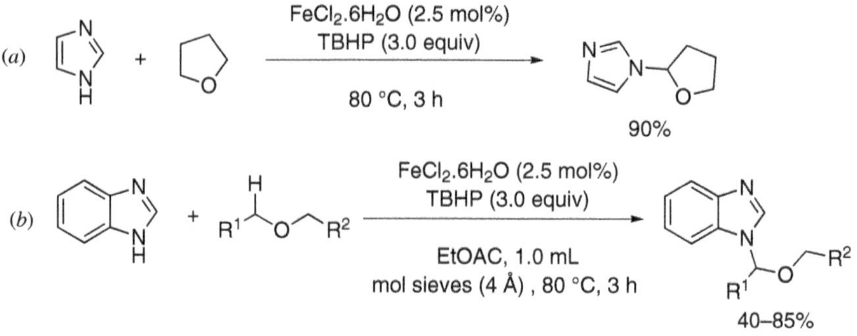

Scheme 139. Iron-catalyzed oxidative C−N bond formation.

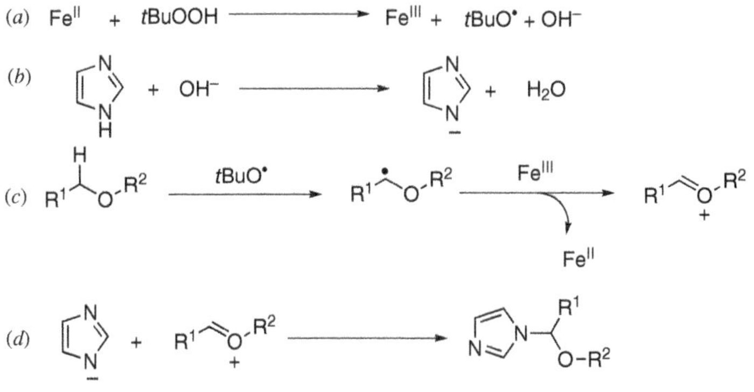

(a) Fe^{II} + tBuOOH → Fe^{III} + $tBuO^{\bullet}$ + OH^-

(b) imidazole + OH^- → imidazolide + H_2O

(c) $R^1CH_2OR^2$ —tBuO•→ $R^1\dot{C}HOR^2$ —Fe^{III} ($\to Fe^{II}$)→ $R^1CH{=}O^+R^2$

(d) imidazolide + $R^1CH{=}O^+R^2$ → N-(R^1CH(O–R^2))-imidazole

Scheme 140.

R¹-azole (NH; X,Y = C or N) + $R^2CH_2N(R^3)R^4$ —$FeCl_2$ (5 mol%), DTBP (3.0 equiv), C_6H_5Cl, 120 °C, 3–6 h→ N-[CH(R^2)N(R^3)R^4]-azole

X,Y = C or N

R^3 = acyl or sulfonyl

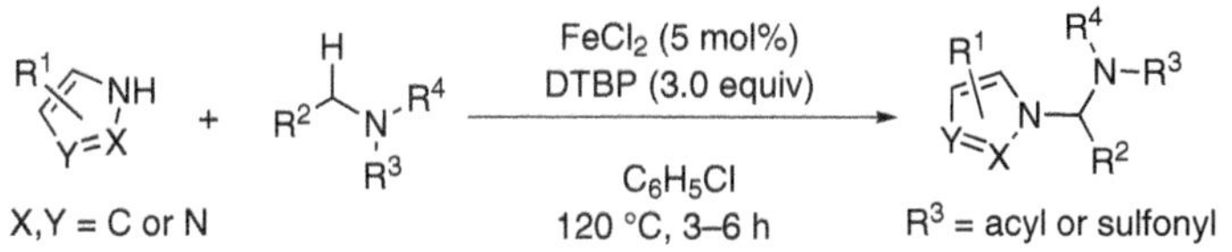

Scheme 141.

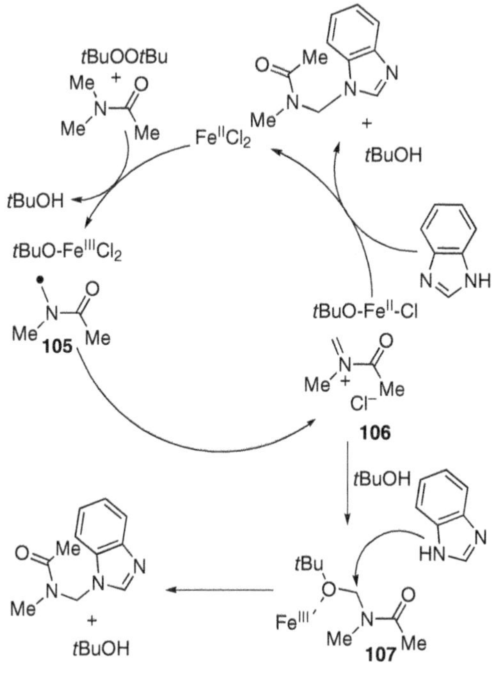

Scheme 142.

TABLE XXXI
Iron-Catalyzed Decarbonylative Amination[a]

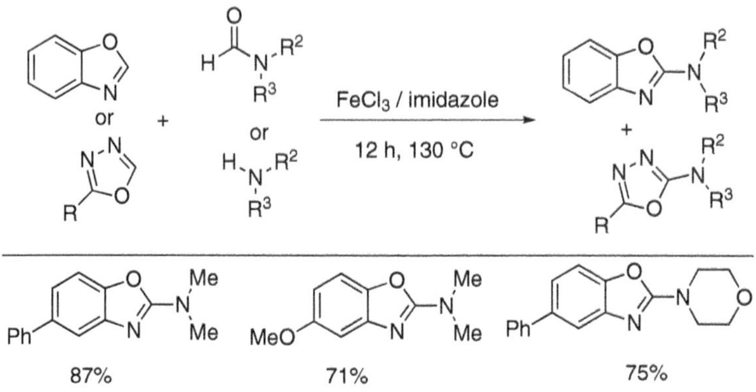

87%	71%	75%

[a]Conditions: DMF (solvent) (2 mL), $FeCl_3$ (0.25 equiv), imidazole (2.0 equiv),130 °C, 12 h, under air. Yields were isolated yields, formamide or amine was the nitrogen source.

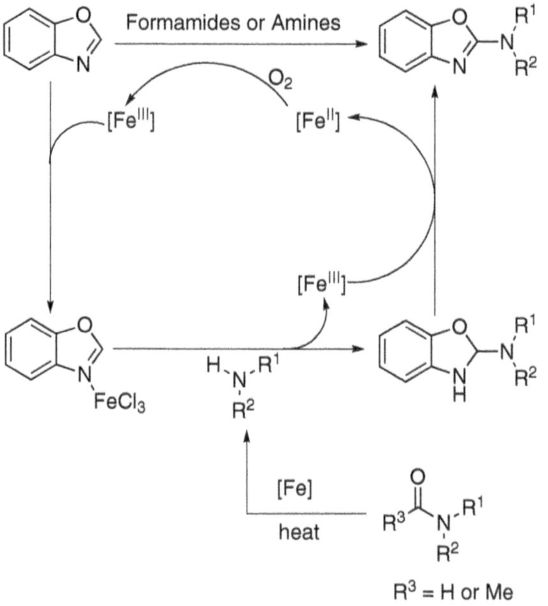

Scheme 143. Mechanism of iron-catalyzed decarbonylative amination.

VII. IRON-CATALYZED AMINATION

Amine compounds are important in the pharmaceuticals industry and different amine-based pesticides are widely used to protect crops. Hence, syntheses of different amine compounds are important.

A. Allylic Aminations

Amine compounds could be synthesized by allyl amination where nitrogen-based compounds added to unsaturated compounds (214). Allylic amination was first reported by Johannsen and Jorgenson (215) and Srivastava and Nicholas (216) independently in 1994 using allylic compounds and a phenylhydroxyl amine in the presence of an iron phthalocyanine complex (method A) or a mixture of $FeCl_2.4H_2O/FeCl_3.6H_2O$ (method B) (Table XXXII).

Olefins generally gave low-to-moderate yields of allyl amines in this reaction via a formal heteroene process (215). Iron–phthalocyanine (FePc) gave an amination product with limited substrates, mainly olefins conjugated to aromatic rings. On the other hand, use of an Fe^{II}/Fe^{III} catalyst system was more beneficial then FePc as it gave good yields with nonterminal acyclic olefins. The later catalytic system (method B), with a mixture of $FeCl_2.4H_2O/FeCl_3.6H_2O$, showed better activity when PhNHOH was changed to 2,4-dinitrophenyl-hydroxylamine (217). Poor results were obtained from both catalyst systems due to the decomposition of phenylhydroxylamine to aniline, azobenzene, and azoxybenzene in the presence of iron complexes (218).

Two different sets of mechanisms were proposed based on the different catalytic systems. Here FePc generally underwent an "–NR" transfer mechanism "off" the metal (Scheme 144). The iron catalyst played an important role in

TABLE XXXII
Allylic Aminations of Olefins[a]

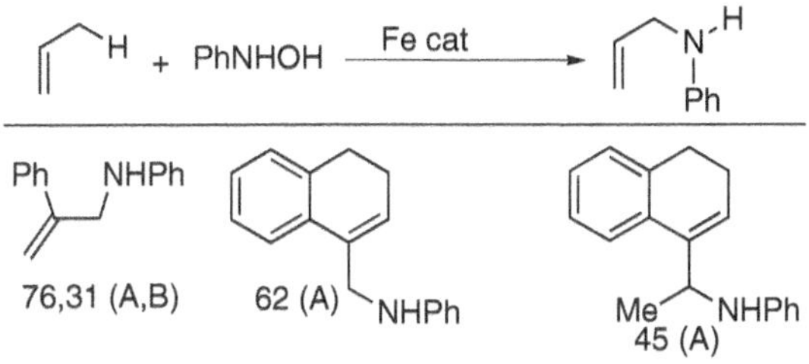

H + PhNHOH —Fe cat→ N-H, Ph

Ph, NHPh	NHPh	Me, NHPh
76,31 (A,B)	62 (A)	45 (A)

[a] Method A: Reaction performed with FePc (5 mol%), olefin (5 equiv), and PhNHOH (1 equiv) in toluene under reflux for 10 h. Method B: Reaction performed with $FeCl_2.4H_2O/FeCl_3.6H_2O$ (9:1, 10 mol%), olefin (1 equiv), and PhNHOH (2 equiv) in dioxane at 80 °C.

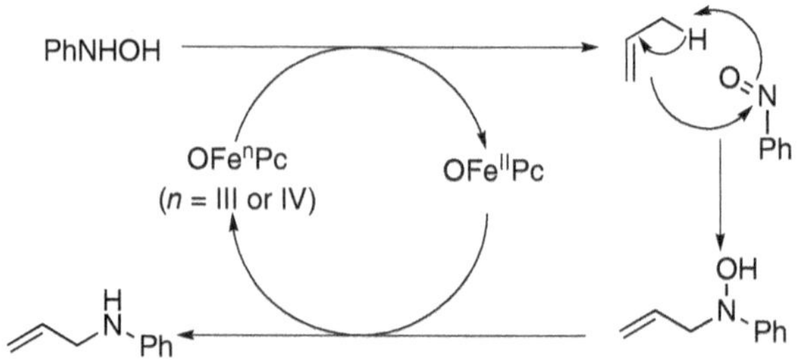

Scheme 144.

forming nitrosobenzene from phenylhydroxylamine and in the formation of an allylic amine from hydroxylamine via a hetero-ene reaction of PhNO with alkene (219).

In the case of a second catalytic system (method B), Nicholas and co-workers (220) showed that reaction occurred via formation of an azo–dioxide complex (**108**) (Scheme 145).

Later, Srivastava and Nicholas (221) reported that nitroarenes could also be used as an aminating reagent in the presence of $[CpFe(CO)_2]_2$ as catalyst under a CO atmosphere at high temperature (Scheme 146).

An iron–cyclopentadienyl–dicarbonyl complex also catalyzed allylation of styrene nitroaryl compounds having electron-withdrawing groups. The reaction temperature and pressure can be reduced (222) under a photoassisted condition.

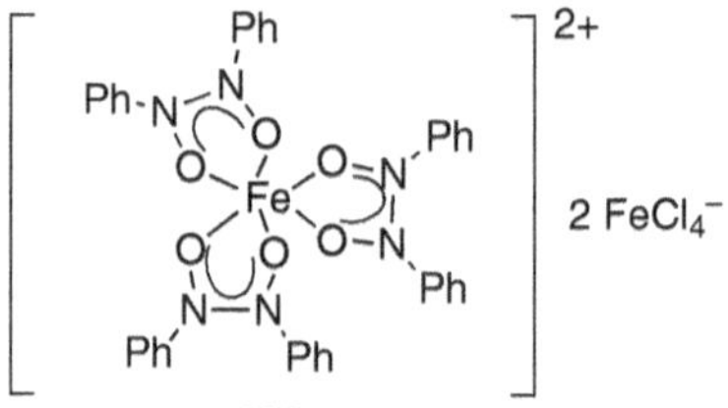

Scheme 145.

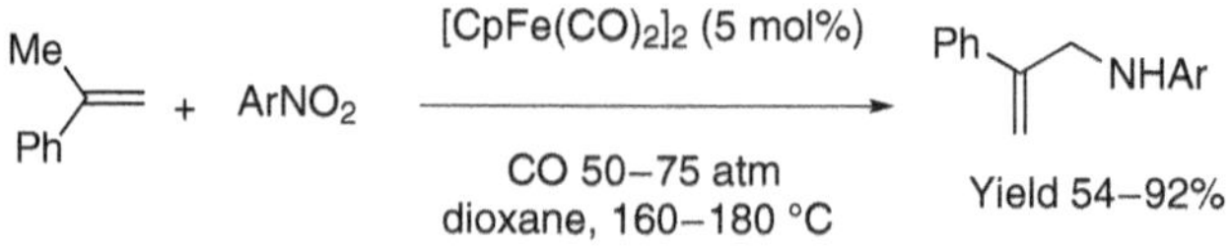

Scheme 146.

TABLE XXXIII
Amination of Allyl Carbonates[a]

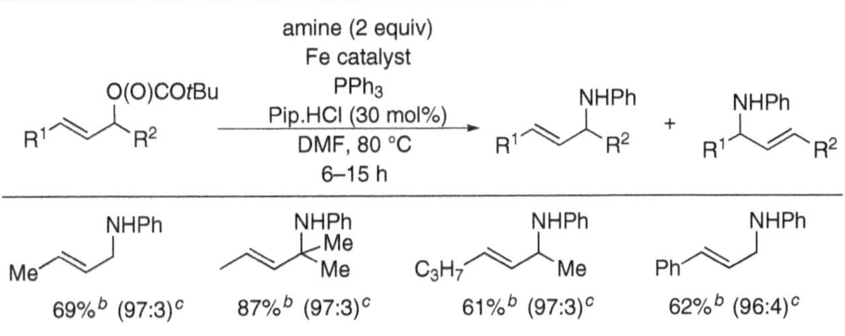

Reaction conditions: allyl carbonate (R¹–CH=CH–CH(O(O)COtBu)–R²), amine (2 equiv), Fe catalyst, PPh_3, Pip.HCl (30 mol%), DMF, 80 °C, 6–15 h → branched allylamine (R¹–CH=CH–CH(NHPh)–R²) + linear allylamine (R¹–CH(NHPh)–CH=CH–R²)

Me–CH=CH–CH₂–NHPh	69%[b]	(97:3)[c]
CH₂=CH–CH=… C(Me)(Me)NHPh	87%[b]	(97:3)[c]
C_3H_7–CH=CH–CH(NHPh)–Me	61%[b]	(97:3)[c]
Ph–CH=CH–CH₂–NHPh	62%[b]	(96:4)[c]

[a] All reactions were performed on a 1-mmol scale.
[b] Yield of isolated product.
[c] Regioselectivity of the crude product according to gas chromatography (GC) integration is given in parenthesis.

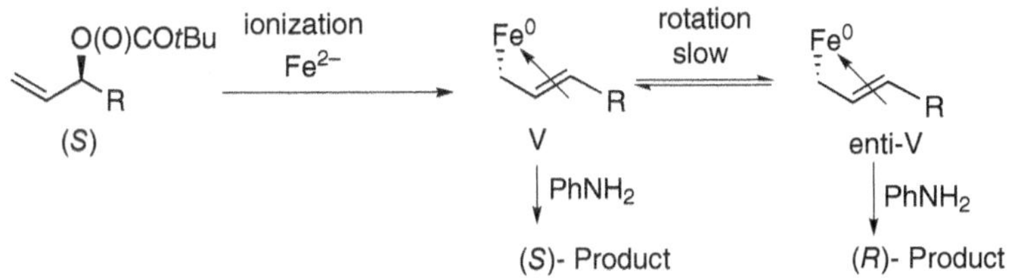

Scheme 147.

Later in 2006, Plietker (223) reported a regio- and stereoselective allylic amination of allyl carbonates with secondary amines using a $[Bu_4N][Fe(CO)_3(NO)]$ and triphenyl phosphine combination as the catalyst. The presence of a catalytic amount of piperidinium chloride (Pip.HCl) as a buffer retarded the decomposition of the catalyst.The variation in carbonate and amine partners provided good-to-excellent yield with high regioselectivity (Table XXXIII).

The reaction occurred via the formation of an σ-allyl metal intermediate (Scheme 147), and therefore provided high regio- and stereospecificity (223).

B. Intramolecular Allylic Amination

In 2011, Bonnamour and Bolm (224) developed an intramolecular C–H amination of azidoacrylates toward the synthesis of indole derivatives using an

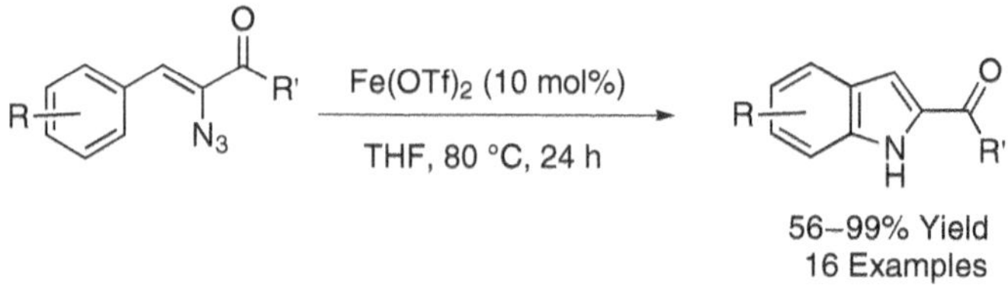

Scheme 148.

iron(II) triflate as catalyst. This reaction tolerated various functional groups including methoxy, alkyl, trifluoromethyl, halo, and phenyl groups at the para position of the aryl ring (Scheme 148).

The field of nitrenoid-based C–H amination was pioneered by Breslow and Gellman (225) by using Fe (TPP). Paradine and White (226) developed a highly selective intramolecular C–H amination by using an inexpensive, nontoxic [Fe^{III}Pc] catalyst, where Pc = phthalocyanine (**109**). This method showed a strong preference for allylic C–H amination over aziridinations (Scheme 149). A selectivity pattern emerged for the C–H amination: allylic > benzylic > ethereal > 3° > 2° >> 1°. In the case of polyolefinic substrates, the selectivity was controlled by an electronic and steric character. As expected, the relative rate of C–H amination inversely varied with bond strength (Scheme 150) (226). Allylic

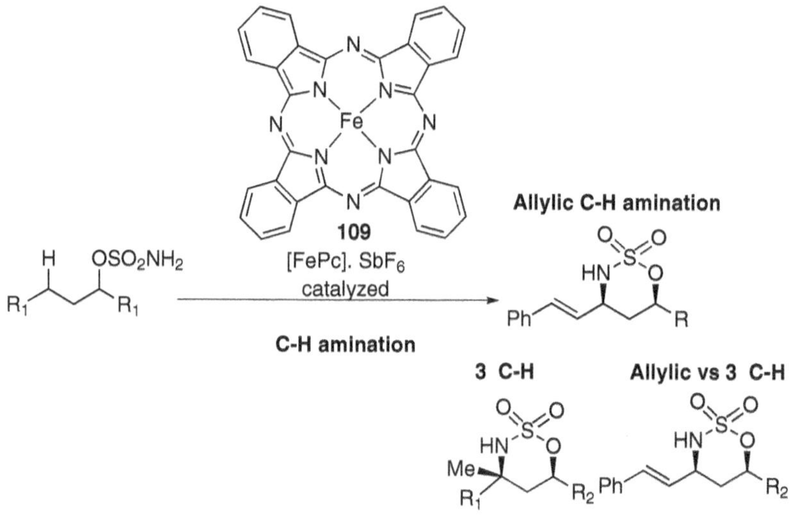

Scheme 149.

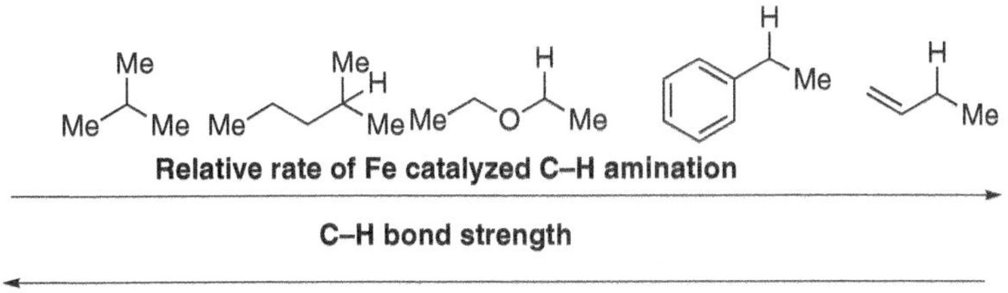

Relative rate of Fe catalyzed C–H amination

C–H bond strength

Scheme 150.

substrates with a wide variety of substituents were successfully employed for such amination reactions (Table XXXIV).

A stepwise mechanism was considered based on experimental observations. However, the stereoretentive nature of C–H amination for 3° aliphatic C–H bonds suggested a rapid radical rebound mechanism.

Later, Sun and co-workers (227) reported an $FeCl_3$ catalyzed intramolecular allylic amination toward the synthesis of dihydroquinolines and quinolones

TABLE XXXIV
Intramolecular C–H Amination

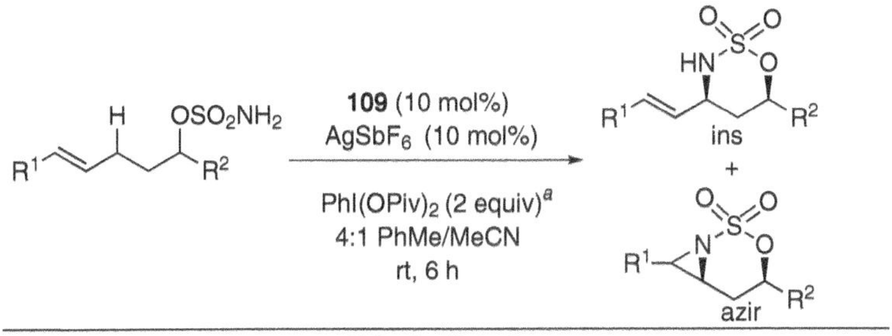

109 (10 mol%), $AgSbF_6$ (10 mol%), $PhI(OPiv)_2$ (2 equiv)[a], 4:1 PhMe/MeCN, rt, 6 h; products: ins + azir

Styrenyl 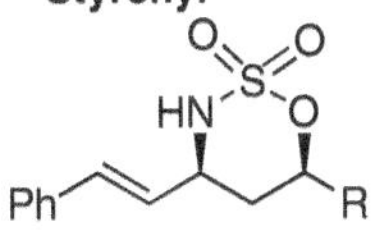	Trisubstituted 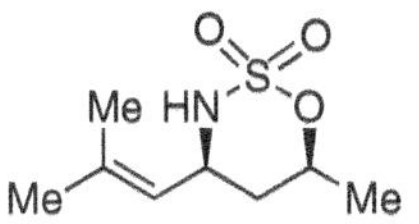	Terminal 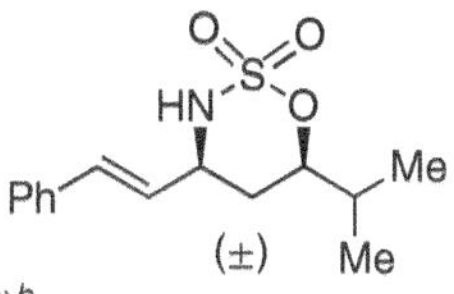
(±)	(±)	(±)
70% yield (0% RSM)[b]	53% yield (<10% RSM)[b]	52% yield (10% RSM)[b]
dr = 3.5:1 syn/anti[c]	dr = 2.7:1 syn/anti	dr = 3:1 syn/anti[c]
(*E*/*Z*) > 20:1	ins/azir > 20:1	ins/azir > 12:1[c]
ins/azir > 20:1[c]		
a = pivolate		

[a] Pivalate = Piv.
[b] Isolated yields (syn + anti); % RSM in parentheses.
[c] All product ratios were determined by ^1H NMR analysis of the crude reaction mixture.

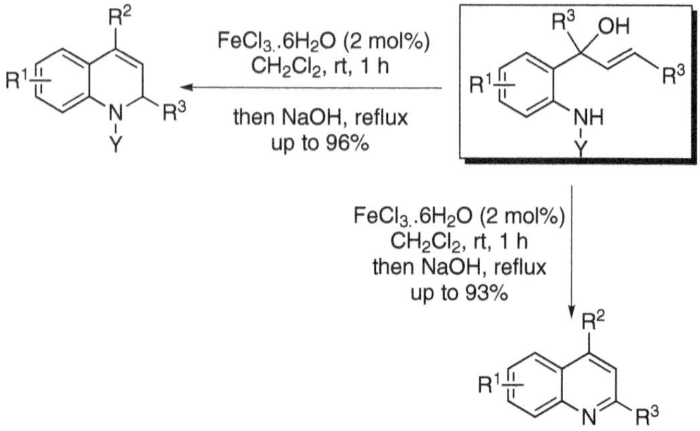

Scheme 151.

(Scheme 151). No stereoretentivity was observed with pure enantiomeric substrates. Hence, it was proposed that the reaction occurred via formation of the carbocationic intermediate (**110**) (Scheme 152).

In 2011, Betley and co-workers (228) employed an iron complex of the dipyrromethane ligand scaffolds bearing large aryl groups for C–H amination. The use of a bulky ligand provided a high-spin iron complex ($S = 2$), which was the catalytically active species. Reaction of iron complexes (**111–113**) with alkyl azides provided C–H amination for toluene and aziridine in the case of styrene

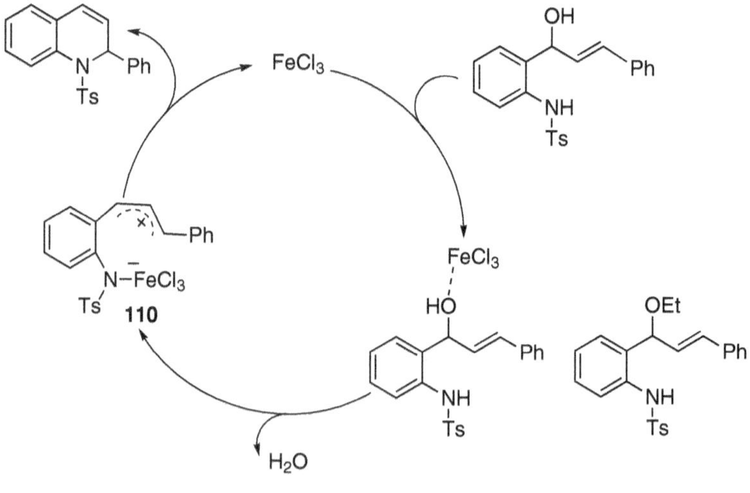

Scheme 152.

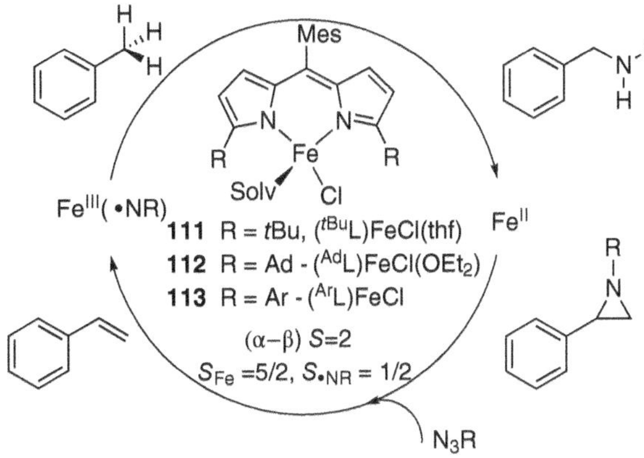

Scheme 153.

(Scheme 153 Ad = adamantyl, Ar = aryl). The complex (^{Ad}L)FeCl(OEt$_2$) (**112**) gave a better yield compared to **111** and **113**. They were able to isolate a high-spin iron complex, (^{Ar}L)FeCl(N(*p*-*t*BuC$_6$H$_4$) during the reaction of (^{Ar}L)FeCl(OEt$_2$) (**113**) with *p*-*t*BuC$_6$H$_4$N$_3$ (228). A radical rebound-like mechanism by hydrogen-atom abstraction was proposed based on their experimental findings.

VIII. SULFOXIDATIONS AND SYNTHESIS OF SULFOXIMINES, SULFIMIDES, AND SULFOXIMIDES

A. Sulfoxidation

Sulfoxides are an important class of compounds since they were used for various ligand syntheses. Selective oxidation of sulfide was reported in the literature by using an iron catalyst in the presence of different oxidants (e.g., H_5IO_6, HNO_3, or other terminal oxidants) (229). Asymmetric sulfoxides were usually synthesized using titanium, vanadium, and manganese complexes. But those were less effective for practical use. One of the earlier methods iron porphyrin in the presence of iodosyl benzene, gave a moderate enantioselectivity (<55% ee) (230). Fontcave and co-workers (231) employed **114** as a ligand for iron-catalyzed sulfoxidation, however, they got low enantioselectivity (ee_{max} = 40%) (Scheme 154).

Legros and Bolm (232) reported the major breakthrough in iron-catalyzed asymmetric sulfoxidation. A highly enantioselective sulfide oxidation with optically purity (ee = 96%) was described under clean reaction condition (Table XXXV). More challenging substrates (e.g., phenyl ethyl and phenyl benzyl sulfides) were

TABLE XXXV
Iron-Catalyzed Asymmetric Sulfoxidation

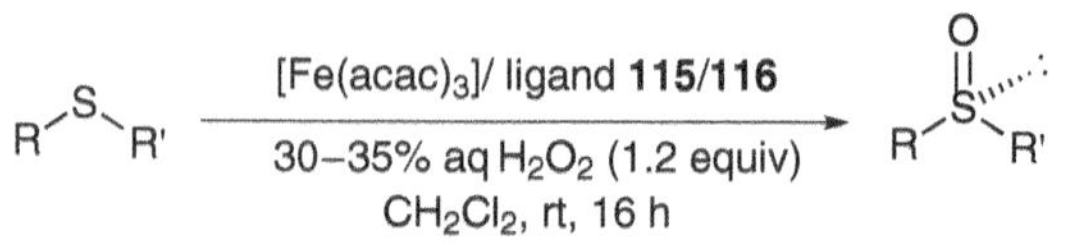

[Fe(acac)3]/ ligand **115/116**
30–35% aq H_2O_2 (1.2 equiv)
CH_2Cl_2, rt, 16 h

Entry	Sulfide	Product	
		Yield (%)	ee (%)
1	Ph-S-Me	63	90
2	Ph-S-Et	56	82
3	Ph-S-CH_2Ph	73	79
4	Ph-S-CH_2-CH=CH_2	63	71

well tolerated with excellent enantioselectivity under their reaction protocol. The reaction was chemoselective since in phenyl allyl sulfide only the sulfur atom was oxidized and the double bond remained unaffected (233). This method was useful for the asymmetric synthesis of sulindac (> 90% ee), a biologically active chiral sulfoxide (234).

An iron(III)–salen complex with H_2O_2 could also catalyze the oxidation of organic sulfides and sulfoxides, in which an iron high-valent oxo (Fe^{IV}=O) species was proposed to be the active species (235).

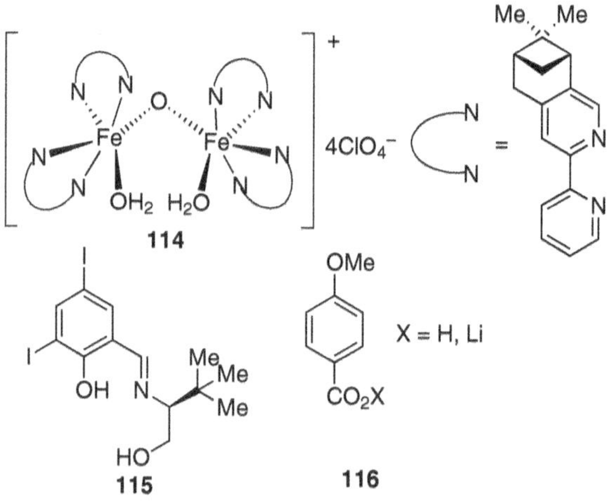

114 **115** **116** X = H, Li

Scheme 154.

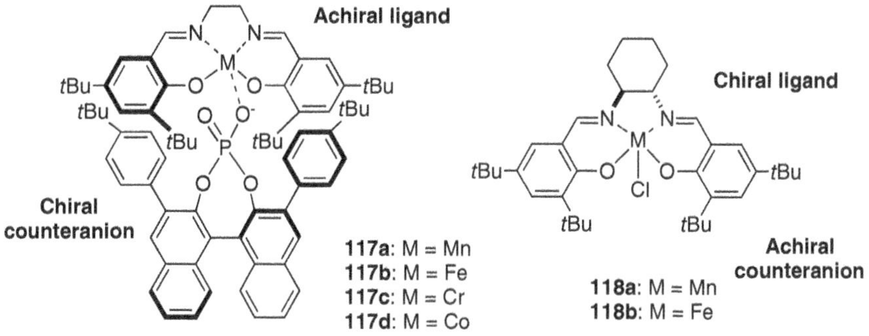

Achiral ligand

Chiral ligand

Chiral counteranion

117a: M = Mn
117b: M = Fe
117c: M = Cr
117d: M = Co

118a: M = Mn
118b: M = Fe

Achiral counteranion

Scheme 155.

117b (1 mol%), PhIO (1.1 equiv), PhH, 15 °C, 8–12 h

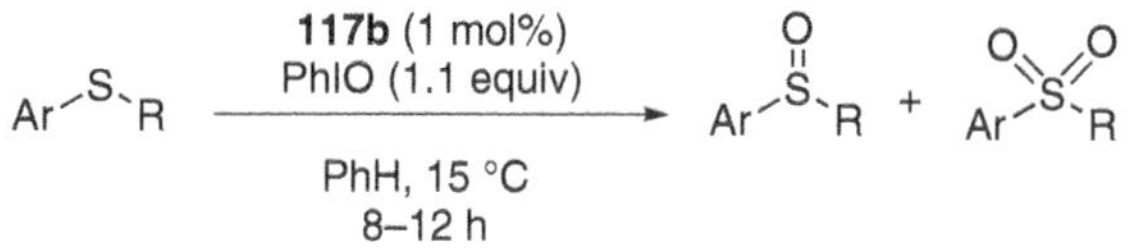

Scheme 156.

Later in 2012, Liao and List (236) reported a highly active and enantioselective sulfoxidation catalyst using an iron(III)–salen cation (**118b**) and a chiral phosphate counteranion (**117a–117d**) (Scheme 155). They used **117b** as the chiral catalyst in combination with PhIO as the terminal oxidant (Scheme 156). This reaction was the first example of an asymmetric counteranion directed catalysis (236).

Their methodology had a potential application toward the enantioselective synthesis of hydroxamic acid (Scheme 157), which is a potent histone deacetylase inhibitor.

Very recently, Chen and co-workers (237) developed another protocol of sulfoxidation using $Fe(acac)_3$ (1 mol%) and polyethylene glycol as an additive and O_2 as the oxidant.

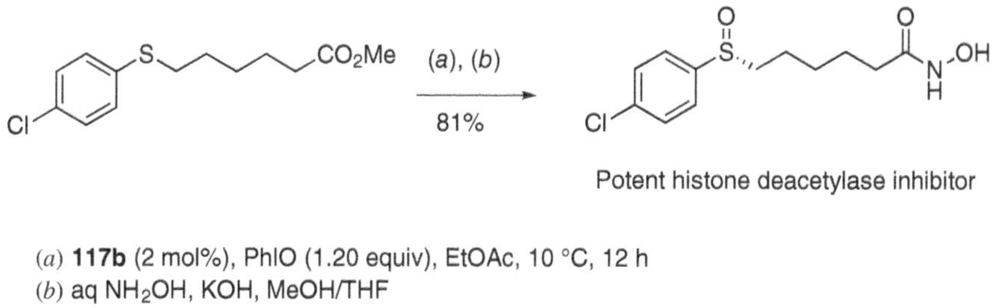

(*a*), (*b*) 81%

Potent histone deacetylase inhibitor

(*a*) **117b** (2 mol%), PhIO (1.20 equiv), EtOAc, 10 °C, 12 h
(*b*) aq NH_2OH, KOH, MeOH/THF

Scheme 157.

B. Synthesis of Sulfoximines, Sulfimides, and Sulfoximides

Nitrene transfer was a key process for the synthesis of various classes of organic compounds (e.g., aziridination and different nitrogen-containing compounds). It was also utilized in the synthesis of N-substituted sulfimides. Bach and Kober (238) first reported the iron-catalyzed method for the synthesis of sulfoximides using $FeCl_3$/*tert*-butyloxycarbonyl azide ($BocN_3$) (Scheme 158). Instead of using $FeCl_3$ in a stoichiometric amount, it could be used in a catalytic amount (25–50 mol%) in their modified condition (Scheme 159). The nitrene could transform sulfoxide and sulfide to sulfoximides and sulfimides, respectively.

These reactions were stereospecific and could be utilized for the synthesis of chiral ligands. Optically pure sulfoxides could be converted into sulfoximides. Subsequently, free NH_2–sulfoximine was obtained after Boc cleavage. Bolm and co-workers (239) synthesized chiral sulfoximines as ligands for enantioselective synthesis. Syntheses of sulfoximines by imidation of sulfides and sulfimides were obtained in moderate-to-good yield in the presence of acetylacetone or DMF (Scheme 160).

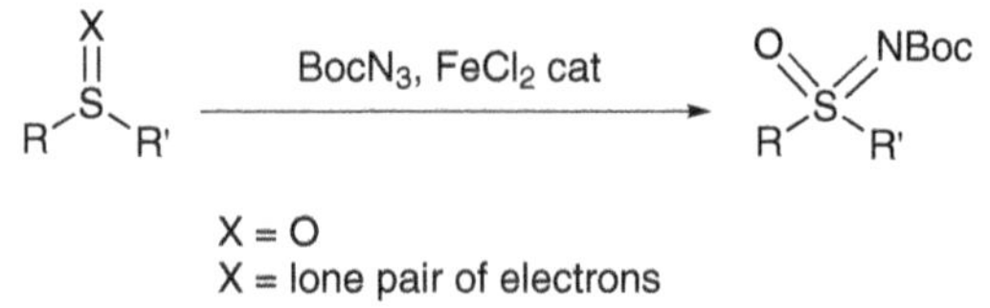

X
S
R R'
$BocN_3$, $FeCl_2$ cat
O NBoc
S
R R'
X = O
X = lone pair of electrons

Scheme 158.

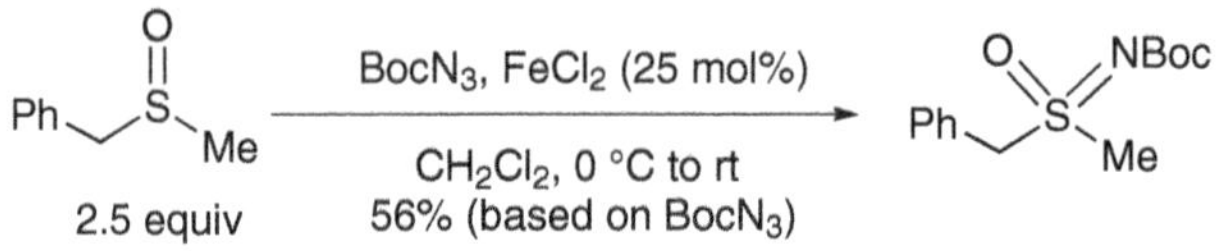

O
Ph S Me
2.5 equiv
$BocN_3$, $FeCl_2$ (25 mol%)
CH_2Cl_2, 0 °C to rt
56% (based on $BocN_3$)
O NBoc
Ph S Me

Scheme 159.

S
R R'
2.5 equiv
R = Ph, Bn
R' = Me, Et
$BocN_3$, $FeCl_2$ (10–25 mol%)
acetylacetone (1.3 equiv)
CH_2Cl_2, 0 °C, 5 h
61–90% based on $BocN_3$
NBoc
S
R R'

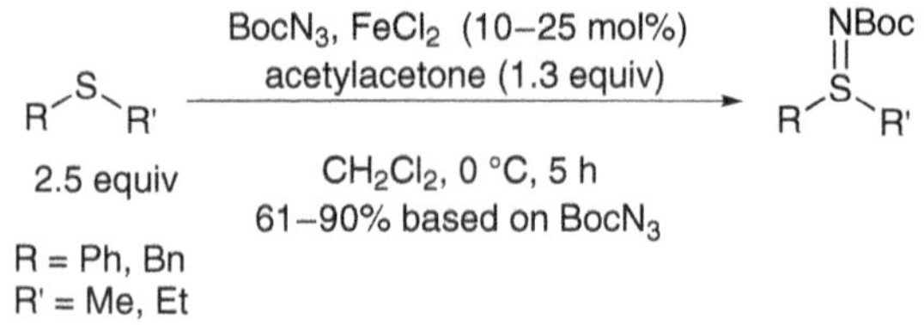

Scheme 160.

Scheme 161.

Later, Mancheno and Bolm (240) developed stereospecific methods for sulfoximine synthesis using $Fe(acac)_3$ as a catalyst (Scheme 161). Unfortunately, the reaction was susceptible toward steric effect.

Subsequently, they reported a modified method, which tolerated challenging substrates including benzyl, sterically demanding alkyl, and heteroaryl substituted sulfoxides at rt (Table XXXVI) (241).

1. Mechanism

An Fe(V)–nitrene complex [(Cl)Fe^{V}=NR] was proposed as a imidation reagent in the catalytic cycle (Scheme 162).

Darcel and co-workers (242) developed a sulfonylimines synthesis under neutral condition starting from an aldehyde (Scheme 163).

TABLE XXXVI
Iron-Catalyzed Sulfoximide Synthesis[a]

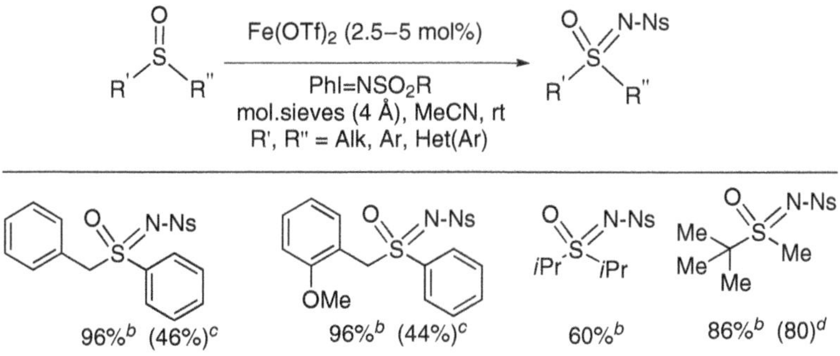

[a]Reaction conditions: sulfoxide (1 equiv), $Fe(OTf)_2$, PhI=NNs (Ns = 2 or 4-nitrobenzenesulfonyl) (1.3 equiv), and molecular sieves 4 Å in acetonitrile (0.1 *M*) at rt.
[b]After column chromatography.
[c]Yield obtained by using 10 mol% of $Fe(acac)_3$ as catalyst.
[d]Reference (240).

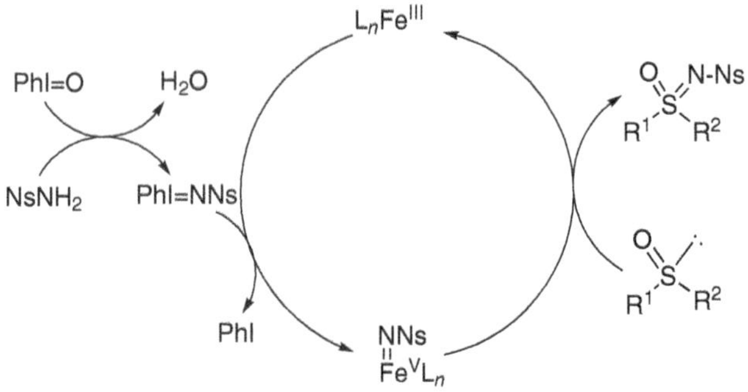

Scheme 162.

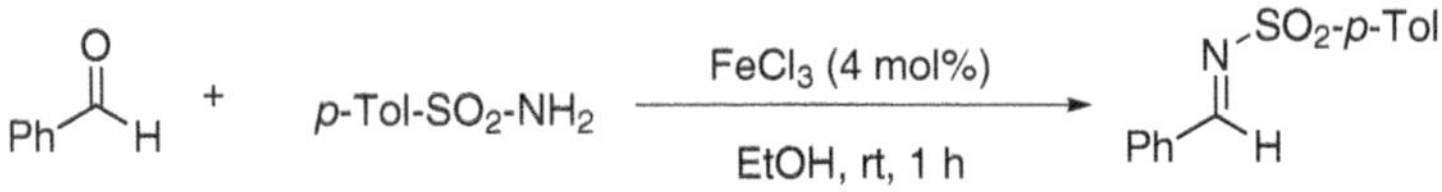

Scheme 163.

IX. REDUCTION REACTIONS

A. Hydrosilylation of Alkenes

Organosilicon compounds can be found in numerous consumer goods that are associated with our daily life (e.g., oil, grease, rubbers, cosmetics, and medicinal compounds). Most of them contain Si–C bonds that need to be synthesized artificially, as they do not exist in nature. Addition of Si-H across an unsaturated double or triple bond provides the most convenient and widely used approach to form such organosilicons. This method is termed as hydrosilylation (HySi) and it requires a transition metal catalyst (Scheme 164).

One of the major problems regarding hydrosilylation of alkenes and alkynes is the selectivity issue, as side reactions (e.g., DHySi and Hy) may decrease the turn

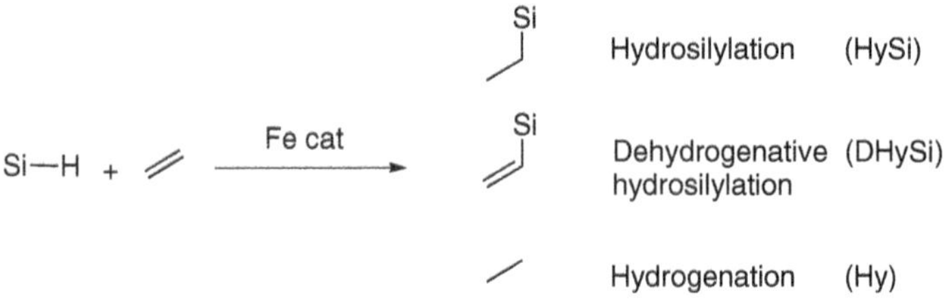

Scheme 164. Iron-catalyzed reactions of alkene and hydrosilane (243).

over number (TON) and can generate unwanted side products in a significant amount (Scheme 164). The success of this method, therefore, lies in discovering highly reactive yet selective reaction conditions that will not only ensure high-catalytic turnover, but also make the process economical and environmentally benign as organosilicons are generated on the millions of tons scale.

Traditionally, for over three decades, among transition metal catalysts, platinum-based Spier's catalyst ($Pt_2([(CH_2{=}CH)SiMe_2]_2\,O)_3$) and Karstedt's catalyst ($H_2PtCl_6.6H_2O/i$PrOH) have been regarded to be very powerful and are often employed for commercial purposes. In 2007, worldwide consumption of platinum by the silicone industry was estimated to be 5.6 metric tons and most of them were not recovered (244). Also, Rh and lanthanide-based catalysts have been reported to affect HySi. The high cost, limited availability, and toxic nature of precious metals imposed a scientific challenge to discover catalytic systems based on earth-abundant first-row transition metals, particularly iron.

Nesmeyanov et al. (244b) reported the first iron-catalyzed HySi reaction in 1962, where $Fe(CO)_5$ served as the catalyst in the reaction of alkenes and tertiary silanes. Products from both HySi and DHySi were generated depending on the substrate and reaction condition. In 1977, Schroeder and Wrighton (244c) disclosed a photoreaction of R_3SiH and alkenes involving the same catalyst, $Fe(CO)_5$. Later, they proposed a mechanism of iron-catalyzed HySi for the $Cp^*Fe(CO)_2R$ catalytic system ($Cp^* = \eta^5$-C_5Me_5, R = alkyl, silyl) that included the insertion of a C=C double bond into an Fe–Si bond.

Completely selective DHySi was realized by Murai and co-workers (245) in 1993 as vinyl silanes were obtained from the reaction of styrenes and Et_3SiH in the presence of $Fe_3(CO)_{12}$ (Scheme 165). Other metals of the same group (Ru, Os) were also effective, but iron exhibited complete selectivity. However, the reaction suffered a serious drawback, since 1-hexene afforded a complex mixture that contained vinylsilane, alkylsilane, and allylsilane.

On the other hand, selective hydrosilylation of alkenes by inexpensive iron catalysts eluded researchers, until Chirik and co-workers (246) made a significant breakthrough in 2004 (Scheme 166). By using a well-characterized iron bis(imino)pyridine dinitrogen complex (**119**), they successfully hydrosilylated a number of unactivated olefins with $PhSiH_3$, whereas with Ph_2SiH_2, a slower reaction rate was observed. The method was also compatible with alkynes and with $PhSiH_3$. However, due to the steric

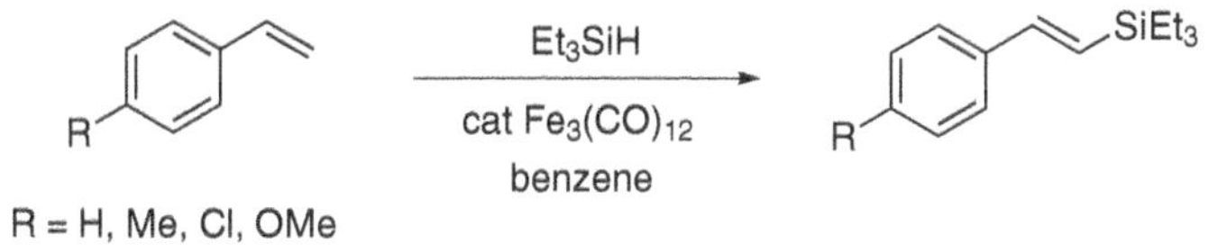

R
Et_3SiH
cat $Fe_3(CO)_{12}$
benzene
$SiEt_3$
R
R = H, Me, Cl, OMe

Scheme 165. Selective DHysi of styrenes catalyzed by triiron decacarbonyl.

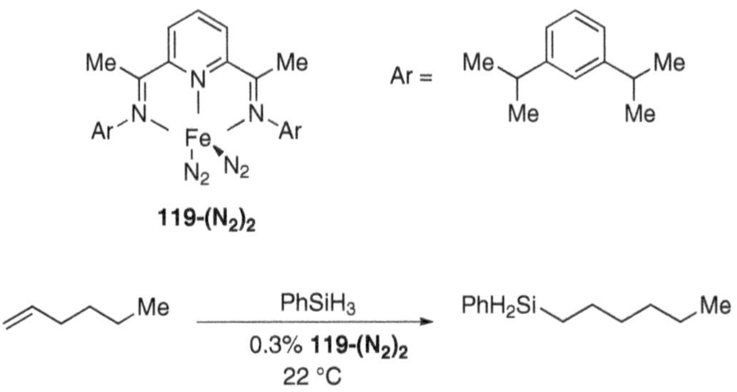

Scheme 166. Selective hydrosilylation of alkenes catalyzed by an iron bis(imino)pyridine dinitrogen complex.

hindrance of the bulky silylalkene, no further reaction was observed even with a large excess of phenylsilane.

Unfortunately, this catalyst was unreactive toward the most commercially relevant silicon hydrides, the tertiary silanes (e.g., Et_3SiH). To overcome this, Chirik and co-workers (244) carried out a series of reactions under solvent-free conditions with iron bis(imino)pyridine complexes containing both linear and bridging nitrogen ligands. By manipulating the size of the 2,6-aryl substituent of the bis(imino)pyridine ligand (*i*Pr, Et) **(120, 121)**, they were successful in employing a number of tertiary silanes as efficient silylating agents under ambient conditions, though Et_3SiH remained unreactive (Scheme 167). With a sterically

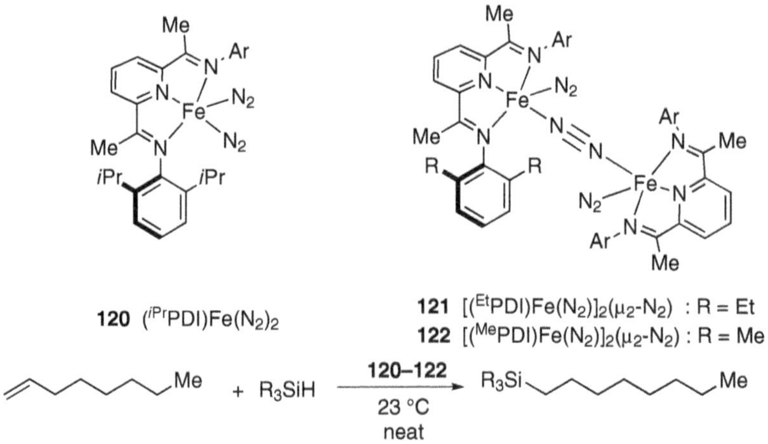

Scheme 167. Iron-catalyzed selective hydrosilylation of alkenes.

less hindered methyl as the substituent (**122**), the most active catalytic system was discovered as a complete conversion of Et_3SiH and 1-octene was observed after 45 min at 23 °C. This method was also applicable to styrene, as well as with amino substituted olefins. Notably, in the case of a platinum catalysts, the amino group often serves as a catalyst poison. Further, this catalytic system was successfully applied in cross-linking of silicone fluids, thus promising a viable alternative to the use of a precious platinum catalysts for such a purpose.

The problem of limited accessibility, as well as the air and moisture sensitivity of iron–bis(imino)pyridine dinitrogen complexes, was overcome in 2012, by the same group. In that year they prepared a series of iron dialkyl complexes containing either bis(imino)pyridine or terpyridine ligands that served as efficient pre-catalysts in the hydrosilylation reaction (247). At the same time, Nakazawa and co-workers (248) independently reported selective single–double hydrosilylation of 1-ocetene by iron complexes containing terpyridine as ancillary ligands.

Meanwhile, Ritter and co-workers (249) reported a very efficient iron-catalyzed hydrosilylation of a 1,4-diene by a well-defined low-valent iron complex (**123**) (Scheme 168). In the presence of redox active imino pyridine ligands, a number of substituted dienes underwent hydrosilylation that provided allylsilanes in excellent yields and selectivities.

In 2012, unprecedented reactivity of $CpFe(CO)_2Me$ was exhibited in the reaction of 1,3-divinyldisiloxane with hydrosilane by Nakazawa and co-workers (241) (Scheme 169). In this reaction, one of the vinyl groups was dehydrogenatively silylated, while the other vinyl group was hydrogenated.

This reaction is a combination of DHySi and Hy, which are regarded as unwanted side reactions in hydrosilylation chemistry, and from the reaction outcome it can very well be termed as HySi. It was observed that two vinyl groups and an oxygen atom connecting the vinylsilyl group were necessary for the success of the reaction. A labeling experiment with $MePh_2SiD$ indicated that hydrogen atoms from the silylated vinyl group and tertiary silane were responsible for hydrogenation of the other vinyl group.

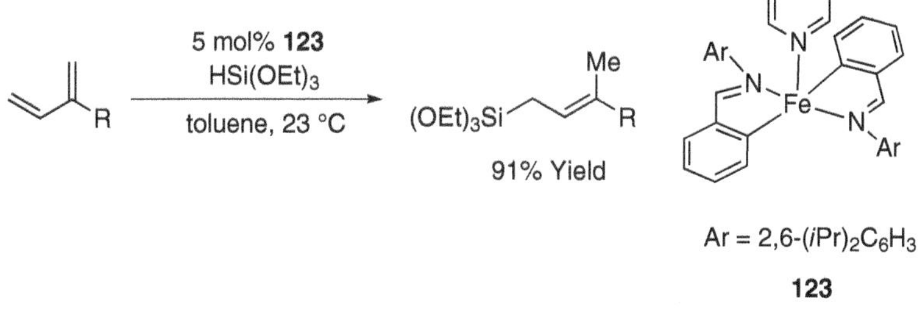

5 mol% **123**
$HSi(OEt)_3$
toluene, 23 °C
Me
R
$(OEt)_3Si$
91% Yield
Ar
N
Fe
Ar = 2,6-(iPr)$_2$C$_6$H$_3$
123

Scheme 168. Iron-catalyzed hydrosilylation of 1,4-diene.

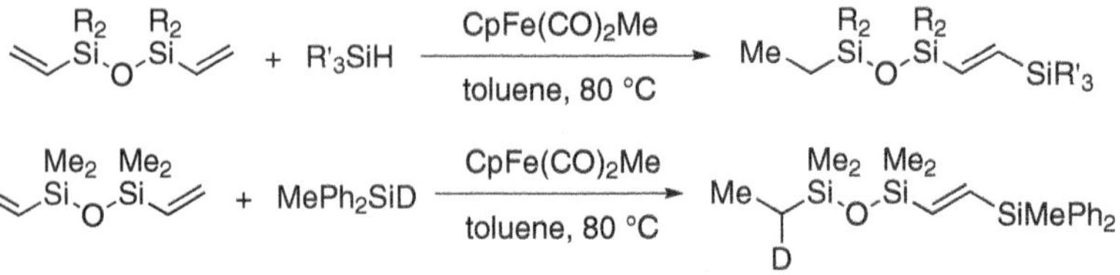

Scheme 169. A different iron-catalyzed "hydrosilylation" reaction.

B. Hydrosilylation of Aldehydes and Ketones

Reduction of aldehydes and ketones to the corresponding alcohols is a fundamental transformation in synthetic chemistry. Although numerous methods are reported for reduction, the lack of chemoselectivity, sensitivity, and toxicity of the reagents available call for the discovery of mild and easy to handle reaction conditions. Hydrosilylation can be envisaged as a useful alternative in this regard. Unfortunately, hydrosilanes by themselves are unreactive toward the carbonyl compounds and a transition metal catalyst is needed for the success of the reaction. In recent decades, several transition metals ranging from Rh, Ru, and Ir to Ti, Zn, Sn, and Cu have been explored. But the cost of the metals, toxicity arising from the waste, and residual toxicity in the product emphasizes the need for more sustainable and greener protocols. Iron due to its relatively lower toxicity, inexpensiveness, and abundance in the earth has drawn significant attention from the scientific community.

In 2007, Nishiyama (250) disclosed a hydrosilylation reaction of aromatic and aliphatic ketones by using readily available $Fe(OAc)_2$/tmeda in conjunction with $(EtO)_2MeSiH$ as the hydride source (Scheme 170). Surprisingly, reaction did not proceed in the absence of ligand; tmeda presumably helped the catalyst system to be homogeneous by facile coordination with iron. By using a chiral ligand (e.g., bopa-*tb* in place of tmeda, up to 79% ee was achieved. But, competing DHySi produced silyl enol ether in a significant amount since in most of the cases starting material was recovered (2–51%) after work up. However, replacing tmeda by thiophene-2-carboxylate as the ligand greatly improved the selectivity, as corresponding alcohols were obtained in excellent yields (> 90%) with no or a trace amount of starting material recovered.

To address the enantioselectivity issue, they prepared an iron catalyst based on NCN type ligands, bis(oxazolinylphenyl), amine) (phebox and bis(oxazolinyl)

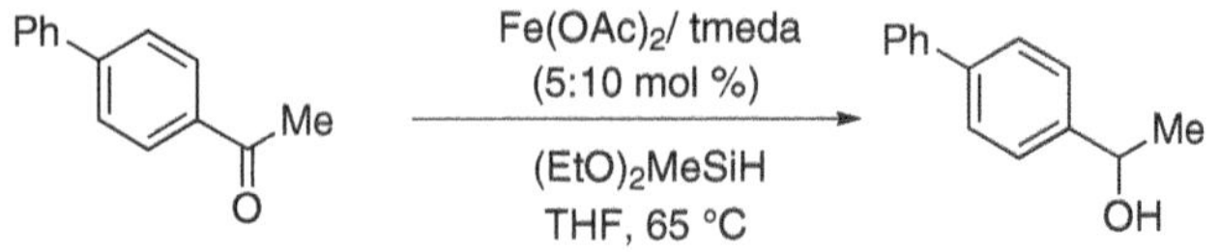

Scheme 170. The $Fe(OAc)_2$/tmeda catalyzed hydrosilylation of ketones.

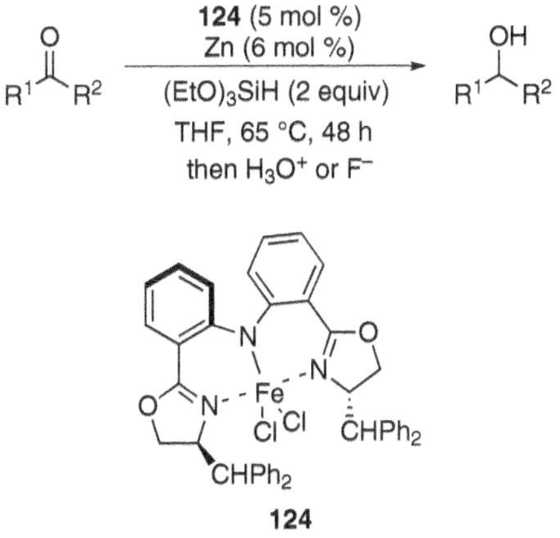

Scheme 171. Enantioselective iron-catalyzed hydrosilylation of ketones.

pyridine (251a), plus $Fe_2(CO)_9$. However, only up to 66% ee with an (*R*) configuration of the product alcohol was obtained with this catalytic system. Note, with the same ligand, an Ru catalyst yielded alcohol of absolute configuration (*S*) with 77% ee. Improved stereoselectivity was observed with bis(oxazolinylphenyl)amine (Bopa) ligand in conjunction with iron (up to 88% ee) (**124**) and cobalt (up to 98% ee). In all these cases, (*R*) products were obtained. Addition of Zn powder to an Fe–Bopa catalytic system dramatically altered the outcome, as alcohol product was obtained with an absolute configuration (*S*) (Scheme 171) (251b).

Electron-rich phosphine ligands (**125**) were applied in conjunction with iron for enantioselective hydrosilylation of ketones (up to 99% ee) (Scheme 172) (252a). Earlier, Beller and co-workers (252b) utilized most inexpensive hydrosilane, PMHS (polymethylhydrosiloxan) to reduce aldehydes by $Fe(OAc)_2/PCy_3$. A broad range of aromatic, aliphatic, and heteroaromatic aldehydes were successfully converted to alcohols with excellent yields under a particularly mild reaction

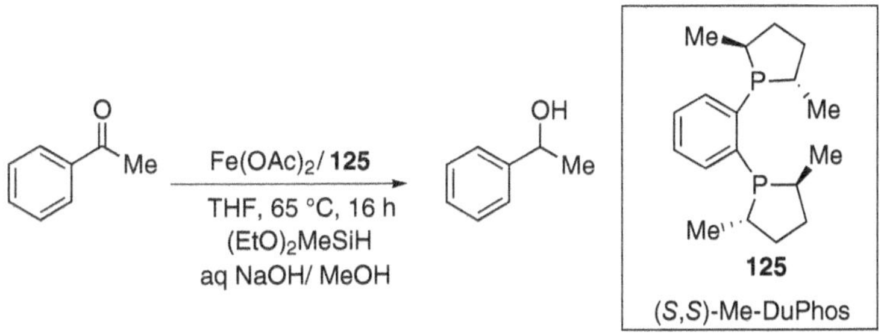

Scheme 172. Highly enantioselective hydrosilylation of ketones by use of an iron–phosphine catalyst [DuPhos = 1,2-bis((2*R*,5*R*)-2,5-di-*i*-isoproplyphospholano)benzene].

condition. Meanwhile, Yang and Tilley (253) made use of a simple iron amide catalyst [$Fe(N(SiMe_3)_2)_2$] to generate alkoxysilanes in high yields from the corresponding aldehydes and ketones.

C. Hydrogenation of C–C Unsaturated Bonds

Iron complexes have high complexing affinity toward the C–C unsaturated bonds, which can cause a problem for hydrogenation due to deactivation of the catalyst by forming a stable complex. In 1965, Frankel et al. (254) reported homogeneous hydrogenation of methyl linoleate using $Fe(CO)_5$ as the precatalyst under a nitrogen (250 psi) and hydrogen atmosphere (400 psi) at 180°C for successive hydrogenation. In the presence of the iron–carbonyl complex, the double-bond isomerizes and forms the conjugated diene leading to the stable complex **126**. High temperature and pressure was used to decompose this catalyst poisoning complex (**126**). Under reaction conditions, an iron carbonyl compound results in the formation of $H_2Fe(CO)_4$, which is considered to be the hydrogenating agent (Scheme 173) (254).

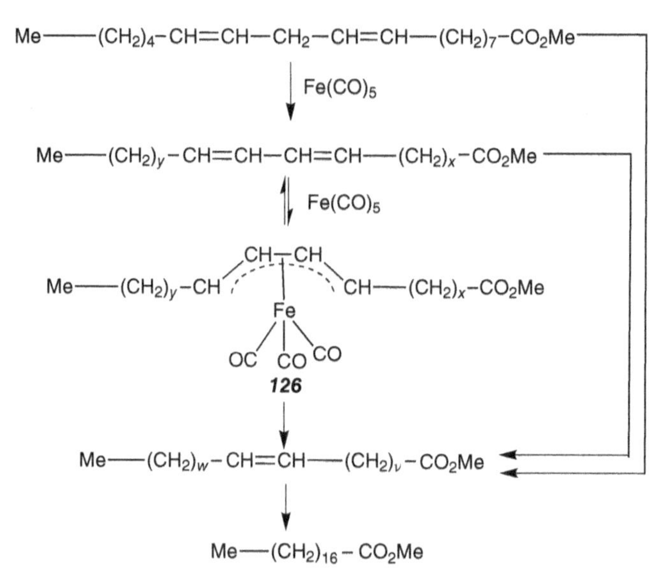

x,y = 4,8; 5,7; 6,6; 7,5; 8,4; 9,3; 10,2.

v,w = 2,12; 3,11; 4,10; 5,9; 6,8; 7,7; 8,6; 9,5; 10,4; 11,3; 12,2; 13,1; 14,0.

$$Fe(CO)_5 + H_2 \longrightarrow H_2Fe(CO)_4 + CO$$

$$2\ \text{Complex X} + 3H_2 \longrightarrow 2\ \text{monoene} + H_2Fe(CO)_4 + 2\ CO + Fe$$

Scheme 173. The $Fe(CO)_5$ catalyzed hydrogenation of olefin.

TABLE XXXVII
Photoassisted Hydrogenation of the olefin by $Fe(CO)_5$

Olefin $\xrightarrow[\text{25°C, near-UV radiation}]{0.011\ M\ Fe(CO)_5 \text{ in benzene or toluene } H_2 \text{ gas (10–14 psi)}}$ Hydrogenated product

Entry	Olefin	Irradiation Time (min)	Product(s)	Conversion (%)
1		60		8.9 7.3
2	Me—Me	60	Me—Me Linear hexenes	30.8 5.2
3	Me—OH	60	Me—O	~100

With fully hydrogenated product, a mixtures of double-bond isomerized monoenoic fatty esters have been observed with considerable cis–trans isomerization (254).

After Frankel's pioneering work in 1976 on hydrogenation under thermal conditions, Schroeder and Wrighton's (255) group carried out the photocatalyzed hydrogenation of the olefin using the same iron–carbonyl precatalyst. In the presence of radiation (UV), $Fe(CO)_5$ performed the hydrogenation of an alkene under a milder condition than before (Table XXXVII). Again, positional isomerization of the double bond is the main limitation of this method. The olefin with the alcohol moiety produced only the corresponding aldehyde.

Schroeder and Wrighton (255) proposed that photoirradiation initiates the dissociation of the pentacarbonyl iron complex to the tetracarbonyl iron complex, which under photolytic or thermal conditions, forms the $H_2Fe(CO)_3$(π-alkene) species (Scheme 174). Under a photoinduced condition, a $H_2Fe(CO)_3$(π-alkene) complex performs hydrogenation through the iron monohydride, $HFe(CO)_3$(alkyl), complex. Surprisingly, in the presence of the 1,3-dienes, the hydrogenation reactions are totally quenched due to the formation of a stable (1,3-diene)$Fe(CO)_3$ complex, which cannot be dissociated under the reaction condition.

Meanwhile, in 1972 Noyori et al. (256) demonstrated another route for the iron pentacarbonyl mediated olefin hydrogenation of α,β-unsaturated ketones (Scheme 175). In the presence of base and protonated solvent, the iron pentacarbonyl

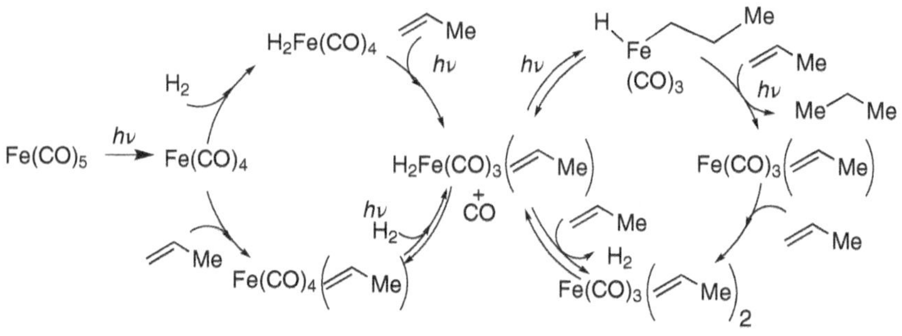

Scheme 174. Mechanism for the $Fe(CO)_5$ catalyzed photoassisted hydrogenation of olefin.

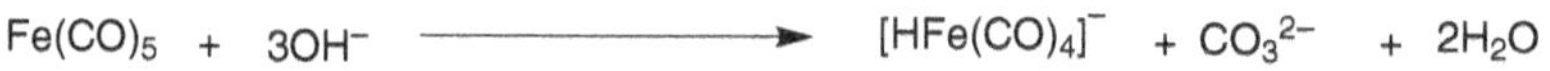

$$Fe(CO)_5 + 3OH^- \longrightarrow [HFe(CO)_4]^- + CO_3^{2-} + 2H_2O$$

Scheme 175.

complex generates a hydrido–iron complex. They also proposed the formation of an iron π-enolate intermediate.

This method showed selectivity toward the olefin moiety in the presence of a conjugated ketone, aldehyde, ester, and nitrile entities. Contrary to earlier reports, no isomerization of the double bond occurred. Steric hindrance near an olefin moiety decreases the reductive efficiency of the complex, for example, yield of the hydrogenation decreases from 2-cyclohexenone to 2-methy-2-cyclohexenone (Table XXXVIII) (256).

Selective hydrogenation of alkyne has also been achieved by using the iron(II)-*cis*-hydride-η^2-H_2 complex. Bianchini and co-workers (257) showed that terminal alkynes are preferentially hydrogenated over a double bond present in the same substrate (Scheme 176). Trimethysilane containing a terminal alkyne gave a low turnover frequency (TOF) value for the desired hydrogenated product due to formation of diene.

According to the reported hypotheses (Scheme 177) (257), one phosphine ligating site of a tetra-phosphine bound iron(II) center unlocks from the metal center to allow alkyne coordination. Subsequent hydride transfer to the alkyne results in the iron–vinyl complex. Subsequently, a proton from dihydrogen to the vinyl moiety is transferred through an intramolecular acid–base reaction as the oxidation of iron(II) to iron(IV) is less likely. In the final step, molecular hydrogen again coordinates with the iron center regenerating the active complex.

Brown and Peters (258) reported a tris(phosphino)borato-ligated iron(III)–imide complex that can perform partial hydrogenation of benzene (Scheme 178). The iron(III) hydride species is formed by a hydrogenolytic cleavage of the imide

TABLE XXXVIII
Substrate Scope for Hydrogenation with a Hydridoiron Complex

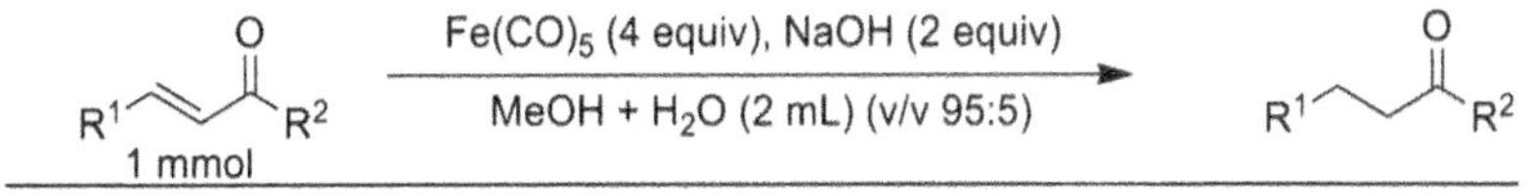

$R^1CH=CHC(O)R^2$ (1 mmol) → [$Fe(CO)_5$ (4 equiv), NaOH (2 equiv); MeOH + H_2O (2 mL) (v/v 95:5)] → $R^1CH_2CH_2C(O)R^2$

Entry	Substrate	Temp. (°C)	Time (h)	Product	Yield (%)
1	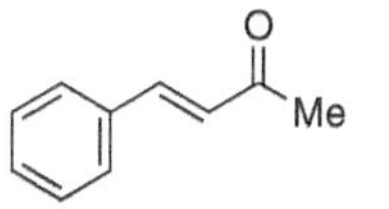 PhCH=CHC(O)Me	20	12	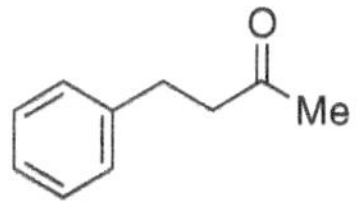 PhCH₂CH₂C(O)Me	>98
2	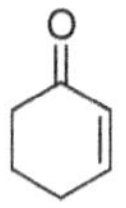 2-cyclohexen-1-one	20	10	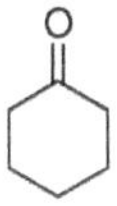 cyclohexanone	96
3	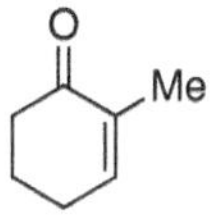 2-methyl-2-cyclohexen-1-one	60	24	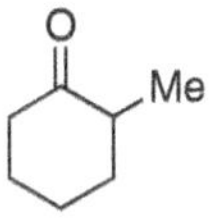 2-methylcyclohexanone	35
4	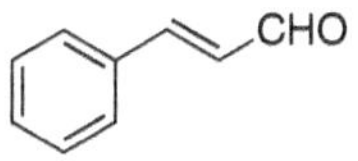 PhCH=CHCHO	20	12	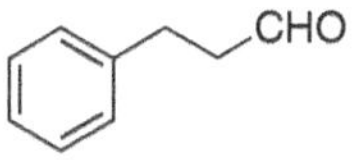 PhCH₂CH₂CHO	98
5	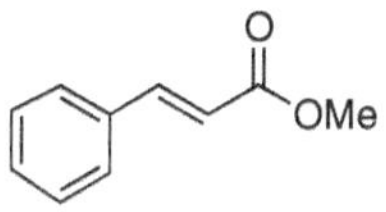 PhCH=CHC(O)OMe	20	48	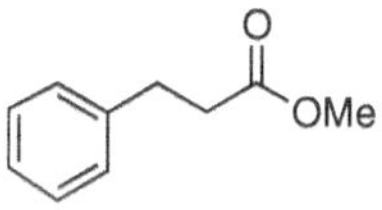 PhCH₂CH₂C(O)OMe	90
6	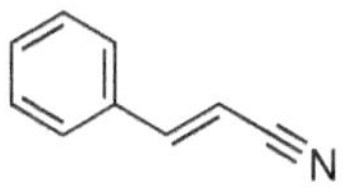 PhCH=CHCN	20	36	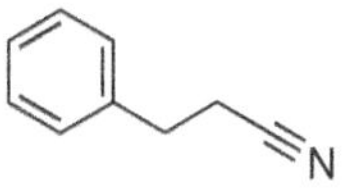 PhCH₂CH₂CN	92

R−C≡C−H (2 mmol) → [$[(PP_3)Fe(H)(N_2)]BPh_4$ (1 mol%), H_2 (1 atm); 63 °C, 2 h, THF (12 mL)] → RCH=CH₂

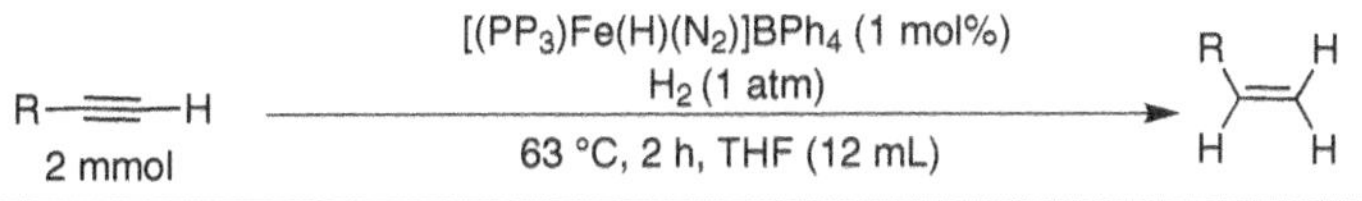

Product	PhCH=CH₂	Me₃SiCH=CH₂	nC_5H_{11}CH=CH₂	MeO−HC=HC−CH=CH₂
TOF (h^{-1})	45.2	5.2	24.5	21.2

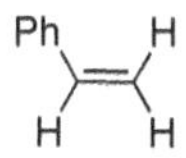

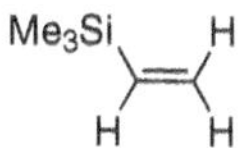

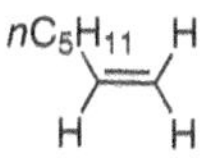

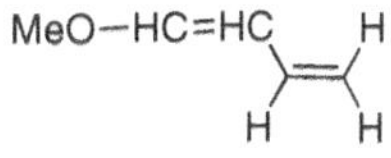

Scheme 176. Hydrogenation of the alkyne iron(II)–cis-hydride complex.

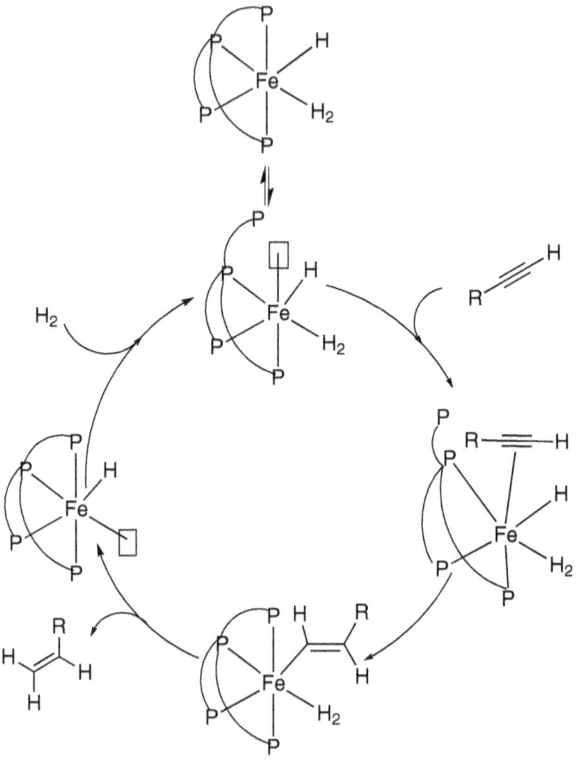

Scheme 177. Catalytic cycle for hydrogenation of alkyne.

complex under a hydrogenation atmosphere and has been suggested to be the active species for hydrogenation. Subsequently, a class of tris(phosphino)borato-ligated iron(III) alkyl and hydride complexes were prepared for olefin and alkyne hydrogenations (Scheme 179) (259).

Simple aromatic and aliphatic alkenes were hydrogenated efficiently following Peter's condition. Catalyst **127** was found to be more active since it is hydrogenated more easily than **128**. Double-bond isomerization of 1-hexene was observed

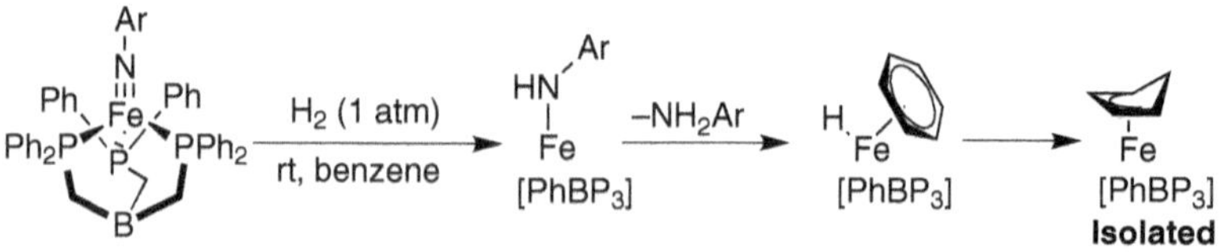

Scheme 178. Hydrogenation of benzene by tris(phosphino)borato-ligated iron(III) hydride.

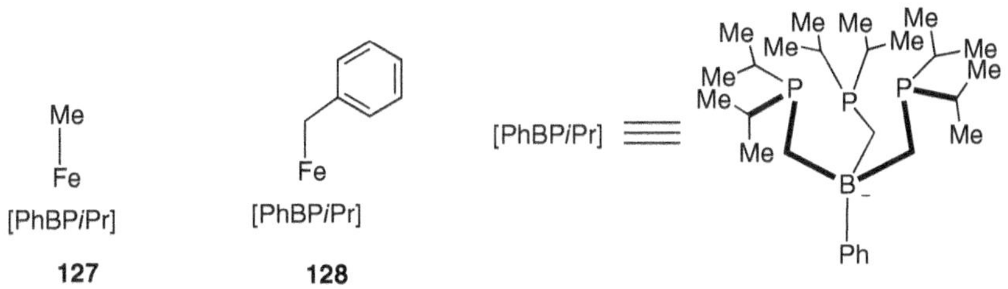

Me
Fe
[PhBP*i*Pr]
127

Fe
[PhBP*i*Pr]
128

[PhBP*i*Pr] ≡

Scheme 179. Catalyst for hydrogenation of unsaturated C–C bond.

during the course of hydrogenation. Alkyne is quantitatively hydrogenated to alkane after successive double hydrogenations (Table XXXIX) (259). The TOF is calculated after >95% conversion of the starting materials.

In their search for a better hydrogenation catalyst, Chirik and co-workers (246) reported bis(imino)pyridine-iron(0)-bis(nitrogen) complex (**129**) (Scheme 180), which can facilitate hydrogenation of a different functionalized alkene and alkyne (Schemes 179–181). The π-acidic imine ligand moiety was chosen to provide stability for the iron(0) oxidation state. During deuteriolysis of norbornene under the same reaction condition, *exo, exo*-2,3-norborane was generated through a cis-addition toward the double bond. Geometrical and stereochemical (cis–trans) isomerizations of the double bond were detected during the course of hydrogenation (246).

A wide range of functionalized alkenes and a disubstituted alkyne were hydrogenated using a catalytic amount of bis(imino)pyridine-iron(0)-bis(nitrogen) complex (**129**) (260). Amino olefins were hydrogenated to amino alkanes without any N-H trans-hydrogenation and the hydrogenation ability was increased with the N-alkylated substrates. Alkenes with various oxygen-containing functional groups and a fluorinated moiety can be tolerated. Di- and trisubstituted alkenes were readily hydrogenated. In the case of an α,β-unsaturated ketone, conversion sharply decreases due to the decomposition of the catalyst. Notably, olefin with a

TABLE XXXIX
Hydrogenation by Tris(phosphino)borato-Ligated Iron(III) Alkyl Complexes

Alkene or Alkyne $\xrightarrow{\textbf{127} \text{ or } \textbf{128} \text{ (10 mol\%), 50°C}}$ Alkane

Entry	Catalyst	Substrate	H_2 (atm)	Time (min)	TOF (h^{-1})
1	**127**	Styrene	4	78	7.7
2	**128**	Ethylene	4	25	24.0
3	**127**	1-Hexene	1	115	5.2
4	**128**	1-Hexene	1	130	4.6
5	**128**	2-Pentyne	1	370	1.6

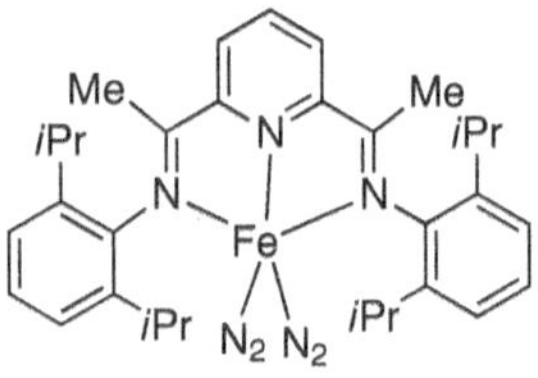

Scheme 180.

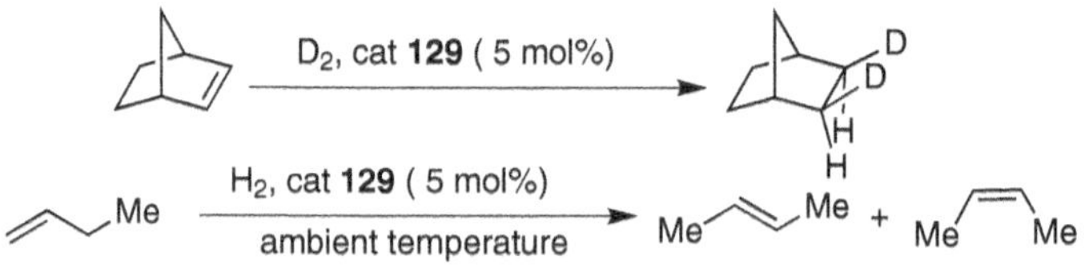

Scheme 181. Bis(imino)pyridine iron(0)-bis(nitrogen) complex for alkene hydrogenation.

conjugated ester moiety was selectively hydrogenated. Functional groups (e.g., amine or carbonyl) coordinate with the iron center prior to hydrogenation and hydrogenation TOF is inversely proportional to the strength of this coordination (Table XL) (260).

In 2012, Beller and co-workers (261) introduced the transfer hydrogenation of the terminal alkyne to the corresponding alkene using formic acid as the hydrogen source (Table XLI). Formic acid releases carbon dioxide and hydrogen, which coordinates to the metal center and performs the hydrogenation. By using deuterated formic acid (DCOOH) in place of HCOOH, deuterium was incorporated into the reduced alkene (261). A number of aliphatic alkynes, heterocyclic alkynes, and aromatic terminal alkynes bearing electron-donating and electron-withdrawing groups have been hydrogenated efficiently and selectively.

D. Hydrogenation of Ketones

Ketones are widely available precursors, for synthesizing enantiopure secondary alcohols. Optically active secondary amines, lactones, and so on, can be prepared from enantiopure alcohols, which are widely used in fragrances, pharmaceuticals, and in the beverage industry. Asymmetric hydrogenation is usually achieved using a chiral environment around the metal center. However, selectivity toward keto functionality in the presence of other unsaturated moieties is a challenging task.

TABLE XL
Olefin Hydrogenation Using Bis(imino)pyridine-Iron(0)-bis(nitrogen) Catalyst

R ⟶ (cat **129** (5 mol %), 4 atm H_2, 23 °C) R–Me

Entry	Substrate	Time (min)	Conversion (%)	TOF (h^{-1})
1	Me O	5	>99	>240
2	O Me O	900	3	0.04
3	O Me O Me	5	>99	>240
4	F F F F F	5	>99	>240
5	O Ph Ph		No conversion	
6	NH_2	1440[a]	20	3
7	H N Me	60[a]	95	320

[a] Catalyst loading 0.3 mol%.

Bianchini et al. (262) explored the iron-catalyzed reduction of ketone in 1993. A trihydride iron(II) catalyst, $[(PP_3)Fe(H)(H_2)]BPh_4$, was used for transfer hydrogenation, where a superstoichiometric (20 equiv) cyclopentanol or isopropyl alcohol was the hydride source. Reduction of ketone in the presence of an olefin is extremely challenging since the C=C bond is more easily reducible than the C=O bond. Increasing the size of the olefin substituent (R_1), decreases the chance of C=C hydrogenation. A large substituent at ketone (R_2) makes it less polarizable and decreases the probability of hydrogenation of the keto-group. Thus, it was possible to keep C=C unaffected even with a conjugated electron-withdrawing group (Table XLII) (262).

TABLE XLI
Iron-Catalyzed Selective Hydrogenation of Terminal Alkynes

R—≡—H → [$Fe(BF_4)_2.6H_2O$, PP_3 (1:1), HCOOH (2 equiv), THF, 40 °C, 5 h] → R(H)C=CH₂

PP_3 = $P(CH_2CH_2PPh_2)_3$ (Ph₂P, PPh₂, PPh₂)

Substrate (product)	Yield, (catalyst loading)
4-Br-styrene (Br)	99%, (0.75)[a]
4-COMe-styrene (COMe)	99%, (1)[a]
4-CO₂Me-styrene (CO_2Me)	99%, (2.5)[a]
3-vinylpyridine (N)	2%, (0.75)[a]
3-vinylthiophene (S)	99%, (1)[a]
1-alkene (Me)	99%, (1)[a]

4-acetylphenylacetylene → [$Fe(BF_4)_2.6H_2O$ (1 mol %), PP_3 (1 mol %), DCOOH, THF, 40 °C, 1 h] → deuterated 4-acetylstyrene (D 0.17, D 0.18, D 0.51)

[a]The values in parentheses are the catalyst loading in mol%.

Further interrogation showed that in the presence of H_2 gas, catalytic activity was totally stopped. This result suggested that hydrogen gas coordinates with the metal center and inhibits the substrate to bind with the metal center. For ketone hydrogenation, a catalytic cycle was proposed (Scheme 182) (262), where (a) at first the ketone coordinates with the metal center (may be in either η^1 or η^2 fashion), (b) then the hydride from the metal center is transferred to the keto carbon center through a four-member transition state, (c) the secondary alcohol coordinates with the metal, breaking the M–O bond of the reduced ketone, and (d) finally, the secondary alcohol donates a hydride to the metal center and departs as the ketone. A similar mechanism also has been proposed for the competing hydrogenation of an olefin. In the case of olefin hydrogenation, the C=C bond first coordinates with the metal center in η^2-fashion and goes through a π-oxa-allyl intermediate. Therefore, any steric hindrance near the keto group prefers olefin hydrogenation more and vice versa.

In 2007, Casey and co-workers (263) reported the first iron-catalyzed efficient chemoselective H_2 hydrogenation of a ketone using the similar catalyst active center of Shvo's ketone hydrogenation catalyst, which replaced ruthenium with iron (**130**) (Table XLIII). Successful hydrogenation with aliphatic and aromatic ketones has proved the utility of this method. Further, the reported reaction condition was milder

TABLE XLII
Chemoselective-Transfer Hydrogenation of α,β-Unsaturated Ketones

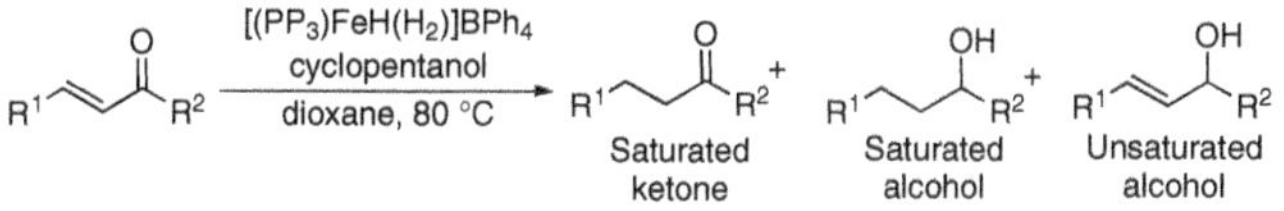

$[(PP_3)FeH(H_2)]BPh_4$, cyclopentanol, dioxane, 80 °C; products: Saturated ketone + Saturated alcohol + Unsaturated alcohol; $PP_3 = P(CH_2CH_2PPh_2)_3$

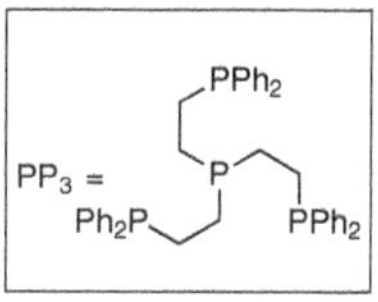

Entry	Substrates	Yield of Saturated Ketone (%)	Yield of Saturated Alcohol (%)	Yield of Unsaturated Alcohol (%)
1	R^1 = Ph; R^2 = Me 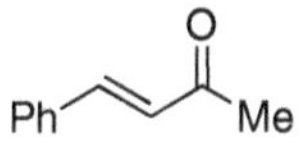	0 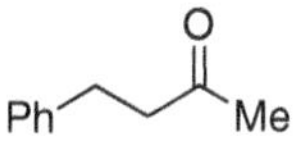	0 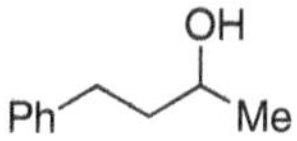	95 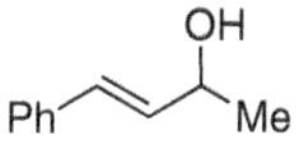
2	R^1 = Ph; R^2 = Ph 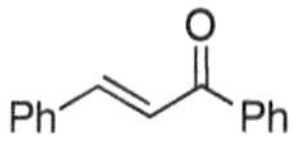	30 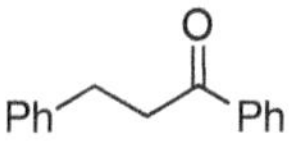	0 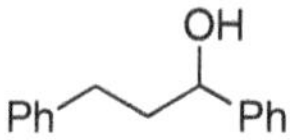	0 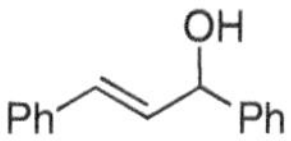
3	R^1 = Me; R^2 = Ph 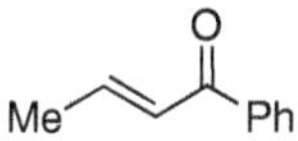	7 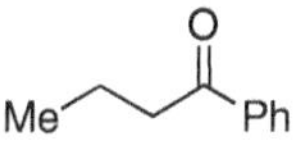	0 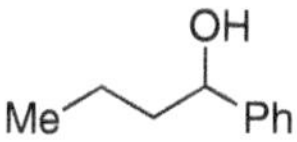	0 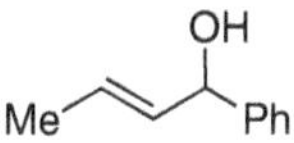
4	R^1 = Me; R^2 = Et 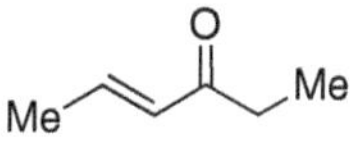	19 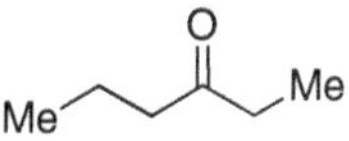	0 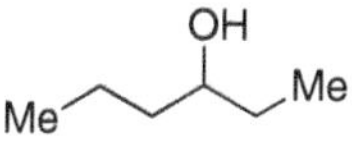	0 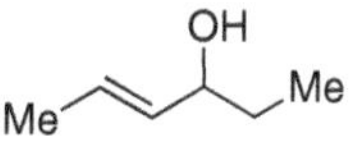
5	R^1 = H; R^2 = Et 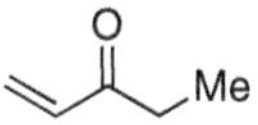	100 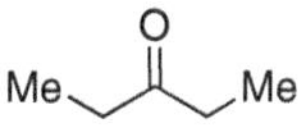	0 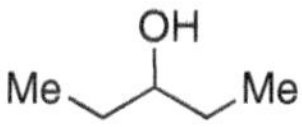	0 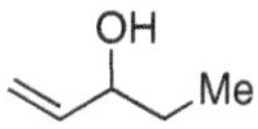
6	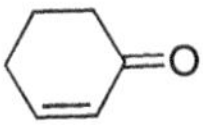 cyclohex-2-enone	0 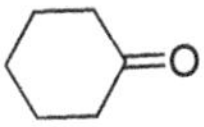	44 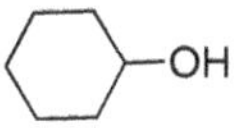	28 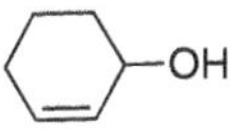
7	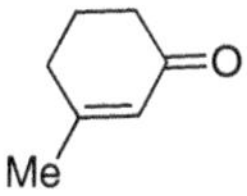 3-methylcyclohex-2-enone	0 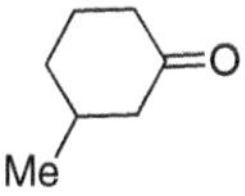	0 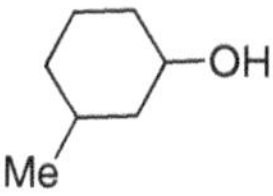	31 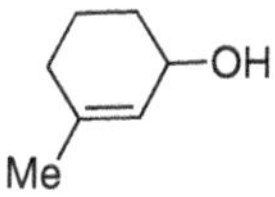
8	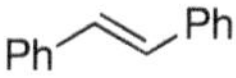 PhCH=CHPh	No reaction	-	-

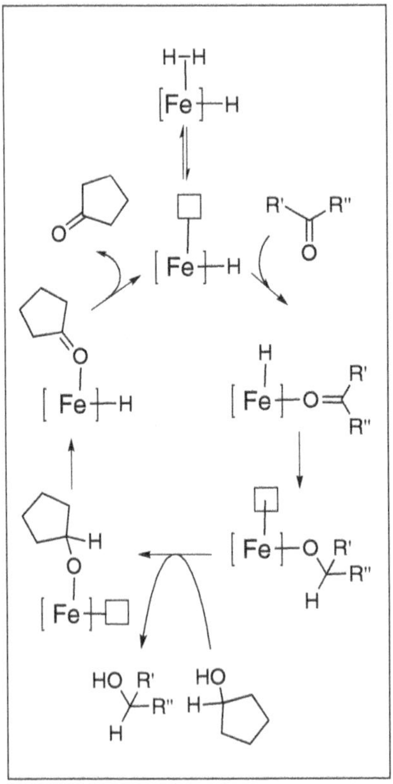

Catalytic cycle for C=O hydrogenation

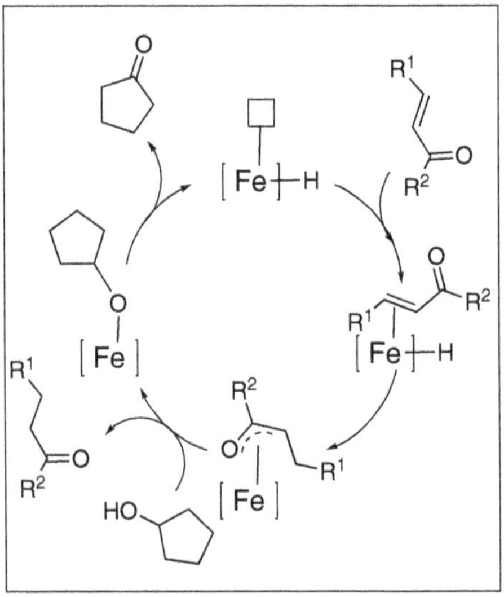

Catalytic cycle for C=C hydrogenation

Scheme 182. Catalytic cycle for transfer hydrogenation of α,β-unsaturated ketones.

than Shvo's reaction condition. Functional groups including a nitro and a double bond at the homoallylic position were well tolerated in this system. A ketone with a pyridine moiety was also hydrogenated efficiently. High diastereoselectivity has also been observed in the case of benzoin (*meso/dl* = 25). Under hydrogenation condition, esters, epoxides, alkynes, and alkenes were survived well. Unfortunately, selective hydrogenation of α,β-unsaturated ketone failed to be impressive.

Casey and co-workers (263b) first established the mechanistic details of hydrogenation with Shvo's catalyst. Detailed studies on the hydrogenation of benzaldehyde revealed first-order dependence with respect to both the aldehyde and the catalyst. In addition, a primary kinetic isotope effect for transfer of both RuD and OD was also observed. Since transfer of ^{13}CO from labeled catalyst has not occurred, they proposed an outer-sphere mechanism as depicted in Scheme 183 (263b).

In the case of an iron-based catalyst, Casey's group found similar experimental results and concluded that the iron catalyst is also going through the same

TABLE XLIII
Substrate Scope for Hydrogenation Explored by Casey's Group

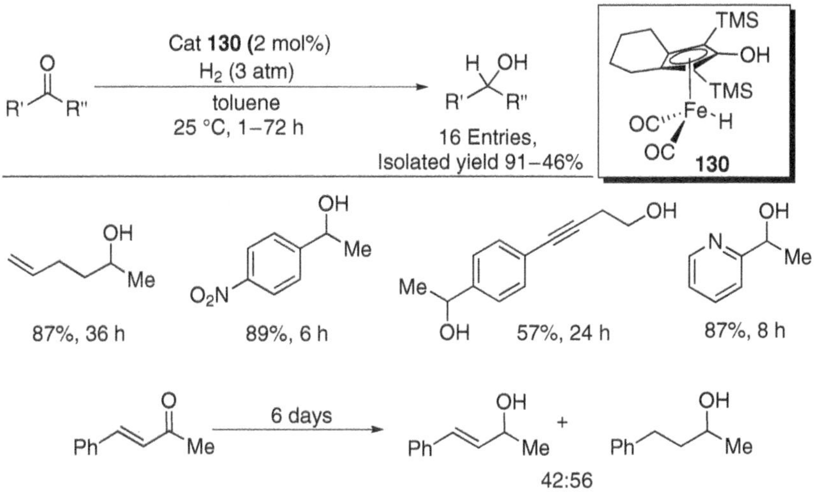

Cat **130** (2 mol%)
H_2 (3 atm)
toluene
25 °C, 1–72 h

16 Entries,
Isolated yield 91–46%

130

87%, 36 h
89%, 6 h
57%, 24 h
87%, 8 h

6 days
42:56

mechanistic pathway as that of ruthenium (Scheme 184) (263a). In the first step, the hydroxyl proton is transferred to the keto-oxygen and hydride is transferred to the keto-carbon. Consequently, hydrogen gas regenerates the catalyst and completes the catalytic cycle. In addition, the proposed intermediate (**131**) was trapped by PPh_3, which provides further support to this hypothesis. Also, there was no rate dependence on the concentration of PPh_3. *In situ*, they quantitatively measured the rate of hydrogenation by ReactIR and concluded that hydrogen transfer from the catalyst is the rate-determining step and first order with respect to the catalyst.

Casey's iron cyclopentadienone catalyst (**130**) with a chiral ligand environment was capable of chiral hydrogenation. In 2012, Berkessel et al. (264) developed a chiral precatalyst (**133**) that replaced one CO of Casey's catalyst with a chiral phosphoramidite ligand under photoirradiation (Scheme 185).

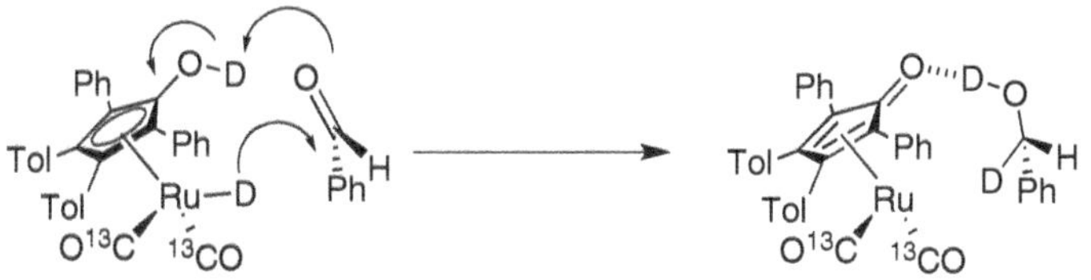

Scheme 183. Labeling experiment on hydrogenation with Shvo' catalyst.

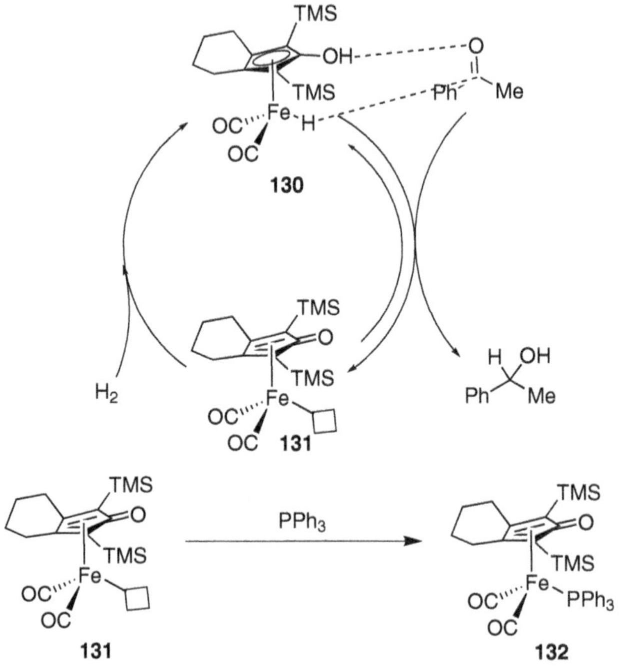

Scheme 184. Outer-sphere mechanism for transfer hydrogenation of ketones.

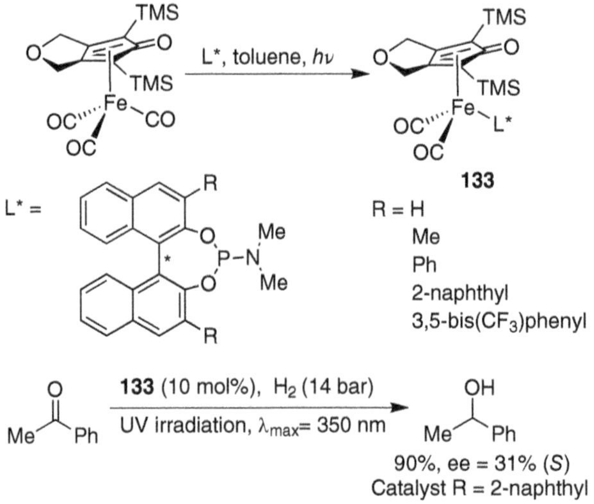

Scheme 185. Enantioselective hydrogenation of a ketone using a chiral phosphoramidite.

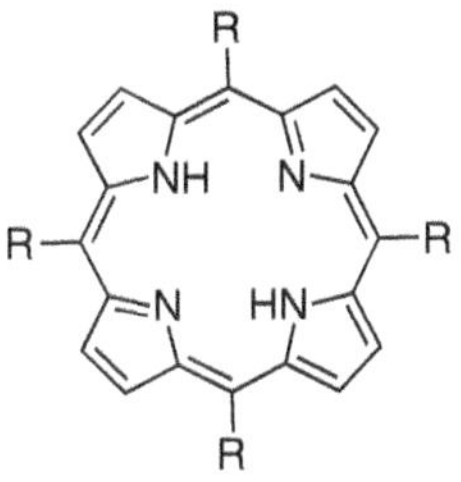

134a R= *p*-Cl-C_6H_4 $E_{1/2}$ = 330 mV, TOF = 2100
134b R= Ph $E_{1/2}$ = 338 mV, TOF = 1500
134c R= C_6F_5 $E_{1/2}$ = 326 mV, TOF = 1150

Scheme 186. Porphyrin ligands used in transfer-hydrogenation reactions.

Iron–porphyrin systems are also known for transfer hydrogenation. In 2008 Beller and co-workers (265) tested a class of porphyrin derivatives (**134a–134c**) and discovered ligand **134a** (Scheme 186), which showed a good catalytic system with $FeCl_2$(Table XLIV) (265). Substitution at the meso-phenylic moiety of the porphyrin system is responsible for tuning the oxidation potential of iron(II) to iron(III). Unfortunately, no clear correlation between the catalytic activity and the oxidation potential can be outlined. Different types of 2-alkoxy- and 2-aryloxy ketones were tested and most of them had shown good-to-moderate yields. Increasing the bulkiness simultaneously at the 2- and 6-positions, yield of hydrogenation reaction decreased. However, a substrate without a six substitution gave a good yield. Unfortunately, other hydroxyl protecting groups like sillyl and acetyl were unsuccessful.

TABLE XLIV
Substrate Scope for Transfer Hydrogenation Using an Iron–Porphyrin Complex

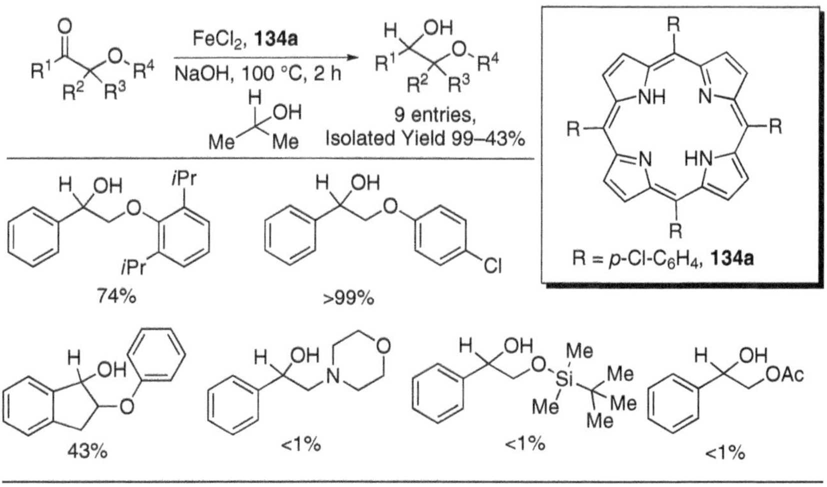

$FeCl_2$, **134a**
NaOH, 100 °C, 2 h
9 entries,
Isolated Yield 99–43%

R = *p*-Cl-C_6H_4, **134a**

74%
>99%
43%
<1%
<1%
<1%

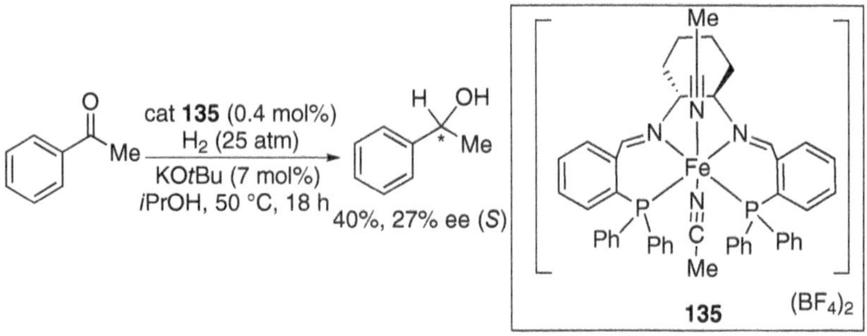

Scheme 187. Asymmetric hydrogenation using iron complexes containing a P–N–N–P based ligand.

Earlier asymmetric hydrogenation of ketones has been reported with a ruthenium-based catalyst containing P–N–N–P based chiral ligands (Gao's catalyst). Using a P–N–N–P chelated iron catalyst (**135**), Morris and co-workers (266) in 2009 reported hydrogenation under a high-pressured hydrogen atmosphere (Scheme 187). Unfortunately, they failed to extend their substrate scope beyond acetophenone.

Catalyst **136** with P–N–N–P and **137** with P–NH–NH–P ligand systems gave a similar result in the hydrogenation of acetophenone (Scheme 188) (266b). Their similar activity suggested that these two systems go through the same active intermediate. It was also assumed that iron-catalyzed reactions are going through the same reaction pathway as the ruthenium based catalyst *trans*-[RuH_2(P-NH-NH-P)]. Density functional theory (DFT) calculations also proved that iron would have similar activation barriers for ruthenium in the catalytic cycle (266b).

An outer-sphere mechanism was proposed for catalyst **135**, where the N–H proton binds with the carbonyl oxygen and the M–H hydride attacks the carbonyl carbon. Therefore, the alcohol was generated along with an unsaturated imido complex, which cleaved the hydrogen molecule heterolytically and regenerated the

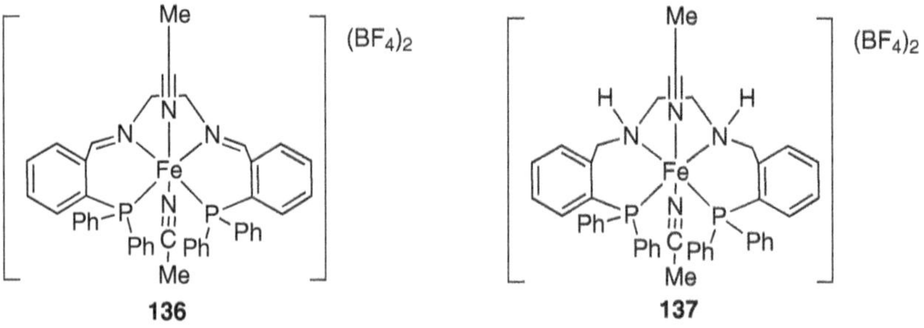

Scheme 188. Iron catalysts for hydrogenation with a P–N–N–P and P–NH–NH–P based ligand.

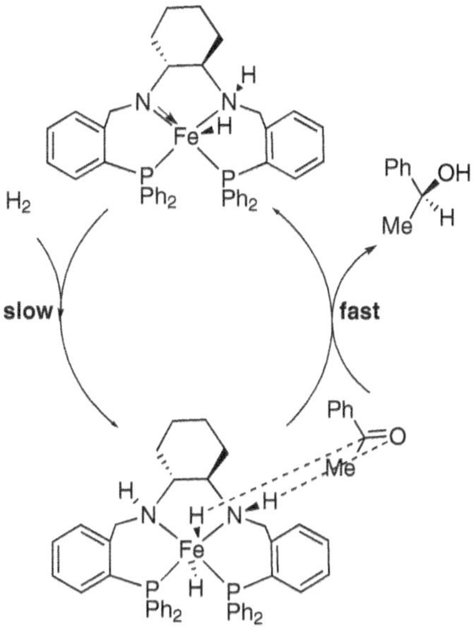

Scheme 189. Iron-catalyzed asymmetric hydrogenation of ketones.

active catalyst. The regeneration step required a high-activation energy and therefore a high-pressured hydrogen atmosphere was needed for smooth functioning of the catalytic cycle (Scheme 189) (266b, 267).

A suitable chiral ligand for iron can promote hydrogenation in an asymmetric fashion. For example, Morris and co-workers (266a, 267–268) found that catalyst **138** was suited for transfer hydrogenation using isopropyl alcohol as the hydride source, as well as the solvent. Functional group variability and moderate-to-good enantioselectivity were notable features of this method. They found that the substituent next to the keto group determined the conversion, as well as the stereoselectivity. From acetophenone to *tert*-butylphenyl ketone, conversion decreased, but selectivity increased. On the other hand, cyclohexylphenyl ketone showed low conversion and less stereoselectivity compared to acetophenone. Unlike the meta position, the π-electron-donating moiety at the para position of the aromatic ketone decreased the stereoselectivity. The method failed to generate the expected product from α,β-unsaturated ketones (Table XLV) (266a). Catalyst **138** followed the same mechanistic pathway as catalyst **135** except for the last step, where a secondary alcohol regenerated the active catalyst (269).

After considerable progress in the development of a phosphorous nitrogen-based iron complex for ketone hydrogenation, Milstein and co-workers (270) reported iron–pincer complexes as effective hydrogenation catalysts (Table XLVI). Even at rt

TABLE XLV

Substrate Scope for Iron-Catalyzed Asymmetric-Transfer Hydrogenation

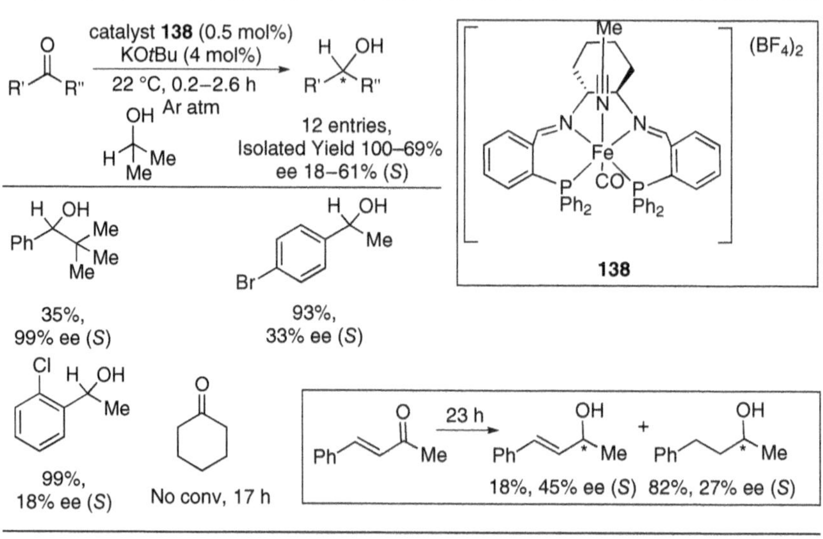

$R'C(O)R''$ → catalyst **138** (0.5 mol%), KO*t*Bu (4 mol%), 22 °C, 0.2–2.6 h, Ar atm, *i*PrOH → $R'CH(OH)R''$	12 entries, Isolated Yield 100–69%, ee 18–61% (*S*)	**138**: [Fe(CO) complex](BF$_4$)$_2$	
PhCH(OH)CMe$_3$: 35%, 99% ee (*S*)	4-BrC$_6$H$_4$CH(OH)Me: 93%, 33% ee (*S*)		
2-ClC$_6$H$_4$CH(OH)Me: 99%, 18% ee (*S*)	Cyclohexanone: No conv, 17 h	PhCH=CHC(O)Me → (23 h) PhCH=CHCH(OH)Me: 18%, 45% ee (*S*)	+ PhCH$_2$CH$_2$CH(OH)Me: 82%, 27% ee (*S*)

(26–28 °C) and at a relatively low hydrogen pressurized atmosphere (4.1 atm) iron–pincer complex [(*i*PrPNP)FeH(CO)Br] [*i*PrPNP = 2,6-bis(diisopropylphosphino-methyl)pyridine] (**139**) showed hydrogenation with a broader substrate scope. Halide substituted aromatic ketones and ketones with a pyridine moiety were well tolerated.

TABLE XLVI

Hydrogenation of Ketones by an Iron–Pincer Complex

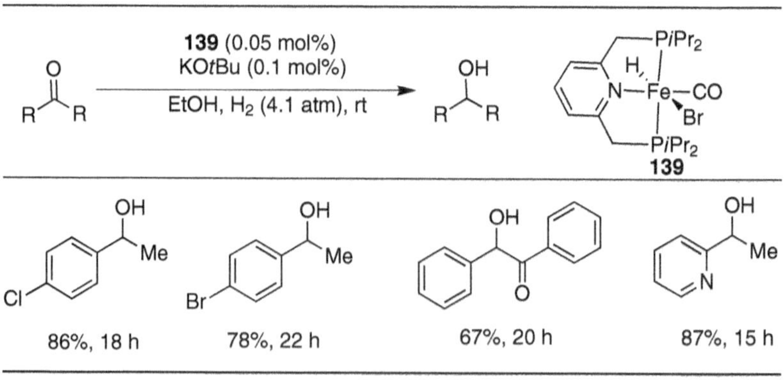

$RC(O)R$ → **139** (0.05 mol%), KO*t*Bu (0.1 mol%), EtOH, H$_2$ (4.1 atm), rt → $RCH(OH)R$		**139**: (P*i*Pr$_2$)$_2$ pyridine pincer Fe(H)(CO)Br	
4-ClC$_6$H$_4$CH(OH)Me: 86%, 18 h	4-BrC$_6$H$_4$CH(OH)Me: 78%, 22 h	PhCH(OH)C(O)Ph: 67%, 20 h	2-PyCH(OH)Me: 87%, 15 h

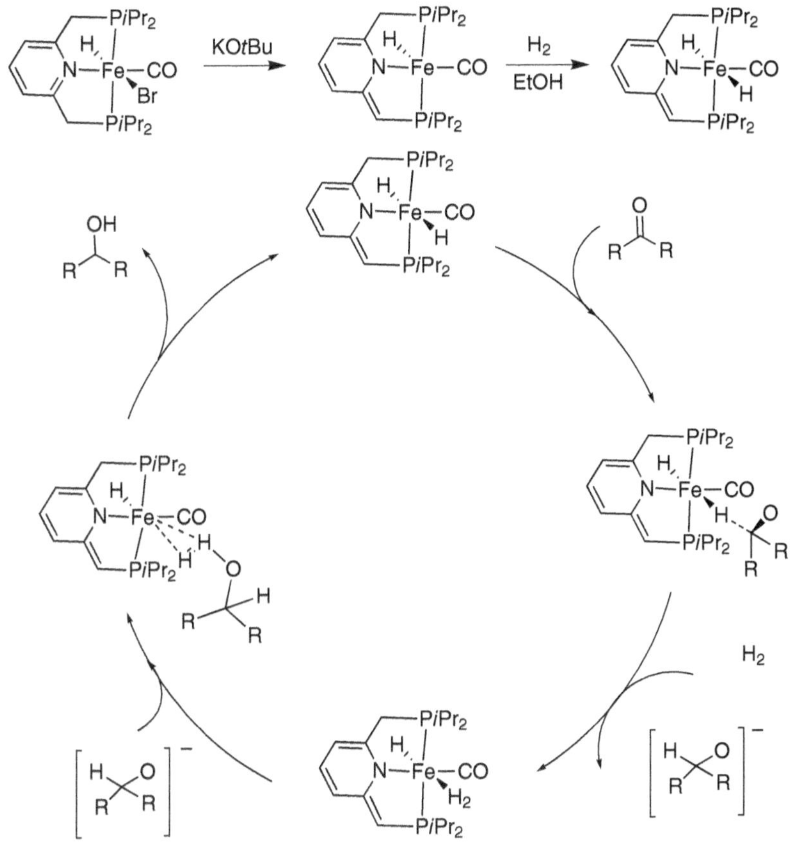

Scheme 190. Catalytic cycle for hydrogenation of ketones with an iron–pincer complex.

The hydrogenation of ketones by the [(*i*PrPNP)FeH(CO)Br] complex putatively underwent a direct reduction mechanism. The DFT studies by Yang (271) showed that the pincer ligand was not involved in hydride transfer to the ketone carbon, as well as the H_2 molecule cleavage by the hydrido-alkoxo species (Scheme 190). Calculations also showed that the choice of EtOH as the solvent is crucial for H_2 cleavage.

Further studies discovered a tetrahydridoborate iron–pincer complex [(*i*PrPNP) FeH(CO)(η^1-BH_4)] that can promote hydrogenation of ketones with almost the same efficiency without any base (272).

Royo and co-workers (273) reported transfer hydrogenation of ketones using a cyclopentadienyl functionalized NHC iron complex (Table XLVII). Their piano-stool type iron complex (**140**) can accomplish hydrogenation in the presence of

TABLE XLVII
Hydrogenation of a Ketone by a Cyclopentadienyl Functionalized NHC Complex

R(C=O)R → 140 (1 mol%), KOH; HO–CH(Me)Me → RCH(OH)R

140

1-Phenylethanol	Diphenylmethanol	Cyclohexanol
86%, 6 h	85%, 18 h	100%, 2 h

isopropyl alcohol as the hydride source. In 2012, Hashimoto et al. (274) applied an iron(II)-N-heterocyclic carbene complexe for transfer hydrogenation. They explored a number of monodentate NHC ligands with an iron(II) complex for catalytic study. Iron NHC complexes $(IMes)_2FeCl_2$, *trans*-$(IMes)_2FeMe_2$, $(IEtPh^*)_2FeCl_2$ [IMes = 1,3-bis(2,4,6-trimethylphenyl)imidazole-2-ylidene] and [IEtPh* = 1,3-bis(*R*)-1'-phenylethyl)imidazole-2-ylidene] were efficient in catalytic-transfer hydrogenation.

Recently, Beller and co-workers (275) published reduction of an aldehyde in a water–gas shift condition using Knolker's iron cyclopentadienone complex (**141**) as catalyst (Table XLVIII). They used K_2CO_3 as the base and water as the cosolvent under a CO atmosphere. In their reduction protocol, aromatic and alphatic aldehyde as well as α,β-unsaturated aldehydes can be employed. Presumably, hydroxide attacked the CO coordinated to the metal center and formed CO_2 and an iron-hydride species.This iron-hydride species was responsible for the reduction.

E. Hydrogenation of Imines

Chiral amine moieties are found in numerous drug and pesticide molecules. The most convenient way of preparing stereochemically active amines is the hydrogenation of imines in an asymmetric way. As expected, plenty of catalytic methods based on transition metals were developed in the past two decades. Iron is most acceptable among other transition metals due to its natural abundance and biorelevance. The reduction of imines is more challenging since they are much more sensitive toward hydrolysis and other side reactions. This problem is settled

TABLE XLVIII
Iron-Catalyzed Reduction of Aldehyde Using a Water–Gas Shift Condition

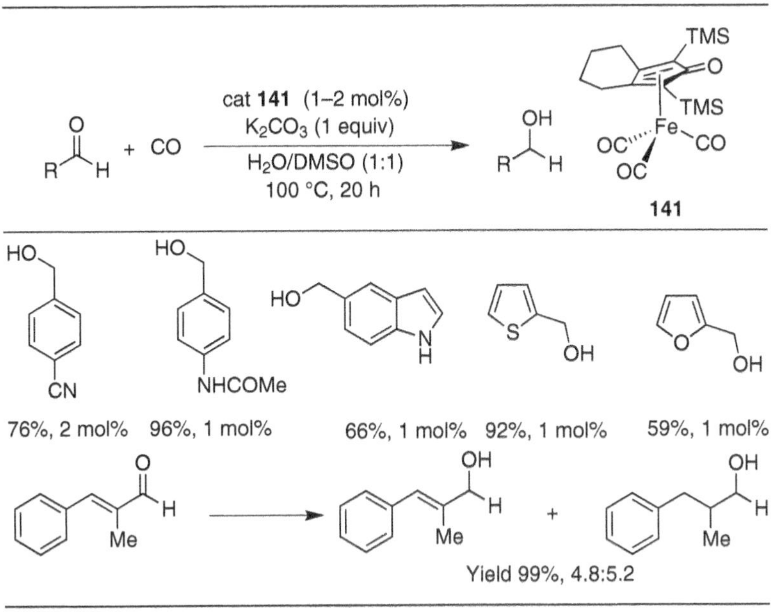

RCHO + CO → RCH₂OH; cat **141** (1–2 mol%), K_2CO_3 (1 equiv), H_2O/DMSO (1:1), 100 °C, 20 h

Product	Yield, catalyst loading
4-Cyanobenzyl alcohol	76%, 2 mol%
4-(Acetylamino)benzyl alcohol (NHCOMe)	96%, 1 mol%
Indol-5-ylmethanol	66%, 1 mol%
2-Thienylmethanol	92%, 1 mol%
2-Furylmethanol	59%, 1 mol%
α-Methylcinnamaldehyde → α-methylcinnamyl alcohol + 2-methyl-3-phenylpropan-1-ol	Yield 99%, 4.8:5.2

by electronic or steric protection (e.g., phosphinoyl, sulfonyl) of imines prior to hydrogenation and then deprotection.

In 2008, Morris and co-workers (266a) showed that the iron–PNNP [P(2)(N2)] ligated complex (**138**) can hydrogenate *N*-benzylideneaniline (Ph-CH=NPh) (Scheme 191). Unfortunately, the same condition completely failed for *N*-(1-phenylethylidene)aniline (Ph-CMe=NPh).

After Morris, Beller and co-workers (276) reported the iron-catalyzed asymmetric hydrogenation of imines (Table XLIX). They used the iron(II) hydride

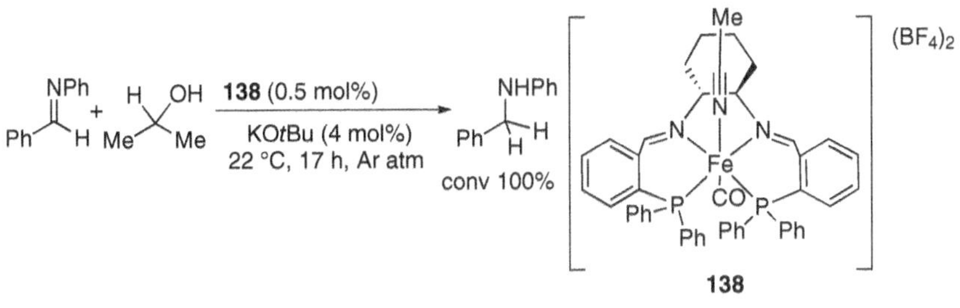

PhCH=NPh + Me_2CHOH → PhCH₂NHPh; **138** (0.5 mol%), KO*t*Bu (4 mol%), 22 °C, 17 h, Ar atm; conv 100%

138: $(BF_4)_2$

Scheme 191. Hydrogenation of imine by an Fe^{II}–PNNP ligated complex.

TABLE XLIX
Assymetric Hydrogenation of *N*-(Diphenylphosphinyl)imines

cat $(Et_3NH)[HFe_3(CO)_{11}]$
142
KOH (5 mol%), *i*PrOH
45 °C, 30 min

142

94%, ee 96% (*R*)
85%, ee 95% (*R*)
85%, ee 95% (*R*)
67%, ee 92% (*R*)
87%, ee 95% (*R*)
29%, ee 35% (*R*)

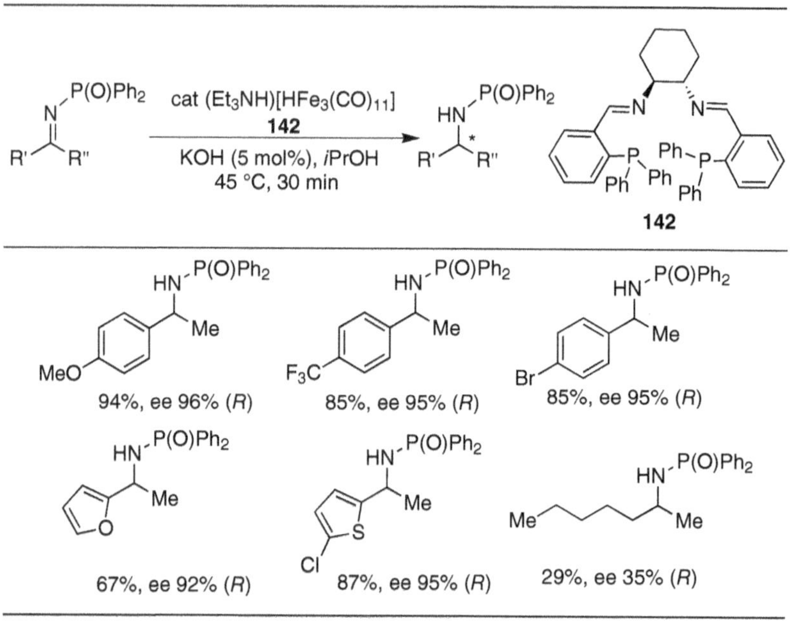

carbonyl complex with the chiral PNNP ligand (**142**) for asymmetric hydrogenation of *N*-(diphenylphosphinyl)imines. Aromatic and heterocyclic imines were efficiently hydrogenated with up to 99% ee. But, aliphatic imines failed to give a meaningful conversion, as well as stereoselectivity.

A year later, the asymmetric hydrogenation reaction of *N*-(1-phenylethylidene) aniline (Ph-CMe=NPh) was demonstrated by Beller and co-workers (277) using the Knolker iron cyclopentadienone complex (**130**) with a chiral phosphate ligand (**143**) (Table L). It is notable that this type of imine was unreactive toward Morris's catalyst. In the current reaction protocol, a class of aromatic ketamines was hydrogenated stereoselectively.

In 2012, Morris and co-workers (278) demonstrated a class of iron–PNNP based precatalyst for asymmetric-transfer hydrogenation of *N*-(diphenylphosphinyl) imines. They observed that the catalytic activity was dependent on the substituent on the phosphorous atom or the type of diamine. Catalyst **148** gave very low hydrogenation conversion due to the presence of an electron-donating ethyl group on the phosphorous atom. However, catalyst **146** was efficient in conversion, as well as in stereoselectivity. Aromatic and heterocyclic *N*-(diphenylphosphinyl)-imines were efficiently hydrogenated with good conversion and excellent stereoselectivity (Scheme 192).

TABLE L
Hydrogenation of Imine by the Knolker Iron Cyclopentadienone Complex[a]

$R^1C(=NR^3)R^2$ → **130** (1 mol%), **143** (5 mol%), H_2 (50 bar), toluene, 65 °C, 24 h → $R^1C^*H(NHR^3)R^2$

130; **143** (S), R' = 2,4,6-$iPr_3C_6H_2$

87%, ee 93% (*S*)	87%, ee 98% (*S*)	91%, ee 93% (*S*)	99%, ee 83% (*S*)

[a] *p*-Methoxyphenyl = PMP.

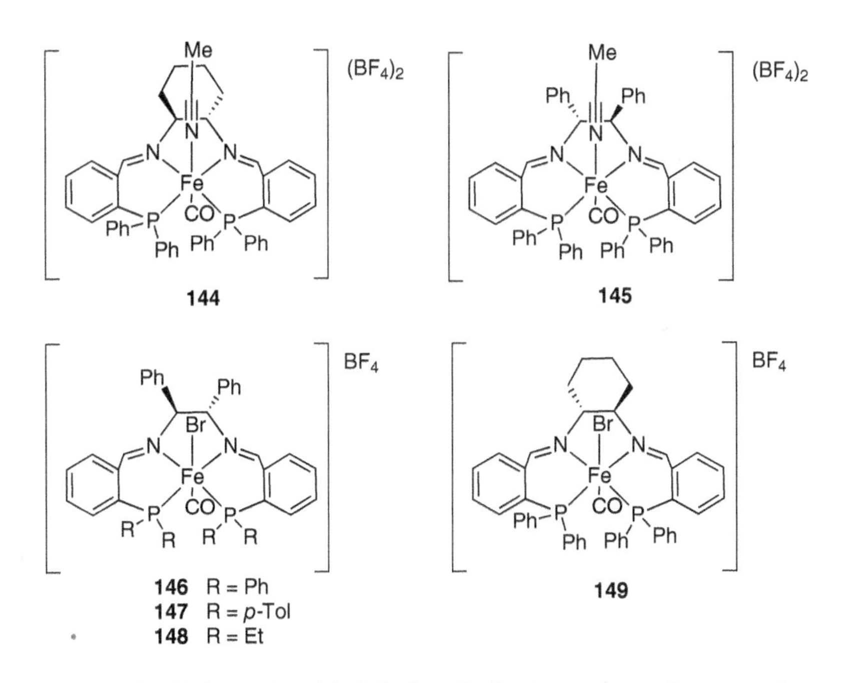

Scheme 192. Hydrogenation of the imine by a Knolker iron–cyclopentadienone complex.

F. Reduction of Nitroarene to Anilines

In synthetic chemistry, reduction of diverse functional groups is an indispensable protocol. Hydrogenation of nitro compounds to the corresponding amines is considered as one of the most important and classical methods for the production of amines. Since amine derivatives are valuable intermediates for the manufacture of many agrochemicals, pharmaceuticals, dyes, and pigments, it is important to develop economically feasible catalytic processes for this transformation.

The selective reduction of the nitro group is often difficult when other reducible groups (e.g., carbon–carbon or carbon–nitrogen double or triple bonds, carbonyl or benzyl groups, and multiple Cl, Br, or I substituents) are present in the same molecule. Continuous progress has been made and is still going on in this area to produce a catalytic method based on iron that can selectively reduce the nitro group to the corresponding amines. In this section, iron-based nitro reduction has been complied.

Bachamp (279a) used iron in a stoichiometric amount for the reduction of nitro compounds to amines with Fe/HCl (Scheme 193). This reaction was originally used to produce large amounts of aniline for industry, but this reaction is the most vigorous reduction method producing merely the amino products. Therefore, if the aromatic moiety contains substituents prone to being reduced (e.g., carbonyl, cyano, azo, or further nitro groups) a significant amount of the byproducts will be produced.

In 1998, Lauwiner et al. (279) applied an iron oxide–hydroxide as a catalyst for the reduction of various substituted nitroarenes to produce the corresponding amines (Table LI). Hydrazine hydrate was applied as the hydrogen source. Various functional groups (e.g., halide, hydroxyl, amide, and acid) were well tolerated. The important finding was that regioselective reduction of 1,3-dinitrobenzene produce the corresponding 3-nitroanilines, but the method was not applicable for 1,2- and 1,4-dinitrobenzene. In most cases, an excellent yield was reported. The reaction kinetics was measured to determine the dependence of the rate of reduction on the nature and position of additional substituents other than the nitro group. It was observed that the rate is enhanced by electron-withdrawing substituents and decreased by electron-donating groups (279).

The same group reported the reduction of a series of monosubstituted 3- and 4-nitrophenylazobenzenes with hydrazine hydrate in the presence of the iron oxide–hydroxide catalyst (Scheme 194) (280). Their study revealed that selectivity for the nitro group reduction versus that of the azo bridge in substituted

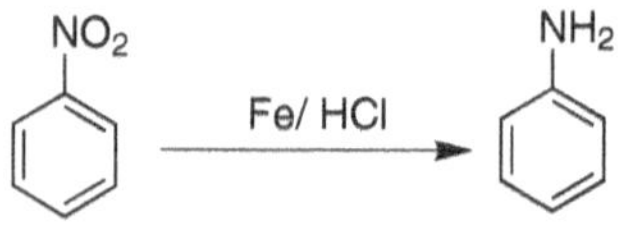

NO_2 —Fe/ HCl→ NH_2

Scheme 193.

TABLE LI
Iron Oxide–Hydroxide Catalyzed Reduction of Substituted Nitroarenes

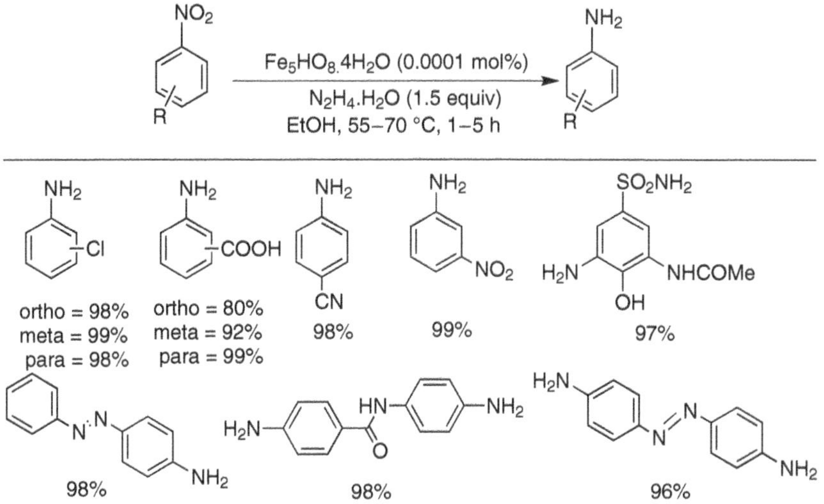

$Fe_5HO_8.4H_2O$ (0.0001 mol%)
$N_2H_4.H_2O$ (1.5 equiv)
EtOH, 55–70 °C, 1–5 h

Product	Yield
Chloroaniline (NH_2, Cl)	ortho = 98%, meta = 99%, para = 98%
Aminobenzoic acid (NH_2, COOH)	ortho = 80%, meta = 92%, para = 99%
4-Aminobenzonitrile (NH_2, CN)	98%
3-Nitroaniline (NH_2, NO_2)	99%
SO_2NH_2, H_2N, NHCOMe, OH substituted benzene	97%
4-Aminoazobenzene	98%
H_2N–C₆H₄–C(O)–NH–C₆H₄–NH_2	98%
H_2N–C₆H₄–N=N–C₆H₄–NH_2	96%

nitrophenylazobenzenes is higher than for the para nitro group than for the azo bridge, than for the corresponding meta counterparts. This selectivity was enhanced with the stronger electron-attracting properties of the substituent. Moreover, the selectivity for the reduction of the nitro group versus that of the azo function can be enhanced by increasing the reaction temperature (280).

Desai et al. (281) applied mild, neutral, ecofriendly, and a useful FeS$-NH_4Cl-$ MeOH$-H_2O$ system for the reduction of nitroarenes to anilines (Scheme 195).

In 2003, Sonavane et al. (282) applied recyclable trivalent iron substituted hexagonal mesoporous aluminnophospahte (FeHMA), for catalytic-transfer hydrogenation of aromatic nitro compounds using isopropyl alcohol as donor (Table LII). High chemo- and regioselectivity was reported in the studied substrates. The catalyst

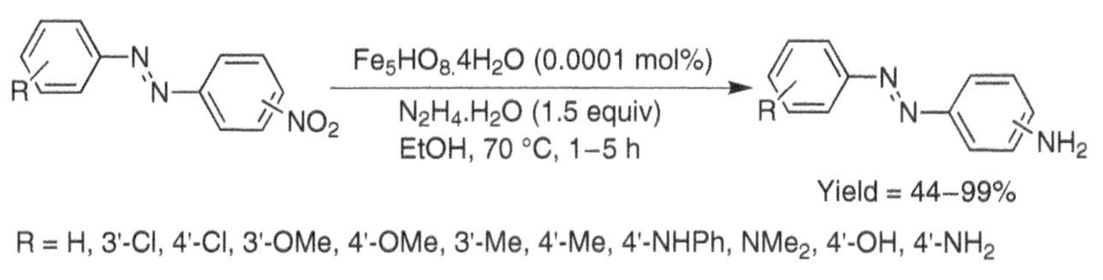

$Fe_5HO_8.4H_2O$ (0.0001 mol%)
$N_2H_4.H_2O$ (1.5 equiv)
EtOH, 70 °C, 1–5 h

Yield = 44–99%

R = H, 3'-Cl, 4'-Cl, 3'-OMe, 4'-OMe, 3'-Me, 4'-Me, 4'-NHPh, NMe_2, 4'-OH, 4'-NH_2

Scheme 194.

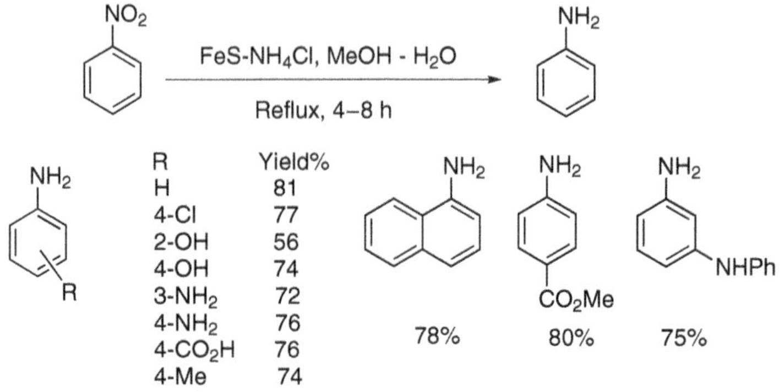

$$NO_2\text{-}C_6H_5 \xrightarrow[\text{Reflux, 4–8 h}]{\text{FeS-NH}_4\text{Cl, MeOH - H}_2\text{O}} NH_2\text{-}C_6H_5$$

R	Yield%
H	81
4-Cl	77
2-OH	56
4-OH	74
3-NH_2	72
4-NH_2	76
4-CO_2H	76
4-Me	74

1-Naphthylamine 78%; methyl 4-aminobenzoate (CO_2Me) 80%; 3-amino-NHPh 75%

Scheme 195.

was recycled up to six times without any loss in selectivity. It was observed that the regioselectivity depended on the reaction time (282).

In 2004, Deshpande et al. (282b) reported chemoselective reduction of substituted nitroarenes that correspond to amines by using water soluble iron ethylenediaminetetraacetic acid disodium salt (Fe^{II}/$EDTANa_2$) with molecular hydrogen (Table LIII). They applied this catalyst for hydrogenation in a biphasic toluene–water system to achieve selectivity and recyclability. Catalyst remained in the aqueous phase, while the nitro substrate and the product formed in the organic layer. A comparision of reactivity of this water soluble iron salt showed little decrease in activity, which may be due to the coordination of $EDTANa_2$ and also due to the fact that catalyst remained in the aqueous phase. Excellent chemoselectivity with a

TABLE LII

The FeHMA Catalyzed Transfer Hydrogenation of Aromatic Nitro Compounds

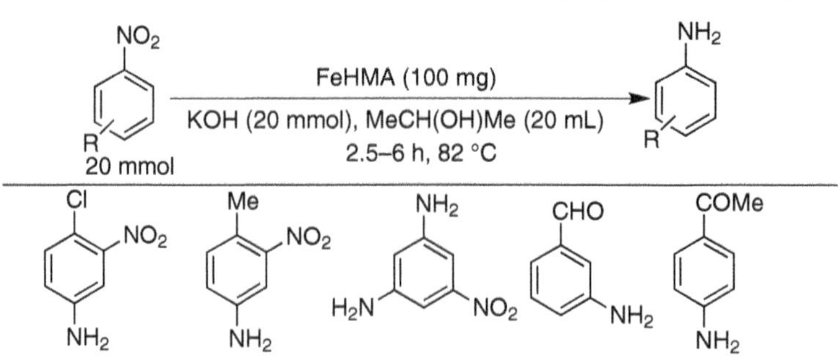

$$R\text{-}C_6H_4\text{-}NO_2 \text{ (20 mmol)} \xrightarrow[\text{KOH (20 mmol), MeCH(OH)Me (20 mL), 2.5–6 h, 82 °C}]{\text{FeHMA (100 mg)}} R\text{-}C_6H_4\text{-}NH_2$$

Product	First run	Six run
4-Amino-2-chloronitrobenzene (Cl, NO_2, NH_2)	82%	81%
4-Amino-2-methylnitrobenzene (Me, NO_2, NH_2)	74%	74%
3,5-Diaminonitrobenzene (NH_2, H_2N, NO_2)	83%	83%
3-Aminobenzaldehyde (CHO, NH_2)	83%	83%
4-Aminoacetophenone (COMe, NH_2)	80%	80%

TABLE LIII
Fe^{II}/EDTANa_2[a] Catalyzed Hydrogenation of Aromatic Nitro Compounds

Fe^{II}/EDTANa_2 (100 mg), H_2 (400 psi), H_2O : C_7H_{10}, 2–8.5 h, 150 °C

Substrate product	Yield	TOF (h^{-1})
2-Me, 4-H_2N aniline	84.6%	134
3-Cl aniline	99.0%	434
4-COMe aniline	99.0%	208
4-COOH aniline	85.8%	393
4-CH_2CN aniline	90.1%	99

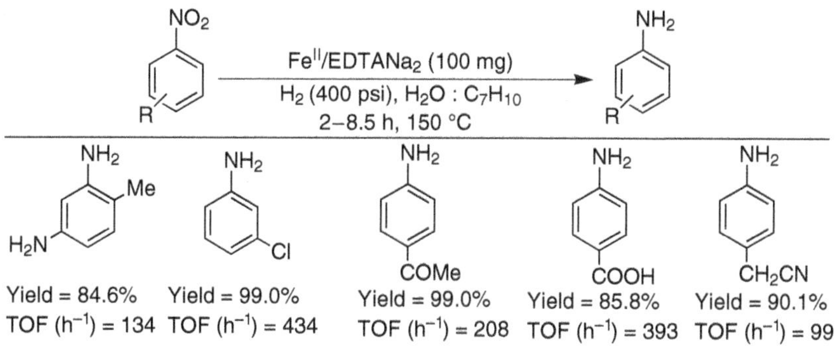

[a] Ethylenediaminetetraacetic acid = EDTA.

good-to-excellent yield was observed in most of the studied substrates with a TOF (h^{-1}) range between 99 and 529.

A solid-supported catalysis has a number of advantages over conventional solution-phase reaction because of the good dispersion of active sites leading to more reactivity (Table LIV). Kumarraja and Pitchumani (283) applied the iron(III) cation exchanged faujasite zeolite for selective reduction of nitro arenes to the corresponding amines with hydrazine hydrate. A nonpolar solvent (e.g., hexane) was used. Polar solvents lead to a decrease in yield by occupying the zeolite cavity and thereby forcing the substrate into the solvent phase.

Kumarraja and Pitchuman (283) author also proposed an electron-transport based mechanism for the reduction (Scheme 196). Hydrazine hydrate leads to the

TABLE LIV
The Iron(III) Cation Exchanged Faujasite Zeolite-Catalyzed Reduction of Aromatic Nitro Compounds

$Fe^{III}Y$ (500 mg), $N_2H_4.H_2O$ (10 equiv), C_6H_{14}, Reflux

Product	Yield
2-Me aniline	91%
2-OH aniline	91%
3-NH_2 aniline	100%
4-OMe aniline	100%
4-Cl aniline	63%
4-COOH aniline	88%

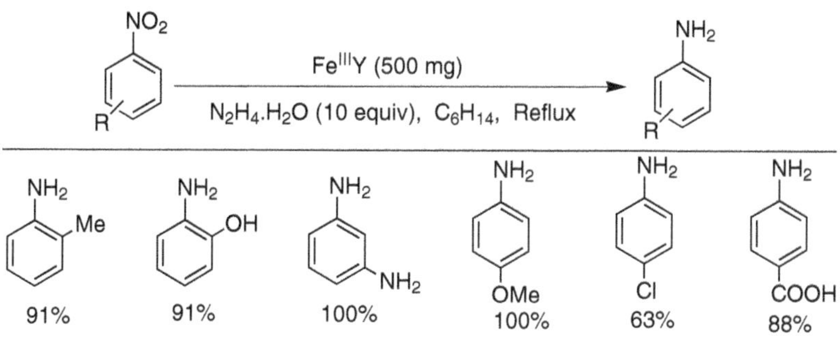

$$2Fe^{III} + NH_2NH_2 \longrightarrow 2Fe^{II} + 2H^+ + N_2H_2$$

$$2F_2^{III} + N_2H_2 \longrightarrow 2Fe^{II} + 2H^+ + N_2$$

$$ArNO_2 \xrightarrow[2H^+]{2e^-} ArNO \xrightarrow[2H^+]{2e^-} ArNHOH \xrightarrow[2H^+]{2e^-} ArNH_2$$

Scheme 196.

one-electron reduction of ferric ion to the ferrous ion and generation of a proton. Finally, oxidation of ferrous ion to ferric ion leads to electron transfer to nitrobenzene. An intial four-electron process leads to hydroxylamine, which was finally reduced to aniline in a two-electron process.

Liu et al. (284) applied activated iron for the reduction of nitroarenes (Table LV). Iron powder was preactivated by hydrochloric acid and then used as a catalyst for the reaction. Other tested acids (e.g, acetic acid, sulfuric acid, and phosphoric acid) did not provide the appropriate yield of the desired amine. Liu et al. (284) had also activated iron with zinc powder in the presence of ammonium chloride. Similar results were obtained as in the case of the acid-activated iron catalyst. Though an excess of iron metal (5 equiv) was used, a sensitive functional group (e.g., cyano, olefin double bond, and carbonyl) were well tolerated.

Shi et al. (285) applied Fe_2O_3–MgO as a catalyst for the reduction of the pharmaceutically important sulfur-containing nitro compounds to amines by using hydrazine hydrate as the hydrogen source (Table LVI). Sulfur-containing compounds are generally considered difficult because of the poisoning of the catalyst, but Shi et al. (285) applied Fe_2O_3–MgO successfully to obtain good yields of the corresponding sulfur-containing amines.

Gamble et al. (286) applied iron powder for selective reduction of nitro compounds under ultrasonic irradiation (35 kHz) (Table LVII). It was proposed

TABLE LV
Activated Iron-Catalyzed Reduction of Nitroarenes

R-C$_6$H$_4$-NO$_2$ → (Fe (5 equiv) / HCl (0.5 equiv); NH$_4$Cl, EtOH, 50–80 °C, 5–6 h) → R-C$_6$H$_4$-NH$_2$

Product	Yield
2-Bromoaniline (NH_2, Br)	90%
4-Aminobenzonitrile (NH_2, CN)	99%
4-Aminophenylacetonitrile (NH_2, CH_2CN)	87%
4-Aminobenzophenone (NH_2, COPh)	98%
4-(Propargyloxy)aniline (NH_2, O-propargyl)	91%
4-Amino-2,1,3-benzothiadiazole (NH_2, N, S, N)	90%
H_2N-benzo[c]chromen-6-one (O, O)	85%

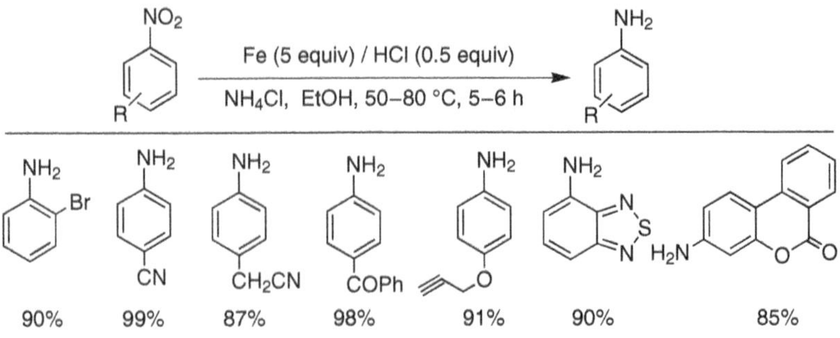

TABLE LVI
The Fe_2O_3–MgO Catalyzed Reduction of Sulfur-Containing Nitro Compounds

Fe_2O_3 - MgO, $N_2H_4.H_2O$
MeCH(OH)Me,1–3.5 h, reflux

Product	Yield
2-(2-hydroxyethylthio)aniline	98%
4-(2-hydroxyethylthio)aniline	95%
4-(methylthio)aniline	99%
3-(methylthio)aniline	87%
4-(benzylthio)aniline	98%

that ultrasonication helps in cleaning and activation of the surface of metal reagents, consequently resulting in the activation of the metal catalyst. Tri- or tetrasubstituted substrates were tested for the reduction. Good yields were reported in most cases.

Iron-catalyzed reduction of nitro compounds to amines was mostly carried out using hydrazine hydrate as the hydrogen source. In 2010, Chandrappa et al. (287) for the first time carried out rapid catalytic transfer hydrogenation using a stoichiometric amount of an Fe–$CaCl_2$ combination in a water: ethanol mixture (Table LVIII). Various functional groups (e.g., cyano, carabonyl, halide, and amide) were well tolerated.

TABLE LVII
Iron Powder Catalyzed Reduction of Nitro Compounds Under Ultrasonic Irradiation

Fe (5 equiv)
EtOH: MeCOOH: H_2O (2 :2: 1),
60 °C, 0.25–2 h, ultrasonication

Product	Yield
4-aminoacetophenone	77%
2-amino-6-bromo-4-methoxyphenol	65%
3-bromo-2,5-dimethoxyaniline	86%
4,6-dichloropyrimidin-5-amine	82%
H_2N-C$_6H_4$-N(CH_2CH_2CN)(CH_2CH_2OAc)	78%

TABLE LVIII
The Fe/$CaCl_2$ Catalyzed Reduction of Nitro Compounds

Fe (3 equiv), $CaCl_2$ (1 equiv)
EtOH : H_2O (20 : 1), 60 °C, 0.5 h

80%	90%	88%	86%	85%

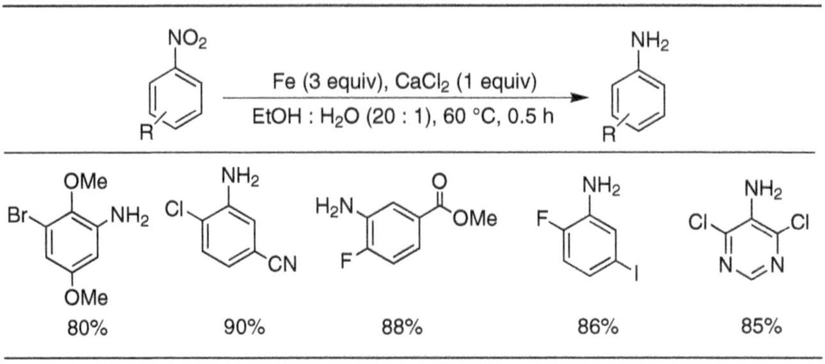

Most importantly, good yields were obtained in the multisubstituted macromolecules as well (Scheme 197*a*). This method was also applicable for the reduction of azo compounds to the corresponding amines (Scheme 197*b*) (287).

In 2010, Beller and co-workers (289) and Lemaire and co-workers (288) independently reported the iron-catalyzed reduction of nitro compounds to the corresponding amines by using silanes as the hydrogen source (Scheme 198). These requests gave the first report on the use of an iron–silane combination for nitro reduction. Beller and co-workers (289) applied ferrous bromide, phosphine

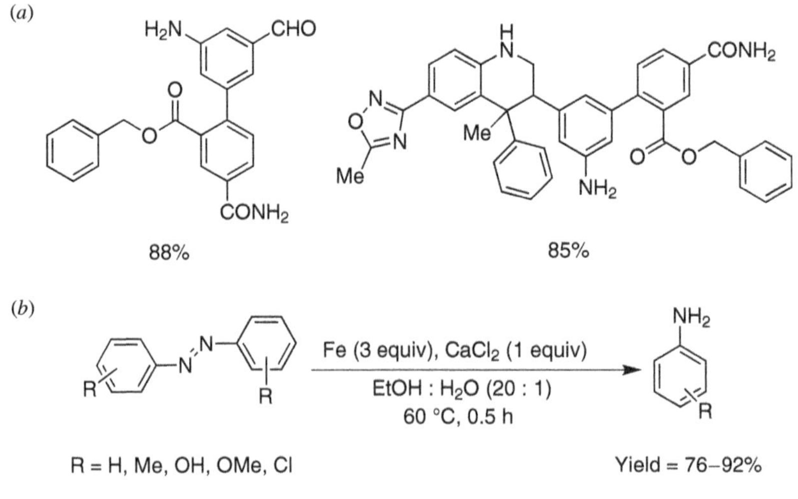

(*a*)

88% 85%

(*b*)

Fe (3 equiv), $CaCl_2$ (1 equiv)
EtOH : H_2O (20 : 1)
60 °C, 0.5 h

R = H, Me, OH, OMe, Cl

Yield = 76–92%

Scheme 197.

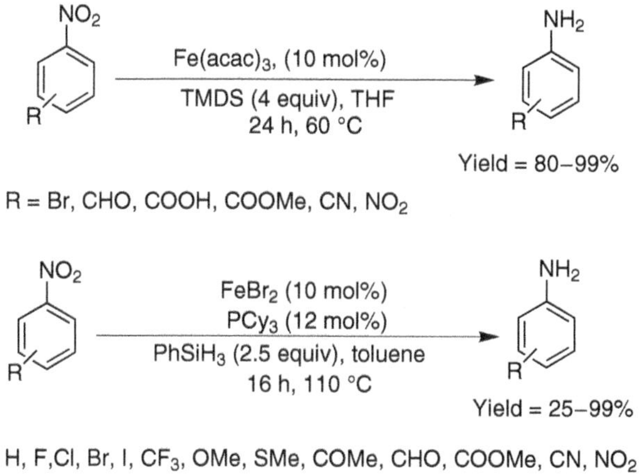

NO_2 → NH_2

$Fe(acac)_3$, (10 mol%)
TMDS (4 equiv), THF
24 h, 60 °C

Yield = 80–99%

R = Br, CHO, COOH, COOMe, CN, NO_2

NO_2 → NH_2

$FeBr_2$ (10 mol%)
PCy_3 (12 mol%)
$PhSiH_3$ (2.5 equiv), toluene
16 h, 110 °C

Yield = 25–99%

R = H, F,Cl, Br, I, CF_3, OMe, SMe, COMe, CHO, COOMe, CN, NO_2

Scheme 198.

ligand and triphenyl silane, whereas Lemaire and co-workers (288) applied Fe $(acac)_3$ and 1,1,3,3-tetramethyldisiloxane (TMDSO) without using any ligand, but the substrate scope was much less.

In 2011, Beller and co-workers (289) applied Fe–phenanthroline–C as a heterogeneous catalyst for highly chemoselective reduction of nitro aromatic compounds (Scheme 199). The combination of $Fe(OAc)_2$–phenanthroline–C provided only 10% conversion even after 24 h, but the pyrolysis of the *in situ* generated $Fe(OAc)_2$–phenanthroline complex supported on carbon led to full conversion of the nitro compounds to the corresponding amines. Challenging functional groups (e.g., halide, cyano, ester, double, and triple bonds were well tolerated [289].

Sharma et al. (290) applied a combination of iron phthalocyanine and an iron salt for a highly chemo- and regioselective reduction of aromatic nitro compounds in a green solvent (water–ethanol) system using hydrazine hydrate as the hydrogen source (Scheme 200). Iron phthalocyanine and iron sulfate were also studied as a separate catalyst, but the combination of both produce a better result than

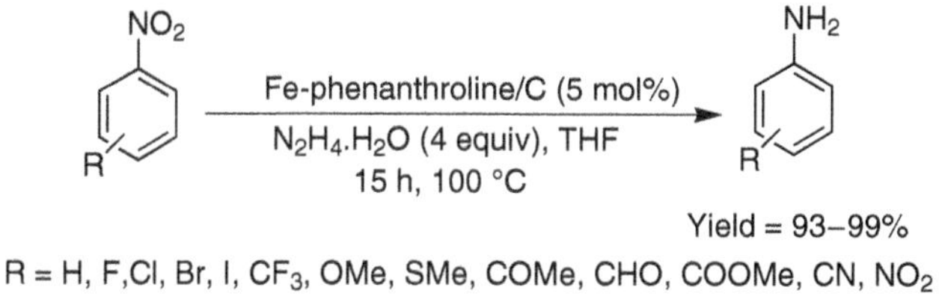

NO_2 → NH_2

Fe-phenanthroline/C (5 mol%)
$N_2H_4.H_2O$ (4 equiv), THF
15 h, 100 °C

Yield = 93–99%

R = H, F,Cl, Br, I, CF_3, OMe, SMe, COMe, CHO, COOMe, CN, NO_2

Scheme 199.

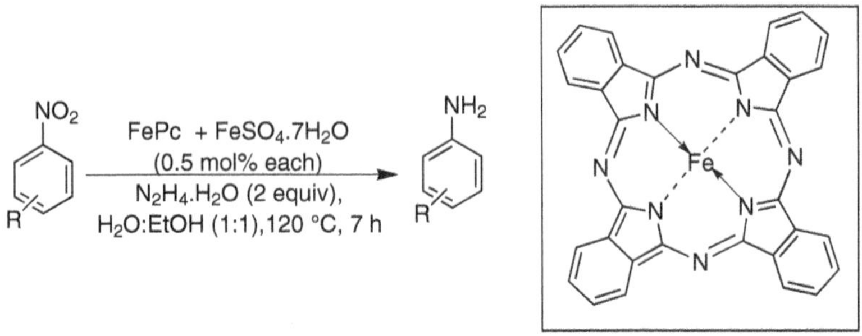

NO_2 → NH_2

FePc + $FeSO_4.7H_2O$ (0.5 mol% each)
$N_2H_4.H_2O$ (2 equiv), H_2O:EtOH (1:1),120 °C, 7 h

R = –F, –Cl, –Br, –I, –OH, –OMe, –COOH, –CN, –COOMe, –$CONH_2$, –NO_2, –$NHCH_2Ph$
Pc = Phthalocyanine

Scheme 200.

individuals do. The reduction sensitive functional group was well tolerated. The most important finding of this method was the regioselective reduction of the nitro group and tolerance of an *N*-benzyl and an *O*-benzyl functional group, which are generally considered acid–base sensitive.

Beller and co-workers (291) further reported the first base-free catalytic-transfer hydrogenation of nitroarenes (Scheme 201). An iron-based catalyst and a phosphine ligand combination was applied in the presence of formic acid as the hydrogen source at ambient temperature in a green solvent (ethanol). This method avoids the use of any base, which is generally required for the catalytic-transfer hydrogenation. High chemoselectivity with good-to-excellent yield was reported for most of the studied substrates (291).

To confirm that the nitro group is reduced by catalytic-transfer hydrogenation and not by hydrogen generated from dehydrogenation of the formic acid, the reduction was performed under a 5-bar hydrogen pressure in the presence of an iron catalyst. No reduction was observed, which confirmed the catalytic-transfer hydrogenation. Beller and co-workers (291) also analyzed the gas phase of the reaction mixture. A 3.2: 1 ratio of carbon monoxide to hydrogen was observed. This ratio showed that only 31% of the consumed formic acid was decomposed to hydrogen showing the necessity of using 4.5 equiv of acid.

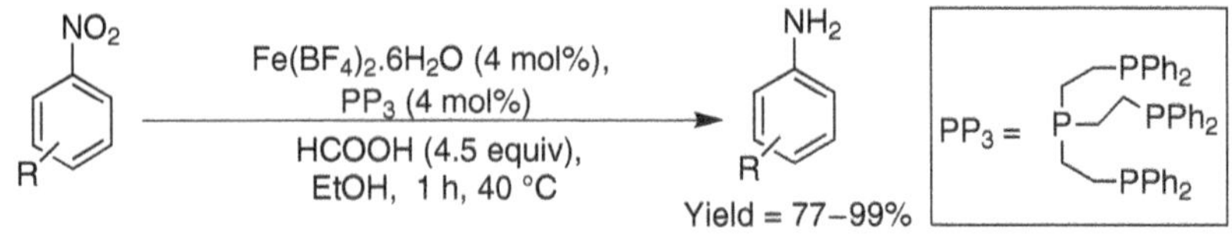

NO_2 → NH_2

$Fe(BF_4)_2.6H_2O$ (4 mol%), PP_3 (4 mol%)
HCOOH (4.5 equiv), EtOH, 1 h, 40 °C

Yield = 77–99%

PP_3 = P(–PPh_2)(–PPh_2)(–PPh_2)

R = F,Cl, Br, I, CF_3, OMe, SMe,COMe, CHO, COOMe, CN, NO_2

Scheme 201.

Reduction of a nitro group to an amine can proceed by two mechanisms: through hydroxylamine or through azobenezene. To confirm which pathway was followed, the authors analyzed the reaction mixture and only a small quantity of azobenzene (<1%) was observed. Further, the direct reduction of azobenzene resulted in only a 48% conversion and 5 % yield of aniline clearly indicating the direct reduction pathway in their reaction. On the basis of these results and isolated iron complexes, it was proposed that the $[FeF(PP2)]^+$ (A) cation act as the active catalyst. This cation coordinated to formate leading to the neutral complex (B). A β-hydride elimination from B resulted in complex C, with the elimination of carbon dioxide. This reaction was followed by protonation of complex C by formic acid leading to the iron dihydride complex D that reduced the nitrobenzene to nitrosobenzene. This cycle was repeated two more times to yield aniline (Scheme 202).

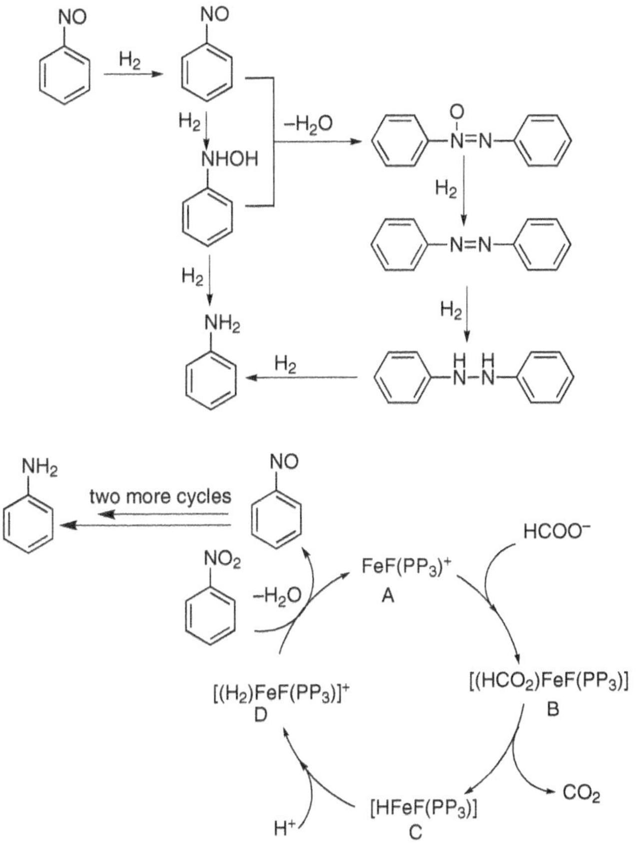

Scheme 202.

TABLE LIX
Iron Nanoparticle-Catalyzed Reduction of Nitro Compounds

NO_2 —[Fe nanoparticles (3 equiv); H_2O, rt, 1–3 h]→ NH_2

NH_2 R = OMe, 86%; CHO, 94%; COOH, 90%; $CONH_2$, 89%; CN, 93%; Br, 90%; I, 89%

NH_2 SePh 86%

NH_2 N O 92%

NH_2 O N H O 90%

NH_2 N 92%

HO N NH_2 92%

N N H NH_2 89%

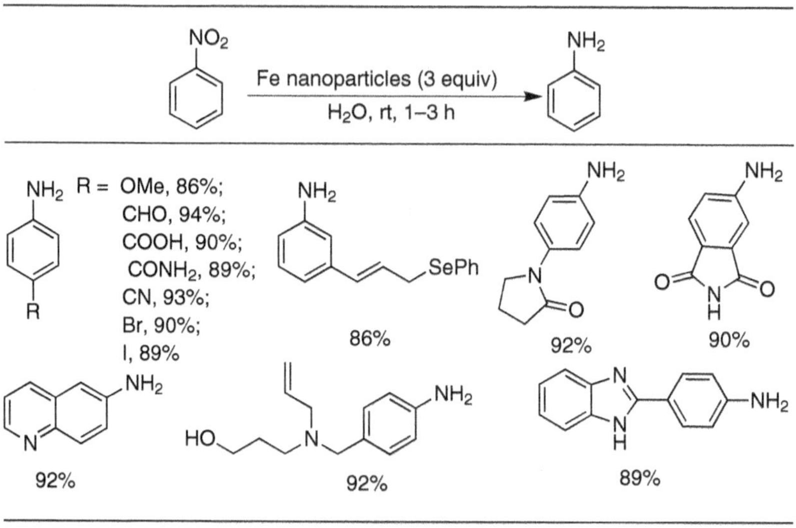

Nanoparticles have unique properties compared to the bulk material due to their high surface-to-volume ratio. Iron nanoparticles have been applied for the selective reduction of aromatic nitro compounds in water without using any additional hydrogen source at rt (Table LIX). It was proposed that the hydrogen coordinated iron nanoparticales are responsible for the hydride atom. This weak hydride acted as a target specific reducing agent for the highly electrophilic nitro group. Good-to-excellent yield was observed with high chemoselectivity in most of the substrates. In this method, an excess amount of iron nanoparticles (i.e., 3 equiv) was applied, which need improvements.

Cantillo et al. (292) applied *in situ* formed iron nanoparticles in a catalytic manner for the reduction of nitro compounds using a microwave. Hydrazine hydrate was used as the hydrogen source and the reaction time was 2–8 min. This method has the advantage of being homogenous as the colloidal nanocrystals remain in the solution. Then nanocrystals start to aggregate forming a precipitate, and therefore making the reaction mixture heterogeneous. After completion, the nanocrystals can be removed easily by using a magnet. Products were purified simply by filtration and excellent yields were observed for all the selected substrates. This method was also applied for the synthesis of amines up to the gram scale using a continuous-flow method.

G. Hydrogenation of Carbon Dioxide and Bicarbonate

Carbon dioxide is a widely available carbon source and also has a greenhouse effect on the environment. Therefore, the reduction of carbon dioxide to formate is a breakthrough froms both the industrial, as well as the environmental, viewpoint. For the same reason, a number of methods involving rhodium, ruthenium, iridium, and palladium were developed in the last 20 years. In 1975, Inoue et al. (293) reported that group 7 transition metal complexes including iron could generate formate from carbon dioxide, hydrogen, and alcohol in the presence of a tertiary amine. The iron complex [$H_2Fe(diphos)_2$] was tested with high pressurized CO_2 and H_2 at 140 °C (Scheme 203, diphos = 1,2-bis(diphenylphosphino)ethane.

In 1978, Evans and Newell (294) reported an anionic iron–hydride complex [HFe $(CO)_4]^-$ for carbon dioxide reduction. Their method suffered from the use of a high-pressure gas and high temperature. Later, Jessop and co-workers (295) designed a high-pressure combinatorial screening process using pressure sustainable apparatus (up to 200 bar). Their process established that a catalytic amount of $FeCl_3$, depe ($Cy_2PCH_2CH_2PCy_2$) and DBU (1,8-diazabicyclo[5.4.0]undec-7-ene) is essential for generation of formate under 100-bar hydrogen and carbon dioxide atmosphere.

Beller and co-workers (296) reduced bicarbonate to formate (and formamide) in the presence of catalytic $Fe(BF_4)_2.6H_2O$ and PP_3 as ligand [PP_3 = $P(CH_2CH_2PPh_2)_3$]. A reasonable TON = 610 was attained under 60 bar H_2 at 80 °C (Scheme 204).

In 2012, Beller and co-workers (296) reported an improved version for hydrogenation of carbon dioxide and bicarbonate using a catalytic amount of $Fe(BF_4)_2.6H_2O$. Their catalytic proposal included formation of [$FeH(H_2)(PP_3)$] from $Fe(BF_4)_2$ in the presence of H_2 gas and base (Scheme 205) (297). Afterward, reaction of carbon dioxide and hydride, generated formate. Both the [$FeH(H_2)$-$(PP_3)]^+$ and [$FeH(CO_2)(PP_3)]^+$ intermediates were characterized by nuclear magnetic resonance (NMR) spectroscopy. In the last step, H_2 gas regenerated the catalytic species and removed formic acid.

A number of pincer complexes were reported with higher row transitional metals (Ir, Ru) for CO_2 hydrogenation. The strong donating ability of the pincer

$$\underset{}{ROH} + \underset{25\ atm}{CO_2} + \underset{25\ atm}{H_2} \xrightarrow[Et_3N, 140\ °C,\ 21\ h]{[H_2Fe(diphos)_2]} HCO_2R + H_2O$$

Yield = (mol of HCO_2Et/ mol of complex) = 2

Scheme 203. Hydrogenation of Carbon Dioxide Using [$H_2Fe(diphos)_2$].

$$CO_2 + H_2 \xrightarrow[MeOH,\ base]{Fe(BF_4)_2.6H_2O\ (cat)} HCO_2H.base \xrightarrow{ROH/NHR'_2} HCOOR/HCONR'_2$$

Scheme 204. Catalytic hydrogenation of CO_2 and formation of formate ester and formamides.

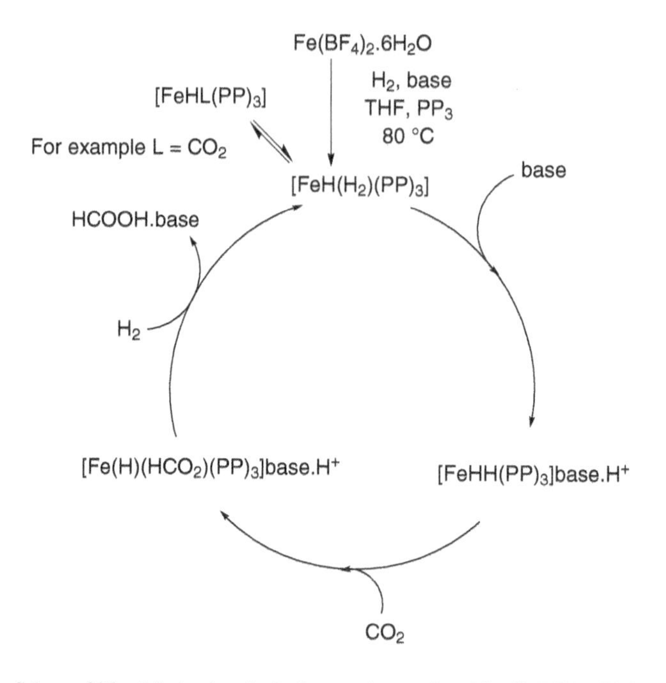

Scheme 205. Mechanism for hydrogenation catalyzed by $Fe(BF_4)_2.6H_2O$.

ligands facilitated the insertion of carbon dioxide into the metal-hydride bonds. In 2011, Milstein and co-workers (298) studied the iron–pincer complex *trans*-[(*t*Bu-PNP)Fe(H)$_2$(CO)] (**150**) for effective hydrogenation of carbon dioxide and bicarbonates (Scheme 206). In their hydrogenation protocol, carbon dioxide was hydrogenated under low-pressure hydrogen (6.66 bar) and a carbon dioxide (3.33 bar) atmosphere with a TOF of 156 h^{-1}. Hydride directly attacked CO_2 generating formate, which was subsequently replaced by the cosolvent water. This water molecule was displaced by H_2 in the next step. In a final step of the catalytic cycle, *trans*-[(*t*Bu-PNP)Fe(H)$_2$(CO)] (**150**) was regenerated through the heterolytic cleavage of the H_2 molecule in the presence of base or dearomatization and succesive proton transfer (Scheme 207) (298).

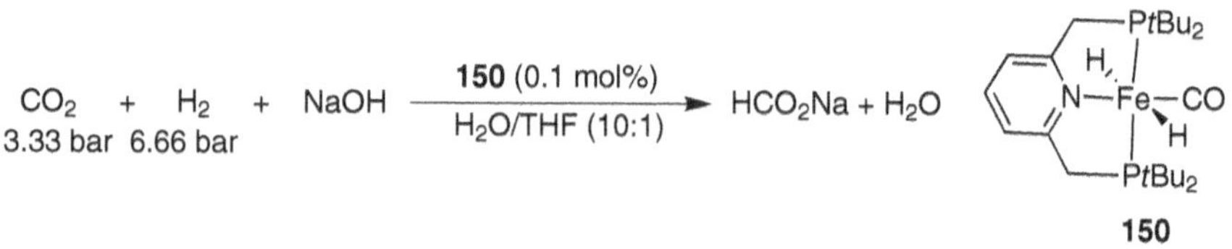

$$\underset{3.33\text{ bar}}{CO_2} + \underset{6.66\text{ bar}}{H_2} + NaOH \xrightarrow[\text{H}_2\text{O/THF (10:1)}]{\textbf{150}\text{ (0.1 mol\%)}} HCO_2Na + H_2O$$

PtBu2 H N—Fe—CO H PtBu2

150

Scheme 206. Hydrogenation of carbon dioxide using an iron–pincer complex.

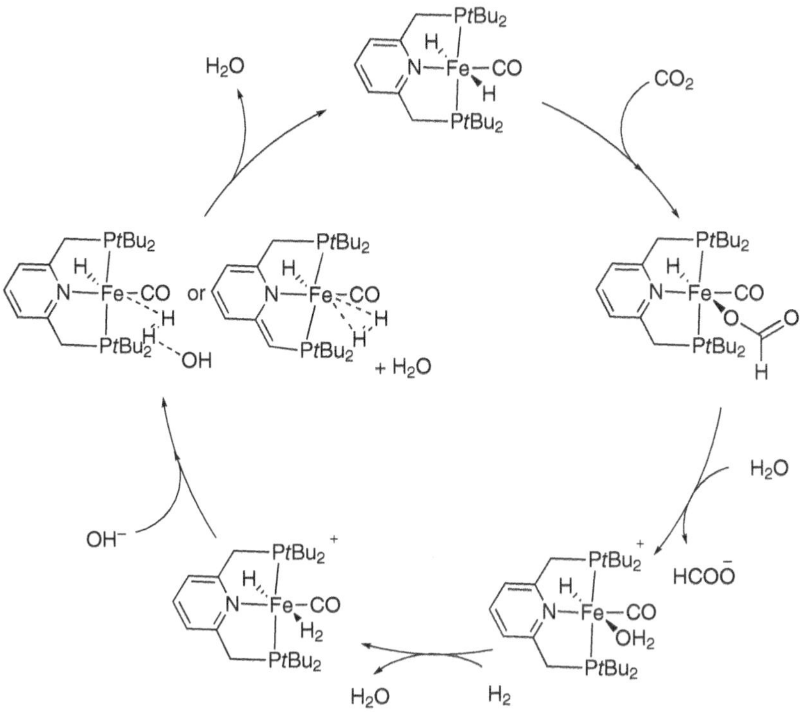

Scheme 207. Hydrogenation of carbon dioxide with an iron–pincer complex.

H. Amide Reduction

Beller and co-workers (299) developed a very convenient and general method of amide reduction for the synthesis of amine derivatives. During their study on iron-catalyzed dehydrations of primary amides in the presence of polymethylhydrosilanes (PMHS) as the reducing agent, a secondary amine was obtained as the byproduct. From their details investigation, they were able to develop a method for amide reduction. They used a $Fe_3(CO)_{12}$ complex as the catalyst in the presence of PMHS as the reducing agent. This finding was the first report of an iron-catalyzed hydrosilylation of amides to an amine (Table LX).

The iron precursor formed an activated species that resulted in hydrosilylations of the carbonyl group in the amide to generate *O*-silylated *N,O*-acetal (**151**). Subsequently, **151** transformed to **152**, which was further reduced to the amine with a second equivalent of activated species (Scheme 208).

TABLE LX
Iron-Catalyzed Reduction of Amides Using PMHS as a Reducing Agent[a]

$R C(O) N R^1 R^2$ —[$Fe_3(CO)_{12}$] (2–10 mol%), PMHS, 100 °C, 24 h→ $R CH_2 N R^1 R^2$

Product	Yield
4-Me-$C_6H_4CH_2N(CH_2Ph)_2$	93%[b]
2-thienyl-$CH_2N(CH_2Ph)_2$	84%[b]
$MeCH_2N(CH_2Ph)_2$	89%[b,c]
4-I-$C_6H_4CH_2N(CH_2Ph)_2$	83%[b,d]

[a] Reaction conditions: Amide (1 mol), [$Fe_3(CO)_{12}$] (2–10 mol%), PMHS (4–8 mmol), $n Bu_2O$ (5 mL).
[b] Yield of isolated product.
[c] Toluene as solvent.
[d] Contained 10% tribenzylamine.

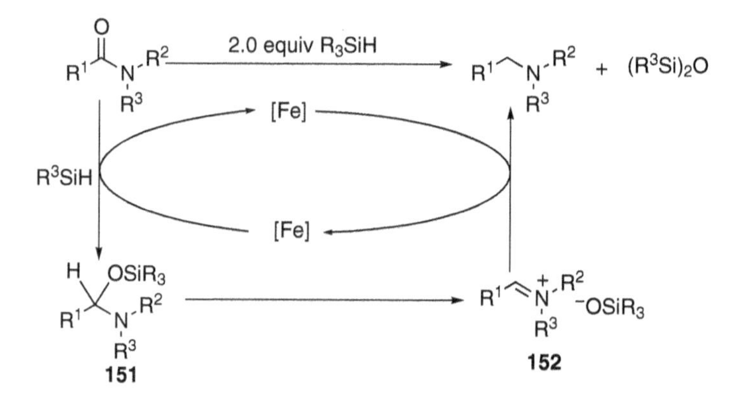

Scheme 208. Mechanism of reduction of amides.

I. Reductive Aminations

Amine compounds are very important for pharmaceuticals and natural product synthesis (300). Reductive amination of carbonyl compounds is one of the alternative protocols for synthesis of amine compounds. Renaud and co-workers (301) developed an iron-catalyzed reductive amination method. They synthesized different iron(II) complexes (**153–158**, (Scheme 209) (266a, 302). Interestingly, full conversion of starting material into the desired product was obtained by using

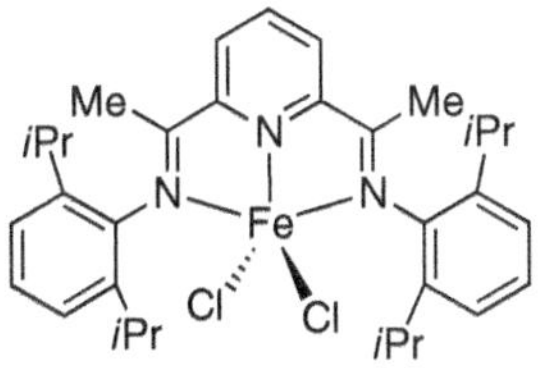

153 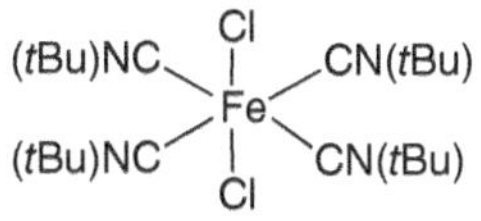 154

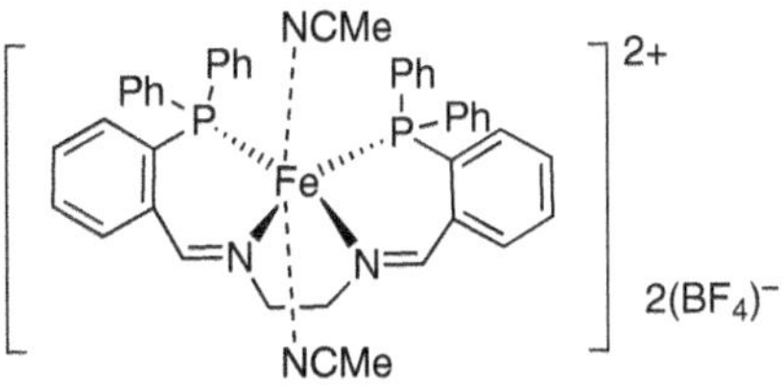

155 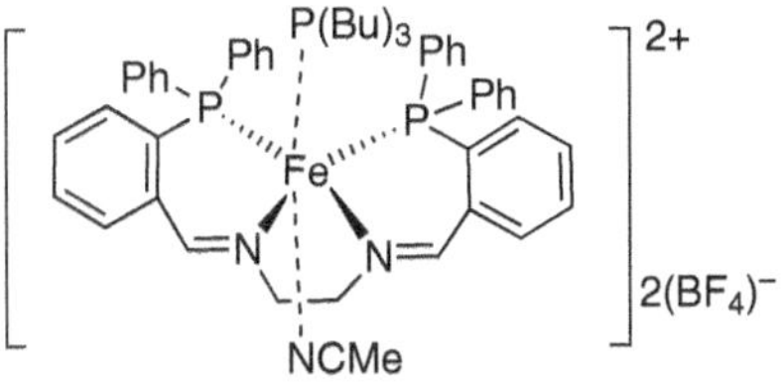 156

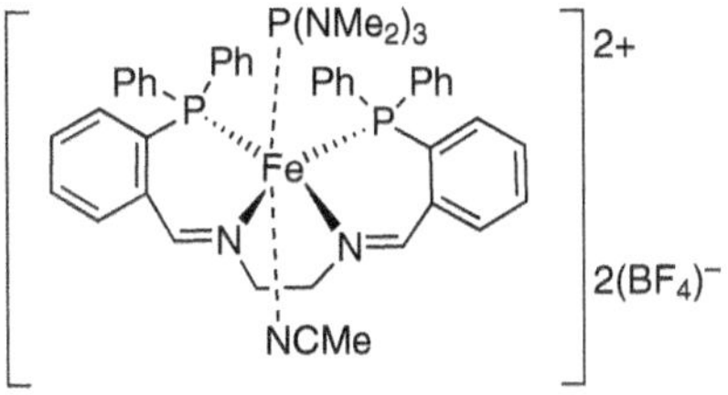

157 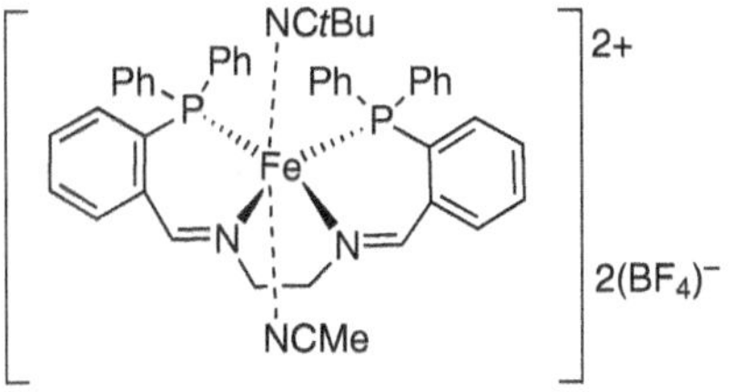 158

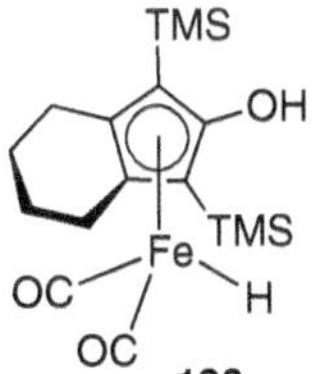

130 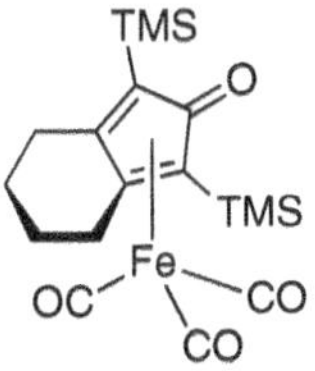 159

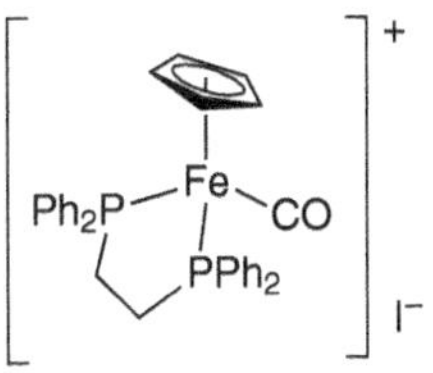

160 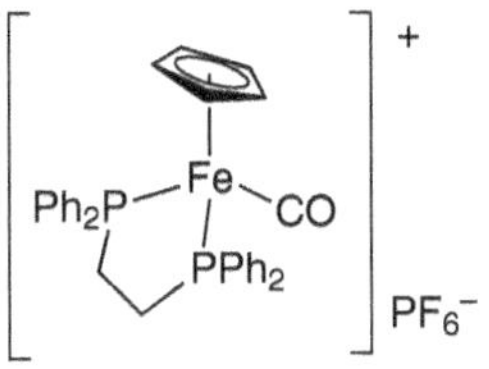 161

Scheme 209.

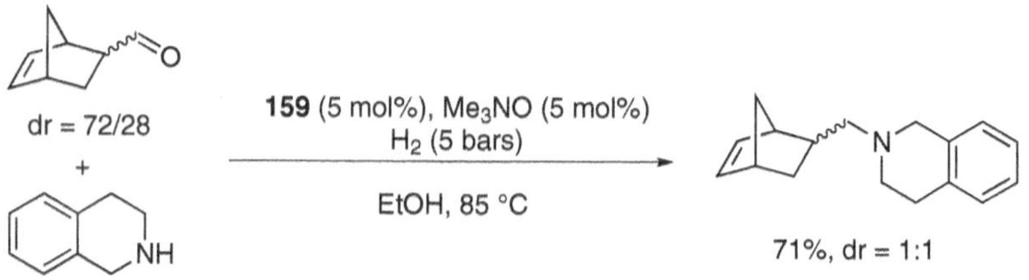

dr = 72/28

+

NH

159 (5 mol%), Me_3NO (5 mol%)
H_2 (5 bars)

EtOH, 85 °C

71%, dr = 1:1

Scheme 210.

the Knolker complex **130** (263a, 264, 277, 303). Complex **130** was air sensitive and decomposed quickly. But, its related complex **159** (Scheme 209, Scheme 210) was air stable but unable to show activity toward reductive amination. Interestingly, when it was treated with trimethylamine *N*-oxide, which oxidatively removed CO ligand from **159** and generated a 16e species (**163**, Scheme 211). Subsequently, **163** reacted with hydrogen to form **130** *in situ*, which showed full conversions with moderate yield and selectivity. Note, other piano-stool complexes, **160** and **161**, failed to give product. The reaction tolerated both primary and secondary amines, as well as both aliphatic and aromatic aldehydes, to provide alkylated amines in good-to-excellent yields (Table LXI). Control experiments between norborene carboxaldehyde and tetrahydroisoquinoline were carried out under standard reaction protocol (Scheme 210). This reaction provided an amine product in 71% yield and 1:1 mixtures of diastereomers. Based on these observations, imine (**165**)/enamine (**166**) equilibrium was proposed in the mechanism (Scheme 211).

TABLE LXI
Substrates Scope of Reductive Amination by a Knolker Complex[a]

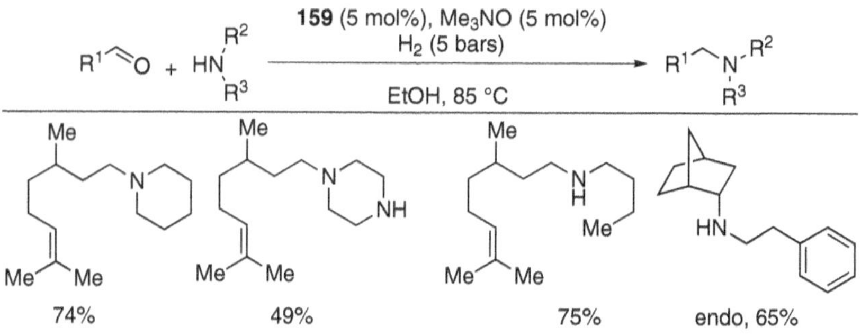

R^1CHO + HNR^2R^3 → $R^1CH_2NR^2R^3$

159 (5 mol%), Me_3NO (5 mol%)
H_2 (5 bars)

EtOH, 85 °C

74%	49%	75%	endo, 65%

[a] The hydrogenation reactions were carried out with aldehyde (1 mmol) and amine (1.2 mmol) in the presence of an iron precatalyst (5 mol%) and trimethyl-*N*-oxide (5 mol%) in ethanol (2 mL) under 5 bar of hydrogen at 85 °C.

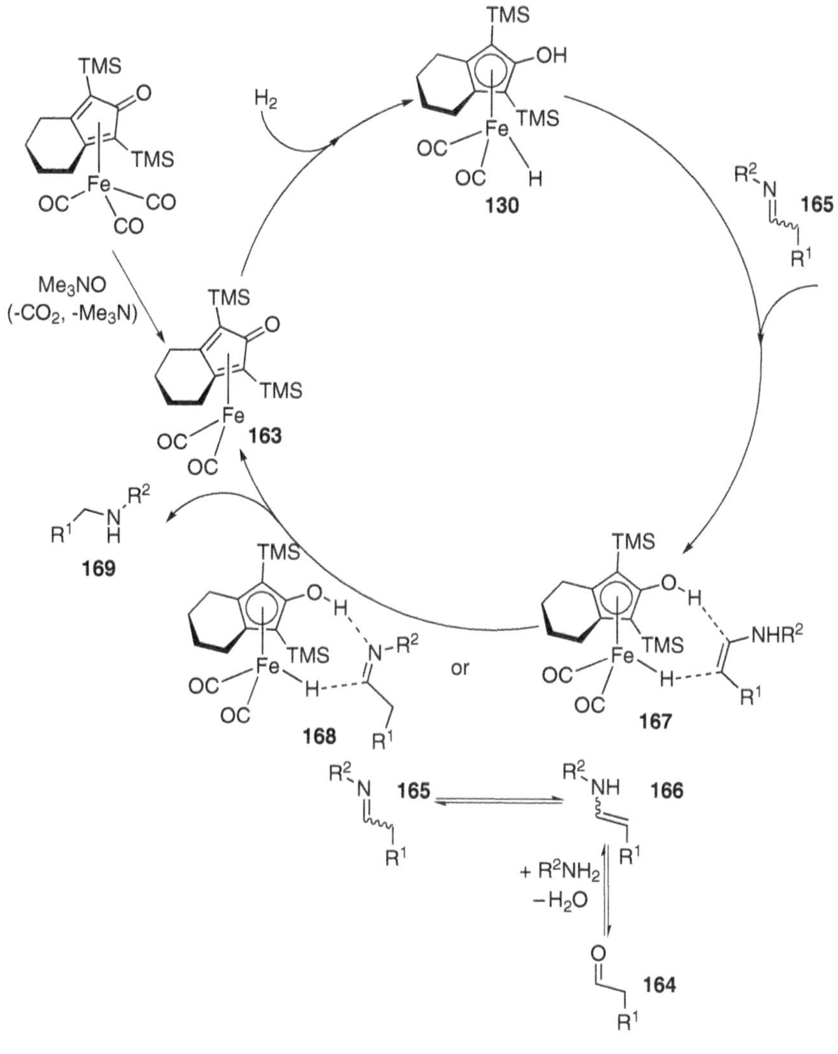

Scheme 211. Mechanism of reductive amination by a Knolker complex.

The *in situ* generated **130** then led to activation of **165** (or **166**) via coordination through the hydroxy group and generated intermediates **167** and **168** (Scheme 211). Subsequently, addition of hydride provided reduced amine and with regeneration of the unsaturated iron(0) complex (**163**).

In 2008, Bhagane and co-workers (304) synthesized amines via reductive amination using an iron(II)/EDTA complex with molecular hydrogen as the

TABLE LXII
Reductive Amination Using $FeCl_3$ as Catalyst[a]

91%[b]	78%[b]	93%[b]

[a]Reaction conditions: $FeCl_3$ (0.12 mmol, 5 mol%), aldehyde (2.4 mmol), amine (2.4 mmol), PMHS (0.5 mL), THF (2.0 mL), 60 °C, 24 h. All products were characterized by GC–MS (mass spectrometry) and ^{1}H NMR.
[b]Isolated yield.

hydrogen source. Later Enthaler reported a very effective method for reductive amination by using $FeCl_3$/PMHS (Table LXII). This reaction tolerated a wide range of aromatics aldehydes (Table LXII) (305).

The reaction was proposed to undergo a Lewis acid catalyzed imine formation followed by the reduction of this imine (Scheme 212).

Later, Beller and co-workers (306) reported another alternative protocol of reductive amination between a ketone–aldehyde and an amine. In Beller's report,

Scheme 212. Mechanism of reductive amination using $FeCl_3$ as catalyst.

TABLE LXIII
Reductive Amination of Ketones and Aldehydes with an Amine in the Presence of H_2[a]

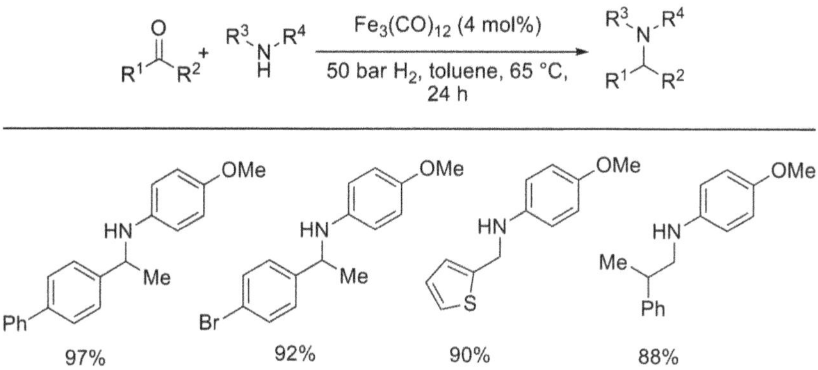

$R^1C(O)R^2 + R^3NHR^4 \xrightarrow[\text{50 bar } H_2\text{, toluene, 65 °C, 24 h}]{Fe_3(CO)_{12}\text{ (4 mol\%)}} R^1CH(NR^3R^4)R^2$

Product	Yield
N-(1-(4-biphenylyl)ethyl)-4-methoxyaniline (Ph, Me, HN, OMe)	97%
N-(1-(4-bromophenyl)ethyl)-4-methoxyaniline (Br, Me, HN, OMe)	92%
N-(thiophen-2-ylmethyl)-4-methoxyaniline (S, HN, OMe)	90%
N-(2-phenylpropyl)-4-methoxyaniline (Me, Ph, HN, OMe)	88%

[a] General reaction condition:0.5-mmol aldehyde, 0.75 mmol amine, 0.02 mmol $Fe_3(CO)_{12}$, 50 bar, H_2, 65 °C, 24 h.

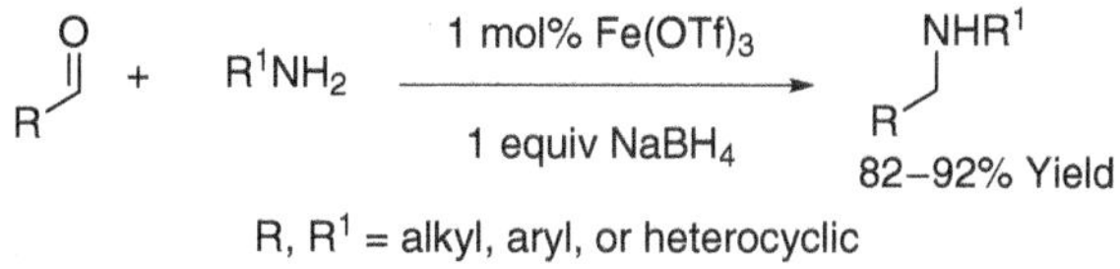

$RCHO + R^1NH_2 \xrightarrow[\text{1 equiv } NaBH_4]{\text{1 mol\% } Fe(OTf)_3} RCH_2NHR^1$ 82–92% Yield

R, R^1 = alkyl, aryl, or heterocyclic

Scheme 213. Reductive amination using $Fe(OTf)_3/NaBH_4$.

the iron carbonyl complex was used as the catalyst in the presence of molecular hydrogen (Table LXIII).

In 2012, Bandichhor and co-workers (307) reported another protocol for reductive amination using $Fe(OTf)_2$ as the catalyst and $NaBH_4$ as a reducing agent (Scheme 213).

X. TRIFLUOROMETHYLATION

The introduction of the trifluoromethyl group ($-CF_3$) into organic molecules is crucial for the discovery of biologically active compounds (308). The CF_3 group can act as a useful biostere of a methyl group and also it does not suffer from metabolic oxidation. In 2012, Buchwald and co-workers (309) developed a facile method for trifluoromethylation of potassium vinyltrifluoroborates using catalytic amounts of $FeCl_2$ and Togni's reagent (**171**) (Table LXIV) as the trifluoromethyl

TABLE LXIV
Trifluoromethylation Reaction with Iron Catalysis[a]

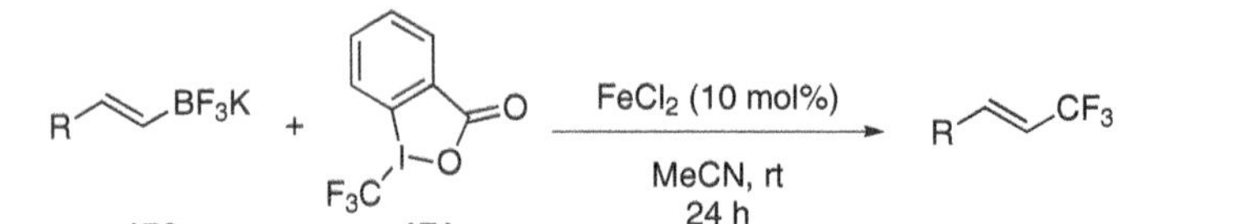

170 1.1 equiv; **171** 1.1 equiv

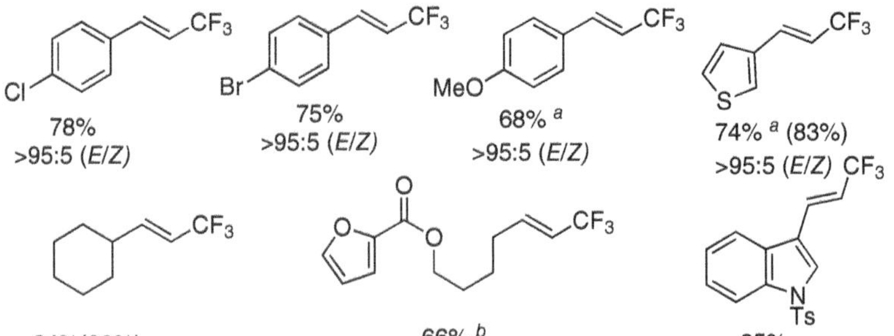

78% >95:5 (*E/Z*)

75% >95:5 (*E/Z*)

68% [a] >95:5 (*E/Z*)

74% [a] (83%) >95:5 (*E/Z*)

34%(80%) 83:17 (*E/Z*)

66% [b] >95:5 (*E/Z*)

65% >95:5 (*E/Z*)

[a] Reaction conditions: compound **170** (1.1–0.55 mmol, 1.1 equiv), **171**(1.0–0.5 mmol, 1.0 equiv), $FeCl_2$ (10 mol%), $[171]_{t=0} = 0.40\,M$. Contains 10–15 mol% of a protodeboronated side product.
[b] Here 15 mol% $FeCl_2$ was used.

source. This method works efficiently for different aryl, heteroaryl, and aliphatic compounds (Table XLIV).

Some trifluoromethylated substrates gave a considerable amount of (*E/Z*) isomers (Scheme 214). According to Parsons et al. (309), such observations

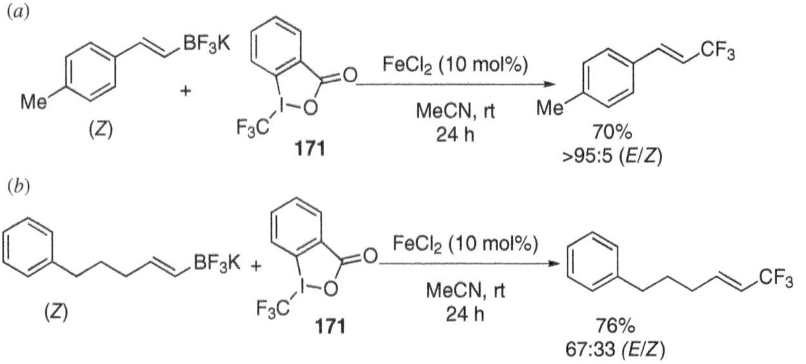

Scheme 214. Selectivity in the formation of trifluoromethylated product.

disfavor transmetalation–reductive elimination type mechanism. Interestingly, $Sn(OTf)_2$ provided a similar selectivity and yields of the desired product. Notably, irrespective of the geometry of the potassium trifluoroborate starting material, same product was obtained. All these observations supported formation of a cationic intermediate by Lewis acid catalysis.

XI. CONCLUSION

Efforts toward the development of sustainable, efficient, and selective processes for the synthesis of fine chemicals and pharmaceuticals are undoubtedly increasing. Designing environmentally benign methods for the construction of highly valuable building blocks from renewable raw materials remain a challenging task. In this context, catalysis based on abundant, inexpensive, and relatively nontoxic metals can play a pivotal role in the chemist's synthetic toolbox. Recently, catalysts based on iron metal have emerged as sustainable alternatives to precious metal-based catalysts in a range of synthetic applications. The wealth of knowledge concerning reaction pathways that is available for transition metal catalyzed reactions will continue to advance the growth of well-defined catalytic systems with iron. Handling of a number of useful "ligand–iron" complexes is often problematic due to their air sensitivity. Expensive and toxic late transition metals are used as an alternative for a given reaction. Consequently, designing and developing air stable, robust, and easy to handle iron catalyst should be a primary goal for aspiring chemists.

Acknowledgments

This work is supported by DST-India. Financial support received from CSIR-India (fellowships to S.R., A.M., and S.M.) and UGC-India (fellowship to T.P.) is gratefully acknowledged.

ABBREVIATIONS

acac	Acetylacetonate
BINAP	2,2′-Bis(diphenylphosphino)-1,1′-binaphthyl
BINOL	1,1′-Bi-2-naphthol
Bn	Benzyl
BNP	Bis-Binaphthylporphyrin
Boc	*tert*-Butyloxycarbonyl
$BocN_3$	*N-tert*-Butyloxycarbonyl azide

Bopa	Bis(oxazolinylphenyl)amine
BPBP	1,1′-Bis(pyridin-2-ylmethyl)-2,2′-bipyrrolidine
BPMCN	*N,N′*-Bis(2-pyridyl-methyl)-*N,N′*-dimethyl-1,2-cyclohexanediamine bpmcn (ligand)
bpmen	*N,N′*-Dimethyl-*N,N′*-bis(pyridin-2-ylmethyl)ethane-1,2-diamine
BPN	Butylphenylnitrone
BQBP	1,1′-Bis(quinolin-2-ylmethyl)-2,2′,-bipyrrolidine
BTA	bis[(trifluoromethyl)sulfonyl]-amide
Bu	Butyl
CDA	Cross-dehydrogenative arylation
CDC	Cross-dehydrogenative coupling
CHT	Cyclohepta-1,3,5-triene
COD	1,5-cyclooctadiene
COX	Acyl halide
Cp	Cyclopentadienyl
Cy	Cyclohexyl
CyH	Cyclohexane
Cyclen	1,4,7,10-Tetraazacyclododecane
dach	1,2-Diaminocyclohexane
dapb	2,6-Bis[1-(benzylemino)ethyl]pyridine
DBFOX	4,6-Dibenzofurandilyl-2,2′-bis(4-phenyloxazoline)
DBM	Dibenzoylmethane (solvent), dbm (ligand)
DBU	1,8-Diazabicyclo[5.4.0]undec-7-ene
DCE	1,2-Dichloroethane
DCIB	1,2-Dichloroisobutane
DCM	Dichloromethane
dppbz	1,2-Bis(diphenylphosphino)benzenedichloride
DDQ	2,3-Dichloro-5,6-dicyano-1,4-benzoquinone
de	Diastereomeric excess
depe	$Cy_2PCH_2CH_3PCy_2$
DFT	Density functional theory
DHySi	Dehydrogenative hydrosilylation
diphos	1,2-Bis(diphenylphosphino)ethane
Dipp	2,6-Diisopropylphenyl
DMAP	4-Dimethylaminopyridine
DMCH	Dimethyl cyclohexane
DME	Dimethoxyethane
dmeda	*N,N′*-Dimethylethylenediamine (solvent) dmeda (ligand)
DMF	*N,N*-Dimethylformamide
DMSO	Dimethyl sulfoxide (solvent), dmso (ligand)
dpaq	2-(Bis(pyridin-2-ylmethyl)amino-*N*-(quinolin-8-yl)acetamide
dppbz	1,2-Bis(diphenylphosphino)benzene

dppe	1,2-Bis(diphenylphosphino)ethane
dppp	1,3-Bis(diphenylphosphino)propane
DTPB	Di-(*tert*-butyl)peroxide
dtbpy	4,4′-Di-*tert*-butyl-2,2′-bipyridine
D_4-TpAP	$H_2[HPhH(dach)_2]$ where 1,2-diaminocyclohexane (dach)
dr	Diastereomeric ratio
DuPhos	(-)-1,2-bis((2S,5S)-2,5-di-*i*-propylphospholano)benzene
EDA	Ethyldiazoacetate
EDTA	Ethylenediaminetetraacetic acid
ee	Enantiomeric excess
emim	Ethyl methyl imidazolium
EWG	Electron-withdrawing group
FTMS	Fourier transform mass spectrometer
GC	Gas chromatography
HMA	Hexagonal mesoporous aluminophosphate
HMDS	Bis(trimethysilyl)amide
^{1}H NMR	Proton nuclear magnetic resonance
HPLC	High-performance liquid chromatography
H_2Pydic	Pyridine-2,6-dicarboxylic acid
HOMO	Highest occupied molecular orbital
Hy	Hydrogenation
HySi	Hydrosilylation
IEtPh	1,3-Bis(*R*)-1′-phenylethyl)imidazole-2-ylidene)
INAS	Intramolecular nucleophilic acyl substitution
IMes	1,3-Bis(2,4,6-trimethylphenyl)imidazole-2-ylidene
Ipr	1,3-Bis(2,6-diisopropylphenyl)-2,3-dehydro-1*H*-imidazole
*i*PrPNP	2,6-Bis(diisopropylphosphinomethyl)pyridine
KHMDS	Potassium bis(trimethylsilyl)amide
KIE	Kinetic isotopic effect
LDA	Lithium diisopropylamide
LiHMDS	Lithium bis(trimethylsilyl)amide
LUMO	Lowest unoccupied molecular orbital
mcpp	N1,N2-dimethyl-N1,N2-bis(pyridin-2-ylmethyl)cyclohexane-1,2-diamine
mep	*N,N′*-Dimethyl-*N,N′*-bis(2-pyridylmethyl)ethane
Mes	2,4,6-Trimethylphenyl
MOM	Methoxymethyl ether
NBS	*N*-Bromosuccinamide
NHC	*N*-Heterocyclic carbene
N-IBN	*N*-(2-Iodophenyl) methyl
NMP	*N*-methylpyridine
NMR	Nuclear magnetic resonance
NTs	Tosylaziridine

OTf	Trifluoromethanesulfonate
OMP	*o*-methoxyphenyl
Pc	Phthalocyanine
PDI	(*N*,*N'*E,*N*,*N'*E)-*N*,*N'*-(1,1′)-(Pyridine-2,6-diyl)bis(ethan-1-yl-1-ylidene))bis(2,6-diisopropylaniline)
PDP	PDP=2-({(S)-2-[(S)-1-(pyridin-2-ylmethyl)pyrrolidin-2-yl] pyrrolidin-1-yl}methyl)pyridine
phebox	Bis(oxazolinyl)phenyl
Phen	1,10-Phenanthroline
$PhNTf_2$	*N*-Phenyl-bis(trifluoromethanesulfonimide)
PhINTs	*N*-Tosyliminobenzyliodinane
Pip	Piperidinium
Piv	Pivaloyl
PMB	4-Methoxybenzyl bromide
PMHS	Polymethylhydrosiloxane
PMP	*p*-Methoxyphenyl
PNNP	[P(2)(N2)] ligad set
Pr	Propyl
PrPNP	2,6-Bis(diisopropylphosphinomethyl)pyridine
pybox	Bis(oxazolinyl)pyridine
PyTACN	Bis(2-pyridylmethyl)-1,4,7-triazacyclononane
rt	Room temperature
RM	Organometalic reagent
RSM	Recovered starting material
salen	*N*,*N'*-Bis(3,5-di-*tert*-butylsalicylidene)-1,2-cyclohexanediamino
SET	Single-electron transfer
SIPr	1,3-Bis(2,6-diisopropylphenyl)-imidazolidinium
TACN	Triazacyclononane
TBA	Tetrabutylammonium
TBDMS	*tert*-Butyldimethylsilyl
TBDPS	*tert*-Butyldiphenylsilane
TBHP	*tert*-Butyl hydrogen peroxide
TBS	*tert*-Butyldimethylsilyl
TEMPO	2,2,6,6-Tetramethyl-piperidin-1-yl)oxyl
tf	Trifluoromethylsulfonyl
TFA	Trifluoroacetic acid
THF	Tetrahydrofuran (solvent), thf (ligand)
THP	Tetrahydropyran
TIP	Tetrakis(imino)pyracene
TIPS	Triisopropylsilyl
TMDSO	1,1,3,3-Tetramethyldisiloxane
TMEDA	*N*,*N*,*N'N'*-Tetramethylethane-1,2-diamine (solvent), tmeda (ligand)

TMP	2,3,6-Trimethylphenol
TMS	Trimethylsilyl
TON	Turn over number
TPA	Tris(2-pyridylmethyl)amine
TPP	Tetraphenylporphyrin
TOF	Turn over frequency
tol	Tolyl
Tosyl	4-Toluenesulfonyl
TTP	*meso*-Tetra-*p*-tolylporphyrin
UV	Ultraviolet
VCP	Vinylcyclopropane

REFERENCES

1. (a) C. Bolm, J. Legros, J. Le Paih, and L. Zani, *Chem. Rev.*, *104*, 6217 (2004); (b) S. Enthaler, K. Junge, and M. Beller, *Angew. Chem. Int. Ed.*, *47*, 3317 (2008); (c) A. Correa, O. G. Mancheno, and C. Bolm, *Chem. Soc. Rev.*, *37*, 1108 (2008); (d) C. L. Sun, B. J. Li, and Z. J. Shi, *Chem. Rev.*, *111*, 1293 (2011); (e) K. Gopalaiah, *Chem. Rev. 113*, 3248 (2013).
2. H. Ohara, T. Itoh, M. Nakamura, and E. Nakamura, *Chem. Lett.*, 624 (2001).
3. H. Ohara, H. Kiyokane, and T. Itoh, *Tetrahedron Lett.*, *43*, 3041 (2002).
4. M. Rosenblum and D. Scheck, *Organometallics*, *1*, 397 (1982).
5. M. W. Bouwkamp, A. C. Bowman, E. Lobkovsky, and P. J. Chirik, *J. Am. Chem. Soc.*, *128*, 13340 (1982).
6. S. K. Russell, E. Lobkovsky, and P. J. Chirik, *J. Am. Chem. Soc.*, *133*, 8858 (2011).
7. A. P. Dieskau, M. S. Holzwarth, and B. Plietker, *J. Am. Chem. Soc.*, *134*, 5048 (2012).
8. J. M. Fan, L. F. Gao, and Z. Y. Wang, *Chem. Commun.*, 5021 (2009).
9. H. S. Wu, B. Wang, H. Q. Liu, and L. Wang, *Tetrahedron*, *67*, 1210 (2011).
10. E. P. Kundig, C. M. Saudan, and F. Viton, *Adv. Synth. Catal.*, *343*, 51 (2002).
11. W. S. Jen, J. J. M. Wiener, and D. W. C. MacMillan, *J. Am. Chem. Soc.*, *122*, 9874 (2000).
12. C. Breschi, L. Piparo, P. Pertici, A. M. Caporusso, and G. Vitulli, *J. Organomet. Chem.*, *607*, 57 (2000).
13. F. Knoch, F. Kremer, U. Schmidt, U. Zenneck, P. LeFloch, and F. Mathey, *Organometallics*, *15*, 2713 (1996).
14. K. Ferre, L. Toupet, and V. Guerchais, *Organometallics*, *21*, 2578 (2002).
15. N. Saino, D. Kogure, and S. Okamoto, *Org. Lett.*, *7*, 3065 (2005).
16. N. Saino, D. Kogure, K. Kase, and S. Okamoto, *J. Organomet. Chem.*, *691*, 3129 (2006).
17. A. Furstner, K. Majima, R. Martin, H. Krause, E. Kattnig, R. Goddard, and C. W. Lehmann, *J. Am. Chem. Soc.*, *130*, 1992 (2008).

18. Y. Yu, J. M. Smith, C. J. Flaschenriem, and P. L. Holland, *Inorg. Chem.*, *45*, 5742 (2006).
19. C. X. Wang, X. C. Li, F. Wu, and B. S. Wan, *Angew. Chem. Int. Ed.*, *50*, 7162 (2011).
20. B. R. D'Souza, T. K. Lane, and J. Louie, *Org. Lett.*, *13*, 2936 (2011).
21. E. J. Corey, N. Imai, and H. Y. Zhang, *J. Am. Chem. Soc.*, *113*, 728 (1991).
22. M. P. Sibi, S. Manyem, and H. Palencia, *J. Am. Chem. Soc.*, *128*, 13660 (2006).
23. N. Khiar, I. Fernandez, and F. Alcudia, *Tetrahedron Lett.*, *34*, 123 (1993).
24. S. Matsukawa, H. Sugama, and T. Imamoto, *Tetrahedron Lett.*, *41*, 6461 (2000).
25. S. Kanemasa, Y. Oderaotoshi, H. Yamamoto, J. Tanaka, E. Wada, and D. P. Curran, *J. Org. Chem.*, *62*, 6454 (1997).
26. S. Kanemasa, Y. Oderaotoshi, S. Sakaguchi, H. Yamamoto, J. Tanaka, E. Wada, and D. P. Curran, *J. Am. Chem. Soc.*, *120*, 3074 (1998).
27. H. Usuda, A. Kuramochi, M. Kanai, and M. Shibasaki, *Org. Lett.*, *6*, 4387 (2004).
28. Y. Shimizu, S. L. Shi, H. Usuda, M. Kanai, and M. Shibasaki, *Angew. Chem. Int. Ed.*, *49*, 1103 (2010).
29. M. E. Bruin and E. P. Kundig, *Chem. Commun.*, 2635 (1998).
30. R. Bakhtiar, J. J. Drader, and D. B. Jacobson, *J. Am. Chem. Soc.*, *114*, 8304 (1992).
31. K. Fujiwara, T. Kurahashi, and S. Matsubara, *J. Am. Chem. Soc.*, *134*, 5512 (2012).
32. (a) G. Wuitschik, M. Rogers-Evans, A. Buckl, M. Bernasconi, M. Marki, T. Godel, H. Fischer, B. Wagner, I. Parrilla, F. Schuler, J. Schneider, A. Alker, W. B. Schweizer, K. Muller, and E. M. Carreira, *Angew. Chem. Int. Ed.*, *47*, 4512 (2008); (b) G. Wuitschik, M. Rogers-Evans, K. Muller, H. Fischer, B. Wagner, F. Schuler, L. Polonchuk, and E. M. Carreira, *Angew. Chem. Int. Ed.*, *45*, 7736 (2006).
33. W. A. Donaldson, *Tetrahedron*, *57*, 8589 (2001).
34. (a) J. Salaun, *Chem. Rev.*, *89*, 1247 (1989); (b) M. P. Doyle and D. C. Forbes, *Chem. Rev.*, *98*, 911 (1998).
35. M. Brookhart and W. B. Studabaker, *Chem. Rev.*, *87*, 411 (1987).
36. P. W. Jolly and R. Pettit, *88*, 5045 (1966).
37. W. J. Seitz, A. K. Saha, and M. M. Hossain, *Organometallics*, *12*, 2604 (1993).
38. (a) S. K. Edulji and S. T. Nguyen, *Organometallics*, *22*, 3374 (2003); (b) S. K. Edulji, and S. T. Nguyen, *Pure. Appl. Chem.*, *76*, 645 (2004).
39. X. Morise, M. L. H. Green, P. Braunstein, L. H. Rees, and I. C. Vei, *New J. Chem.*, *27*, 32 (2003).
40. J. R. Wolf, C. G. Hamaker, J. P. Djukic, T. Kodadek, and L. K. Woo, *J. Am. Chem. Soc.*, *117*, 9194 (1995).
41. G. D. Du, B. Andrioletti, E. Rose, and L. K. Woo, *Organometallics*, *21*, 4490 (2002).
42. I. Nicolas, T. Roisnel, P. Le Maux, and G. Simonneaux, *Tetrahedron Lett.*, *50*, 5149 (2009).
43. B. Morandi and E. M. Carreira, *Science*, *335*, 1471 (2012).
44. B. Morandi, A. Dolva, and E. M. Carreira, *Org. Lett.*, *14*, 2162 (2012).

45. (a) M. Schlosser, *Angew. Chem. Int. Ed.*, *45*, 5432 (2006); (b) C. Isanbor and D. O'Hagan, *J. Fluorine Chem.*, *127*, 303 (2006).
46. B. Morandi and E. M. Carreira, *Angew. Chem. Int. Ed.*, *49*, 938 (2010).
47. B. Morandi, J. Cheang, and E. M. Carreira, *Org. Lett.*, *13*, 3080 (2011).
48. (a) I. D. G. Watson, L. L. Yu, and A. K. Yudin, *Acc. Chem. Res.*, *39*, 194 (2006); (b) B. Zwanenburg and P. ten Holte, *Top. Curr. Chem.*, *216*, 93 (2001); (c) W. McCoull and F. A. Davis, *Synthesis*, 1347 (2001); (d) S. Lociuro, L. Pellacani, and P. A. Tardella, *Tetrahedron Lett.*, *24*, 593 (1983); (e) S. Fioravanti, M. A. Loreto, L. Pellacani, and P. A. Tardella, *J. Org. Chem.*, *50*, 5365 (1985); (f) D. A. Evens, M. M. Faul, and M. T. Bilodeau, *J. Am. Chem. Soc.*, *116*, 2742 (1994); (g) K. Surendra, N. S. Krishnaveni, and K. R. Rao, *Tetrahedron Lett.*, *46*, 4111 (2005); (h) M. S. Reddy, M. Narender, and K. R. Rao, *Tetrahedron Lett.*, *46*, 1299 (2005).
49. M. Nakanishi, A. F. Salit, and C. Bolm, *Adv. Synth. Catal.*, *350*, 1835 (2008).
50. A. C. Mayer, A. F. Salit, and C. Bolm, *Chem. Commun.*, 5975 (2008).
51. A. Marti, L. Richter, and C. Schneider, *Synlett*, 2513 (2011).
52. (a) A. M. Caporusso, L. Lardicci, and G. Giacomelli, *Tetrahedron Lett.*, 4351 (1977); (b) A. M. Caporusso, G. Giacomelli, and L. Lardicci, *J. Chem. Soc., Perkin Trans. 1*, 3139 (1979).
53. M. Nakamura, A. Hirai, and E. Nakamura, *J. Am. Chem. Soc.*, *122*, 978 (2000).
54. M. Hojo, Y. Murakami, H. Aihara, R. Sakuragi, Y. Baba, and A. Hosomi, *Angew. Chem. Int. Ed.*, *40*, 621 (2001).
55. D. H. Zhang and J. M. Ready, *J. Am. Chem. Soc.*, *128*, 15050 (2006).
56. E. Shirakawa, T. Yamagami, T. Kimura, S. Yamaguchi, and T. Hayashi, *J. Am. Chem. Soc.*, *127*, 17164 (2005).
57. E. Shirakawa, D. Ikeda, S. Masui, M. Yoshida, and T. Hayashi, *J. Am. Chem. Soc.*, *134*, 272 (2012).
58. T. Yamagami, R. Shintani, E. Shirakawa, and T. Hayashi, *Org. Lett.*, *9*, 1045 (2007).
59. E. Shirakawa, D. Ikeda, T. Ozawa, S. Watanabe, and T. Hayashi, *Chem. Commun.*, 1885 (2009).
60. L. Adak and N. Yoshikai, *Tetrahedron 68*, 5167 (2012).
61. S. Ito, T. Itoh, and M. Nakamura, *Angew. Chem. Int. Ed.*, *50*, 454 (2011).
62. J. Cabral, P. Laszlo, L. Mahe, M. T. Montaufier, and S. L. Randriamahefa, *Tetrahedron Lett.*, *30*, 3969 (1989).
63. (a) J. Christoffers, *Chem Commun*, 943 (1997); (b) J. Christoffers, *Tetrahedron Lett.*, *39*, 7083 (1998).
64. (a) J. Christoffers, *Eur. J. Org. Chem.*, 1259 (1998); (b) S. Pelzer, T. Kauf, C. van Wullen, and J. Christoffers, *J. Organomet. Chem.*, *684*, 308 (2003).
65. J. Christoffers, *Eur. J. Org. Chem.*, 759 (1998).
66. J. Christoffers and A. Mann, *Eur. J. Org. Chem.*, 1977 (2000).
67. (a) J. Christoffers and A. Mann, *Eur. J. Org. Chem.*, 1475 (1999); (b) J. Christoffers, *J. Prak Chem-Chem Ztg.*, *341*, 495 (1999).

68. K. Shimizu, M. Miyagi, T. Kan-no, T. Kodama, and Y. Kitayama, *Tetrahedron Lett.*, *44*, 7421 (2003).
69. G. A. Molander and C. R. Harris, *Chem. Rev.*, *96*, 307 (1996).
70. (a) G. A. Molander and J. B. Etter, *Tetrahedron Lett.*, *25*, 3281 (1984); (b) G. A. Molander and J. B. Etter, *J. Org. Chem.*, *51*, 1778 (1986).
71. G. A. Molander and J. A. McKie, *J. Org. Chem.*, *56*, 4112 (1991).
72. G. A. Molander and J. A. McKie, *J. Org. Chem.*, *58*, 7216 (1993).
73. G. A. Molander and S. R. Shakya, *J. Org. Chem.*, *59*, 3445 (1994).
74. M. S. Kharasch, P. S. Skell, and P. Fisher, *J. Am. Chem. Soc.*, *70*, 1055 (1948).
75. J. Elzinga and H. Hogeveen, *J. Org. Chem.*, *45*, 3957 (1980).
76. (a) T. Susuki and J. Tsuji, *J. Org. Chem.*, *35*, 2982 (1970); (b) R. Davis, J. L. A. Durrant, N. M. S. Khazal, and T. E. Bitterwolf, *J. Organomet. Chem.*, *386*, 229 (1990).
77. L. Forti, F. Ghelfi, and U. M. Pagnoni, *Tetrahedron Lett.*, *37*, 2077 (1996).
78. L. Forti, F. Ghelfi, E. Libertini, U. M. Pagnoni, and E. Soragni, *Tetrahedron*, *53*, 17761 (1997).
79. (a) T. K. Hayes, A. J. Freyer, M. Parvez, and S. M. Weinreb, *J. Org. Chem.*, *51*, 5501 (1986); (b) T. K. Hayes, R. Villani, and S. M. Weinreb, *J. Am. Chem. Soc.*, *110*, 5533 (1988); (c) G. M. Lee, M. Parvez, and S. M. Weinreb, *Tetrahedron*, *44*, 4671 (1988); (d) G. M. Lee and S. M. Weinreb, *J. Org. Chem.*, *55*, 1281 (1990).
80. J. Norinder, A. Matsumoto, N. Yoshikai, and E. Nakamura, *J. Am. Chem. Soc.*, *130*, 5858 (2008).
81. N. Yoshikai, A. Matsumoto, J. Norinder, and E. Nakamura, *Angew. Chem. Int. Ed.*, *48*, 2925 (2009).
82. L. Ilies, J. Okabe, N. Yoshikai, and E. Nakamura, *Org. Lett.*, *12*, 2838 (2010).
83. N. Yoshikai, A. Mieczkowski, A. Matsumoto, L. Ilies, and E. Nakamura, *J. Am. Chem. Soc.*, *132*, 5568 (2010).
84. L. D. Tran and O. Daugulis, *Org. Lett.*, *12*, 4277 (2010).
85. J. Wang, S. Wang, G. Wang, J. Zhang, and X. Q. Yu, *Chem. Commun.*, *48*, 11769 (2012).
86. F. Vallee, J. J. Mousseau, and A. B. Charette, *J. Am. Chem. Soc.*, *132*, 1514 (2010).
87. W. Liu, H. Cao, and A. W. Lei, *Angew. Chem. Int. Ed.*, *49*, 2004 (2010).
88. S. L. Buchwald and C. Bolm, *Angew. Chem. Int. Ed.*, *48*, 5586 (2009).
89. (a) H. Chen, S. Schlecht, T. C. Semple, and J. F. Hartwig, *Science*, *287*, 1995 (2000); (b) V. Ritleng, C. Sirlin, and M. Pfeffer, *Chem. Rev.*, *102*, 1731 (2002); (c) T. Naota, H. Takaya, and S.-I. Murahashi, *Chem. Rev.*, *98*, 2599 (1998); (d) B. A. Arndtsen, R. G. Bergman, T. A. Mobley, and T. H. Peterson, *Acc. Chem. Res.*, *28*, 154 (1995); (e) A. S. Goldman, *Nature* (*London*), *366*, 514 (1993).
90. C.-J. Li, *Acc. Chem. Res.*, *42*, 335 (2009).
91. Z. Li, L. Cao, and C.-J. Li, *Angew. Chem., Int. Ed.*, *46*, 6505 (2007).
92. S.-I. Murahashi, N. Komiya, and H. Terai, *Angew. Chem., Int. Ed.*, *44*, 6931 (2005).
93. Z. Li, R. Yu, and H. Li, *Angew. Chem., Int. Ed.*, *47*, 7497 (2008).

94. H. Li, Z. He, X. Guo, W. Li, X. Zhao, and Z. Li, *Org. Lett.*, *11*, 4176 (2009).
95. Y. Zhang and C.-J. Li, *Eur. J. Org. Chem.*, *2007*, 4654 (2007).
96. Y.-Z. Li, B.-J. Li, X.-Y. Lu, S. Lin, and Z.-J. Shi, *Angew. Chem., Int. Ed.*, *48*, 3817 (2009).
97. C.-X. Song, G.-X. Cai, T. R. Farrell, Z.-P. Jiang, H. Li, L.-B. Gan, and Z.-J. Shi, *Chem. Commun.*, 6002 (2009).
98. X. Guo, R. Yu, H. Li, and Z. Li, *J. Am. Chem. Soc.*, *131*, 17387 (2009).
99. C. M. R. Volla and P. Vogel, *Org. Lett.*, *11*, 1701 (2009).
100. (a) R. Shang, Y. Fu, Y. Wang, Q. Xu, H.-Z. Yu, and L. Liu, *Angew. Chem. Int. Ed.*, *48*, 9350 (2009); (b) Z.-M. Sun and P. Zhao, *Angew. Chem. Int. Ed.*, *48*, 6726 (2009); (c) F. Zhang and M. F. Greaney, *Angew. Chem. Int. Ed.*, *49*, 2768 (2010).
101. H.-P. Bi, W.-W. Chen, Y.-M. Liang, and C.-J. Li, *Org. Lett.*, *11*, 3246 (2009).
102. S.-Y. Zhang, Y.-Q. Tu, C.-A. Fan, F.-M. Zhang, and L. Shi, *Angew. Chem., Int. Ed.*, *48*, 8761 (2009).
103. P. S. Baran, B. D. Hafensteiner, N. B. Ambhaikar, C. A. Guerrero, and J. D. Gallagher, *J. Am. Chem. Soc.*, *128*, 8678 (2006).
104. M. P. DeMartino, K. Chen, and P. S. Baran, *J. Am. Chem. Soc.*, *130*, 11546 (2008).
105. J. M. Richter, Y. Ishihara, T. Masuda, B. W. Whitefield, T. s. Llamas, A. Pohjakallio, and P. S. Baran, *J. Am. Chem. Soc.*, *130*, 17938 (2008).
106. H. Egami, K. Matsumoto, T. Oguma, T. Kunisu, and T. Katsuki, *J. Am. Chem. Soc.*, *132*, 5886 (2010).
107. (a) I. Fridovich, *Science*, *201*, 875 (1978); (b) S. H. Han, B. Zheng, D. G. Schatz, E. Spanopoulou, and G. Kelsoe, *Science*, *274*, 2094 (1996); (c) P. L. Roach, I. J. Clifton, C. M. H. Hensgens, N. Shibata, C. J. Schofield, J. Hajdu, and J. E. Baldwin, *Nature* (London), *387*, 827 (1997); (d) P. C. Ford, B. O. Fernandez, and M. D. Lim, *Chem. Rev.*, *105*, 2439 (2005).
108. (a) D. H. R. Barton and D. Doller, *Acc. Chem. Res.*, *25*, 504 (1992); (b) D. H. R. Barton, *Chem. Soc. Rev.*, *25*, 237 (1996); (c) D. H. R. Barton, *Tetrahedron*, *54*, 5805 (1998); (d) P. Stavropoulos, R. Celenligil-Cetin, and A. E. Tapper, *Acc. Chem. Res.*, *34*, 745 (2001); (e) M. J. Perkins, *Chem. Soc. Rev.*, *25*, 229 (1996); (f) D. T. Sawyer, A. Sobkowiak, and T. Matsushita, *Acc. Chem. Res.*, *29*, 409 (1996); (g) C. Walling, *Acc. Chem. Res.*, *31*, 155 (1998); (h) P. A. MacFaul, D. D. M. Wayner, and K. U. Ingold, *Acc. Chem. Res.*, *31*, 159 (1998); (i) S. Goldstein and D. Meyerstein, *Acc. Chem. Res.*, *32*, 547 (1999); (j) M. Fontecave, S. Menage, and C. Duboc-Toia, *Coord. Chem. Rev.*, *178*, 1555 (1998); (k) M. Costas, K. Chen, and L. Que, Jr., *Coord. Chem. Rev.*, *200*, 517 (2000); (l) M. Costas, M. P. Mehn, M. P. Jensen, and L. Que, *Chem. Rev.*, *104*, 939 (2004); (m) E. Y. Tshuva and S. J. Lippard, *Chem. Rev.*, *104*, 987 (2004).
109. M. Nakanishi and C. Bolm, *Adv. Synth. Catal.*, *349*, 861 (2007).
110. K. Moller, G. Wienhofer, K. Schroder, B. Join, K. Junge, and M. Beller, *Chem. Eur. J.*, *16*, 10300 (2010).
111. (a) F. Shi, M. K. Tse, and M. Beller, *J. Mol. Catal. A: Chem.*, *270*, 68 (2007); (b) F. Shi, M. K. Tse, and M. Beller, *Adv. Synth. Catal.*, *349*, 303 (2007); (c) G. Wienhofer,

K. Schroder, K. Moller, K. Junge, and M. Beller, *Adv. Synth. Catal.*, *352*, 1615 (2010); (d) M. Periasamy and M. V. Bhatt, *Tetrahedron Lett.*, 4561 (1978); (e) A. Bohle, A. Schubert, Y. Sun, and W. R. Thiel, *Adv. Synth. Catal.*, *348*, 1011 (2006); (f) T. Takai, E. Hata, and T. Mukaiyama, *Chem. Lett.*, 885 (1994); (g) R. Song, A. Sorokin, J. Bernadou, and B. Meunier, *J. Org. Chem.*, *62*, 673 (1997).

112. C. Kim, K. Chen, J. H. Kim, and L. Que, Jr., *J. Am. Chem. Soc.*, *119*, 5964 (1997).
113. K. Chen and L. Que, Jr., *J. Am. Chem. Soc.*, *123*, 6327 (2001).
114. (a) G. J. P. Britovsek, J. England, and A. J. P. White, *Inorg Chem*, *44*, 8125 (2005); (b) J. England, G. J. P. Britovsek, N. Rabadia, and A. J. P. White, *Inorg. Chem.*, *46*, 3752 (2007); (c) J. England, C. R. Davies, M. Banaru, A. J. P. White, and G. J. P. Britovsek, *Adv. Synth. Catal.*, *350*, 883 (2008); (d) J. England, R. Gondhia, L. Bigorra-Lopez, A. R. Petersen, A. J. P. White, and G. J. P. Britovsek, *Dalton Trans.*, 5319 (2009); (e) O. Y. Lyakin, K. P. Bryliakov, G. J. P. Britovsek, and E. P. Talsi, *J. Am. Chem. Soc.*, *131*, 10798 (2009).
115. (a) M. Merkx, D. A. Kopp, M. H. Sazinsky, J. L. Blazyk, J. Muller, and S. J. Lippard, *Angew. Chem. Int. Ed.*, *40*, 2782 (2001); (b) M. M. Abu-Omar, A. Loaiza, and N. Hontzeas, *Chem. Rev.*, *105*, 2227 (2005).
116. L. Que and W. B. Tolman, *Nature* (*London*), *455*, 333 (2008).
117. A. Company, L. Gomez, M. Guell, X. Ribas, J. M. Luis, L. Que, and M. Costas, *J. Am. Chem. Soc.*, *129*, 15766 (2007).
118. A. Company, L. Gomez, X. Fontrodona, X. Ribas, and M. Costas, *Chem. Eur. J.*, *14*, 5727 (2008).
119. B. Retcher, J. S. Costa, J. K. Tang, R. Hage, P. Gamez, and J. Reedijk, *J. Mol. Catal. A: Chem.*, *286*, 1 (2008).
120. M. S. Chen and M. C. White, *Science*, *318*, 783 (2007).
121. N. A. Vermeulen, M. S. Chen, and M. C. White, *Tetrahedron*, *65*, 3078 (2009).
122. M. S. Chen and M. C. White, *Science*, *327*, 566 (2010).
123. L. Gomez, I. Garcia-Bosch, A. Company, J. Benet-Buchholz, A. Polo, X. Sala, X. Ribas, and M. Costas, *Angew. Chem. Int. Ed.*, *48*, 5720 (2009).
124. Y. Hitomi, K. Arakawa, T. Funabiki, and M. Kodera, *Angew. Chem. Int. Ed.*, *51*, 3448 (2012).
125. (a) K. A. Jorgensen, Wiley-VCH, Weinheim 2, 1998; (b) U. Sundermeier, C. Döbler, and M. Beller, *Modern Oxidation Methods*, Wiley-VCH, Weinheim, 2004.
126. (a) Y. Shi, *Acc. Chem. Res.*, *37*, 488 (2004); (b) E. N. Jacobsen and M. H. Wu, *Comprehensive Asymmetric Catalysis*, Vol. 2, 1999;. (c) T. Katsuki, *Comprehensive in Asymmetric Catalysis*, Vol. 2, Springer, Berlin, 1999.
127. M. C. White, A. G. Doyle, and E. N. Jacobsen, *J. Am. Chem. Soc.*, *123*, 7194 (2001).
128. G. Dubois, A. Murphy, and T. D. P. Stack, *Org. Lett.*, *5*, 2469 (2003).
129. J. Y. Ryu, J. Kim, M. Costas, K. Chen, W. Nam, and L. Que, *Chem. Commun.*, 1288 (2002).
130. R. Mas-Balleste and L. Que, *J. Am. Chem. Soc.*, *129*, 15964 (2007).
131. G. Anilkumar, B. Bitterlich, F. G. Gelalcha, M. K. Tse, and M. Beller, *Chem. Commun.*, 289 (2007).

132. B. Bitterlich, G. Anilkumar, F. G. Gelalcha, B. Spilker, A. Grotevendt, R. Jackstell, M. K. Tse, and M. Beller, *Chem. Asian J.*, *2*, 521 (2007).

133. (a) F. G. Gelalcha, G. Anilkumar, M. K. Tse, A. Brueckner, and M. Beller, *Chem. Eur. J.*, *14*, 7687 (2008); (b) B. Bitterlich, K. Schroder, M. K. Tse, and M. Beller, *Eur. J. Org. Chem.*, 4867 (2008); (c) S. Enthaler, K. Schroder, S. Inoue, B. Eckhardt, K. Junge, M. Beller, and M. Driess, *Eur. J. Org. Chem.*, 4893 (2010); (d) K. Schroder, K. Junge, A. Spannenberg, and M. Beller, *Catal. Today*, *157*, 364 (2010); (e) K. Schroder, S. Enthaler, B. Join, K. Junge, and M. Beller, *Adv. Synth. Catal.*, *352*, 1771 (2010).

134. K. Hasan, N. Brown, and C. M. Kozak, *Green Chem.*, *13*, 1230 (2012).

135. F. G. Gelalcha, B. Bitterlich, G. Anilkumar, M. K. Tse, and M. Beller, *Angew. Chem. Int. Ed.*, *46*, 7293 (2007).

136. K. Schroder, B. Join, A. J. Amali, K. Junge, X. Ribas, M. Costas, and M. Beller, *Angew. Chem. Int. Ed.*, *50*, 1425 (2011).

137. Y. Nishikawa and H. Yamamoto, *J. Am. Chem. Soc.*, *133*, 8432 (2011).

138. (a) D. J. Ferraro, L. Gakhar, and S. Ramaswamy, *Biochem. Biophys. Res. Commun.*, *338*, 175 (2005); (b) S. Beil, B. Happe, K. N. Timmis, and D. H. Pieper, *Eur. J. Biochem.*, *247*, 190 (1997).

139. K. Suzuki, P. D. Oldenburg, and L. Que, Jr., *Angew. Chem. Int. Ed.*, *47*, 1887 (2008).

140. Y. Feng, C. Y. Ke, G. Q. Xue, and L. Que, Jr., *Chem. Commun.*, 50 (2009).

141. (a) F. Dietrich and P. J. Stang, *Metal-Catalyzed Cross-Coupling Reactions*, Wiley–VCH, Weinhem, 1998; (b) N. Miyaura, *Cross-Coupling Reaction. A Practical Guide*, Vol. 219, Springer, Berlin, 2002.

142. M. Beller, A. Zapf, and W. Magerlein, *Chem. Eng. Technol.*, *24*, 575 (2001).

143. (a) L. Acemoglu and J. M. J. Williams, *in Handbook of Organopalladium Chemistry for Organic Synthesis*, John Wiley & Sons, Inc., New York, 2002; (b) J. Tsuji, *Palladium Reagents and Catalysts: Innovation in Organic Synthesis*, John Wiley & Sons, Inc., New York, 1996; (c) B. M. Trost, and T. R. Verhoeven, Vol. 8, *Pergamon, Oxford, 1982.*

144. (a) J. P. Corriu and J. P. Masse, *J. Chem. Soc. Chem. Commun.*, 144 (1972); (b) K. Tamao, K. Sumitani, and M. Kumada, *J. Am. Chem. Soc.*, *94*, 4374 (1972).

145. (a) M. Tamura and J. Kochi, *J. Am. Chem. Soc.*, *93*, 1487 (1971); (b) M. Tamura and J. K. Kochi, *Bull. Chem. Soc. Jpn.*, *44*, 3063 (1971); (c) M. Tamura and J. Kochi, *Synthesis*, 303 (1971); (d) J. K. Kochi, *Acc. Chem. Res.*, *7*, 351 (1974); (e) S. M. Neumann and J. K. Kochi, *J. Org. Chem.*, *40*, 599 (1975); (f) R. S. Smith and J. K. Kochi, *J. Org. Chem.*, *41*, 502 (1976); (g) J. K. Kochi, *J. Organomet. Chem.*, *653*, 11 (2002).

146. (a) G. A. Molander, B. J. Rahn, D. C. Shubert, and S. E. Bonde, *Tetrahedron Lett.*, *24*, 5449 (1983); (b) H. Felkin, P. J. Knowles, and B. Meunier, *J. Organomet. Chem.*, *146*, 151 (1978).

147. V. Fiandanese, G. Miccoli, F. Naso, and L. Ronzini, *J. Organomet. Chem.*, *312*, 343 (1986).

148. (a) J. L. Fabre, M. Julia, and J. N. Verpeaux, *Tetrahedron Lett.*, *23*, 2469 (1982); (b) J. L. Fabre, M. Julia, and J. N. Verpeaux, *Bull. Soc. Chim. Fr.*, 772 (1985); (c) E. Alvarez, T. Cuvigny, C. H. Dupenhoat, and M. Julia, *Tetrahedron*, *44*, 111 (1988); (d) E. Alvarez, T. Cuvigny, C. H. Dupenhoat, and M. Julia, *Tetrahedron*, *44*, 119 (1988).
149. (a) L. Jin, M. Julia, and J. N. Verpeaux, *Synlett*, 215 (1994); (b) I. Daub, A. K. Habermann, A. Hobert, and M. Julia, *Eur. J. Org. Chem.*, 163 (1999).
150. H. M. Walborsky and R. B. Banks, *J. Org. Chem.*, *46*, 5074 (1981).
151. (a) W. Dohle, F. Kopp, G. Cahiez, and P. Knochel, *Synlett*, 1901 (2001); (b) G. Cahiez and H. Avedissian, *Synthesis*, 1199 (1998).
152. (a) G. Cahiez and S. Marquais, *Tetrahedron Lett.*, *37*, 1773 (1996); (b) A. Furstner and H. Brunner, *Tetrahedron Lett.*, *37*, 7009 (1996); (c) G. Cahiez and S. Marquais, *Pure Appl. Chem.*, *68*, 53 (1996).
153. (a) A. Furstner and R. Martin, *Chem. Lett.*, *34*, 624 (2005); (b) B. Scheiper, M. Bonnekessel, H. Krause, and A. Furstner, *J. Org. Chem.*, *69*, 3943 (2004); (c) K. Itami, S. Higashi, M. Mineno, and J. Yoshida, *Org. Lett.*, *7*, 1219 (2005).
154. (a) A. Furstner and P. Hannen, *Chem. Eur. J.*, *12*, 3006 (2006); (b) G. Berthon-Gelloz and T. Hayashi, *J. Org. Chem.*, *71*, 8957 (2006).
155. (a) A. Furstner and A. Leitner, *Angew. Chem. Int. Ed.*, *41*, 609 (2002); (b) A. Furstner, A. Leitner, M. Mendez, and H. Krause, *J. Am. Chem. Soc.*, *124*, 13856 (2002).
156. G. Seidel, D. Laurich, and A. Furstner, *J. Org. Chem.*, *69*, 3950 (2004).
157. B. Scheiper, F. Glorius, A. Leitner, and A. Furstner, *Proc. Ind. Natl. Sci. Acad.*, *101*, 11960 (2004).
158. A. Furstner, A. Leitner, *Angew. Chem. Int. Ed.*, *42*, 308 (2003).
159. T. Hatakeyama and M. Nakamura, *J. Am. Chem. Soc.*, *129*, 9844 (2007).
160. M. Hocek and H. Dvorakova, *J. Org. Chem.*, *68*, 5773 (2003).
161. E. Colacino, H. Benakki, F. Guenoun, J. Martinez, and F. Lamaty, *Synth. Commun*, *39*, 1583 (2009).
162. A. L. Silberstein, S. D. Ramgren, and N. K. Garg, *Org. Lett.*, *14*, 3796 (2012).
163. O. M. Kuzmina, A. K. Steib, D. Flubacher, and P. Knochel, *Org. Lett.*, *14*, 4818 (2012).
164. A. C. Frisch and M. Beller, *Angew. Chem. Int. Ed.*, *44*, 674 (2005).
165. M. Tamura and J. Kochi, *J. Organomet. Chem.*, *31*, 289 (1971).
166. (a) M. Nakamura, K. Matsuo, S. Ito, and B. Nakamura, *J. Am. Chem. Soc.*, *126*, 3686 (2004); (b) M. Nakamura, S. Ito, K. Matsuo, and E. Nakamura, *Synlett*, 1794 (2005).
167. T. Nagano and T. Hayashi, *Org. Lett.*, *6*, 1297 (2004).
168. (a) A. Guerinot, S. Reymond, and J. Cossy, *Angew. Chem. Int. Ed.*, *46*, 6521 (2007); (b) G. Cahiez, C. Duplais, and A. Moyeux, *Org. Lett.*, *9*, 3253 (2007).
169. K. G. Dongol, H. Koh, M. Sau, and C. L. L. Chai, *Adv. Synth. Catal.*, *349*, 1015 (2007).
170. Y. Yamaguchi, H. Ando, M. Nagaya, H. Hinago, T. Ito, and M. Asami, *Chem. Lett.*, *40*, 983 (2011).

171. Z. B. Mo, Q. Zhang, and L. Deng, *Organometallics*, *31*, 6518 (2012).
172. R. Martin and A. Furstner, *Angew. Chem. Int. Ed.*, *43*, 3955 (2004).
173. K. Jonas, L. Schieferstein, C. Kruger, and Y. H. Tsay, *Angew. Chem. Int. Ed.*, *18*, 550 (1979).
174. (a) R. B. Bedford, D. W. Bruce, R. M. Frost, and M. Hird, *Chem. Commun.*, 4161 (2005); (b) K. Bica and P. Gaertner, *Org. Lett.*, *8*, 733 (2006); (c) R. R. Chowdhury, A. K. Crane, C. Fowler, P. Kwong, and C. M. Kozak, *Chem. Commun.*, 94 (2008).
175. M. Jin and M. Nakamura, *Chem. Lett.*, *40*, 1012 (2011).
176. G. Cahiez, V. Habiak, C. Duplais, and A. Moyeux, *Angew. Chem. Int. Ed.*, *46*, 4364 (2007).
177. W. C. Percival, R. B. Wagner, and N. C. Cook, *J. Am. Chem. Soc.*, *75*, 3731 (1953).
178. F. Babudri, A. Dettole, V. Fiandanese, G. Marchese, and F. Naso, *J. Organomet. Chem.*, *405*, 53 (1991).
179. (a) B. D. Sherry and A. Furstner, *Acc. Chem. Res.*, *41*, 1500 (2008); (b) C. Cardellicchio, V. Viandanese, G. Marchese, and L. Ronzini, *Tetrahedron Lett.*, *26*, 3595 (1985); (c) C. Duplais, F. Bures, I. Sapountzis, T. J. Korn, G. Cahiez, and P. Knochel, *Angew. Chem. Int. Ed.*, *43*, 2968 (2004).
180. V. Fiandanese, G. Marchese, and F. Naso, *Tetrahedron Lett.*, *29*, 3587 (1988).
181. (a) A. Furstner, D. De Souza, L. Parra-Rapado, and J. T. Jensen, *Angew. Chem. Int. Ed.*, *42*, 5358 (2003); (b) K. Lehr and A. Furstner, *Tetrahedron*, *68*, 7695 (2012).
182. M. Taillefer, N. Xia, and A. Ouali, *Angew. Chem. Int. Ed.*, *46*, 934 (2007).
183. A. Correa and C. Bolm, *Angew. Chem. Int. Ed.*, *46*, 8862 (2007).
184. A. Correa, M. Carril, and C. Bolm, *Angew. Chem. Int. Ed.*, *47*, 2880 (2008).
185. Y. Y. Lin, Y. J. Wang, C. H. Lin, J. H. Cheng, and C. F. Lee, *J. Org. Chem.*, *77*, 6100 (2012).
186. O. Bistri, A. Correa, and C. Bolm, *Angew. Chem. Int. Ed.*, *47*, 586 (2008).
187. J. C. Mao, G. L. Xie, J. M. Zhan, Q. Q. Hua, and D. Q. Shi, *Adv. Synth. Catal.*, *351*, 1268 (2009).
188. (a) T. Mizoroki, K. Mori, and A. Ozaki, *Bull. Chem. Soc. Jpn.*, *44*, 581 (1971); (b) R. F. Heck and J. P. Nolley, *J. Org. Chem.*, *37*, 2320 (1972).
189. R. Loska, C. M. R. Volla, and P. Vogel, *Adv. Synth. Catal.*, *350*, 2859 (2008).
190. T. Hatakeyama, Y. Kondo, Y. I. Fujiwara, H. Takaya, S. Ito, E. Nakamura, and M. Nakamura, *Chem. Commun.*, 1216 (2009).
191. R. B. Bedford, M. Huwe, and M. C. Wilkinson, *Chem. Commun.*, 600 (2009).
192. S. Ito, Y. Fujiwara, E. Nakamura, and M. Nakamura, *Org. Lett.*, *11*, 4306 (2009).
193. T. Hatakeyama, T. Hashimoto, Y. Kondo, Y. Fujiwara, H. Seike, H. Takaya, Y. Tamada, T. Ono, and M. Nakamura, *J. Am. Chem. Soc.*, *132*, 10674 (2010).
194. T. Hashimoto, T. Hatakeyama, and M. Nakamura, *J. Org. Chem.*, *77*, 1168 (2012).
195. T. Hatakeyama, T. Hashimoto, K. Kathriarachchi, T. Zenmyo, H. Seike, and M. Nakamura, *Angew. Chem. Int. Ed.*, *51*, 8834 (2012).
196. S. Kawamura, T. Kawabata, K. Ishizuka, and M. Nakamura, *Chem. Commun.*, *48*, 9376 (2012).

197. M. Carril, A. Correa, and C. Bolm, *Angew. Chem. Int. Ed.*, *47*, 4862 (2008).
198. B. Bogdanovic and M. Schwickardi, *Angew. Chem. Int. Ed.*, *39*, 4610 (2000).
199. A. Furstner, R. Martin, H. Krause, G. Seidel, R. Goddard, and C. W. Lehmann, *J. Am. Chem. Soc.*, *130*, 8773 (2008).
200. B. Holzer and R. W. Hoffmann, *Chem. Commun.*, 732 (2003).
201. D. Noda, Y. Sunada, T. Hatakeyama, M. Nakamura, and H. Nagashima, *J. Am. Chem. Soc.*, *131*, 6078 (2009).
202. (a) J. Kleimark, A. Hedstrom, P. F. Larsson, C. Johansson, and P. O. Norrby, *Chemcatchem*, *1*, 152 (2009); (b) A. Hedstrom, U. Bollmann, J. Bravidor, and P. O. Norrby, *Chem. Eur. J.*, *17*, 11991 (2011).
203. C. J. Adams, R. B. Bedford, E. Carter, N. J. Gower, M. F. Haddow, J. N. Harvey, M. Huwe, M. A. Cartes, S. M. Mansell, C. Mendoza, D. M. Murphy, E. C. Neeve, and J. Nunn, *J. Am. Chem. Soc.*, *134*, 10333 (2012).
204. R. B. Bedford, E. Carter, P. M. Cogswell, N. J. Gower, M. F. Haddow, J. N. Harvey, D. M. Murphy, E. C. Neeve, and J. Nunn, *Angew. Chem. Int. Ed.*, *51*, 5435 (2012).
205. M. D. Greenhalgh and S. P. Thomas, *J. Am. Chem. Soc.*, *134*, 11900 (2012).
206. (a) R. Chinchilla, C. Najera, *Chem. Rev.*, *107*, 874 (2007); (b) H. Doucet and J. C. Hierso, *Angew. Chem. Int. Ed.*, *46*, 834 (2007).
207. T. Hatakeyama, Y. Yoshimoto, T. Gabriel, and M. Nakamura, *Org. Lett.*, *10*, 5341 (2008).
208. (a) T. E. Muller and M. Beller, *Chem. Rev.*, *98*, 675 (1998); (b) R. N. Salvatore, C. H. Yoon, and K. W. Jung, *Tetrahedron*, *57*, 7785 (2001); (c) P. Muller and C. Fruit, *Chem. Rev.*, *103*, 2905 (2003).
209. (a) J. L. Liang, J. S. Huang, X. Q. Yu, N. Y. Zhu, and C. M. Che, *Chem. Eur. J.*, *8*, 1563 (2002); (b) Y. Cui and C. He, *J. Am. Chem. Soc.*, *125*, 16202 (2003); (c) Z. G. Li, D. A. Capretto, R. Rahaman, and C. A. He, *Angew. Chem. Int. Ed.*, *46*, 5184 (2007).
210. Z. Wang, Y. M. Zhang, H. Fu, Y. Y. Jiang, and Y. F. Zhao, *Org. Lett.*, *10*, 1863 (2008).
211. S. G. Pan, J. H. Liu, H. R. Li, Z. Y. Wang, X. W. Guo, and Z. P. Li, *Org. Lett.*, *12*, 1932 (2010).
212. Q. Q. Xia and W. Z. Chen, *J. Org. Chem.*, *77*, 9366 (2012).
213. J. A. Wang, J. T. Hou, J. Wen, J. Zhang, and X. Q. Yu, *Chem. Commun.*, *47*, 3652 (2011).
214. (a) M. Johannsen and K. A. Jorgensen, *Chem. Rev.*, *98*, 1689 (1998); (b) Y. Tamaru, and M. Kimura, *Synlett*, 749 (1997).
215. M. Johannsen and K. A. Jorgensen, *J. Org. Chem.*, *59*, 214 (1994).
216. R. S. Srivastava and K. M. Nicholas, *Tetrahedron Lett.*, *35*, 8739 (1994).
217. S. Singh and K. M. Nicholas, *Synth. Commun*, *31*, 3087 (2001).
218. (a) J. E. Kmiecik, *J. Org. Chem.*, *30*, 2014 (1965); (b) D. Mulvey and W. A. Waters, *J. Chem. Soc., Perkin Trans. 2*, 1868 (1977).
219. M. Johannsen and K. A. Jorgensen, *J. Org. Chem.*, *60*, 5979 (1995).

220. (a) R. S. Srivastava, M. A. Khan, and K. M. Nicholas, *J. Am. Chem. Soc.*, *118*, 3311 (1996); (b) R. S. Srivastava and K. M. Nicholas, *J. Am. Chem. Soc.*, *119*, 3302 (1997).
221. (a) R. S. Srivastava and K. M. Nicholas, *Chem. Commun.*, 2705 (1998); (b) M. K. Kolel-Veetil, M. A. Khan, and K. M. Nicholas, *Organometallics*, *19*, 3754 (2000).
222. R. S. Srivastava, M. Kolel-Veetil, and K. M. Nicholas, *Tetrahedron Lett.*, *43*, 931 (2002).
223. B. Plietker, *Angew. Chem. Int. Ed.*, *45*, 6053 (2006).
224. J. Bonnamour and C. Bolm, *Org. Lett.*, *13*, 2012 (2011).
225. (a) R. Breslow and S. H. Gellman, *J. Chem. Soc. Chem. Commun.*, 1400 (1982); (b) R. Breslow and S. H. Gellman, *J. Am. Chem. Soc.*, *105*, 6728 (1983).
226. S. M. Paradine and M. C. White, *J. Am. Chem. Soc.*, *134*, 2036 (2012).
227. Z. Wang, S. Li, B. Yu, H. Wu, Y. Wang, and X. Sun, *J. Org. Chem.*, *77*, 8615 (2012).
228. E. R. King, E. T. Hennessy, and T. A. Betley, *J. Am. Chem. Soc.*, *133*, 4917 (2011).
229. (a) S. S. Kim, K. Nehru, S. S. Kim, D. W. Kim, and H. C. Jung, *Synthesis*, 2484 (2002); (b) A. R. Suarez, A. M. Baruzzi, and L. I. Rossi, *J. Org. Chem.*, *63*, 5689 (1998).
230. (a) J. T. Groves and P. Viski, *J. Org. Chem.*, *55*, 3628 (1990); (b) E. Baciocchi, M. F. Gerini, and A. Lapi, *J. Org. Chem.*, *69*, 3586 (2004).
231. (a) C. Duboc-Toia, S. Menage, R. Y. N. Ho, L. Que, C. Lambeaux, and M. Fontecave, *Inorg. Chem.*, *38*, 1261 (1999); (b) Y. Mekmouche, H. Hummel, R. Y. N. Ho, L. Que, V. Schunemann, F. Thomas, A. X. Trautwein, C. Lebrun, K. Gorgy, J. C. Lepretre, M. N. Collomb, A. Deronzier, M. Fontecave, and S. Menage, *Chem. Eur. J.*, *8*, 1196 (2002).
232. (a) J. Legros and C. Bolm, *Angew. Chem. Int. Ed.*, *42*, 5487 (2003); (b) J. Legros and C. Bolm, *Angew. Chem. Int. Ed.*, *43*, 4225 (2004).
233. J. Legros and C. Bolm, *Chem. Eur. J.*, *11*, 1086 (2005).
234. (a) A. Korte, J. Legros, and C. Bolm, *Synlett*, 2397 (2004); (b) J. Legros, J. R. Dehli, and C. Bolm, *Adv. Synth. Catal.*, *347*, 19 (2005).
235. A. M. I. Jayaseeli and S. Rajagopal, *J. Mol. Catal. A: Chem.*, *309*, 103 (2009).
236. S. H. Liao and B. List, *Adv. Synth. Catal.*, *354*, 2363 (2012).
237. B. Li, A. H. Liu, L. N. He, Z. Z. Yang, J. Gao, and K. H. Chen, *Green Chem.*, *14*, 130 (2012).
238. (a) T. Bach and C. Korber, *Tetrahedron Lett.*, *39*, 5015 (1998); (b) T. Bach and C. Korber, *Eur. J. Org. Chem.*, 1033 (1999).
239. (a) C. Bolm, K. Muniz, N. Aguilar, M. Kesselgruber, and G. Raabe, *Synthesis*, 1251 (1999); (b) H. Okamura and C. Bolm, *Chem. Lett.*, *33*, 482 (2004); (c) M. Reggelin and C. Zur, *Synthesis*, 1 (2000).
240. O. G. Mancheno and C. Bolm, *Org. Lett.*, *8*, 2349 (2006).
241. O. G. Mancheno, J. Dallimore, A. Plant, and C. Bolm, *Org. Lett.*, *11*, 2429 (2009).
242. X. F. Wu, C. Vovard-Le Bray, L. Bechki, and C. Darcel, *Tetrahedron*, *65*, 7380 (2009).
243. R. N. Naumov, M. Itazaki, M. Kamitani, and H. Nakazawa, *J. Am. Chem. Soc.*, *134*, 804 (2012).

244. (a) A. M. Tondreau, C. C. H. Atienza, K. J. Weller, S. A. Nye, K. M. Lewis, J. G. P. Delis, and P. J. Chirik, *Science*, *335*, 567 (2012); (b) A. N. Nesmeyanov, R. Kh. Freedline, E. C. Chukovskaya, R. G. Petrova, A. B. Belyarsky, *Tetrahedron*, *17*, 61 (1962); (c) M. A. Schroeder and M. S. Wrighton, *J. Organoment. Chem.*, *128*, 345 (1977).
245. F. Kakiuchi, Y. Tanaka, N. Chatani, and S. Murai, *J. Organomet. Chem.*, *456*, 45 (1993).
246. S. C. Bart, E. Lobkovsky, and P. J. Chirik, *J. Am. Chem. Soc.*, *126*, 13794 (2004).
247. A. M. Tondreau, C. C. H. Atienza, J. M. Darmon, C. Milsmann, H. M. Hoyt, K. J. Weller, S. A. Nye, K. M. Lewis, J. Boyer, J. G. P. Delis, E. Lobkovsky, and P. J. Chirik, *Organometallics*, *31*, 4886 (2012).
248. K. Kamata, A. Suzuki, Y. Nakai, and H. Nakazawa, *Organometallics*, *31*, 3825 (2012).
249. J. Y. Wu, B. N. Stanzl, and T. Ritter, *J. Am. Chem. Soc.*, *132*, 13214 (2010).
250. H. Nishiyama and A. Furuta, *Chem. Commun.*, 760 (2007).
251. (a) S. Hosokawa, J. Ito, and H. Nishiyama, *Organometallics*, *29*, 5773 (2010): (b) T. Inagaki, A. Ito, J. Ito, and H. Nishiyama, *Angew. Chem. Int. Ed.*, *49*, 9384 (2010).
252. (a) N. S. Shaikh, S. Enthaler, K. Junge, and M. Beller, *Angew. Chem. Int. Ed.*, *47*, 2497 (2008); (b) N. S. Shaikh, K. Junge, and M. Beller, *Org. Lett.*, *9*, 5429 (2007).
253. J. A. Yang and T. D. Tilley, *Angew. Chem. Int. Ed.*, *49*, 10186 (2010).
254. E. N. Frankel, E. A. Emken, and V. L. Davison, *J. Org. Chem.*, *30*, 2739 (1965).
255. M. A. Schroeder and M. S. Wrighton, *J. Am. Chem. Soc.*, *98*, 551 (1976).
256. R. Noyori, I. Umeda, and T. Ishigami, *J. Org. Chem.*, *37*, 1542 (1972).
257. C. Bianchini, A. Meli, M. Peruzzini, P. Frediani, C. Bohanna, M. A. Esteruelas, and L. A. Oro, *Organometallics*, *11*, 138 (1992).
258. S. D. Brown and J. C. Peters, *J. Am. Chem. Soc.*, *126*, 4538 (2004).
259. E. J. Daida and J. C. Peters, *Inorg. Chem.*, *43*, 7474 (2004).
260. R. J. Trovitch, E. Lobkovsky, E. Bill, and P. J. Chirik, *Organometallics*, *27*, 1470 (2008).
261. G. Wienhofer, F. A. Westerhaus, R. V. Jagadeesh, K. Junge, H. Junge, and M. Beller, *Chem. Commun.*, *48*, 4827 (2012).
262. C. Bianchini, E. Farnetti, M. Graziani, M. Peruzzini, and A. Polo, *Organometallics*, *12*, 3753 (1993).
263. (a) C. P. Casey and H. R. Guan, *J. Am. Chem. Soc.*, *129*, 5816 (2007); (b) C. P. Casey, S. E. Beetner, and J. B. Johnson, *J. Am. Chem. Soc.*, *130*, 2285 (2008).
264. A. Berkessel, S. Reichau, A. von der Hoh, N. Leconte, and J. M. Neudorfl, *Organometallics*, *30*, 3880 (2011).
265. S. Enthaler, B. Spilker, G. Erre, K. Junge, M. K. Tse, and M. Beller, *Tetrahedron*, *64*, 3867 (2008).
266. (a) C. Sui-Seng, F. Freutel, A. J. Lough, and R. H. Morris, *Angew. Chem. Int. Ed.*, *47*, 940 (2008); (b) C. Sui-Seng, F. N. Haque, A. Hadzovic, A. M. Puetz, V. Reuss, N. Meyer, A. J. Lough, M. Z. D. Iuliis, and R. H. Morris, *Inorg. Chem.*, *48*, 735 (2009).
267. R. H. Morris, *Chem. Soc. Rev.*, *38*, 2282 (2009).

268. (a) A. Mikhailine, A. J. Lough, and R. H. Morris, *J. Am. Chem. Soc.*, *131*, 1394 (2009); (b) N. Meyer, A. J. Lough, and R. H. Morris, *Chem. Eur. J.*, *15*, 5605 (2009).
269. (a) P. O. Lagaditis, A. J. Lough, and R. H. Morris, *Inorg. Chem.*, *49*, 10057 (2010); (b) P. E. Sues, A. J. Lough, and R. H. Morris, *Organometallics*, *30*, 4418 (2011); (c) D. E. Prokopchuk, J. F. Sonnenberg, N. Meyer, M. Z. D. Iuliis, A. J. Lough, and R. H. Morris, *Organometallics*, *31*, 3056 (2012); (d) D. E. Prokopchuk and R. H. Morris, *Organometallics*, *31*, 7375 (2012).
270. R. Langer, G. Leitus, Y. Ben-David, and D. Milstein, *Angew. Chem. Int. Ed.*, *50*, 2120 (2011).
271. X. Z. Yang, *Inorg. Chem.*, *50*, 12836 (2011).
272. R. Langer, M. A. Iron, L. Konstantinovski, Y. Diskin-Posner, G. Leitus, Y. Ben-David, and D. Milstein, *Chem. Eur. J.*, *18*, 7196 (2012).
273. V. Kandepi, J. M. S. Cardoso, E. Peris, and B. Royo, *Organometallics*, *29*, 2777 (2010).
274. T. Hashimoto, S. Urban, R. Hoshino, Y. Ohki, K. Tatsumi, and F. Glorius, *Organometallics*, *31*, 4474 (2012).
275. A. Tlili, J. Schranck, H. Neumann, and M. Beller, *Chem. Commun.*, *18*, 15935 (2012).
276. S. L. Zhou, S. Fleischer, K. Junge, S. Das, D. Addis, and M. Beller, *Angew. Chem. Int. Ed.*, *49*, 8121 (2010).
277. S. L. Zhou, S. Fleischer, K. Junge, and M. Beller, *Angew. Chem. Int. Ed.*, *50*, 5120 (2011).
278. A. A. Mikhailine, M. I. Maishan, and R. H. Morris, *Org. Lett.*, *14*, 4638 (2012).
279. (a) A. Bechamp, *Annal. Chim.*, *42*, 186 (1854); (b) M. Lauwiner and P. Rys, J. Wissmann, *Appl. Catal., A*, *172*, 141 (1998).
280. M. Lauwiner, R. Roth, and P. Rys, *Appl. Catal., A*, *177*, 9 (1999).
281. D. G. Desai, S. S. Swami, S. K. Dabhade, and M. G. Ghagare, *Synth. Commun.*, *31*, 1249 (2001).
282. (a) S. U. Sonavane, S. K. Mohapatra, R. V. Jayaram, and P. Selvam, *Chem. Lett.*, *32*, 142 (2003); (b) R. M. Deshpande, A. N. Mahajan, M. M. Diwakar, P. S. Ozarde, and R. V. Chaudhari, *J. Org. Chem.*, *69*, 4835 (2004).
283. M. Kumarraja and K. Pitchumani, *Appl. Catal., A*, *265*, 135 (2004).
284. Y. G. Liu, Y. S. Lu, M. Prashad, J. Repic, and T. J. Blacklock, *Adv. Synth. Catal.*, *347*, 217 (2005).
285. Q. X. Shi, R. W. Lu, Z. X. Zhang, and D. F. Zhao, *Chin. Chem. Lett.*, *17*, 441 (2006).
286. A. B. Gamble, J. Garner, C. P. Gordon, S. M. J. O'Conner, and P. A. Keller, *Synth. Commun.*, *37*, 2777 (2007).
287. S. Chandrappa, K. Vinaya, T. Ramakrishnappa, and K. S. Rangappa, *Synlett*, 3019 (2010).
288. L. Pehlivan, E. Metay, S. Laval, W. Dayoub, P. Demonchaux, G. Mignani, and M. Lemaire, *Tetrahedron Lett.*, *51*, 1939 (2010).
289. R. V. Jagadeesh, G. Wienhofer, F. A. Westerhaus, A. E. Surkus, M. M. Pohl, H. Junge, and K. Junge, and M. Beller, *Chem. Commun.*, *47*, 10972 (2011).

290. U. Sharma, P. K. Verma, N. Kumar, V. Kumar, M. Bala, and B. Singh, *Chem. Eur. J.*, *17*, 5903 (2011).
291. G. Wienhofer, I. Sorribes, A. Boddien, F. Westerhaus, K. Junge, H. Junge, R. Llusar, and M. Beller, *J. Am. Chem. Soc.*, *133*, 12875 (2011).
292. D. Cantillo, M. Baghbanzadeh, and C. O. Kappe, *Angew. Chem. Int. Ed.*, *51*, 10190 (2012).
293. Y. Inoue, Y. Sasaki, and H. Hashimoto, *J. Chem. Soc., Chem. Commun.*, 718 (1975).
294. G. O. Evans and C. J. Newell, *Inorg. Chim. Acta*, *31*, L387 (1978).
295. C. C. Tai, T. Chang, B. Roller, and P. G. Jessop, *Inorg. Chem.*, *42*, 7340 (2003).
296. C. Federsel, A. Boddien, R. Jackstell, R. Jennerjahn, P. J. Dyson, R. Scopelliti, G. Laurenczy, and M. Beller, *Angew. Chem. Int. Ed.*, *49*, 9777 (2010).
297. C. Ziebart, C. Federsel, P. Anbarasan, R. Jackstell, W. Baumann, A. Spannenberg, and M. Beller, *J. Am. Chem. Soc.*, *134*, 20701 (2012).
298. R. Langer, Y. Diskin-Posner, G. Leitus, L. J. W. Shimon, Y. Ben-David, and D. Milstein, *Angew. Chem. Int. Ed.*, *50*, 9948 (2011).
299. S. Zhou, K. Junge, D. Addis, S. Das, and M. Beller, *Angew. Chem. Int. Ed.*, *48*, 9507 (2009).
300. B. Merla and N. Risch, *Synthesis*, 1365 (2002).
301. A. Pagnoux-Ozherelyeva, N. Pannetier, M. D. Mbaye, S. Gaillard, and J. L. Renaud, *Angew. Chem. Int. Ed.*, *51*, 4976 (2012).
302. (a) P. O. Lagaditis, A. J. Lough, and R. H. Morris, *J. Am. Chem. Soc.*, *133*, 9662 (2011); (b) A. Mikhailine, A. J. Lough, and R. H. Morris, *J. Am. Chem. Soc.*, *131*, 1394 (2009); (c) C. P. Casey, and H. R. Guan, *J. Am. Chem. Soc.*, *131*, 2499 (2009).
303. J. P. Hopewell, J. E. D. Martins, T. C. Johnson, J. Godfrey, and M. Wills, *Org. Biol. Chem.*, *10*, 134 (2012).
304. M. D. Bhor, M. J. Bhanushali, N. S. Nandurkar, and B. M. Bhanage, *Tetrahedron Lett.*, *49*, 965 (2008).
305. S. Enthaler, *Chemcatchem*, *2*, 1411 (2010).
306. S. Fleischer, S. L. Zhou, K. Junge, and M. Beller, *Chem. Asian J.*, *6*, 2240 (2011).
307. N. U. Kumar, B. S. Reddy, V. P. Reddy, and R. Bandichhor, *Tetrahedron Lett.*, *53*, 4354 (2012).
308. K. Muller, C. Faeh, and F. Diederich, *Science*, *317*, 1881 (2007).
309. A. T. Parsons, T. D. Senecal, and S. L. Buchwald, *Angew. Chem. Int. Ed.*, *51*, 2947 (2012).

A New Paradigm for Photodynamic Therapy Drug Design: Multifunctional, Supramolecular DNA Photomodification Agents Featuring Ru(II)/Os(II) Light Absorbers Coupled to Pt(II) or Rh(III) Bioactive Sites

JESSICA D. KNOLL AND KAREN J. BREWER

Department of Chemistry, Virginia Tech, Blacksburg, VA

CONTENTS

Progress in Inorganic Chemistry, Volume 59, First Edition. Edited by Kenneth D. Karlin.

I. INTRODUCTION

A. Scope and Limitations

This chapter focuses on Ru(II) or Os(II) light absorbers coupled to Pt(II) or Rh(III) bioactive sites. These bioactive sites interact with deoxyribonucleic acid (DNA) through thermal and/or photoinduced binding and photocleavage via oxygen dependent or independent mechanisms designed as a new paradigm for photodynamic therapy drug design. A brief introduction into targeting DNA in cancer therapy research, photodynamic therapy (PDT), cisplatin and related platinum-based chemotherapy drugs, and Rh(III)–polyazine and Rh_2(II,II)–DNA modification agents is given to set the stage for the discussion of supramolecular complexes containing PDT active light absorbing units coupled to Pt(II) or Rh(III) centers. In the context of this work, a supramolecular complex is defined as an assembly of molecular components that individually perform a specific task and are coupled through coordinate covalent bonds to function together for a complex function. Properties and functions can be tuned systematically through component modification to afford a diverse assortment of supramolecules that hold promise in the general field of DNA modification and PDT drug development.

B. Cancer

The quest for effective anti-cancer treatments is an important research focus spanning many fields, thus requiring an interdisciplinary approach. Cancer is defined as a group of diseases in which abnormal, uncontrolled cells form a tumor that interferes with the functioning of nearby tissues and organs (1, 2). Cancer cells lack a mechanism to control replication. Abnormal cells can separate from the tumor and migrate through the body to affect healthy organs through metastasis (2, 3). Current cancer treatments are often plagued with negative side effects and additional risk; surgery is invasive and provides a risk of infection, chemotherapy suffers from poor selectivity–specificity and is often toxic to the system. Radiation can cause adverse reactions and the potential for new tumors. The ideal cancer treatment should exhibit high selectivity and specificity to only cancerous tumors, be noninvasive, and have a minimal impact on healthy cells.

C. Deoxyribonucleic Acid as a Target

Deoxyribonucleic acid contains the genetic information in cells utilized in development and functioning (2, 3). This macromolecule is composed of nucleotides that contain four nucleobases [adenine (A), cytosine (C), guanine (G), and thymine (T)], sugar (deoxyribose), and phosphates that form the backbone. Base pairs are formed between two strands, where A pairs with T and C pairs with G via hydrogen bonding to form a double helical structure, Fig. 1.

Deoxyribonucleic acid is an ideal biological target for cancer treatment to inhibit cell replication, possessing many properties that allow targeting by metal complexes. Metal complexes can bind to DNA at the polyanionic phosphate backbone, the nucleophilic bases, and the major and minor grooves (3–12). Cationic compounds, (e.g., many metal complexes) can interact through ionic binding to the polyanionic DNA backbone. Deoxyribonucleic acid modification agents can also interact through covalent modification via Lewis acidic metal

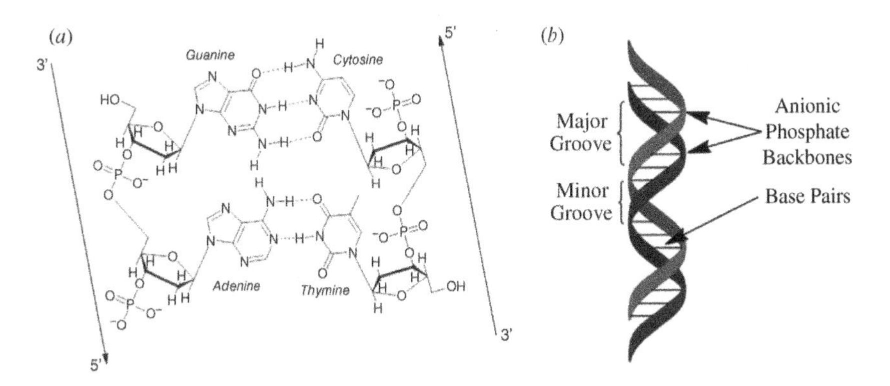

Figure 1. Structural representation of Watson–Crick base pairs (*a*) and a schematic representation of the DNA double helix highlighting the anionic phosphate backbones, base pairs, and major and minor grooves (*b*). (See the color version of this figure in color plates section.)

binding to Lewis basic sites on the nitrogen bases or phosphate backbone. Rigid, planar molecules can associate between the base pairs through intercalation to localize the molecule at DNA. Finally, molecules of an appropriate size and shape can bind in the major or minor groove of the DNA double helix.

II. PHOTODYNAMIC THERAPY

Photodynamic therapy is an expanding area of cancer treatment that was first approved by the Food and Drug Administration (FDA) in 1995 to treat bladder, brain, ovarian, lung, and esophageal cancers (13). This therapy has, at the heart of the design, inactive forms of drugs that can be locally or systematically delivered, and the highly active forms of drugs are generated at the tumor site via excitation with light in the photodynamic therapeutic window, 600–900 nm (wavelengths of light that can pass through the skin). Using light to selectively kill cancer cells is an interesting and promising arena. Many new systems are being explored for application in this arena (13–15).

A. Requirements

Three components are required for PDT via the most commonly employed oxygen-dependent pathway: a light absorber (LA), light of appropriate energy to excite the photosensitizer, and molecular oxygen (16, 17). Individually, these components are nontoxic to both cancerous and healthy cells; however, a combination of these components causes photochemical reactions targeted at a tumor with light, which are toxic to the cancerous cells. Successful candidates for PDT should exhibit low dark toxicity, absorption at low energy (~600–900 nm, which is known as the therapeutic or PDT window), selective accumulation in the tumor, and amphiphilicity to facilitate movement through the cell membrane (17). A general scheme for the excited-state processes of the photosensitizer and its interactions with molecular oxygen ($^3O_2$) is presented in Fig. 2 (18).

The LA in its ^{1}GS is first excited with visible light to populate a ^{1}ES, which can then deactivate back to the ^{1}GS through fluorescence or nonradiative decay or can undergo intersystem crossing (isc) to populate a ^{3}ES. This ^{3}ES can deactivate to the ^{1}GS radiatively through phosphorescence or nonradiatively through thermal deactivation. Photoreactions can occur from the ^{3}ES of the LA: Type I, II, or III. Type I involves electron transfer to H_2O or O_2 to form reactive oxygen species (ROS). Type II involves energy transfer to oxygen to produce highly reactive $^1O_2$. Type III involves direct interaction with a cellular target (18). Reactive oxygen species are known to target mitochondria, the plasma membrane, tumor blood vessels, or DNA (12). Type III is the only mode that acts independent of oxygen. Oxygen independence is important in aggressive, hypoxic tumors (19).

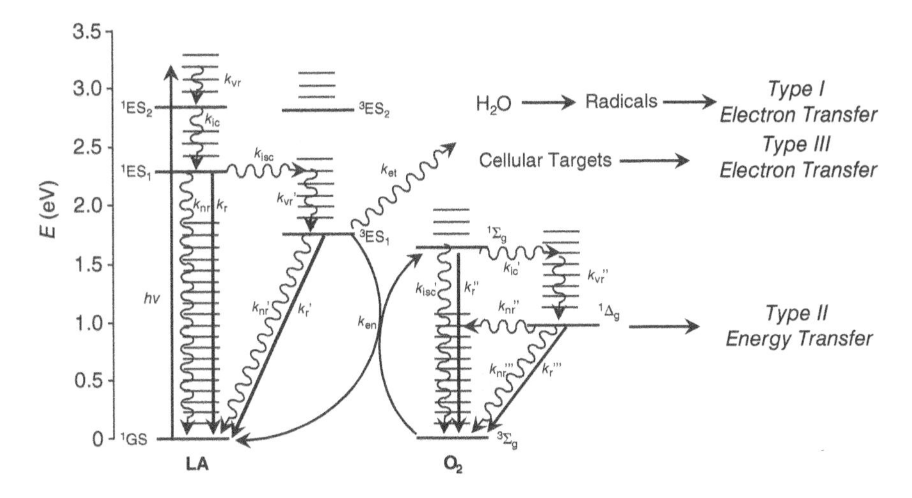

Figure 2. State diagram for a general PDT scheme demonstrating Type I, II, and III photoreactions. Here ^{1}GS = singlet ground state, ^{1}ES = singlet excited state, ^{3}ES = triplet excited state, $^3\Sigma_g$ = molecular oxygen ground state, $^1\Sigma_g$ = singlet oxygen highest energy excited state, $^1\Delta_g$ = singlet oxygen lowest energy excited state, k_r = rate constant for radiative decay, k_{nr} = rate constant for nonradiative decay, k_{vr} = rate constant for vibrational relaxation, k_{isc} = rate constant for intersystem crossing, k_{ic} = rate constant for internal conversion, k_{en} = rate constant for energy transfer, and k_{et} = rate constant for electron transfer.

B. Traditional PDT Agents

The first FDA approved PDT agent, Photofrin® (HpD, hematoporphyrin derivative, Fig. 3), is used to treat esophageal and endobronchial cancers. It consists of a heterogeneous mixture of monomers, dimers, and polymers of the substituted porphyrin (20). Photofrin absorbs at 400 nm due to Soret band

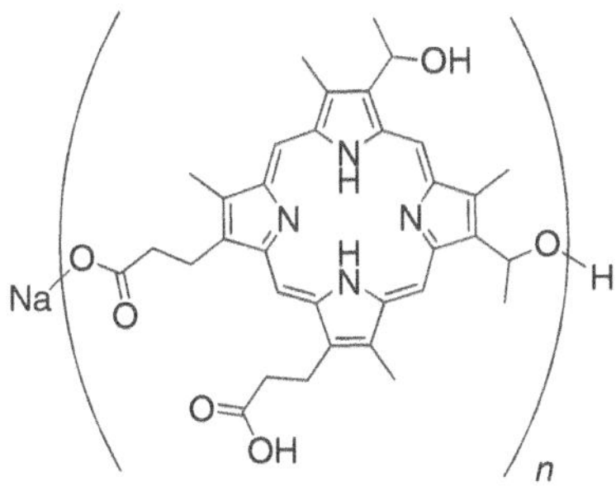

Figure 3. Structural representation of Photofrin (HpD).

absorption, as well as low-energy absorption in the 500–650-nm range due to Q band absorptions. These characteristics are of the extended π system provided by the porphyrin ring. A drawback of this PDT agent arises from non-unity population of the ^{3}ES [quantum yield (Φ) = 0.83], which limits the efficiency of $^1O_2$ generation to only $\Phi = 0.65$ (20) and limits the molar absorptivity (ε) between 620 and 650 nm to 3500 $M^{-1}cm^{-1}$. A benefit to this drug is that the porphyrin mixture tends to accumulate in rapidly growing tissue to aid in selectively targeting tumors. Photofrin is often used in combination therapy with chemotherapy, radiation, or surgery. Phthalocyanines and chlorins are also used as clinical PDT agents (21).

C. Ruthenium(II) Light Absorbers as PDT Agents

Ruthenium(II)–polyazine complexes are of interest in PDT applications due to their rich and efficient light absorbing properties, unity population of relatively long-lived ^{3}MLCT (metal-to-ligand charge transfer) emissive excited states, tunable properties, and efficient $^1O_2$ generation (4, 22–27). The cationic nature of these complexes also provides an affinity for negatively charge biomolecules (i.e., DNA). The prototypical inorganic LA $[Ru(bpy)_3]^{2+}$ (Fig. 4), where bpy = 2,2′-bipyridine, absorbs in the ultraviolet (UV) and visible (vis) regions with high molar absorptivities due to bpy π→π* transitions and Ru(dπ)→bpy(π*) charge-transfer (CT) transitions (27). Excitation of $[Ru(NN)_3]^{2+}$ complexes, where NN = bidentate polyazine ligand, such as bpy, phen (1,10-phenanthroline), or Ph_2phen (4,7-diphenyl-1,10-phenanthroline) (Fig. 4), provide unity population of the lowest lying ^{3}MLCT state.

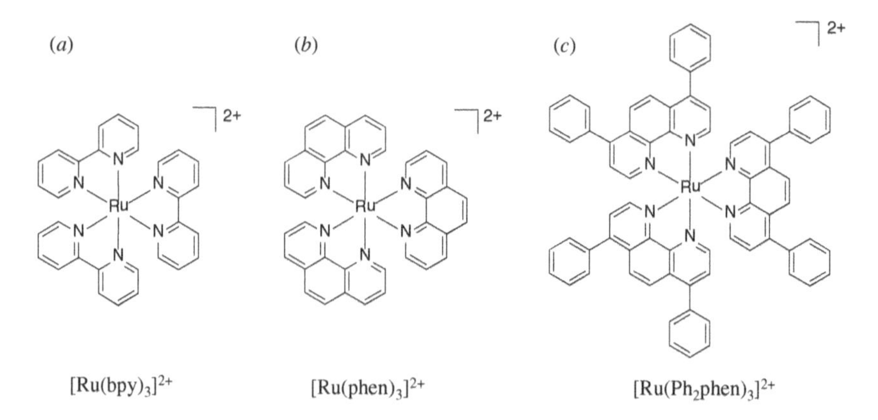

Figure 4. Structural representations of $[Ru(bpy)_3]^{2+}$ (*a*), $[Ru(phen)_3]^{2+}$ (*b*), and $[Ru(Ph_2phen)_3]^{2+}$ (*c*).

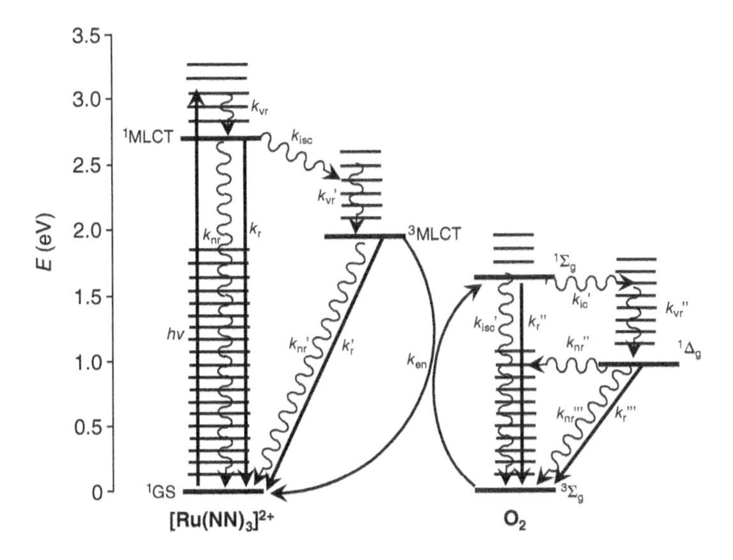

Figure 5. Simplified state diagram for a $^1O_2$ generation using the light absorber $[Ru(NN)_3]^{2+}$, where NN = a bidentate polyazine ligand. ^{1}GS = singlet ground state, ^{1}MLCT = singlet metal-to-ligand charge transfer excited state, ^{3}MLCT = triplet metal-to-ligand charge transfer excited state.

A simplified state diagram depicting the photoprocesses involved in $^1O_2$ generation with a $[Ru(NN)_3]^{2+}$ complex (28) is displayed in Fig. 5. Population of a ^{1}MLCT excited state is followed by intersystem crossing (isc) to populate a ^{3}MLCT excited state with unit efficiency, typical of Ru(II)–polyazine complexes (29). In the case of $[Ru(bpy)_3]^{2+}$ in methanol, this excited state is relatively long lived ($\tau = 0.765\ \mu s$) (28) and will deactivate back to the ^{1}GS by radiative (r) or nonradiative (nr) decay, or in the presence of molecular oxygen, it can undergo efficient energy transfer to form $^1O_2$ and the GS species, $[Ru(bpy)_3]^{2+}$. Replacing the bpy ligand with phen or Ph_2phen (Fig. 4) tunes the light absorbing and excited state properties of the Ru(II) complex. The excited-state lifetimes in methanol of $[Ru(phen)_3]^{2+}$ and $[Ru(Ph_2phen)_3]^{2+}$ are reported as 0.313 and 5.34 μs, respectively (28). The differences in τ are reflected in the quantum yields of $^1O_2$ generation in air saturated methanol, where $\Phi_{^1O_2} = 0.73$, 0.54, and 0.97 for $[Ru(bpy)_3]^{2+}$, $[Ru(phen)_3]^{2+}$, and $[Ru(Ph_2phen)_3]^{2+}$, respectively (18, 25). The $[Ru(bpy)_3]^{2+}$ complex photocleaves pBR322 DNA when excited with visible light ($\lambda^{irr} \geq 450$ nm) through the Type II mechanism of $^1O_2$ generation (24, 30, 31). The need for O_2 to cleave DNA as well as the requirement for the complex to localize near the biomolecule to facilitate efficient $^1O_2$ reactions are limiting factors for the use of Ru(II)–polyazine complexes for PDT. Complexes that couple a Ru(II) based LA to a bioactive site (BAS) that binds to DNA to afford localization of the singlet oxygen generation site at DNA have the potential to overcome these limitations.

III. PLATINUM AND RHODIUM CENTERS AS BIOACTIVE SITES

The study of smaller complexes that exhibit DNA binding is important for understanding the interactions of larger complexes that combine light activated PDT agents with DNA targeting BASs. A variety of Pt and Rh complexes have been reported to interact with DNA, and a few pertinent examples are provided herein.

A. Platinum(II) Based Chemotherapeutics

1. Cisplatin

The reactivity of cisplatin, *cis*-$[PtCl_2(NH_3)_2]$, with biomolecules was discovered in 1965 by Rosenberg et al. (32). While studying the effect of electric fields on the bacteria *Escherichia coli* using a Pt electrode, a compound released from the electrode was found to inhibit replication of the bacteria. The active compound, later discovered to be cisplatin, also inhibits replication of mammalian cells and causes apoptosis in cancerous cells (33, 34). It was approved by the FDA as an anti-cancer drug in 1978 and is used to treat testicular, ovarian, bladder, head, and neck tumors, often in combination therapies (35).

The mechanism of cisplatin activation is well studied. Cisplatin remains intact in the blood stream where the Cl^- concentration is high (~100 m*M*). Upon moving into the cell where the Cl^- concentration is much lower (~4 m*M*), the complex undergoes sequential aquation of the chlorides via ligand exchange with one Cl^- ligand replaced with water to form the mono-aquated *cis*-$[(NH_3)_2PtCl(OH_2)]^{2+}$. The complex then undergoes a second ligand exchange to afford the bis-aquated *cis*-$[(NH_3)_2Pt(OH_2)]^{2+}$ complex (8). Structures of cisplatin and the mono- and bis-aquated species are given in Fig. 6. The aquation of cisplatin provides a positively charged metal complex with higher water solubility capable of ionic and covalent binding to DNA. The mono- and bis-aquated complexes are the active species that

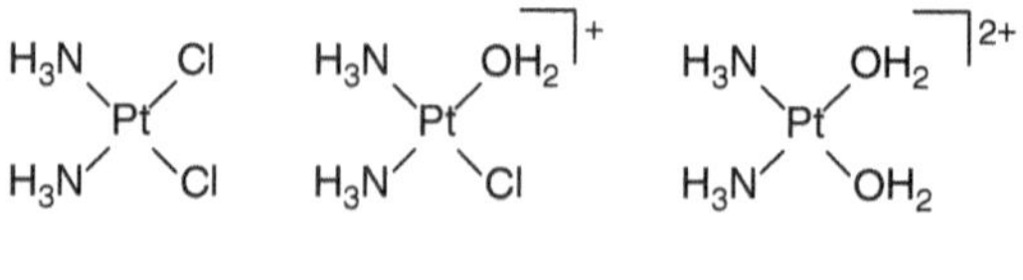

Figure 6. Structural representations of *cis*-$[PtCl_2(NH_3)_2]$ (cisplatin), *cis*-$[(NH_3)_2PtCl(OH_2)]^+$, and *cis*-$[(NH_3)_2Pt(OH_2)_2]^{2+}$.

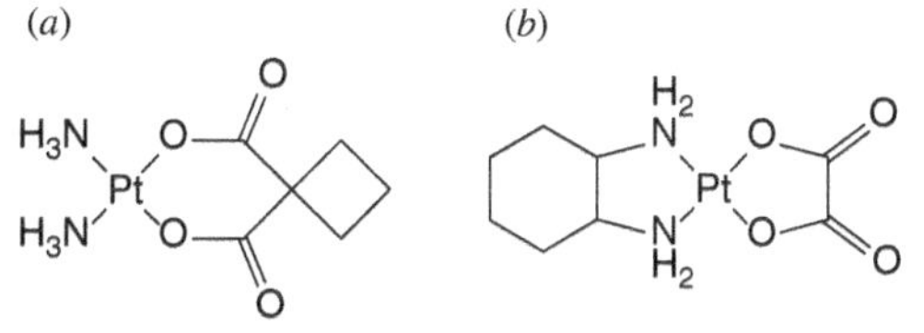

Figure 7. Structural representations of second and third generation Pt(II) drugs carboplatin (*a*) and oxaliplatin (*b*).

permeate the nuclear membrane and bind to DNA through the primary active mode, 1,2-interstrand cross-links. These species primarily access DNA via the major groove, covalently binding the N^7 of the guanine and adenine bases due to the accessibility and high nucleophilicity of these sites (8). This Pt(II) drug causes cancer cell death through an apoptotic mechanism (controlled cell death). The limitations of cisplatin arise from its inability to target proliferating cancer cells, its toxic side effects, and intrinsic or acquired resistance (11, 35, 36).

2. *Second and Third Generation Pt(II) Drugs*

Second and third generation Pt(II) drugs (Fig. 7) were developed in an effort to improve upon the successes and overcome the limitations of cisplatin (37). Carboplatin, FDA approved in 1989, exhibits a lower toxicity to the body than cisplatin while maintaining the efficacy; however, it displays the same problems with resistance. Oxaliplatin, approved by the FDA in 2002, does not exhibit cisplatin cross-resistance. Oxaliplatin has improved water solubility and increased lipophilicity (37). While great strides have been made in the realm of Pt(II) based anti-cancer drugs, effects of systemic delivery and drug resistance has fostered much research aimed at developing more effective anti-cancer agents.

B. Rhodium as a Bioactive Site

1. *Complexes of* $[Rh^{III}(NN)_2X_2]^+$

Octahedral Rh(III) complexes, $[Rh^{III}(NN)_2X_2]^+$ (where NN = bidentate polyazine ligand and X = halide) are known to bind to DNA (38). The complex $[Rh(phen)_2Cl_2]^+$ (Fig. 8) was designed to possess a cis-dihalide moiety similar to cisplatin, providing labile ligands to undergo dissociation followed by covalent binding of the metal center to DNA (39). An advantage of using $[Rh(phen)_2Cl_2]^+$ for DNA modification is its photoactivity. The complex absorbs in the UV and near-UV regions, with transitions centered at 334 nm and 351 nm assigned to intraligand (IL) transitions and a transition centered at 380 nm assigned to a ligand

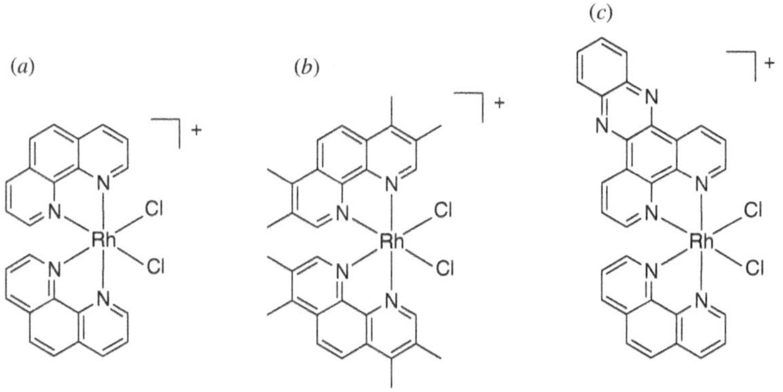

(a) $[Rh(phen)_2Cl_2]^+$ (b) $[Rh(Me_4phen)_2Cl_2]^+$ (c) $[Rh(phen)(dppz)Cl_2]^+$

Figure 8. Structural representations of (a) $[Rh(phen)_2Cl_2]^+$, (b) $[Rh(Me_4phen)_2Cl_2]^+$, and (c) $[Rh(phen)(dppz)Cl_2]^+$.

field (LF) transition (40). Excitation of an aqueous solution of $[Rh(phen)_2Cl_2]^+$ into the LF excited state results in photolabilization of a Cl^-, which is substituted by a water molecule to form the monoaquated $[Rh(phen)_2Cl(OH_2)]^{2+}$ species (39). Additionally, the positive charge gives $[Rh(phen)_2Cl_2]^+$ the ability to associate in the ground state with DNA through electrostatic interactions (38). Photolysis of the complex and calf thymus DNA with $\lambda^{irr} > 330$ nm results in the formation of covalent adducts with moderate efficiency ($\Phi \sim 10^{-3}$) (38). The complex binds preferentially to the purine bases with guanine as the major target, much like in the case of cisplatin (38, 41, 42). The complex does not bind in the dark, and photolysis of the complex to form the bis-aquated $[Rh(phen)_2(OH_2)_2]^{3+}$ that was then added to DNA did not result in binding. Presence of O_2 has little impact on the degree of covalent binding (38). Binding of the complex is initiated by reductive quenching of the excited state, where an electron is transferred from the base to the complex in a ^{3}LF excited state. This process is then followed by Cl^- loss and subsequent binding to DNA (41).

The photochemistry and photobiology of this complex is impacted by variation of the bidentate ligands. Replacing the phen ligands in $[Rh(phen)_2Cl_2]^+$ with Me_4phen ligands (Me_4phen = 3,4,7,8-tetramethyl-1,10-phenanthroline) to give $[Rh(Me_4phen)_2Cl_2]^+$ (Fig. 8) results in a larger hydrophobicity to enable the complex to pass through the cell membrane and undergo more efficient ground-state association with DNA (43). Methylation results in minor shifts in the electronic absorption and emission spectroscopy of this motif; however, photoaquation to form the $[Rh(NN)_2Cl(OH_2)]^{2+}$ complex is greatly enhanced with 347 nm excitation with NN = Me_4phen compared to phen ($\Phi = 0.63$ and 0.03, respectively). This difference is attributed to the greater σ-donating ability of Me_4phen, resulting in enhanced stabilization of the pentacoordinate complex formed upon chloride dissociation when the complex is directly excited into a LF excited state. Significant uptake in human KB cells was observed with $[Rh(Me_4phen)_2Cl_2]^+$, a great improvement on the lack of uptake of $[Rh(phen)_2Cl_2]^+$ and the Me_2phen (4,7-dimethyl-1,10-phenanthroline) derivative. The $[Rh(Me_4phen)_2Cl_2]^+$ [Fig. 8(*b*)] complex exhibits only a moderate phototoxicity in KB and M109 tumor cells with 55 μM metal complex and $\lambda^{irr} = 311$ nm. Increased concentration improves the phototoxicity, but an undesired increase in dark toxicity is also observed.

Substituting one phen ligand in $[Rh(phen)_2Cl_2]^+$ [Fig. 8(*a*)] for a dppz (dipyrido [3,2-*a*-2′,3′-*c*]phenazine) ligand to form the complex $[Rh(phen)(dppz)Cl_2]^+$ [(Fig. 8 (*c*)] results in a species with enhanced ground state association with DNA through intercalation of the dppz ligand and more efficient photoaquation when excited with near UV light (44). Utilization of the dppz ligand results in a ~30 nm red shift in the absorption of the ^{1}LF state compared to the phen analogue, extending the absorption to ~400 nm. At 77 K, $[Rh(phen)(dppz)Cl_2]^+$ [Fig. 8(*c*)] exhibits dual emission at 554 and 710 nm assigned as the dppz $^3\pi{\rightarrow}\pi^*$ and ^{3}LF excited states, respectively. The emission of $[Rh(phen)(dppz)Cl_2]^+$ (Fig. 8*c*) is blue shifted compared to $[Rh(phen)_2Cl_2]^+$ [Fig. 8(*a*)], a result of the increased ligand strength of dppz, which enhances photosolvolysis in aqueous solution. The presence of O_2 decreases photoaquation in $[Rh(phen)(dppz)Cl_2]^+$ [Fig. 8(*c*)], which is attributed to the possibility for electron transfer from the dppz $^3\pi{\rightarrow}\pi^*$ excited state to O_2; however, O_2 does not impact the degree of photobinding or photocleavage with calf thymus DNA. The two complexes exhibit similar degrees of photobinding to DNA, but $[Rh(phen)(dppz)Cl_2]^+$ (*c*) also photocleaves DNA into several smaller fragments. Binding and cleavage of DNA by Rh(III)–polyazine monometallic complexes provide a means toward efficient Type III PDT.

2. *Dirhodium(II,II) Complexes*

Dihodium(II,II) complexes are known to bind to DNA and provide mechanisms for DNA modification. Metal–metal bonded dirhodium(II,II) complexes with carboxylate bridges are an important class of metal-based chemotherapeutic drugs that have been thoroughly studied for DNA photocleavage and anti-tumor activity since 1978 (45). A variety of these complexes have been reported, such as substituted tetra-μ-carboxylato-dirhodium(II), $[Rh_2(O_2CR)_4]$, as well as complexes with one or more polyazine ligand to impart a cationic charge and intercalating ligands to modify interactions with DNA.

The dirhodium(II,II) complex $[Rh_2(O_2CEt)_4]$ and its oxidized form $[Rh_2(O_2CEt)_4]^+$ were reported in 1978 to exhibit anti-tumor activity when tested against L1210 and Ehrlich ascites tumors in mice (45). Complexes of this type

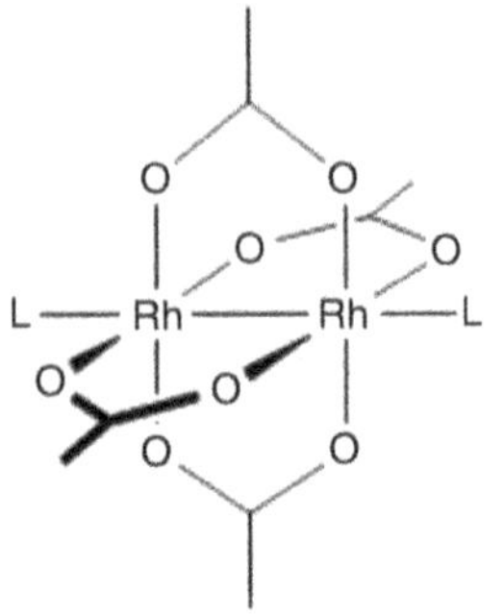

Figure 9. Structural representation of $[Rh_2(O_2CCH_3)_4(L)_2]$, where L = solvent or other Lewis base.

readily coordinate solvent molecules or other species with oxygen, nitrogen, sulfur, or phosphorous donor atoms at the axial positions to form the structure in Fig. 9 (46, 47).

The DNA photocleavage activity of the carboxylate-bridged complex $[Rh_2(\mu\text{-}O_2CCH_3)_4(H_2O)_2]$ was studied with pUC18 DNA (47). The complex has a long-lived excited state (3.5–5.0 μs), making it a suitable photoactivated DNA cleaving agent (48). Interestingly, the excited state of the complex ($\lambda^{irr} \geq 395$ nm) causes no cleavage of the plasmid; however, when excited in the presence of an electron acceptor (e.g., py^+, 3-cyano-1-methylpyridinium tetrafluoroborate, or AQ^{2-}, 1,8-anthraquinone disulfonate) to form the stable Rh_2(III,II) species, photocleavage is observed after 10 min of irradiation (48). This process is more efficient when under N_2 rather than air, owing to the deactivation of the excited state by O_2. Substitution of the labile H_2O axial ligands by py (pyridine) or PPh_3 (triphenylphosphine) results in no photocleavage, suggesting that labile axial ligands are necessary to provide an open coordination site for DNA binding. Matrix-assisted laser desorption/ionization (MALDI) mass spectrometry and enzymatic digestion experiments with the dirhodium complex and small oligonucleotides containing a single AA or GG binding site provide evidence that such complexes are able to bind adenine or guanine bases while the dirhodium core and two coordinated acetate bridges remain intact (49).

The cationic complex *cis*-$[Rh_2(\mu\text{-}O_2CCH_3)_2(phen)_2]Cl_2$ (Fig. 10) was designed to possess the same structural architecture of the previously discussed $[Rh_2(\mu\text{-}O_2CCH_3)_4]$ complex yet more efficiently interact with the anionic DNA double helix by virtue of the positive charges (50). This complex exhibits a binding constant 100 times larger with calf thymus DNA than the neutral $[Rh_2(\mu\text{-}O_2CCH_3)_4]$. Transcription inhibition was reported for *cis*-$[Rh_2(\mu\text{-}O_2CCH_3)_2(phen)_2]Cl_2$ at a lower concentration than the activated cisplatin complex $[Pt(NH_3)_2(H_2O)_2]Cl_2$. This inhibition was found to be a result of the Rh_2(II,II) complex directly interacting with the T7-RNA (ribonucleic acid) polymerase

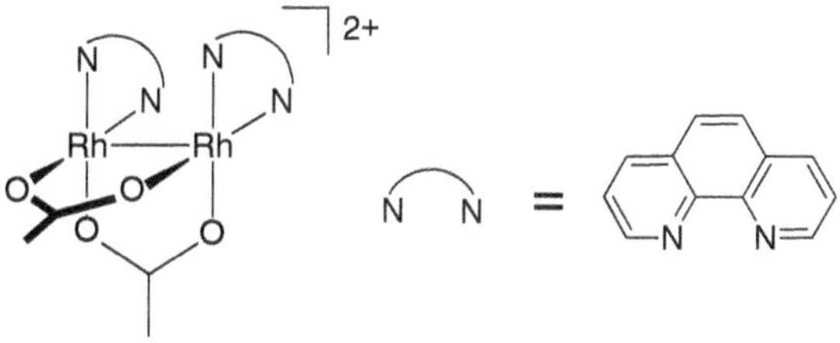

Figure 10. Structural representation of *cis*-$[Rh_2(\mu\text{-}O_2CCH_3)_2(phen)_2]^{2+}$.

enzyme, unlike cisplatin, which was reported to inhibit transcription through interaction with DNA (50).

A dirhodium(II,II) complex was synthesized to increase interactions of the complex with DNA by incorporating a dppz ligand on one Rh(II) center to afford *cis*-$[Rh_2(\mu\text{-}O_2CCH_3)_2(\eta^1\text{-}O_2CCH_3)(MeOH)(dppz)]^+$ (Fig. 11), which is expected to bind with DNA through intercalation or groove binding (51). The complex absorbs UV–vis light in H_2O with a lowest energy transition occurring at 590 nm ($\varepsilon = 350\,M^{-1}\,cm^{-1}$), assigned as a Rh–Rh$\pi^*$→Rh–Rh$\sigma^*$ transition. A more intense Rh–Rhπ^*→Rh–Oσ^* absorption is observed at 428 nm ($\varepsilon = 3180\,M^{-1}cm^{-1}$). An enhanced binding constant (K_b) to calf thymus DNA is observed for *cis*-$[Rh_2(\mu\text{-}O_2CCH_3)_2(\eta^1\text{-}O_2CCH_3)(MeOH)(dppz)]^+$ compared to $[Rh_2(\mu\text{-}O_2CCH_3)_4]$ with $K_b = 4.6\times10^{-2}$ and $1.8\times10^{-5}\,M^{-1}$, respectively. The binding constant, as well as hypochromicity resulting from titration of *cis*-$[Rh_2(\mu\text{-}O_2CCH_3)_2(\eta^1\text{-}O_2CCH_3)\text{-}(MeOH)(dppz)]^+$ with calf thymus DNA, suggests intercalation of the dppz ligand or groove binding as the major binding mode. Irradiation with $\lambda \geq 395$ nm in both the absence and presence of molecular oxygen results in DNA photocleavage without the need for an electron acceptor. The oxygen-independent DNA photocleavage occurs through dppz based DNA targeting and a reactive excited state (51). Coordination of a second dppz ligand to afford *cis*-$[Rh_2(\mu\text{-}O_2CCH_3)_2(dppz)_2]^{2+}$ (Fig. 11) results in a decrease in DNA

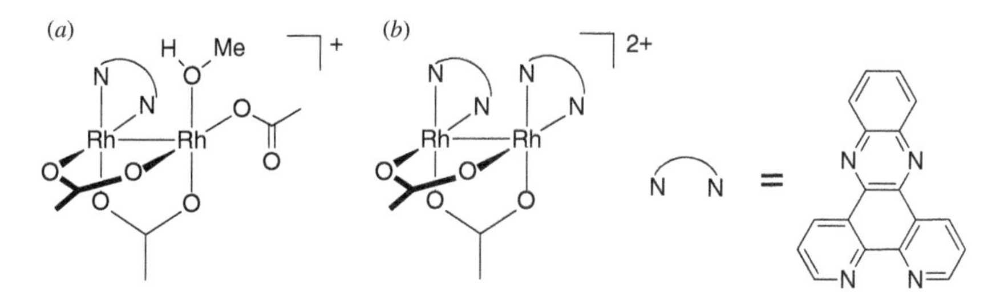

Figure 11. Structural representations of *cis*-$[Rh_2(\mu\text{-}O_2CCH_3)_2(\eta^1\text{-}O_2CCH_3)(MeOH)(dppz)]^+$ (*a*) and *cis*-$[Rh_2(\mu\text{-}O_2CCH_3)_2(dppz)_2]^{2+}$ (*b*).

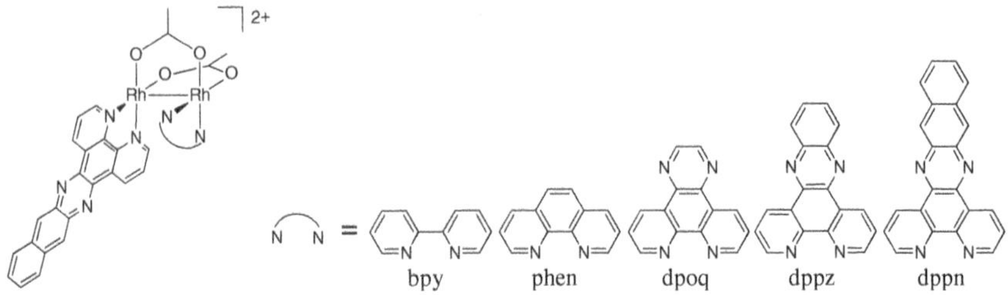

Figure 12. Structural representation of *cis*-$[Rh_2(\mu\text{-}O_2CCH_3)_2(dppn)(L)]^{2+}$, where L = bpy, phen, dpoq, dppz, and dppn.

photocleavage compared to the single dppz complex, as the distance between the two dppz π-systems is not appropriate for intercalation (52). Instead, the complex interacts with DNA through surface aggregation due to electrostatic interactions. However, the lower dark cytotoxicity of *cis*-$[Rh_2(\mu\text{-}O_2CCH_3)_2(dppz)_2]^{2+}$ compared to *cis*-$[Rh_2(\mu\text{-}O_2CCH_3)_2(\eta^1\text{-}O_2CCH_3)(MeOH)(dppz)]^+$ makes it a potential PDT agent (53). Strong binding to DNA by *cis*-$[Rh_2(\mu\text{-}O_2CCH_3)_2(\eta^1\text{-}O_2CCH_3)\text{-}(MeOH)(dppz)]^+$ in the dark increases dark cytotoxicity.

A series of dirhodium(II,II) complexes of the type *cis*-$[Rh_2(\mu\text{-}O_2CCH_3)_2(dppn)(L)]^{2+}$, where dppn = benzo[*i*]dipyrido[3,2-*a*:2′,3′-*h*]quinoxaline and L = bpy, phen, dpoq (dipyrido[3,2-*f*:2′,3′-*h*]quinoxaline), dppz, or dppn, Fig. 12, were synthesized and the photophysical properties, DNA photocleaving activity, and photocytotoxicity were studied (54). The electronic absorption spectra exhibit ligand-centered (LC) transitions between 250 and 430 nm in water. The MLCT transitions, as well as metal-centered (MC) rhodium transitions, are expected to occur but are likely hidden by overlapping LC transitions. Each complex has a nonemissive, long-lived excited state observed by transient absorption spectroscopy that is dppn LC $\pi \rightarrow \pi^*$ in nature, with lifetimes of 2.7, 2.4, 2.4, 3.5, and 4.1 μs in deoxygenated dimethyl sulfoxide (solvent) (DMSO) for L = bpy, phen, dpoq, dppz, and dppn, respectively. In air-saturated methanol, the complexes produce $^1O_2$ with quantum yields of 0.7, 0.9, 0.8, 0.4, and 0.4, respectively.

The ligand choice plays an important role in the activity of these complexes. The complexes with smaller L (bpy and phen) were found to bind with pUC18 DNA through intercalation, and the complexes with larger L (dppz and dppn) were found to aggregate on the DNA through electrostatic interactions (54). When L = dpoq, the complex both intercalates and aggregates. Upon photoexcitation in the presence of pUC18 plasmid DNA, the L = bpy, phen, dpoq, and dppz complexes cleave DNA through an O_2 mediated mechanism, while *cis*-$[Rh_2(\mu\text{-}O_2CCH_3)_2(dppn)_2]^{2+}$ cleaves DNA through a mostly O_2 independent mechanism. The photocytotoxicity of all five complexes are significantly higher toward Hs-27 human skin fibroblasts compared to the dark cytotoxicity (54). The *cis*-$[Rh_2(\mu$-$O_2CCH_3)_2(dppn)(L)]^{2+}$ complexes were also reported to kill human cancer cells HeLa and COLO-316, in which apoptosis is triggered by the complex when L = bpy, phen, dpoq, and dppz, while necrosis is triggered when L = dppn. A correlation was observed between the complex's lipophilicity and its cytotoxicity, demonstrating that increased lipophilicity results in enhanced activity. Dirhodium (II,II) complexes possess interesting reactivity with biomolecules and exhibit many properties that make them potential anti-cancer agents.

IV. SUPRAMOLECULAR COMPLEXES AS DNA PHOTOMODIFICATION AGENTS

A. Supramolecular Chemistry

A supramolecular complex, as defined by Balzani et al. (55) in 1987, is an assembly of components that individually perform a specific role, but act together to execute a complex function. When the function is initiated by light, the assembly is known as a photochemical molecular device (PMD). Supramolecular chemistry is a promising arena for successful PDT agents as it allows for development of multifunctional agents that provide combination therapy in a single molecule. This is accomplished by the coupling of LAs to promote light activated DNA damage with a BAS to prelocalize the LA in close proximity to the DNA double helix through covalent binding (56). The LA with two bidentate polyazine terminal ligands (TLs) or one tridentate TL and one monodentate TL is coupled to the BAS through a bidentate or tridentate polyazine bridging ligand (BL). Tuning of redox and spectroscopic properties, as well as control of the size, stereoisomerization, lipophilicity, and water solubility of the supramolecule, can be achieved by component variation, providing the possibility for a wide array of PMDs to assay the effects of varied structures and orbital energetics. This ability to easily modify structure can be exploited to avoid drug resistance in application of these supramolecules as PDT agents. Discussed in Sections IV.B and IV.C are PMDs that utilize Ru(II)– or Os(II)–polyzine LAs coupled to a Pt(II) or Rh(III) BAS. The modulation of redox, spectroscopic, and photophysical properties are described in the context of the PMDs' ability to bind to or to cleave DNA and application as PDT agents.

B. Light Absorbers with Ru(II)/Os(II) Coupled to Pt(II) Bioactive Sites

A variety of Ru(II),Pt(II) and Os(II),Pt(II) supramolecular complexes, discussed herein, have been reported to undergo reactivity with DNA initiated thermally or photochemically. These complexes were designed to possess a

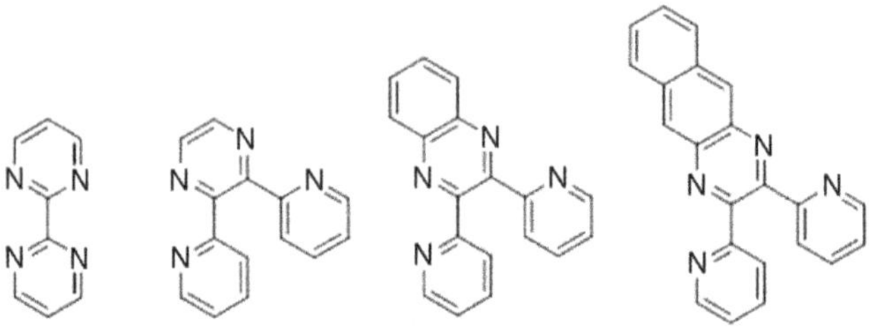

Figure 13. Bidentate polyazine bridging ligands. [bpm = 2,2′-bipyrimidine and dpp = 2,3-bis(2-pyridyl)pyrazine].

cis-$PtCl_2$ BAS similar to cisplatin to afford binding to DNA and achieve efficient cleavage of the DNA backbone through $^1O_2$ production following localization of the drug at the DNA target. This provides for more efficient DNA photocleavage even at reduced oxygen concentrations.

1. *Bimetallic Complexes Ru,Pt and Os,Pt With Bidentate Ligands*

Bidentate polyazine BLs allow for $[Ru(bpy)_3]^{2+}$ type complexes to be coupled into larger supramolecular systems through replacement of a bpy TL with a BL (e.g., dpp, dpq, dpb, or bpm Fig. 13). These BLs possess two bidentate sites for remote metal coordination, providing for construction of complex systems (57). Importantly, $(TL)_2Ru^{II}(BL)$ motifs undergo MLCT excitation to move an excited electron to the BL, toward the coupled BAS.

a. Complex $[(bpy)_2M(dpb)PtCl_2]^{2+}$ [M = Ru(II) or Os(II)]. The first bimetallic complexes featuring Ru(II) or Os(II) LA units bridged to a *cis*-$PtCl_2$ unit that were reported to interact with DNA were of the supramolecular architecture $[(bpy)_2M(dpb)PtCl_2]^{2+}$ (58, 59), where M = Ru(II) or Os(II), Fig. 14. These complexes were designed to feature two structural motifs: the planar benzoquinoxaline portion of the BL that may provide strong groove binding or intercalation to DNA, and the Pt BAS for covalent binding to DNA.

The electrochemistry of $[(bpy)_2M(BL)PtCl_2](PF_6)_2$ features a M(II,III) oxidation occurring at 1.61 and 1.05 V vs Ag/AgCl for M = Ru and Os, respectively, consistent with the higher energy dπ orbitals of Os(II) compared to Ru(II) (58). The first reduction process of each complex is assigned to $dpb^{0/-}$ at −0.11 and −0.22 V for M = Ru and Os, respectively. The redox properties suggest a M(II) based highest occupied molecular orbital (HOMO) and a dpb based lowest unoccupied molecular orbital (LUMO). This result is observed in the electronic absorption spectroscopy with a lowest lying M(dπ)$\rightarrow$dpb(π*) ^{1}MLCT transition centered at

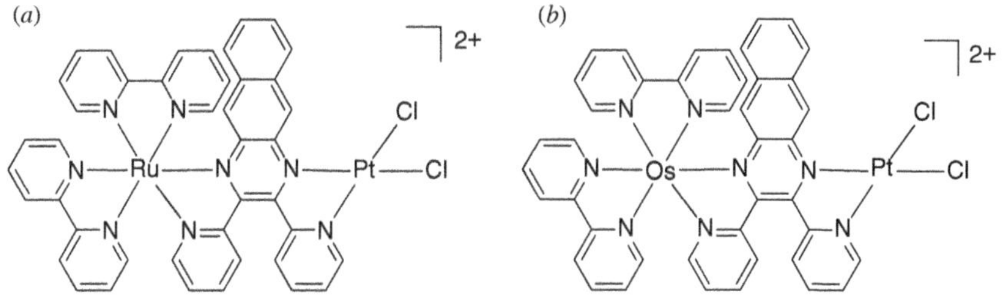

(*a*) $[(bpy)_2Ru(dpb)PtCl_2]^{2+}$

(*b*) $[(bpy)_2Os(dpb)PtCl_2]^{2+}$

Figure 14. Structural representations of $[(bpy)_2Ru(dpb)PtCl_2]^{2+}$ (*a*) and $[(bpy)_2Os(dpb)PtCl_2]^{2+}$ (*b*).

630 (M = Ru) and 638 nm (M = Os). No emission is observed for either of the bimetallic complexes, owing to the low-energy absorption resulting in lower energy emission beyond the PMT detection limit.

Concentration- and time-dependent thermal-binding studies of the $[(bpy)_2M$-$(dpb)PtCl_2]Cl_2$ bimetallic complexes with pBluescript KS+ linear plasmid DNA were performed in comparison to the known DNA binder cisplatin (59). Incubation of each $[(bpy)_2M(dpb)PtCl_2]Cl_2$ complex with DNA at 37 °C results in enhanced migration retardation in gel electrophoresis compared to cisplatin. This result along with comparison to the lack of retarded migration of the plasmid that was incubated with $[(bpy)_2Ru(dpb)]Cl_2$ suggests the primary binding mode of $[(bpy)_2M(dpb)$-$PtCl_2]Cl_2$ is covalent binding through the Pt site. The more dramatically retarded migration using the bimetallic complex compared to cisplatin is a result of the overall decreased negative charge of DNA upon binding of a more positively charged cation and the increased size or change in the DNA three-dimensional structure upon metal complex binding. Denaturing gel electrophoresis was utilized to study whether the $[(bpy)_2M(dpb)PtCl_2]Cl_2$ complexes bind through intrastrand or interstrand cross-linking (60). The primarily intrastrand cross-linker cisplatin and the primarily interstrand cross-linker $[\{t\text{-}PtCl(NH_3)_2\}_2\{\mu\text{-}H_2N(CH_2)_4HN_2\}]Cl_2$ were studied as controls. Results indicate that $[(bpy)_2M(dpb)PtCl_2]Cl_2$ exhibits ~90% intrastrand and 10% interstrand cross-linking. The light absorbing properties in the low-energy visible region make these complexes potential PDT agents; however, no photolysis studies were reported.

b. Complex $[(Ph_2phen)_2Ru(BL)PtCl_2]^{2+}$ (BL = dpp or dpq). Two Ru(II), Pt(II) PMDs were designed to couple the enhanced $^1O_2$ generation of $[Ru(Ph_2phen)_3]^{2+}$ with the DNA binding ability of cisplatin (61). These complexes with the architecture $[(Ph_2phen)_2Ru(BL)PtCl_2]^{2+}$ (BL = dpp or dpq), Fig. 15, possess redox, spectroscopic, and photophysical properties that make them potential PDT (*a*) $[(Ph_2phen)_2Ru(dpp)PtCl_2]^{2+}$ (*b*) $[(Ph_2phen)_2Ru(dpq)PtCl_2]^{2+}$

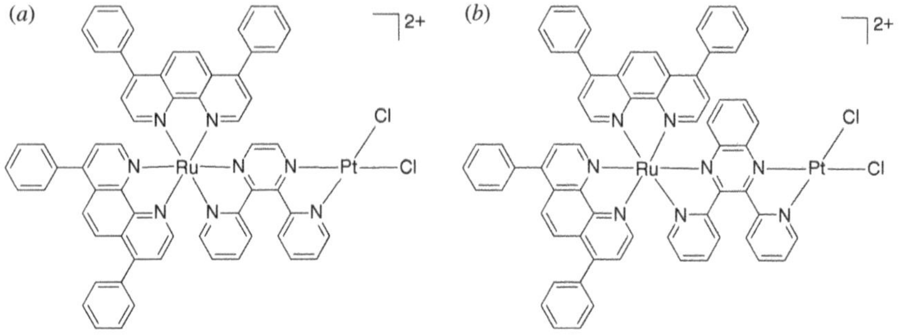

Figure 15. Structural representations of $[(Ph_2phen)_2Ru(dpp)PtCl_2]^{2+}$ (*a*) and $[(Ph_2phen)_2Ru(dpq)PtCl_2]^{2+}$ (*b*).

agents. The electrochemistry predicts a Ru based HOMO, with Ru(II,III) oxidation occurring at 1.60 and 1.61 V vs Ag/AgCl for BL = dpp and dpq, respectively. An irreversible shoulder appears on the oxidation couple at 1.50 V for both complexes, assigned to a Pt(II,IV) oxidation. The LUMO is predicted to be BL based, with $BL^{0/-}$ reductions occurring at −0.48 and −0.21 V for BL = dpp and dpq, respectively. The electronic absorption spectra of $[(Ph_2phen)_2Ru(dpp)PtCl_2](PF_6)_2$ and $[(Ph_2phen)_2Ru(dpq)PtCl_2](PF_6)_2$ in MeCN are given in Fig. 16. The difference in the lowest energy absorption between the two bimetallic complexes is consistent with the varied HOMO–LUMO gaps from electrochemistry, with the Ru($d\pi$)→BL (π^*) CT transition centered at 517 nm ($\varepsilon = 11{,}400\ M^{-1}cm^{-1}$) and 600 nm ($\varepsilon = 9{,}800\ M^{-1}cm^{-1}$) for BL = dpp and dpq, respectively. Utilizing Ph_2phen TL has been

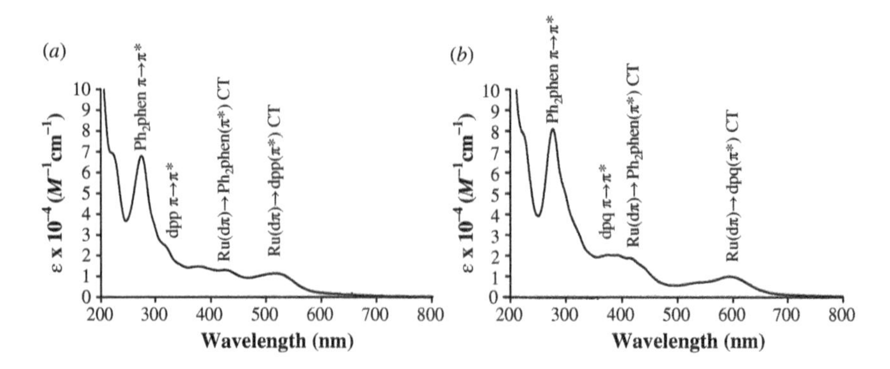

Figure 16. Electronic absorption spectra of (*a*) $[(Ph_2phen)_2Ru(dpp)PtCl_2](PF_6)_2$ and (*b*) $[(Ph_2phen)_2Ru(dpq)PtCl_2](PF_6)_2$ at room termperature (rt) in MeCN.

reported to enhance absorptivity in the UV and visible regions for transitions involving Ph_2phen π^* orbitals. This trend is also observed in the $[(Ph_2phen)_2Ru(BL)PtCl_2](PF_6)_2$ complexes, with the $Ru(d\pi)\rightarrow Ph_2phen(\pi^*)$ CT transition centered at 424 nm with high absorptivity of $18{,}000\,M^{-1}cm^{-1}$, compared to $\varepsilon = 9{,}300$ $M^{-1}cm^{-1}$ for the $Ru(d\pi)\rightarrow bpy(\pi^*)$ CT transition of $[(bpy)_2Ru(dpp)PtCl_2](PF_6)_2$ (62). The Ph_2phen TL selection results in fewer spectral gaps, making the PMD a more efficient visible light absorber.

The emissive nature of $[(Ph_2phen)_2Ru(dpp)PtCl_2](PF_6)_2$ provides a probe into the excited state dynamics and photoreactivity with O_2 (61). At rt in MeCN solution, the complex emits at 740 nm with the quantum yield of emission $(\Phi^{em}) = 4.1 \times 10^{-4}$ and lifetime $(\tau) = 44$ ns when deoxygenated with Ar, and $\Phi^{em} = 3.6 \times 10^{-4}$ and $\tau = 40$ ns in air saturated solution. No emission was observed for $[(Ph_2phen)_2Ru(dpq)PtCl_2](PF_6)_2$, consistent with the stabilized ^3MLCT excited state and expected red-shifted weaker emission. Because Φ^{em} and τ are sensitive to the presence of molecular oxygen, the quantum yield for $^1O_2$ production $(\Phi_{^1O_2})$ was measured in MeOH against the reference $[Ru(Ph_2phen)_3]^{2+}$ $(\Phi_{^1O_2} = 0.97)$. The $\Phi_{^1O_2}$ values for $[(Ph_2phen)_2Ru(BL)PtCl_2](PF_6)_2$ were measured as 0.067 and 0.033 for BL = dpp and dpq, respectively. The lower efficiencies compared to the reference are likely due to the shortened excited state lifetimes of the bimetallic complexes; however, the results suggest that DNA photocleavage may occur through an O_2 facilitated pathway.

The thermal-binding abilities of the $[(Ph_2phen)_2Ru(BL)PtCl_2]Cl_2$ complexes were probed with pUC18 plasmid DNA and assayed by gel electrophoresis (Fig. 17). When incubated at 37 °C in the dark, both complexes retard migration through the gel with base pair/metal complex (BP/MC) ratios as high as 20:1. This result is similar to the activity of cisplatin under the same conditions, suggesting that the bimetallic complexes are efficient DNA binding agents (61).

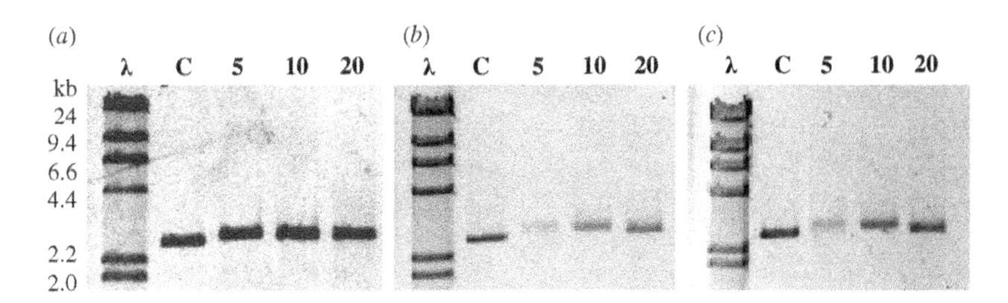

Figure 17. Gel electrophoresis assays of (*a*) cisplatin, (*b*) $[(Ph_2phen)_2Ru(dpp)PtCl_2]Cl_2$, and (*c*) $[(Ph_2phen)_2Ru(dpq)PtCl_2]Cl_2$ incubated at 37 °C in the dark for 1 h with linear plasmid pUC18 DNA. Lane λ is a lambda molecular weight marker, lane *c* is linear pUC18 control, lane 5 is 5:1 BP/MC, lane 10 is 10:1 BP/MC, and lane 20 is 20:1 BP/MC (61). [Reprinted with permission from S.L.H. Higgins, T. A. White, B.S.J. Winkel. *Inorg. Chem.*, *50*, 463 (2010). Copyright © 2010 American Chemical Society.]

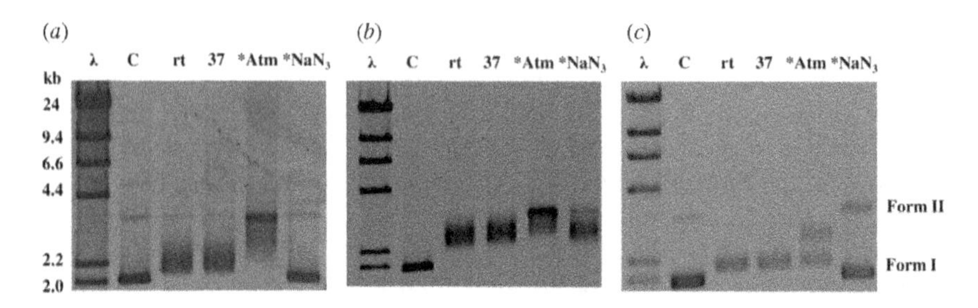

Figure 18. Gel electrophoresis assays of $[(Ph_2phen)_2Ru(dpp)PtCl_2]Cl_2$ (*a* and *b*) and $[(Ph_2phen)_2Ru(dpq)PtCl_2]Cl_2$ (C) with circular plasmid pUC18 DNA. Lane λ is lambda molecular weight marker, lane *c* is circular pUC18 control, lane rt is 20:1 BP/MC incubated at rt for 1 h, lane 37 is 20:1 BP/MC incubated at 37 °C for 1 h, lane *Atm is 20:1 BP/MC photolyzed at 455 nm for 1 h in air saturated solution, lane *NaN_3 is 20:1 BP/MC photolyzed at 455 nm for 1 h in the presence of the $^1O_2$ quencher NaN_3, and lane *FPT is 20:1 BP/MC photolyzed for 1 h at 455 nm following six FPT cycles (61). [Reprinted with permission from S.L.H. Higgins, T.A. White, B.S.J. Winkel, *Inorg. Chem.*, *50*, 463, (2010). Copyright © 2010 American Chemical Society.]

Remarkably, this new motif displays an unprecedented reactivity for Ru,Pt complexes: DNA photobinding. The DNA photocleavage of circular pUC18 DNA imparted by $[(Ph_2phen)_2Ru(BL)PtCl_2]Cl_2$ was assayed by gel electrophoresis studies in which a solution of 20:1 BP/MC was photolyzed at 455 nm. In the presence of O_2, the supercoiled DNA (SC, Form I) was converted to open circular (OC, Form II) for BL = dpp, evidenced by slowed migration (Fig. 18). When BL = dpq, a band corresponding to double-strand cleavage to produce the linear form is observed. The addition of $^1O_2$ quencher, sodium azide (NaN_3), results in most of the plasmid remaining in the SC form, suggesting that $^1O_2$ causes the DNA cleavage converting SC DNA to relaxed, cleaved DNA. This assay is complicated by the potential Pt binding of the Lewis basic azide anion. When the solution is degassed by multiple freeze–pump–thaw (FPT) cycles, the photocleavage is inhibited.

Further studies of the interactions of $[(Ph_2phen)_2Ru(dpp)PtCl_2]Cl_2$ with DNA suggests a new mechanism of DNA modification with the metal complex binding induced by the MLCT excited state. In addition, this photoreactivity uses red light in the therapeutic window (63). Through MLCT excitation, an electron is moved to the μ-dpp, decreasing the Lewis acidity of the Pt center to provide for photolabilization of Cl^- and covalent binding to DNA. Gel electrophoresis studies using $\lambda^{irr} = 455$ nm and ≥ 590 nm to photolyze 5:1 BP/MC samples at time intervals up to 60 min along with DNA selective precipitation studies show that metal complex binding to DNA occurs rapidly when excited at 455 nm (Fig. 19). When excited with $\lambda \geq 590$ nm, the process is slower due to decreased absorptivity in this spectral window. The red light photobinding of $[(Ph_2phen)_2Ru(BL)PtCl_2]Cl_2$ to DNA provides a mode of targeting this complex to DNA for applications in PDT. The

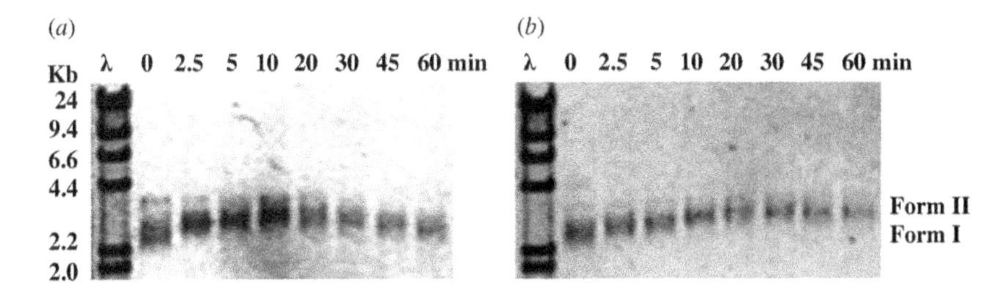

Figure 19. Gel electrophoresis assays of $[(Ph_2phen)_2Ru(dpp)PtCl_2]Cl_2$ and circular pUC18 DNA photolyzed at *(a)* 455 nm or *(b)* ≥ 590 nm. Lane λ corresponds to lambda molecular weight marker, and lanes 0, 2.5, 5, 10, 20, 30, 45, and 60 correspond to 5:1 BP/MC solutions photolyzed for 0, 2.5, 5, 10, 20, 30, 45, and 60 min, respectively. (63). [Reproduced by permission of the Royal Society of Chemistry.]

$[(Ph_2phen)_2Ru(dpp)PtCl_2]Cl_2$ complex exhibits rapid photobinding to DNA, and the use of lower energy light in the therapeutic window is important in PDT drug development.

c. Complex $[(bpy)_2Ru\{bpy-[CONH-(CH_2)_3NH_2]_2\}PtCl_2]^{2+}$. The Ru(II), Pt(II) bimetallic complex reported by Sakai, et al. (64, 65) was designed to couple a $[Ru(bpy)_3]^{2+}$ type LA to a *cis*-$PtCl_2$ BAS through aliphatic amines to afford the complex $[(bpy)_2Ru\{bpy-[CONH-(CH_2)_3NH_2]_2\}PtCl_2]^{2+}$, Fig. 20. The use of aliphatic amines rather than α-diimines more closely mimics the coordination environment in cisplatin. Additionally, use of N heterocycles typically has a major impact on the spectroscopic properties of the LA once Pt is coordinated, so aliphatic amines were used to maintain the photophysical properties of $[Ru(bpy)_3]^{2+}$. The electronic absorption spectroscopy of the monometallic $[(bpy)_2Ru\{bpy-[CONH-(CH_2)_3NH_2]_2\}]^{2+}$ and the platinated bimetallic are quite similar as a result of the intervening aliphatic linkers between the chromophore and the Pt unit (64). The enhanced emission at ~660 nm and a longer lived excited state observed upon Pt(II) coordination (244 ns for the monometallic and 518 ns for the bimetallic) is a result of the formation of a rigid metallocycle that decreases the nonradiative decay processes

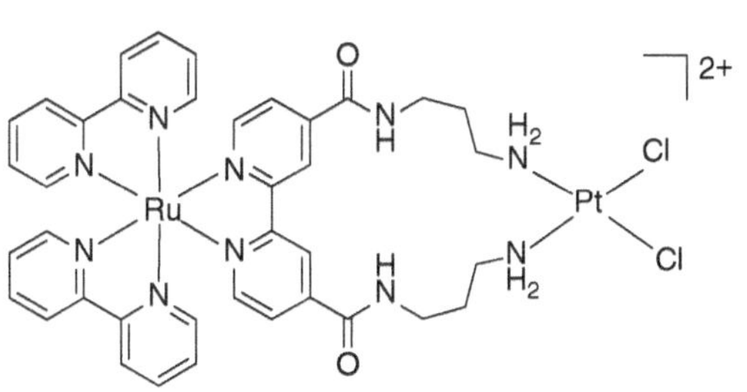

Figure 20. Structural representation of $[(bpy)_2Ru\{bpy-[CONH-(CH_2)_3NH_2]_2\}PtCl_2]^{2+}$.

that are promoted when the aliphatic ligands are flexible in the monometallic. The addition of DNA or 5′-GMP (guanosine-5′-phosphate disodium salt) also enhances the emission of the complex, which is suggested to result in less structural flexibility and provide the previously reported "light switch effect" (6).

This Ru,Pt bimetallic motif is an active DNA photocleavage agent. When the complex is photolyzed at 470 nm in the presence of pBR322 plasmid DNA and air, photocleavage to form the OC plasmid, as well as a small amount of double nicked linear form, is observed (64). Dark 37 °C incubation at 5:1, 10:1, 20:1, and 50:1 BP/MC ratios without photolysis resulted in no cleavage, indicating that light is necessary. Photoinduced electron transfer to cleave DNA is suggested as a mechanism; however, no O_2 free studies were reported to support this process.

d. Tetrametallic Ru,Pt Complex. The tetrametallic supramolecule, $[\{(bpy)_2Ru(dpp)\}_2Ru(dpp)PtCl_2]^{6+}$, Fig. 21, features a three-ruthenium centered LA coupled to a *cis*-$PtCl_2$ unit through a dpp BL on the central Ru (66). The movement of the Ru LAs, which possess the lowest lying ^{3}MLCT state, away from the Pt BAS provides for much longer lived excited states to enhance photoreactivity. The two peripheral Ru centers oxidize at 1.58 V vs Ag/AgCl, prior to the central Ru oxidation that is expected to occur outside of the electrochemical window due to coordination of three μ-BLs. The first reduction is assigned to the dpp, which bridges the central Ru and Pt (−0.40 V vs Ag/AgCl) due to the common observation that Pt(II) coordination stabilizes a BL acceptor orbital to a greater extent than Ru(II) coordination (67). The two dpp ligands bridging the Ru(II) centers are reduced at −0.60 and −0.71 V.

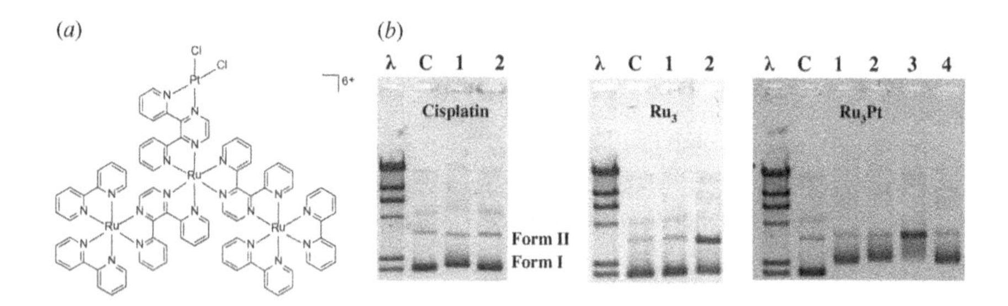

Figure 21. (*a*) Structural representation of $[\{(bpy)_2Ru(dpp)\}_2Ru(dpp)PtCl_2]^{6+}$. (*b*) Gel electrophoresis assays of binding and photocleavage studies with cisplatin, $[\{(bpy)_2Ru(dpp)\}_2Ru(dpp)]^{6+}$ (**Ru_3**), and $[\{(bpy)_2Ru(dpp)\}_2Ru(dpp)PtCl_2]^{6+}$ (**Ru_3Pt**). In all cases, lane λ is lambda molecular weight marker and lane C is pUC18 DNA control with no metal complex. In the **Cisplatin** study, incubation at 37 °C in the dark was carried out for 1 h, and lane 1 is 5:1 BP/MC and lane 2 is 20:1 BP/MC. For **Ru_3**, lane 1 is 20:1 BP/MC for 1 h at 37 °C and lane 2 is 20:1 BP/MC photolyzed with 450–1000 nm light for 1 h in atmospheric conditions. For **Ru_3Pt**, lane 1 is 20:1 BP/MC in the dark at rt, lane 2 is 20:1 BP/MC incubated in the dark at 37 °C for 1 h, lane 3 is 20:1 BP/MC photolyzed with 450–1000-nm light for 1 h in atmospheric conditions, and lane 4 is 20:1 BP/MC photolyzed with 450–1000 nm light for 1 h in deoxygenated conditions. [Adapted from (66).]

The complex exhibits rich light absorbing properties due to the many overlapping Ru(dπ)→TL(π^*) CT and Ru(dπ)→BL(π^*) CT transitions in the visible region. The Ru→bpy CT transitions have a maximum absorption at 416 nm ($\varepsilon = 25{,}000\ M^{-1}cm^{-1}$) and the Ru→dpp CT transitions have a maximum at 542 nm ($\varepsilon = 35{,}000\ M^{-1}cm^{-1}$). Upon visible light excitation, the complex emits at 750 nm from a peripheral Ru→dpp ^{3}MLCT excited state with $\Phi^{em} = 3.2 \times 10^{-4}$ and $\tau = 100$ ns.

Deoxyribonucleic acid binding and photocleavage experiments using $[\{(bpy)_2Ru(dpp)\}_2Ru(dpp)PtCl_2]^{6+}$ were performed in comparison to the known DNA binder cisplatin, as well as to the trimetallic $[\{(bpy)_2Ru(dpp)\}_2Ru(dpp)]^{6+}$, to observe the impact of the *cis*-$PtCl_2$ unit, Fig. 21. The trimetallic complex $[\{(bpy)_2Ru(dpp)\}_2Ru(dpp)]^{6+}$ does not exhibit thermal binding, consistent with the lack of BAS (66). In the presence of O_2 and visible light, a significant band corresponding to cleaved DNA is observed, indicating that the complex is able to produce $^1O_2$; however, supercoiled (SC, Form I) DNA persists, potentially due to the inability of the LA to remain in close enough proximity for efficient photocleavage. Thermal binding is observed for the tetrametallic complex $[\{(bpy)_2Ru(dpp)\}_2Ru(dpp)PtCl_2]^{6+}$, and the enhanced retardation of migration compared to cisplatin is consistent with the larger size and higher positive charge. Exposure to O_2 upon photolysis results in efficient conversion of SC (Form I) DNA to open circular (OC, Form II), a result of the Pt BAS covalently bound to prelocalize the LA close to the DNA for efficient photocleavage. In the absence of molecular oxygen, photocleavage is not observed, suggesting that an O_2 dependent mechanism is operative. This multifunctional DNA modification agent is the first of its architecture to be reported in this arena.

2. *Bimetallic and Trimetallic Complexes of Ru,Pt With Tridentate Ligands*

Tridentate TLs and BLs are of interest in supramolecular complexes for DNA modification as they limit the number of optical and/or geometric isomers, providing the potential for additional methods of analysis. Isolating one isomer of a complex can also give more detailed information about how the complex interacts with DNA. The commonly employed $(TL)_2Ru^{II}(dpp)$ bidentate bridged LA has Λ and Δ stereoisomers, as well as new isomers often introduced by the attachment of the AB chelating dpp to the BAS.

a. Complex $[(tpy)Ru(PEt_2Ph)(BL)PtCl_2]^{2+}$ (BL = dpp or bpm). Two Ru(II),Pt(II) bimetallic complexes featuring a tridentate tpy (2,2′:6′,2″-terpyridine) TL, a monodentate PEt_2Ph TL, and a bidentate dpp or bpm BL, Fig. 22, were designed to limit the number of stereoisomers with the tpy and bpm ligands and provide a ^{31}P NMR tag for structure and reactivity studies by incorporation of the phosphine ligand (68, 69).

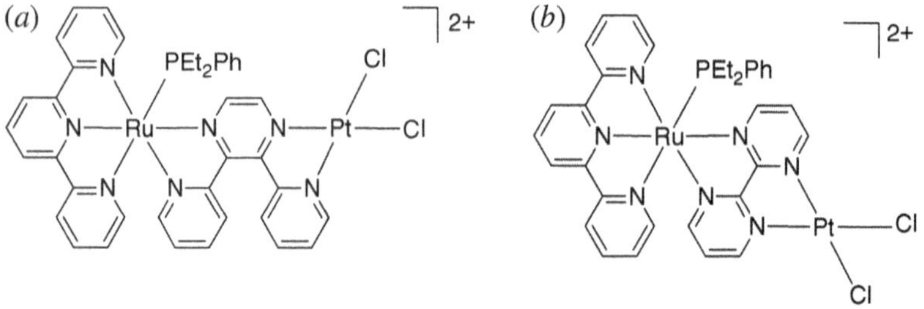

$[(tpy)Ru(PEt_2Ph)(dpp)PtCl_2]^{2+}$

$[(tpy)Ru(PEt_2Ph)(bpm)PtCl_2]^{2+}$

Figure 22. Structural representations of (*a*) $[(tpy)Ru(PEt_2Ph)(dpp)PtCl_2]^{2+}$ and (*b*) $[(tpy)Ru(PEt_2Ph)(bpm)PtCl_2]^{2+}$.

The redox and spectroscopic properties of this motif provides information into the orbital energetics. The electrochemistry of the $[(tpy)Ru(PEt_2Ph)(BL)PtCl_2](PF_6)_2$ complexes in MeCN exhibits a Ru(II,III) first oxidation (relating to the HOMO) and $BL^{0/-}$ first reduction (relating to the LUMO) (69). The reversible Ru based oxidations occur at 1.63 and 1.55 V vs Ag/AgCl for BL = bpm and dpp, respectively. An irreversible oxidation following the Ru(II,III) couple is assigned as a Pt(II,IV) couple at 1.82 and 1.70 V for BL = bpm and dpp, respectively. The first reduction potential is less negative when bpm is selected instead of dpp (−0.34 vs −0.50 V), owing to the lower energy π^* orbitals of bpm relative to dpp. The smaller HOMO–LUMO gap of $[(tpy)Ru(PEt_2Ph)(bpm)PtCl_2](PF_6)_2$ relative to the dpp analogue is observed in the electronic absorption spectrum where the $Ru(d\pi)\rightarrow BL(\pi^*)$ CT transition is centered at 560 nm with BL = bpm and 506 nm with BL = dpp. While no detectable emission is observed for $[(tpy)Ru(PEt_2Ph)(bpm)PtCl_2](PF_6)_2$, the dpp analogue emits at 750 nm with $\Phi^{em} = 1.8 \times 10^{-4}$ and $\tau = 56$ ns.

Photocleavage studies were not reported for the $[(tpy)Ru(PEt_2Ph)(BL)PtCl_2]^{2+}$ motif; however, thermal-binding studies were reported and compared to cisplatin (68). A concentration-dependent study was performed in which solutions of metal complex and pBluescript DNA (5:1, 10:1, 20:1, 100:1, 200:1, and 300:1 BP/MC ratios) were incubated at 37 °C in the dark for 4 h (Fig. 23). Binding with cisplatin becomes apparent at a lowest relative metal complex concentration of 20:1 BP/MC, while the $[(tpy)Ru(PEt_2Ph)(BL)PtCl_2](PF_6)_2$ complexes appear to be more avid DNA binders with binding apparent at a low concentration of 100:1 BP/MC. Slightly enhanced thermal binding is observed with BL = dpp compared to the bpm analogue. The efficient binding through the Pt(II) BAS to bring the LA in close proximity to the DNA target makes these complexes interesting potential PDT agents.

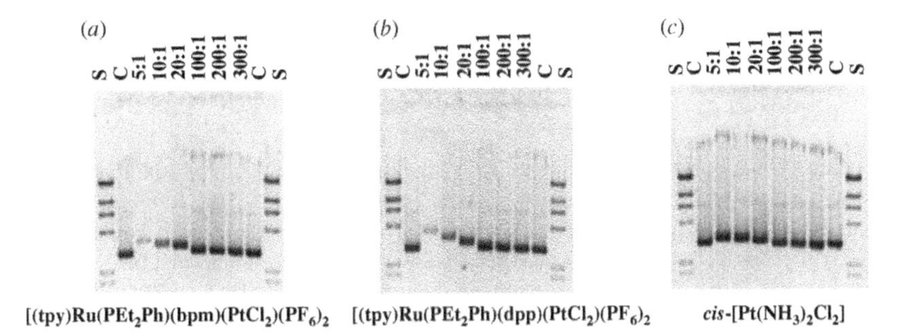

Figure 23. Gel electrophoresis images of DNA–metal binding studies for (*a*) $[(tpy)Ru(PEt_2Ph)(bpm)PtCl_2](PF_6)_2$, (*b*) $[(tpy)Ru(PEt_2Ph)(dpp)PtCl_2](PF_6)_2$, and (*c*) cisplatin incubated at 37 °C in the dark for 4 h. Lane S corresponds to a lambda molecular weight marker, lane C corresponds to DNA control with no metal complex, and the 5:1, 10:1, 20:1, 100:1, 200:1, and 300:1 lanes corresponds to the BP/MC ratio. [From (68) with permission of Elsevier.]

b. Complex $[(tpy)RuCl(BL)PtCl_2]^{2+}$ (BL = dpp, dpq, or dpb). A series of tpy terminated Ru(II),Pt(II) bimetallic complexes were designed to study the effect of BL on bioactivity. The $[(tpy)RuCl(BL)PtCl_2]^+$ complexes with BL = dpp, dpq, and dpb (70) are shown below in Fig. 24. These complexes have been studied to demonstrate DNA biding at the Pt BAS.

The BL identity impacts the orbital energetics in this motif, and the electrochemistry gives information about these effects. Typical of Ru(II)–polyazine complexes, the Ru(II,III) oxidation occurs at 1.14, 1.10, and 1.12 V vs Ag/AgCl for BL = dpp, dpq, and dpb, respectively. These potentials are less positive than the previously discussed complexes due to the coordination of a σ-donating chloride ligand and consequential destabilization of the Ru(dπ) orbitals. The LUMO energy is greatly affected by BL selection, with the $BL^{0/-}$ reduction occurring at −0.50, −0.32, and −0.20 V for BL = dpp, dpq, and dpb,

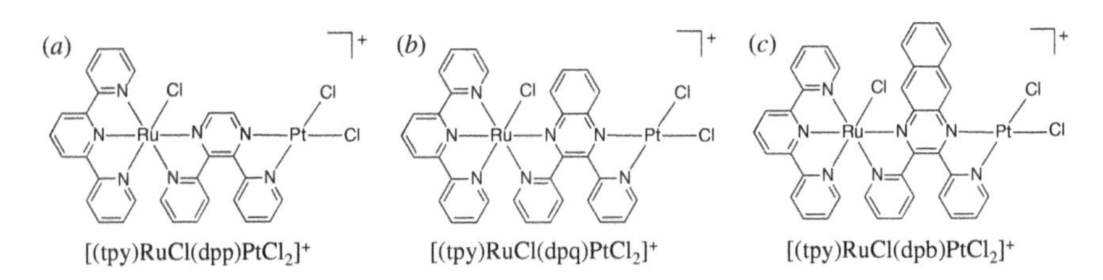

Figure 24. Structural representations of (*a*) $[(tpy)RuCl(dpp)PtCl_2]^+$, (*b*) $[(tpy)RuCl(dpq)PtCl_2]^+$, and (*c*) $[(tpy)RuCl(dpb)PtCl_2]^+$.

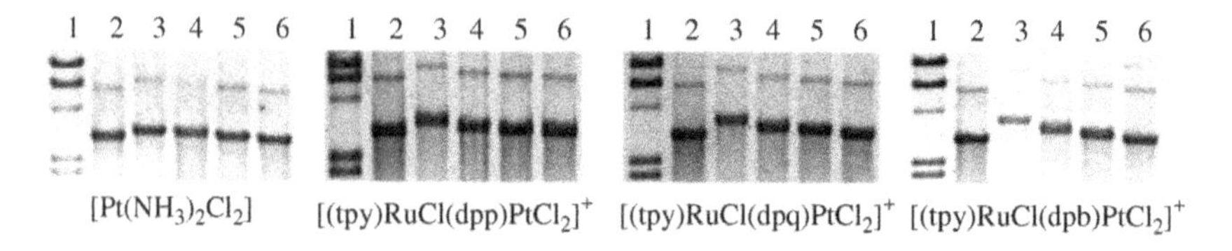

Figure 25. Gel electrophoresis assay of DNA binding studies with cisplatin and $[(tpy)RuCl(BL)PtCl_2]^+$ (BL = dpp, dpq, and dpb) incubated in the presence of linearized pBluescript DNA at 37 °C in the dark for 4 h. Lane 1 is a lambda molecular weight marker, lane 2 is a DNA control with no metal complex, lane 3 is a 5:1 BP/MC ratio, lane 4 is a 10:1 BP/MC ratio, lane 5 is a 20:1 BP/MC ratio, and lane 6 is a 100:1 BP/MC ratio. Figure from (70). [Reprinted with permission from R.J Williams, H.N. Toft, and K.J. Brewer. *Inorg. Chem.*, *42*, 4394 (2003). Copyright © 2003 American Chemical Society.]

respectively. The Ru(dπ)→BL(π*) CT transition observed in electronic absorption spectroscopy of the three complexes in MeCN correlates to the HOMO–LUMO gaps predicted by electrochemistry, with the transitions centered at 544 nm (BL = dpp), 632 nm (BL = dpq), and 682 nm (BL = dpb). Utilization of stabilized BL and σ-donating chloride coordinated to a Ru(II) center provides for low-energy absorption extending into the therapeutic window.

This structural motif, $[(tpy)RuCl(BL)PtCl_2]^+$, was reported to thermally bind to the DNA target. Thermal-binding studies of the complexes compared to cisplatin were performed by incubating solutions of the metal complex and linearized pBluescript DNA (5:1, 10:1, 20:1, and 100:1 BP/MC ratios) at 37 °C in the dark for 4 h, Fig. 25. Enhanced slowing of migration through the gel of the metal complex modified DNA was observed for the three Ru,Pt complexes compared to cisplatin. This result suggests more avid binding or a more substantial change in the three-dimensional structure upon metal complex binding for the Ru,Pt supramolecules.

The chromophoric properties of the $[(tpy)RuCl(BL)PtCl_2]^+$ complexes were exploited to spectrophotometrically probe the degree of metal complex binding to calf thymus DNA. A 10:1 BP/MC solution was incubated at 37 °C in the dark, and at time points during incubation, an aliquot was removed in which the DNA was precipitated and the concentration of remaining unbound metal complex was determined by the absorption of the solution. The amount of complex bound to the DNA decreases with increasing BL size (80% for BL = dpp, 70% for BL = dpq, and 45% for BL = dpb), suggesting that the steric bulk imparted by the BL requires larger binding sites for the BAS to bind to DNA.

The complex $[(tpy)RuCl(dpp)PtCl_2]^+$ was also reported to inhibit *in vivo* growth of *E. coli* as a direct result of *cis*-$PtCl_2$ coordination, as the $[(tpy)RuCl(dpp)]^+$ synthon exhibits no antibacterial properties (71). The bimetallic complex $[(tpy)RuCl(dpp)PtCl_2]^+$ more avidly binds to pBluescript DNA than to cisplatin. Concentration studies comparing $[(tpy)RuCl(dpp)PtCl_2]^+$ and cisplatin showed complete *E. coli* growth inhibition upon treatment with 0.2, 0.4, and 0.6 m*M* (*a*) $[(tpy)RuCl(dpp)PtCl_2]^+$ (*b*) $[(MePhtpy)RuCl(dpp)PtCl_2]^+$ (*c*) $[(t\text{-}Bu_3tpy)RuCl(dpp)PtCl_2]^+$

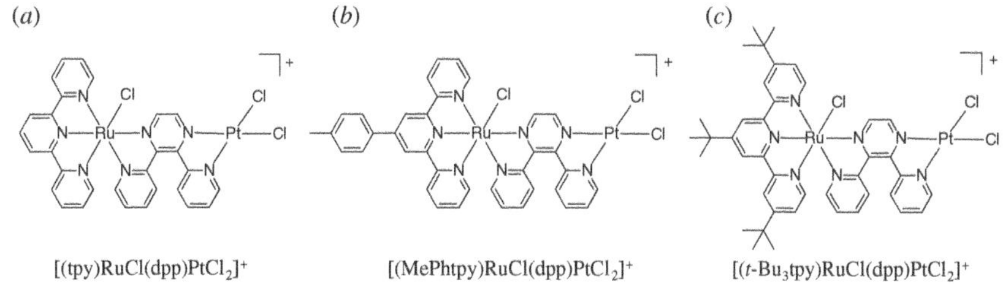

Figure 26. Structural representations of (*a*) $[(tpy)RuCl(dpp)PtCl_2]^+$, (*b*) [(MePhtpy)RuCl(dpp)-$PtCl_2]^+$, (*c*) and $[(t\text{-}Bu_3tpy)RuCl(dpp)PtCl_2]^+$.

cisplatin, while 0.2 m*M* $[(tpy)Ru(dpp)PtCl_2]^+$ had no effect on growth and at 0.4 and 0.6 m*M* $[(tpy)RuCl(dpp)PtCl_2]^+$, growth inhibition was observed, but not to the same extent as cisplatin. The differences in bacterial growth inhibition may be due to different rates of cellular uptake of the metal complexes. Additionally, the bimetallic complex may form an adduct with DNA that less effectively inhibits replication relative to cisplatin. This *in vivo* study is unusual for supramolecular complexes as typically only DNA gel shift assays are performed.

c. Complex $[(TL)RuCl(dpp)PtCl_2]^+$ (TL = tpy, MePhtpy, *t*-Bu$_3$tpy). Two analogues of the previously discussed $[(tpy)RuCl(dpp)PtCl_2]^+$ complex were synthesized by using substituted tpy ligands to yield the complexes [(MePh-tphy)RuCl(dpp)$PtCl_2]^+$ and $[(t\text{-}Bu_3tpy)RuCl(dpp)PtCl_2]^+$, where MePhtpy = 4′-(4-methylphenyl)-2,2′:6′,2″-terpyridine and *t*-Bu$_3$tpy = 4,4′,4″-tri-*tert*-butyl-2,2′:6′,2″-terpyridine, Fig. 26 (72). The purpose of TL variation was to vary the partition coefficient, log *P*, a physicochemical property that estimates the ability for a drug to permeate the cell membrance due to hydrophilicity or lipophilicity.

The electrochemistry of these complexes predicts that the energy of the Ru(II) based HOMO is finely tuned by the TL selection. The Ru(II,III) couples occur at 1.10, 1.10, and 1.01 V vs Ag/AgCl for TL = tpy, MePhtpy, and *t*-Bu$_3$tpy, respectively. The LUMO energy is minimally impacted by TL identity as well, with $dpp^{0/-}$ occurring at −0.50, −0.55, and −0.59 V for TL = tpy, MePhtpy, and *t*-Bu$_3$tpy, respectively. The three complexes exhibit a lowest energy ^{1}MLCT absorption, Ru(dπ)→dpp(π*) in nature, at ~545 nm. The complexes are efficient light absorbers throughout the visible region, making them candidates for DNA photocleavage agents.

The presence of the *cis*-$PtCl_2$ unit suggests that these complexes may exhibit DNA thermal binding similar to that of cisplatin. Incubation of the metal complex with linear pUC18 DNA at 37 °C in the dark with varied BP/MC ratios (5:1, 10:1, and 20:1) results in the TL = tpy and MePhtpy complexes binding to DNA in a similar fashion to cisplatin. However, a smaller degree of binding is observed for

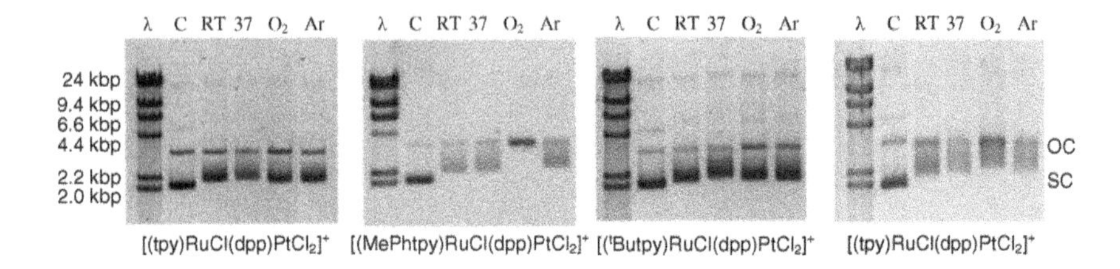

Figure 27. Gel electrophoresis assays of DNA photocleavage studies with $[(TL)RuCl(dpp)PtCl_2]^+$ (TL = tpy, MePhtpy, and t-Bu_3tpy) and circular pUC18 DNA. Lane λ is a lambda molecular weight marker, lane C is a DNA control with no metal complex, lane RT is a 20:1 BP/MC ratio incubated at RT in the dark for 2 h, lane 37 is 20:1 BP/MC ratio incubated at 37 °C in the dark for 2 h, lane O_2 is a 20:1 BP/MC ratio photolyzed with 450–1000-nm light for 2 h in air saturated solution, and lane Ar is a 20:1 BP/MC ratio photolyzed with 450–1000-nm light for 2 h in deoxygenated solution. The gel on the right is $[(tpy)RuCl(dpp)PtCl_2]^+$ and pUC18 incubated/photolyzed for 4 h. Figure from (72). [Reprinted with permission from A. Jain, J. Wang, E. R. Mashack, B. S. J. Winkel, and K. J. Brewer, *Inorg. Chem.*, *48*, 9077 (2009). Copyright © 2009 American Chemical Society.]

TL = t-Bu_3tpy, evidenced by less retardation of DNA migration through the gel. This result may be due to the increased steric bulk from the *tert*-butyl groups. Photocleavage studies of circular pUC18 DNA were performed with the three complexes by photolyzing 20:1 BP/MC solutions with 450–1000 nm light for 2 h in atmospheric and deoxygenated conditions (Fig. 27). In 2 h, the SC DNA is fully converted to the OC form in the presence of O_2 when TL = MePhtpy. Minimal photocleavage is observed after 2 h photolysis with TL = tpy or t-Bu_3tpy, and an increased photolysis time of 4 h results in enhanced photocleavage with TL = tpy, but still very little cleavage with TL = t-Bu_3bpy. This difference is attributed to the greater binding of the tpy complex. To assay the lipophilicity of the Ru(II),Pt(II) complexes, the partition coefficient (log P) values were measured in water and octanol with values of −2.00, −0.39, and 4.00 for TL = tpy, MePhtpy, and t-Bu_3tpy, respectively. The ideal log P for metal-based drugs is in the 0.5–2.5 range to allow for water solubility and cell membrane permeability (73). The variation of TL dramatically impacts lipophilicity in this series, greater providing a means to tune this property. The ability to tune the DNA binding and photocleaving ability of a supramolecular architecture through TL substitution is important in the field of PDT.

d. Complexes $[(tpy)Ru(tppz)PtCl]^{3+}$ and $[ClPt(tppz)Ru(tppz)PtCl]^{4+}$. The complexes $[(tpy)Ru(tppz)PtCl]^{3+}$ and $[ClPt(tppz)Ru(tppz)PtCl]^{4+}$, Fig. 28, where tppz = 2,3,5,6-tetrakis(2-pyridyl)pyrazine, were synthesized using only tridentate TLs and BLs, creating molecules that are stereochemically defined, allowing proton nuclear magnetic resonance spectroscopy (^{1}H NMR) characterization (74). The bimetallic complex $[(tpy)Ru(tppz)PtCl]^{3+}$ features one BAS with a labile

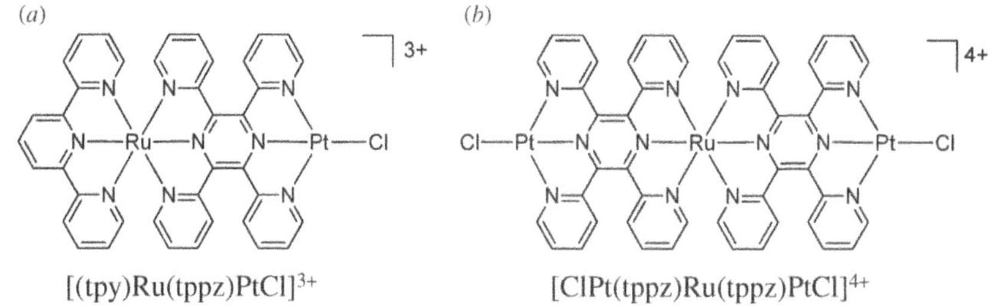

(*a*)

$[(tpy)Ru(tppz)PtCl]^{3+}$

(*b*)

$[ClPt(tppz)Ru(tppz)PtCl]^{4+}$

Figure 28. Structural representations of (*a*) $[(tpy)Ru(tppz)PtCl]^{3+}$ and (*b*) $[ClPt(tppz)Ru(tppz)PtCl]^{4+}$.

chloride ligand coordinated to Pt(II), while the trimetallic complex $[ClPt(tppz)Ru(tppz)PtCl]^{4+}$ possesses two BAS of the same nature, spatially separated and oriented at a 180° angle. Typical of Ru(II)–polyazine complexes, the electrochemistry predicts a Ru(II) based HOMO with the Ru(II,III) couple occurring at 1.63 and 1.83 V vs Ag/AgCl for $[(tpy)Ru(tppz)PtCl]^{3+}$ and $[ClPt(tppz)Ru(tppz)PtCl]^{4+}$, respectively. The larger oxidation potential for $[ClPt(tppz)Ru(tppz)PtCl]^{4+}$ is due to the presence of two Pt(II) centers coordinated on either side of the Ru center, stabilizing the Ru(dπ) orbitals. The LUMO in both complexes is expected to be tppz based, with the first reduction occurring at −0.16 and −0.03 V vs Ag/AgCl for the bimetallic and trimetallic complex, respectively. The bimetallic $[(tpy)Ru(tppz)PtCl]^{3+}$ exhibits a lowest energy Ru(dπ)→tppz(π*) CT transition centered at 530 nm, while this transition is slightly red shifted in $[ClPt(tppz)Ru(tppz)PtCl]^{4+}$ to 538 nm due to stabilization of the tppz π* orbitals upon PtCl coordination. Only the trimetallic complex exhibits a detectable emission at rt ($\lambda^{max} = 754$ nm, $\Phi^{em} = 5.4 \times 10^{-4}$, $\tau = 80$ ns).

Concentration- and temperature-dependent thermal-binding studies of the complexes with one and two BAS were performed and compared to cisplatin (75). Both supramolecular complexes show enhanced modification of DNA migration upon binding compared to cisplatin, with enhanced retardation of migration observed at larger BP/MC ratios. Complexes $[ClPt(tppz)Ru(tppz)PtCl]^{4+}$ and $[(tpy)Ru(tppz)PtCl]^{3+}$ exhibit comparable degrees of slowed migration at 10:1 and 5:1, respectively, consistent with 2 vs 1 BAS available for DNA binding. The gel electrophoresis assays of the temperature-dependent binding studies are provided in Fig. 29. The number of BAS were held constant, so cisplatin and $[(tpy)Ru(tppz)PtCl]^{3+}$ were performed at 5:1 BP/MC and $[ClPt(tppz)Ru(tppz)PtCl]^{4+}$ was performed at 10:1 BP/MC. In agreement with previously reported cisplatin–DNA interactions, DNA migration is increasingly retarded as temperature increases, with the greatest retardation occurring at 37 °C when the DNA has completely unwound. This finding is followed by faster migration at 50 °C, which is attributed to rewinding of the DNA in the form of

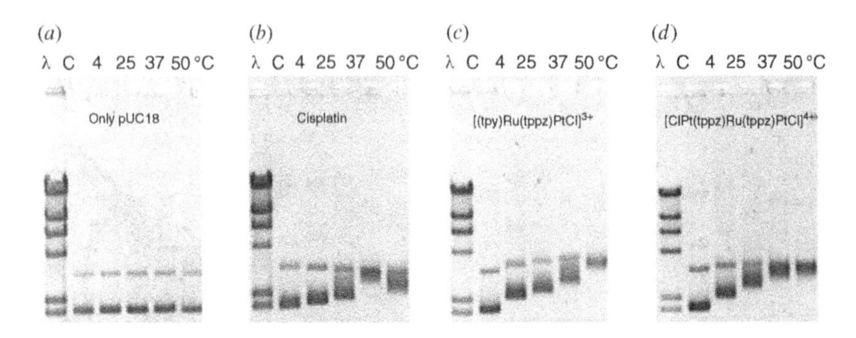

Figure 29. Temperature-dependent DNA binding assays of pUC18 with (*a*) no metal complex, (*b*) 5:1 BP/MC cisplatin, (*c*) 5:1 BP/MC $[(tpy)Ru(tppz)PtCl]^{3+}$, (*d*) 10:1 BP/MC $[ClPt(tppz)Ru(tppz)PtCl]^{4+}$ incubated in the dark. Lane λ is a lambda molecular weight marker, lane C is pUC18 control, lanes 4, 25, 37, and 50 are incubation at 4, 25, 37, and 50 °C, respectively. [From (75) with permission of Elsevier.]

negative supercoils with enhanced metal cross-linking. This effect is not observed at 50 °C with the bimetallic and trimetallic complexes although the migration continues to retard as temperature increases, suggesting that the type of binding in these complexes is different than in cisplatin. Cisplatin causes bifunctional DNA adducts as a result of the two labile Cl^- ligands, whereas $[(tpy)Ru(tppz)PtCl]^{3+}$ only allows monofunctional DNA adducts due to the single labile ligand on Pt(II), and $[ClPt\text{-}(tppz)Ru(tppz)PtCl]^{4+}$ allows bifunctional adducts at a longer distance and in a different orientation than cisplatin.

Both Ru,Pt and Os,Pt supramolecular complexes couple one or more $^1O_2$ generating LA to a DNA binding Pt BAS for prelocalization of the drug to aid in more efficient DNA cleavage. The complexes were studied by a variety of methods (e.g., electrochemistry, electronic absorption spectroscopy, steady-state and time-resolved emission spectroscopy, and gel electrophoresis) to understand how the properties impact biological reactivity, and these properties are summarized in Table I. Similar to cisplatin, ligand labilization occurs thermally to remove chloride ligands coordinated to Pt and provide an open coordination site for DNA binding. The TL variation significantly impacts the bioreactivity of these supramolecules. Use of Ph_2phen in the $[(TL)_2Ru(BL)PtCl_2]^{2+}$ motif provides enhanced visible light absorption and $^1O_2$ generation, resulting in more efficient DNA photocleavage compared to bpy analogues, and enables a new type of photobinding resulting from MLCT excitation. The lipophilicity of the metal complex in the $[(TL)RuCl(dpp)PtCl_2]^+$ motif can be tuned by variation of the substituents on the TL, and the complex $[(TL)RuCl(dpp)PtCl_2]^+$ exhibits unusual antibacterial activity. The BL identity tunes the acceptor orbital energy, providing a means to extend light absorption to lower energy toward the therapeutic window. Utilization of a larger LA in the tetrametallic motif $[\{(bpy)_2Ru(dpp)\}_2Ru(dpp)PtCl_2]^{6+}$

TABLE I
Electrochemical, Spectroscopic, and Photophysical Properties and Biological Activity of Supramolecular Complexes

Complex[a]	Electrochemistry[b]		Electronic Absorption Spectroscopy[c]			Emission Spectroscopy[c]		Biological Activity	References
	$E_{1/2}$ (V)	Assignment	λ_{max}^{abs} (nm)	ε ($M^{-1}cm^{-1}$)	Assignment	λ_{max}^{em} (nm)	τ (ns)		
$[(bpy)_2Ru(dpb)PtCl_2]^{2+}$	1.61	$Ru^{III/II}$	290	55,000	bpy π→π*	[d]	[d]	Thermal binding	58
	−0.11	$dpb^{0/-}$	385	32,000	dpb π→π*				
	−0.75	$dpb^{-/2-}$	415 (sh)	17,000	Ru→bpy CT				
			630	10,000	Ru→dpb CT				
$[(bpy)_2Os(dpb)PtCl_2]^{2+}$	1.05	$Os^{III/II}$	290	55,000	bpy π→π*	[d]	[d]	Thermal binding	58
	−0.22	$dpb^{0/-}$	380	30,000	dpb π→π*				
	−0.88	$dpb^{-/2-}$	415 (sh)	18,000	Os→bpy CT				
			638	11,000	Os→dpb CT				
$[(Ph_2phen)_2Ru(dpp)$-	1.60	$Ru^{III/II}$	274	90,000	Ph_2phen π→π*	740	40	λ^{irr} = 455 nm,	61, 63
$PtCl_2]^{2+}$	1.50	$Pt^{IV/II}$ [x]	310 (sh)	35,000	dpp π→π*			photocleavage with	
	−0.48	$dpp^{0/-}$	424	18,00	Ru→Ph_2phen CT			O_2, thermal and	
	−1.06	$dpp^{-/2-}$	517	11,400	Ru→dpp CT			photobinding	
	−1.37	$Ph_2phen^{0/-}$							
$[(Ph_2phen)_2Ru(dpq)$-	1.61	$Ru^{III/II}$	274	90,000	Ph_2phen π→π*	[d]	[d]	λ^{irr} = 455 nm,	61
$PtCl_2]^{2+}$	1.50	$Pt^{IV/II}$ [x]	320 (sh)	34,000	dpq π→π*			photocleavage with	
	−0.21	$dpq^{0/-}$	424	18,000	Ru→Ph_2phen CT			O_2, thermal and	
	−0.99	$dpq^{-/2-}$	600	9,800	Ru→dpq CT			photobinding	
	−1.37	$Ph_2phen^{0/-}$							
$[(bpy)_2Ru\{bpy\text{-}[CONH\text{-}$	[e]	[e]	470	[e]	Ru→bpy CT	660	518	λ^{irr} = 470 nm,	64
$(CH_2)_3NH_2]_2\}PtCl_2]^{2+}$								photocleavage with O_2	
$[\{(bpy)_2Ru(dpp)\}_2Ru$-	1.58	$Ru^{III/II}$	290	100,000	bpy π→π*	750	100	$\lambda^{irr} \geq$ 470 nm,	66
$(dpp)PtCl_2]^{6+}$	−0.40	$dpp^{0/-}$	320 (sh)	55,000	dpp π→π*			photocleavage with	
	−0.60	$dpp^{0/-}$	416	25,000	Ru→bpy CT			O_2, thermal binding	
	−0.71	$dpp^{0/-}$	542	35000	Ru→dpp CT				
	1.70	$Pt^{IV/II}$	276	20,900	tpy π→π*	750	56	Thermal binding	69

(continued)

TABLE I
(*Continued*)

Complex[a]	Electrochemistry[b]		Electronic Absorption Spectroscopy[c]			Emission Spectroscopy[c]		Biological Activity	References
	$E_{1/2}$ (V)	Assignment	λ_{max}^{abs} (nm)	ε ($M^{-1}cm^{-1}$)	Assignment	λ_{max}^{em} (nm)	τ (ns)		
[(tpy)Ru(PEt_2Ph)(dpp)-	1.55	$Ru^{III/II}$	308	24,100	tpy $\pi\rightarrow\pi^*$,				
$PtCl_2]^{2+}$	−0.50	$dpp^{0/-}$			Ru→tpy CT				
	−1.15	$dpp^{-/2-}$	336	22,200	tpy $\pi\rightarrow\pi^*$				
	−1.40	$tpy^{0/-}$	424 (sh)	5,300	Ru→dpp CT				
			506	10,500	Ru→dpp CT				
[(tpy)Ru(PEt_2Ph)(bpm)-	1.82	$Pt^{IV/II}$	274	47,100	tpy $\pi\rightarrow\pi^*$	[d]	[d]	Thermal binding	69
$PtCl_2]^{2+}$	1.63	$Ru^{III/II}$	304	40,100	tpy $\pi\rightarrow\pi^*$,				
	−0.34	$bpm^{0/-}$			Ru→tpy CT				
	−1.04	$bpm^{-/2-}$	336	31,400	tpy $\pi\rightarrow\pi^*$				
	−1.45	$tpy^{0/-}$	416	15,200	Ru→tpy CT,				
			560	6,100	Ru→bpm CT				
					Ru→bpm CT				
[(tpy)RuCl(dpp)$PtCl_2]^+$	1.14	$Ru^{III/II}$	272	23,900	tpy $\pi\rightarrow\pi^*$	[d]	[d]	$\lambda^{irr} \geq 450$ nm,	70, 72
	−0.50	$dpp^{0/-}$	316	35,600	tpy $\pi\rightarrow\pi^*$			photocleavage with	
	−1.05	$dpp^{-/2-}$	362 (sh)	15,000	dpp $\pi\rightarrow\pi^*$			O_2, thermal binding	
	−1.43	$tpy^{0/-}$	460 (sh)	4,680	Ru→tpy CT				
			544	14,600	Ru→dpp CT				
[(tpy)RuCl(dpq)$PtCl_2]^+$	1.10	$Ru^{III/II}$	272	26,100	tpy $\pi\rightarrow\pi^*$	[d]	[d]	Thermal binding	70
	−0.32	$dpq^{0/-}$	316	36,900	tpy $\pi\rightarrow\pi^*$				
	−0.91	$dpq^{-/2-}$	362 (sh)	13,600	dpq $\pi\rightarrow\pi^*$				
	−1.50	$tpy^{0/-}$	460 (sh)	4,610	Ru→tpy CT				
			632	10,000	Ru→dpq CT				
[(tpy)RuCl(dpb)$PtCl_2]^+$	1.12	$Ru^{III/II}$	272	29,600	tpy $\pi\rightarrow\pi^*$	[d]	[d]	Thermal binding	70
	−0.20	$dpb^{0/-}$	316	38,300	tpy $\pi\rightarrow\pi^*$				
	−0.81	$dpb^{-/2-}$	380	27,000	dpb $\pi\rightarrow\pi^*$				
	−1.51	$tpy^{0/-}$	460 (sh)	5,510	Ru→tpy CT				
			682	9,860	Ru→dpb CT				

[(MePhtpy)RuCl(dpp)-$PtCl_2$]$^{+}$	1.10	$Ru^{III/II}$	227	46,700	MePhtpy $\pi\rightarrow\pi^*$	[d]	[d]	$\lambda^{irr} \geq 450$ nm, photocleavage with O_2, thermal binding	72
	−0.55	$dpp^{0/-}$	318	48,200	MePhtpy $\pi\rightarrow\pi^*$				
	−1.15	$dpp^{-/2-}$	354 (sh)	19,000	dpp $\pi\rightarrow\pi^*$				
	−1.44	$MePhtpy^{0/-}$	464 (sh)	6,700	Ru→MePhtpy CT				
			548	20,900	Ru→dpp CT				
[($t$$Bu_3$tpy)RuCl(dpp)-$PtCl_2$]$^{+}$	1.01	$Ru^{III/II}$	230	39,200	t-Bu_3tpy $\pi\rightarrow\pi^*$	[d]	[d]	$\lambda^{irr} \geq 450$ nm, minimal photocleavage with O_2, thermal bidning	72
	−0.59	$dpp^{0/-}$	314	35,300	t-Bu_3tpy $\pi\rightarrow\pi^*$				
		$dpp^{-/2-}$	354 (sh)	14,200	dpp $\pi\rightarrow\pi^*$				
		$tBu_3tpy^{0/-}$	462 (sh)	4,900	Ru→ $t$$Bu_3$tpy CT				
			545	15,400	Ru→dpp CT				
[(tpy)Ru(tppz)PtCl]$^{3+}$	1.63	$Ru^{III/II}$	530	28,000	Ru→tppz CT	[d]	[d]	Thermal binding	74
	−0.16	$tppz^{0/-}$							
	−0.70	$tppz^{-/2-}$							
[ClPt(tppz)Ru(tppz)-PtCl]$^{4+}$	1.83	$Ru^{III/II}$	538	30,000	Ru→tppz CT	754	80	Thermal binding	74
	−0.03	$tppz^{0/-}$							
	−0.17	$tppz^{0/-}$							
[{$(bpy)_2$Ru-(dpp)}$_2$$RhCl_2$]$^{5+}$	1.61	$Ru^{III/II}$	284	99,000	bpy $\pi\rightarrow\pi^*$	776	38	$\lambda^{irr} \geq 475$ nm, photocleavage without O_2, photobinding, PDT with Vero cells	76, 77
	−0.35	$Rh^{III/II/I}$	338 (sh)	41,400	dpp $\pi\rightarrow\pi^*$				
	−0.76	$dpp^{0/-}$	416	16,400	Ru→bpy CT				
	−1.01	$dpp^{0/-}$	518	26,100	Ru→dpp CT				
[{$(bpy)_2$Ru-(bpm)}$_2$$RhCl_2$]$^{5+}$	1.70	$Ru^{III/II}$	278	90,000	bpy $\pi\rightarrow\pi^*$	800	10	Not active	76, 78
	−0.13	$bpm^{0/-}$	412	37,000	Ru→bpy CT,				
	−0.26	$bpm^{0/-}$	|||594	|||9,900	Ru→bpm CT				
	−0.78	$Rh^{III/II/I}$			Ru→bpm CT				
[{$(bpy)_2$Os-(dpp)}$_2$$RhCl_2$]$^{5+}$	1.21	$Os^{III/II}$	284	120,000	bpy $\pi\rightarrow\pi^*$	[d]	[d]	$\lambda^{irr} \geq 475$ nm, photocleavage without O_2, photobiding, PDT with Vero cells	76
	−0.39	$Rh^{III/II/I}$	336	45,000	dpp $\pi\rightarrow\pi^*$				
	−0.76	$dpp^{0/-}$	412	22,000	Os→bpy CT				
	−1.00	$dpp^{0/-}$	534	36,000	Os→dpp CT				
			798	6,100	Os→dpp ^{3}CT				
[{(tpy)RuCl-(dpp)}$_2$$RhCl_2$]$^{3+}$	1.12	$Ru^{III/II}$	265	48,000	tpy $\pi\rightarrow\pi^*$	[d]	[d]	$\lambda^{irr} \geq 475$ nm, photocleavage without O_2, photobinding	76
	−0.47	$Rh^{III/II/I}$	310	60,000	tpy $\pi\rightarrow\pi^*$				
	−0.87	$dpp^{0/-}$	350 (sh)	30,000	dpp $\pi\rightarrow\pi^*$				

(continued)

TABLE I
(*Continued*)

Complex[a]	Electrochemistry[b]		Electronic Absorption Spectroscopy[c]			Emission Spectroscopy[c]		Biological Activity	References
	$E_{1/2}$ (V)	Assignment	λ_{max}^{abs} (nm)	ε ($M^{-1}cm^{-1}$)	Assignment	λ_{max}^{em} (nm)	τ (ns)		
	−1.20	$dpp^{0/-}$	460 (sh)	10,000	Ru→tpy CT				
			530	30,000	Ru→dpp CT				
$[(bpy)_2Os(dpp)$-$RhCl_2(phen)]^{3+}$	1.20	$Os^{III/II}$	284	70,000	bpy $\pi\rightarrow\pi^*$	[d]	[d]	$\lambda^{irr} \geq 645$ nm, photocleavage without O_2, photobinding, amplification inhibition	79, 80
	−0.46	$Rh^{III/II}$	336	28,000	dpp $\pi\rightarrow\pi^*$				
	−0.75	$Rh^{II/I}$	413	10,000	Os→bpy CT				
	−0.98	$dpp^{0/-}$	521	18,000	Os→dpp CT				
			750	2,900	Os→dpp ^{3}CT				
$[(bpy)_2Ru(dpp)$-$RhCl_2(phen)]^{3+}$	1.61	$Ru^{III/II}$	279	71,300	bpy $\pi\rightarrow\pi^*$, phen $\pi\rightarrow\pi^*$	786	32	$\lambda^{irr} = 455$ nm, photocleavage without O_2, photobinding	15, 81
	−0.39	$Rh^{III/II}$							
	−0.74	$Rh^{II/I}$	338 (sh)	23,400	dpp $\pi\rightarrow\pi^*$				
	−0.98	$dpp^{0/-}$	418	8,100	Ru→bpy CT				
			515	14,700	Ru→dpp CT				
$[(bpy)_2Ru(bpm)$-$RhCl_2(phen)]^{3+}$	1.76	$Ru^{III/II}$	276	78,400	bpy $\pi\rightarrow\pi^*$, phen $\pi\rightarrow\pi^*$, bpm $\pi\rightarrow\pi^*$	[d]	[d]	$\lambda^{irr} = 455$ nm, photocleavage without O_2, photobinding	15, 81
	−0.14	$bpm^{0/-}$							
	−0.77	$Rh^{III/II}$							
	−0.99	$Rh^{II/I}$	393 (sh)	14,200	Ru→bpm CT				
			412	16,600	Ru→bpy CT				
			581	4,000	Ru→bpm CT				

[a] 2,3-Bis(2-pyridyl)benzoquinoxaline = dpb.

[b] Potentials in V vs Ag/AgCl (3 *M* NaCl). Standard reference vs Ag/AgCl: NHE = −0.209 V, SCE = −0.035 V, $FeCp_2^{0/-} = 0.461$ V.

[c] Recorded in rt MeCN.

[d] No detectable emission at rt.

[e] Values not reported.

provides for enhanced UV and visible light absorption and a relatively long-lived excited state. The trimetallic complex $[ClPt(tppz)Ru(tppz)PtCl]^{4+}$ features two BAS for more active binding to DNA. These Ru,Pt and Os,Pt supramolecules provide significant insight into the properties impacting reactivity with biomolecules (e.g., DNA).

C. Light Absorbers with Ru(II)/Os(II) Coupled to Rh(III) Bioactive Sites

While *cis*-$Rh^{III}Cl_2$ BAS are structurally similar to *cis*-$Pt^{II}Cl_2$ BAS, $Rh^{III}-Cl$ bonds are not labile at rt in contrast to $Pt^{II}-Cl$ bonds. This result provides a unique forum to develop light activated halide labilization in Rh based systems. In addition, the presence of Ru^{II}(BL) LAs provides for lower energy excitation of these coupled Rh BAS. Also unique to this Ru,Rh or Os,Rh forum is the ability to display O_2 independent light induced reactivity with biomolecules including DNA.

1. *Trimetallic Complexes of Ru,Rh,Ru and Os,Rh,Os*

A supramolecular architecture that couples two Ru(II)- or Os(II)-based LAs to a *cis*-Rh(III)Cl_2 BAS was studied through component variation to understand the effects of TL, LA metal, and BL variation on the orbital energetics and DNA reactivity on this motif (76, 82, 83). The four complexes, $[\{(bpy)_2Ru(dpp)\}_2RhCl_2]^{5+}$, $[\{(bpy)_2Ru(bpm)\}_2RhCl_2]^{5+}$, $[\{(bpy)_2Os(dpp)\}_2RhCl_2]^{5+}$, and $[\{(tpy)RuCl(dpp)\}_2RhCl_2]^{3+}$ are pictured in Fig. 30.

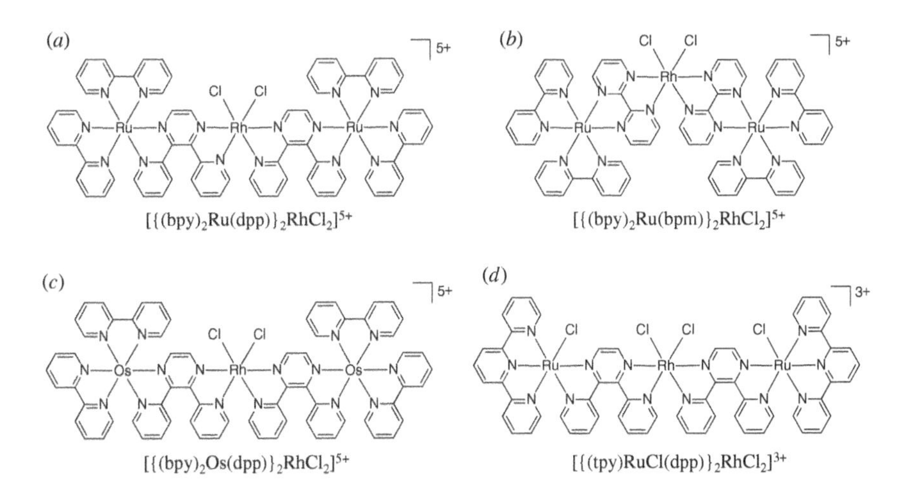

Figure 30. Structural representations of (*a*) $[\{(bpy)_2Ru(dpp)\}_2RhCl_2]^{5+}$, (*b*) $[\{(bpy)_2Ru(bpm)\}_2RhCl_2]^{5+}$, (*c*) $[\{(bpy)_2Os(dpp)\}_2RhCl_2]^{5+}$, (*d*) and $[\{(tpy)RuCl(dpp)\}_2RhCl_2]^{3+}$.

Electrochemical analysis of the four Rh(III)-centered trimetallic complexes exhibits Ru or Os oxidations and Rh and ligand reductions. The Ru(II,III) oxidation potentials for $[\{(bpy)_2Ru(dpp)\}_2RhCl_2]^{5+}$ and $[\{(bpy)_2Ru(bpm)\}_2RhCl_2]^{5+}$ occur at 1.60 and 1.70 V, respectively, typical of Ru(II) centers coordinated to three bidentate polyazine ligands (76). The first oxidation potential is decreased by replacing Ru with Os to afford $[\{(bpy)_2Os(dpp)\}_2RhCl_2]^{5+}$ [1.21 V for Os(II,III)] and by replacing the two bpy ligands with a tpy TL and a σ-donating Cl^- ligand to afford $[\{(tpy)RuCl(dpp)\}_2RhCl_2]^{3+}$ [1.12 V for Ru(II,III)]. The first reduction observed for $[\{(bpy)_2Ru(dpp)\}_2RhCl_2]^{5+}$, $[\{(bpy)_2Os(dpp)\}_2RhCl_2]^{5+}$, and $[\{(tpy)RuCl(dpp)\}_2RhCl_2]^{3+}$ is an irreversible couple at −0.39, −0.39, and −0.47 V vs Ag/AgCl, respectively, assigned as a Rh(III/II/I) reduction. The $dpp^{0/-}$ couples occur after the Rh reduction around −0.80 V. The complex with bpm BL, $[\{(bpy)_2Ru(bpm)\}_2RhCl_2]^{5+}$, exhibits two reversible reductions at −0.13 and −0.26 V, assigned as $bpm^{0/-}$ reductions. Following the BL reductions is an irreversible $Rh^{III/II/I}$ couple at −0.78 V. The orbital inversion placing the LUMO on the bpm BL is designed as bpm acceptor orbitals are typically stabilized compared to dpp (76). Orbital energy diagrams for the four complexes are pictured in Fig. 31. From the electrochemistry, a lowest lying Ru(dπ) or Os(dπ)→Rh(dσ*) MMCT (metal-to-metal charge transfer) excited state is predicted for $[\{(bpy)_2Ru(dpp)\}_2RhCl_2]^{5+}$, $[\{(bpy)_2Os(dpp)\}_2RhCl_2]^{5+}$, and $[\{(tpy)RuCl(dpp)\}_2RhCl_2]^{3+}$; a lowest lying Ru(dπ)→bpm(π*) MLCT excited state is predicted for $[\{(bpy)_2Ru(bpm)\}_2RhCl_2]^{5+}$.

The electronic absorption spectroscopy of these Rh(III)-centered trimetallic complexes provides efficient coverage of the UV and visible regions due to many overlapping transitions (76, 78). The UV region is dominated by TL and BL π→π* transitions and the visible region is dominated by Ru(dπ) or Os(dπ)→TL(π*) or BL(π*) CT transitions. The complexes $[\{(bpy)_2Ru(dpp)\}_2RhCl_2]^{5+}$, $[\{(bpy)_2Os-$

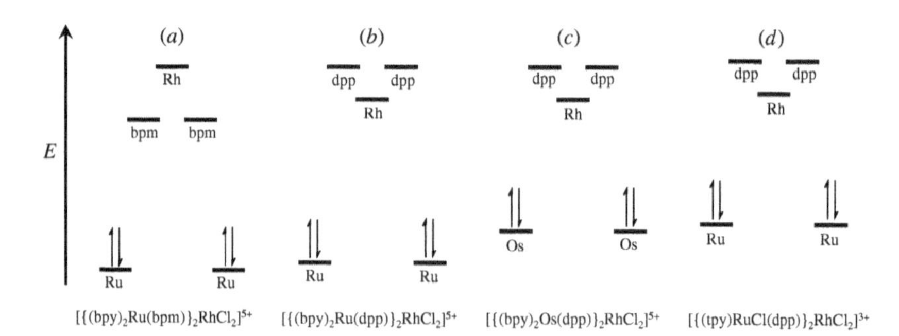

Figure 31. Simplified orbital energy diagrams depicting the frontier molecular orbitals for the trimetallic complexes (*a*) $[\{(bpy)_2Ru(bpm)\}_2RhCl_2]^{5+}$, (*b*) $[\{(bpy)_2Ru(dpp)\}_2RhCl_2]^{5+}$, (*c*) $[\{(bpy)_2Os(dpp)\}_2RhCl_2]^{5+}$, and (*d*) $[\{(tpy)RuCl(dpp)\}_2RhCl_2]^{3+}$.

(dpp)}$_2$RhCl$_2$]$^{5+}$, and [{(tpy)RuCl(dpp)}$_2$RhCl$_2$]$^{3+}$ absorb strongly with a lowest lying ^{1}MLCT transition centered at ~530 nm. The lowest lying ^{1}MLCT transition for [{(bpy)$_2$Ru(bpm)}$_2$RhCl$_2$]$^{5+}$ is centered at 594 nm, consistent with the stabilized bpm acceptor orbitals relative to dpp. An advantage of using an Os(II) LA is enhanced near-infrared (NIR) absorption from direct excitation into the ^{3}MLCT excited state, a result of spin–orbit coupling typically observed in Os(II)–polyazine complexes. This NIR absorption is important as it coincides with the therapeutic window for PDT applications.

Unlike most complexes that utilize Ru MLCT light absorbers, the Rh centered systems display oxygen independent photoreactions with DNA. This design constraint is for application of PDT in aggressive, hypoxic tumors. Photocleavage studies with the four Rh centered trimetallic complexes were performed to monitor the impact of TL, BL, and LA metal variation on the ability to photomodify pUC18 and pBluescript DNA when excited with irradiation wavelength (λ^{irr}) $\geq$ 475 nm in the absence of molecular oxygen (76). The systems that use dpp as the BL photocleave DNA via an oxygen-independent mechanism. When photolyzed in the presence of DNA plasmid, [{(bpy)$_2$Ru(bpm)}$_2$RhCl$_2$]$^{5+}$ does not cause cleavage. This result is rationalized by the inaccessible Rh(dσ*) orbitals in this motif; photoexcitation results in electron localization on the bpm π* orbitals. The gel electrophoresis images from photocleavage studies of [{(bpy)$_2$Ru(dpp)}$_2$RhCl$_2$]$^{5+}$, [{(bpy)$_2$Os(dpp)}$_2$RhCl$_2$]$^{5+}$, and [{(tpy)RuCl(dpp)}$_2$RhCl$_2$]$^{3+}$ are shown below in Fig. 32. All three complexes exhibit rapid photocleavage of both pUC18 and pBluescript DNA plasmids in the absence of O_2 when irradiated with wavelengths longer than 475 nm. Excitation is red shifted via the Ru or Os→dpp MLCT excitation with stabilized μ-dpp(π*) acceptor orbitals. Each of these complexes possess a Rh(III)-based LUMO and has the ability to populate a lower lying Ru or Os→Rh ^{3}MMCT excited state following MLCT excitation. Photocleavage results from reactivity of DNA with the newly generated Rh(II) site upon excitation in the Ru→Rh MMCT photoreactive state. No change in migration is observed when the complexes are incubated with DNA, demonstrating that these complexes likely ionically bind or groove bind DNA prior to excitation and photocleavage. This activity is expected to result in low dark toxicity of this new class of potential PDT drugs.

The trimetallic complexes [{(bpy)$_2$Ru(dpp)}$_2$RhCl$_2$]$^{5+}$ and [{(bpy)$_2$Os(dpp)}$_2$RhCl$_2$]$^{5+}$ were reported to inhibit cell growth when Vero cells (African green monkey kidney epithelial cells) were treated with metal complex and photolyzed at $\lambda \geq$ 460 nm with a focused beam for 4 min (82, 83). The micrographs of cells treated with [{(bpy)$_2$Ru(dpp)}$_2$RhCl$_2$]$^{5+}$ showing cell death where the sample was illuminated are pictured on the left in Fig. 33. Plots of cell replication inhibition vs metal complex concentration are depicted on the right in Fig. 33. Cells treated with metal complex, but not irradiated with visible light, demonstrated normal growth at all concentrations studied. Upon irradiation, cell

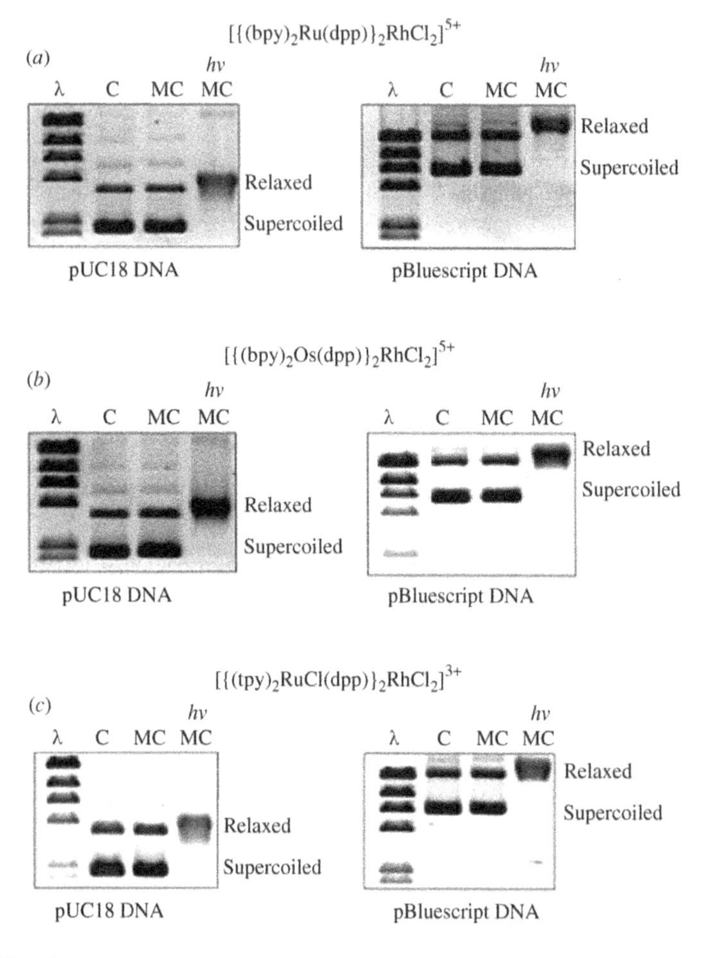

Figure 32. Gel electrophoresis assays of the photocleavage of pUC18 (left) and pBluescript (right) plasmids by (*a*) $[\{(bpy)_2Ru(dpp)\}_2RhCl_2]^{5+}$, (*b*) $[\{(bpy)_2Os(dpp)\}_2RhCl_2]^{5+}$, and (*c*) $[\{(tpy)RuCl(dpp)\}_2RhCl_2]^{3+}$. All experiments were performed in oxygen-free conditions. Lane λ is a lambda molecular weight marker, lane C is a plasmid control, lane MC is a dark control incubated at 37 °C for 2 h with a 5:1 BP/MC ratio, and lane *hν* MC is a 5:1 BP/MC ratio photolyzed with $\lambda \geq 475$ nm for 20 min. [Adapted from (76).]

replication was inhibited by the Os complex, and light-induced cell death occurred with the use of $[\{(bpy)_2Ru(dpp)\}_2RhCl_2]^{5+}$. The ability to inhibit cell replication is an important criterion for light activated anti-cancer agents. The unique activity of these complexes upon photolysis using an oxygen-independent pathway and lack of dark toxicity makes these systems promising for PDT drug development.

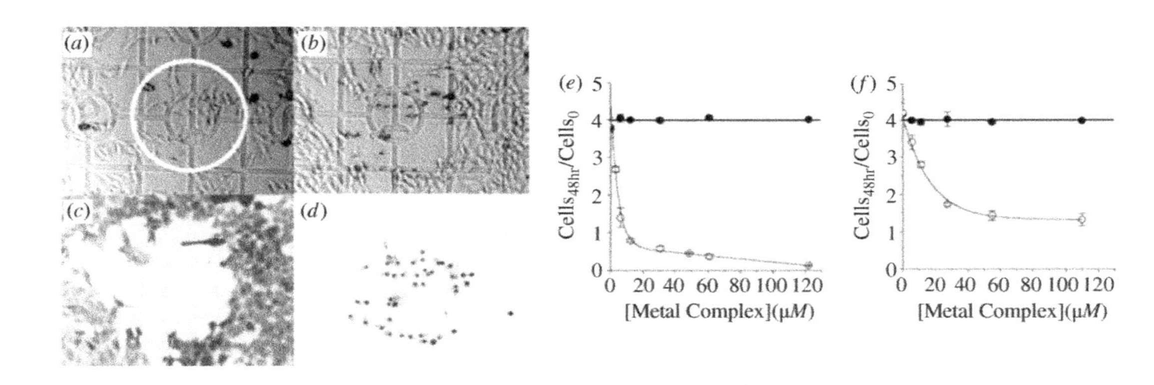

Figure 33. Micrograph images of Vero cells following exposure to 122-μ*M* $[\{(bpy)_2Ru(dpp)\}_2RhCl_2]Cl_5$ and irradiation for 4 min with a focus beam of ≥ 460 nm. (*a*) Micrograph after photolysis. The white circle represents the border of the irradiation spot. (*b*) Micrograph after a 48-h growth period in the dark. (*c*) Micrograph after the 48-h growth period with green live cell visualization using calcein AM (acetmethyloxy) fluorescence. (*d*) Micrograph after the 48-h growth period with red dead cell visualization using ethidium homodimer-1 fluorescence. Cell replication inhibition after exposure to (*e*) $[\{(bpy)_2Ru(dpp)\}_2RhCl_2]Cl_5$ or (*f*) $[\{(bpy)_2Os(dpp)\}_2RhCl_2]Cl_5$ with (○) or without (•) irradiation with ≥ 460-nm light as a function of [metal complex]. [Adapted from (82).] (See the color version of this figure in color plates section.)

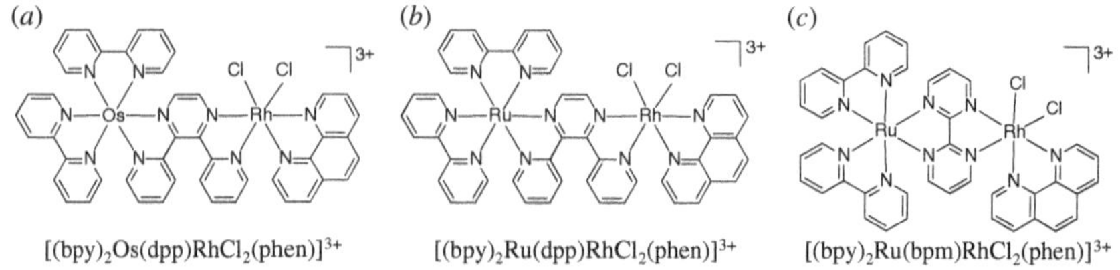

(*a*) $[(bpy)_2Os(dpp)RhCl_2(phen)]^{3+}$

(*b*) $[(bpy)_2Ru(dpp)RhCl_2(phen)]^{3+}$

(*c*) $[(bpy)_2Ru(bpm)RhCl_2(phen)]^{3+}$

Figure 34. Structural representations of (*a*) $[(bpy)_2Os(dpp)RhCl_2(phen)]^{3+}$, (*b*) $[(bpy)_2Ru(dpp)\text{-}RhCl_2(phen)]^{3+}$, and (*c*) $[(bpy)_2Ru(bpm)RhCl_2(phen)]^{3+}$.

2. *Bimetallic Complexes of Ru,Rh and Os,Rh*

Smaller analogues to the Ru,Rh,Ru and Os,Rh,Os trimetallics that may exhibit better cell-membrane permeability, more efficient binding to DNA, provide a lower cationic charge, reduce the number of stereoisomers, and provide an easily tunable TL on the Rh(III) center are desired. A new class of PMDs were designed that couple one Ru(II)- or Os(II)-based LA to a *cis*-$RhCl_2$(TL) BAS (79, 81). The structural representations of the $[(bpy)_2M(BL)RhCl_2(phen)]^{3+}$ complexes, where M = Ru(II) or Os(II) and BL = dpp or bpm, are depicted in Fig. 34. These new two-metal systems are designed to display the same O_2 independent DNA photocleavage ability while possessing a more sterically accessible Rh site to enhance photoreactivity. In fact, these new motifs display a new photoreactivity: MMCT facilitated DNA photobinding. This result seems to photolocalize the molecules at the DNA target, providing enhanced DNA photocleavage.

Electrochemistry allows the determination of the nature and energy of the frontier orbitals. The oxidative electrochemistry of the Os,Rh and Ru,Rh complexes is similar to the previously discussed Ru,Rh,Ru and Os,Rh,Os trimetallic complexes, where the complexes oxidize at 1.20, 1.61, and 1.76 V vs Ag/AgCl for $[(bpy)_2Os(dpp)RhCl_2(phen)]^{3+}$, $[(bpy)_2Ru(dpp)RhCl_2(phen)]^{3+}$, and $[(bpy)_2Ru\text{-}(bpm)RhCl_2(phen)]^{3+}$, respectively. The bimetallic complexes display complicated reductive electrochemistry; however, the LUMO is localized on the Rh center with BL = dpp and localized on the BL with BL = bpm. The steric accessibility of the Rh site in the new Os,Rh and Ru,Rh bimetallics leads to slower halide loss at the *cis*-$Rh^{III}Cl_2$ site following generation of the Rh(II) state. This complicates the electrochemistry of this motif (83). The reductive couples for $[(bpy)_2Os(dpp)RhCl_2(phen)]^{3+}$ are observed at −0.46 V ($Rh^{III/II}$), −0.75 V ($Rh^{II/I}$), and −0.98 V ($dpp^{0/-}$). Similarly, $[(bpy)_2Ru(dpp)RhCl_2(phen)]^{3+}$ reduces at −0.39 V ($Rh^{III/II}$), −0.74 V ($Rh^{II/I}$), and −0.98 V ($dpp^{0/-}$). The bpm bridged complex, $[(bpy)_2Ru(bpm)RhCl_2(phen)]^{3+}$, reduces at −0.14 V ($bpm^{0/-}$), −0.77 V ($Rh^{III/II}$), and −0.91 V ($Rh^{II/I}$).

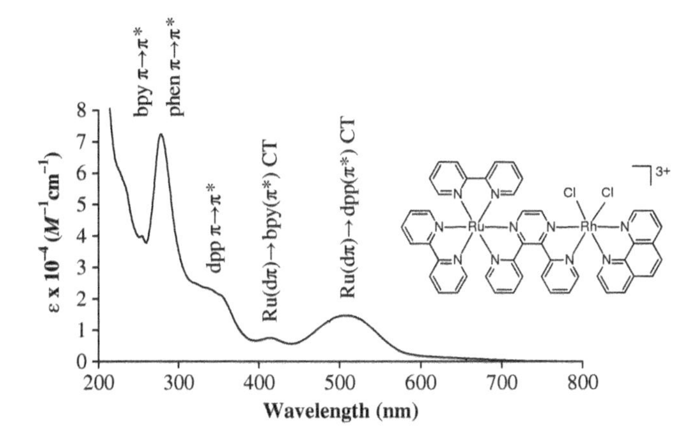

Figure 35. Electronic absorption spectroscopy of $[(bpy)_2Ru(dpp)RhCl_2(phen)](PF_6)_3$ in rt MeCN.

The Ru,Rh and Os,Rh bimetallic complexes exhibit light absorbing properties similar to the analogous trimetallic complexes. The electronic absorption spectroscopy of the complexes is in agreement with the orbital energetics from electrochemical analysis. Either Ru(dπ) or Os(dπ)→bpy(π*) CT transitions are observed at ~410 nm, and the Ru(dπ) or Os(dπ) → BL(π*) CT are observed at 521, 515, and 592 nm for $[(bpy)_2Os(dpp)RhCl_2(phen)]^{3+}$, $[(bpy)_2Ru(dpp)RhCl_2(phen)]^{3+}$, and (*c*) $[(bpy)_2Ru(bpm)RhCl_2(phen)]^{3+}$, respectively. Additionally, the Os,Rh complex exhibits a relatively intense Os→dpp ^{3}MLCT absorption centered at 750 nm (~3000 $M^{-1}cm^{-1}$), facilitating absorption in the PDT window (81). The electronic absorption spectrum of $[(bpy)_2Ru(dpp)RhCl_2(phen)](PF_6)_3$ in MeCN is shown in Fig. 35.

The Ru,Rh complexes display O_2 independent DNA photocleavage characteristic of the Ru,Rh,Ru motifs, as well as a new DNA photobinding reaction made possible by MMCT labilization of the chloride and a sterically accessible Rh BAS. Photocleavage and photobinding studies were performed with pUC18 DNA and 455-nm excitation of the 5:1 BP/MC solutions using $[(bpy)_2Ru(dpp)RhCl_2(phen)]^{3+}$ and $[(bpy)_2Ru(bpm)RhCl_2(phen)]^{3+}$ in the absence of O_2 to analyze the impact of Rh or BL based LUMO, Fig. 36 (81). Binding of the metal complexes to DNA is observed due to the smaller size, leading to a sterically accessible Rh BAS. Conversion of SC DNA to the OC form is observed more rapidly when BL = dpp compared to BL = bpm. Emission studies of $[(bpy)_2Ru(dpp)RhCl_2(phen)]^{3+}$ reveal that the complex emits at 786 nm with a lifetime of 30 ns. No emission is detectable for the bpm analogue. The emission is expected to occur at lower energy, which should result in a shorter excited-state lifetime. The slower photocleavage using (*c*) $[(bpy)_2Ru(bpm)RhCl_2(phen)]^{3+}$ can be attributed

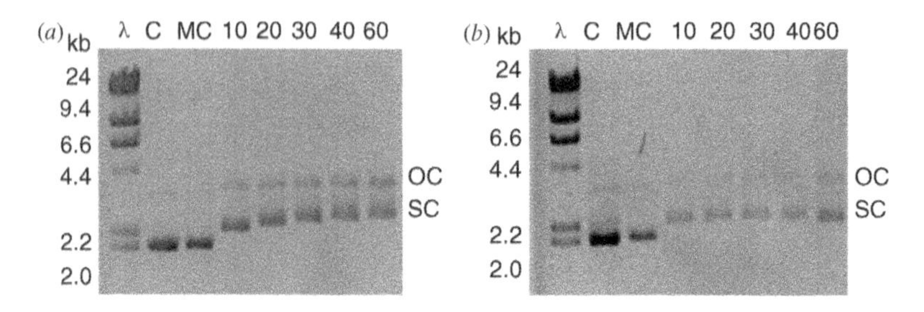

Figure 36. Gel electrophoresis images depicting photobinding and photocleavage of pUC18 DNA with (*a*) $[(bpy)_2Ru(dpp)RhCl_2(phen)]^{3+}$ and (*b*) $[(bpy)_2Ru(bpm)RhCl_2(phen)]^{3+}$. Lane λ is a lambda molecular lane marker, lane C is a pUC18 control, lane MC is an incubation in the dark at rt for 2 h with 5:1 BP/MC, lanes 10, 20, 30, 40, and 60 are irradiation at 455 nm of a 5:1 BP/MC solution at 455 nm under Ar for 10, 20, 30, 40, and 60 min. [From (81) with permission of Elsevier.]

to the complex remaining in the excited state for less time relative to the dpp analogue. The activity with DNA does not require molecular oxygen, which is an important property of potential anti-cancer agents.

The new DNA photobinding reaction observed for $[(bpy)_2Ru(dpp)RhCl_2(phen)]^{3+}$ via a unique MMCT state prompted a study of the photosubstitution chemistry of this system. Photolabilization of the chloride ligands coordinated to the Rh center was observed when $[(bpy)_2Ru(dpp)RhCl_2(phen)]^{3+}$ (0.1 m*M*) and NaH_2PO_4 buffer (10 m*M*) were photolyzed with 455-nm light for 30 min (81). A small blue shift of ~15 nm in the Ru→dpp CT absorption was observed, and $[(bpy)_2Ru(dpp)Rh(HPO_4)(phen)]^{3+}$, $[(bpy)_2Ru(dpp)Rh(OH)(HPO_4)(phen)]^{2+}$, $[(bpy)_2Ru(dpp)Rh(H_2O)(HPO_4)(phen)]^{3+}$, and $[(bpy)_2Ru(dpp)Rh(H_2O)_2(phen)]^{3+}$ were found by electrospray ionization–mass spectrometry (ESI–MS). This experiment provides more information into the photolabilization of chloride upon MLCT excitation and conversion to the lower lying MMCT state. To better understand the mode of interaction between pUC18 DNA and the metal complex, an experiment was performed where the plasmid and $[(bpy)_2Ru(dpp)RhCl_2(phen)]^{3+}$ in a 5:1 BP/MC ratio were photolyzed together for 20 min (lane 1 in Fig. 37). The metal complex was also photolyzed with 10-m*M* phosphate buffer for 20 min followed by addition of pUC18 DNA (lane 2 in Fig. 37). The gel shift assay indicates that when the bimetallic complex loses a chloride and subsequently binds species (e.g., H_2O, OH^-, and HPO_4^-), the Rh center will not bind to DNA at rt. Rather, photoexcitation in the presence of DNA is needed due to the high thermal stability of rhodium–ligand bonds.

Osmium light absorbers can be used to provide for supramolecules that absorb red light in the therapeutic window, as previously discussed, while maintaining photoreactivity with DNA. For $[(bpy)_2Os(dpp)RhCl_2(phen)]^{3+}$, this is followed by internal conversion to the Os→Rh ^{3}MMCT state, which should lead to DNA

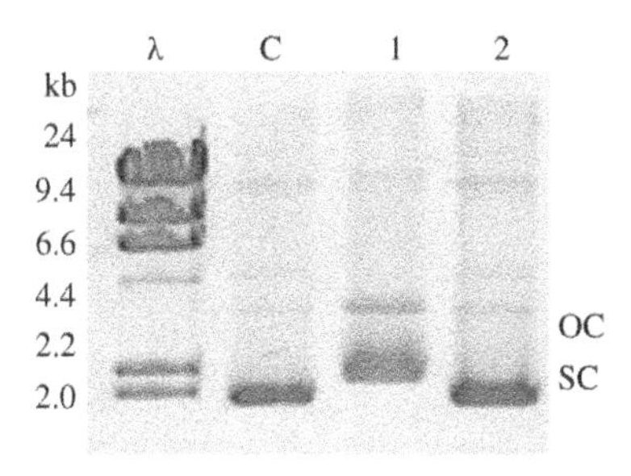

Figure 37. Gel electrophoresis assay of $[(bpy)_2Ru(dpp)RhCl_2(phen)]^{3+}$ and pUC18 DNA. Lane λ is the lambda molecular weight marker, lane C is the pUC18 control, lane 1 is a 5:1 BP/MC ratio photolyzed at 455 nm for 20 min in the presence of pUC18, and lane 2 is the same concentration photolyzed at 455 nm for 20 min in 10-m*M* phosphate buffer with pUC18 added after photolysis. [From (81) with permission of Elsevier.]

photobinding and photocleavage. Red light photobinding and O_2 independent photocleavage were demonstrated with the Os complex $[(bpy)_2Os(dpp)$-$RhCl_2(phen)]^{3+}$ and pUC18 DNA in deoxygenated solutions (81). Assay of the 5:1 BP/MC solutions irradiated at either $\lambda \geq 590$ or $\lambda \geq 645$ nm by gel electrophoresis shows that photobinding occurs within 60 min of photolysis and complete conversion from SC to OC DNA was achieved within 240 min (Fig. 38). The ability of this complex to both photobind to and photocleave DNA with low-energy excitation in the PDT window without the need for molecular O_2 makes this PMD a promising candidate for PDT drug development and light activated anticancer therapy.

The $[(bpy)_2Os(dpp)RhCl_2(phen)]^{3+}$ complex inhibits DNA replication and amplification in polymerase chain reaction (PCR) studies when excited with red, low-energy visible light in the therapeutic window without the need for molecular oxygen (84). This unusual property is desirable in the development of new types of PDT agents for the treatment of disease. A 5:1 BP/MC ratio of the Os, Rh complex and a 670 BP fragment of pUC 18 excited with low-energy visible light (either $\lambda^{irr} \geq 590$ or ≥ 645 nm) for 60 min in the absence of O_2 inhibits DNA amplification. In the corresponding gel shift assay with pUC18 photolyzed and $[(bpy)_2Os(dpp)RhCl_2(phen)]^{3+}$ previously discussed in Fig. 38, significant binding is observed after 60 min of red light irradiation with little photocleavage. The metal complex–DNA adduct prevents replication and amplification in the PCR experiment. The ability of this Os,Rh complex to impede DNA replication using red light without the need for molecular oxygen is unique, providing a platform to develop a new type of PDT agent. The effect of a lower concentration of metal complex on PCR was also studied with $\lambda^{irr} \geq 590$ nm. The imaged gel shift assay with pUC18 plasmid and the Os,Rh complex in a 50:1 BP/MC ratio [Fig. 39(*a*)]

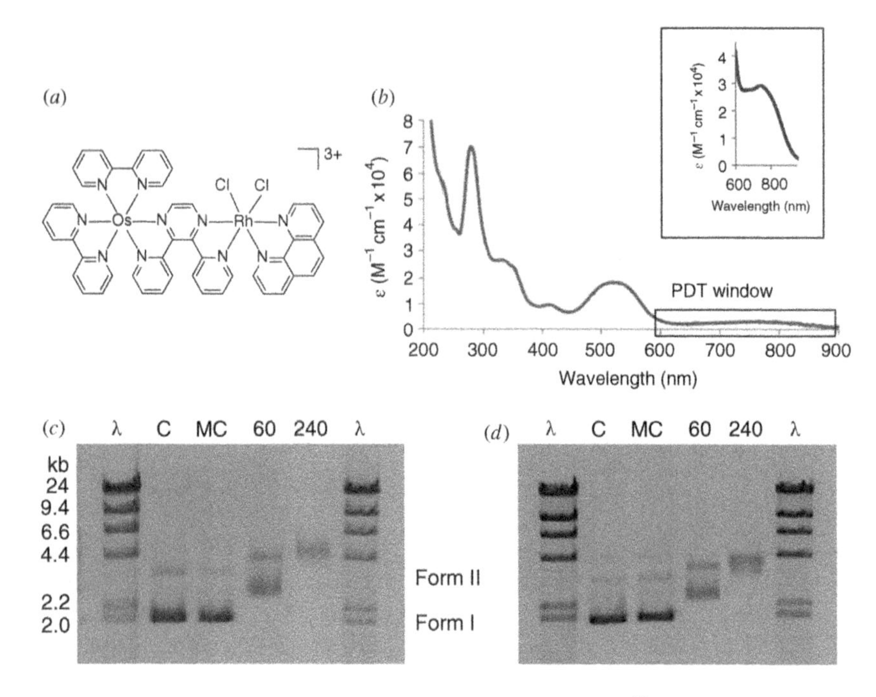

Figure 38. (*a*) Structural representation of $[(bpy)_2Os(dpp)RhCl_2(phen)_2]^{3+}$. (*b*) Electronic absorption spectrum of the complex in MeCN. Inset is a zoom-in on the absorptivity in the PDT window. Bottom: Gel electrophoresis assays of the interactions of $[(bpy)_2Os(dpp)RhCl_2(phen)_2]^{3+}$ with pUC18 DNA in the absence of O_2 and irradiate with (*c*) $\lambda \geq 590$ nm or (*d*) $\lambda \geq 645$ nm. Lane λ is the lambda molecular weight marker, lane C is the pUC18 control, lane MC is a 5:1 BP/MC solution incubated in the dark for 120 min, lanes 60 and 240 are 5:1 BP/MC solutions photolyzed for 60 or 240 min, respectively, under Ar. [Adapted from (79).]

shows that the smaller amount of metal complex photobound to DNA leads to minor changes in DNA migration in the gel shift assay. Very little photocleavage is observed at this low metal loading. Remarkably, the complex fully inhibits DNA replication and amplification after just 60 min of irradiation, demonstrating that a small number of photobound MC–DNA adducts is sufficient to prohibit DNA replication by this Os,Rh supramolecule. The PCR experiment requires 35 cycles of 94, 58, and 72 °C incubation over a time period of 180 min. The adduct formed between the DNA and $[(bpy)_2Os(dpp)RhCl_2(phen)]^{3+}$ is quite stable to withstand such conditions, providing support that this motif may serve as a new paradigm for PDT drug development. The results obtained in this PCR model for *in vivo* DNA replication demonstrates the promise of these new motifs for light activated treatment of diseases that can be targeted by inhibition of DNA including cancer.

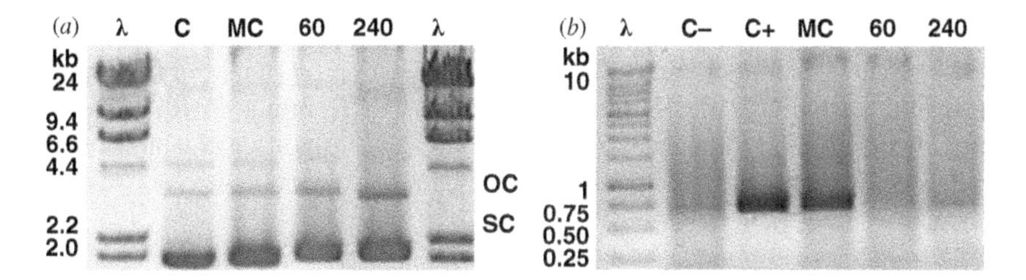

Figure 39. (*a*) Imaged agarose gel electrophoresis assay of the light-induced interaction of pUC18 DNA and $[(bpy)_2Os(dpp)RhCl_2(phen)_2]^{3+}$ irradiated with $\lambda \geq 590$ nm. Lane λ is the lambda molecular weight marker, lane C is the pUC18 control, lane MC is a 50:1 BP/MC solution incubated in the dark for 240 min under Ar, lanes 60 and 240 are 50:1 BP/MC solutions photolyzed for 60 or 240 min, respectively, under Ar. (*b*) The PCR amplification assay with $[(bpy)_2Os(dpp)RhCl_2(phen)]^{3+}$ irradiated with $\lambda \geq 590$ nm. Lane λ is the lambda molecular weight marker, lane C− is a negative control using deionized H_2O, lane C+ is a positive control using pUC18 template, lane MC is the PCR product using pUC18 incubated with the complex at a 50:1 BP/MC ratio for 240 min in the dark under Ar, lanes 60 and 240 are the PCR results using pUC18 and the complex irradiated for 60 or 240 min, respectively (80). [From J. Wang et al., *Angew. Chem. Inter. Ed. 125*, 1300 (2013) Copyright © 2014 by John Wiley & Sons, Inc. Reprinted by permission of John Wiley & Sons, Inc.]

Coupling Ru- or Os-polyazine light absorbers to a *cis*-$Rh^{III}Cl_2$ BAS affords supramolecules that undergo O_2 independent photobinding and photocleavage of DNA. This O_2 independent pathway (Type III PDT) is of great importance in developing light-activated anti-cancer drugs to treat aggressive, hypoxic tumors. The series of complexes discussed above were analyzed by electrochemistry, electronic absorption spectroscopy, steady-state and time-resolved emission spectroscopy, and gel electrophoresis to gain insight into the effects of these properties on biological reactivity, and these properties are summarized in Table I. Selection of BL in the $[\{(bpy)_2Ru(BL)\}_2RhCl_2]^{5+}$ motif is important in designing systems with appropriate orbital energetics to populate a Rh→Rh MMCT excited state, photolabilize the chloride ligands, and bind to DNA. This effect is also observed in the bimetallic $[(bpy)_2Ru(BL)RhCl_2(phen)]^{3+}$, where BL impacts the rate of photocleavage. The visible light absorption of the bimetallic complex $[(bpy)_2Os(dpp)RhCl_2(phen)]^{3+}$ extends into the therapeutic window with relatively intense direct ^{3}MLCT absorption due to spin–orbit coupling. This complex photocleaves DNA in the absence of O_2 in the therapeutic window ($\lambda^{irr} \geq 645$ nm), and recent studies highlight its ability to inhibit DNA replication and amplification in PCR studies using low-energy visible light. The characteristics of the Os,Rh bimetallic system, which can photobind to DNA and photocleave DNA without the need for molecular oxygen using red light excitation, are quite unusual and a result of the progressive study of the bioreactivity of these supramolecular motifs. The further demonstration that photomodification of the DNA by the Os,Rh complex using red light leads to complete inhibition of DNA replication even at relatively low loading provides support for the development of new drug therapies based on these motifs. These promising PDT agents featuring *cis*-$Rh^{III}Cl_2$ BAS provide much insight into the role of orbital energetics, sterics, and labilization of metal-halide bonds in photomodification of biomolecules.

V. CONCLUSIONS

The development of effective anti-cancer agents that are selective, targeted, noninvasive, and exhibit low toxicity to healthy cells can be transformational in disease treatment. Many cancer therapies are cross-resistant and toxic to the body, making PDT particularly attractive in this forum. Cancer is identified as uncontrolled cell growth due to uncontrolled DNA replication. Compounds that damage DNA through cleavage or binding can inhibit that uncontrolled replication to impede cancerous cell growth. The well-known anti-cancer drug, cisplatin, and its second and third generation derivatives, carboplatin and oxaliplatin, are currently used to treat cancers, such as testicular, ovarian, bladder, head, and neck. Major drawbacks of many chemotherapeutics, including platinum drugs, include a lack of selectivity to cancerous cells leading to damage of healthy cells and intrinsic or acquired cross-resistance.

Photodynamic therapy offers a selective treatment for cancer using highly toxic materials present only upon optical excitation at the tumor site. An ideal PDT drug is localized at a tumor, nontoxic in the dark, and highly toxic upon direct excitation with visible light in the therapeutic window (600–900 nm, wavelengths of light that can pass through the skin). Many PDT agents undergo energy transfer to molecular oxygen to produce $^1O_2$, a ROS. This ROS can damage DNA by causing cleavage of the phosphate backbone. Photofrin is a porphyrin-based PDT agent used clinically to treat esophageal and endobronchial tumors, and it tends to accumulate in rapidly growing tissue. A drawback of this treatment is the non-unity population of the ^{3}ES that leads to an efficiency of $^1O_2$ generation of only 0.65 and a need for significant oxygen concentrations to be active. Coordination complexes with Ru(II) or Os(II) centers and polypyridyl TLs provide a new method of PDT. Tris-homoleptic complexes of the form $[Ru(TL)_3]^{2+}$, where TL = phen, bpy, and Ph_2phen, generate $^1O_2$ with efficiencies of 0.54, 0.73, and 0.97, respectively. This trend can be correlated to the ^{3}MLCT excited-state lifetimes of 0.313, 0.765, and 5.34 μs in methanol, respectively. Longer lived excited states are desirable as they provide more time for $^3O_2$ to interact with the excited state and undergo energy transfer. A common drawback to using $[Ru(TL)_3]^{2+}$ complexes is the lack of absorbance in the therapeutic window, transport through cell and nuclear membranes, and little specificity of localization near DNA or other biomolecules.

Supramolecular complexes, which combine the characteristic traits of individual building blocks into larger assemblies that perform a complex function through contributions of each building block, offer the ability to couple multifunctional units into one molecule providing a type of targeting not possible otherwise and combination-type therapy in one molecular architecture. Early approaches couple one or more $^1O_2$ generating PDT agents or LA to a covalent DNA binder through a BL to target ROS generation to the DNA biomolecule. Platinum and rhodium complexes are reported to thermally or photochemically bind to DNA, making them attractive to incorporate into supramolecular complexes for light-activated DNA modification. Properties of supramolecular complexes can be controlled by subunit variation. Subunits (e.g., LA metal, TL, BL, and BAS) each play a specific and vital role in the functioning of supramolecular complexes for DNA modification and tuning excited-state reactions in these new classes of potential PDT drugs.

Selecting the LA metal center is important in the photophysical properties of PDT agents. The Ru(II)–polypyridyl LAs are typically used as they provide rich UV and visible light absorption, relatively long-lived excited states, and photostability. Their ^{3}MLCT excited states undergo energy transfer in the presence of molecular oxygen to form ROS that are potent in cleavage of the DNA phosphate backbone. Substitution of Ru(II) with Os(II) maintains the rich light absorption and extends it into the therapeutic window, a result of direct excitation into the spin-forbidden ^{3}MLCT excited state, a transition that is much weaker in the Ru(II) analogues. This property is very important in anti-cancer therapy, as it allows photoactivation of the complex through the skin. The Os,Rh bimetallic complex $[(bpy)_2Os(dpp)RhCl_2(phen)]^{3+}$ efficiently photobinds to and photocleaves DNA with red light excitation due to this ^{3}MLCT absorption, a rare property reported for such complexes. Most recently photomodification of DNA with this Os,Rh complex has been shown to completely inhibit DNA replication via an oxygen independent pathway using red light, even during 90 °C thermal cycling in PCR, demonstrating a very stable metal–DNA adduct. The promising reactivity of the Os,Rh complex, photobinding to and photocleaving DNA using red light excitation via an oxygen independent pathway is unusual and illustrates the promise of these motifs for a new paradigm for PDT drug development. The development of this system follows from the detailed analysis and long-term study of a variety of supramolecular complexes, allowing the basic science knowledge base to develop systems with unusual and promising properties. The previously discussed Os,Pt bimetallic complex $[(bpy)_2Os(dpb)PtCl_2]^{2+}$ has the potential for similar red light induced interactions with DNA, yet photolysis studies have not been reported with this complex. Incorporation of Os(II) in place of Ru(II) in other previously reported supramolecular complexes may also result in interesting interactions with DNA in the PDT window. This result provides a promising direction for mixed-metal DNA photomodification agents.

The selection of TL and BL in these supramolecular architectures impacts properties, such as light absorption, photophysics, and lipophilicity of the metal complex. The use of Ph_2phen as the TL in the bimetallic motif $[(TL)_2Ru(BL)$-$PtCl_2]^{2+}$ enhances lipophilicity as expected based on studies with the $[Ru(TL)_3]^{2+}$ complexes. Enhanced ^3MLCT lifetimes and $^1O_2$ generation are observed despite the formally Ru→BL CT nature of the photoactive state. This finding suggests that the TL may have an unexpectedly significant contribution to that Ru→BL CT excited state. Studies of the bimetallic complexes with tridentate TLs $[(TL)RuCl$-$(dpp)PtCl_2]^+$, where TL = tpy, MePhtpy, and *t*-Bu_3tpy, give information into the impacts of TL on the metal complex's ability to permeate the cell membrane. The lipophility of the complex is enhanced with the use of MePhtpy or *t*-Bu_3tpy compared to tpy. The complex $[(t\text{-}Bu_3tpy)RuCl(dpp)PtCl_2]^+$ is expected to permeate the lipophilic cell membrane more efficiently than the MePhtpy and tpy analogues due to its log P value of 4.00 compared to −0.39 (MePhtpy) and −2.00 (tpy). The steric bulk of *t*-Bu_3tpy may impede DNA binding at the Pt BAS. Use of MePhtpy may result in an appropriate lipophilicity that will both facilitate permeation of the cell membrane and association with DNA.

The effects of subunit variation on orbital energetics, spectroscopy, and DNA interactions were summarized in the report of the Rh(III) centered trimetallic complexes (M = Ru or Os). The trimetallic complex $[\{(bpy)_2Ru(dpp)\}_2RhCl_2]^{5+}$ acts as a starting point to analyze the impact of LA metal variation to Os(II) to afford $[\{(bpy)_2Os(dpp)\}_2RhCl_2]^{5+}$, the impact of TL variation to a tridentate and a monodentate ligand to produce $[\{(tpy)RuCl(dpp)\}_2RhCl_2]^{3+}$, and the influence of BL from dpp to bpm to give the complex $[\{(bpy)_2Ru(bpm)\}_2RhCl_2]^{5+}$. Use of Os(II) as the LA metal center extends light absorption into the therapeutic window due to the optically accessible $^1GS \rightarrow {}^3MLCT$ absorption; however, red light studies were not reported. The Os based HOMO in this motif is destabilized compared to the Ru based HOMO in the analogous complex, however, this does not appear to have a major impact on the DNA interaction. Use of tpy and Cl in place of two bpy ligands also destabilizes the HOMO, but it also does not impact the DNA interaction compared to the bpy analogue. The striking difference in this trimetallic motif is observed when the bpm BL is used in place of dpp. This BL variation places the LUMO on the bpm ligand, preventing the population of a Ru→Rh ^3MMCT excited state that is observed in the other three complexes. This ^3MMCT excited state plays an important role in the ability of this architecture to bind to and cleave DNA with visible light irradiation.

A new motif for metal complex binding to DNA reported with the bimetallic complex $[(Ph_2phen)_2Ru(dpp)PtCl_2]^{2+}$ provides interesting insight into the interactions of supramolecular complexes with biomolecules. Excitation into the Ru→dpp MLCT excited state provides enhanced electron density on the dpp ligand and thereby the Pt site, decreasing Lewis acidity and facilitating Cl^- loss followed by hydrolysis and covalent binding of the Pt center to DNA. This process was shown to occur in the therapeutic window when exciting at the low-energy absorption tail of the Ru→dpp MLCT transition. Photobinding was faster when excited at 455 nm where the complex strongly absorbs visible light. This report is important in understanding how such molecules interact with biomolecules and uncovering a new paradigm for targeting: MLCT facilitated DNA photobinding.

The ability to photocleave DNA in the absence of O_2 is a critical goal in developing successful anti-cancer drugs as O_2 concentration is low in aggressive, hypoxic tumors. The Rh(III) centered bimetallic and trimetallic complexes have been reported to promote DNA photocleavage in deoxygenated conditions. The ability of the Rh complexes $[(bpy)_2M(BL)RhCl_2(phen)_2]^{3+}$ [M = Ru(II) or Os(II), BL = dpp or bpm], $[\{(bpy)_2M(dpp)\}_2RhCl_2]^{5+}$ [M = Ru(II) or Os(II)], and $[\{(tpy)RuCl(dpp)\}_2RhCl_2]^{3+}$ to populate an unusual Ru(dπ)→Rh(dσ*) ^{3}MMCT excited state to promote halide loss and enable DNA binding is central to bioactivity of this new molecular architecture. The complex $[(bpy)_2Os(dpp)RhCl_2(phen)]^{3+}$ has recently been shown to inhibit DNA replication following red light activated DNA photobinding. Combining the benefits of O_2 independent photobinding and photocleavage and low-energy excitation in the therapeutic window make the bimetallic complex $[(bpy)_2Os(dpp)RhCl_2(phen)]^{3+}$ a very interesting candidate for *in vitro* and *in vivo* studies to learn how these properties translate to biological studies. Ruthenium-centered supramolecular complexes offer an exciting and promising stride toward potent anti-cancer drugs.

The ability to tune the properties of supramolecular complexes to afford a molecular device that can target biomolecules to prelocalize a LA or PDT agent close to DNA, efficiently absorb low-energy visible light that can pass through the skin, cleave DNA through formation of ROS or through an O_2 independent pathway, and exhibit very low dark toxicity to healthy cells, while selectively killing cancer cells when excited, is a forum that holds much potential. Much progress has recently be reported in this new field. Ardent research and dedication in the field must continue in order to build and improve upon the successes of currently utilized drugs for disease treatment including cancer and to develop PDT agents and realize the potential of these supramolecular light-activated anti-cancer drugs.

ACKNOWLEDGMENTS

Acknowledgment is made to the past and present members of the Brewer Group who have contributed to this field. Special acknowledgment is given to Prof. Brenda Winkel, Prof. John Robertson, Dr. Travis White, Dr. Jing Wang, Dr. Samantha Higgins, Ms. Hannah Mallalieu, and Mr. Roberto Padilla for assistance with this manuscript and our work in this field. Acknowledgment is also made to the National Science Foundation and the Institute of Critical Technology and Applied Science for their generous funding of our research in this field and to the Virginia Tech Graduate School Dissertation Fellowship for support provided to JDK.

ABBREVIATIONS

$^1\Delta_g$	Singlet oxygen lowest energy excited state
$^1\Sigma_g^+$	Singlet oxygen highest energy excited state
$^3\Sigma_g^-$	Molecular oxygen ground state
5′-GMP	Guanosine-5′-phosphate disodium salt
A	Adenine
AM	Acetomethoxy
AQ^{2-}	1,8-Anthraquinone disulfonate
ATM	Under atmospheric conditions
BAS	Bioactive site
BL	Bridging ligand
BP/MC	Base pair/metal complex ratio
bpm	2,2′-Bipyrimidine
bpy	2,2′-Bipyridine
C	Cytosine
CT	Charge transfer
DMSO	Dimethyl sulfoxide (solvent)
DNA	Deoxyribonucleic acid
dpb	2,3-Bis(2-pyridyl)benzoquinoxaline
dpoq	Dipyrido[3,2-*f*:2′,3′-*h*]quinoxaline
dpp	2,3-Bis(2-pyridyl)pyrazine
dppn	Benzo[*i*]dipyrido[3,2-*a*:2′,3′-*h*]quinoxaline
dppz	Dipyrido[2,3-*a*:2′,3′-*c*]phenazine
dpq	2,3-Bis(2-pyridyl)quinoxaline
ES	Excited state
ESI	Electro spray ionization
ε	Absorptivity
FDA	Food and Drug Administration
FPT	Freeze–pump–thaw
G	Guanine
GS	Ground state
^{1}H NMR	Proton nuclear magnetic resonance spectroscopy
HOMO	Highest occupied molecular orbital
HpD	Hematoporphyrin derivative
IL	Intraligand
ISC	Inersystyem crossing
k_b	Binding constant
k_{en}	Rate constant for energy transfer
k_{et}	Rate constant for electron transfer
k_{ic}	Rate constant for internal conversion
k_{isc}	Rate constant for intersystem crossing
k_{nr}	Rate constant for nonradiative decay

k_r	Rate constant for radiative decay
k_{vr}	Rate constant for vibrational relaxation
λ^{im}	Irradiation wavelength
LA	Light absorber
LC	Ligand centered
LF	Ligand field
log *P*	Partition coefficient
LUMO	Lowest unoccupied molecular orbital
MALDI	Matrix-assisted laser desorption/ionization
MC	Metal centered
Me_2phen	4,7-Dimethyl-1,10-phenanthroline
Me_4phen	3,4,7,8-Tetramethyl-1,10-phenanthroline
MePhtpy	4′-(4-Methylphenyl)-2,2′:6′,2″-terpyridine
MLCT	Metal-to-ligand charge transfer
MMCT	Metal-to-metal charge transfer
MS	Mass spectrometry
NIR	Near-infrared
NN	Bidentate polyazine ligand
OC	Open circular
PCR	Polymerase chain reaction
PDT	Photodynamic therapy
PPh_3	Triphenylphosphine
Ph_2phen	4,7-Diphenyl-1,10-phenanthroline
phen	1,10-Phenanthroline
PMD	Photochemical molecular device
py^+	3-Cyano-1-methylpyridinium tetrafluoroborate
RNA	Ribonucleic acid
ROS	Reactive oxygen species
rt	Room temperature
SC	Supercoiled
T	Thymine
t-Bu_3tpy	4,4′,4″-Tri-*tert*-butyl-2,2′:6′,2″-terpyridine
TL	Terminal ligand
tppz	2,3,5,6-Tetrakis(2-pyridyl)pyrazine
tpy	2,2′:6′,2″-Terpyridine
UV	Ultraviolet
vis	Visible
λ^{irr}	Irradiation wavelength
τ	Lifetime
Φ	Quantum yield
$\Phi_{^1O_2}$	Quantum yield of singlet oxygen generation
Φ^{em}	Quantum yield of emission

REFERENCES

1. R. Siegel, E. Ward, O. Brawley, and A. Jemal, *CA: A Cancer J. Clin. 61*, 212 (2011).
2. B. Alberts, A. Johnson, J. Lewis, M. Raff, K. Roberts, and P. Walter, *Molecular Biology of the Cell.* 5th ed., Garland Science, Taylor & Francis Group, LLC, NY, 2008.
3. M. J. Hannon, *Chem. Soc. Rev. 36*, 280 (2007).
4. B. Armitage, *Chem. Rev. 98*, 1171 (1998).
5. L. J. K. Boerner and J. M. Zaleski, *Curr. Opin. Chem. Biol. 9*, 135 (2005).
6. A. E. Friedman, J. C. Chambron, J. P. Sauvage, N. J. Turro, and J. K. Barton, *J. Am. Chem. Soc. 112*, 4960 (1990).
7. C. V. Kumar, J. K. Barton, and N. J. Turro, *J. Am. Chem. Soc. 107*, 5518 (1985).
8. R. B. Martin, *Cisplatin* Zurich: Verlag Helvetica Chimica Acta; Weinheim, New York: Wiley-VCH, 1999, pp. 111–134.
9. A. D. Richards and A. Rodger, *Chem. Soc. Rev. 36*, 471 (2007).
10. S. E. Sherman and S. J. Lippard, *Chem. Rev. 87*, 1153 (1987).
11. D. B. Zamble and S. J. Lippard, *Trends Biochem. Sci. 20*, 435 (1995).
12. R. Zhao, R. Hammitt, R. P. Thummel, Y. Liu, C. Turro, and R. M. Snapka, *Dalton Trans.* 10926 (2009).
13. R. A. Hsi, D. I. Rosenthal, and E. Glatstein, *Drugs 57*, 725 (1999).
14. L. D. Via and S. M. Magno, *Curr. Med. Chem. 8*, 1405 (2001).
15. D. F. Zigler and K. J. Brewer, in *Metal Complex–DNA Interactions*, John Wiley & Sons, Ltd., Chickester, West Sussex, UK, 2009, pp. 235–272.
16. I. J. MacDonald and T. J. Dougherty, *J. Porphyrins Phthalocyanines 5*, 105 (2001).
17. K. Szaciłowski, W. Macyk, A. Drzewiecka-Matuszek, M. Brindell, and G. Stochel, *Chem. Rev. 105*, 2647 (2005).
18. M. C. DeRosa and R. J. Crutchley, *Coord. Chem. Rev. 233–234*, 351 (2002).
19. L. Wyld, M. W. Reed, and N. J. Brown, *Br. J. Cancer 77*, 1621 (1998).
20. G. Stochel, A. Wanat, E. Kuliś, and Z. Stasicka, *Coord. Chem. Rev. 171*, 203 (1998).
21. K. F. M. Fan, C. Hopper, P. M. Speight, G. A. Buonaccorsi, and S. G. Bown, *Int. J. Cancer 73*, 25 (1997).
22. C. P. Anderson, D. J. Salmon, T. J. Meyer, and R. C. Young, *J. Am. Chem. Soc. 99*, 1980 (1977).
23. J. N. Demas, D. Diemente, and E. W. Harris, *J. Am. Chem. Soc. 95*, 6864 (1973).
24. J. W. Dobrucki, *J. Photochem. Photobiol., B. 65*, 136 (2001).
25. D. Garcìa-Fresnadillo, Y. Georgiadou, G. Orellana, A. M. Braun, and E. Oliveros, *Helv. Chim. Acta 79*, 1222 (1996).
26. A. Hergueta-Bravo, M. E. Jiménez-Hernández, F. Montero, E. Oliveros, and G. Orellana, *J. Phys. Chem. B. 106*, 4010 (2002).
27. A. Juris, V. Balzani, F. Barigelletti, S. Campagna, P. Belser, and A. von Zelewsky, *Coord. Chem. Rev. 84*, 85 (1988).

28. J. N. Demas, E. W. Harris, and R. P. McBride, *J. Am. Chem. Soc. 99*, 3547 (1977).
29. A. Juris, V. Balzani, F. Barigelletti, S. Campagna, P. Belser, and A. von Zelewsky, *Coord. Chem. Rev. 84*, 85 (1988).
30. J. M. Kelly, A. B. Tossi, D. J. McConnell, and C. OhUigin, *Nucleic Acids Res. 13* 6017 (1985).
31. A. B. Tossi and J. M. Kelly, *Photochem. Photobiol. 49*, 545 (1989).
32. B. Rosenberg, L. Van Camp, and T. Krigas, *Nature (London) 205*, 698 (1965).
33. B. Rosenberg, *Interdiscipl. Sci. Rev. 3*, 134 (1978).
34. B. Rosenberg, L. Vancamp, J. E. Trosko, and V. H. Mansour, *Nature (London) 222*, 385 (1969).
35. K. Barabas, R. Milner, D. Lurie, and C. Adin, *Vet. Comp. Oncol. 6*, 1 (2008).
36. J. Reedijk, *Chem. Commun.*, 801 (1996).
37. I. Kostova, *Recent Pat Anticancer Drug Discov 1*, 1 (2006).
38. R. E. Mahnken, M. A. Billadeau, E. P. Nikonowicz, and H. Morrison, *J. Am. Chem. Soc. 114*, 9253 (1992).
39. M. M. Muir, and W.-L. Huang, *Inorg. Chem. 12*, 1831 (1973).
40. G. A. Crosby and J. N. Demas, *J. Am. Chem. Soc. 92*, 7262 (1970).
41. M. A. Billadeau, K. V. Wood, and H. Morrison, *Inorg. Chem. 33*, 5780 (1994).
42. H. L. Harmon and H. Morrison, *Inorg. Chem. 34*, 4937 (1995).
43. D. Loganathan and H. Morrison, *Photochem. Photobiol. 82*, 237 (2006).
44. E. L. Menon, R. Perera, M. Navarro, R. J. Kuhn, and H. Morrison, *Inorg. Chem. 43*, 5373 (2004).
45. K. M. Kadishi, K. Das, R. Howard, A. Dennis, and J. L. Bear, *Bioelectroch. Bioener. 5*, 741 (1978).
46. K. Das, K. M. Kadish, and J. L. Bear, *Inorg. Chem. 17*, 930 (1978).
47. P. K. L. Fu, P. M. Bradley, and C. Turro, *Inorg. Chem. 40*, 2476 (2001).
48. P. M. Bradley, B. E. Bursten, and C. Turro, *Inorg. Chem. 40*, 1376 (2001).
49. J. M. Asara, J. S. Hess, E. Lozada, K. R. Dunbar, and J. Allison, *J. Am. Chem. Soc. 122*, 8 (2000).
50. K. Sorasaenee, P. K. L. Fu, A. M. Angeles-Boza, K. R. Dunbar, and C. Turro, *Inorg. Chem. 42*, 1267 (2003).
51. P. M. Bradley, A. M. Angeles-Boza, K. R. Dunbar, and C. Turro, *Inorg. Chem. 43*, 2450 (2004).
52. A. M. Angeles-Boza, P. M. Bradley, P. K. L. Fu, S. E. Wicke, J. Bacsa, K. R. Dunbar, and C. Turro, *Inorg. Chem. 43*, 8510 (2004).
53. A. M. Angeles-Boza, P. M. Bradley, P. K. L. Fu, M. Shatruk, M. G. Hilfiger, K. R. Dunbar, and C. Turro, *Inorg. Chem. 44*, 7262 (2005).
54. L. E. Joyce, J. D. Aguirre, A. M. Angeles-Boza, A. Chouai, P. K. L. Fu, K. R. Dunbar, and C. Turro, *Inorg. Chem. 49*, 5371 (2010).
55. V. Balzani, L. Moggi, and F. Scandola, V. Balzani, Ed., Reidel, Dordrecht, 1987, pp. 1 and references cited therein.

56. S. Swavey, R. L. Williams, Z. Fang, M. Milkevitch, and K. J. Brewer, *Society of Photo-Optical Instrumentation Engineers (SPIE) Conference Series*, W. B. Spillman, Ed., Vol. *4512*, Bellingham, WA, pp. 75–83, 2001.
57. S. Swavey and K. J. Brewer, in *Comprehensive Coordination Chemistry II*, T. J. Meyer, Ed. Pergamon, Oxford, 2003, pp. 135–157.
58. M. Milkevitch, E. Brauns, and K. J. Brewer, *Inorg. Chem. 35*, 1737 (1996).
59. M. Milkevitch, H. Storrie, E. Brauns, K. J. Brewer, and B. W. Shirley, *Inorg. Chem. 36*, 4534 (1997).
60. M. Milkevitch, B. W. Shirley, and K. J. Brewer, *Inorg. Chim. Acta 264*, 249 (1997).
61. S. L. H. Higgins, T. A. White, B. S. J. Winkel, and K. J. Brewer, *Inorg. Chem. 50*, 463 (2010).
62. V. W.-W. Yam, V. W.-M. Lee, and K.-K. Cheung, *J. Chem. Soc., Chem. Commun.*, 2075 (1994).
63. S. L. H. Higgins, A. J. Tucker, B. S. J. Winkel, and K. J. Brewer, *Chem. Commun. 48*, 67 (2012).
64. K. Sakai, H. Ozawa, H. Yamada, T. Tsubomura, M. Hara, A. Higuchi, and M.-a. Haga, *Dalton Trans.* 3300 (2006).
65. K. Sakai and T. Tsubomura, *J. Inorg. Biochem. 67*, 349 (1997).
66. R. Miao, M. T. Mongelli, D. F. Zigler, B. S. J. Winkel, and K. J. Brewer, *Inorg. Chem. 45*, 10413 (2006).
67. R. Sahai and D. P. Rillema, *J. Chem. Soc., Chem. Communs.*, 1133 (1986).
68. Z. Fang, S. Swavey, A. Holder, B. Winkel, and K. J. Brewer, *Inorg. Chem. Commun. 5*, 1078 (2002).
69. S. Swavey, Z. Fang, and K. J. Brewer, *Inorg. Chem. 41*, 2598 (2002).
70. R. L. Williams, H. N. Toft, B. Winkel, and K. J. Brewer, *Inorg. Chem. 42*, 4394 (2003).
71. A. Jain, B. S. J. Winkel, and K. J. Brewer, *J. Inorg. Biochem. 101*, 1525 (2007).
72. A. Jain, J. Wang, E. R. Mashack, B. S. J. Winkel, and K. J. Brewer, *Inorg. Chem. 48*, 9077 (2009).
73. C. Hansch, J. P. Björkroth, and A. Leo, *J. Pharm. Sci. 76*, 663 (1987).
74. S. Zhao, S. M. Arachchige, C. Slebodnick, and K. J. Brewer, *Inorg. Chem. 47*, 6144 (2008).
75. A. J. Prussin II, S. Zhao, A. Jain, B. S. J. Winkel, and K. J. Brewer, *J. Inorg. Biochem. 103*, 427 (2009).
76. A. A. Holder, S. Swavey, and K. J. Brewer, *Inorg. Chem. 43*, 303 (2004).
77. S. Swavey and K. J. Brewer, *Inorg. Chem. 41*, 6196 (2002).
78. G. N. A. Nallas, S. W. Jones, and K. J. Brewer, *Inorg. Chem. 35*, 6974 (1996).
79. J. Wang, S. L. H. Higgins, B. S. J. Winkel, and K. J. Brewer, *Chem. Commun. 47*, 9786 (2011).
80. J. Wang, J. Newman, S. L. H. Higgins, K. M. Brewer, B. S. J. Winkel, and K. J. Brewer, *Angew. Chem., Inter. Ed. 125*, 1300 (2013).

81. J. Wang, D. F. Zigler, N. Hurst, H. Othee, B. S. J. Winkel, and K. J. Brewer, *J. Inorg. Biochem.*, *116*, 135 (2012).
82. A. A. Holder, D. F. Zigler, M. T. Tarrago-Trani, B. Storrie, and K. J. Brewer, *Inorg. Chem. 46*, 4760 (2007).
83. B. Storrie, A. Holder, and K. J. Brewer, in *Society of Photo-Optical Instrumentation Engineers (SPIE) Conference Series* D. Kessel, Ed., Vol. *6139*, Bellingham, WA, pp. 336–342, 2006.
84. T. A. White, J. Wang, H. E. Mallalieu, and K. J. Brewer, *Chem.-A Eur. J.*, in press (2014).

Selective Binding of Zn^{2+} Complexes to Non-Canonical Thymine or Uracil in DNA or RNA

KEVIN E. SITERS, STEPHANIE A. SANDER, AND JANET R. MORROW

Department of Chemistry, University at Buffalo, State University of New York, Amherst, NY

CONTENTS

Progress in Inorganic Chemistry, Volume 59, First Edition. Edited by Kenneth D. Karlin.

I. INTRODUCTION

A. Interactions of Zn^{2+} With Nucleobases in Nucleic Acids

Late transition metal ions in the divalent state typically bind to the nucleobases of deoxyribonucleic acid (DNA) or ribonucleic acid (RNA) rather than to the phosphate ester backbone. The most well-known examples are Pt^{2+} anticancer drugs that bind primarily to the N7 position of purine nucleobases in duplex DNA, especially guanine (G) (1). The N3 position of thymine (T) is the primary binding site for Hg^{2+} ions. This site selectivity has been utilized for catalytic DNA molecules for detection of Hg^{2+} (2). These nucleic acid sensors are activated by Hg^{2+} stabilization of T–T mismatches in duplexes. A final example is the oxidation of guanine nucleobases catalyzed by Ni^{2+} complex binding to the N7 position. This selective reaction has been applied for structural analysis by marking exposed guanines in RNA or DNA (3). These examples show that the nucleobase binding selectivity of transition metal ions is quite remarkable and can be applied in research areas ranging from the design of anticancer drugs, to metal ion sensors and probes of nucleic acid structure.

In this chapter, we discuss Zn^{2+} macrocyclic complexes that bind to DNA and RNA with an emphasis on the recognition of non-canonical DNA and RNA structures. Zn^{2+} is a post-transition metal ion grouped with the transition elements based on its similar chemical properties. The Zn^{2+} complexes presented here selectively bind to thymine or uracil nucleobases. The interaction of Zn^{2+} complexes of L1 and their derivatives with simple nucleosides, single-stranded oligonucleotides, and duplex DNA were reviewed several years ago (4). However, macrocycles other than L1 derivatives were not reported previously and binding to nucleobases in non-canonical nucleic acid structures was not discussed. Our focus here is on the recognition of unpaired thymine or uracil (U) nucleobases in bulges or abasic sites with the goal of designing Zn^{2+} complexes of macrocycles as probes for specific DNA or RNA secondary structures.

There are many advantages of using Zn^{2+} macrocyclic complexes as recognition agents for nucleic acids. First, Zn^{2+} is a strongly Lewis acidic divalent metal ion with borderline hard–soft acid–base characteristics (4). This promotes binding to available ligating oxygen- or nitrogen-containing groups in nucleic acids. The coordination bond to Zn^{2+} is stronger than the typical hydrogen bonds formed by the purely organic heterocyclic recognition agents for nucleobase binding. Second, ligand exchange reactions of Zn^{2+} are typically rapid, so that nucleic acid binding events are at equilibrium after a relatively short period of time. A third advantage of Zn^{2+} is that there are no other readily accessible oxidation states such that Zn^{2+} is not redox active under biologically relevant conditions. Redox active metal ions may damage DNA through promoting the formation of reactive oxygen species (3, 5). Finally, because Zn^{2+} is a biologically relevant metal ion, the development of Zn^{2+} based recognition agents for nucleic acids may be useful for biologically interesting sensing applications. Macrocyclic complexes of Zn^{2+} are useful in this regard because the macrocycle binds tightly to the Zn^{2+} ion in water, making it feasible to use the complexes under biologically relevant conditions (6, 7). The amine groups of these macrocycles are readily functionalized with groups that further enhance selective binding (4). As described here, the key to developing the Zn^{2+} complexes as recognition agents for RNA–DNA secondary structures is combining organic–inorganic components to build multifunctional agents.

Prior to describing binding of metal ions to non-canonical nucleic acids, it is useful to examine binding to double-stranded DNA or RNA. Binding of divalent metal ion complexes to double-stranded nucleic acids normally does not involve the groups that are involved in Watson–Crick bond formation (8). Pyrimidine bases (e.g., C, T, and U, see Scheme 1) in double-stranded structures have all heteroatoms except for the carbonyl of thymine or uracil and the nitrogen attached to the sugar involved in Watson–Crick hydrogen bonds. In contrast, purines including G and adenine (A) have N7 exposed in the major groove. The N7 of purines, especially G, is a common metal ion binding site. The N3 site of purines is exposed in the minor groove, but this site does not typically bind transition metal ions in duplex DNA or RNA.

Metal ion binding to nucleobases in non-canonical structures is generally quite different from binding to nucleobases involved in double-stranded nucleic acids because non-canonical bases have more accessible ligating groups for binding metal ions (9). These binding sites include the carbonyl oxygens of G, C, U, and T. In contrast, the exocyclic amino groups of G, A, and C do not generally participate in metal ion binding because the amine lone pair is not very basic. Amide groups

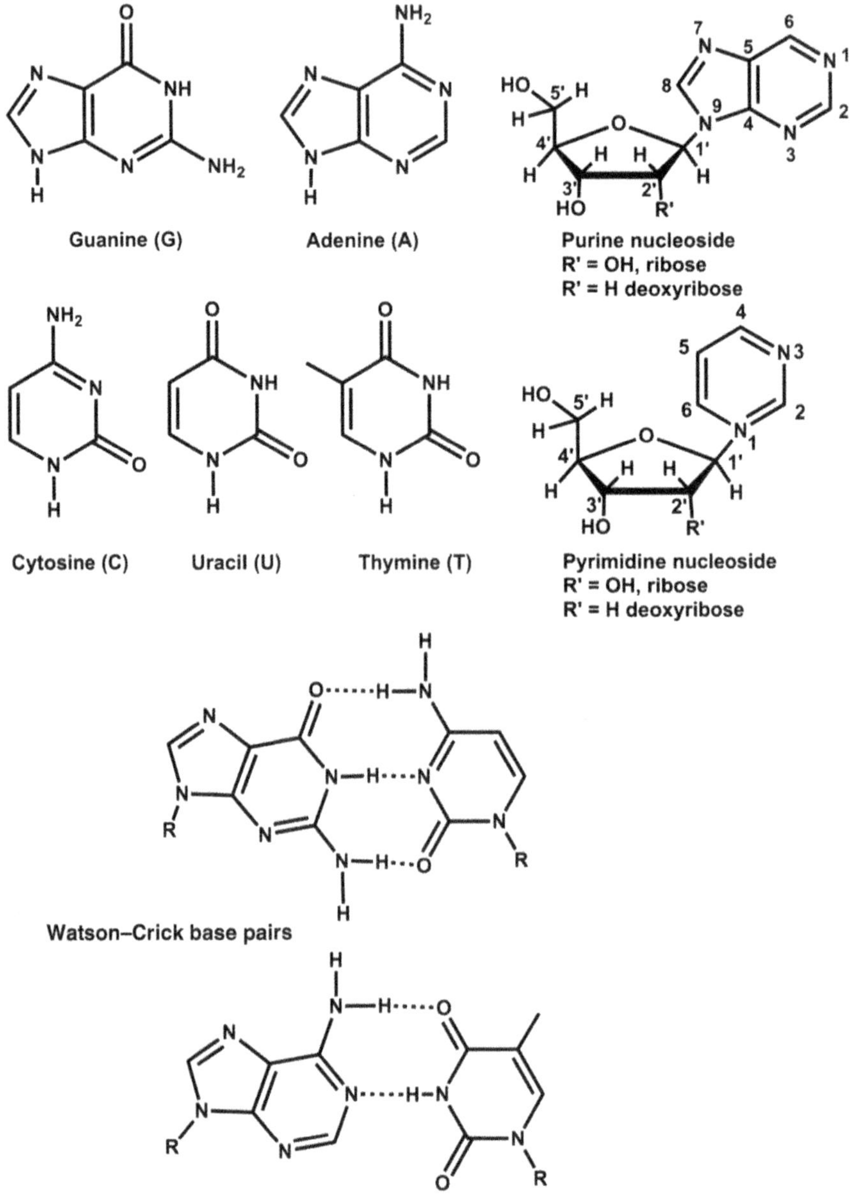

Guanine (G)

Adenine (A)

Purine nucleoside
R' = OH, ribose
R' = H deoxyribose

Cytosine (C)

Uracil (U)

Thymine (T)

Pyrimidine nucleoside
R' = OH, ribose
R' = H deoxyribose

Watson–Crick base pairs

Scheme 1. Nucleobases, nucleosides, and Watson–Crick base pairing.

(e.g., N3 in pyrimidines and N1 in guanine), typically bind transition metal ions in the deprotonated form of the NH group (10, 11). For purines, the amide NH (N1) of guanosine nucleoside with 3′,5′- bis (phosphate) diester substitutents has a pK_a of 9.6, whereas the N1 of corresponding adenosine nucleoside is 3.8 (12). Uridine and thymidine amides are especially important metal ion binding sites because of the strong basicity of the nitrogen in the amide NH group (acid dissociation constant, pK_a for NH = 9.6 and 10.1, respectively). In comparison, the pK_a of the NH of cytidine (N3) nucleoside with 3′,5′- bis phosphate diester substitutents is 4.4 (12). Metal ion–aquo complexes coordinate to N-amide and may also interact with the carbonyls of the pyridimides through hydrogen bonding (10). Macrocyclic Zn^{2+} complexes interact with the N-amide and hydrogen bond to the carbonyls through NH of the macrocyclic amines. In comparison to uracil and thymine, cytosine does not typically bind metal ions (e.g., Zn^{2+}) strongly because of the lowered basicity of the amide nitrogen (11).

Steric factors are also important in metal ion binding. Steric factors have been calculated for six-coordinate first-row transition metal ion complexes. These calculations show that the greater steric restrictions on the N1 of guanine may be responsible for weaker binding to this site in comparison to the N3 of uracil or thymine (13). Guanine bases typically bind Zn^{2+} complexes more strongly through the N7 than at the deprotonated amide N1 group (6).

B. Structures That Contain Thymine or Uracil in Non-Canonical DNA or RNA

There are several types of secondary structures that contain T or U in non-canonical contexts. Removal of the complementary base to form an abasic site in DNA is conceptually the simplest. In this type of structure, the Watson–Crick face of the thymine is accessible for Zn^{2+} macrocycle binding because the complementary base is absent. The base opposite the abasic site (apurinic or apyrimidinic sites) can be either extrahelical or intrahelical; it is possible, however, for both to exist in equilibrium. Studies suggest that a purine base in an apyrimidinic site tends to be stacked into the helix. The Patel group observed an adenine nucleobase stacking into the duplex toward an apyrimidinic site (14). However, introducing an abasic site by using an unnatural mimic [e.g., tetrahydrofuran (THF), acyclic propanyl, or acyclic ethanyl moieties] can disrupt the local structure around the abasic site. In these examples, downfield shifts of phosphate esters observed with ^{31}P nuclear magnetic resonance (NMR) indicate a structure that is not B-form. The structure of the abasic site is dependent on the abasic site mimic (natural or unnatural) and the flanking base. These alterations can cause the duplex to lose its B-form around the abasic site (15). In contrast, using a natural aldehydic site causes an adenine or guanine opposite an apyrimidinic site to stack back into the duplex. In addition, introducing an aldehydic abasic site minimizes structural changes, conserving the right-handedness and B-form of the duplex (16).

The nucleobases opposing apurinic sites, including thymines, may be either extrahelical or intrahelical, depending primarily on flanking base pairs and temperature. Coppel et al. (17) reported a structure in which the unpaired thymine stacks back into the duplex. In this example, a bifurcated hydrogen bond between O4 on the flipped-in thymine and amino groups on a flanking C appears to help stabilize the intrahelical conformation. In a study reported by Singh et al. (18), a thymine residue adjacent to an apurinic site, deoxyribose, can be extrahelical or intrahelical. The study shows that the intrahelical conformation is favored by 68.1 kcal mol^{-1}. The extrahelical conformation is most likely driven by π–π stacking between guanines that flank the abasic site, but is disfavored by desolvation penalties of both the thymine residue and abasic site. The conformation of the thymine adjacent to the abasic site is temperature dependant; at $\geq$15 °C the intrahelical to extrahelical thymine equilibrium is shifted toward the extrahelical conformation. In these studies, it was also determined that in an apurinic site, the pyrimidine is more likely to be flipped out due to favorable stacking of neighboring base pairs (19). A study to determine the structure of an abasic site formed by topoisomerase II cleavage shows that a cytidine residue opposite an apurinic site is flipped-out of the duplex toward the minor groove (20).

Abasic sites are biologically important in that they are found during repair processes upon excision of an incorrect base. Recognition agents for abasic sites are typically flat aromatic compounds that can form hydrogen bonds to the unpaired base. These compounds include naphthyridine derivatives studied by Teramae. 2-Amino-7-methyl-1,8-naphthyridine (AMND) was shown to stabilize abasic sites in aqueous media. The AMND binds to an abasic site with an opposing cytosine that exhibits fluorescent quenching and increases the duplex melting temperature by 13.7 °C. However, AMND also increases the thermal melting temperature of abasic sites with opposing thymine (11.2 °C), Guanine (4.9 °C) and adenine (3.0 °C) nucleobases (21). The AMND was modified by adding methyl groups to 5,6,7 positions to increase selectivity for pyrimidines over purines (22). Attempts to improve selectivity for thymine over cytosine by addition of a trifluoromethyl group were not successful (23). An example of the interaction between an AMND scaffold and cytosine or thymine residues is shown in Scheme 2. Abasic sites have also been used as aptamers for biologically relevant molecules (24–26).

Single thymine or uracil base bulges consist of an extra base in one strand. The other neighboring bases on the same strand and on the opposing strand form Watson–Crick base pairs in a helical structure. Single base bulges may have the nucleobase flipped away from the helix or tucked back into the helical stack. In the former case, the Watson–Crick face of the nucleobase is accessible for Zn^{2+} complex binding. Deoxyribonucleic acid bulges typically form as intermediates in

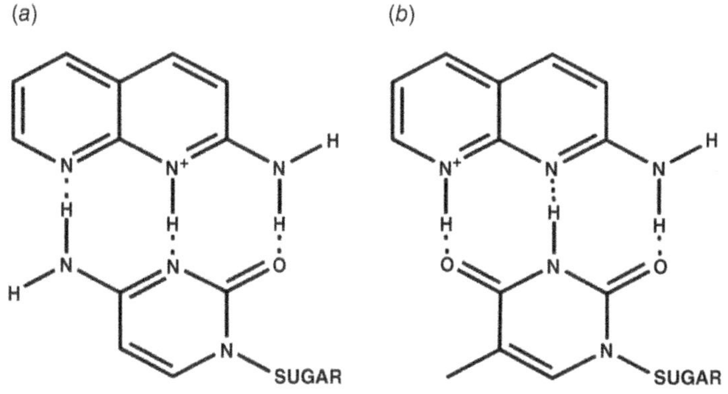

Scheme 2. Hydrogen-bonding interactions between AMNDH^{+} and (*a*) cytosine and (*b*) thymine. [Adapted from (22, 23).]

slipped DNA synthesis involving expanded DNA trinucleotide repeats or arise during errors in replication. In contrast, bulges are a commonly occurring structural motif in RNA.

Bulge structure and position are both dynamic. Temperature and flanking base pairs may dictate the bulge location and stability. For example, Patel showed that the structure of a thymine bulge flanked by guanines (GTG) is very temperature dependent. At 5 °C, the thymine appears to be extrahelical; however, as the temperature is increased to 35 °C, the thymine intercalates back into the duplex and becomes intrahelical. The structure of a thymine bulge flanked by cytosine (CTC) is less temperature dependent (27). Work done by LeBlanc and Morden (28) showed that pyrimidine bulges (T or C bulge) are more stable than purine bulges (A or G bulges). Nuclear magnetic resonance spectroscopy experiments were used to show that a thymine bulge with flanking adenines is extrahelical at temperatures below its melting temperature (29).

Bulges that are flanked by pyrimidines are less stable than bulges flanked by purines, as determined through thermal-melting experiments (28). Specifically, one study showed that incorporating a single base bulge into a duplex could destabilize the oligonucleotide by as much as 10.1 kcal mol^{-1}. Bulges flanked by cytosine (CNC, where N = A, T, C, G) showed the largest overall destabilization with a decrease in melting temperature (8.8–15.6 °C) and a less favorable $\Delta G°$ for duplex formation by 3.2–10.1 kcal mol^{-1}. However, when a bulge is positioned between two purines (GNG, where N = A, T, C, G), destabilization is generally less pronounced. Thermal-melting temperatures decrease by 8.2–12.4 °C and $\Delta G°$ for duplex formation is less favorable by 2.6–7.7 kcal mol^{-1}. Notably, thymine bulges are 2.6 kcal mol^{-1} more stable with guanine flanking bases than with cytosine flanking bases (30).

Recognition of thymine bulges typically capitalizes on interaction of the recognition agent with the nucleobase, generally flat aromatic organic compounds that form hydrogen bonds to exposed nucleobases. For example, Zimmerman and co-workers (31) reported on several aromatic compounds that form hydrogen bonds to the thymine. In some cases, a bifunctional agent that also contains a stacking moiety is included to increase the strength of the interaction. However, nucleobase selectivity is not high. Only five-fold selectivity of thymine bulges over cytosine bulges was observed. Additionally, the compounds reported do not exhibit a preference for bulges with pyrimidine or purine flanking bases. Most of the compounds reported contain a naphthyl group that increases interactions with the bulged duplex, and this might contribute to decreased selectivity. The proposed hydrogen bonds that form in these interactions are shown in Scheme 3.

Nakatani and co-workers (32, 33) reported a series of aromatic compounds, similar to Zimmerman, that utilize hydrogen bonding to recognize bulges. These compounds intercalate into the duplex and form hydrogen bonds with the nucleobase in the adjacent bulge. One such series of compounds contain naphthyride scaffolds as shown for 2,7-diamino-1,8-naphthyridine (DANP). Similar to the compounds discussed above, poor nucleobase selectivity is observed. Thymine and cytosine are especially difficult to discriminate by using recognition agents that rely primarily on hydrogen bonding. This finding may be attributed to a protonated species of DANP scaffold (DANPH$^+$) that can form hydrogen bonding interactions with both C and T residues as shown in Scheme 4. Thus, while this class of compounds recognizes nucleobases through hydrogen bonding and intercalation and stabilizes secondary structures (e.g., bulges and abasic sites), nucleobase selectivity is not high.

Goldberg and co-workers (34) developed a series of spirocyclic molecules that mimic a molecule that cleaves bulges: the neocarzinostatin chromophore (NCS-chrom). Neocarzinostatin chromophore is an antitumor antibiotic that

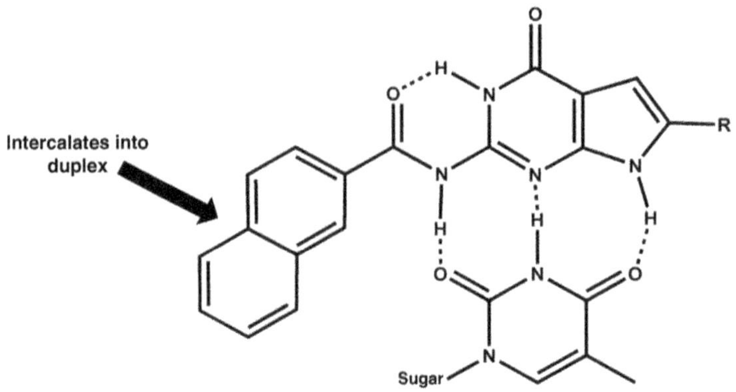

Scheme 3. Interaction of 7-deazagunaine scaffold with thymine. [Adapted from (31).]

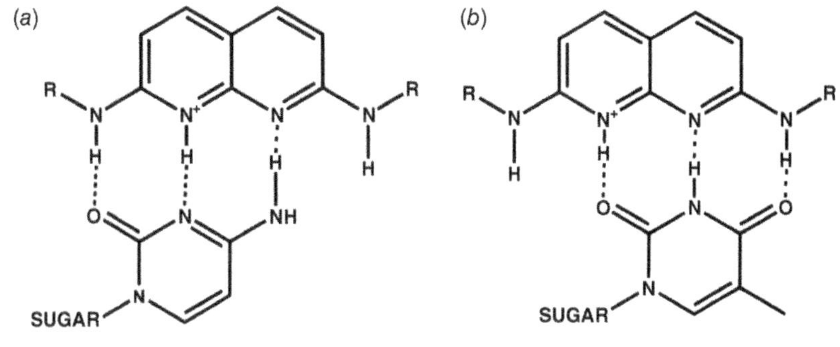

Scheme 4. The DANP^{+} scaffold interacting with a C residue (*a*) and a T residue (*b*) via hydrogen bonding. [Adapted from (32).]

cleaves DNA bulges via a radical mechanism. In the absence of bulged DNA, NCS-chrom decomposes into a spirolactam metabolite (NSCi-gb) that binds specifically to two-nucleotide bulged DNA. Several molecules were synthesized that conserved this structure. It was proposed that maintaining a wedge shape and intramolecular 35° twist would allow for the strongest binding to bulges because the compounds would mimic the natural metabolite. Compounds with *N*-methylfucosamine as the sugar moiety in α-glycosidic linkage bind similarly to bulges and show preferential binding to two-nucleotide bulges. Additionally, these compounds were shown to prevent duplex cleavage by NCS-chrom. The metabolite analogue, SCA-α2, exhibited a K_i value of 62 μM (35). Structural studies show that SCA-α2 intercalates into the major groove and ejects the nucleobases in the bulge toward the minor groove. Alternatively, compounds with a larger aromatic surface (e.g., NCSi-gb) intercalate into the minor grove and direct the nucleobases in the bulge into the major groove. However, SCA-α2, does not perturb the structure of the duplex as markedly as does NCSi-gb (36). Meyer and Hergenrother (37) developed a series of analogous compounds lacking a sugar appendage. These compounds show strong binding to RNA bulges, especially two-nucleotide RNA bulges. Using an ethidium bromide (EtBr) displacement assay, K_d values of 2.3 and 9.4 μM were measured. The nucleobase selectivity of these compounds appears to be based on bulge structure, rather than on the recognition of the nucleobase primarily through hydrogen bonds.

A series of rhodamine metalloinsertors that selectively bind to abasic or to bulge sites was reported by Barton and co-workers (38). In these examples, the Rh^{3+} center organizes the aromatic groups that may include 2,2′-bipyridine (bpy) and chrysene-5,6-quinone diimine (chrysi) or benzo[*a*]phenazine-5,6-quinone diimine (phzi). The complex $[Rh(bpy)_2(chrysi)]^{3+}$ shows preferential binding to abasic sites with an observed K_a of $1.2–3.9 \times 10^6 M^{-1}$. Single-base bulges are not recognized as strongly by the chrysi complex when compared to abasic sites.

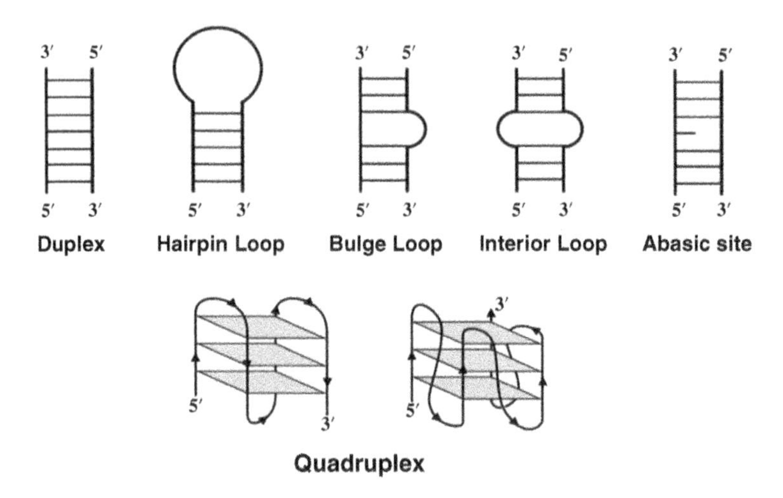

Scheme 5. Secondary structures with unpaired and paired thymine residues.

A preference for bulges with pyrimidine over purine flanking bases is observed. Despite strong binding to bulged and abasic sites, the Rh^{3+} complexes destabilize the oligonucleotides by ≤11 °C for abasic sites and ≤ 9.5 °C in bulges (39). The complexes, $[\Delta\text{-Rh(bpy)}_2\text{(chrysi)}]^{3+}$ and $[\Delta\text{-Rh(bpy)}_2\text{(phzi)}]^{3+}$, interact as metalloinsertors in the presence of both bulges and abasic sites.

Other commonly occurring DNA or RNA structures contain thymine or uracil in loop structures such as those found in hairpins or as internal loops. These structures are shown in Scheme 5. Notably nucleobases in loops are not involved in canonical base pairing and thus may be sites for metal ion interaction. However, nucleobases in loops generally interact with other bases through intrastrand stacking or hydrogen bonding. Higher order structures (e.g., triplexes or quadruplexes) are also of interest for the design of recognition agents. The G-quadruplex structures in particular frequently contain long stretches of thymines in loops that connect the guanines that form the tetrad the G-quadruplex. Interactions of Zn^{2+} complexes with these structures will not be discussed here, but are a topic of investigation in our laboratory.

II. INTERACTION OF Zn^{2+} MACROCYCLIC COMPLEXES WITH NUCLEOSIDES AND SINGLE-STRANDED OLIGONUCLEOTIDES

Several groups have studied Zn^{2+} macrocyclic complex interactions with nucleobases, especially thymine or uracil, in simple nucleosides, nucleotides, and unstructured oligonucleotides. This research was initiated largely by Aoki and

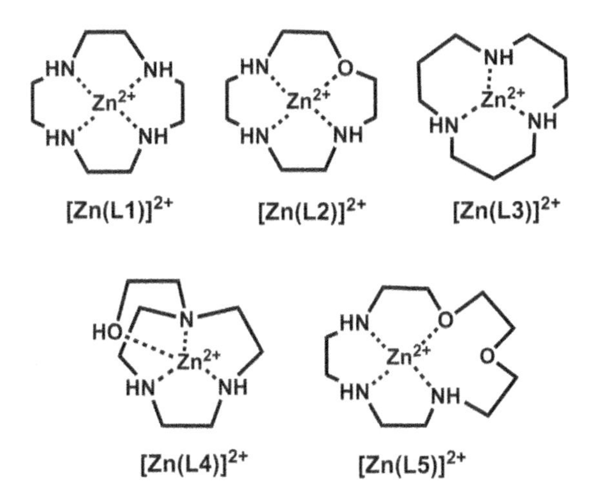

Scheme 6. The Zn^{2+} complexes of simple aza- and aza–oxa -macrocycles.

Kimura (4), a pioneer in the study of Zn^{2+} complexes for molecular recognition of anions with an interest in applying these complexes to practical medicinal chemistry applications. His work focused primarily on Zn^{2+} complexes of the tetraazamacrocycle, cyclen (L1), and its derivatives (Scheme 6). Early research in his laboratory on mononuclear $[Zn(L1)]^{2+}$ complexes (11, 40) was later expanded to multinuclear Zn^{2+} complexes (41–45) and to Zn^{2+} complexes with pendent aromatic groups (6, 46–50). Some highlights of Kimura's work will be given here, but the reader is referred to a 2004 review for further details (4). In this section, we will also present Zn^{2+} complexes of triazamacrocycles and mixed aza–oxa macrocycles that bind to thymine and uracil groups as studied in the laboratories of Morrow and co-workers (7, 51), Lönnberg and co-workers (52, 53), Martell and co-workers (54), Spiccia and co-workers (55), and Valtancoli and co-workers (56).

This section will focus on binding of Zn^{2+} macrocyclic complexes to nucleobases in relatively unstructured nucleic acids. Of special interest is the basis for the selectivity in nucleobase binding by these complexes. Both the Zn^{2+} center and additional groups (e.g., pendent aromatic groups or multiple Zn^{2+} centers) contribute to binding strength. This analysis will lay the groundwork for recognition in more structurally complex DNA and RNA sequences.

Two types of binding constants will be cited in this chapter. The first is an effective or apparent dissociation constant ($K_{d\ (app)}$), which is typically measured at near neutral pH. This constant ($K_{d\ (app)}$) is a practical measure of the interaction of Zn^{2+} macrocyclic complex with nucleoside or with DNA/RNA under the reported conditions. The K_d is, of course, the inverse of the association constant for binding of the Zn^{2+} macrocyclic complex to nucleic acid. The most common techniques used to obtain effective binding constants are ultraviolet–visible (UV–vis) and fluorescence spectroscopy. In these methods, a change in a spectroscopic property under the condition of interest is monitored as an index of complex formation (45).

The second type of binding constant is reported for a particular equilibrium step between specified solution species. The most important of these is the binding of the Zn^{2+} complex to deprotonated thymine (or uracil) groups in order to compare the strength of the interaction for different Zn^{2+} complexes (K_1 in Eq. 1) (7). Notably, both thymine groups and Zn^{2+} macrocyclic complexes exist in different states of protonation that depend on solution pH. In Eq. 1, dT^- represents the N3H deprotonated form of deoxythymidine (thymidine, dT), which has a pK_a of 9.8 ($I = 0.1\ M$, $NaNO_3$) (11) or 9.2 for uridine ($I = 0.1\ M$, $NaNO_3$) (11) and thus forms predominantly at alkaline pH. The complex $[Zn(L)(OH_2)]^{2+}$ designates the cationic form of the complex with one bound water molecule. (Note that water molecules are not generally included in the formula, or in drawings and that the number of water ligands is not always known.) A water ligand is displaced from the Zn^{2+} complex upon binding anions (e.g., dT^- or U^-) (11). An additional equilibrium is the ionization of $[Zn(L)(OH_2]^{2+}$ to form the hydroxide complex $[Zn(L)(OH)]^+$. The pK_a values for this ionization range from 7.9 to 9.3 ($I = 0.100$, NaCl), depending on the macrocycle (7). Given that $[Zn(L)(OH)]^+$ does not appreciably bind thymidine or uridine, there is an optimal pH for formation of $[Zn(L)(dT^-)]^+$ under conditions where the dT^- is available to compete with formation of the hydroxide complex. The binding constant K_1 (or log K_1) is obtained from measurements that also monitor several other equilibria in solution including the ionization of thymine and $[Zn(L)(OH_2)]^{2+}$ (7, 11). The most common way to monitor these equililbria is to conduct pH potentiometric titrations of the Zn^{2+} macrocyclic complex with and without thymine or uracil containing substrates. Several equilibrium expressions are used to fit the data, which are collected as a function of pH with added base. An advantage of this method is that once the different equilibrium constants are determined, effective binding constants under different conditions can be calculated from these data.

$$[Zn(L1)]^{2+} + dT^- = \left([Zn(L1)]^{2+}\right)(dT^-) \tag{1}$$

$$K_1 = \frac{\left[\left([Zn(L1)]^{2+}\right)(dT^-)\right]}{\left[\left([Zn(L1)]^{2+}\right) + (dT^-)\right]}$$

A. Cyclen Zn^{2+} Complexes

In 1993, Kimura and co-workers (11) reported that $[Zn(L1)]^{2+}$ binds thymidine, uridine, and related derivatives through the deprotonated amide N3 (Scheme 7). A crystal structure of $[Zn(L1)(AZT)]^+$ (3′-azido-3′-deoxythmidine = AZT) showed

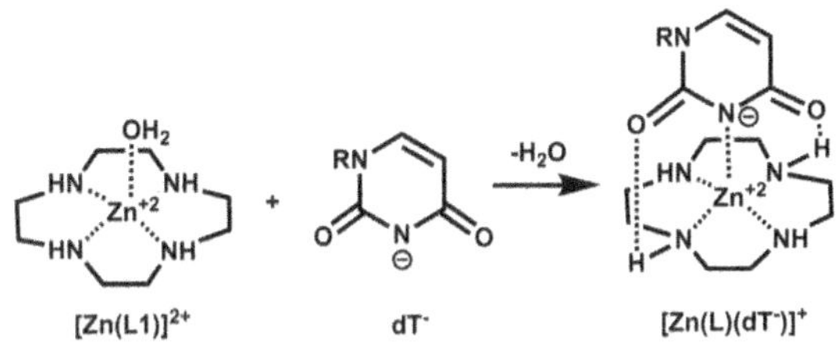

Scheme 7. Interaction of $[Zn(L1)]^{2+}$ and dT^-, where R = deoxyribose.

that the deprotonated N3 of the thymine group was coordinated to the Zn^{2+} in a complex with distorted square-pyramidal geometry. The amine protons of the cyclen (1,4,7,10-tetraa zacyclododecane) backbone form hydrogen bonds to the carbonyl groups of the thymine, consistent with delocalization of the negative charge on the oxygen atoms. The pH dependence of binding to thymine is consistent with the requirement of deprotonation of the N3 amide in solution. The pH potentiometric titrations of all four DNA nucleosides, deoxyguanosine (dG), deoxycytidine (dC), deoxyadenosine (dA), and dT show that only dT binds detectably to $[Zn(L1)]^{2+}$ at milimolar concentrations. The apparent dissociation constant ($K_{d\ (app)}$) for $[Zn(L1)]^{2+}$ with dT is 0.75 m*M* at pH 7.5 or log K_1 of 5.6. Other heterocycles that contain similar amide groups including uridine, AZT, inosine and riboflavin also bind to $[Zn(L1)]^{2+}$. Binding constants strengthen with increasing basicity of the amide nitrogen, consistent with the formation of a strong coordination bond as a driving factor for selective complexation. In more recent work, the Spiccia group attached ferrocene to $[Zn(L1)]^{2+}$ to convert the complex into an electrochemical sensor for dT derivatives (55).

B. Complexes of Zn^{2+} Triazamacrocycles and Mixed Aza/oxa-Macrocycles

The complex $[Zn(L1)]^{2+}$ is not the only macrocyclic complex of Zn^{2+} that binds to thymine or uracil groups. Any Zn^{2+} center with an available coordination site might be expected to bind to the deprotonated thymine. Macrocycles with different ring sizes and numbers of amine groups form the Zn^{2+} complexes shown in Scheme 6 in aqueous solution. For $[Zn(L1)]^{2+}$ - $[Zn(L5)]^{2+}$, the strength of binding to uridine increases as the Lewis acidity of the Zn^{2+} complex increases (7). Here Lewis acidity is calibrated through the water ligand pK_a, a value related to the magnitude of hydroxide binding. Essentially, the lower the bound water pK_a, the higher the affinity of hydroxide for the Zn^{2+} complex. The Zn^{2+} complexes that bind uridine the most tightly of the mononuclear complexes are $[Zn(L3)]^{2+} \gg [Zn(L1)]^{2+}$ and $[Zn(L2)]^{2+}$ with the latter two being nearly equal. For example, the binding constant of $[Zn(L4)]^{2+}$ for uridine is nearly two orders of magnitude weaker than $[Zn(L1)]^{2+}$ at 25 °C, 100 m*M* NaCl, correlating to the high pK_a of

$[Zn(L4)]^{2+}$ (9.2) compared to $[Zn(L1)]^{2+}$ (8.1) in 0.1 *M* NaCl, 25 °C. The fact that uridine binding correlates to hydroxide anion binding strength suggests that the strength of the coordination bond to Zn^{2+} and not hydrogen bonds to the macrocycle are the primary energetic components driving recognition (Scheme 7).

C. Dinuclear and Trinuclear Zn^{2+} Complexes

Linking two Zn^{2+} complexes together produces a bifunctional receptor for interaction with two distinct groups on the nucleic acid (Scheme 8). For example, dinuclear Zn^{2+} complexes (e.g., $[Zn_2(L6)]^{4+}$ or $[Zn_2(L7)]^{4+}$) bind to phosphorylated thymidine (e.g., 5′-pT) 100-fold more tightly than to thymidine, whereas $[Zn(L1)]^{2+}$ binds thymidine and 5′-pT nearly equally as well (41, 42). The dinuclear complexes bind the thymine N3 and phosphate through use of both Zn^{2+} centers. These terminal phosphate groups are much stronger ligands than are the phosphate diesters in DNA, however. Multinuclear Zn^{2+} complexes also bind

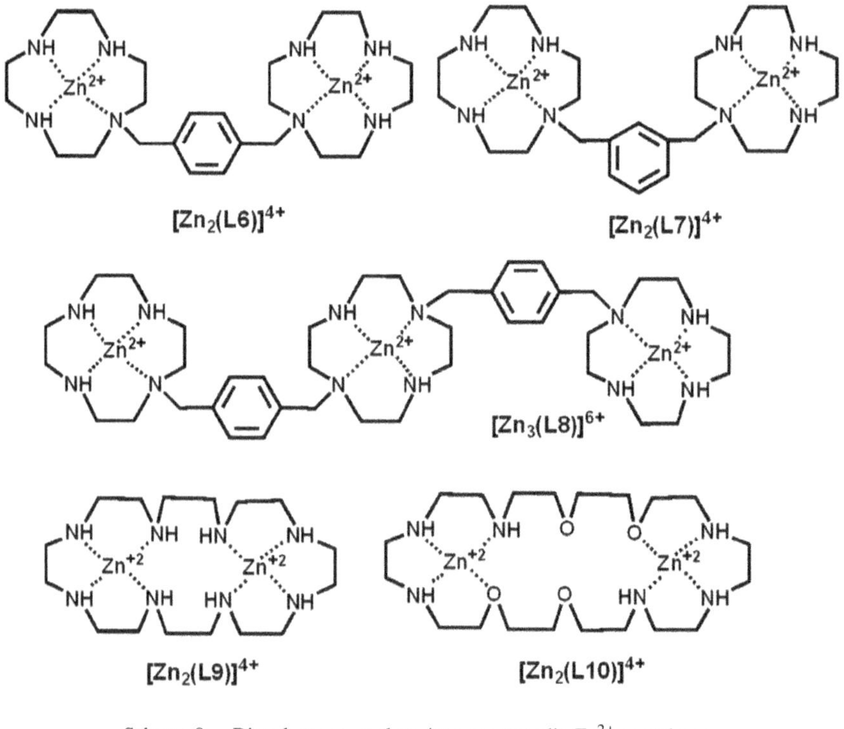

$[Zn_2(L6)]^{4+}$ $[Zn_2(L7)]^{4+}$

$[Zn_3(L8)]^{6+}$

$[Zn_2(L9)]^{4+}$ $[Zn_2(L10)]^{4+}$

Scheme 8. Dinuclear aza- and aza/oxo-macrocyclic Zn^{2+} complexes.

more tightly to dinucleotides or oligonucleotides with adjacent T groups. For example, thymidylylthymidine (TT) binds to $[Zn_2(L6)]^{4+}$ with a dissociation constant of 0.6 μ*M* at pH 7.5, ($I = 0.1\ M$, $NaNO_3$), 25 °C (45). This interaction that involves two Zn^{2+} centers is 1000-fold stronger than that of mononuclear $[Zn(L1)]^{2+}$ with either dT or 5′-pT. Other dinucleotides are bound much less strongly to the dinuclear complex. The complex $[Zn_2(L6)]^{4+}$ binds GT with a dissociation constant of 13 μ*M*, 20-fold weaker than TT, but stronger than dT alone (45). This result is consistent with interaction of the Zn^{2+} center with the N7 of the guanine base in GT. Both CT and AT have $K_d > 100\ \mu M$, suggesting that cytosine or adenine groups do not bind the second Zn^{2+} center. Addition of a third Zn^{2+} cyclen center to give $[Zn_3(L8)]^{6+}$ shows even stronger interactions with oligonucleotides containing multiple thymine groups. Binding of the trinuclear complex with thymidylylthymidylylthymidine (TTT) has a K_d of 0.8 n*M* under similar conditions (45).

A related example of multinuclear Zn^{2+} binding to consecutive U groups is the interaction of $[Zn_3(L8)]^{6+}$ with human immunodeficiency virus (HIV-1) messenger RNA (mRNA) containing the trans-activation responsive (TAR) structure (57) that contains a three nucleotide uracil bulge. The HIV-1 regulatory protein trans-activator of transcription (TAT) binds to the bulge region. Inhibition of TAT binding to HIV-1 TAR RNA is attributed to competition of $[Zn_3(L8)]^{6+}$ with TAT for the uridine bulge. However, the authors did not include structural details on the binding of the multinuclear Zn^{2+} complexes to the bulge, so it is not known how binding of the Zn^{2+} complex affects the stability of the RNA structure. For this reason, we include this example in the section on unstructured RNA.

Notably, it is not necessary to link Zn^{2+} cyclen centers through aromatic groups for binding of adjacent thymines. Dinuclear Zn^{2+} complexes of ligands with even larger macrocyclic ring size ($[Zn_2(L9)]^{4+}$ and $[Zn_2(L10)]^{4+}$) bind two thymidines or uridines (Scheme 8) (54, 56). $[Zn_2(L9)]^{4+}$ also binds TT, albeit not as strongly as two individual thymidines, suggesting that the two Zn^{2+} centers are not optimally oriented for binding both thymine N3 sites in TT.

D. Multinuclear Zn^{2+} Complexes for Cleavage of Phosphate Esters Containing Uracil

Complexes of Zn^{2+} for the neutral macrocycles L1–L3 contain a Lewis acidic Zn^{2+} center for anion binding including hydroxide or deprotonated thymine. Such Zn^{2+} complexes are also catalysts for the cleavage of RNA through transesterification of the phosphate ester (7, 51, 58, 59). These "hydrolytic" cleavage reactions proceed by attack of the 2′-hydroxyl and stabilization of the anionic phosphorane transition state by the metal ion catalyst (Scheme 9). It has long been noted that cleavage is base-sequence selective for both Zn^{2+} and Cu^{2+} complex catalysts, even for short oligonucleotides that do not form secondary structures in

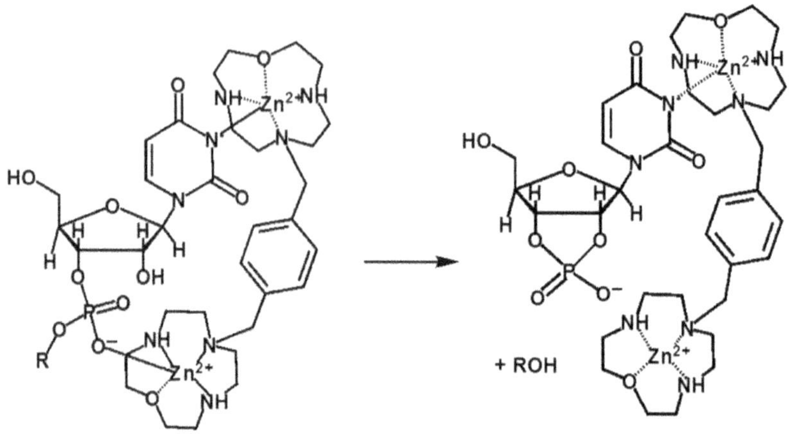

Scheme 9. Cleavage of a phosphate diester by a dinuclear catalyst that binds a uracil nucleobase.

solution (60, 61). Binding of the metal ion to the nucleobase most likely produces this sequence selective cleavage. Nucleobase binding by the multinuclear catalysts might be either productive or inhibitory. For example, binding of a neighboring base by one metal ion center might bring the phosphate ester close to a second metal ion center in a productive interaction for catalytic cleavage, as shown in Scheme 9 (62). Thus, cleavage of dinucleotides containing uridine or guanine is enhanced by calixarenes containing multinuclear metal ion centers as catalysts (61). An inhibitory interaction might occur if nucleobase binding to the catalytic center blocks interaction with the phosphate ester. An example of this has been observed for uridylyluridine (UU) sequences that bind two Zn^{2+} centers in dinucleotides and in longer sequences, resulting in inhibition of cleavage (53).

Several dinuclear Zn^{2+} complexes have been studied for cleavage of phosphate esters or RNA containing uracil groups. Two linked L2 macrocyclic ligands were used to form $[Zn_2(L11)]^{4+}$ for the study of RNA cleavage (62). The L11 macrocycle is used rather than the L6 macrocycle to capitalize on the more effective catalytic properties of $[Zn(L2)]^{2+}$ for RNA cleavage in comparison to $[Zn(L1)]^{2+}$. The complex $[Zn_2(L11)]^{4+}$ cleaves the uridine-containing phosphate ester, uridine-3′-4-nitrophenylphosphate (UPNP), more rapidly than does the mononuclear catalyst, $[Zn(L2)]^{2+}$. The complex $[Zn_2(L12)]^{4+}$ contains two triazacyclononane macrocycles and a bridging alkoxide to maintain the two Zn^{2+} centers in close proximity for interaction with the phosphate ester in RNA cleavage. The binding of $[Zn_2(L12)]^{4+}$ to UU and UPNP, as measured by kinetic experiments, strengthens as the number of uracil groups increase, consistent with an interaction between the Zn^{2+} centers and the uracil groups (59).

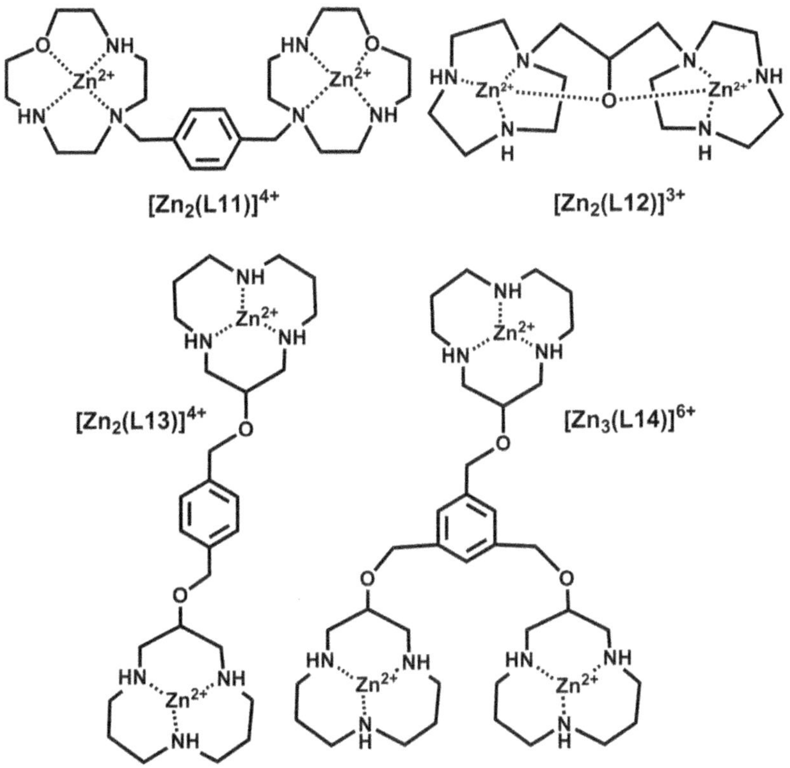

Scheme 10. Multinuclear Zn^{2+} complexes used for RNA cleavage.

Multinuclear Zn^{2+} complexes of linked triazacyclododecane macrocycles have been designed with the goal of obtaining sequence selective cleavage of RNA containing uridine (Scheme 10). The complex $[Zn(L3)]^{2+}$ has the advantage that it binds strongly to uridine in comparison to other macrocyclic complexes, but has the disadvantage of having a relatively weak formation constant of the macrocycle to Zn^{2+}, such that the complex has a high degree of dissociation at low millimolar concentrations (52, 53). The complex $[Zn_2(L13)]^{4+}$ and derivatives with different aromatic linkers cleave dinucleotides that contain a single uracil (AU, UA) 10-fold more rapidly than nucleosides that do not contain uracil (AA). This finding suggests that one of the Zn^{2+} centers binds to the uracil while the other interacts with the phosphate ester that undergoes cleavage. Uridylyluridine is cleaved slowly, suggesting that the two Zn^{2+} centers bind both uracils and cannot interact with the phosphate ester moiety to promote cleavage. The trinuclear complex, $[Zn_3(L14)]^{6+}$, shows a different base sequence selectivity for cleavage than the dinuclear complexes. Dinucleotides AU, UA, and UU are all cleaved with similar rate constants (52). This result is consistent with the third Zn^{2+} center promoting cleavage. Similar base-sequence selectivities are observed in unstructured short oligonucleotides for these catalysts (53, 63).

E. Macrocyclic Complexes of Zn^{2+} Containing Aromatic Pendents

Attachment of aromatic pendent groups to Zn^{2+} macrocycles substantially increases the binding constant of the complex to uracil or thymine groups (Scheme 11). For example $[Zn(L14)(dT)]^+$, which contains an acridine pendent group, has a $K_{d\ (app)}$ of 8 μM at pH 7.4 which is 100–fold tighter than that of $[Zn(L1)(dT)]^+$ under similar conditions (6). The log K_1 for this interaction is 7.2. Kimura and co-workers (6) proposed that interactions of the acridine pendent

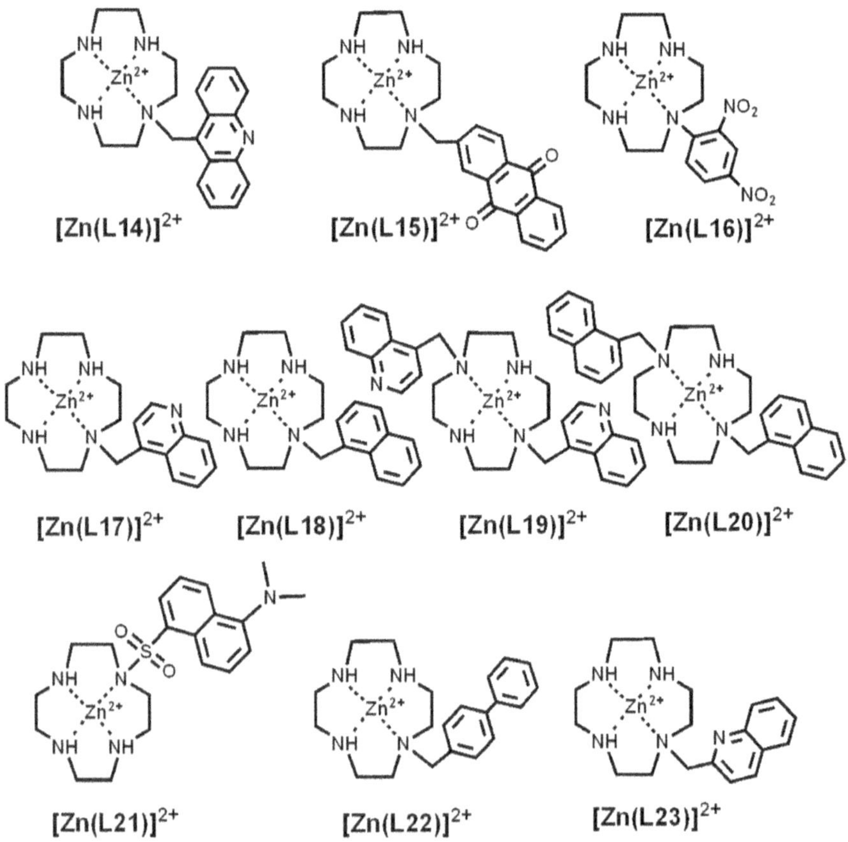

$[Zn(L14)]^{2+}$ $[Zn(L15)]^{2+}$ $[Zn(L16)]^{2+}$

$[Zn(L17)]^{2+}$ $[Zn(L18)]^{2+}$ $[Zn(L19)]^{2+}$ $[Zn(L20)]^{2+}$

$[Zn(L21)]^{2+}$ $[Zn(L22)]^{2+}$ $[Zn(L23)]^{2+}$

Scheme 11. Complexes of Zn^{2+} with aromatic pendent groups.

group through π–π stacking interactions on the thymine contributed to the more favorable binding.

A crystal structure of 1-methyl-thymine bound to $[Zn(L14)]^{2+}$ shows the acridine group stacked on the T in a parallel fashion with an interplane separation of 3.28–3.42 Å (Fig. 1) (6). Proton NMR spectra of the complex show shifted aromatic protons for acridine and thymine, consistent with a π–π stacking interaction in solution. The acridine group also increased the extent of binding to guanosine from an undetectable value for $[Zn(L1)]^{2+}$ to a log K of 4.1 for $[Zn(L14)]^{2+}$ binding of the neutral guanosine at the N7 position. The complex $[Zn(L14)(dG)]^{2+}$ is the predominant form of the guanosine-containing complex at neutral pH. A crystal structure of $[Zn(L14)(dG)]^{2+}$ shows stacking of the acridine on the N7-bound guanosine. At high pH values (>8), this complex deprotonates at N1 of guanosine. However, under physiologically relevant conditions, $[Zn(L14)]^{2+}$ is specific for thymine over guanosine. Solutions containing 1 mM of each nucleoside and Zn^{2+} complex at pH 7.6 and 25 °C would form $[Zn(L14)]^{2+}$ complexes with 71% bound dT and 24% bound dG (6).

The complex $[Zn(L14)]^{2+}$ binds selectively to dinucleotides and oligonucleotides containing thymine groups with the binding mode dependent on the neighboring nucleobases (6). For example, NT (N = G, C, A, T) bound a single $[Zn(L14)]^{2+}$ with similar binding constants ($K_d \approx 10\ \mu M$ at pH 8.0). However, a second $[Zn(L14)]^{2+}$ bound to TT with a binding constant that is 20-fold larger than the first binding event. Stacking of the two acridine linkers in $[(Zn(L14))_2(TT)]^{2+}$

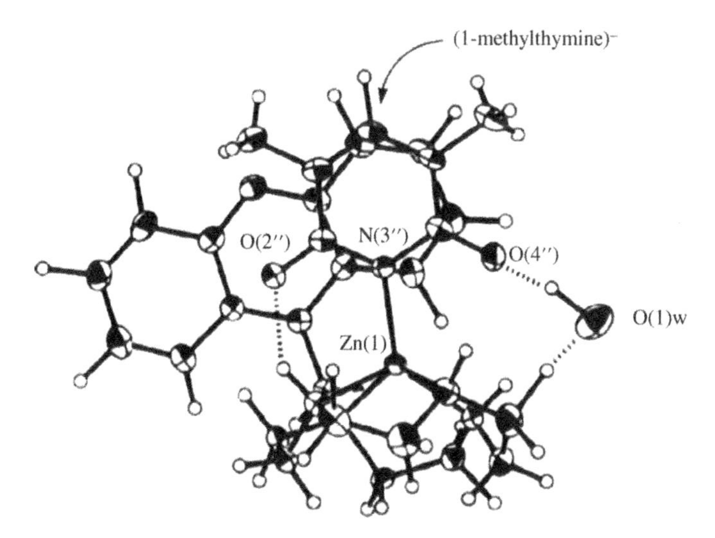

Figure 1. Crystal structure of the complex cation of $[Zn(L14)(1\text{-}Me\text{-}T^-)]ClO_4^- \cdot H_2O$ showing stacking of the acridine group on methyl-thymine. [Reprinted with permission from M. Shionoya and E. Kimura, JACS, *115*, 6730 (1993) (6). Copyright © (1994) American Chemical Society.]

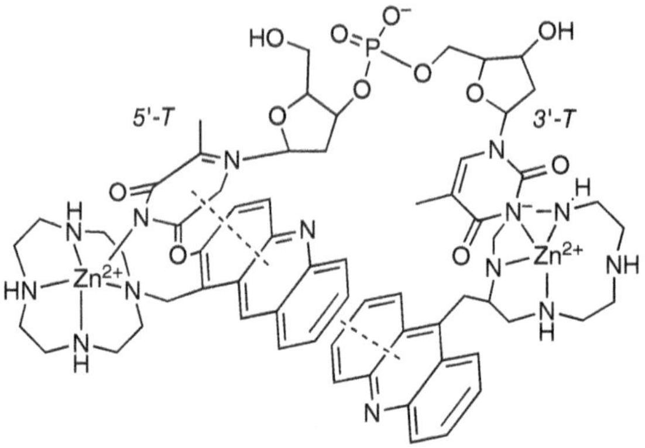

Scheme 12. Interaction of 2:1 $[Zn(L14)]^{2+}$ and TT. [Reprinted with permission from E. Kimura et al., *JACS*, *122*, 4668 (200) (49). Copyright © (2000) American Chemical Society.]

is proposed in order to account for the favorable formation of this complex (Scheme 12). Deoxyribonucleic acid containing GT sequences also form complexes with two bound $[Zn(L14)]^{2+}$.

Other pendents include anthraquinone ($[Zn(L15)]^{2+}$) which binds dT with log K_1 of 6.6 (64). Interestingly, complexes with a simple aromatic ring (e.g., a benzyl group) did not markedly change the binding constant in comparison to $[Zn(L1)]^{2+}$, whereas a Zn^{2+} complex with 2,4-dinitrophenyl-cyclen ($[Zn(L16)]^{2+}$) did bind dT strongly (log $K = 6.9$) (50). This result suggests that the benzyl group does not interact strongly with the thymine ring. The stronger interaction of the dinitrophenyl group with the bound thymine may be due to the effect of the electron-withdrawing substituents on the π–π interaction or, alternatively, the direct attachment of the aromatic group to the cyclen ring, rather than through a methylene linker. A third factor is that direct alkylation of an amine in cyclen decreases the basicity of the amine and increases the Lewis acidity of the Zn^{2+} center for anions (e.g., deprotonated thymine groups) (50).

Complexes with aromatic pendents containing two fused six-membered rings $[Zn(L17)]^{2+}$–$[Zn(L21)]^{2+}$ have been studied for binding dT. The complex $[Zn(L17)]^{2+}$ binds dT with a log K_1 of 6.8 or apparent K_d of 25 μM at pH 7.5 100 mM NaCl (48, 65). $[Zn(L18)]^{2+}$ binds dT with log K_1 of 6.3 and apparent K_d of 63 μM under similar conditions. Two aromatic pendent groups, for example, in $[Zn(L19)]^{2+}$, produce a complex that has slightly stronger binding of dT (log $K_1 = 7.7$, Eq. 1), consistent with stacking of both pendents in a sandwich-like interaction around the dT^-.

Binding of Zn^{2+} complexes to pentanucleotides was studied (66). The complex $[Zn(L21)]^{2+}$ binds to a thymine group in a pentanucleotide (CCTCC) with a dissociation constant of 55 μM (100 mM NaCl). Additionally, $[Zn(L21)]^{2+}$ binds to the pentanucleotide AATAA with a dissociation constant of 50 μ*M* (100 m*M* NaCl), implying the flanking base pairs do not substantially affect binding. No fluorescence change was observed upon addition of 30-fold excess AAAAA to 5 u*M* (100 m*M* NaCl) of $[Zn(L21)]^{2+}$, suggesting that the thymidine is important for micromolar range oligonucleotide binding. A small fluorescent change (17% relative increase) was observed upon treatment of $[Zn(L21)]^{2+}$ with CCCCC to give a $K_{d\ (app)}$ of ~ 200 μ*M* (100 m*M* NaCl). The complex $[Zn(L17)]^{2+}$ binds to CCTCC with a similar apparent dissociation constant of 25 μ*M* (100 m*M* NaCl). In contrast, $[Zn(L14)]^{2+}$ binds to CCTCC 10-fold more tightly than does $[Zn(L21)]^{2+}$, with a $K_{d\ (app)}$ of 5.2 μ*M* (100 m*M* NaCl) (67). Although the binding mode has not been investigated, it is possible that the increased accessible solvent surface area of $[Zn(L14)]^{2+}$ yields tighter binding upon stacking of the acridine pendent to the thymine. The complex $[Zn(L14)]^{2+}$ does not bind appreciably to AAAAA as shown by a lack of change in its intrinsic fluorescence upon the addition of 100-fold excess AAAAA to 5 μ*M* (100 m*M* NaCl) $[Zn(L21)]^{2+}$. The data suggests that aromatic pendents function synergistically with the Zn^{2+} center to bind thymine groups. O'Neil and Wiest (69) proposed a model for $[Zn(L21)]^{2+}$ binding to the single-stranded pentanucleotide CCTCC (68) that shows possible binding modes.

It is intriguing that Zn^{2+} macrocyclic complexes containing pendents with two fused aromatic rings ($[Zn(L17)]^{2+}$, $[Zn(L18)]^{2+}$, and $[Zn(L21)]^{2+}$) bind to thymine in simple oligonucleotides with binding constants that are similar to that of $[Zn(L14)]^{2+}$, which contains an acridine pendent with three fused rings (48). Naively, it would seem that stacking interactions of the acridine on the bound thymine would be substantially more favorable. What is known about these stacking interactions? Notably, the stacking interaction between acridine and thymine in a crystal structure of $[Zn(L14)]^{2+}$ (5-1-Me-T^-) features parallel stacked aromatic rings. Solution NMR data supports this interaction (6). By contrast, the crystal structure of $[Zn(L18)]^{2+}$ (5-1-MeT^-) shows the quinoline pendents splayed away from the bound thymine ring (48). However, solution proton NMR (^1H NMR) studies in this report support stacking of the quinoline pendents on the thymine groups in aqueous media.

The importance of the aromatic pendent is unquestionable given the stronger binding that is produced in Zn^{2+} complexes containing pendent aromatic groups. Parsing the free energy of binding into energetic factors, as discussed further in Section V, suggests that the aromatic group is likely to contribute to stronger binding through more favorable hydrophobic transfer energies and through stronger molecular interactions with the DNA by stacking of the pendent on the thymine. However, the nature of the stacking interaction in solution has not been well defined. What orientations of the aromatic groups with respect to thymine are favorable? Crystal structures and solution structures derived from NMR spectroscopy of Zn^{2+} complexes show stacking interactions that involve

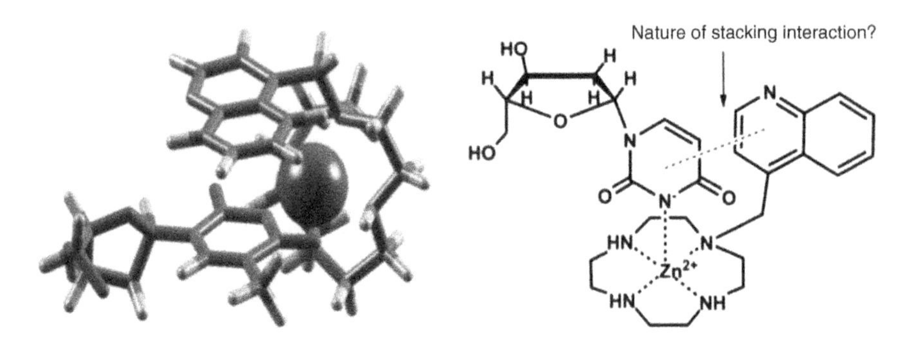

Figure 2. Stacking of quinoline in $[Zn(L17)]^{2+}$ on thymine in bulged structure. [Reprinted with permission from I. M. A. del Mundo et al., *Chem. Commun.*, *47*, 8566 (2011) (70). Copyright © (2011) American Chemical Society.] (See the color version of this figure in color plates section.)

face-to-face orientations of aromatic pendents and thymine (Fig. 2) (66, 70). However, recent discussions on the nature of stacking interactions between aromatic groups make it important to consider that face-to-face π–π stacking of aromatic groups is not always the most favorable interaction (71, 72). Off-center stacking and edge-to-face stacking are also important interactions. Thus pendent groups might twist and interact with the thymine or uracil groups in an edge-to-face interaction. Notably, different linkers that connect the pendent aromatic group to the macrocycle influence binding interactions of the Zn^{2+} complexes with thymine groups. For example, most complexes presented here have methylene linkers, but some complexes have direct connections to the macrocycle ($[Zn(L16)]^{2+}$) and one complex ($[Zn(L21)]^{2+}$) has a sulfone linker. These linkers will influence the range of conformations of the aromatic pendent group that are available for interaction with the thymine groups. Further work to delineate the aromatic interactions in Zn^{2+} complexes with thymine or uracil would be useful in the application of these complexes in nucleic acid recognition. As discussed in Section III, the aromatic linker may play a role in the recognition of thymine or uracil groups in more complicated nucleic acid structures, thus it would be beneficial to elucidate this role.

III. INTERACTION OF Zn^{2+} MACROCYCLIC COMPLEXES WITH DOUBLE-STRANDED NUCLEIC ACIDS

The mode of binding of Zn^{2+} complexes to double-helical nucleic acids containing thymine is more difficult to elucidate than Zn^{2+} binding to single-stranded analogues that contain an exposed thymine. Pairing of the thymine to adenine in double-helical nucleic acids buries the binding site for the Zn^{2+}

complex. Yet, there is a pathway that enables Zn^{2+} complexes to bind to thymine or uracil bases in duplexes as shown by several types of experiments described below. For this to occur there must be some partial or complete denaturation of the double-stranded helix in order to expose the Watson–Crick face of thymine or uracil for recognition. A brief summary of the effect of base sequence on duplex stability and conformation is given in this section to better understand binding pathways.

The nucleic acid sequence is an important factor in duplex stability. A duplex composed of poly(dA)·poly(dT) is less stable than a similar duplex composed of poly(dG)·poly(dC); thermal melting temperature (T_m) ~40 vs >95 °C, respectively (43, 65). This stability difference is in part due to the A-T pair having two hydrogen bonds in comparison to the three hydrogen bonds in G-C pairs. When considering mixed nucleobase sequences, an increase in the G-C content increases the overall stability of the duplex, with some helical stabilization arising from stacking interactions of the adjacent base pairs. The G-G stacking typically has the most favorable interaction with $\Delta G = -11.3\ \mathrm{kcal\ mol^{-1}}$ (73). Notably, A-T base pairs do not necessarily compromise overall stability; on the contrary, it has been shown that certain types of A-T tracts may promote duplex stability. These A-T runs cause a bend in the helix and promote a propeller twist conformation (74, 75). The propeller twist increases purine–purine stacking interactions and enables additional hydrogen-bond formation. This occurs at the major groove side of each base, pushing the N6 of adenine toward the O4 of thymine. This conformation allows for a diagonal non-Watson–Crick hydrogen bond to form across the major groove. For this to truly increase stability, an A-T tract of at least three is necessary (74). Thus, the base sequence dictates the overall structure and stability of the double-stranded helix (76, 77).

To date, there are only a few metal complexes that selectively bind to duplex DNA and compromise the overall structure, one of these being cisplatin, *cis*-$Pt(NH_3)_2Cl_2$ (65). Another metal complex that perturbs duplex structure is the binuclear ruthenium complex, [μ-(11,11′-bis(dipyrido[3,2-*a*:2′,3′-*c*]phenazinyl) (1,10-phenanthroline)$_4$Ru$_2$]$^{4+}$. This complex initially acts as an A-T selective groove binder, then is able to insert into the helix, described as a "threading" binding mode. The threading of this complex is rapid with A-T tracts that contain sections of non-canonical DNA, but slow with canonical duplex DNA (78). Such metal ion complexes are of interest to modulate duplex structure with the goal of inhibiting gene expression. Inhibition can occur through promoting structural changes of the duplex, or by blocking certain enzymes and proteins necessary for gene expression (65). In this regard, the biological importance of T and U tracts in DNA or RNA, such as those found in TATA boxes, has made it interesting to study the binding of Zn^{2+} complexes to these sequences. In this section, we will review studies on binding of Zn^{2+} macrocyclic complexes to duplex DNA. A major question we will attempt to address is the mechanism whereby the Zn^{2+} complex binds to T or U in duplex DNA.

A. Role of Nucleic Acid Sequence and Structure

Understanding the binding mode of the Zn^{2+} complexes to duplex DNA is challenging because a case can be made for more than one mode of binding for certain Zn^{2+} complexes. At different ratios of complex to DNA, the Zn^{2+} complexes interact differently with the DNA and perhaps also interact with each other in a concentration-dependent fashion. Comparison to reference compounds with known binding interactions is one way to provide binding information. Distamycin A and 4,6-diamidino-2-phenylindole (DAPI) are known minor groove binders and have been used for many studies to better understand the groove preference of small molecules (Scheme 13) (79).

One property of duplex behavior that is important for certain Zn^{2+} binding modes is the propensity of the duplex to "breathe" (Scheme 14). Deoxyribonucleic acid breathing occurs by "unzipping" of the duplex to create internal loops. At lower temperatures, the DNA rehybridizes; as the temperature increases, it promotes extension of the internal loop, which can ultimately lead to unwinding and dissociation of the duplex (80). This dissociation is followed by Watson–Crick interactions re-forming and the duplex closing itself. Dissociation is more prone to occur for long A-T tracts since this pair is weaker than the G-C pair. This unzipping mechanism may facilitate binding of Zn^{2+} complexes to the face of nucleobases, even though they are fully paired most of the time. The threading Ru^{2+} complex mentioned above is thought to take advantage of this pathway in order to insert itself through the helix that ultimately affects the helical structure (78). Several types of experiments that yield information on the binding mode are given Section III.B.2 and III.B.3.

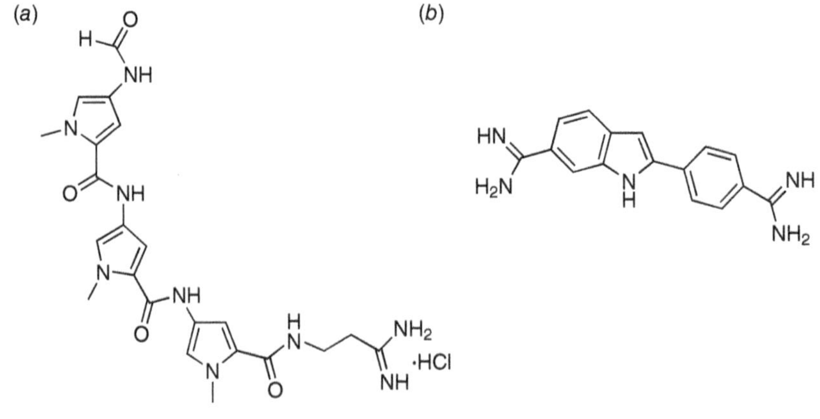

(*a*) (*b*)

Scheme 13. The dA-dT selective minor groove binders, (*a*) Distamycin A and (*b*) DAPI [79].

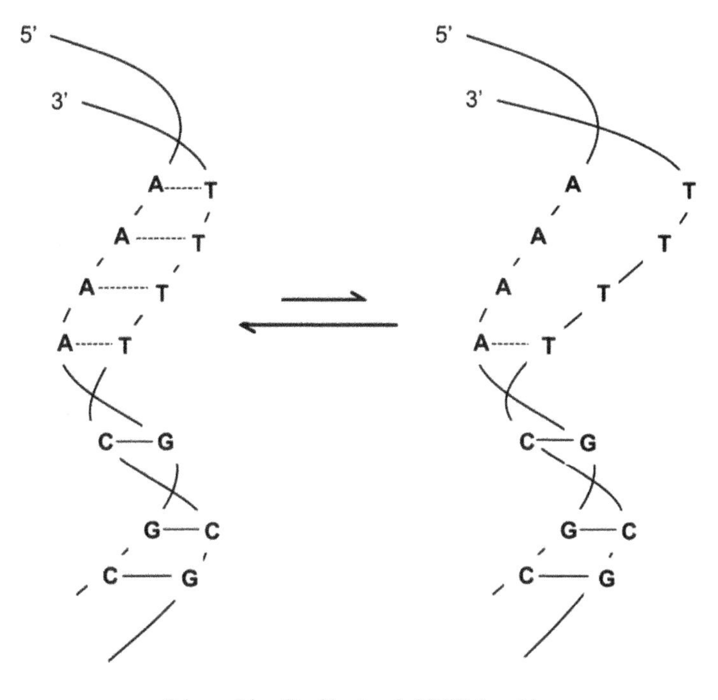

Scheme 14. Double-stranded DNA breathing.

B. Optical Studies

1. *UV–Vis*

Binding of compounds containing chromophores, such as the Zn^{2+} complexes containing pendent aromatic groups (Scheme 11), to nucleic acids is readily monitored by using UV–vis spectroscopy. These studies are useful for determining binding constants of Zn^{2+} complexes to nucleic acids, but give little information on the binding mode. Here we will highlight the importance of the Zn^{2+} center and its central role in binding to nucleobases.

Binding to all DNA sequences studied was tighter for $[Zn(L14)]^{2+}$ than for the respective Zn^{2+} free ligand. The strongest binding was observed for $[Zn(L14)]^{2+}$ to poly(dA)·poly(dT). The Zn^{2+} complex bound 31-fold more tightly to this sequence than did the Zn^{2+} free ligand. The spectrum for the $[Zn(L14)]^{2+}$ titration with poly(dA)·poly(dT) had a complicated dependence on Zn^{2+} complex, consistent with two different modes of binding. The single strands of this duplex were also studied to better understand this interaction. The poly(dA) titration with the Zn^{2+} complex had a different spectrum than that of the duplex, while the poly(dT) titration had a very similar spectrum to that of the duplex. The fact that the spectrum of poly(dT) titrated with $[Zn(L14)]^{2+}$ matched that of the duplex, is consistent with denaturation of the duplex. Concentration-dependent titrations showed that when the concentration of DNA was low compared to $[Zn(L14)]^{2+}$ (20 μ*M*) concentration, the change in the appearance of the absorbance spectrum upon addition of DNA was complicated and lacked an isosbestic point. In contrast, when high DNA concentrations (>80 μ*M*) were used compared to the concentration of $[Zn(L14)]^{2+}$ (<10 μ*M*), an isosbestic point was observed. These results suggest a bimodal binding interaction. If the concentration of Zn^{2+} complex is low, a single binding mode is observed that likely involves intercalation of the acridine pendent group into the intact helical DNA. As the concentration of Zn^{2+} complex increases, interaction of the Zn^{2+} center with the face of thymine dominates, causing a disruption in the A-T duplex (46).

The complex $[Zn(L15)]^{2+}$ contains a distinct type of pendent group that lends the complex a slight preference for dT over dG. Titrations were carried out with double-stranded poly(dG)·poly(dC), poly(dG-dC)$_2$, poly(dA)·poly(dT), and poly (dA-dT)$_2$. Binding of $[Zn(L15)]^{2+}$ to double-stranded DNA containing GC base pairs showed a single type of binding mode as supported by UV–vis studies that produced data with a single isosbestic point. The binding mode is not well defined, but may involve binding of the Zn^{2+} complex to N7 of the guanine bases, similar to that observed for guanosine. This type of binding to G bases is also supported by nuclease footprinting assays discussed in Section III.C. In contrast, the poly(dA)·poly(dT)·titration with $[Zn(L15)]^{2+}$ gave results that were more complex, most likely due to multiple-binding modes.

2. *Circular Dichroism*

Circular dichroism (CD) studies were useful to identify whether the helical structure is compromised upon addition of Zn^{2+} complexes. Circular dichroism spectra of proteins and nucleic acids are a result of the chirality of the biopolymers, creating a signature spectrum for different helical forms. Nucleic acids form different types of helical structures with distinct characteristics, including the A, Z and B-form with the latter being the most common. This spectroscopic technique gives qualitative information on the perturbation of helical structure upon interaction with small molecules. Linear dichroism (LD) is a technique capable of defining parallel or perpendicular orientation of a ligand in respect to the nucleic acid, protein, or enzyme (among others) being studied. The major difference between the two is that CD can be used as a fingerprint for structural properties, while LD is used to define orientation within a system (81, 82). Circular dichroism is primarily used in the studies discussed here because of the focus on structural changes due to the interactions of the Zn(II) complexes with nucleic acids.

The poly(A)·poly(U) duplex gave a significantly different CD spectrum upon addition of the Zn^{2+} complex. Each of the strands was also studied separately to better understand the observed differences. The spectrum for poly(U)·$[Zn(L14)]^{2+}$ was virtually identical to that of poly(A)·poly(U)·$[Zn(L14)]^{2+}$ while poly(A)·

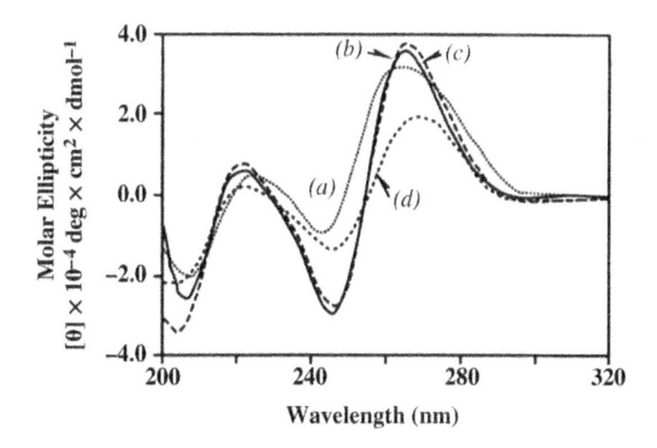

Figure 3. The CD spectra (*a*) poly(A)·poly(U), (*b*) poly(A)·poly(U) in the presence of 40 μ*M* [Zn(L14)]$^{2+}$, (*c*) spectrum of poly(U) in the presence of 40 μ*M* [Zn(L14)]$^{2+}$ combined with the spectrum of poly (A), and (*d*) spectrum of poly(A) in the presence of 40 μ*M* [Zn(L14)]$^{2+}$ combined with the spectrum of poly (U) at 25 °C in 5 m*M* Tris-HCl buffer (pH 7.6) and 10 m*M* NaCl. The concentration of each polynucleotide was kept constant at 50 μ*M*(P). [Reprinted with permission [65]. Copyright © (1998) Springer.]

[Zn(L14)]$^{2+}$ had a completely different spectrum (Fig. 3). This data suggests that [Zn(L14)]$^{2+}$ denatures poly(A)·poly(U) to give [Zn(L14)]$^{2+}$ bound to the uracil nucleobase of poly(U). In contrast, [Zn(L14)]$^{2+}$ does not bind strongly with the adenine nucleobase of poly(A) (65).

Additional studies were conducted with poly(dG)·poly(dC). The CD spectra changed when compared to the original duplex spectrum as the concentration of [Zn(L14)]$^{2+}$ increased (Fig. 4). This alteration of the spectrum supports a structural

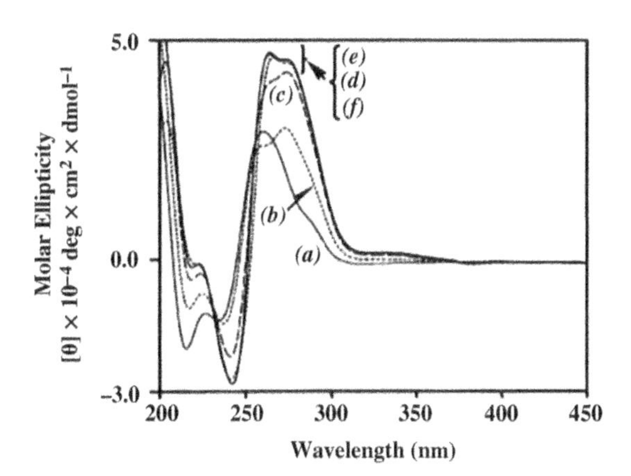

Figure 4. The CD spectra of poly(dG)·(dC) in the absence and presence of [Zn(L14)]$^{2+}$. (*a*) 100 μ*M* (P) of poly(dG)·poly(dC), (*b*) *a*+10 μ*M* of [Zn(L14)]$^{2+}$, (*c*) *a*+20 μ*M* of [Zn(L14)]$^{2+}$, (*d*) *a*+30 μ*M* of [Zn(L14)]$^{2+}$, (*e*) *a*+40 μ*M* of [Zn(L14)]$^{2+}$, (*e*) *a*+40 μ*M* of [Zn(L14)]$^{2+}$ at 25.0 ± 0.5 °C in 5 m*M* Tris-HCl buffer (pH 7.6) and 10 m*M* NaCl. [Reprinted with permission [65]. Copyright © (1998) Springer.]

change that may be attributed to binding of $[Zn(L14)]^{2+}$ to dG. Although binding is stronger to thymine or uracil nucleobases than to the guanine nucleobase, $[Zn(L14)]^{2+}$ also interacts with guanine through N7 recognition and stacking interactions of the acridine on the nucleobase, as discussed in Section II and III. B.3 (65, 79). Interaction with poly(dG)·poly(dC) may involve similar binding modes or may simply involve intercalation.

3. *Thermal Melting Experiments*

Small molecules that selectively bind to nucleic acids affect the overall stability of the helix. Optical thermal melting studies are typically used to monitor an increase or decrease in stability (4). Intercalators (e.g., ethidium bromide and actinomycin) slide in between the base pairs increasing both the stacking interactions and helix stability and therefore increases the T_m (65). Other small molecules that can increase the melt temperatures are groove binders. For example, the minor groove binder distamycin A stacks parallel with the helix and binds to a series of A-T(U) bases. This stabilizes the helix by rigidifying the DNA. Major groove binders act in a similar way to minor groove binders, also increasing thermal melt temperatures. Metal ions are known to interact with nucleic acids in different ways and can cause a change in melt temperatures. The ions Mg^{2+}, Na^{+}, and Ba^{2+} are known to neutralize the negatively charged phosphodiester backbone and increase helical stabilization. In contrast, metal ions (e.g., Cu^{2+}, Cd^{2+}, and Pb^{2+}) are thought to disrupt the hydrogen bonds between base pairs and replace them with coordination bonds, therefore destabilizing the helix and lowering the melt temperature (65). This destabilization is the anticipated mechanism when considering Zn^{2+} macrocycles and their effect on the double-stranded helix. However, for the bifunctional complexes discussed here, it is difficult to predict which factor will dominate. If the Zn^{2+} center were the primary source of interaction, helical destabilization would be expected. Alternatively, the aromatic pendents might act as intercalators or groove binders that stack on nucleobases and increase helical stability to give an increase in melting temperature.

The $[Zn(L1)]^{2+}$ complex interactions with poly(A)·poly(U), poly(dA)·poly(dT), and poly(dA-dT)$_2$ were studied. It was expected that the helix would be denatured, leading to lowering of the T_m of poly(A)·poly(U), poly(dA)·poly(dT), and poly(dA-dT)$_2$ due to binding of the Zn^{2+} complex to the thymine or uracil N3-imide (Scheme 7 and Section I). This was in fact the case, confirming destabilization upon binding of the metal complex to the homopolymer duplex [Fig. 5(*a*)] (65). These results were compared to those of $[Zn(L14)]^{2+}$ to better understand the effect of the pendent groups in binding of the complexes to RNA and DNA.

Stabilization of the duplex was observed by an increase of the T_m as $[Zn(L14)]^{2+}$ was added to poly(A)·poly(U). This stabilizing effect was observed at $[Zn(L14)]^{2+}$/nucleotide ratios of 0.2 for poly(A)·poly(U) and for poly(dA-dT)$_2$.

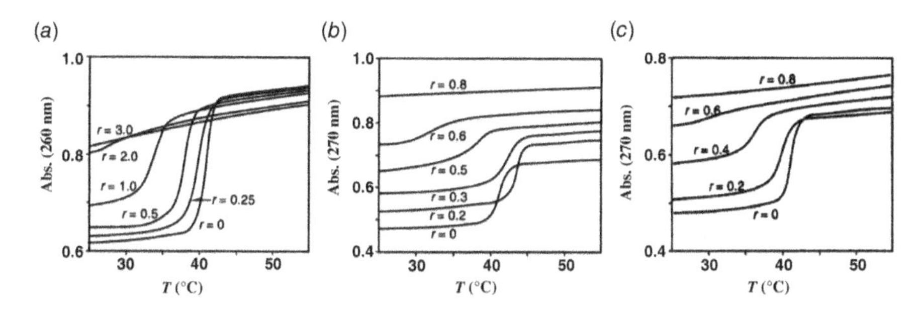

Figure 5. Melting curves for 100 *M* (P) poly(A)·poly(U) with increasing concentrations (*r*) of (*a*) $[Zn(L1)]^{2+}$ (*b*) $[Zn(L14)]^{2+}$, and (*c*) $[Zn(L18)]^{2+}$ [47, 65].

For poly(dA)·poly(dT), the stabilization was observed up to a ratio of 0.1. As the concentration of $[Zn(L14)]^{2+}$ rose above these ratios, the T_m began to decrease, indicating a destabilization of the helix [Fig. 5(*b*)]. This behavior can be explained by the effect of the pendent group dominating at low concentrations and the Zn^{2+} center at higher concentrations. Stacking interactions of the acridine pendent on the nucleobases, likely through an intercalation mode, initially predominate and give rise to stabilization of the helix. At higher concentrations of Zn^{2+} complex, thymine binding to the Zn^{2+} center is accompanied by disruption of Watson–Crick base pairing (Scheme 15). These studies support the different binding modes assigned for $[Zn(L1)]^{2+}$ and $[Zn(L14)]^{2+}$ as shown by optical titrations discussed above. The unusual dependence of T_m on the concentration of the Zn^{2+} complex is consistent with binding of multiple Zn^{2+} complexes to the partially denatured duplex with exposed thymines, resulting in an unzipping of the helix and domination of the thymine-binding mode.

In contrast to the $[Zn(L14)]^{2+}$ complex that contains a potential intercalator, $[Zn(L18)]^{2+}$, which contains a naphthyl aromatic pendent group, monotonically decreases the melting temperature of poly(A)·poly(U) [Fig. 5(*c*)] (47). The Zn^{2+} center in $[Zn(L18)]^{2+}$ appears to dictate the binding interaction at all ratios of metal ion complex to DNA. The simple two-ring nature of the $[Zn(L18)]^{2+}$ pendent could contribute to the minimal interactions observed with the helix by not forming favorable intercalating interactions. The naphthyl pendent could also play a less important role in this interaction due to the lack of a heteroatom in the aromatic ring system, which is thought to be important for stacking interactions (72).

The complexes $[Zn(L1)]^{2+}$ and $[Zn(L14)]^{2+}$ were also titrated with the poly(dG)·poly(dC) duplex. This duplex is so stable that it naturally melts above 95 °C, which makes thermal melting experiments impractical. In order to decrease the melting temperature, studies were done in 50% (v/v) formamide solution. Under these conditions, the $[Zn(L1)]^{2+}$ complex increased the melting temperature

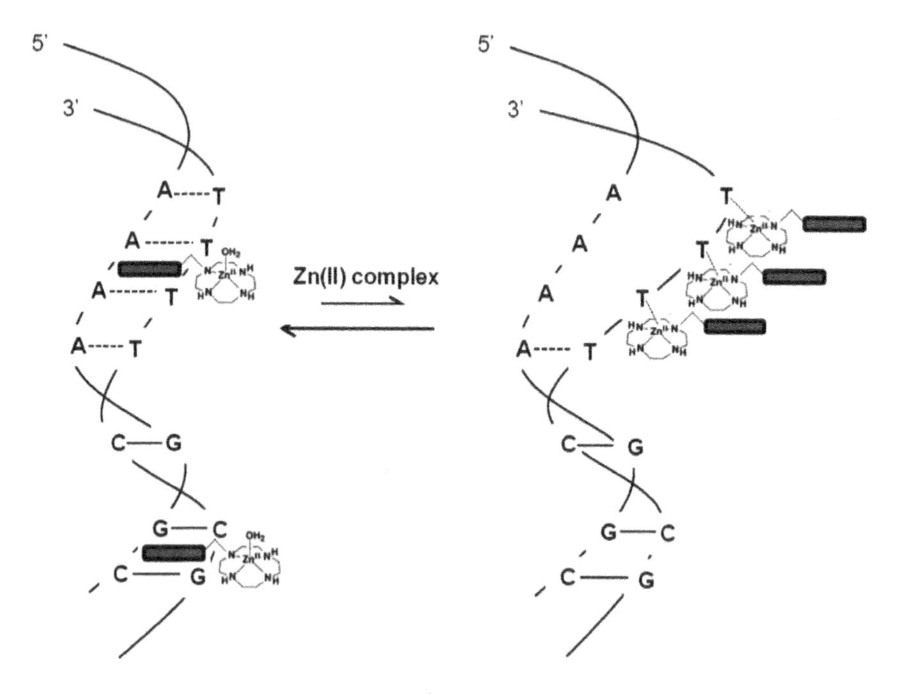

Scheme 15. Bimodal interaction of $[Zn(L14)]^{2+}$: as Zn^{2+} complex concentration increases, binding mode goes from pendent dominating to Zn^{2+} dominating.

by ~8 °C. This slight increase in helical stabilization resulting from $[Zn(L1)]^{2+}$ interaction with the poly(dG)·poly(dC) duplex is a different trend than what was observed with the poly(dA)·poly(dT), poly(A)·poly(U), and poly(dA-dT)$_2$ duplexes, where $[Zn(L1)]^{2+}$ had a destabilizing effect. Similarly, $[Zn(L14)]^{2+}$ increased the melting temperature of the poly(dG)·poly(dC) duplex by > 20 °C. Thus both $[Zn(L1)]^{2+}$ and $[Zn(L14)]^{2+}$ interact with and stabilize the poly(dG)·poly(dC) duplex. The complex $[Zn(L14)]^{2+}$ most likely interacts with the duplex through intercalation of the acridine pendent and possibly by binding to the N7 of guanine nucleobases.

To better elucidate binding modes, both L1 and L14 (Zn^{2+} free) were studied and shown to increase helical stability. For the poly(A)·poly(U) duplex, the L1 ligand at an L1/nucleotide ratio of 0.1 raised the T_m by 6.0 °C. For the same duplex, the L14 compound at a ratio of 0.8 raised the T_m by 18.5 °C. These results shed light on the effect of the Zn^{2+} center in the titrations described above. Thus, L1 showed the opposite trend to $[Zn(L1)]^{2+}$, confirming that the Zn^{2+} center is in fact disrupting the helix and causing denaturation to occur. The L14 versus $[Zn(L14)]^{2+}$ comparison shows that $[Zn(L14)]^{2+}$ displays both stabilizing and destabilizing effects, unlike the free ligand that only exhibits stabilizing effects.

Acridine derivatives are known intercalators so this trend can be explained by intercalation of the acridine along with favorable electrostatic interactions of the DNA with the protonated macrocycle leading to an increase in T_m. This also supports the role of the Zn^{2+} center in the concentration-dependent binding mode for $[Zn(L14)]^{2+}$. The acridine pendent dominates binding at low concentrations ($[Zn(L14)]^{2+}$/nucleotide ratio ~ 0.2), but as the ratio increases > 0.2 the Zn^{2+} center dominates binding, leading to helical denaturation. To further confirm the importance of the Zn^{2+} center for the disruption of the helix, $[Cu(L1)]^{2+}$ was tested against the poly(A)·poly(U) duplex. This complex did not disrupt the helix at all; rather, a stabilizing effect was observed through an increasing melt temperature. When the ratio of $[Cu(L1)]^{2+}$ to nucleotide was 0.1, the T_m increased +6.5 °C (65).

C. Nuclease Footprinting Studies With Zn^{2+} Complexes

In order to further assess the binding mode of the Zn^{2+} complexes, nuclease footprinting assays were carried out. A 150 base pair sequence from the plasmid, pUC19 was analyzed. The binding modes of the Zn^{2+} complexes were compared to A-T specific minor groove binders, DAPI and distamycin A, and to echinomycin as a reference compound for G-C selective binding. Not only were reference compounds used, several different nucleases were used as well including DNase I and micrococcal nuclease.

DNase I is a nuclease that cleaves the phosphodiester linkage on the 5′ end of a pyrimidine and cleaves both strands. The complexes $[Zn(L1)]^{2+}$, $[Zn(L14)]^{2+}$, and $[Zn(L16)]^{2+}$ all showed specificity for the A-T tracts in the plasmid DNA upon digestion of the DNA with this nuclease. The digestion of the DNA was protected on both strands of the helix. Digestion was similar to controls showing that both DAPI and distamycin A protected the A-T tracts of DNA from DNase I digestion. The A-T sequences on pUC19 that were protected by Zn^{2+} complex binding almost overlapped with the protected areas from DAPI and distamycin A. Comparison of these results with those of the micrococcal nuclease studies showed distinct variations that allowed for a better understanding of the Zn^{2+} complex binding mode.

Micrococcal nuclease footprinting experiments showed digestion was protected with DAPI and distamycin A in virtually an identical manner to that with DNase I. However, when the Zn^{2+} complexes were analyzed, only the poly(dT) tract on a single strand was protected, while its pair, the poly(dA) strand was cleaved by the nuclease. This nuclease shows sequence specific cleavage by cleaving only T−A bonds, and not G−C bonds. It also capitalizes on the breathing mechanism of DNA and binds and cleaves a single strand of the duplex, allowing for more selective cleavage (43).

The combination of these nuclease studies showed that the Zn^{2+} complexes specifically recognize dA-dT tracts in duplex DNA and effectively only bind to the

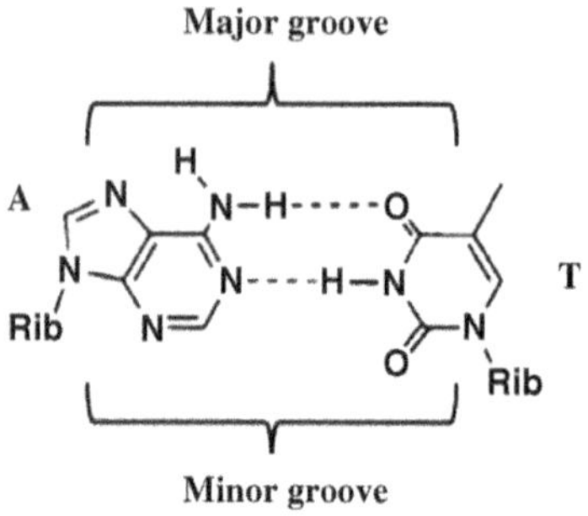

Scheme 16. The dA-dT base pair (46).

thymine, and leave its nucleobase pair, adenine, intact. Further studies suggest that the complexes initially bind in the minor groove prior to disrupting dT-dA and A-U nucleobase pairs. To elucidate which groove $[Zn(L14)]^{2+}$ binds into, a displacement assay was run with distamycin and methyl green as minor and major groove binders, respectively. Approximately 50% dissociation of $[Zn(L14)]^{2+}$ and calf thymus DNA was observed when 6 μM distamycin was added to a mixture containing 10.5 μM $[Zn(L14)]^{2+}$ and 96.0 μM (nucleobase concentration) calf thymus DNA. In contrast, for methyl green, 50% dissociation occurred at 30 μM (46). This finding is consistent with the importance of the minor groove for $[Zn(L14)]^{2+}$ binding. This result is attributed to the presence of a carbonyl oxygen on T or U (O4) that is available for Zn^{2+} binding in the minor groove (Scheme 16) (65).

Binding constants derived from footprinting information showed that DAPI and distamycin A bound the strongest, followed by $[Zn(L14)]^{2+} > [Zn(L16)]^{2+} >> [Zn(L1)]^{2+}$. This analysis confirms the requirement of the pendent group to improve not only recognition, but binding affinity.

The complex $[Zn(L15)]^{2+}$ was studied by using the DNase I assay on DNA that contained a sequence rich in G-C boxes that are located upstream from a TATA box region. The TATA box region is rich in A-T base pairs while the G-C box region is rich in G-C base pairs. These regions are biologically relevant in that they provide a binding domain for replication proteins (43, 83). When $[Zn(L15)]^{2+}$ was added to the DNA, there was protection from nuclease cleavage in the region where the G-C boxes were located, but none where the TATA box was located. Not only was no cleavage present in the G-C rich region, it was evident that the protection was localized to the strand containing 5′-GGGCGGG-3′, whereas the complementary strand, 5′-CCCGCCC-3′ was exposed to nuclease activity. This result is consistent with preferential binding of $[Zn(L15)]^{2+}$ to guanine in double-stranded DNA. Studies on binding to simple nucleosides show that this pendent promotes binding to N7 of the guanine nucleobase.

D. Mode of Binding

Data from optical spectroscopic titrations, CD spectroscopy, T_m experiments, and nuclease footprinting experiments all support binding to duplex DNA or RNA through multiple modes that initially may involve stacking or intercalation of the pendent aromatic group of the Zn^{2+} complex. Most Zn^{2+} complexes discussed here eventually denature DNA or RNA containing a strand with a run of consecutive thymine or uracil nucleobases, respectively, albeit at relatively high Zn^{2+} complex/nucleotide concentration ratios (>0.2). $[Zn(L1)]^{2+}$, $[Zn(L14)]^{2+}$, $[Zn(L17)]^{2+}$- $[Zn(L20)]^{2+}$ preferentially interact with thymine or uracil in double helical DNA in at least partially denatured DNA to expose the thymine or uracil bases as shown by nuclease protection experiments. The exposed face of the thymine or uracil base in duplexes is important to the denaturation of the duplex; in fact destabilization is promoted by this pivotal interaction. Duplex DNA, however, is not the only source for exposed faces of nucleobases; many non-canonical secondary motifs have incomplete pairing of bases that can leave the imide (N3) face of thymine or uracil exposed. The natural exposure of the nucleobases can be utilized for recognition, and denaturation is less significant with these interactions. In fact, non-canonical nucleobase recognition with Zn^{2+} complexes may stabilize certain structures. These interactions are discussed in Section IV with emphasis on the non-canonical bulged secondary structure.

IV. RECOGNITION OF THYMINE OR URACIL BULGES AND OTHER NON-CANONICAL STRUCTURES

Binding of thymine or uracil groups in nucleic acids containing secondary structural motifs (e.g., bulges, loops, and abasic sites) provides additional challenges for recognition agent design. Factors that could perturb binding to thymine or uracil in structured nucleic acids include the involvement of these nucleobases in additional interactions with other groups in the nucleic acid. Such interactions may be disrupted upon Zn^{2+} binding. In addition, binding of the Zn^{2+} complex may alter the conformation of neighboring bases or even affect more remote parts of the oligonucleotide structure. In more complex structures, other proximal groups may affect π–π stacking interactions between the Zn^{2+} complex pendent aromatic group and the pyrimidine face of thymine. Therefore, the requirements for Zn^{2+} complex recognition of a thymine or uracil nucleobase in a secondary structure are more stringent than in a relatively flexible single-stranded oligonucleotide. In this sense, the recognition of a uracil or a thymine group in a larger folded nucleic acid is analogous to fitting a small molecule into a well-defined binding pocket. Additional considerations of the important energetic factors contributing to binding are discussed in Section V.

In this section, we describe the recognition of two common types of structures that involve non-canonical thymine or uracil. Nucleic acids with single nucleobase bulges have been a popular target based on their importance in the biology of DNA repair and their ubiquitous nature in RNA. Single nucleobase bulges have been the subject of many recognition studies given the relative simplicity of their structures. For pyrimidines in particular, the bulged nucleobase is relatively accessible because it generally protrudes away from the helix (Section I.B). Thus, one simple approach has been to use planar aromatic groups that hydrogen bond to the Watson–Crick face of the bulged nucleobase. Yet even this approach must take into consideration the interaction of the recognition agent with the rest of the nucleic acid because aromatic groups are not likely to protrude into the surrounding aqueous environment. Similarly, Zn^{2+} complex binding to uridine or thymine bulges must take into account the position of the bulged nucleobase with respect to the rest of the nucleic acid and the possible reorganization of the nucleic acid structure upon binding.

Abasic sites are also discussed here, although more briefly. Thymine or uracil groups that are opposite an abasic site are typically intrahelical, placing a greater restriction on the recognition agent. Perhaps for this reason, the type of molecules that function well are planar aromatic groups that fit into the space normally occupied by the missing nucleobase. This type of site would seem to be less suited for Zn^{2+} macrocyclic complexes. The major application for Zn^{2+} complex binding to abasic sites is, in fact, to measure flipping of the T base into the extrahelical form.

A. Structure of DNA Containing a Thymine Base Bulge

Factors that are important for binding of Zn^{2+} complexes to a DNA secondary structure (e.g., a bulge) include aromatic base stacking, interaction of the Zn^{2+} with the N3 of thymine, stacking of neighboring nucleobases and interaction of the bulged moiety with the helical portion of the nucleic acid. The nature of the environment surrounding the bulged thymine is also important. For example, thymine bulges are most stable with purine flanking base pairs (29). As discussed in Section I.B, the extrahelical nature of the T may be dependent on flanking nucleobases, temperature, and pH (27, 28). Given the lack of structural information on thymine bulges, Morrow and co-workers (66) produced a NMR structure of a DNA hairpin containing a single thymine bulge.

The structure of the TggX bulge (Fig. 6) was determined by using NMR spectroscopy. Upfield shifts of C6 H4′, H5′, and H5″ are consistent with a hairpin loop, and the number of proton resonances suggest that there is one stable secondary structure (hairpin) under the conditions of the experiment. Two-Dimensional Nuclear Overhauser Effect Spectroscopy (2D NOESY) showed Nuclear Overhauser Effects (NOEs) between G8 H1′ and G10 H8, implying the two bases are stacked on each other and the thymine is looped out from the

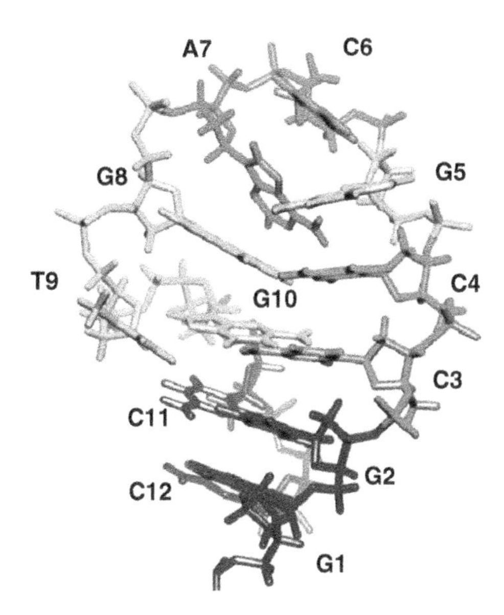

Figure 6. The NMR structure of TggX. Here T9 is flipped away from the stem and located in the major groove. [Reprinted with permission from (66) I.M.A. del Mundo et al. *Inorg. Chem.*, *51*, 5444 (2012). Copyright © (2012) American Chemical Society.] (See the color version of this figure in color plates section.)

stack. Additionally, weak or absent NOEs between T9 H6 and G8 H1 or G10 H8 and T9 H1′ are consistent with the T being extrahelical and oriented away from the flanking base pairs. For a complete list of NOE contacts, the reader should refer to a 2012 publication (66). The resulting structure, utilizing the NMR data presented above, illustrates that the thymine is extrahelical and directed toward the major groove of the hairpin stem. The model also shows the flanking G–C base pairs stacking on each other. However, G8 remains partially unstacked and displaced toward the major groove. The structure (Fig. 6) illustrates that the deprotonated amide, N3, is positioned to coordinate with the Zn^{2+} center of the macrocyclic complex. In order to determine how the Zn^{2+} complex might interact with this site, the NMR structure of a Zn^{2+} complex, $[Zn(L17)]^{2+}$, with a similar DNA bulge was determined, as described further in Section IV.B.

B. Interactions of Zn^{2+} Complexes With DNA Bulges

Binding of Zn^{2+} complexes to a series of thymine bulges was studied to determine selectivity for the bulged T in comparison to other nucleobase bulges and to study the affect of neighboring bases on binding. Complexes $[Zn(L17)]^{2+}$, $[Zn(L18)]^{2+}$, $[Zn(L21)]^{2+}$, and $[Zn(L23)]^{2+}$ studied by Morrow and

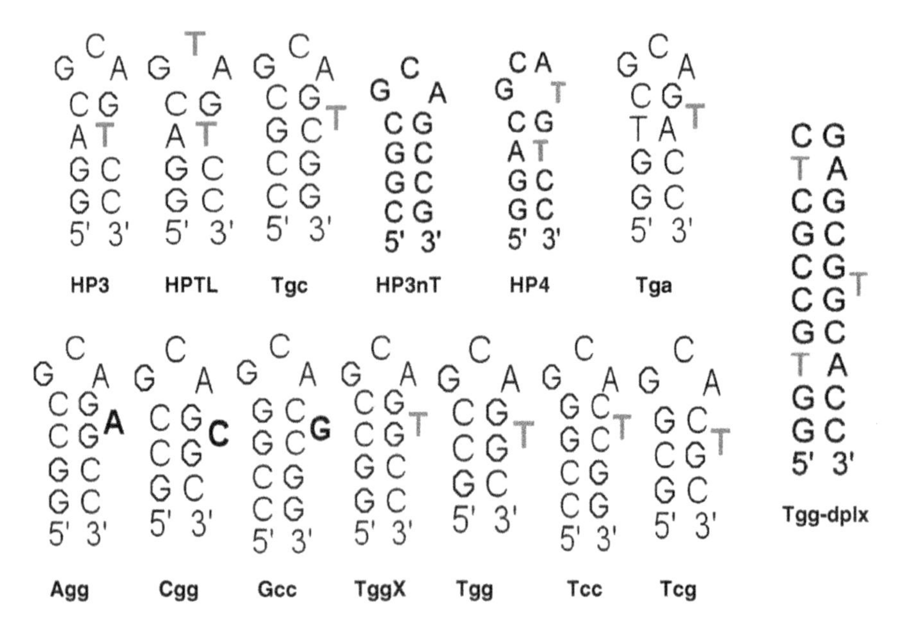

Scheme 17. Hairpins and Bulges used to study Zn^{2+} complex interactions.

co-workers (66, 67) for bulges and $[Zn(L21)]^{2+}$ by O'Neil and Wiest (68, 69) for abasic sites capitalize on the observation that two-ring pendent groups are not effective intercalators and do not bind indiscriminately to double-stranded DNA. In contrast, $[Zn(L14)]^{2+}$ that contains an acridine pendent binds tightly to duplex DNA (46, 67).

Morrow and co-workers (66,70) investigated $[Zn(L17)]^{2+}$, $[Zn(L18)]^{2+}$, $[Zn(L21)]^{2+}$, and $[Zn(L23)]^{2+}$ complexes for their affinity to single-thymine bulges. Selectivity of $[Zn(L17)]^{2+}$ was greatest toward single-thymine bulges. The bulges that were studied (Scheme 17) differed in stem length and flanking base pairs. The $K_{d\ (app)}$ and K_d values were obtained by using an indirect fluorescent displacement assay with EtBr. An indirect fluorescent assay was used because the photophysical properties of $[Zn(L17)]^{2+}$ were not sufficiently different from DNA to use direct fluorescence measurements. Instead, binding was quantified by observing a decrease in EtBr fluorescence. As $[Zn(L17)]^{2+}$ binds to DNA, the EtBr is displaced and a decrease of fluorescence is observed (Scheme 18).

The $[Zn(L17)]^{2+}$ complex showed strong binding to the majority of T bulges with a preference for bulges containing an adjacent 5′ purine base (Table I). The $K_{d\ (app)}$ values were between 1.2 and 2.2 μM to bulges containing an adjacent 5′ purine base. Bulges containing flanking pyridimines that were 5′ relative to the thymine bulge showed weaker binding as demonstrated by Tcc ($K_{d\ (app)}$ 22 μM at

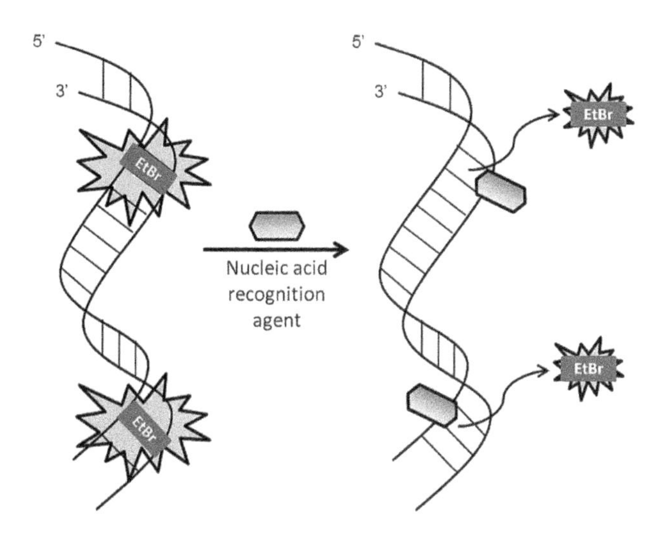

Scheme 18 Illustration of ethidium bromide displacement assay used. (See the color version of this scheme in color plates section.)

pH 7.5, 100 m*M* NaCl). Due to the nature of the experiment, the apparent K_d values were corrected for the affinity of EtBr to DNA by using Eq. 2 (66). The values for EtBr binding to the oligonucleotide types are as follows: $K_{EtBr\ (DNA\ Bulge)}$: 1.33×10^5, $K_{EtBr\ (RNA\ Bulge)}$: 8.00×10^5, $K_{EtBr\ (DNA\ Stem)}$: 7.41×10^4 and $K_{EtBr\ (RNA\ Stem)}$: 1.32×10^4 (70). A summary of $[Zn(L17)]^{2+}$ binding constants, $K_{d\ (app)}$ and K_d, to T bulges is shown in Table I. Note that bulges in Table I are identified by

TABLE I
The $[Zn(L17)]^{2+}$ Dissociation Constants Binding to Oligonucleotides[a]

Sequence	$K_{d\ (app)}$ (μ*M*)	K_d (μ*M*)[b]	Sequence	$K_{d\ (app)}$ (μ*M*)	K_d (μ*M*)[b]
Tgg	2.2 ± 0.2[c]	3.8	Tcg	3.8 ± 1.4	6.5
Agg	>200	>400	TggX	1.2 ± 0.4	2.1
Cgg	185 ± 40	315	Tgg-dplx	1.5 ± 0.1	2.6
Gcc	108 ± 1	184	HPTL	127 ± 16	571.3[d]
Tgc	1.2 ± 0.8	3.4	HP3	142 ± 50	435
Tga	1.3 ± 0.1	2.2	HP4	104 ± 22	468[d]
Tcc	22.2 ± 0.2	37.8	HP3nT	435 ± 10	1334

$[Zn(L17)]^{2+}$

[a] At pH 7.5 100 m*M* NaCl [66].
[b] From Eq. 2; $K_{EtBr\ (DNA\ Bulge)}$: 1.33×10^5, $K_{EtBr\ (RNA\ Bulge)}$: 8.00×10^5, $K_{EtBr\ (DNA\ Stem)}$: 7.41×10^4, and $K_{EtBr\ (RNA\ Stem)}$: 1.32×10^4.
[c] From [70].
[d] Data not published.

using lowercase letters to denote the flanking base pairs 5′ to 3′. Importantly, $[Zn(L17)]^{2+}$ shows at least 100-fold selectivity toward T bulges, exhibited by K_d values of 244, 185, and 106 μ*M* for adenine, cytosine and guanosine bulges, respectively, at pH 7.5 and in 100 m*M* NaCl (66, 70).

$$K_d = \frac{K_{d\,(app)}}{(K_{EtBr}[EtBr])} \tag{2}$$

The $[Zn(L17)]^{2+}$ complex bound to Tgc (see Scheme 17) has a $K_{d\,(app)}$ of 1.2 μ*M* at pH 7.5, 100 m*M* NaCl, the strongest affinity observed between $[Zn(L17)]^{2+}$ and thymine bulges to date (70). Interestingly, fluorescence increases when $[Zn(L17)]^{2+}$ binds to Tgc-EtBr adducts. This suggests that $[Zn(L17)]^{2+}$ binding to Tgc–EtBr is cooperative with ethidium bromide because EtBr fluorescence is enhanced. Although the exact mechanism is not currently known, it is possible that the quinoline moiety stacks on the EtBr or binds to a structure stabilized by the EtBr and elicits an increase in fluorescence. The EtBr assay is useful to quantify binding, however, care must be taken in the interpretation of the K_d values generated.

To understand the basis for the selective interaction of Zn^{2+} complexes, it is important to study how the bound Zn^{2+} complexes affect the bulge structure. An increased T_m suggests that the complex stabilizes the bulge. The extent of stabilization is, however, difficult to predict based on the dissociation constant. For example, $[Zn(L17)]^{2+}$ binds tightly to both Tgg ($K_{d\,(app)}$: 2.2 μ*M*) and to Tcg ($K_{d\,(app)}$: 3.8 μ*M*) yet, $[Zn(L17)]^{2+}$ stabilizes Tgg by 5 °C and destabilizes Tcg by 6.1 °C with 1 equiv and an additional 3 °C with 2 equiv (66). Thus, despite the similarity in $K_{d\,(app)}$ of Tgg and Tcg, their interaction with $[Zn(L17)]^{2+}$ must be different. As discussed in this section, this may be due to either different binding pockets for the two bulges, or other energetic factors, such as reorganization of the nucleic acid structure.

Circular dichroism spectroscopy was utilized to better understand how $[Zn(L17)]^{2+}$ impacted the structural integrity of a single-thymine bulge. The complex $[Zn(L17)]^{2+}$ binds to Tgg to produce a CD signal with a minimum at 240 nm as the complex concentration is increased up to 1 equiv [Fig. 7(*a*)]. Inducing a change in CD spectra implies a structural change. It is interesting that $[Zn(L17)]^{2+}$ does not change the helical structure of the DNA as it retains B-form DNA. However, at higher concentrations (3 equiv) of $[Zn(L17)]^{2+}$, more structural changes occur as evidenced by a disappearance of the 240-nm minimum. At these concentrations, the bulge structure has not been disrupted significantly; the maximum at 280 nm remains [Fig. 7(*b*)] (84, 85). Not surprisingly, at very high concentrations (13 equiv) of $[Zn(L17)]^{2+}$ the bulge structure is now compromised as observed by loss of most of the CD signal [Fig. 7(*c*)].

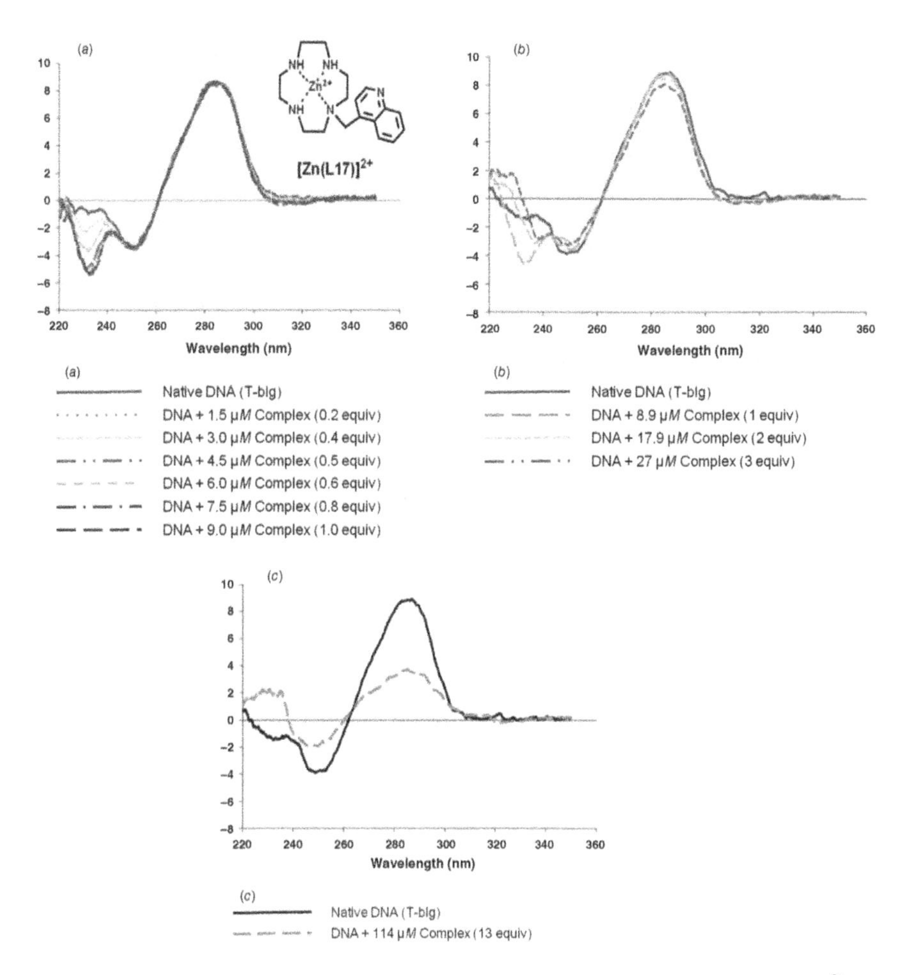

Figure 7. Circular dichroism spectra of Tgg(8.9 μ*M*) in increasing concentrations of $[Zn(L17)]^{2+}$ at pH 7.5 100 m*M* NaCl. (*a*) Low concentrations of $[Zn(L17)]^{2+}$. (*b*) Moderate concentrations of $[Zn(L17)]^{2+}$. (*c*) High concentrations of $[Zn(L17)]^{2+}$. *Inset*: $[Zn(L17)]^{2+}$. [Adapted from 86.] (See the color version of this figure in color plates section.)

NMR experiments were utilized to better understand the binding mode of the Zn^{2+} complex to the T bulge. Proton NMR experiments with Tgg at 2 m*M* at pH 7.5 100 m*M* NaCl, showed multiple sets of resonances, consistent with two separate conformations. The multiple conformations are most likely due to equilibrium between the bulged sequence and a minor non-hairpin conformation (Fig. 8).

However, when 1 equiv of $[Zn(L17)]^{2+}$ is introduced, the correct number of proton resonances for a single secondary structure is observed. This NMR

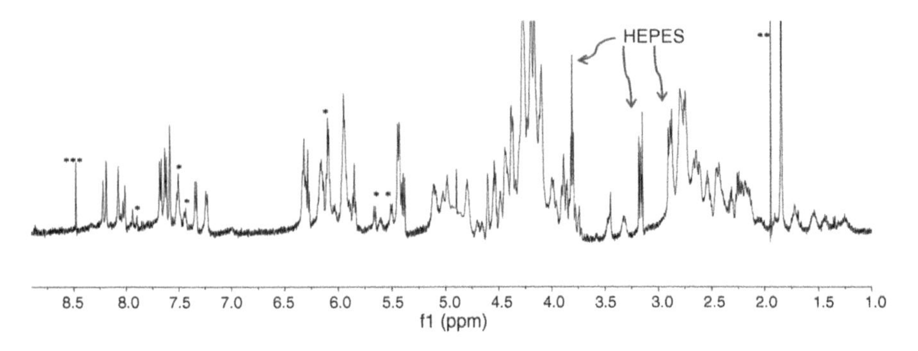

Figure 8. Proton-NMR spectra of Tgg, 2 m*M* at pH 7.5 100 m*M* NaCl. **Me_3COO^- peak, ***impurity, *extra peaks due to more than one conformation of the DNA (HEPES = 2-[4-(2-hydroxyethyl)-piperazin-l-yl]ethane sulforic acid). [Adapted from (86).]

spectrum supports the ability of $[Zn(L17)]^{2+}$ to stabilize the bulged hairpin Tgg. Interestingly, a stoichometry of 1:1 $[Zn(L17)]^{2+}$: Tgg leads to a decrease in T-methyl peak intensity at 1.7 ppm with the increase of a new resonance at 2.1 ppm (66). Shifting of the thymine methyl resonance is consistent with the quinoline moiety of $[Zn(L17)]^{2+}$ stacking on the face of the bulged thymine. Further supporting this interaction is a shift in five proton resonances of the quinoline pendent as well as the resonances from the bulged thymine. These are observed as new resonances that grow in the 6.3–7 ppm region of the ^{1}H NMR between Tgg and $[Zn(L17)]^{2+}$. These resonances are C4H1′, T9H4′ from the thymine, q-H8, q-H3, and the methylene linker from $[Zn(L17)]^{2+}$ (66). The similarity of the proton resonances between 6.3–7.0 ppm in ^{1}H NMR titrations of $[Zn(L17)]^{2+}$ between Tgg or with Tggx suggest that $[Zn(L21)]^{2+}$ interacts similarly with Tgg and Tggx (Fig. 9).

The NMR structure of the $[Zn(L17)]^{2+}$ (Tgg) adduct is shown in Fig. 10. This structure provides a great deal of information on the interaction of the Zn^{2+} macrocycle with quinoline pendent to the T bulge. In the model shown in Fig. 10, the $[Zn(L17)]^{2+}$ forms a coordination bond between the Zn^{2+} center and the deprotonated amide, N3, of the bulged thymine. More importantly, the quinoline moiety stacks face-to-face with the thymine residue. The model places the quinoline moiety 3–8 Å above the thymine.

The quinoline moiety is slightly skewed with respect to the thymine with the q-H2, q-H3, q-H7, and q-H8 protons above the thymine ring plane. Two-dimensional NOESY experiments revealed an NOE between q-H8 and the adjacent G8-H2″ sugar proton. Quinoline stacking on thymine is confirmed with observed NOEs between q-H8 and T9H4′, H5′, and H5″. An NOE was observed between q-H3 and the T-methyl. This suggests the quinoline moiety is positioned toward the major groove (Figs. 10 and 11) (66). The models presented in Figs. 10 and 11 suggest that

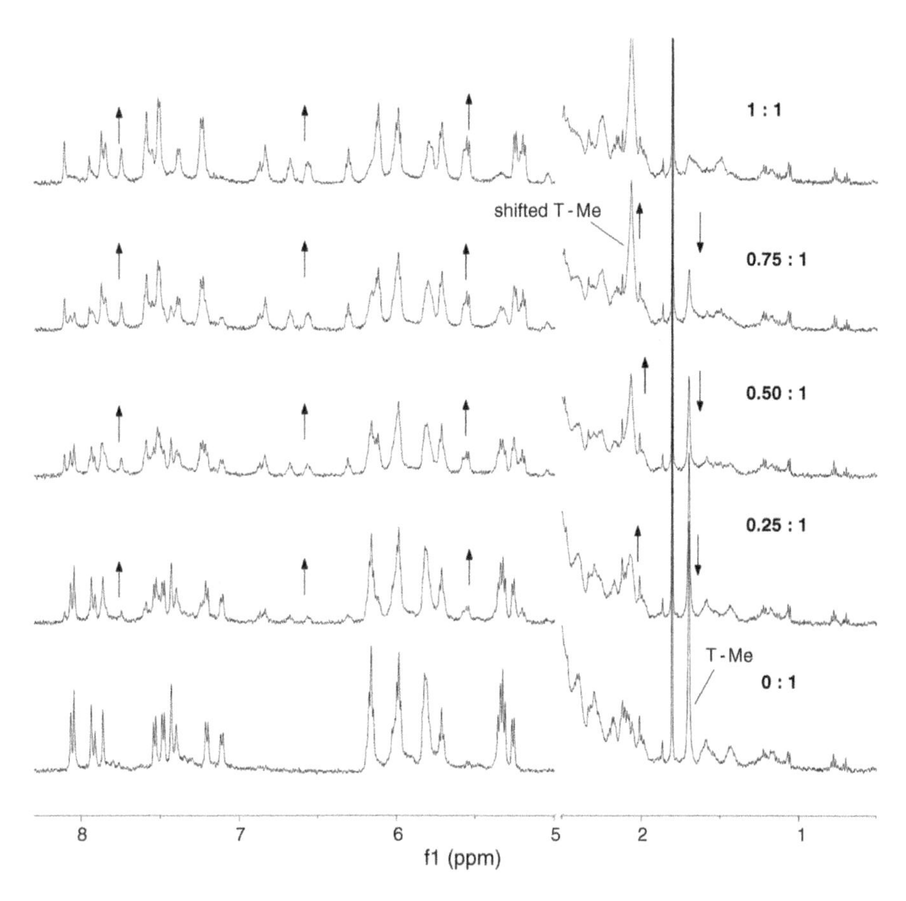

Figure 9. Proton NMR titration of $[Zn(L17)]^{2+}$ into Tgg (2 m*M*) at varying concentrations $[Zn(L17)]^{2+}$: Tgg. Arrows denote new resonances. [Adapted from (86).]

Tgg and similar thymine bulges produce a binding pocket for the Zn^{2+} complex. The thymine is oriented away from the hairpin stem in the major groove. However, the aromatic pendent must be compact and planar to fit between the bulged thymine and hairpin stem. Additionally, a flexible linker may be needed to facilitate stacking on the extrahelical thymine. The planar quinoline moiety can fit into the binding pocket observed in Tgg and TggX; this suggests a rationale for low micromolar binding constants and stabilization of Tgg and TggX by $[Zn(L17)]^{2+}$. Additionally, the two flanking guanines are more closely spaced, and presumably stacked more strongly on each other in the structure containing the bound Zn^{2+} complex.

The Morrow group studied a series of compounds that showed both weaker binding and lower selectivity to thymine bulges (Table II). In these complexes, the

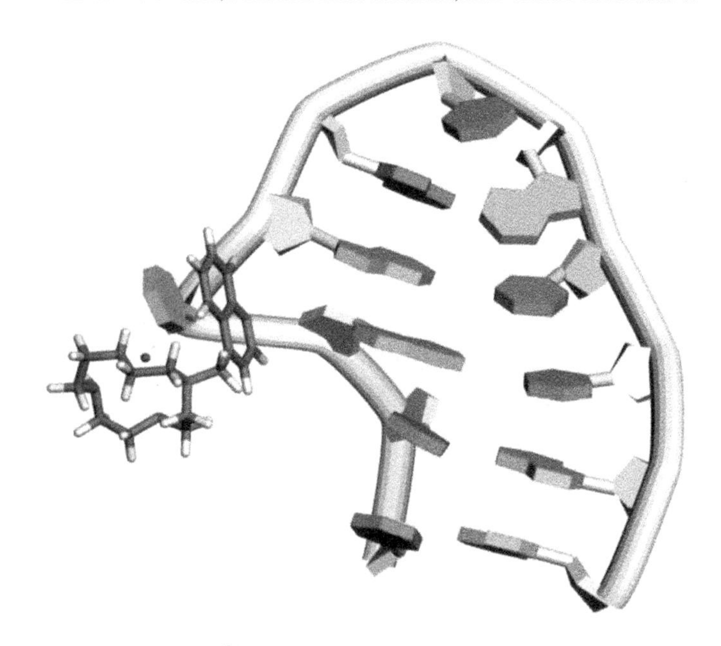

Figure 10. Model of $[Zn(L17)]^{2+}$ binding to TggX. Quinoline moiety is stacked on thymine and directed toward the major groove. (See the color version of this figure in color plates section.)

TABLE II

Binding Constants of Zn^{2+} Complexes to Thymine Bulge and Control Hairpins at pH 7.5 100 mM NaCl[a]

	K_{app} ($\times 10^{-6}$ M)			
Sequence	$[Zn(L14)]^{2+}$ [b,c]	$[Zn(L21)]^{2+}$ [c]	$[Zn(L22)]^{2+}$ [d]	$[Zn(L23)]^{2+}$ [d]
Tgg	≤ 1	250 ± 34	9 ± 2	171 ± 25
HPTL	0.27 ± 0.02	>1000	75 ± 26	459 ± 35
HP3	1.0 ± 0.2	>1000	93 ± 18	427 ± 77
HP4	nd[e]	nd[e]	79 ± 21	193 ± 91
HP3nT	nd[e]	nd[e]	118 ± 7	421 ± 56

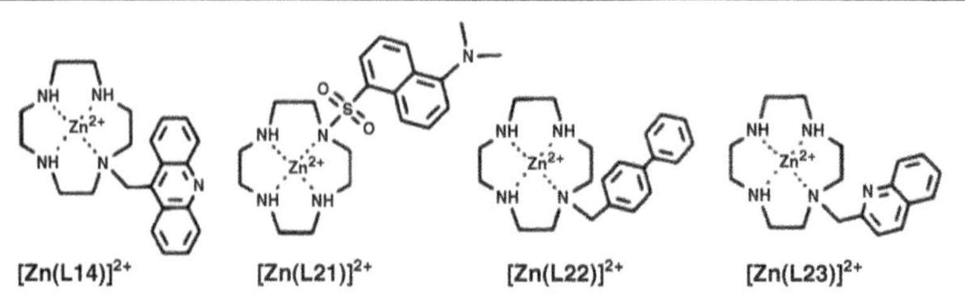

[a] See (66).

[b] Data not published.

[c] Direct fluorescence used to determine binding.

[d] Ethylbromide used to determine binding.

[e] Not determined = nd.

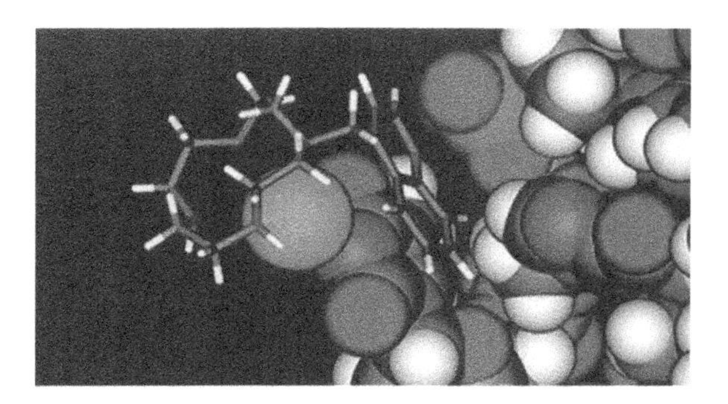

Figure 11. Close-up of T bulge and adjacent stem. Bulge binding pocket can be observed. [Reprinted with permission from I.M.A. del Mundo et al. *Inorg. Chem.*, *51*, 5444 (2012) [66]. Copyright © (2012) American Chemical Society.] (See the color version of this figure in color plates section.)

Zn^{2+} cyclen macrocycle was conserved and the aromatic pendents were varied to give $[Zn(L14)]^{2+}$ and $[Zn(L21)]^{2+}$-$[Zn(L23)]^{2+}$. Complex $[Zn(L22)]^{2+}$ showed relatively strong binding to the Tgg bulge, with an observed $K_{d\ app}$ value of 9 μ*M* at pH 7.5 100 m*M* NaCl (66). However, unlike $[Zn(L17)]^{2+}$, $[Zn(L22)]^{2+}$ showed more moderate selectivity toward control hairpins. For example, HP3, HP4, and HPTL were found to have $K_{d\ app}$ values of 93, 79, and 75 μ*M*, respectively, at pH 7.5 100 m*M* NaCl (66). The $[Zn(L23)]^{2+}$ complex, while similar to $[Zn(L17)]^{2+}$, displayed weak binding to all oligonucleotides it was tested against. This weak binding is attributed to binding of the 2-quinoline pendent to the Zn^{2+} center to inhibit thymine binding. This result was shown by both ^{1}H NMR studies and by the high pK_a (11.2) of the Zn^{2+}-H_2O of $[Zn(L23)]^{2+}$ indicating suppression of $[Zn^{2+}\text{-}OH]^-$ formation and thus blocking of the available coordination site. The pendent group binding to the Zn^{2+} center thus prevents binding to the deprotonated amide N3 of thymine.

Notably, the fluorescence properties of dansyl pendent of $[Zn(L21)]^{2+}$ allows for direct monitoring of DNA binding. As the complex binds to a thymine in the loop of a hairpin, for example, the environment around the dansyl pendent will become more hydrophobic and fluorescence will increase (87). However, $[Zn(L21)]^{2+}$ shows little increase in fluorescence and binds very weakly to thymine bulges (Tgg). The $K_{d\ (app)}$ for $[Zn(L21)]^{2+}$ (Tgg) adduct at pH 7.5, 100 m*M* NaCl is 250 μ*M*. Nevertheless, $[Zn(L21)]^{2+}$ shows still weaker binding to the control hairpins HP3 and HPTL. In fact, only approximate binding constants were assigned due to the weak binding: >1 m*M* (66).

Several factors dictate bulge binding. The $[Zn(L14)]^{2+}$ complex contains an acridine pendent group with good aromatic stacking properties, yet it is unselective. This complex binds strongly to both Tgg as well as to control hairpins HPTL and HP3 with $K_{d\ (app)}$ of 0.27 and 1.0 μ*M*, respectively. As discussed in

Section III.B.3, the acridine pendent of $[Zn(L14)]^{2+}$ intercalates into DNA. This relatively nonspecific interaction is likely to compete with the mode observed for simple thymines that has the Zn^{2+} center interacting with the deprotonated amide, N3, and the acridine stacked on the thymine.

Weak binding to the thymine bulge by $[Zn(L21)]^{2+}$ can be attributed to several factors. As seen in Figs. 10 and 11, a tight binding pocket is formed by the bulged thymine and the major groove of the hairpin stem. A binding pocket as presented in Fig. 11 can be compared to one found in a protein. Similar to protein receptor interactions, a certain shape is required for proper binding. The data presented in Table II regarding $[Zn(L21)]^{2+}$ suggests that the dansyl pendent prevents favorable bulge binding. The dimethylamine group pf the dansyl pendent makes it nonplanar, unlike the quinoline pendent in $[Zn(L17)]^{2+}$. This dimethylamine substituent may sterically prevent $[Zn(L21)]^{2+}$ from stacking on the extrahelical thymine in Tgg. Additionally, $[Zn(L21)]^{2+}$ differs from the complexes previously discussed because of its sulfone (SO_2) linker. The sulfone linker, connecting the aromatic pendent to the Zn^{2+} center, is slightly more rigid than the methylene (CH_2) linker observed in the remainder of the complexes discussed in this section. The more rigid linker of $[Zn(L21)]^{2+}$ may not align the aromatic pendent in a manner that is optimal for bulge binding (66). This difference in bulge binding is contrasted to the similar binding constants of $[Zn(L21)]^{2+}$ and $[Zn(L17)]^{2+}$ to thymines in unstructured nucleosides or oligonucleotides. The unstructured oligonucleotides are more flexible and thus are expected to have less specific requirements for the shape of the pendent group. Further energetic factors are discussed in Section V.

C. Interactions of Zn^{2+} Complexes With RNA Bulges

The complexes $[Zn(L17)]^{2+}$, $[Zn(L22)]^{2+}$, and $[Zn(L23)]^{2+}$ were studied as recognition agents for uracil bulges in RNA by using an EtBr displacement assay. Equation 2 was used to calculate K_d from $K_{d\ (app)}$. $[Zn(L23)]^{2+}$, due to pendent group coordination, shows very poor binding to U bulges (Table III) (66). The complex $[Zn(L22)]^{2+}$ also shows very poor binding to U bulges and has very poor selectivity toward the control RNA stem. Sizeable differences in $K_{d\ (app)}$ and K_d are rationalized by the weak affinity of EtBr to RNA hairpins (rHP) and stronger affinity for RNA bulges (Ugg), relative to DNA.

The complex $[Zn(L17)]^{2+}$ shows the strongest binding and selectivity toward uridine bulges. A 7 °C increase in thermal melting temperature is indicative of a stabilization of the uridine bulge, Ugg (66). The ability of modified Zn^{2+} complexes to selectively bind RNA and DNA verifies their versatility as nucleic acid probes for thymine or uridine.

TABLE III

Binding constants for Zn^{2+} Complexes to Thymine Bulge and Control Hairpins at pH 7.5 100 m*M* NaCl[a]

Sequence	$K_{d\ (app)}$ ($\times 10^{-6}$ *M*)	K_d ($\times 10^{-6}$ *M*)[b]
Zn(L17)		
Ugg	12.3	3.5
rHP	20	343
Zn(L22)		
Ugg	285	81[c]
rHP	102	1756[c]
Zn(L23)		
Ugg	334	95[c]
rHP	299	5148[c]

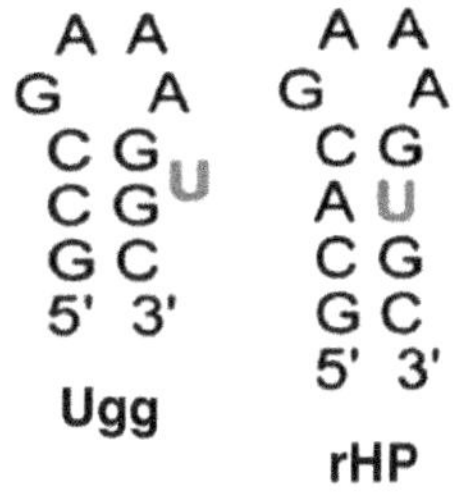

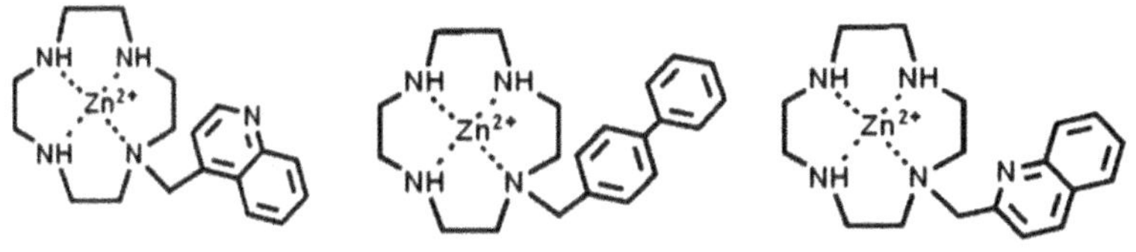

Ugg

rHP

[Zn(L17)]$^{2+}$ [Zn(L22)]$^{2+}$ [Zn(L23)]$^{2+}$

[a] See (66).

[b] From Eq. 2; $K_{EtBr\ (DNA\ Bulge)}$: 1.33×10^5, $K_{EtBr\ (RNA\ Bulge)}$: 8.00×10^5, $K_{EtBr\ (DNA\ Stem)}$: 7.41×10^4, and $K_{EtBr\ (RNA\ Stem)}$: 1.32×10^4.

[c] Data not published

D. Recognition of Abasic Sites

The complex [Zn(L21)]$^{2+}$ showed poor binding toward extrahelical thymine bulges, attributed to steric hindrance within the thymine bulge-induced binding pocket. However, the fluorescent nature of [Zn(L21)]$^{2+}$ makes it a very useful tool for detecting thymines across from abasic sites in DNA. O'Neil and Wiest (69) have taken advantage of the dansyl pendent's solvachromatic nature. The complex [Zn(L21)]$^{2+}$ was observed to bind selectively to an abasic thymine. Binding of [Zn(L21)]$^{2+}$ to an abasic thymine is facilitated when the thymine flips out of the duplex. However, an abasic site provides more room for the complex to bind. O'Neil and Wiest observed a $K_{d\ (app)}$ value of 0.72 μ*M* for [Zn(L21)]$^{2+}$ binding to an abasic thymine, in 100 m*M* pH 7.0 phosphate buffer. They observed a significant increase in fluorescence to saturation as [Zn(L21)]$^{2+}$ bound to abasic thymine. As [Zn(L21)]$^{2+}$ binds to abasic thymine, the relative normalized fluorescence increases to 1.0 (69). Conversely, as [Zn(L21)]$^{2+}$ binds to Tgg a relative fluorescent increase of 0.35 is observed. Despite weak binding to bulged nucleic acids, tight binding to abasic sites justifies [Zn(L21)]$^{2+}$ as a useful nucleic acid probe because of its intrinsic fluorescent nature.

V. ENERGETIC FACTORS IN DNA BINDING: TOWARD MORE SELECTIVE Zn^{2+} RECOGNITION AGENTS

A. Parsing Free Energy Contributions for Binding of Small Molecules to DNA

In this chapter, we focused on using structural information from NMR spectroscopy to rationalize the selectivity of binding of monomeric Zn^{2+} complexes to thymine bulges in DNA. X-ray crystal structures of Zn^{2+} complexes bound to thymidine also provide useful molecular details. However, structural information representing molecular interactions is only one component in the energetics of binding. Here we consider the underlying thermodynamics of binding of small molecules to DNA (88, 89). This approach takes into consideration that there are several energetic factors that may substantially contribute to binding strength. Information on the magnitude of these contributions can lead to a better understanding of the driving force for binding and ultimately guide the design of more selective recognition agents. Unfortunately, experimental data or theoretical computational work on the contributions of different energetic factors to Zn^{2+} complex binding to DNA or RNA is scarce. For example, isothermal calorimetry (ITC) experiments are especially useful in giving insight into certain thermodynamic contributions (88), but detailed studies are lacking for Zn^{2+} complexes.

Free energies of binding of small molecules to DNA have been divided into several different contributions, as shown in Eq. 3, with the assumption that the free energies from the different contributions are additive (88, 89). Understanding the magnitude of the different energetic contributions to binding is useful for defining the recognition process for different classes of binding agents (90). For example, the free energy of binding of classical intercalators and groove binders to DNA has been analyzed (89). Notable differences include structural rearrangement of DNA to create a larger spacing between base pairs at the intercalation site, which gives rise to an unfavorable energetic term. In comparison, there is little reported DNA reorganization upon interaction with groove binders (e.g., Hoechst derivatives). By comparison, there are no studies to the best of our knowledge, on the different contributions to the free energy of Zn^{2+} macrocyclic complex binding to DNA or RNA. Here, we discuss the application of this approach to Zn^{2+} complex binding to DNA, with an emphasis on complexes with aromatic pendent groups.

$$\Delta G_{\text{obs}} = \Delta G_{\text{conf}} + \Delta G_{\text{l+r}} + \Delta G_{\text{hyd}} + \Delta G_{\text{pc}} + \Delta G_{\text{mol}} \tag{3}$$

The free energy terms in Eq. 3 represent the following contributions as proposed by Chaires (88). The observed free energy, ΔG_{obs}, of binding is obtained from the association constant (K) and the relationship $\Delta G = -RT \ln K$. The free energy contribution, ΔG_{conf}, arises from DNA conformational changes that occur upon binding of the small molecule. The free energy contribution, ΔG_{l+r}, is associated with loss of translational and rotational degrees of freedom upon binding of the small molecule to DNA. This term is generally unfavorable and increases with the number of bonds in the molecule. The free energy contribution, ΔG_{hyd}, is associated with hydrophobic transfer of the small molecule from water into its DNA binding site. The polyelectrolyte contribution, ΔG_{pc}, arises from the release of counterions from DNA upon binding of charged molecules (e.g., the Zn^{2+} complexes). The free energy contribution, ΔG_{mol}, is from molecular interactions between the recognition agents and the DNA. Molecular interaction would include stacking interactions, hydrogen bonding, and Zn^{2+} coordination to nucleobases or other ligating groups.

B. Thermodynamic Contributions of Zn^{2+} Complex Interactions With Thymine Bulges

Our focus in this section is on binding interactions of Zn^{2+} complexes that contain a pendent two-ring aromatic group. These complexes have the highest selectivity for binding to non-canonical thymine in DNA or uracil in RNA bulges (66, 70). Comparison of the potential differences in energetic contributions for Zn^{2+} complexes is instructive. The identical overall complex charge and similar size of $[Zn(L17)]^{2+}$, $[Zn(L18)]^{2+}$, $[Zn(L21)]^{2+}$, and $[Zn(L22)]^{2+}$ suggest that ΔG_{pc} and ΔG_{l+r} may be close in magnitude for these complexes. In contrast, ΔG_{hyd} will differ between the complexes based on the number and placement of heteroatoms in the aromatic pendent, as well as the solvent accessible surface area of the pendent group. The parameter ΔG_{hyd} is an important energetic contribution that is a large favorable driving force for DNA interactions for many drugs containing aromatic groups (88). Some of the pendent groups present different shapes, such as the twisted biphenyl in $[Zn(L22)]^{2+}$, compared to the planar naphthyl group in $[Zn(L18)]^{2+}$, which would be anticipated to give different ΔG_{hyd}. The ΔG_{hyd} for $[Zn(L17)]^{2+}$ and $[Zn(L18)]^{2+}$ may be comparable given the similar surface area of the two planar pendent groups. Two additional terms that are important and would likely differ between complexes and DNA sequences are ΔG_{mol} and $\Delta G_{conf.}$ The magnitude of the contribution of ΔG_{conf} for Zn^{2+} complex binding to bulged DNA is not known at this juncture, but reorganization in base stacking as shown by NMR spectroscopy experiments suggests a nonzero contribution. The parameter ΔG_{mol} depends on molecular variations in the aromatic pendent groups and their interaction with the pocket created by the DNA bulge. To date, we have T bulge binding constants for four Zn^{2+} complexes with two-ring pendent aromatics that give us insight into the role of pendent group molecular interactions in ΔG_{mol}.

Prior to examining energetic contributions to bulge binding, it is instructive to examine binding of $[Zn(L17)]^{2+}$ and $[Zn(L21)]^{2+}$ to single-stranded DNA. Binding to CCTCC does not differ markedly for the two complexes despite their different aromatic pendent group (25 and 55 μM, respectively) (66). This result suggests that either the Zn^{2+} interactions with N3 are similar and that the aromatic group stacking interactions do not differ markedly or that the two factors compensate. For example, $[Zn(L21)]^{2+}$ is a better Lewis acid, as shown by the lower water ligand pK_a and should thus bind more tightly to the N3 of thymine. This more favorable interaction of the Zn^{2+} center is probably negated by the poorer stacking properties of the dansyl group. The flexibility of the single-stranded CCTCC that permits accommodation of the stacking interaction may be important. Also, for single-stranded DNA, the aromatic group is unlikely to be completely solvent free so that ΔG_{hyd} may not be fully realized. Note that single-stranded DNA containing multiple thymines are in another class because of the cooperative binding of multiple Zn^{2+} complexes that feature aromatic pendents stacked on each other (6).

The binding strength of the macrocyclic complexes $[Zn(L17)]^{2+}$ and $[Zn(L21)]^{2+}$ for bulged DNA is quite different than for single-stranded DNA (66). Whereas $[Zn(L17)]^{2+}$ shows a 50- to 200-fold enhancement for binding to bulged thymine compared to duplex DNA, $[Zn(L21)]^{2+}$ binding is approximately 4-fold stronger to the thymine bulge and binding is weak to all DNA sequences studied. The complex $[Zn(L17)]^{2+}$ binds 100-fold more strongly to the Tgg bulge than does $[Zn(L21)]^{2+}$. This most likely reflects a difference in the molecular interactions of $[Zn(L21)]^{2+}$ with the DNA bulge, due to steric interactions within the bulge binding pocket that abuts the major groove. The dansyl group is unlikely to be able to stack compactly on top of the T base of the bulge. Another potential difference is that of the free energy of hydrophobic transfer because the dansyl group contains an exocyclic dimethyl amino group. By contrast, $[Zn(L18)]^{2+}$, which contains a pendent aromatic group that lacks a heteroatom, binds to Tgg 40-fold more weakly than does the related $[Zn(L17)]^{2+}$ (91). The naphthyl pendent group in $[Zn(L18)]^{2+}$ is compact and similarly shaped to the quinoline group, but the lack of the nitrogen in the ring is expected to lead to less favorable stacking interactions (91). Heterocycles containing nitrogen have relatively electronegative π-systems and generally show stronger stacking interactions than aromatics lacking nitrogen heteroatoms (72).

The energetic contribution of DNA reorganization upon binding of the Zn^{2+} complexes, $\Delta G_{conf,}$ may be significant for DNA containing thymine bulges. The NMR structural data for Tgg bound to $[Zn(L17)]^{2+}$ suggests that the two flanking guanines have increased stacking interactions in comparison to the Zn^{2+} free structure (66). This result may contribute to the higher stability of the Zn-Tgg adduct and give a favorable energetic contribution to the free energy of binding. Other neighboring base pairs undergo changes in stacking interactions upon binding of $[Zn(L17)]^{2+}$ as well as shown by NMR spectroscopy experiments. At this juncture, the contribution to binding free energy is unknown.

An additional consideration is that selectivity toward non-canonical thymine or uracil may be improved by decreasing binding strength of Zn^{2+} complexes to double-stranded DNA or RNA, respectively. Aoki and Kimura (4) studies showed that duplex nucleic acids are denatured by Zn^{2+} complexes of L1 and derivatives if they contain long runs of T or U. However, the requirement of high concentrations of Zn^{2+} complex (1 equiv per phosphate for $[Zn(L1)]^{2+}$) suggests that this interaction is not competitive with thymine or uracil bulge binding at low concentrations of Zn^{2+} complex (4). The effective denaturation of these poly(dA)·poly(dT) or poly(A)·poly(U) duplexes by Zn^{2+} complexes with aromatic pendent groups at ratios of 0.3–0.5 Zn^{2+} complex to phosphate is consistent with the stronger interaction of these complexes with uracil or thymine in comparison to $[Zn(L1)]^{2+}$. It is likely that the aromatic group plays a role in the denaturation, perhaps by promoting binding of multiple Zn^{2+} complexes to neighboring thymine or uracil groups (6). These postulated interactions, as shown in Scheme 15, may be similar to that observed for TT with two $[Zn(L17)]^{2+}$ complexes. Thus, it would be useful to elucidate the importance of aromatic pendent group stacking interactions on binding of Zn^{2+} complexes to long sequences of adjacent thymine or uracil nucleobases. The sequence length of consecutive T or U nucleobases required to enable effective Zn^{2+} complex induced denaturation is unknown.

VI. SUMMARY

Macrocyclic complexes of Zn^{2+} show promise for the recognition of non-canonical thymines in DNA or uracil in RNA. A better understanding of the magnitude and corresponding importance of the energetic factors involved in binding of the Zn^{2+} complexes to nucleic acids will be useful in the design of the next generation agents. In particular, the role of the aromatic group in interaction with DNA or RNA and the extent of structural reorganization of the nucleic acid upon binding are important topics for further study. Elucidation of the sequence requirements in Zn^{2+} complex denaturation of double-stranded DNA would also be useful. This information will contribute toward the design of more specific DNA and RNA recognition agents based on Zn^{2+} complexes.

ACKNOWLEDGMENTS

JRM thanks the National Science Foundation (CHE-0911375) for support of the Zn^{2+} promoted nucleic acid cleavage and recognition research in her laboratory.

ABBREVIATIONS

2D NOESY	Two-dimensional nuclear Overhauser effect spectroscopy
A	Adenine
AMND	2-amino-7-methyl-1,8-naphthyridine
AZT^-	N3 deprotonated 3′-azido-3′-deoxythmidine
bpy	2,2′-Bipyridine
C	Cytosine
CD	Circular dichroism
chrysi	Chrysene-5,6-quinone diimine
cyclen	1,4,7,10-Tetraazacyclododecane
dA	Deoxyadenosine
dC	Deoxycytidine
dG	Deoxyguanosine
dT	Deoxythymidine (thymidine)
DANP	2,7-Diamino-1,8-naphthyridine
DAPI	4,6-Diamidino-2-phenylindole
DNA	Deoxyribonucleic acid
EtBr	Ethidium bromide
G	Guanine
HEPES	2-[4-(2-Hydroxyethyl)piperazin-1-yl]ethanesulfonic acid
HIV	Human immunodeficiency virus
^{1}H NMR	Proton NMR
ITC	Isothermal calorimetry
$K_{d\ (app)}$	Apparent dissociation constant
K_d	Dissociation constant
LD	Linear dichroism
mRNA	Messenger RNA
NCS-chrom	Neocarzinostatin chromophore
nd	Not determined
NSCi-gb	Spirolactam metabolite
NMR	Nuclear magnetic resonance
NOE	Nuclear overhauser effects
pK_a	Acid dissociation constant
phzi	Benzo[*a*]phenazine-5,6-quinone diimine
RNA	Ribonucleic acid
T	Thymine
5-pT	5-Thymidine monophosphate
TAR	Trans-activation responsive
TAT	Trans-activator of transcription
THF	Tetrahydrofuran (solvent)
T_m	Thermal melting temperature

TT	Thymidylylthymidine
TTT	Thymidylylthymidylylthymidine
U	Uracil
UPNP	Uridine-3′-4-nitrophenylphosphate
UU	Uridylyluridine
UV	Ultraviolet
vis	Visible

REFERENCES

1. J. Reedijk, *Pure Appl. Chem.*, *83*, 1709 (2011).
2. S.-F. Torabi and Y. Lu, *Faraday Discuss.*, *149*, 125 (2011).
3. P. Ghude, M. A. Schallenberger, A. M. Fleming, J. G. Muller, and C. J. Burrows, *Inorg. Chim. Acta*, *369*, 240 (2011).
4. S. Aoki and E. Kimura, *Chem. Rev.*, *104*, 769 (2004).
5. J. G. Muller, L. A. Kayser, S. J. Paikoff, V. Duarte, N. Tang, R. J. Perez, S. E. Rokita, and C. J. Burrows, *Coord. Chem. Rev.*, *185*, 761 (1999).
6. M. Shionoya, T. Ikeda, E. Kimura, and M. Shiro, *J. Am. Chem. Soc.*, *116*, 3848 (1994).
7. C. S. Rossiter, R. A. Mathews, and J. R. Morrow, *Inorg. Chem.*, *44*, 9397 (2005).
8. V. J. DeRose, S. Burns, N. K. Kim, and M. Vogt, in *Comprehensive Coordination Chemistry II*, J. A. McCleverty and T. J. Meyer, Eds. Oxford University Press, Oxford, 2003, UK, pp. 787–813.
9. E. Freisinger and R. K. O. Sigel, *Coord. Chem. Rev.*, *251*, 1834 (2007).
10. B. Knobloch, W. Linert, and H. Sigel, *Proc. Natl Acad. Sci. USA*, *102*, 7459 (2005).
11. M. Shionoya, E. Kimura, and M. Shiro, *J. Am. Chem. Soc.*, *115*, 6730 (1993).
12. S. Chatterjee, W. Pathmasiri, O. Plashkevych, D. Honcharenko, O. P. Varghese, M. Maiti, and J. Chattopadhyaya, *Org. Biomol. Chem.*, *4*, 1675 (2006).
13. E. Yuriev and J. D. Orbell, *Inorg. Chem.*, *35*, 7914 (1996).
14. M. W. Kalnik, C. N. Chang, A. P. Grollman, and D. J. Patel, *Biochemistry*, *27*, 924 (1988).
15. M. W. Kalnik, C. N. Chang, F. Johnson, A. P. Grollman, and D. J. Patel, *Biochemistry*, *28*, 3373 (1989).
16. J. M. Withka, J. A. Wilde, P. H. Bolton, A. Mazumder, and J. A. Gerlt, *Biochemistry*, *30*, 9931 (1991).
17. Y. Coppel, N. Berthet, C. Coulombeau, C. Coulombeau, J. Garcia, and J. Lhomme, *Biochemistry*, *36*, 4817 (1997).
18. M. P. Singh, G. C. Hill, D. Peoc'h, B. Rayner, J.-L. Imbach, and J. W. Lown, *Biochemistry*, *33*, 10271 (1994).
19. P. Cuniasse, G. V. Fazakerley, W. Guschlbauer, B. E. Kaplan, and L. C. Sowers, *J. Mol. Biol.*, *213*, 303 (1990).

20. S. D. Cline, W. R. Jones, M. P. Stone, and N. Osheroff, *Biochemistry*, *38*, 15500 (1999).
21. K. Yoshimoto, S. Nishizawa, M. Minagawa, and N. Teramae, *J. Am. Chem. Soc.*, *125*, 8982 (2003).
22. Y. Sato, S. Nishizawa, K. Yoshimoto, T. Seino, T. Ichihashi, K. Morita, and N. Teramae, *Nucleic Acids Res.*, *37*, 1411 (2009).
23. Y. Sato, Y. Zhang, T. Seino, T. Sugimoto, S. Nishizawa, and N. Teramae, *Org. Biomol. Chem.*, *10*, 4003 (2012).
24. Z. Xu, Y. Sato, S. Nishizawa, and N. Teramae, *Chem. Eur. J.*, *15*, 10375 (2009).
25. Y. Sato, S. Nishizawa, and N. Teramae, *Chem. Eur. J.*, *17*, 11650 (2011).
26. S. Nishizawa, S. Yusuke, X. Zhiai, K. Morita, L. Minjie, and N. Teramae, *Supramol. Chem.*, *22*, 467 (2010).
27. M. W. Kalnik, D. G. Norman, B. F. Li, P. F. Swann, and D. J. Patel, *J. Biol. Chem.*, *265*, 636 (1990).
28. D. A. LeBlanc and K. M. Morden, *Biochemistry*, *30*, 4042 (1991).
29. K. M. Morden, B. M. Gunn, and K. Maskos, *Biochemistry*, *29*, 8835 (1990).
30. C. A. S. A. Minetti, D. P. Remeta, R. Dickstein, and K. J. Breslauer, *Nucleic Acids Res.*, *38*, 97 (2010).
31. H. C. Ong, J. F. Arambula, S. Rao Ramisetty, A. M. Baranger, and S. C. Zimmerman, *Chem. Commun.*, 668 (2009).
32. H. Suda, A. Kobori, J. Zhang, G. Hayashi, and K. Nakatani, *Biorg. Med. Chem.*, *13*, 4507 (2005).
33. A. Kobori, T. Murase, H. Suda, I. Saito, and K. Nakatani, *Biorg. Med. Chem. Lett.*, *14*, 3431 (2004).
34. G. B. Jones, Y. Lin, Z. Xiao, L. Kappen, and I. H. Goldberg, *Biorg. Med. Chem.*, *15*, 784 (2007).
35. L. S. Kappen, Y. Lin, G. B. Jones, and I. H. Goldberg, *Biochemistry*, *46*, 561 (2006).
36. N. Zhang, Y. Lin, Z. Xiao, G. B. Jones, and I. H. Goldberg, *Biochemistry*, *46*, 4793 (2007).
37. S. T. Meyer, and P. J. Hergenrother, *Org. Lett.*, *11*, 4052 (2009).
38. B. M. Zeglis, J. A. Boland, and J. K. Barton, *J. Am. Chem. Soc.*, *130*, 7530 (2008).
39. B. M. Zeglis, J. A. Boland, and J. K. Barton, *Biochemistry*, *48*, 839 (2009).
40. M. Shionoya, M. Sugiyama, and E. Kimura, *J. Chem. Soc., Chem. Commun.*, 1747 (1994).
41. S. Aoki, and E. Kimura, *J. Am. Chem. Soc.*, *122*, 4542 (2000).
42. S. Aoki, and E. Kimura, *Rev. Mol. Biotechnol.*, *90*, 129 (2002).
43. E. Kikuta, S. Aoki, and E. Kimura, *J. Biol. Inorg. Chem.*, *7*, 473 (2002).
44. E. Kikuta, S. Aoki, and E. Kimura, *J. Am. Chem. Soc.*, *123*, 7911 (2001).
45. E. Kimura, M. Kikuchi, H. Kitamura, and T. Koike, *Chem. Eur J.*, *5*, 3113 (1999).
46. E. Kikuta, N. Katsube, and E. Kimura, *J. Biol. Inorg. Chem.*, *4*, 431 (1999).
47. E. Kimura, T. Ikeda, and M. Shionoya, *Pure &App. Chern*, *69*, 2187 (1997).

48. E. Kimura, N. Katsube, T. Koike, M. Shiro, and S. Aoki, *Supramol. Chem.*, *14*, 95 (2002).
49. E. Kimura, H. Kitamura, K. Ohtani, and T. Koike, *J. Am. Chem. Soc.*, *122*, 4668 (2000).
50. T. Koike, T. Gotoh, S. Aoki, E. Kimura, and M. Shiro, *Inorg. Chim. Acta*, *270*, 424 (1998).
51. C. S. Rossiter, R. A. Mathews, and J. R. Morrow, *J. Inorg. Biochem.*, *101*, 925 (2007).
52. M. Laine, K. Ketomaki, P. Poijarvi-Virta, and H. Lonnberg, *Org. Biomol. Chem.*, *7*, 2780 (2009).
53. Q. Wang, E. Leino, A. Jancsó, I. Szilágyi, T. Gajda, E. Hietamäki, and H. Lönnberg, *ChemBioChem*, *9*, 1739 (2008).
54. J. Gao, J. H. Reibenspies, and A. E. Martell, *Org. Biomol. Chem.*, *1*, 4242 (2003).
55. G. Gasser, M. J. Belousoff, A. M. Bond, Z. Kosowski, and L. Spiccia, *Inorg. Chem.*, *46*, 1665 (2007).
56. C. Bazzicalupi, A. Bencini, E. Berni, S. Ciattini, A. Bianchi, C. Giorgi, P. Paoletti, and B. Valtancoli, *Inorg. Chim. Acta*, *317*, 259 (2001).
57. E. Kikuta, S. Aoki, and E. Kimura, *J. Am. Chem. Soc.*, *123*, 7911 (2001).
58. R. A. Mathews, C. S. Rossiter, J. R. Morrow, and J. P. Richard, *Dalton Trans.* 3804 (2007).
59. A. O'Donoghue, S. Y. Pyun, M.-Y. Yang, J. R. Morrow, and J. P. Richard, *J. Am. Chem. Soc.*, *128*, 1615 (2006).
60. M. Komiyama, S. Kina, K. Matsumura, J. Sumaoka, S. Tobey, V. M. Lynch, and E. Anslyn, *J. Am. Chem. Soc.*, *124*, 13731 (2002).
61. R. Cacciapaglia, A. Casnati, L. Mandolini, A. Peracchi, D. N. Reinhoudt, R. Salvio, A. Sartori, and R. Ungaro, *J. Am. Chem. Soc.*, *129*, 12512 (2007).
62. C. S. Rossiter, R. A. Mathews, I. M. A. del Mundo, and J. R. Morrow, *J. Inorg. Biochem.*, *103*, 64 (2009).
63. Q. Wang, and H. Lonnberg, *J. Am. Chem. Soc.*, *128*, 10716 (2006).
64. J. H. R. Tucker, M. Shionoya, T. Koike, and E. Kimura, *Bull. Chem. Soc. Jpn.*, *68*, 2465 (1995).
65. E. Kimura, T. Ikeda, S. Aoki, and M. Shionoya, *J. Biol. Inorg. Chem.*, *3*, 259 (1998).
66. I. M. A. del Mundo, K. E. Siters, M. A. Fountain, and J. R. Morrow, *Inorg. Chem.*, *51*, 5444 (2012).
67. K. E. Siters and J. R. Morrow (unpublished results) (2013).
68. L. L. O'Neil and O. Wiest, *Org. Biomol. Chem.*, *6*, 485 (2008).
69. L. L. O'Neil and O. Wiest, *J. Am. Chem. Soc.*, *127*, 16800 (2005).
70. I. M. A. del Mundo, M. A. Fountain, and J. R. Morrow, *Chem. Commun.*, *47*, 8566 (2011).
71. C. R. Martinez and B. L. Iverson, *Chem. Sci.*, *3*, 2191 (2012).
72. C. Janiak, *Dalton Trans.* 3885 (2000).
73. J. Šponer, J. Leszczyński, and P. Hobza, *J. Phys. Chem.*, *100*, 5590 (1996).

74. H. C. M. Nelson, J. T. Finch, B. F. Luisi, and A. Klug, *Nature (London)*, *330*, 221 (1987).
75. R. A. Hunt, M. Munde, A. Kumar, M. A. Ismail, A. A. Farahat, R. K. Arafa, M. Say, A. Batista-Para, D. Tevis, D. W. Boykin, and W. D. Wilson, *Nucleic Acids Res.*, *39*, 4265 (2011).
76. G. M. Blackburn, in *Nucleic Acids in Chemistry and Biology;* 2nd ed. G. M. Blackburn, and M. J. Gait, Eds., Oxford University Press, UK, 1996, pp. 17–70.
77. P. R. N. Kamya and H. M. Muchall, *J. Phys. Chem. A.*, *115*, 12800 (2011).
78. M. Kogan, B. Nordén, P. Lincoln, and P. Nordell, *ChemBioChem*, *12*, 2001 (2011).
79. E. Kikuta, M. Murata, N. Katsube, T. Koike, and E. Kimura, *J. Am. Chem. Soc.*, *121*, 5426 (1999).
80. M. Bandyopadhyay, S. Gupta, and D. Segal, *Phys. Rev. E*, *83*, 031905 (2011).
81. B. M. Bulheller, A. Rodger, and J. D. Hirst, *Phys. Chem. Chem. Phys.*, *9*, 2020 (2007).
82. N. J. Wheate, C. R. Brodie, J. G. Collins, S. Kemp, and J. R. Aldrich-Wright, *Mini-Rev. Med. Chem.*, *7*, 627 (2007).
83. E. Kikuta, R. Matsubara, N. Katsube, T. Koike, and E. Kimura, *J. Inorg. Biochem.*, *82*, 239 (2000).
84. K. Nakamoto, M. Tsuboi, and G. D. Strahan, *Drug-DNA Interactions: Structures and Spectra;* John Wiley & Sons, Hoboken, NJ, 2008, pp. 40–42, 312.
85. D. H. Turner, *Curr. Opin. Struct. Biol.*, *2*, 334 (1992).
86. I. M. A. del Mundo, Ph.D. Thesis, Thymine/uracil recognition in non-canonical nucleic acid structures by Zn (II) macrocycle complexes University at Buffalo (2011).
87. J. R. Lakowicz, *Principles of Fluorescence Spectroscopy;* Springer: New York, 1999, pp. 203–205.
88. J. B. Chaires, *Biopolymers*, *44*, 201 (1998).
89. I. Haq, T. C. Jenkins, B. Z. Chowdhry, J. Ren, and J. B. Chaires, *Methods Enzymol.*, *323*, 373 (2000).
90. J. B. Chaires, *Arch. Biochem. Biophys.*, *453*, 26 (2006).
91. K. E. Siters, S. Gardina, and J. R. Morrow, *manuscript in preparation* (2014).

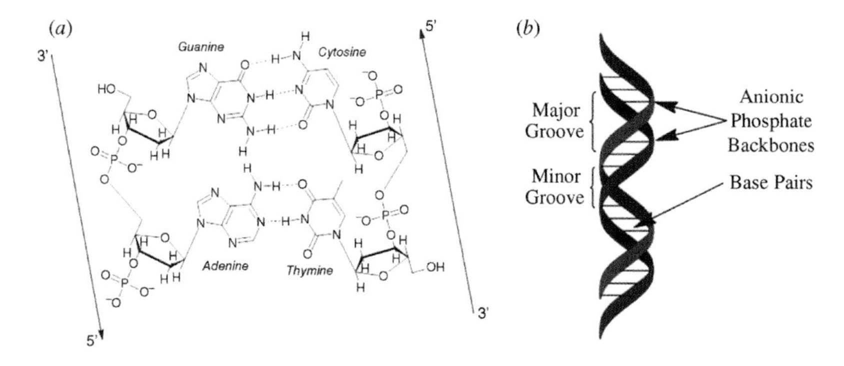

Chapter 2, Figure 1. Structural representation of Watson–Crick base pairs (*See text for full caption.*)

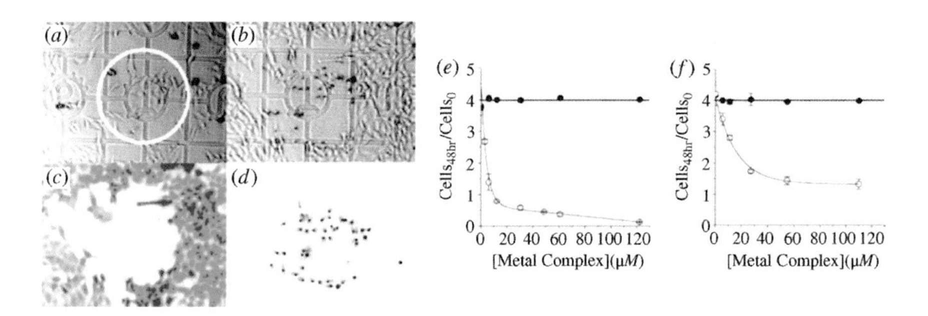

Chapter 2, Figure 33. Micrograph images of Vero cells following exposure to 122-μ*M* $[\{(bpy)_2Ru(dpp)\}_2RhCl_2]Cl_5$ and irradiation for 4 min with a focus beam of ≥ 460 nm. (*See text for full caption.*)

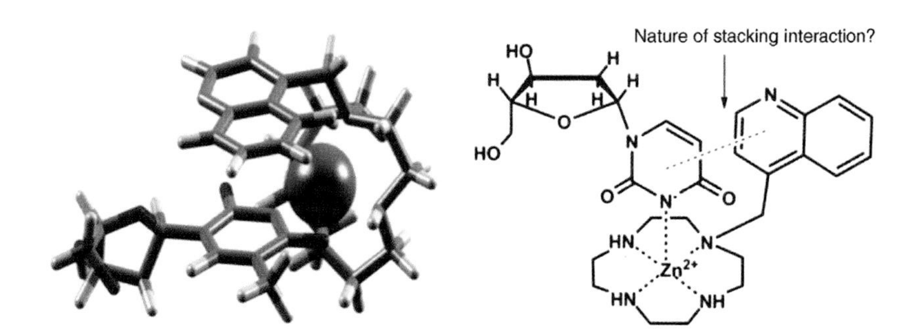

Chapter 3, Figure 2. Stacking of quinoline in $[Zn(L17)]^{2+}$ on thymine in bulged structure. (*See text for full caption.*)

Progress in Inorganic Chemistry, Volume 59, First Edition. Edited by Kenneth D. Karlin.

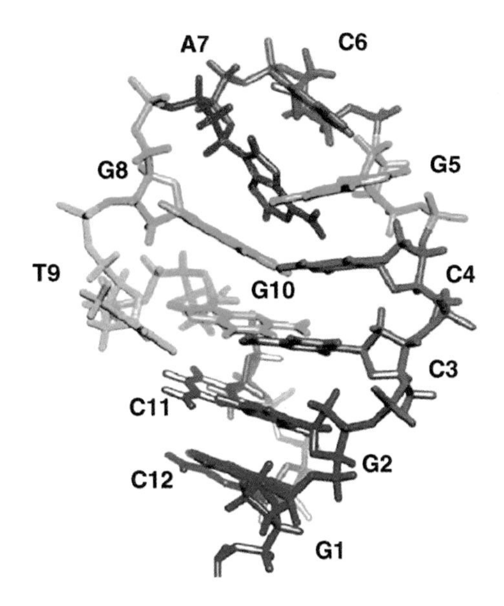

Chapter 3, Figure 6. The NMR structure of TggX. Here T9 is flipped away from the stem and located in the major groove. (*See text for full caption.*)

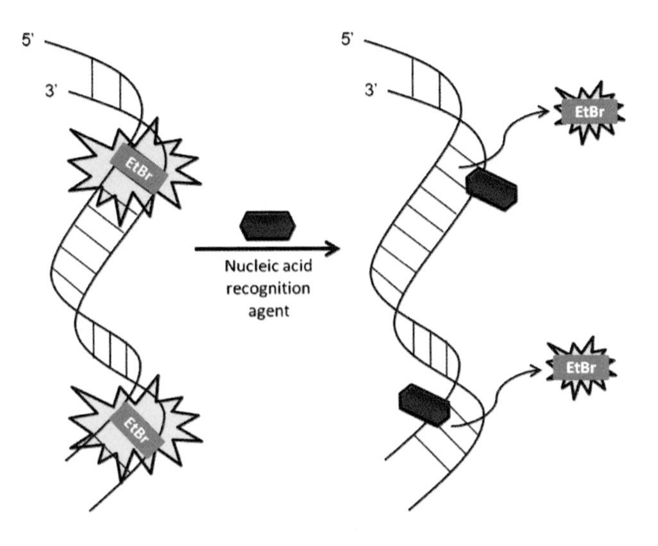

Chapter 3, Scheme 18. Illustration of ethidium bromide displacement assay used.

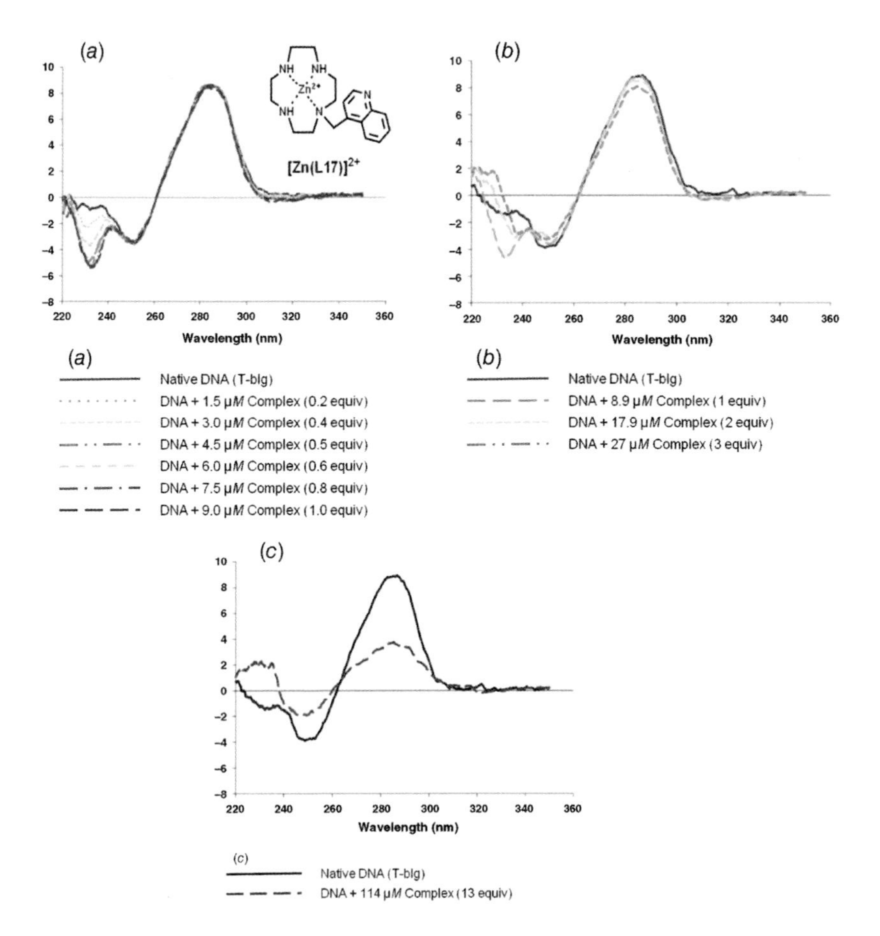

Chapter 3, Figure 7. Circular dichroism spectra of Tgg(8.9 μ*M*) in increasing concentrations of $[Zn(L17)]^{2+}$ at pH 7.5 100 m*M* NaCl. (*See text for full caption.*)

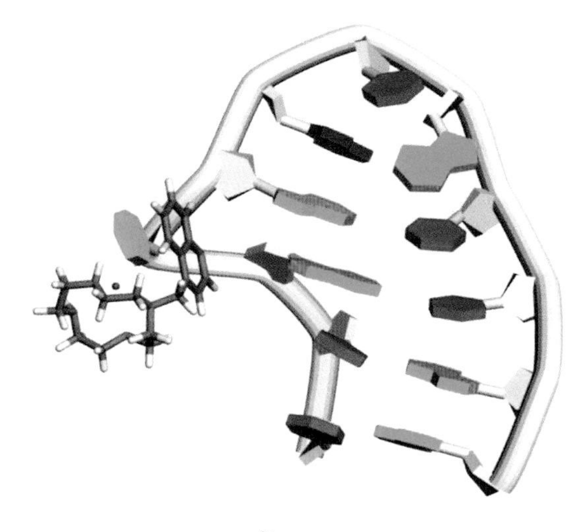

Chapter 3, Figure 10. Model of $[Zn(L17)]^{2+}$ binding to TggX. Quinoline moiety is stacked on thymine and directed toward the major groove.

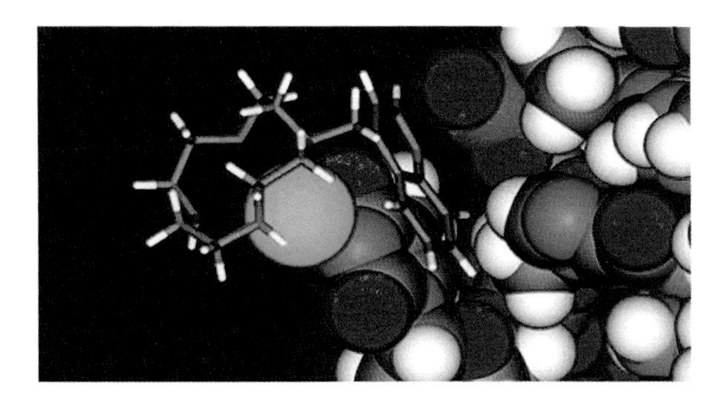

Chapter 3, Figure 11. Close-up of T bulge and adjacent stem. Bulge binding pocket can be observed. [Reprinted with permission from I.M.A. del Mundo et al. *Inorg. Chem.*, *51*, 5444 (2012) [66]. Copyright © (2012) American Chemical Society.]

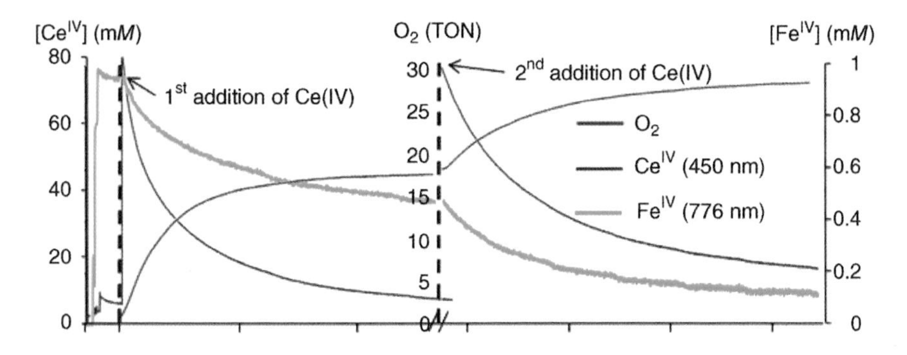

Chapter 7, Scheme 59. Water oxidation reaction monitored by UV–vis spectroscopy and a pressure sensor for the observation of the Ce(IV) consumption (blue), the complex degradation (green), and the oxygen evolution (red). Fe^{IV}=O was fully formed by addition of CAN (6 equiv) over [Fe(CF$_3$SO$_3$)$_2$(^{Me,H}Pytacn)] (1 m*M* in MiliQ water). Oxygen was not detected and the Fe^{IV}=O species were stable for >2 h. A second addition of oxidant (75 equiv) allowed the oxygen production.

Progress Toward the Electrocatalytic Production of Liquid Fuels from Carbon Dioxide

JOEL ROSENTHAL

Department of Chemistry and Biochemistry, University of Delaware, Newark, DE

CONTENTS

Progress in Inorganic Chemistry, Volume 59, First Edition. Edited by Kenneth D. Karlin.

I. INTRODUCTION

A. Reduction of CO_2 for Solar Energy Storage and Distribution

The energy needs of our society are massive and continue to grow each year. Conservative estimates predict that the rate of global energy consumption will nearly double by the middle of this century, reaching 25–30 TW by the year 2050 (1). Over 80% of global energy presently consumed is generated by fossil fuels, which drives atmospheric emissions of ~7 Gt (1 Gt = 10^{12} kg) of carbon per annum in the form of CO_2 (2). If society's dependence on fossil fuels is not alleviated over the next 40 years, the concentration of atmospheric CO_2 will more than double by 2050 to >750 ppm (3, 4). There exists a general consensus in the scientific community that the large additional quantities of CO_2 emitted into the atmosphere contribute to the greenhouse effect (5, 6) and global climate change (7, 8). While the consequences of this increase cannot be predicted precisely, most agree that we are perturbing the planet on an unprecedented scale, particularly with respect to the biosphere (9), water resources (10–12), and human health (13–16).

If the world's global energy portfolio is to grow by at least 15 TW over the next several decades, the use of renewable energy resources will need to be emphasized in order to avoid enormous increases in atmospheric CO_2 concentrations (17). Solar energy is the prime candidate to fill this void, as terrestrial solar insolation can fulfill the energy needs of our technologically advanced society in perpetuity (18). There remains, however, a major barrier to implementation of solar resources on a large scale. Given that mankind relies on a continuous energy supply, the diurnal and intermittent nature of sunlight requires that efficient, inexpensive, and scalable energy storage mechanisms be developed in order for solar energy to become a major contributor to our energy infrastructure (19).

An attractive method for solar energy storage is the generation of chemical fuels via artificial photosynthesis (20). Solar fuels are an appealing medium for energy storage, as the confinement of two electrons in the miniscule volume of a chemical bond leads to generation of an energy carrier with exceptionally high energy density (21). Solar fuel production also provides a clear route to energy distribution. The production of liquid fuels from sunlight is particularly important in addressing energy needs associated with transportation, which accounts for > 20% of US energy demand (1). Moreover, liquid fuels are compatible with existing infrastructure for energy supply and distribution. As such, the development of new methods for the generation of liquid fuels from CO_2 and sunlight is critical to the storage and distribution of solar resources (22, 23).

The societal importance and economic value of liquid fuel resources distinguishes CO_2 activation and reduction chemistry as a critical area of focus in the fields of renewable energy storage and molecular energy conversion. The ability to use CO_2 as a substrate for production of high-energy species (e.g., methanol, methane, and higher hydrocarbons) allows for energy to be stored within the many C–H and C–C bonds of these fuels. There are several strategies to enable the reduction of CO_2, which include (1) hydrogenation or chemical reduction of the gas using thermal methods (24–26); (2) photochemical activation of CO_2 using organic or inorganic photocatalysts (27, 28); and (3) the direct electrochemical reduction of CO_2 in organic or aqueous solution (29).

B. Challenges to CO_2 Conversion

The principle challenge to CO_2 reduction chemistry is that it is an energetically uphill (endergonic) process. As such, the catalytic conversion of CO_2 to more energy-rich species requires either a direct photochemical or electrochemical input of energy to drive the reaction. The standard potentials for several common CO_2 reduction processes are outlined by Eq. 1–6. Each of these potentials is referenced to the reversible hydrogen electrode (RHE).

$$CO_2 + e^- \rightarrow CO_2^{\bullet-} \qquad E^\circ = -1.48\ \text{V} \qquad (1)$$

$$CO_2 + 2H^+ + 2e^- \rightarrow HCO_2H \qquad E^\circ = -0.19\ \text{V} \qquad (2)$$

$$CO_2 + 2H^+ + 2e^- \rightarrow CO + H_2O \qquad E^\circ = -0.11\ \text{V} \qquad (3)$$

$$CO_2 + 4H^+ + 4e^- \rightarrow H_2CO + H_2O \qquad E^\circ = -0.06\ \text{V} \qquad (4)$$

$$CO_2 + 6H^+ + 6e^- \rightarrow MeOH + H_2O \qquad E^\circ = 0.03\ \text{V} \qquad (5)$$

$$CO_2 + 8H^+ + 8e^- \rightarrow CH_4 + 2H_2O \qquad E^\circ = 0.17\ \text{V} \qquad (6)$$

The standard potentials for CO_2 reduction span nearly 1.7 V in potential, however, the majority of the processes highlighted above are grouped between −0.19 and 0.17 V vs RHE (Eqs. 2–6). The main outlier is the single-electron reduction of CO_2 to $CO_2^{\bullet-}$, which takes place at $E^\circ = -1.48$ V. The energy input required to effect the formation of $CO_2^{\bullet-}$ is much greater than that for the other reduction processes due in part to the sizable reorganization energy attendant with conversion of the linear CO_2 molecule to the bent radical anion $CO_2^{\bullet-}$ (30, 31). Equally if not more important, is that the lowest unoccupied molecular orbital (LUMO) of CO_2 is extremely high in energy. This molecular orbital ($2\pi_u$) is doubly degenerate and is comprised of two orthogonal antibonding π orbitals. Accordingly, injection of a single electron into this high-energy orbital is energetically prohibitive.

The CO_2 reduction reactions listed in Eqs. 2–6 all take place at more modest potentials. The first reason for this is that each of these reactions involves the transfer of multiple reducing equivalents and as a result, circumvents the formation of the high-energy $CO_2^{\bullet-}$ intermediate. Additionally, each of these multielectron reduction processes is associated with the transfer of an equivalent number of protons. These multielectron proton-coupled electron transfer (PCET) reactions involve the formation of more stable intermediates than the single-electron reduction of CO_2. Moreover, four of these PCET reactions (Eqs. 3–6) result in the formation of at least 1 equiv of water in addition to the reduced carbon-containing species. Water is a very stable molecule and its formation helps to offset the energy needed to drive these CO_2 reduction processes.

Although the standard potentials for the multielectron PCET reactions (Eqs. 2–6) are roughly −1.3 to −1.65 V less negative than formation of $CO_2^{\bullet -}$, each of the CO_2 conversion processes highlighted above face significant kinetic challenges. For instance, although the direct conversion of CO_2 to methane occurs at a relatively modest 0.17 V vs RHE, this reaction requires the controlled delivery of eight protons in concert with an equivalent number of reducing equivalents. As such, electrochemical generation of highly reduced carbon-containing species is extremely challenging, as multielectron reductions and PCET processes are inherently slower than single electron-transfer reactions (32–35). From a practical standpoint, the standard potentials for the five PCET reactions (Eq. 2–6) are clustered between −0.19 and 0.17 V (Fig. 1). Since each of these CO_2 reduction

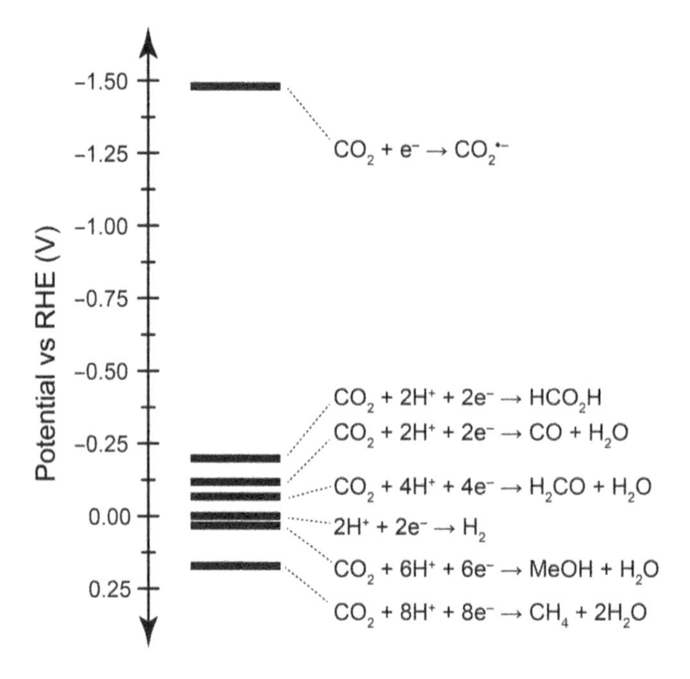

Figure 1. Standard potentials for various fuel-forming reactions in water.

processes takes place within a narrow potential range of <0.4 V, it can be difficult to target a given CO_2 reduction product from a thermodynamic standpoint. For instance, electrochemical reduction of CO_2 to formaldehyde occurs at −0.06 V. Any reaction that targets this product must be run at a potential at which production of methanol and methane is also thermodynamically feasible. This matter is further complicated by the fact that each of the PCET reactions requires protons. Since the thermochemical potential for reduction of protons to hydrogen gas ($E° = 0.0$ V vs RHE) is interdigitated among the other PCET reactions of Fig. 1, this process can occur as a side reaction and lower the selectivity and overall Faradaic efficiency (FE) for a desired CO_2 reduction process. Given the close separation between $E°$ values for the various fuel-forming reactions of Fig. 1, development of catalysts that can control the relative rates of the various PCET reactions is critical in driving the selective reduction of CO_2 to a given species with high faradaic and energy efficiency.

C. Electrochemical Fuel Synthesis

Direct solar-to-fuel conversion can be achieved by combining electrocatalysts that promote energy storing half-reactions with a light harvesting material, such as a photovoltaic (PV) device. Five of the six CO_2 reduction reactions of Fig. 1 directly yield high-energy species that could form the basis for electrochemical fuel synthesis (36). Several of these products are well suited for direct integration into existing fuel infrastructure. For instance, the $8e^-/8H^+$ and $6e^-/6H^+$ reduction of CO_2 to yield methane and methanol, respectively, provide routes to useful fuels via direct electrochemical processes. Although the direct electrochemical reduction of CO_2 to methane or methanol are attractive energy-storing reactions, the kinetic hurdles associated with these multielectron PCET reactions are very large, which complicates such reactions. In contrast, the $2e^-/2H^+$ reduction of CO_2 to CO is another energetically uphill half-reaction that delivers a versatile and energy-rich chemical that forms a basis for fuels production (37–39). The multifaceted utility of CO is depicted in Scheme 1, in which CO_2 is reduced by $2e^-$ and $2H^+$ to generate CO and H_2O. The CO can be funneled through a host of subsequent processes, such as the industrial production of commodity chemicals including methanol (40, 41), acetic acid (42), and some plastics (43) Moreover, CO can react with H_2O via the water–gas shift (WGS) reaction to generate H_2 and CO_2 (44, 45). Perhaps most importantly, CO is the principal carbon-based reactant for Fischer–Tropsch (FT) processes that yield synthetic petroleum and liquid fuels (46–48) Accordingly, Scheme 1 illustrates how the electrochemical reduction of CO_2 to CO (using an electrocatalyst denoted by the "?") provides a route to the renewable generation of liquid fuels, when a PV device provides the energy input for the CO_2 reduction reaction.

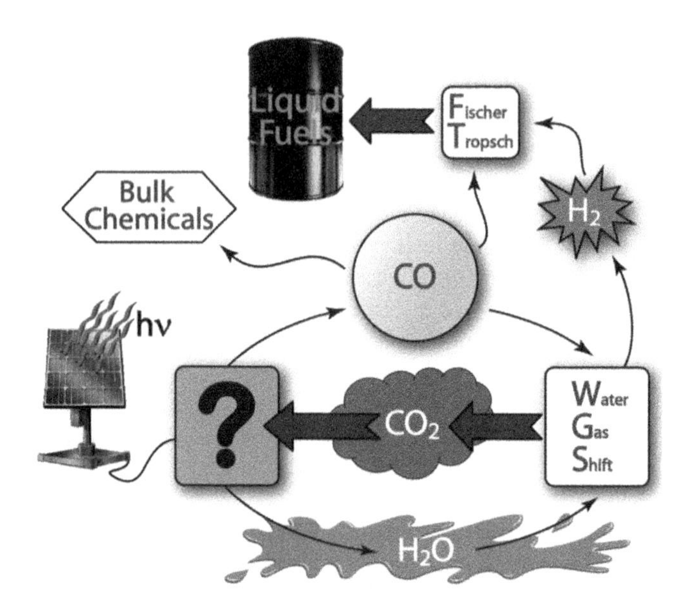

Scheme 1. Cycle for H_2 and liquid fuel generation driven by reduction of CO_2 by $2e^-/2H^+$ to generate CO.

II. ELECTROCATALYTIC REDUCTION OF CO_2

If the electrosynthesis of fuels from CO_2 is to be accomplished, the development of robust and efficient electrocatalysts that can be interfaced with PVs or other light-harvesting devices will be critical. An electrocatalyst is a redox active species that facilitates a chemical reaction upon either oxidation or reduction. For an electrocatalyst to be efficient, it must display both fast electron-transfer kinetics and significantly reduce the activation energy barrier associated with the reaction of interest. Moreover, for a CO_2 reduction catalyst to function with high-energy efficiency it must be reduced at a thermodynamic potential that is close to that for the energy-storing reaction of interest and operate with fast kinetics–high turnover frequency (TOF) (49).

An electrocatalyst can either be a heterogeneous material (i.e., an electrode) that is poised at a given potential and reacts directly with a substrate of interest (i.e., CO_2) or a molecular species that is dissolved within the electrolyte and participates in electron transfer (ET) with a subsequent thermal reaction near the electrode surface. Figure 2 illustrates these two limiting cases. For the case of heterogeneous electrocatalysis [(Fig. 2*a*)], CO_2 dissolved in the electrolyte diffuses and binds to the surface of the cathode, which is polarized at a negative potential. In this process, the cathode material directly activates and reduces CO_2. By contrast, a

(a) **Heterogeneous Electrocatalysis**

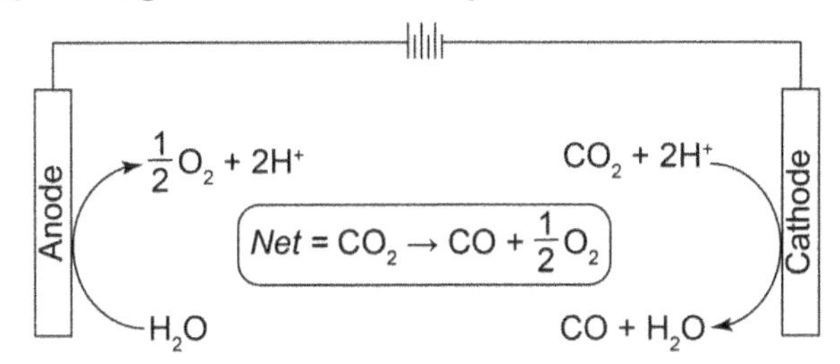

(b) **Homogeneous Electrocatalysis**

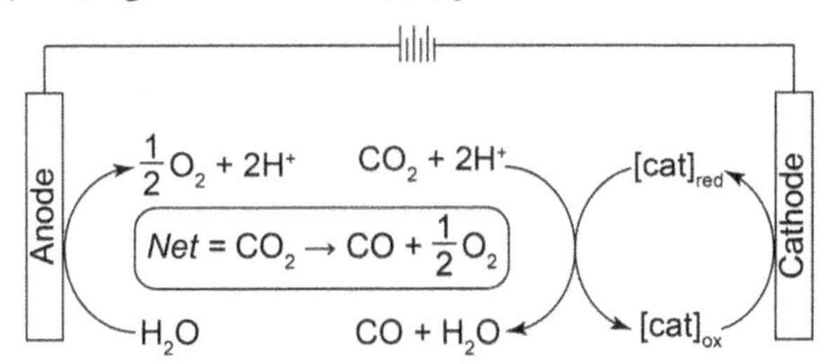

Figure 2. Heterogeneous vs homogeneous CO_2 reduction pathways. Here $[cat]_{ox}$ = oxidized electrocatalyst and $[cat]_{red}$ = reduced electrocatalyst.

prominent pathway by which homogeneous CO_2 reduction takes place [(Fig. 2*b*)] involves a system in which the catalyst primarily exists in an inactive, oxidized (ox) redox state ($[cat]_{ox}$). Upon diffusion to and interaction with the cathode, this species is then reduced (red) by one or more electrons to generate $[cat]_{red}$, which can then bind to CO_2 and drive its conversion to reduced products. Upon reacting with CO_2, the active $[cat]_{red}$ species is converted back to $[cat]_{ox}$. For studies involving the reduction of CO_2, the oxidation of water to generate O_2 often comprises the anodic half-reaction. In this way, oxidation of H_2O provides the electrons needed to drive the production of CO from CO_2. For convenience, the anodic half-reaction is generally carried out using an auxiliary platinum electrode as opposed to a molecular catalyst.

Much effort has been dedicated to the use of CO_2, for energy storage (50, 51), particularly with respect to development of heterogeneous materials and electrocatalysts for the reduction of CO_2 to CO and H_2O (52). Such systems have traditionally relied on precious metals [e.g., Ag (53, 54), Au (55, 56), Pt (57, 58), and nanoparticles] thereof deposited on various conducting supports (59). In addition to the high costs associated with the fabrication of these materials, heterogeneous catalysts have historically been difficult to study in detail and the precise path to efficient and selective CO_2 reduction remains ill-defined, which complicates the systematic improvement of such systems (60). In contrast, homogeneous catalysts offer several important advantages in studying molecular energy conversion reactions: (1) each metal atom is a catalytically active site, as opposed to just those that are surface accessible for a heterogeneous catalyst; (2) all metal atoms are in identical environments, which can increase reaction specificity and activity; (3) reaction selectivities can be fine tuned by attenuating the supporting ligands or solvents, and most importantly, (4) it is much easier to conduct mechanistic studies, which afford a better understanding and control of the reaction (61–63).

The scale and complexity of the CO_2 remediation–conversion problem has distinguished the efficient conversion of CO_2 to fuels as a critical area of research and has spurred the development of both hetero- and homogeneous electrocatalysts. In particular, the study of molecular complexes that can promote CO_2 activation has been the traditional purview of inorganic chemists and several excellent reviews of CO_2 coordination chemistry have been published (64–66). This field was launched by the first report of a structurally characterized metal–CO_2 adduct by Aresta et al. in 1975 (67). This first metal–CO_2 complex $[Ni(Pcy_3)_2(\eta^2\text{-}CO_2)]$ (cy = cyclohexyl) involved η^2-CO_2 binding and similar complexes have since been reported (68–71). Given that CO_2 is an amphoteric molecule, however, there are several other modes by which it can coordinate to transition metal centers (72). The Lewis acidic carbon atom is the electrophilic center of CO_2, whereas the Lewis basic oxygen atoms are nucleophilic in character. These electrophilic and nucleophilic centers can contribute to metal binding in addition to the two π-bonds of the molecule. Accordingly, CO_2 can form adducts with mononuclear metal centers via the three distinct modes shown in Fig. 3 and monometallic η^1-CO_2 (73, 74) and η^1-OCO (75–77) adducts are known.

Over the course of the last 35 years, much emphasis has been devoted to the development of molecular electrocatalysts for the activation of CO_2 (29, 78). Given that the activation pathway of small molecule substrates by transition metal complexes is dictated by the mode of binding, and that the conversion of CO_2 to CO is a formal two-electron reduction, molecular electrocatalysts typically are comprised of low-valent electron replete metals that are nucleophilic in character. Many of these systems are capable of promoting the production of CO in either organic or aqueous electrolyte. These systems primarily span groups 7–10 of the periodic table and can be divided into three main classes that are distinguished by

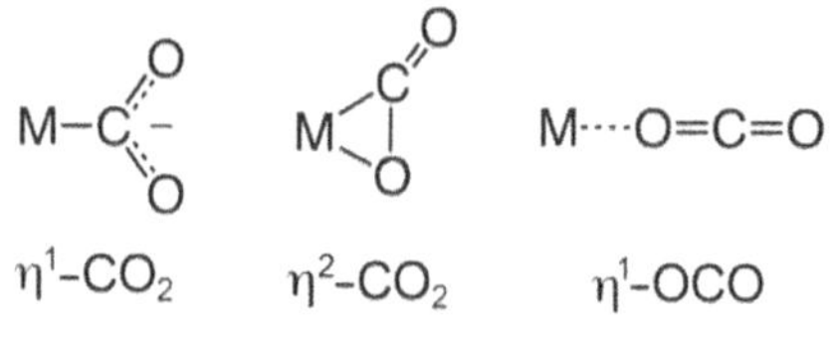

η^1-CO_2 $\quad$ η^2-CO_2 $\quad$ η^1-OCO

Figure 3. Modes of CO_2 binding to a single transition metal center.

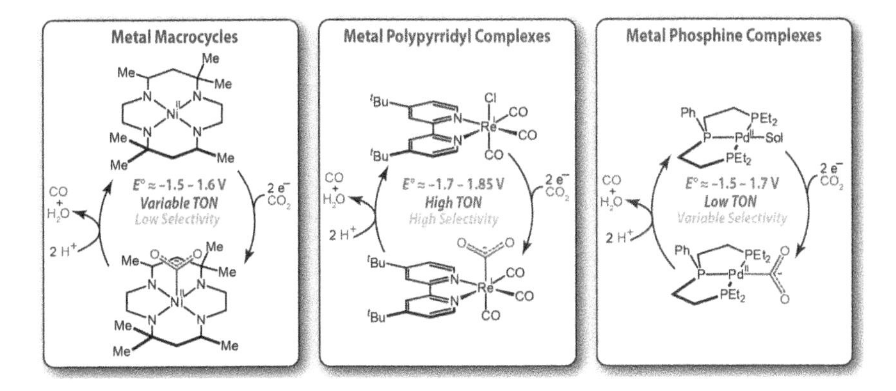

Figure 4. Some of the most thoroughly studied homogeneous electrocatalysts for reduction of CO_2 to CO include Ni supported by macrocyclic ligands, Re polypyridyl complexes, and Pd phosphine complexes (TON = turnover number).

the structure of the supporting ligand. These systems include (1) metal complexes supported by macrocycles; (2) metal polypyridyl complexes; and (3) complexes supported by phosphines. Each of these general frameworks has implicit strengths and weaknesses and examples of each are illustrated in Fig. 4. The following sections will detail the properties of specific molecular electrocatalysts for CO generation and solar-fuel production.

III. METAL COMPLEXES SUPPORTED BY MACROCYCLES

A. Complexes Supported by Tetrapyrrole Ligands

One of the first reports of CO_2 reduction using a molecular electrocatalyst involved nickel and cobalt phthalocyanines that had been immobilized on a graphite support (79). This work did not identify the reduced carbon products, current efficiencies, or TONs for the electrocatalysis. It made clear, however, that enhanced current was observed in the presence of CO_2 upon applying a potential more negative than roughly -1.5 V vs saturated calomel electrode (SCE). Most importantly, this early work clearly established that metal tetrapyrrole complexes could serve as electrocatalysts for CO_2 reduction under reducing conditions. Roughly a decade later, Kapusta and Hackerman (80) showed that formate and methanol were the primary products formed from reduction of CO_2 using a glassy carbon electrode modified by cobalt phthalocyanine (Pc). It was also demonstrated that the identity of the metal bound within the Pc dramatically impacts the overall effectiveness of the catalyst. The relative order of activity was found to be Co(Pc) > Ni(Pc) ≫ Fe(Pc) ~ Cu(Pc) (80). These complexes (**1–4**) are shown in Fig. 5.

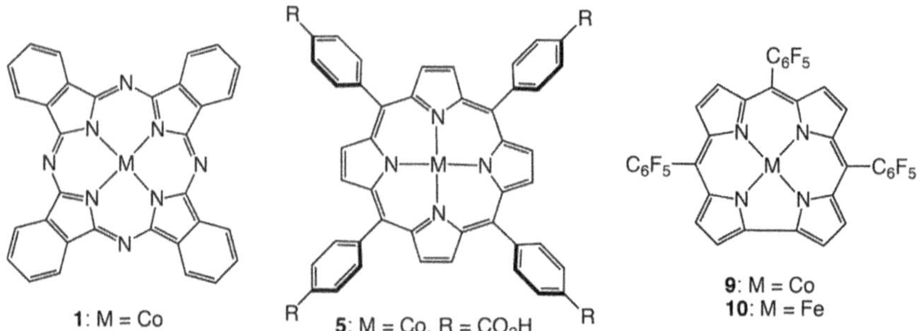

1: M = Co
2: M = Ni
3: M = Fe
4: M = Cu

5: M = Co, R = CO_2H
6: M = Co, R = SO_3H
7: M = Co, R = H
8: M = Fe, R = H

9: M = Co
10: M = Fe

Figure 5. Phthalocyanine-, porphyrin-, and corrole-based electrocatalysts for CO production from CO_2.

Around the same time, Lewis showed that Co(Pc) (**1**) that had been deposited on a carbon cloth electrode could catalyze the reduction of CO_2 to CO at −1.15 V vs SCE using an aqueous citrate electrolyte (pH 5) (81). During the course of these experiments, Lewis found that hydrogen gas was also formed from reduction of protons at the carbon cloth electrode. This side reaction limited the Faradaic efficiency (FE) for CO production by the immobilized Co(Pc) to roughly 50%. This work clearly established, however, that the CO_2 reduction efficiency and product distribution could be dramatically impacted by solution pH, electrolyte composition, and the identity of the solid conducting support. As such, considerable effort has been devoted to the development of interfaces and membranes to tailor the activity of immobilized Co(Pc). Kaneko and co-workers (82, 83) demonstrated that Co(Pc) interfaced with poly-4-vinylpyridine modified carbon and indium tin oxide (ITO) electrodes are more robust and much more selective for electrochemical production of CO over H_2, especially at high Pc loadings. Pulse radiolysis and photochemical experiments carried out by Neta and co-workers (84) helped to identify the electrocatalytically active complex in such systems. This work has demonstrated that the Co(Pc) platform must be reduced by two electrons in order to induce CO_2 binding and activation. The active species has been purported to be $[Co^{I}(Pc^{\bullet -})]^{2-}$, which is a complex in which both the Co center and Pc ligand are reduced by a single electron.

Other tetrapyrrole complexes have also been studied as CO_2 reduction catalysts. One of the first descriptions of electrocatalytic CO_2 activation using metalloporphyrins was published in 1979. This work compared the ability of water-soluble porphyrins containing ancillary carboxylate (TCPP = tetracarboxyphenyl porphyrin) or sulfonate (TPPS tetrasulfonatophenyl porphyrin) groups to effect CO_2 reduction (Fig. 5). This study surveyed freebase, cobalt, iron and copper porphyrin complexes in phosphate buffer and demonstrated that only Co(TCPP) (**5**) and

Co(TSPP) (**6**) were able to activate CO_2 (85). For these systems, the primary product formed upon electrolysis at potentials more negative than −1.4 vs SCE was formic acid. Much like the early Pc studies, however, current efficiencies and TONs were not disclosed. In subsequent work, Aramata and co-workers (86) showed that a cobalt 5,10,15,20-tetraphenylporphyrin (TPP) containing an axial pyridine ligand that had been grafted onto a glassy carbon electrode was able to catalyze the reduction of CO_2 at potentials 100 mV more positive than the water-soluble Co(II) porphyrins. Moreover, these assemblies selectively generated CO as the reduced carbon species with FE >50% and TONs on the order of 10^5. More recently, Hamers and co-workers (87) applied a similar strategy to develop "Smart" electrodes in which a cobalt porphyrin complex was covalently linked to a conducting diamond substrate using Huisgen click chemistry. Conversion of CO_2 to CO was confirmed for these systems by Fourier transforms infrared (FTIR) analysis.

Neta and co-workers (88) showed that cobalt porphyrins can also serve as electrocatalysts for CO_2 reduction under homogenous conditions in organic solution. Cyclic voltammetric (CV) analysis of Co(TPP) (**7**, TPP = 5,10,15,20-tetraphenylporphyrin) in acetonitrile (MeCN) or butyronitrile containing tetrabutylammonium perchlorate as an electrolyte showed two reversible redox waves centered at roughly −0.85 V and −2.0 V vs SCE. Spectroscopic studies demonstrated that unlike the Co(Pc) systems detailed above, neither of the Co(TPP) redox events are due to reduction of the porphyrin macrocycle and instead correspond to Co(II/I) and Co(I/0) couples, respectively. Electrochemical and photochemical experiments indicated that the electrogenerated $[Co^0(TPP)]^{2-}$ complex binds and activates CO_2. Cobalt porphyrins containing electron-withdrawing fluorine substituents on the macrocycle periphery are reduced and catalyze CO_2 reduction at less negative potentials, but are less active as compared to Co(TPP) (88). Similar studies have been carried out for cobalt corrins, which are highly saturated porphyrin homologues containing only six double bonds about the tetrapyrrole periphery. The reduction potentials for cobalt corrins (e.g., hydroxocobalamin, cyanocobalamin, and cobinamide) are similar to those for cobalt porphyrins, however, these saturated complexes are more robust catalysts for CO_2 reduction (89). Mechanistic studies have indicated that the pathway by which Co corrins activate CO_2 is identical to that observed for the cobalt porphyrins described above.

Iron porphyrins have also been shown to function as homogeneous electrocatalysts for conversion of CO_2 to CO at −1.8 V vs SCE in *N,N*-dimethylformamide (DMF) (90). Reduction of Fe(TPP) (**8**) generates $[Fe^0(TPP)]^{2-}$, which is the active form of the catalyst (91). Savéant and co-workers (90) demonstrated that this system is not robust and that the catalyst can decompose either by carboxylation or hydrogenation of the porphyrin ring after a few turnovers. Addition of Lewis acidic Mg^{2+} ions resulted in a marked improvement in the reactivity and stability of the

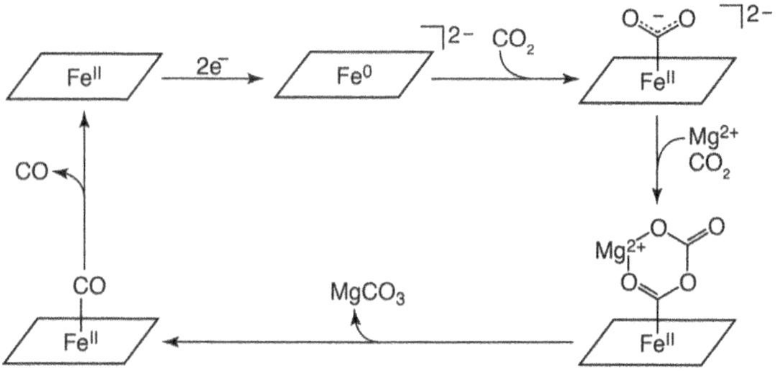

Scheme 2. Proposed electrocatalytic pathway for Mg^{2+} assisted conversion of CO_2 to CO by an iron porphyrin.

iron porphyrin electrocatalyst. Mechanistic studies showed that the Mg^{2+} ions help drive C−O bond cleavage from the CO_2 adduct of Fe^0(TPP) to form the carbonyl complex of Fe^{II}(TPP) and $MgCO_3$ (Scheme 2). As such, this work represents an important example of bimetallic CO_2 activation in which an electron-rich (Fe^0) and an electron-deficient (Mg^{2+}) center act in concert to drive an energy-storing catalytic reaction. Bhugun and Savéant (92) also showed that this "push–pull" catalysis is observed for other Lewis acids including Ca^{2+}, Ba^{2+}, Li^+, and Na^+ and that the order of activity roughly parallels the oxophilicity of the metal cation.

Savéant and co-workers (93) also showed that weak Brønsted acids can enhance the ability of iron porphyrins to reduce CO_2 to CO. Proton donors [e.g., 1-propanol, 2-pyrrolidine, and 2,2,2-trifluoroethanol (TFE)] significantly increase the kinetics and TONs for CO_2 activation by Fe^0(TPP). Notably, the electrogenerated Fe^0(TPP) does not react with these Brønsted acids to generate H_2 as an unwanted side product. However, these studies necessitated a mercury pool electrode in order to avoid unwanted hydrogen generation and fouling of the electrode surface. Since the reduction of CO_2 to CO is a PCET reaction, the proton donor shifts the potential of the catalytic wave for CO_2 reduction to more positive potentials. In the best case, Savéant and co-workers (94) reported efficient CO production at −1.5 V vs SCE with a TOF approaching $350\,h^{-1}$.

Neta, Fujita, and Gross and co-workers (95) have also probed the ability of cobalt and iron corroles (**9** and **10**) to serve as electrocatalysts for CO_2 reduction (Fig. 2). These studies employed corroles with electron-withdrawing aryl substituents [e.g., 5,10,15-tris(pentafluorophenyl)corrole (TPFC) (95). These systems showed a catalytic wave with onset of current at approximately −1.7 V vs SCE and primarily produced CO as the reduced carbon product. Photochemical studies showed that the active state of the catalyst was $[M^I(TPFC)]^{2-}$. The fact that the Fe(I) and Co(I) states of the cobalt and iron corroles can react with CO_2 is in contrast to the case of the respective porphyrins (see above), which do not react with CO_2 until they are reduced down to the M^0 state. Since the corrole ligand is trianionic and in general, is more electron donating than typical porphyrins, such as those described above, this observation is not completely surprising.

B. Complexes Supported by Non-Porphyrinic Macrocycles

Transition metal complexes with non-porphyrinic macrocyclic ligands have also been extensively studied as catalysts for CO_2 activation. In fact, the first description of homogeneous CO_2 electrocatalysis mediated by transition metal complexes displaying high FEs and TONs involved tetraazamacrocyles of cobalt and nickel (Fig. 6). In this pioneering work, Fishu and Eisenberg (96) demonstrated that complexes **11–15** were able to reduce CO_2 to CO at potentials in the range of −1.3 to −1.6 V vs SCE. These experiments employed a water–acetonitrile mixture as the electrolyte, and consequently, proton reduction was observed as a side reaction for most of the systems studied. Complex **14** produced the highest FE for CO production (~60%), albeit with relatively sluggish kinetics (TOF < 10 h^{-1}).

Some of the most thoroughly studied electrocatalysts for CO_2 reduction are Ni^{II}(cyclam) derivatives (cyclam = 1,4,8,11-tetraazacyclotetradecane) of the type shown in Fig. 7. Sauvage and co-workers (97) first reported that $[Ni^{II}(cyclam)]^{2+}$ (**16**) was a robust and selective electrocatalyst for conversion of CO_2 to CO in 1984. This system operated with a FE of ~95% under aqueous conditions, but required the use of a mercury pool electrode. Subsequent work established that the high electrocatalytic activity is due to adsorption of electrogenerated $[Ni^{I}(cyclam)]^{+}$ on the surface of the mercury electrode (98). It was found that deposition of $[CO{-}Ni^{II}(cyclam)]^{2+}$ can passivate the mercury cathode (99). Sauvage and co-workers (100) also demonstrated that electrolyte composition impacted the kinetics of CO_2 electrocatalysis, with weakly coordinating anions (e.g., NO_3^- and ClO_4^-) showing the highest activity.

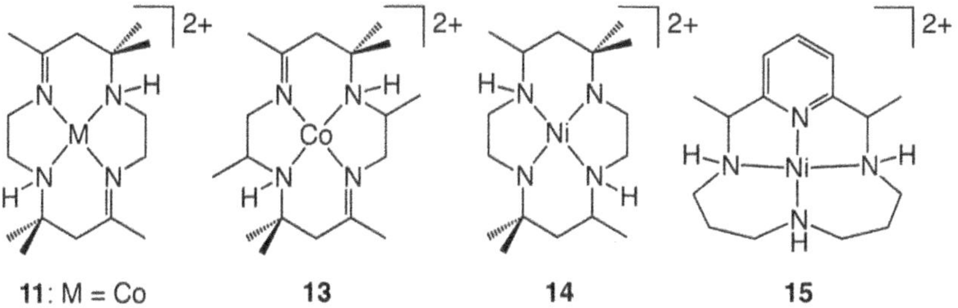

Figure 6. Early non-porphyrinoid-based transition metal complexes for CO_2 reduction.

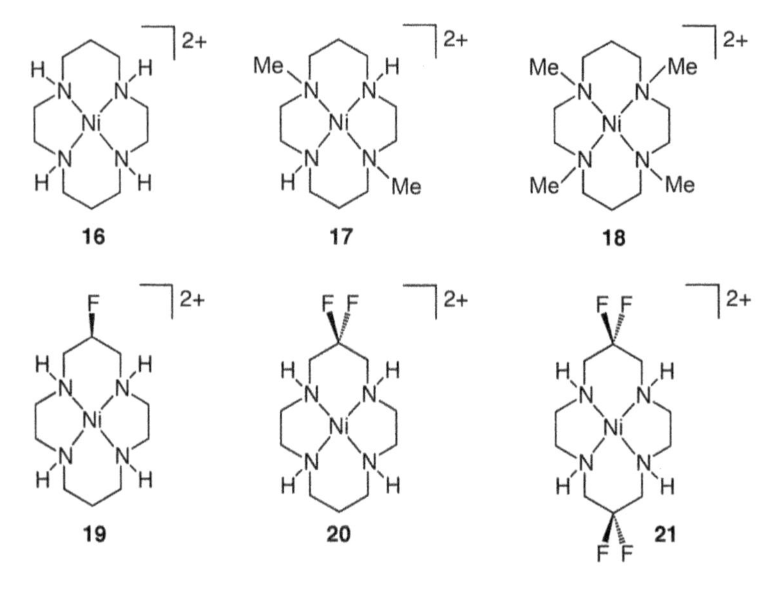

Figure 7. Nickel cyclam derivatives that have been studied as electrocatalysts for CO_2 reduction.

More recently, Froehlich and Kubiak (101) probed the effect of substitution of the cyclam amines on the ability of this platform to activate CO_2. This work demonstrated that $[Ni^{II}(cyclam)]^{2+}$ (**16**) could function as an electrocatalyst for CO production using a glassy carbon cathode and aqueous acetonitrile-containing tetrabutylammonium hexafluorophosphate as the electrolyte. The current densities observed under these conditions were roughly 10–20% those observed using a mercury cathode. Most significantly, this work demonstrated that the $[Ni^{II}(cyclam)]^{2+}$ platform could function under homogenous conditions with a FE of ~90% for CO generation at −1.2 V vs normal hydrogen electrode (NHE) (101). Moreover, this work conclusively established that CO_2 reduction was significantly attenuated by substitution of methyl groups on the amines on the cyclam ring. This difference in reactivity may be due to the reduced nickel complexes of 1,8-dimethylcyclam (**17**) and 1,4,8,11-tetramethylcyclam (**18**) derivatives having insufficient reductive power to activate CO_2, as CV experiments established that both these complexes are easier to reduce than unsubstituted $[Ni^{II}(cyclam)]^{2+}$ (**16**) (101).

Nickel cyclam derivatives with substituents on the carbon skeleton of the macrocycle have also been studied as CO_2 reduction electrocatalysts. For example, a series of mono-, di-, and tetrafluorinated $[Ni^{II}(cyclam)]^{2+}$ derivatives (**19–21**) were prepared by Kimura and co-workers (102) and assayed for electrocatalytic conversion of CO_2 to CO (102). It was demonstrated that the Ni(II/I) reduction potential shifts positively with increased fluorine substitution and that the efficiency and selectivity of CO production is dependent on the number of fluorine atoms on the ligand. In general, increased fluorine substitution decreased both the TOF and selectivity for CO generation.

In related work, it was shown that substitution of alkyl groups on the cyclam periphery can perturb the conformation and geometry of these metal complexes. Fujita et al. (103) showed that (*RRSS*)-$[Ni^{II}(2,3,9,10\text{-tetramethylcyclam})]^{2+}$ (**22**) and (*RR*)-$[Ni^{II}(5,12\text{-dimethylcyclam})]^{2+}$ (**23**) are more active electrocatalysts for CO_2 reduction on a mercury cathode as compared to $[Ni^{II}(\text{cyclam})]^{2+}$ (**16**). Both these systems (Fig. 8) are nearly completely selective for formation of CO over other carbon-containing products and H_2. By contrast, the peak current density of (*RSSR*)-$[Ni^{II}(2,3,9,10\text{-tetramethylcyclam})]^{2+}$ (**24**), which is a geometric isomer of **22**, is lower than that of $[Ni^{II}(\text{cyclam})]^{2+}$ (**16**). This discrepancy has been rationalized by noting that the axial methyl groups of complex **24** may sterically impede binding of CO_2 to the nickel center or adsorption of the complex onto the mercury cathode. This work was recently expanded to include complexes **25**–**27**, which are shown in Fig. 8. In addition to **22** and **23**, complex **25** is a particularly efficient platform for activation of CO_2 and allows for the electrochemical production of CO with high current densities and a FE of ~80% when used in combination with a mercury pool electrode (104).

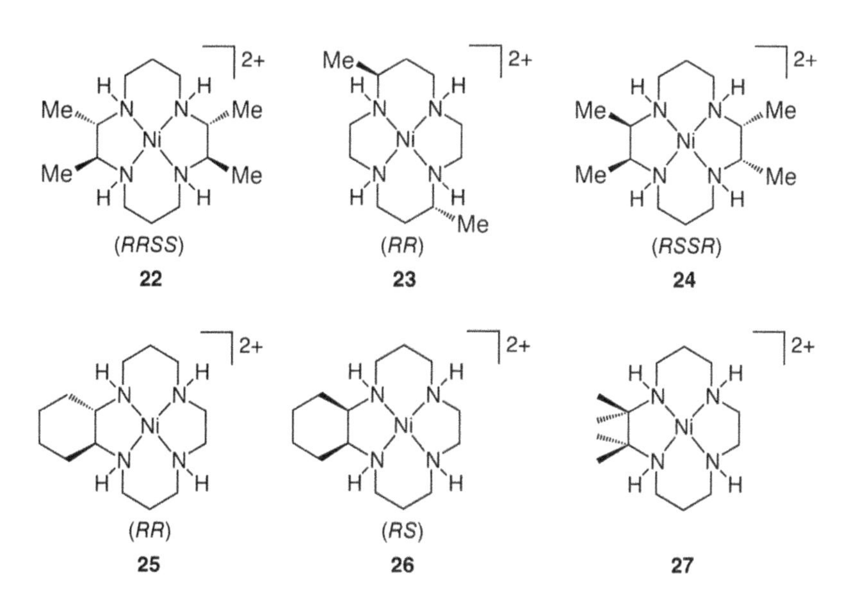

Figure 8. Chiral nickel cyclam derivatives that catalyze reduction of CO_2 to CO.

IV. METAL POLYPYRIDYL COMPLEXES

Metal polypyridyl complexes represent another important and thoroughly studied class of electrocatalyst for reduction of CO_2 to CO. An early example of such a system is Re(bpy)$(CO)_3$Cl (bpy = 2,2-bipyridine) (**28**), which was first reported by Hawecker and Lehn (105) in 1984. The structure of this complex is shown in Fig. 9. Lehn's studies established this platform as an efficient electrocatalyst for CO_2 reduction by showing that the complex can selectively convert CO_2 to CO at roughly −1.5 V vs SCE in aqueous DMF. The FE for CO production was nearly quantitative when the solvent contained only 10% water, but cogeneration of H_2 became more problematic as this percentage was increased (106).

While Lehn's catalyst displayed high selectivity and FE for CO production, the TOF for this system was relatively low (~$20\ h^{-1}$). Efforts to improve the kinetics for this system were recently undertaken by Kubiak and co-workers. These researchers described the synthesis and electrochemical analysis of a family of rhenium bpy complexes with varied substituents at the 4,4′-positions of the ligand. In particular, complexes with *tert*-butyl, methyl, methoxy, and carboxylic acid substituents (**29–32**) were screened for CO_2 electrocatalysis. This work established that certain groups on the bpy ligand lead to enhanced catalytic activity. Of the systems considered, the best catalyst was found to be **29**, which showed a second-order rate constant approaching $1000\ M^{-1}\ s^{-1}$ and a FE of 99%. This marked a vast improvement in rate of catalysis over the original Lehn catalyst (**28**). The *tert*-butyl appended catalyst also showed excellent long-term stability, displaying no significant loss in activity over the course of a 5 h controlled potential electrolysis.

The mechanism by which the rhenium bpy catalysts operate also has been studied. These complexes undergo two one-electron reductions that are centered at roughly −1.4 and −1.8 V vs SCE. The first reduction is reversible and is believed to correspond to reduction of the bpy ligand, while the second redox event corresponds to the irreversible $Re^{I/0}$ couple (107, 108). Notably, the roughly 50 fold increase in activity of **29** as compared to **28** is not believed to be due to

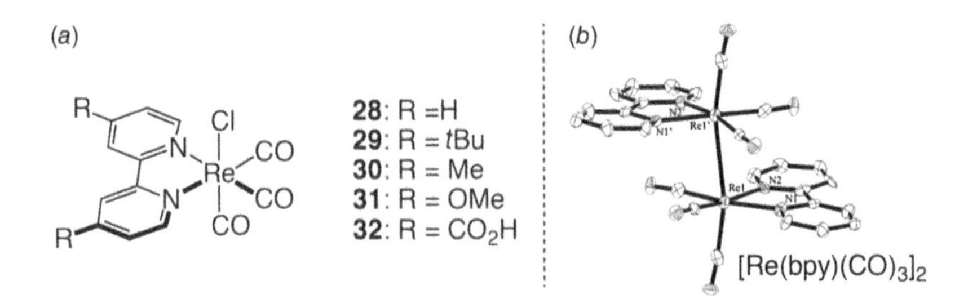

Figure 9. (*a*) Library of rhenium bpy complexes for electrocatalytic CO_2 reduction and (*b*) molecular structure of $[Re(bpy)(CO)_3]_2$ formed from the electrochemical dimerization of Re(bpy)$(CO)_3$Cl.

enhanced electronic donation by the 4,4′-substituents on the bpy ligand. Structural studies show that the rhenium bpy catalyst is deactivated by dimerization of the singly electron reduced Re(I) complex to form the neutral bimetallic complex shown in Fig. 9(*b*) (109). Formation of this dimer is an unproductive side reaction that decreases the total available active species that can react with CO_2. The increased catalytic current observed for complex **29** may be due to a steric barrier that prohibits dimer formation and results in a larger proportion of rhenium complex in the electrochemical double-layer being reduced to $[Re^{0}(bpy)(CO)_3]^{-}$. The Re(0) intermediate is believed to be the active species for CO_2 activation (Scheme 3). Electrocatalytic studies performed with the addition of Brønsted acids revealed a primary H/D kinetic isotope effect, indicating that proton transfer to the Re$-CO_2$ adduct helps to drive CO evolution and is involved in the rate-limiting step of the catalytic process (110).

The *tert*-butyl substituted rhenium bpy complex (**29**) also has been used to drive the photoelectrochemical reduction of CO_2. A hydrogen-terminated p-type silicon substrate was used as a photocathode for the generation of CO in the presence of **29**, which served as the electrocatalyst for CO_2 reduction (111). By using a PV in conjunction with the rhenium electrocatalyst (homogeneous in MeCN), the

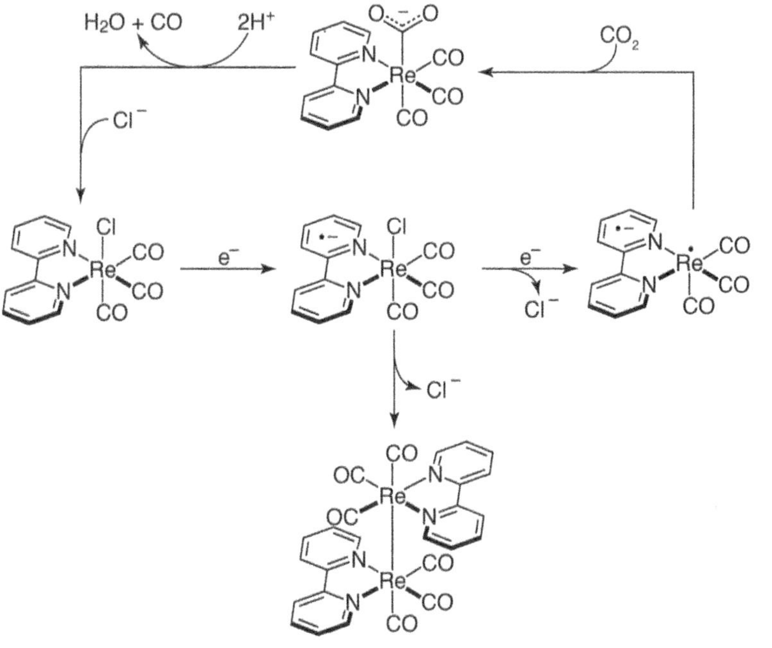

Scheme 3. Proposed electrocatalytic cycle for reduction of CO_2 by $Re(bpy)(CO)_3Cl$.

conversion of CO_2 to CO was achieved at a potential that was 600 mV more positive than that required with a bare Pt electrode. Moreover, a short-circuit quantum efficiency of ~60% for light-to-chemical energy conversion was observed for reduction of CO_2 to CO using this approach (111). In subsequent work, it was shown that by addition of water to the electrolyte solution, the light-assisted cogeneration of syngas (H_2 and CO) could be carried out with an overall FE of nearly 100%. This process involved homogeneous CO_2 reduction by **29** concomitant with heterogeneous proton reduction directly at the surface of the hydrogen-terminated silicon photocathode (112). By varying the concentration of water and rhenium electrocatalyst in the electrolyte the H_2/CO product ratio could be tuned from 0:1 to 2:1. Lastly, Kubiak and co-workers (110) showed that modification of the silicon electrode surface impacts the kinetics of interfacial electron transfer (ET) between the semiconductor and rhenium catalyst. A styrene modified p-Si surface showed a threefold increase in photocatalytic current density compared to hydrogen or hexane terminated surfaces. These studies suggest that the properties of the electrode surface are an important consideration when developing artificial photosynthetic systems and other energy conversion platforms (113).

Ruthenium polypyridyl complexes also have been investigated as electrocatalysts for CO_2 reduction. Some of the earliest such studies were carried out in the late 1980s by Tanaka and co-workers (114), who reported that $[Ru(bpy)_2(CO)_2]^{2+}$ (**33**) and $[Ru(bpy)_2(CO)Cl]^{+}$ (**34**) were competent electrocatalysts for reduction of CO_2 in water containing 10% DMF. Similar results were also reported using $[Ru(bpy)(tpy)(CO)]^{2+}$ (**35**, tpy = terpyridine) in aqueous ethanol (115). For each of these systems, the structures of which are shown in Fig. 10, the effect of solution pH was found to be very important in determining the product distribution of electrocatalysis. Although Tanaka's systems were limited by poor stabilities and sluggish kinetics, these studies helped map the reaction pathways by which CO_2 is converted to reduced products at −1.5 V vs SCE.

The general mechanism by which these systems operate is shown in Scheme 4. Reduction of the starting Ru(II) complex by two electrons leads to loss of CO to generate a five-coordinate 18-electron Ru^0 intermediate. This neutral complex can

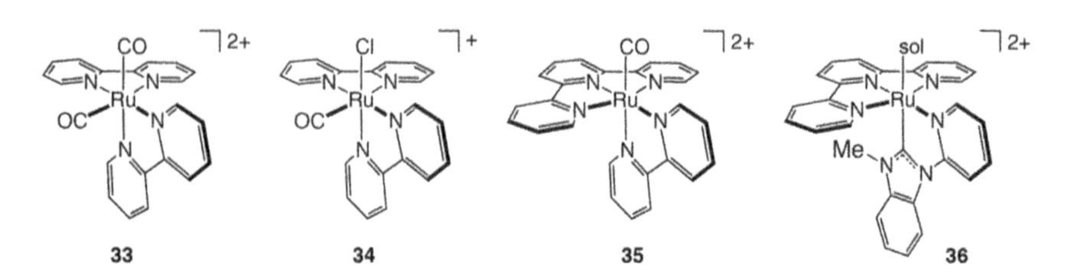

Figure 10. Ruthenium polypyrridyl complexes that function as electrocatalysts for CO_2 reduction.

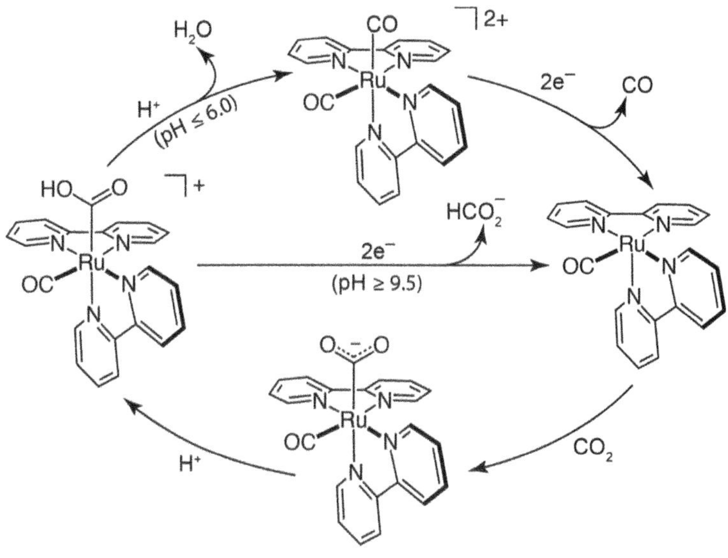

Scheme 4. Proposed electrocatalytic cycle for reduction of CO_2 by $[Ru(bpy)_2(CO)_2]^{2+}$.

bind CO_2 to form the corresponding η^1–CO_2 adduct, which is protonated to yield a Ru(II) metallocarboxylic acid. The catalytic pathway can then diverge depending on the pH of the solution. Under acidic conditions ($pH \leq 6.0$) protonation of an oxygen atom on the metallocarboxylic acid induces dehydration to generate $[Ru(bpy)_2(CO)_2]^{2+}$. Subsequent reduction of the dicarbonyl leads to release of CO. In contrast, if protons are not available to drive this process ($pH \geq 9.5$), a two-electron PCET reaction can take place, which results in loss of formate to regenerate the five-coordinate $Ru^0(bpy)_2(CO)$ species. This process is likely mechanistically distinct from that demonstrated by Darensbourg et al. (116) for the chemical interconversion of CO_2 and formate using Group 6 (117–121) and Group 10 metals hydrides.

The ability of a ruthenium polypyridyl complex containing an axial carbene ligand to activate CO_2 also has been ascertained. Meyer and co-workers (122, 123) described the synthesis of a ruthenium analogue in which the bpy ligand was replaced by a 2-pyridylcarbene (**36**). The ability of this system to activate CO_2 was compared to that observed for the corresponding bpy complex. It was determined that both these complexes operate by similar mechanisms, but that the carbene derivative is much more active for CO_2 reduction (124). The improved kinetics are likely due to enhanced electron donation of the carbene ligand over bpy. This system is selective for CO production (FE ~85%) in an acetonitrile electrolyte. Complex **36** was also shown to be effective as an electrocatalyst for splitting CO_2 to $CO + \frac{1}{2} O_2$ in a two-compartment electrochemical cell (125).

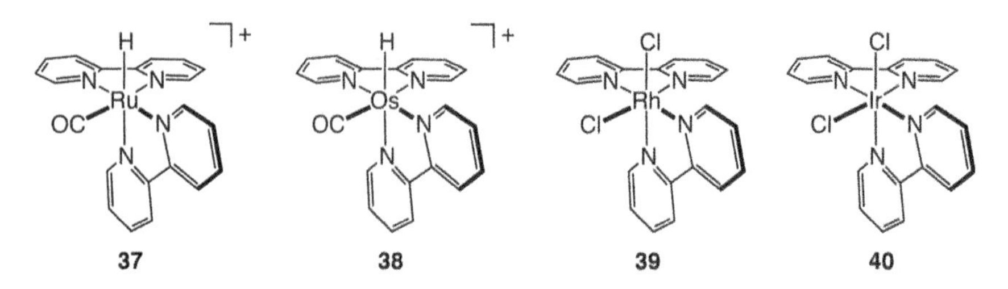

Figure 11. Metal polypyrridyl complexes studied by Meyer and co-workers (122, 123) as electrocatalysts for CO_2 reduction.

Meyer and co-workers (126, 127) also studied analogous hydride complexes of ruthenium and osmium. This work demonstrated that under strictly anhydrous conditions, both $[Ru(bpy)_2(CO)H]^+$ (**37**) and $[Os(bpy)_2(CO)H]^+$ (**38**) are electrocatalysts for reduction of CO_2 to CO. Introduction of water to these systems results in a FE of ~25% for formate production as a side product. The ability of rhodium and iridium polypyridyl complexes to electrochemically activate CO_2 also has been assessed (Fig. 11). Both *cis*-$[Rh(bpy)_2Cl_2]$ (**39**) and *cis*-$[Ir(bpy)_2Cl_2]$ (**40**) can reduce CO_2 at −1.6 V vs SCE (128). Although these systems are structurally similar to **37** and **38**, CO is not observed as a CO_2 reduction product. These rhodium complexes display low TONs and only generate formate and hydrogen with FEs of roughly 65 and 10%, respectively.

All of the complexes described so far in this section have made use of heavy and relatively expensive metals. However, there are also several examples of metal–pyridine complexes containing earth abundant metals. Kubiak and co-workers (129) reported a dinuclear copper complex containing 6-(diphenylphosphino)-2,2′-bipyridyl ligands. This dicopper complex (**41**), which is shown in Fig. 12, could be reduced via two sequential single-electron transfers in MeCN at −1.35 and −1.53 V vs SCE. Both these redox events correspond to reduction of the bpy ligands and are required to drive the reduction of CO_2 to CO. This system was found to be robust over the course of a 24 h electrolysis experiment with a TOF for CO production of $\sim 2\,h^{-1}$.

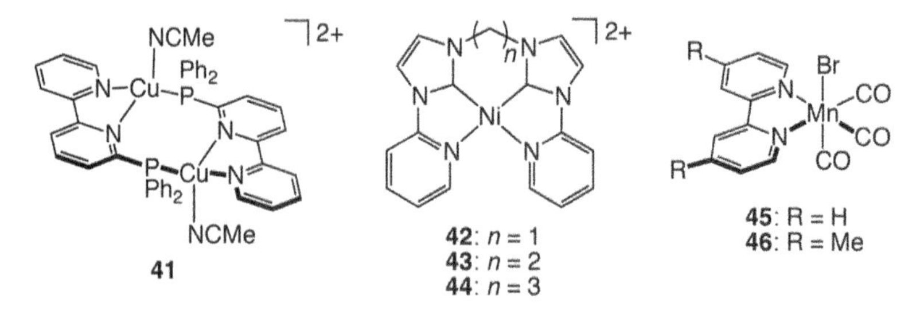

Figure 12. Pyrridyl supported electrocatalysts for CO_2 reduction based upon earth-abundant metals.

More recently, Thoi and Chang (130) described a homologous series of nickel complexes (**42–44**) supported by *N*-heterocyclic carbene pyridine ligands bridged by linkers of one to three carbons in length. Each of these systems exhibited high selectivity for reduction of CO_2 to CO over water at roughly −1.6 V vs SCE. Complexes **42–44** display modest kinetics with TOF ranging from 3.9 to 5.9 h^{-1}, but are not robust, and readily decompose due to hydrogenation or carboxylation of the ligand frameworks.

One final study that involves inexpensive metal–pyridyl complexes for CO_2 reduction was also described in 2011. Manganese variants of the rhenium complexes described above have been studied. $Mn(bpy)(CO)_3Br$ (**45**) was shown to be an electrocatalyst for reduction of CO_2 to CO. Such complexes are attractive since manganese is inexpensive and is the third most abundant transition metal, comprising 0.1% of the Earth's crust (131). The manganese homologue of compound **30**, which contains a 4,4′-dimethyl-2,2′-bipyridyl ligand, was also prepared (**46**). Both **45** and **46** are excellent molecular catalysts for the selective electrochemical reduction of CO_2 to CO (FE ~85%) under mild conditions at roughly −1.85 V vs Ag/AgCl. It will be interesting to see how further development of this platform leads to additional advances in molecular CO_2 electrocatalysis.

V. METAL PHOSPHINE COMPLEXES

Metal complexes that are supported by phosphine ligands have been explored as platforms for CO_2 activation. The first such study was reported by Slater and Wagenknecht (132) in 1984. This work detailed the ability of $Rh(dppe)_2Cl$ [dppe = 1,2-bis(diphenylphosphino)ethane] to activate CO_2 in acetonitrile solution at roughly −1.3 V versus a silver wire quasireference electrode. The electrochemistry of this system was complex, as reduction of the $Rh(dppe)_2Cl$ complex led to hydrogen abstraction from the MeCN solvent to yield the corresponding rhodium hydride. Insertion of CO_2 into this hydride ultimately led to formate production with FEs of 20–40%, depending on the length of the electrolysis experiment.

Significant emphasis has been placed on the development of multimetallic complexes that can activate CO_2. One of the earliest such examples was reported by Kubiak's group in 1987. This study detailed the preparation and electrochemical properties of a binuclear "Cradle" complex of Ni(0), $[Ni_2(\mu\text{-CNMe})(CNMe)_2\text{-}(dppm)_2]^{2+}$ (**47**) (dppm = diphenylphosphinomethane). This compound displays two reversible redox waves at −0.51 and −0.83 V vs Ag/AgCl in THF containing 0.1 *M* tetrabutylammonium tetrafluoroborate (133). Following reduction, complex **47** reacts with CO_2 by an electrochemical (EC) mechanism to generate an adduct in which CO_2 binding and activation occurs at the bridging isocyanide ligand. Labeling experiments showed that this adduct undergoes oxygen-atom transfer from CO_2 to the isonitrile to yield the corresponding complex with a bridging carbonyl

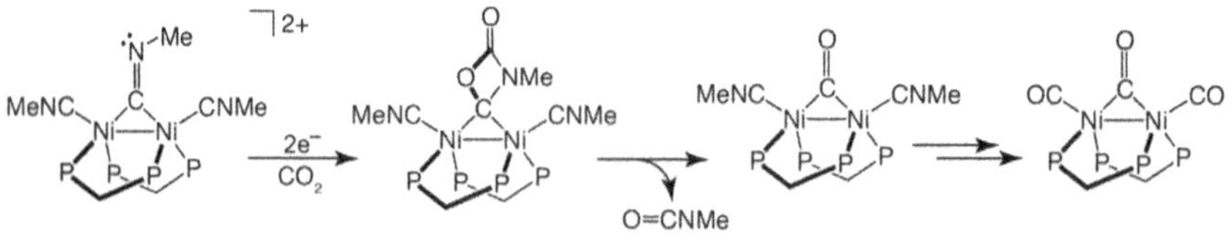

Scheme 5. Proposed pathway for CO_2 activation by $[Ni_2(\mu\text{-CNMe})(CNMe)_2(dppm)_2]^{2+}$ (**47**).

(Scheme 5) (133). The carbonyl complex can drive the production of CO from CO_2, however, the CO is trapped by the nickel complex to generate $[Ni_2(\mu\text{-CO})(CO)_2(dppm)_2]$ and limits the utility of this system. This work was recently extended to include a family of binuclear nickel complexes supported by dppa [dppa = bis (diphenylphosphino)amine] ligands of the general form $[Ni_2(dppa)_2(\mu\text{-CNR})(CNR)_2]$ (**48–50**) (134). These complexes (Fig. 13) demonstrated a similar reactivity profile and were also inhibited by the CO produced upon CO_2 activation.

Trinuclear clusters of nickel have also been studied as electrocatalysts for CO_2 reduction. One of the first such systems was the cationic cluster $[Ni_3(\mu_3\text{-I})(\mu_3\text{-CNMe})(\mu_2\text{-dppm})]^+$ (**51**) (135). This complex exhibits a reversible single-electron reduction at −1.1 V vs Ag/AgCI to give the corresponding mixed-valent species. In the presence of CO_2, reduction of **51** leads to reductive disproportionation to generate CO and CO_3^{2-} (136) via an apparent EC′ electrochemical mechanism. In the presence of even trace amounts of a proton donor (e.g., water), only formate was observed as a CO_2 reduction product.

An entire family of trinuclear electrocatalysts for CO_2 reduction has been reported. This set of complexes includes systems containing bridging halide (137), telluride (138), carbonyl, and a varied array of isocyanide capping ligands. The set of isocyanide complexes has been particularly well studied and the kinetics of CO_2 activation by these clusters (**51–57**) have been reported (139). Each of the systems shown in Fig. 14 catalyzes the disproportionation of CO_2 to give CO and CO_3^{2-} in dry MeCN. Notably, no oxalate or other single-electron reduction products are observed following the electrolysis experiment. Analytical electrochemistry experiments showed that the CO_2 reduction process is first order in both nickel

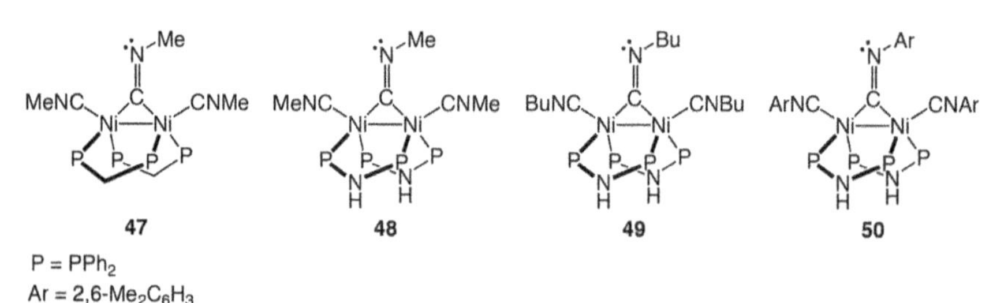

Figure 13. Binuclear nickel isocyanide complexes that electrochemically activate CO_2.

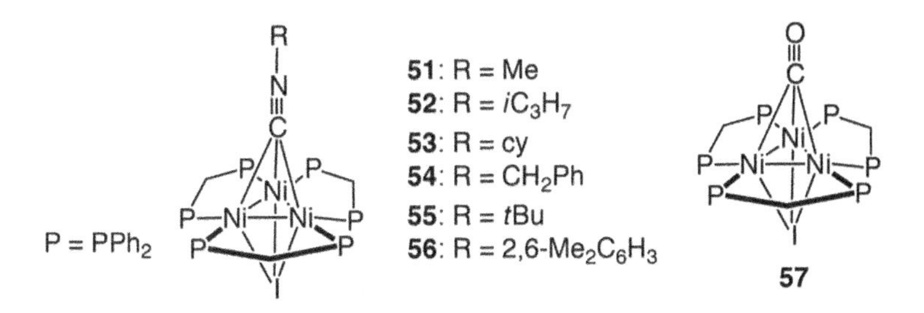

Figure 14. Trinuclear nickel clusters that activate CO_2.

cluster and CO_2. Moreover, the relative rates of reaction of the trinuclear nickel clusters with CO_2 depend on the identity of the capping ligand. The relative activity of the various isocyanide and carbonyl bridged clusters was found to be CNMe (**51**) ~ CN(i-C_3H_7) (**52**) > CN–cy (**53**) > $CNCH_2Ph$ (**54**) > CO (**57**) > CN(tBu) (**55**) > CN(2,6-$Me_2C_6H_3$) (**56**) (139). It was also found that small differences in the redox potentials between the clusters dramatically affect the rates of reaction with CO_2, even though these values fall into a very narrow range (−1.08 to −1.18 V vs SCE). The size and orientation of the isonitrile substituents plays a much smaller role in dictating the kinetics of CO_2 reduction.

Unlike the case for dinuclear nickel complexes (**47–50**) Kubiak and co-workers (137) proposed that interaction of CO_2 with the reduced trinuclear clusters takes place at the cluster core, as opposed to the bridging isocyanide ligand. This pathway was favored based upon molecular orbital calculations. It was determined that the LUMO for the oxidized cluster is almost entirely metal centered. Upon reduction, this orbital becomes a singly occupied molecular orbital (SOMO) and is of appropriate symmetry to interact with the LUMO of CO_2. The electrochemical studies also indicated that the energy of the SOMO is well correlated with CO_2 reduction. Finally, the secondary steric effect observed in the kinetics study suggested that CO_2 binding occurs on the more congested face of the clusters.

One of the most successful classes of electrocatalysts for CO_2 reduction are palladium phosphine complexes (140). Dubois (141) demonstrated that properly substituted complexes of palladium that are supported by tridentate phosphine ligands (Fig. 15) can display fast kinetics for CO_2 reduction with good selectivity for CO production (141). Systems of this type were first reported in 1987 (142). It was shown that a palladium triphos complex [Pd(etpC)(MeCN)]$(BF_4)_2$ (**58**) [etpC = bis[(dicyclohexylphosphino)ethyl]phenylphosphine] could catalyze the reduction of CO_2 to CO in acetonitrile solutions containing HBF_4 (143). This system was selective for CO (FE = 85%) over other reduced carbon products and produced a small amount of hydrogen gas (FE = 15%) (143). Kinetic experiments showed that at acid concentrations > 1.0×10^{-2} M, the reaction was first order in catalyst, first order in CO_2, and independent of acid concentration. At acid

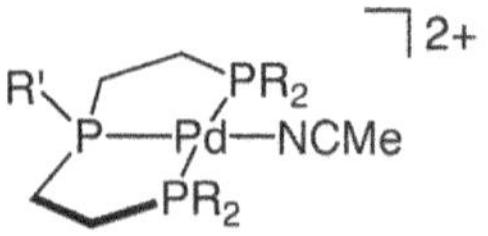

58: R = cy, R' = Ph
59: R = Et, R' = Ph
60: R = Ph, R' = Ph
61: R = cy, R' = Me
62: R = Et, R' = Me
63: R = *t*Bu, R' = neopentyl
64: R = neopentyl, R' = neopentyl

Figure 15. A library of palladium triphos complexes studied for electrocatalytic conversion of CO_2 to CO.

concentrations < $4.0 \times 10^{-3}\,M$, the catalytic rate was shown to be first order in catalyst, second order in acid, and independent of CO_2 (143).

The substituents on the triphos ligand dramatically impact the observed electrocatalysis. Steric effects were shown to play a particularly important role. The R′ group on the ligand backbone dramatically effected the selectivity of the catalyst, as systems in which this group was a phenyl substituent (**58–60**) were much more selective for reduction of CO_2 to CO over proton reduction. In contrast, platforms in which this R′ group was a methyl substituent (**61** and **62**) show FEs for H_2 production of 90%. Systems containing bulky *tert*-butyl (**63**) or neopentyl (**64**) groups showed similarly low FEs for CO production. The R substituents had a more subtle effect on the observed electrocatalysis, as comparison of **58–60** shows that bulky alkyl substituents lead to a much higher selectivity for CO production. In addition to being selective for production of CO, complexes **58** and **59** also exhibited relatively fast kinetics for CO_2 activation of roughly 45–50 $M^{-1}\,s^{-1}$.

Despite the excellent kinetics and selectivity for CO production that was displayed by some of these complexes, the instability of the Pd(triphos) (triphos = 1,1,1-tris(diphenylphosphinomethyl)ethane) platform under electrocatalytic conditions proved to be a critical limitation of these systems. The electrochemical mechanism by which the Pd(triphos) architectures are believed to activate CO_2 is shown in Scheme 6 (140, 141). In this cycle, the starting Pd(II) complex is reduced by a single electron at the cathode to generate a distorted Pd(I) species, which binds CO_2 to form a five-coordinate η^1–Pd–CO_2 intermediate. Subsequent reduction and proton transfer to the CO_2 adduct ultimately leads to CO formation. The labile nature of the flexible phosphine ligand, coupled with the general lability of low-valent Pd centers, introduces the possibility for side reactions. Rearrangement of the phosphine ligand and dimerization of two Pd(I) centers results in the formation of a very stable Pd^I–Pd^I dimer. This dimer is the thermodynamic sink for this system and is not an active electrocatalyst for CO_2 conversion. Accordingly, formation of this bridged dimer siphons palladium from the catalytic cycle and compromises catalyst lifespan. As a result, despite the fact that the best Pd(triphos) architectures reported by Dubois et al. (143) demonstrate second-order kinetics for CO_2 activation on the order of 180 $M^{-1}\,s^{-1}$, the optimal TONs observed for these catalysts are usually in the range of 5–10. Indeed, even for the most sterically encumbered Pd(triphos) systems that have been reported to date, the highest TONs observed for conversion of CO_2 to CO are < 150 (141, 143).

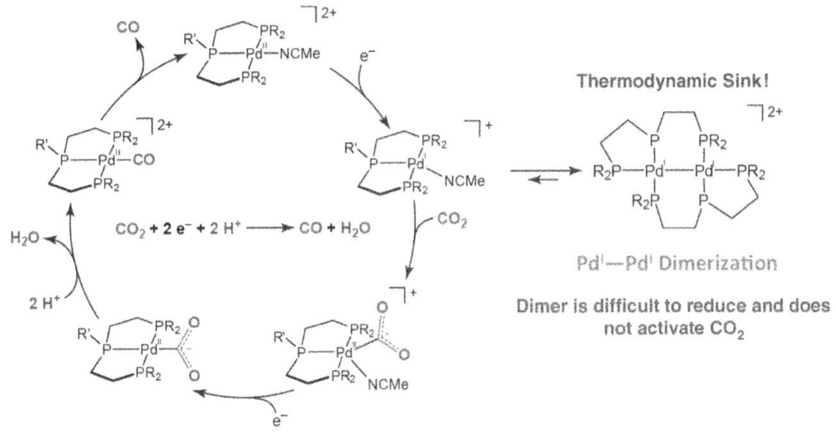

Scheme 6. Proposed pathway for CO_2 activation by and deactivation of palladium triphos complexes.

The design of the triphos pincer framework was amenable to synthetic alteration of the central donating atom of the chelating ligand. Dubois and co-workers (144) described the synthesis and study of the complexes shown in Fig. 16, which contain C (**65**), N (**66** and **67**), S (**68**), and As (**69**) atoms as the central donor. This study showed that the phosphorus-substituted ligands support complexes that have the highest activities and selectivities for CO generation, while the other complexes largely promote proton reduction to H_2.

Dubois and co-workers has also developed water soluble (145) and cationic (146) palladium phosphorus complexes for CO_2 activation. Additional work has centered on the study of complexes containing multiple Pd triphos centers. Initial systems consisted of dendrimers containing up to 15 phosphorus atoms (147). These systems catalyzed the reduction of CO_2 to CO, but showed decreased activity and selectivity compared to the best monomeric Pd systems. In subsequent work, a methylene spaced bis-Pd(triphos) platform (**70**) was described (148). This system showed vastly enhanced activity for CO_2 binding and conversion to CO compared to the previously studied Pd(triphos) monomers, with a second order

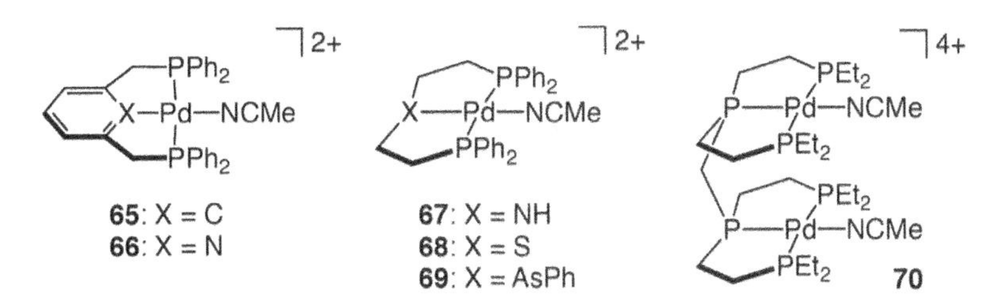

Figure 16. Palladium complexes with varying ligand substituents and number of metal centers.

rate constant for CO_2 reduction that was estimated to be ~$25 \times 10^3 M^{-1}s^{-1}$. This rate enhancement is roughly three orders of magnitude larger than the typical values obtained for mononuclear Pd(triphos) systems (e.g., **58** and **59**). These vastly accelerated kinetics were attributed to a cooperative mechanism in which both palladium centers can bridge and activate CO_2 to enhance C–O bond cleavage. A *meta*-phenylene bridged bimetallic system also has been described (149). but this system does not exhibit the cooperative reactivity that was observed for **70**. Despite the improved kinetics observed for complex **70**, this bimetallic species proved to be highly unstable under the electrocatalytic conditions, presumably due to formation of an intramolecular $Pd^I–Pd^I$ dimer (Section VI).

VI. FUTURE DEVELOPMENT OF CO_2 REDUCTION CATALYSTS: NATURE AS INSPIRATION

Despite the progress that has been made over the last 35 years, there are several issues that must be overcome in order to effect the efficient electrochemical conversion of CO_2 to fuel. For example, none of the molecular CO_2 reduction catalysts described above display efficiencies or stabilities necessary for the practical generation of CO (150, 151). Furthermore, while the preceding sections have outlined several examples of molecular catalysts that can selectively convert CO_2 to CO, very few of these, if any, can drive this transformation at a reasonable rate while operating at low overpotentials (i.e., $\eta < 0.3$ V). Indeed, the main hurdles that must be overcome in advancing CO_2 catalysis to the point of technological relevance are the high overpotentials that are required to drive CO_2 reduction with high current densities and reasonable kinetics (152). The major reason for this shortcoming is that the electrocatalysts reported to date do not minimize the large nuclear reorganization energy associated with binding and reduction of CO_2 upon interaction with the electron-rich metal center.

As illustrated in the preceding sections, the most common mode by which CO_2 binds to electrocatalyst complexes that generate CO is via η^1–CO_2 coordination. The formation of η^1–CO_2 adducts is a direct consequence of nucleophilic attack by an electron-rich metal center onto the Lewis acidic carbon atom of CO_2. In general, this process has a sizeable activation barrier due to the large reorganization energy attendant with conversion of the linear CO_2 molecule to a bent η^1–CO_2 adduct (30, 31). As such, the kinetic barrier associated with CO_2 binding and activation by molecular electrocatalysts often necessitates that such reactions be run at sizeable electrochemical overpotentials, which significantly compromises the energy efficiency of the CO_2 reduction reaction (49, 153).

In contrast, natural systems are astoundingly successful at activating CO_2 with minimal energy input. One of the major pathways through which CO_2 is fixed by Nature is the Wood–Ljungdahl pathway (154). This process involves the reduction of CO_2 to generate CO and H_2O at the Ni–S–Fe active site within the C-cluster of carbon monoxide dehydrogenase (CODH) (155). As discussed in Section I, this transformation is a $2e^- + 2H^+$ multielectron PCET process and is thermodynamically uphill by slightly > 0.5 V at pH 7.0 (Eq. 3). The ability to efficiently promote this PCET process at low overpotential allows the Ni–Fe–S form of CODH to carry out the interconversion of $CO_2 + 2e^-/2H^+ \rightleftarrows CO + H_2O$ at close to the thermodynamic limit, with a staggering TOF of roughly 31,000 s^{-1} (156). Moreover, Shin et al. (157) demonstrated that CODH can be driven electrochemically using methyl viologen as a redox mediator at pH 6.3. Under these conditions, CO generation occurs with a FE of nearly 100% and TOF of ~700 h^{-1} at an applied potential of −0.57 V vs NHE. These conditions represent an overpotential of < 100 mV for CO production.

Carbon monoxide dehydrogenase derives its spectacular ability to convert CO_2 to CO from the structure and electronic properties of the enzyme's C-Cluster, which is the site of catalytic activity. The C-Cluster contains a complex $NiFe_4S_4$ center, which has been resolved crystallographically in three distinct redox states. Activation of CO_2 is believed to take place at the Ni(II) center, which is coordinated by three sulfur ligands that form a distorted T-shaped geometry (Scheme 7.I). Electron injection into the distorted $NiFe_4S_4$ cubane generates the reduced state of the enzyme (Scheme 7.II), which can bind CO_2. This binding completes the coordination around the Ni atom, which acts as a nucleophile and binds CO_2 to produce a square-planar Ni(II) center (Scheme 7.III). Additionally, the partial negative charge on the oxygen atoms of the reduced CO_2 adduct is stabilized by interaction with three discrete Lewis acidic sites, which include: (1) a dangling Fe(II) metal center, (2) a hydrogen bond to His_{93}, and (3) a hydrogen bond to Lys_{563}. These interactions are important in that they work to stabilize the reduced CO_2 adduct and greatly reduce the large reorganization energy associated with its formation. It is believed that the ability of CODH to catalyze the conversion of CO_2 to CO with virtually zero overpotential is attributable to the concerted action of these three Lewis acid–base interactions (158).

Given that CODH catalyzes the $2e^- + 2H^+$ reduction of CO_2 to CO and H_2O at the thermodynamic limit with unparalleled kinetics and high selectivity, this enzyme should serve as inspiration for the rational design of molecular electrocatalysts, capable of carrying out this energy-storing transformation. In keeping with this, salient features of the Ni–Fe–S center of CODH are common to some of the most successful CO_2 reduction catalysts that have been highlighted in this chapter. For instance, bimetallic cooperativity has been shown to be an important element for activation of CO_2. Savéant and co-workers (92–94) demonstrated the efficacy of this reactivity mode in the mid-1990s by detailing the extent to which Lewis acidic metal cations can enhance the ability of a nucleophilic Fe(0) porphyrin complex to active CO_2 (Scheme 2). The synergistic action of a nucleophilic electron-rich metal working in concert with an oxophilic metal cation

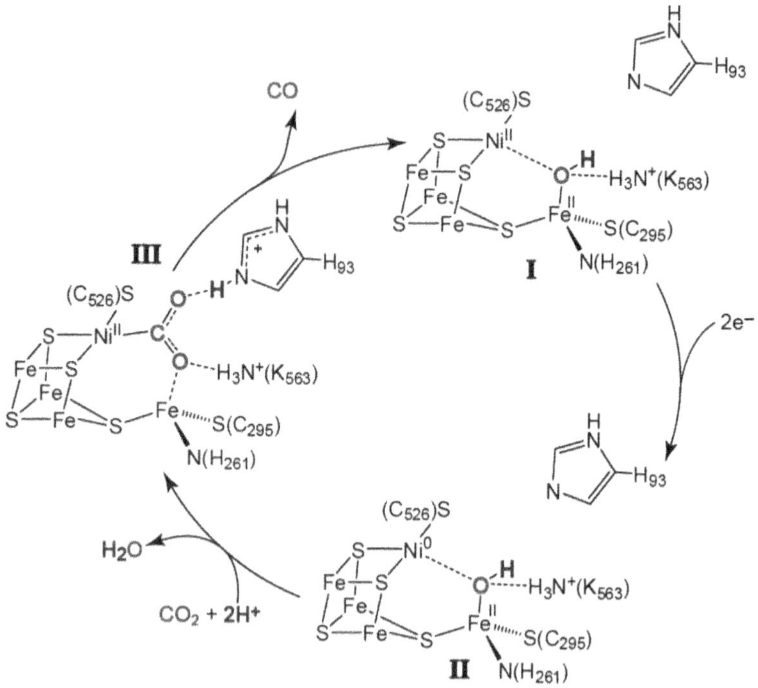

Scheme 7. Mechanism for CO_2 binding and reduction by the C-Cluster of CODH.

to bind CO_2 and drive C–O bond cleavage to liberate CO and H_2O is a powerful strategy for the development of new electrocatalysts that can carry out this reaction with fast kinetics and high selectivity.

Dubois' dinuclear palladium catalyst is another example in which two metals work in concert to successfully activate CO_2. Although complex **70** is decidedly abiotic, it does share several structural and electronic features with the active site of CODH. For example, although Dubois' systems are comprised of palladium as opposed to nickel, both these transition metals are members of group 10 of the periodic table, and therefore display similar properties and reactivity in many respects (159). Furthermore, although the palladium centers of **70** are not supported by sulfur-based ligands, the chelating triphos platform does provide a strongly donating ligand that reproduces the distorted T-shaped coordination environment of the natural system. Most notably, however, is the nature of the intermediate formed upon reduction of this complex in the presence of CO_2. As was described in Section V, CO_2 binds to complex **70** to generate the bridged structure shown in Scheme 8. This structure is analogous to that proposed for the adduct that is formed upon reaction of CO_2 at the $NiFe_4S_4$ cluster of CODH

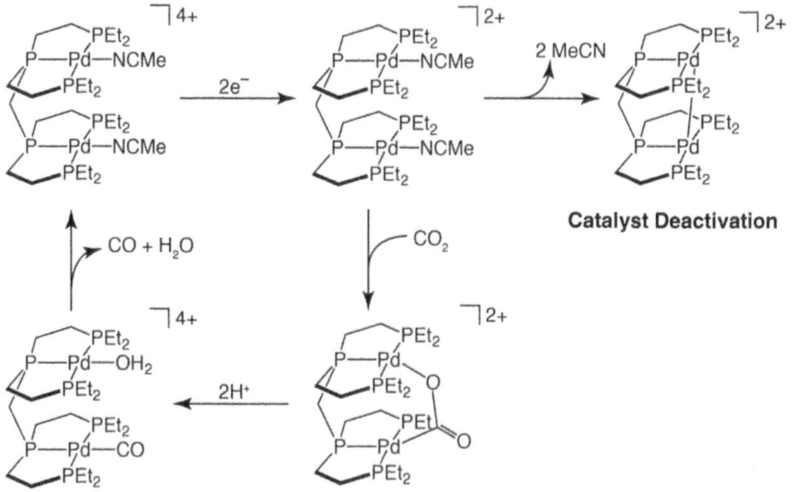

Scheme 8. Mechanism for CO_2 activation by and decomposition of Dubois' dipalladium electrocatalyst.

(Scheme 7.III). The two palladium centers work to weaken the C–O bond that bridges the bimetallic assembly and facilitate extrusion of CO.

The major limitation of complex **70** is that it displays a very low TON for reduction of CO_2 to CO. Since complex **70** presents two palladium centers within identical coordination environments, both metals will be reduced at similar potentials. The generation of two Pd(I) metalloradicals in proximity leads to Pd^I–Pd^I bond formation and catalyst deactivation (Scheme 8). One strategy to prevent intramolecular M–M dimerization is incorporation of two different metals and/or coordination environments that have different reduction potentials into a bimetallic assembly. Such a system would avoid the formation of metalloradicals at similar potentials. Toward this end, Holm and co-workers (160–163) described several bimetallic C-Cluster models that contain both nickel and iron centers. Although the ability of these systems to catalyze the reduction of CO_2 has not yet been demonstrated, the electrochemical formation of M–M bonds has also not been observed. We await future efforts in this area to produce molecular electrocatalysts that mimic the impressive activity and stability of CODH.

In addition to stabilizing CO_2 binding at the Ni–Fe–S center of CODH, the hydrogen-bonding interactions highlighted in Scheme 7 also serve to control the delivery of protons to the reduced CO_2 adduct bound within the C-Cluster. Intimate control of the proton inventory and the timing of H^+ and e^- delivery during catalytic turnover is critical to the high selectivity that CODH displays for CO production versus other CO_2 reduction products (e.g., formate, oxalate, and formaldehyde). This selectivity is particularly noteworthy given that, other than formic acid, all other C_1 compounds that can be derived from CO_2 are formed at lower standard potentials than CO (Section I). Furthermore, the standard potential for proton reduction to H_2 occurs at −0.41 V vs NHE at pH 7, which is also below the standard potential for the conversion of CO_2 to CO. Accordingly, from a purely thermodynamic standpoint, CODH operates at a potential that is sufficiently negative to enable the reduction of adventitious protons to H_2 and convert CO_2 to $H_2C{=}O$, MeOH, or CH_4. The ability of CODH or any other CO_2 reduction catalyst to selectively promote formation of CO over other reduced carbon products stems from the ability of the catalyst to slow the kinetics associated with generation of hydrogen and other more highly reduced carbon-containing products.

The installation of hydrogen-bonding functionalities onto the ligand framework of CO_2 reduction catalysts is another important area of investigation. Prior developments in this field have already helped to demonstrate the validity of this approach. For instance, Sauvage and co-workers (100) supposed that hydrogen-bonding interactions were important to the function of one of the earliest studied electrocatalysts for CO_2 reduction. The compound Ni(cyclam) (**16**) is believed to derive its tremendous activity and selectivity for reduction of CO_2 to CO from the structure of the macrocyclic complex. The ligand geometry produces a highly accessible metal center that can readily bind CO_2 and form the η^1–CO_2 complex shown in Scheme 9(*a*). It has been suggested that hydrogen bonding between the N–H protons and the oxygen atoms of the CO_2 adduct increases the selectivity for binding of CO_2 to the reduced nickel center over protons. In addition to stabilizing the formation of a η^1–CO_2 complex, the hydrogen-bonding

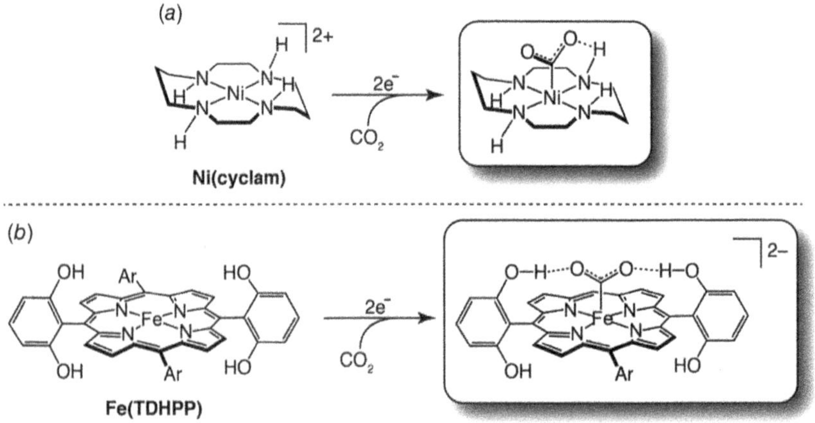

Scheme 9. Formation and proposed structures of η^1–CO_2 adducts stabilized by intramolecular hydrogen bonds for (*a*) Ni(cyclam) and (*b*) an iron porphyrin derivative [Fi(TDHPP)] = iron 5,10,15,20-tetrakis(2′,6′-dihydroxylphenyl)porphyrin].

interactions shown in Scheme 9(*a*) may also provide a direct pathway for protonation of the CO_2 adduct and help to drive the C–O bond scission pathway associated with generation of CO.

Additional support for the hydrogen-bonding interactions of Scheme 9(*a*) exists. For example, Kubiak's and co-workers (101) demonstration that methylation of the macrocycle amines compromises the ability of the Ni(cyclam) platform to activate CO_2 further bolsters this mechanistic picture. Moreover, in 1991, Fujita et al. (164) studied the reaction of various cobalt tetraazamacrocycles with CO_2. These researchers showed that the orientation of alkyl and N–H groups on the ligand periphery dramatically affects the ability of such complexes to bind CO_2. In subsequent work, Fujita et al. (164) made use of variable temperature nuclear magnetic resonance (VT NMR) and FTIR methods to unambiguously demonstrate that intramolecular hydrogen bonds are important to the formation and stabilization of the η^1–CO_2 adducts that are implicated in the catalytic cycle for CO_2 reduction by metal macrocycles (165). Together, these studies clearly indicate the importance of properly positioned proton donors in driving the efficient activation of CO_2.

The importance of PCET to the reduction of CO_2 to CO has also been demonstrated for porphyrin-based electrocatalysts. Building on his earlier studies that established the effect of Brønsted acids on CO_2 reduction, Savéant and co-workers (93, 94) recently showed that a porphyrin complex containing proton donors appended to the meso-aryl groups are extremely active electrocatalysts for conversion of CO_2 to CO in DMF containing 2 M H_2O (166). Indeed, the complex shown in Scheme 9(*b*) catalyzed the production of CO with a FE > 90% and exhibited a TON in excess of 50 million over 4 h. Moreover, this system operated at low overpotential ($\eta = 0.46$–0.56 V) with a TOF in the range of 10^2–10^4 s^{-1}. Comparison between the porphyrin shown in Scheme 9(*b*) and a homologue that lacks the phenol functionalities clearly showed that the proton donors decrease the operating potential of the iron porphyrin catalyst by nearly 0.5 V while increasing the rate of catalysis by one to two orders of magnitude. Although the exact mechanism by which the pendant phenols enhance catalysis has not been studied in detail, it was presumed that the reaction proceeds via the same type of "push–pull" mechanism that was proposed for enhanced CO_2 reduction in the presence of external proton donors (93, 94). The enhanced catalytic activity of Fe(TDHPP) was shown to be related to the high local concentration of phenolic protons; the eight phenolic hydroxyl groups being equivalent to a 150 *M* phenol concentration. As such, the introduction of protonic functionalities into the secondary coordination sphere of a CO_2 reduction catalyst dramatically amplifies the proton-donating effect as compared to use of external acid. A similar effect is almost certainly at play for the C-cluster of CODH. The future development of molecular electrocatalysts that can convert CO_2 to CO and other valuable products, while operating with reasonable kinetics at low overpotential, will likely need to incorporate similar structural and electronic motifs.

VII. SUMMARY AND CONCLUSIONS

The electrochemical conversion of CO_2 to CO and H_2O is a promising strategy for solar energy storage, as this $2e^- + 2H^+$ reaction produces a versatile and energy-rich chemical. This process also provides a means for solar energy distribution since CO can be readily converted into gasoline, diesel, and jet fuel using existing FT methods and infrastructure. To this end, the ability to drive the reduction of CO_2 to CO using a photoelectrochemical cell is an attractive proposition, and the development of new materials and molecular complexes that can promote this reaction continues to attract great interest.

This chapter has detailed progress made over the last 35 years in the fields of molecular energy conversion and CO_2 electrocatalysis. Particular emphasis has been placed on the development and mechanistic study of synthetic platforms for production of CO. Although significant progress has been made in these areas over the last four decades, each of the catalysts described in the preceding sections pales in comparison to CODH, which is an enzymatic system that promotes CO_2 reduction chemistry. The natural system derives its exquisite activity, selectivity, and energy efficiency by managing the *multielectron proton coupled electron transfer* events that drive the C−O bond-breaking reactions that are attendant with production of CO from CO_2. Whereas some of the molecular electrocatalysts we have discussed are almost completely selective for CO, none can match the TOF of CODH at modest potentials. If CO_2 reduction chemistry is to permit the storage of large amounts of solar energy in an economically viable way, meaningful progress will need to be made in the area of CO_2 reduction catalysis. Cues from the C-Cluster of CODH and some of the more successful molecular electrocatalysts for CO_2 reduction suggest that future emphasis will need to be placed on the development of ditopic catalyst platforms that incorporate multiple metal sites and/or engineered hydrogen-bonding functionalities, that engender efficient CO_2 binding and activation. Incorporation of these constructs onto robust electrode assemblies with tailored interfacial and electronic properties will also be important if the photoelectrochemical conversion of CO_2 to CO is to be carried out on a practical scale.

ACKNOWLEDGMENTS

Our efforts in the area of CO_2 electrocatalysis have been supported by a Ralph E. Powe Junior Faculty Enhancement Award from Oak Ridge Associated Universities, and a DuPont Young Professor Award. Additional financial support for our work in this area has been provided by the American Chemical Society Petroleum Research Fund and the University of Delaware Research Foundation.

ABBREVIATIONS

bpy	2,2′-Bipyridine
CODH	Carbon monoxide dehydrogenase
CV	Cyclic voltammetry
cy	Cyclohexyl
cyclam	1,4,8,11-Tetraazacyclotetradecane
DMF	*N*,*N*-Dimethylformamide
dppa	Bis(diphenylphosphino)amine
dppe	1,2-Bis(diphenylphosphino)ethane
dppm	Diphenylphosphinomethane
EC	Electrochemical
ET	Electron transfer
etpC	Bis[(dicyclohexylphosphino)ethyl]phenylphosphine)
FE	Faradaic efficiency
Fe(TDHPP)	Iron 5,10,15,20-tetrakis(2′,6′-dihydroxylphenyl)porphyrin
FT	Fischer-Tropsch
FTIR	Fourier transform infrared
ITO	Indium tin oxide
LUMO	Lowest unoccupied molecular orbital
MeCN	Acetonitrile
NHE	Normal hydrogen electrode
ox	Oxidized
Pc	Phthalocyanine
PCET	Proton-coupled electron transfer
PV	Photovoltaic
red	Reduced
RHE	Reversible hydrogen electrode
SCE	Saturated calomel electrode
SOMO	Singly occupied molecular orbital
TCPP	Tetracarboxyphenyl porphyrin
TFE	2,2,2-Trifluoroethanol
TOF	Turnover frequency
TON	Turnover number
TPFC	5,10,15-Tris(pentafluorophenyl)corrole
TPP	5,10,15,20-Tetraphenylporphyrin
TPPS	Tetrasulfonatophenyl porphyrin
tpy	2,2′,2″-Terpyridine
triphos	1,1,1-Tris(diphenylphosphinomethyl)ethane
WGS	Water–gas shift
VT NMR	Variable temperature nuclear magnetic resonance

REFERENCES

1. *International Energy Outlook 2011*. US Energy Information Administration (September 2011).
2. *Annual Energy Review 2011*. US Energy Information Administration, (September 2012).
3. T. M. L. Wigley, R. Richels, and J. A. Edmonds, *Nature (London)*, *379*, 240 (1996)
4. M. I. Hoffert, K. Caldeira, A. K. Jain, E. F. Haites, L. D. Harvey, S. D. Potter, M. E. Schlesinger, T. M. L. Wigley, and D. J. Wuebbles, *Nature (London)*, *395*, 881 (1998).
5. A. A. Lacis, G. A. Schmidt, D. Rind, and R. A. Ruedy, *Science*, *330*, 356 (2011).
6. J. D. Shakun, Clark, P. U. He, F. Marcott, S. A.; Mix, A. C.; Liu, Z.; Otto-Biesner, B.; Schmittner, A.; and Bard, E. *Nature (London)*, *484*, 49 (2012).
7. S. Solomon, G.-K. Plattner, R. Knutti, and P. Friedlingstein, *Proc. Natl. Acad. Sci.*, *106*, 1704 (2009).
8. J. Hansen, M. Sato, and R. Ruedy, *Proc. Natl. Acad. Sci. U. S. A.*, *109*, E2415 (2012).
9. A. D. Barnosky, E. A. Hadly, J. Bascompte, E. L. Berlow, J. H. Brown, and M. Fortelius, et al. *Nature (London)*, *486*, 52 (2012).
10. R. A. Kerr, *Science*, *336*, 405 (2012).
11. N. W. Arnell, *Global Environ. Change*, *9*, S31 (1999).
12. N. W. Arnell, D. P. van Vuuren, and M. Isaac, *Global Environ. Change*, *21*, 592 (2011).
13. J. A. Patz, D. Campbell-Lendrum, and T. Holloway, J. A. Foley, *Nature (London)*, *438*, 310 (2005).
14. A. J. McMichael, R. E. Woodruff, and S. Hales, *Lancet*, *367*, 859 (2006).
15. A. Costello, M. Abbas, A. Allen, S. Ball, S. Bell, and R. Bellamy, et al. *Lancet*, *373*, 1693 (2009)
16. D. G. Nocera, *Chem. Soc. Rev.*, *38*, 13 (2009).
17. *Climate Change 2007: Mitigation of Climate Change.* IPCC Working Group III Fourth Assessment Report; Intergovernmental Panel on Climate Change: Geneva, 2007.
18. N. S. Lewis, *MRS Bull.*, *32*, 808 (2007).
19. T. R. Cook, D. K. Dogutan, S. Y. Reece, Y. Surendranath, T. S. Teets, and D. G. Nocera, *Chem. Rev.*, *110*, 6474 (2010).
20. D. G. Nocera, *Acc. Chem. Res.*, *45*, 767 (2012).
21. D. G. Nocera, *Inorg. Chem.*, *48*, 10001 (2009).
22. G. A. Olah, A. Goeppert, and G. K. S. Prakash, *J. Org. Chem.*, *74*, 487 (2009).
23. G. A. Olah, G. K. S. Prakash, and A. Goeppert, *J. Am. Chem. Soc.*, *133*, 12881 (2011).
24. G. C. v P. J. Denny, J. R. Jennings, M. S. Spencer, and K. C. Waugh, *Appl. Catal.*, *36*, 1 (1988).
25. Y. Yang, J. Evans, J. A. Rodriguez, M. G. White, and P. Liu, *Phys. Chem. Chem. Phys.*, *12*, 9909 (2010).

26. J. Choudhury, *ChemCatChem*, *4*, 609 (2012).
27. A. J. Morris, G. J. Meyer, and Fujita, E. *Acc. Chem. Res.*, *42*, 1983 (2009).
28. E. B. Cole and A. B. Cocarsly, in *Carbon Dioxide as Chemical Feedstock*, M. Aresta Ed., (pp. 291–316), Wiley–VHC, Weinheim, Germany, 2010.
29. J. M. Savéant, *Chem. Rev.*, *108*, 2348 (2008).
30. G. L. Gutsev, R. J. Bartlett, and R. N. Compton, *J. Chem. Phys.*, *108*, 6756 (1998).
31. R. N. Compton, P. R. Reinhardt, and R. N. Cooper, *J. Chem. Phys.*, *63*, 3821 (1975).
32. N. H. Damrauer, J. M. Hodgkiss, J. Rosenthal, and D. G. Nocera, *J. Phys. Chem. B*, *108*, 6315 (2004).
33. J. Rosenthal, J. M. Hodgkiss, E. R. Young, and D. G. Nocera, *J. Am. Chem. Soc.*, *128*, 10474 (2006).
34. J. M. Hodgkiss, N. H. Damrauer, S. Pressé, J. Rosenthal, and D. G. Nocera, *J. Phys. Chem. B*, *110*, 18853 (2006).
35. E. R. Young, J. Rosenthal, J. M. Hodgkiss, and D. G. Nocera, *J. Am. Chem. Soc.*, *131*, 7678 (2009).
36. E. J. Maginn, *J. Phys. Chem. Lett.*, *1*, 3478 (2010).
37. R. Eisenberg and D. G. Nocera, *Inorg. Chem.*, *44*, 6799 (2005).
38. M. Jacoby, *Chem. Eng. News*, *83* (22), 35 (2005).
39. Z. Jiang, T. Xiao, V. L. Kuznetsov, and P. P. Edwards, *Phi. Trans. Royal Soc. A*, *368*, 3343 (2010).
40. W.-.H. Cheng and H. H. Kung, Eds., 1994, *Methanol Production and Use*, Marcel Dekker, New York.
41. E. Fiedler, G. Grossmann, D. B. Kersebohm, G. Weiss, C. Witte, and Methanol, in *Ullmann's Encyclopedia of Industrial Chemistry*, 6th ed, Vol 21.
42. J. A. Rabo, Catalysis: past, present and future, *Proceedings of the 10th International Congress on Catalysis*, Budapest, Hungary, July 19–24, (1993).
43. G. A. Olah, A. Goeppert, and G. K. Surya Praksh, in *The Methanol Economy*, 2nd ed., Wiley–VHC, Weinheim, Germany, 2009.
44. N. S. Lewis and D. G. Nocera, *Proc. Natl. Acad. Sci.*, *103*, 15729 (2006).
45. P. C. Ford, *Acc. Chem. Res.*, *14*, 31 (1981).
46. G. C. Bond, in *Catalysis by Metals*, Academic Press London, 1962, pp. 353–370.
47. P. N. R. Vennestrøm, C. M. Osmundsen, C. H. Christensen, and E. Taarning, *Angew. Chem. Int. Ed.*, *50*, 10502 (2011).
48. C. K. Rofer-DePoorter, *Chem. Rev.*, *81*, 447 (2012).
49. A. J. Santhrum and C. P. Kubiak, *J. Phys. Chem. Lett.*, *2*, 2372 (2011).
50. R. Williams, R. S. Crandall, and A. Bloom, *Appl. Phys. Lett.*, *33*, 381 (1978).
51. Z. Jiang, T. Xiao, V. L. Kuznetsov, and P. P. Edwards, *Phils. Trans. R. Soc. A*, *368*, 3343 (2010).
52. Y. Hori, H. Wakebe, T. Tsukamoto, and O. Koga, *Electrochim. Acta*, *39*, 1833 (1994).
53. Y. Hori, K. Kikuchi, and S. Suzuki, *Chem. Lett.*, *14*, 1695 (1985).

54. H. Noda, S. Ikeda, Y. Oda, K. Imai, M. Maeda, and K. Ito, *Bull. Chem. Soc. Jpn.*, *63*, 2459 (1990).
55. Y. Hori, A. Murata, K. Kikuchi, and S. Suzuki, *Chem. Commun.* 728 (1987).
56. Y. Chen, C. W. Li, and M. W. Kanan, *J. Am. Chem. Soc.*, *134*, 19969 (2012).
57. J. Giner, *Electrochim. Acta*, *8*, 857 (1963).
58. M. W. Breiter, *Electrochim. Acta*, *12*, 1213 (1967).
59. L. Wenzhen, in *ACS Symposium Series*, Vol. 1056, American Chemical Society, Washington, D.C., 2010, pp. 55–76.
60. Y. Hori, in *Modern Aspects of Electrochemistry*, C. G. Vayenas, R. E. White, and M. E. Gamboa-Aldeco Eds., Vol. *42*, Springer, New York, 2002, pp. 89–189.
61. B. M. Trost, *Angew. Chem. Int. Ed. Engl.*, *34*, 259 (1995).
62. P. W. N. M. Van Leeuwen, *Homogeneous Catalysis: Understanding the Art*, Springer-Verlag, New York, (2005).
63. G. W. Parshall and S. D. Ittel, in *Homogeneous Catalysis: The Applications and Chemistry of Catalysis by Soluble Transition Metal Complexes*, 2nd ed, John Wiley & Sons, Inc., N.Y., 1992.
64. D. H. Gibson, *Chem. Rev.*, *96*, 2063 (1996).
65. X. Yin and J. R. Moss, *Coord. Chem. Rev.*, *181*, 27 (1999).
66. D. Walther, M. Ruben, and S. Rau, *Coord. Chem. Rev.*, *182*, 67 (1999).
67. M. Aresta, C. F. Novile, V. G. Albano, E. Forni, and M. Manassero, *Chem. Commun.* 636 (1975).
68. J. S. Anderson, V. M. Lluc, and G. L. Hillhouse, *Inorg. Chem.*, *49*, 10203 (2010).
69. S. Gambarotta, C. Floriani, A. Chiesi-Villa, and C. Guastini, *J. Am. Chem. Soc.*, *107*, 2985 (1985).
70. G. S. Bristow, P. B. Hitchcock, and M. F. Lappert, *Chem. Commun.* 1145 (1981).
71. R. Alvarez, E. Carmona, E. Gutierrez-Puebla, J. M. Marin, A. Monge, and M. L. Poveda, *Chem. Commun.* 1326 (1984).
72. C. H. Lee, D. S. Laitar, P. Mueller, and J. P. Sadighi, *J. Am. Chem. Soc.*, *129*, 13802 (2007).
73. J. C. Calabrese, T. Herskovitz, and J. B. Kinney, *J. Am. Chem. Soc.*, *105*, 5914 (1983).
74. J. A. Treadway and T. Meyer, *J. Inorg. Chem.*, *38*, 2267 (1999).
75. Castro-Rodriguez, I. Nakai, H. Zakharov, L. N. Rheingold, A. L. and Meyer, K. A. *Science*, *305*, 1757 (2004).
76. C.-C. Chang, M.-C. Liao, T.-H. Chang, S.-M. Peng, and G.-H. Lee, *Angew. Chem. Int. Ed.*, *44*, 7418 (2005).
77. G. Gao, F. Li, L. Xu, X. Liu, and Y. Yang, *J. Am. Chem. Soc.*, *130*, 10838 (2008).
78. J. Schneider, H. Jia, J. T. Muckerman, and E. Fujita, *Chem. Soc. Rev.*, *41*, 2036 (2012).
79. S. Meshitsuka, M. Ichikawa, and K. Tamaru, *Chem. Commun.*, *5*, 158 (1974).
80. S. Kapusta and N. Hackerman, *J. Electrochem. Soc.*, *131*, 1511 (1984).
81. C. M. Lieber and N. S. Lewis, *J. Am. Chem. Soc.*, *106*, 5033 (1984).

82. T. Yoshida, K. Kamato, M. Tsukamoto, T. Iida, D. Schlettwein, D. Wöhrle, and M. Kaneko, *J. Electroanal. Chem.*, 385, 209 (1995).
83. T. Abe, T. Yoshida, S. Tokita, F. Taguchi, H. Imaya, and M. Kaneko, *J. Electroanal. Chem.*, *412*, 125 (1996).
84. J. Grodkowski, T. Dhanasekaran, P. Neta, P. Hambright, B. S. Brunschwig, K. Shinozaki, and E. Fujita, *J. Phys. Chem. A*, *104*, 11332 (2000).
85. K. Takahashi, K. Hiratsuka, H. Sasaki, and S. Toshima, *Chem. Lett.* 305 (1979).
86. T. Atoguchi, A. Aramata, A. Kazusaka, and M. Enyo, *Chem. Commun.* 156 (1991).
87. S. A. Yao, R. E. Ruther, L. Zhang, R. A. Franking, R. J. Hamers, and J. F. Berry, *J. Am. Chem. Soc.*, *134*, 15632 (2012).
88. D. Behar, T. Dhanasekaran, P. Neta, C. M. Hosten, D. Ejeh, P. Hambright, and E. Fujita, *J. Phys. Chem. A*, *102*, 2870 (1998).
89. J. Grodkowski and P. Neta, *J. Phys. Chem. A*, *104*, 1848 (2000).
90. M. Hammouche, D. Lexa, M. Momenteau, and J. M. Savéant, *J. Am. Chem. Soc.*, *113*, 8455 (1991).
91. J. Grodkowski, D. Behar, P. Neta, and P. Hambright, *J. Phys. Chem. A*, *101*, 248 (1997).
92. I. Bhugun, D. Lexa, and J. M. Savéant, *J. Phys. Chem.*, *100*, 19981 (1996).
93. I. Bhugun, D. Lexa, and J. M. Savéant, *J. Am. Chem. Soc.*, *116*, 5015 (1994).
94. I. Bhugun, D. Lexa, and J. M. Savéant, *J. Am. Chem. Soc.*, *118*, 1769 (1996).
95. J. Grodkowski, P. Neta, E. Fujita, A. Mahammed, L. Simkhovich, and Z. Gross, *J. Phys. Chem. A*, *106*, 4772 (2002).
96. B. Fisher and R. Eisenberg, *J. Am. Chem. Soc.*, *102*, 7361 (1980).
97. M. Beley, J.-P. Collin, R. Ruppert, and J.-P. Sauvage, *Chem. Commun.* 1315 (1984).
98. G. B. Balazs and F. C. Anson, *J. Electroanal. Chem.*, *322*, 325 (1992).
99. G. B. Balazs and F. C. Anson, *J. Electroanal. Chem.*, *361*, 149–157 (1993).
100. M. Beley, J.-P. Collin, R. Ruppert, and J.-P. Sauvage, *J. Am. Chem. Soc.*, *108*, 7461 (1986).
101. J. D. Froehlich and C. P. Kubiak, *Inorg. Chem.*, *51*, 3932 (2012).
102. M. Shionoya, E. Kimura, and Y. Iitaka, *J. Am. Chem. Soc.*, *112*, 9237 (1990).
103. E. Fujita, J. Haff, R. Sanzenbacher, and H. Elias, *Inorg. Chem.*, *33*, 4627 (1994).
104. J. Schneider, H. Jia, K. Kobiro, D. E. Cabelli, J. T. Muckerman, and E. Fujita, *Energy Environ. Sci.*, *5*, 9502 (2012).
105. J. Hawecker, J.-M. Lehn, and R. Ziessel, *Chem. Commun.* 328 (1984).
106. J. Hawecker, J.-M. Lehn, and R. Ziessel, *Helv. Chim. Acta*, *69*, 1990 (1986).
107. B. P. Sullivan, C. M. Bolinger, D. Conrad, W. J. Vining, and T. J. Meyer, *Chem. Commun.* 1414 (1985).
108. F. P. A. Johnson, M. W. George, F. Hartl, and J. J. Turner, *Organometallics*, *15*, 3374 (1996).
109. E. E. Benson and C. P. Kubiak, *Chem. Commun.*, *48*, 7374 (2012).

110. J. M. Smieja, E. E. Benson, B. Kumar, K. A. Grice, C. S. Seu, A. J. M. Miller, J. M. Mayer, and C. P. Kubiak, *Proc. Natl. Acad. Sci. USA*, *109*, 15646 (2012).
111. B. Kumar, J. M. Smieja, and C. P. Kubiak, *J. Phys. Chem. C*, *114*, 14220 (2010).
112. B. Kumar, J. M. Smieja, A. F. Sasayama, and C. P. Kubiak, *Chem. Commun.*, *48*, 272 (2011).
113. A. A. S. Gietter, R. C. Pupillo, G. P. A. Yap, T. P. Beebe, J. Rosenthal, and D. A. Watson, *Chem. Sci.*, *4*, 437 (2013).
114. H. Ishida, K. Tanaka, and T. Tanaka, *Organometallics*, *6*, 181 (1987).
115. H. Nagao, T. Mizukawa, and K. Tanaka, *Inorg. Chem.*, *33*, 3415 (1994).
116. D. J. Darensbourg, P. W. Wiegreffe, and C. G. Riordan, *J. Am. Chem. Soc.*, *112*, 5759 (1990).
117. D. J. Darensbourg M. B. Fischer, R. E. Schmidt, and B. J. Baldwin, *J. Am. Chem. Soc.*, *103*, 1297 (1981).
118. D. J. Darensbourg and A. Rokicki, *J. Am. Chem. Soc.*, *104*, 349 (1982).
119. D. J. Darensbourg and A. Rokicki, *Organometallics*, *1*, 1685 (1982).
120. D. J. Darensbourg, H. P. Wiegreffe, and P. W. Wiegreffe, *J. Am. Chem. Soc.*, *112*, 9252 (1990).
121. D. J. Darensbourg, *Inorg. Chem.*, *49*, 10765 (2010).
122. K. J. Takeuchi, M. S. Thompson, D. W. Pipes, and T. J. Meyer, *Inorg. Chem.*, *23*, 1845 (1984).
123. J. J. Concepcion, J. W. Jurss, M. R. Norris, Z. Chen, J. L. Templeton, and T. J. Meyer, *Inorg. Chem.*, *49*, 1277 (2010).
124. Z. Chen, C. Chen, D. R. Weinberg, P. Kang, J. J. Concepcion, D. P. Harrison, M. S. Brookhart, and T. J. Meyer, *Chem. Commun.*, *47*, 12607 (2011).
125. Z. Chen, J. J. Concepcion, M. K. Brennaman, P. Kang, M. R. Norris, P. G. Hoertz, and T. J. Meyer, *Proc. Natl. Acad. Sci. USA*, *109*, 15606 (2012).
126. M. R. M. Bruce, E. Megehee, B. P. Sullivan, H. Thorp, T. R. O'Toole, A. Downard, and T. J. Meyer, *Organometallics*, *7*, 238 (1988).
127. M. R. M. Bruce, E. Megehee, B. P. Sullivan, H. H. Thorp, T. R. O'Toole, A. Downard, J. R. Pugh, and T. J. Meyer, *Inorg. Chem.*, *31*, 4864 (1992).
128. C. M. Bolinger, N. Story, B. P. Sullivan, and T. J. Meyer, *Inorg. Chem.*, *27*, 4582 (1988).
129. R. J. Haines, R. E. Wittrig, and C. P. Kubiak, *Inorg. Chem.*, *33*, 4723 (1994).
130. V. S. Thoi and C. J. Chang, *Chem. Commun.*, *47*, 6578 (2011).
131. J. Elmsley, *The elements*, Oxford University Press , Oxford, UK, 1989, pp. 218–219.
132. S. Slater and J. H. Wagenknecht, *J. Am. Chem. Soc.*, *106*, 5367 (1984).
133. D. L. DeLaet, R. Del Rosario, P. E. Fanwick, and C. P. Kubiak, *J. Am. Chem. Soc.*, *109*, 754 (1987).
134. E. Simón-Manso and C. P. Kubiak, *Organometallics*, *24*, 96 (2005).
135. K. S. Ratliff, R. E. Lentz, and C. P. Kubiak, *Organometallics*, *11*, 1986 (1992).
136. C. Amatore and J. M. Savéant, *J. Am. Chem. Soc.*, *103*, 5021 (1981).

137. D. A. Morgenstern, G. M. Ferrence, J. Washington, J. I. Henderson, L. Rosenhein, J. D. Heise, P. E. Fanwick, and C. P. Kubiak, *J. Am. Chem. Soc.*, *118*, 2198 (1996).
138. G. M. Ferrence, P. E. Fanwick, and C. P. Kubiak, *Chem. Commun.*, *13*, 1575 (1996).
139. R. E. Wittrig, G. M. Ferrence, J. Washington, and C. P. Kubiak, *Inorg. Chim. Acta*, *270*, 111 (1998).
140. M. R. Dubois and D. L. Dubois, *Acc. Chem. Res.*, *42*, 1974 (2009).
141. D. L. Dubois, *Comments Inorg. Chem.*, *19*, 307 (1997).
142. D. L. Dubois and A. Miedaner, *J. Am. Chem. Soc.*, *109*, 113 (1987).
143. D. L. Dubois, A. Miedaner, and R. C. Haltiwagner, *J. Am. Chem. Soc.*, *113*, 8753 (1991).
144. B. D. Steffey, A. Miedaner, M. L. Maciejewski-Farmer, P. R. Bernatis, A. M. Herring, V. S. Allured, V. Carperos, and D. L. Dubois, *Organometallics*, *13*, 4844 (1994).
145. A. M. Herring, B. D. Steffey, A. Miedaner, S. A. Wander, and D. L. Dubois, *Inorg. Chem.*, *34*, 1100 (1995).
146. A. Miedaner, B. C. Noll, and D. L. Dubois, *Organometallics*, *16*, 5779 (1997).
147. A. Miedaner, C. J. Curtis, R. M. Barkley, and D. L. Dubois, *Inorg. Chem.*, *33*, 5482 (1994).
148. B. D. Steffey, C. J. Curtis, and D. L. Dubois, *Organometallics*, *14*, 4937 (1995).
149. J. W. Raebiger, J. W. Turner, B. C. Noll, C. J. Curtis, A. Miedaner, B. Cox, and D. L. Dubois, *Organometallics*, *25*, 3345 (2006).
150. E. E. Benson, C. P. Kubiak, A. J. Sathrum, and J. M. Smieja, *Chem. Soc. Rev.*, *38*, 89 (2009).
151. D. L. DuBois, *Encyclopedia of Electrochemistry*, Vol. 7a, Wiley-VCH: Weinheim, Germany, 2006, pp. 202–225.
152. D. T. Whipple and P. J. A. Kenis, *J. Phys. Chem. Lett.*, *1*, 3451 (2010).
153. C. Costentin, S. Drouet, M. Robert, and J. M. Saveant, *J. Am. Chem. Soc.*, *134*, 11235 (2012).
154. S. W. Ragsdale and E. Pierce, *Biochim Biophys Acta. 1784*, 1873 (2008).
155. S. W. Ragsdale, *Crit. Rev. Biochem. Mol. Biol.*, *39*, 165 (2004).
156. J. H. Jeoung and H. Dobbek, *Science*, *318*, 1461 (2007).
157. W. Shin, S. H. Lee, J. W. Shin, S. P. Lee, and Y. Kim, *J. Am. Chem. Soc.*, *125*, 14688 (2003).
158. J. C. Fontecilla-Camps, P. Amara, C. Cavazza, Y. Nicolet, and A. Volbeda, *Nature (London)*, *460*, 814 (2009).
159. A. Earnshaw and N. Greenwood, in *Chemistry of the Elements*, 2nd ed., Butterworth-Heinemann, 1997.
160. Huang, D., and Holm, R. H. *J. Am. Chem. Soc.*, *132*, 4693 (2010).
161. D. Huang, O. V. Makhlynets, L. L. Tan, Lee, S. C.; E. V. Rybak-Akimova, and R. H. Holm, *Inorg. Chem.*, *50*, 10070 (2011).
162. X. Zhang, D. Huang, Y.-S. Chen, and R. H. Holm, *Inorg. Chem.*, *51*, 11017 (2012).

163. D. Huang, O. V. Makhlynets, L. L. Tan, S. C. Lee, E. V. Rybak-Akimova, and R. H. Holm, *Proc. Natl. Acad. Sci. U.S.A.*, *108*, 1222 (2011).
164. E. Fujita, C. Creutz, N. Sutin, and D. J. Szalda, *J. Am. Chem. Soc.*, *113*, 343 (1991).
165. E. Fujita, C. Creutz, N. Sutin, and B. S. Brunschwig, *Inorg. Chem.*, *32*, 2657 (1993).
166. C. Costentin, S. Drouet, M. Robert, and J. M. Savéant, *Science*, *338*, 90 (2012).

Monomeric Dinitrosyl Iron Complexes: Synthesis and Reactivity

CAMLY T. TRAN, KELSEY M. SKODJE, AND EUNSUK KIM

Department of Chemistry, Brown University, Providence, RI

CONTENTS

Progress in Inorganic Chemistry, Volume 59, First Edition. Edited by Kenneth D. Karlin.

I. INTRODUCTION

Nitric oxide (NO) is among the simplest of molecules and chemists have studied its structure and reaction chemistry for several years. Nitric oxide was originally viewed as only a poisonous and toxic gas (1). It has been more than two decades since the discovery of NO's important role in physiological functions [e.g., blood pressure regulation (2), neutrotransmission (3), and immune response (4)]. Most of the NO chemistry found in biology involves redox-active metal centers, where NO can bind directly to the metal center or nitrosylate of the amino acid side chain (5). Relatively recently, it has been shown that the reaction of NO with various non-heme iron-containing biomolecules (e.g., iron–sulfur proteins) results in the formation of a new type of mononuclear dinitrosyl iron complexes (DNICs) that are vastly different from the simple protein–NO adducts (6, 7). This chapter focuses on the synthesis, physical properties, and chemical reactivity of the synthetic analogues of such dinitrosyl iron species.

Mononuclear dinitrosyl complexes are the most abundant NO derived cellular adducts (8). They were initially discovered in the 1960s by their distinctive electron paramagnetic resonance (EPR) signal at $g = 2.03$ in yeast cells and animal tissues (9–11). Subsequently, synthetic DNICs have been prepared and structurally characterized in an effort to assign the $g = 2.03$ species as a stable four-coordinate DNIC, $[Fe(NO)_2(SR)_2]^-$, where SR^- may be cysteine residues of the protein backbone or small biomolecules (e.g., glutathione or free cysteine, Fig. 1) (12). Although cysteine residues have been proposed to be the major thiol components of DNICs *in vivo*, there are three different kinds of potential coordinated ligands: S-donor (6, 13–18), N-donor (14, 19), and O-donor (6, 13–18, 20) residues. They can be generated from two different cellular iron sources. The interactions between NO and either the chelatable iron pool (21, 22) or iron–sulfur proteins (6, 13–18, 23) lead to the rapid formation of DNICs. Although the exact function of DNICs has yet to be determined, recent evidence shows DNICs are capable of storing, transporting, and delivering NO or iron to generate biological activity (24–26). Dinitrosyl iron complex formation may be an indication of NO cytotoxicity (15, 27), or an active genetic switch in signal transductions (6, 18, 28–30).

The complexity of metal nitrosyl chemistry stems from the redox noninnocent character of NO. Depending on the way in which NO binds to a metal center, three

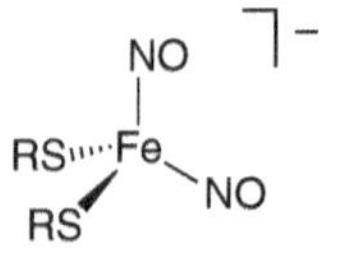

NO
RS Fe NO
RS

Figure 1. General form of a four-coordinate DNIC with thiolate ligands.

formal redox states of NO are possible with their own significance and reactivity; NO^+, $NO^•$, and NO^-/HNO. Unfortunately, assigning formal oxidation states to the metal and NO in metal nitrosyls is not a trivial task and the majority of the reported metal nitrosyl complexes use a simplified formalism, the Enemark–Feltham (E–F) notation (31). This formalism treats the metal nitrosyl as a single entity, represented as $\{M(NO)_x\}^n$, where n is the total number of electrons associated with the metal d and π^* (NO) orbitals. For example, $[Cl_2Fe(NO)_2]^-$ can be described as an $\{Fe(NO)_2\}^9$ species in the E–F formalism without further differentiation of the redox states, such as $[Cl_2Fe^I(•NO)_2]^-$, $[Cl_2Fe^{III}(NO^-)_2]^-$, or others. This notation will be used throughout this chapter.

Though monomeric DNICs have gained interest due to their biological relevance, there are various types of iron nitrosyls that have historical or chemical significance. Probably the three best-known non-heme [Fe–S] cluster nitrosyls are the anions discovered by French chemist Roussin in 1858 (12), almost a century before the first biological DNICs were observed. They are famously known as Roussin's black and red salts and the neutral red ester, Fig. 2. Roussin's black salt possesses the formula $[Fe_4(NO)_7S_3]^-$ with seven nitric oxide ligands positioned at a strongly coupled iron–sulfur core (32). In spite of its complex structure, the synthesis of RBS is straightforward. Reacting iron(II) sulfate with a mixture of sodium nitrite and ammonium sulfide will result in RBS, which is stable up to 120 °C (12, 33). Though the mechanism of this reaction is not well known, it is the most stable of all known [Fe–S] nitrosyls (12). Roussin's black salt is known to act as an antimicrobial agent that has been used for >100 years to inhibit the growth of gram-positive and gram-negative bacteria (33) where the toxicity of RBS stems from its release of NO (34, 35). Similarly, NO can be released from RRS, a bitetrahedron with a pair of fully coupled $Fe(NO)_2$ units through the bridging sulfides containing the formula, $[Fe_2(NO)_4S_2]^{2-}$, Fig. 2 (32, 34, 35). Both RBS and RRS photochemically release NO, which allows them to be potential NO delivering candidates to biological targets (34, 36–38). The red salt may be easily

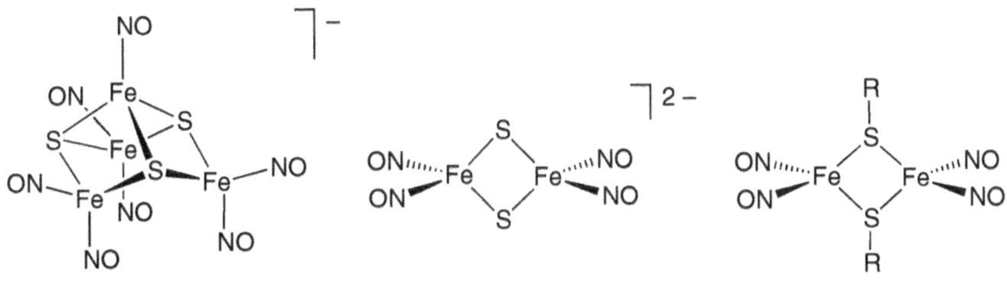

Figure 2. Structures of RBS, RRS, and RRE.

obtained from reacting RBS with sodium hydroxide. Red salt is found to be less toxic to bacteria than the black salt and is rather unstable (34), which makes RRS unlikely to have practical application in complex systems compared to RBS.

Roussin's red esters share a similar bitetrahedron structure like RRS, but the bridging sulfide ligands are replaced with bridging thiolate ligands. Typical RREs are diamagnetic and EPR silent due to the antiferromagnetic coupling between the two iron centers (39, 40). Roussin's red esters may be prepared by alkylation of RRS with an alkyl halide (41, 42) or other methods, such as reacting $[(CO)_2Fe(NO)_2]$ with thiolates (43) or reacting $Fe_2(CO)_9$ with nitrosothiol (RSNO) (44). These esters are important due to their potential as a NO delivery agent for medical applications, similar to monomeric dinitrosyl iron complexes (45). A thorough mechanistic investigation by Ford and co-workers (38, 46) indicated that esters like $Fe_2(\mu\text{-}SMe)_2(NO)_4$, $Fe_2(\mu\text{-}SCH_2Me)_2(NO)_4$, $Fe_2(\mu\text{-}CH_2Ph)_2(NO)_4$, and $Fe_2(\mu\text{-}SCH_2CH_2OH)_2(NO)_4$ are more stable than RRS while remaining active toward NO release and have the potential to be efficient photochemical NO generators. These early discoveries of RBS, RRS, and RREs have provided a strong foundation for iron–nitrosyl chemistry. Chemistry and biology of iron–sulfur cluster nitrosyls (e.g., RBS, RRS, and RRE) have been thoroughly reviewed by Butler and Megson (12) and will not be covered here in detail.

There are excellent recent reviews on the biological aspects of DNICs by Vanin (47), Richardson and co-workers (26, 48), and Lewandowska et al. (49). On the other hand, synthetic DNICs have not been reviewed since the early 2000s (12, 50), despite the considerable number of synthetic DNICs that have been reported in the past decade. Over 80 synthetic DNICs exist in the literature today (Table I). Many of these synthetic DNICs give us an understanding of their unique chemistry, which is crucial to explaining their biological roles, both well known and as of yet uncovered. Throughout this chapter, the DNICs discussed will be considered four-coordinate analogues (Fig. 1) unless specified otherwise. The aim of this chapter is to highlight the synthesis, structure, and reactivity of various known discrete DNICs currently in the literature, which can complement the existing reviews of their biological roles.

II. ANIONIC DINITROSYL IRON COMPLEXES

A. General Overview

Dinitrosyl iron complexes are most commonly found as anionic $\{Fe(NO)_2\}^9$ four-coordinate species (Compounds **1–42** in Table I, Fig. 3). Along with anionic $\{Fe(NO)_2\}^9$ DNICs, stable anionic $\{Fe(NO)_2\}^{10}$ DNICs also have been reported (Compounds **43–45** in Table I, Fig. 4), but these are extremely rare. Anionic $\{Fe(NO)_2\}^9$ DNICs may be ligated by anionic S-, N-, or O-donor ligands, where

TABLE I

Physical Properties of Some Dinitrosyl Iron Model Complexes

Complex	EPR g-value[a]	ν_{NO} (cm^{-1})	<Fe–N–O	Fe–N(O)	N–O	References
Anionic $\{Fe(NO)_2\}^9$ DNICs						
Halides						
$[I_2Fe(NO)_2]^-$ (**1**)	2.072	1719, 1778				51
$[Br_2Fe(NO)_2]^-$ (**2**)	2.049	1710, 1780				51
$[Cl_2Fe(NO)_2]^-$ (**3**)	2.033	1708, 1781				51
Thiolates						
$[(StBu)_2Fe(NO)_2]^-$ (**4**)	2.029	1690, 1739	165.9(2)	1.6824(18)	1.184(2)	52, 53
	2.039^b, 2.027^b, 2.013^b		169.4(3)	1.661(6)	1.183(4)	
$[(SEt)_2Fe(NO)_2]^-$ (**5**)	2.028	1680, 1722	170.4(6)	1.673(6)	1.188(7)	54
	2.042^b, 2.027^b, 2.014^b		173.0(6)	1.674(6)	1.194(6)	
$[(2,4,5\text{-}Cl_3C_6H_2S)_2Fe(NO)_2]^-$ (**6**)		1713, 1760				55
$[(SPh)_2Fe(NO)_2]^-$ (**7**)	2.03	1709, 1744		$1.662(4)^c$	$1.177(5)^c$	56, 57
$[(p\text{-}ClC_6H_4S)_2Fe(NO)_2]^-$ (**8**)		1702, 1748				55
$[(S\text{-}p\text{-}tolyl)_2Fe(NO)_2]^-$ (**9**)[d]	2.029^e	1691, 1732	166.0	1.7096	1.1958	58
			170.1	1.7210	1.1992	
$[(SC_6H_4\text{-}o\text{-}NHCOMe)_2Fe(NO)_2]^-$ (**10**)	2.038	1705, 1752	168.1(2)	1.681(2)	1.166(2)	55
			168.5(2)	1.682(2)	1.174(3)	
$[(SC_6H_4\text{-}o\text{-}NHCOPh)_2Fe(NO)_2]^-$ (**11**)	2.029	1705, 1752				40
$[(2\text{-}SC_7H_4NS)_2Fe(NO)_2]^-$ (**12**)		1716, 1766	169.4(5)	1.680(5)	1.173(6)	55
			170.4(5)	1.687(6)	1.174(6)	
$[(2\text{-}SC_4H_3S)_2Fe(NO)_2]^-$ (**13**)	2.027	1698, 1743	168.28	1.6832	1.178(2)	55
$[(SCH_2CONHMe)_2Fe(NO)_2]^-$ (**14**)	2.028	1700, 1751	168.2(2)	1.661(3)	1.182(3)	59
	2.039^b, 2.029^b, 2.014^b		168.7(3)	1.670(3)	1.192(3)	
$[(SCH_2CONMe_2)_2Fe(NO)_2]^-$ (**15**)	2.027	1686, 1730	164.84^c	1.6870^c	1.1715^c	59
	2.041^b, 2.032^b, 2.012^b					

(continued)

TABLE I
(*Continued*)

Complex	EPR *g*-value[a]	ν_{NO} (cm^{-1})	<Fe–N–O	Fe–N(O)	N–O	References
$[S_5Fe(NO)_2]^-$ (**16**)	2.03	1695, 1739	165.9(2)	1.670(3)	1.177(3)	60
	2.014[b], 2.0270[b], 2.0485[b]		172.8(2)	1.686(2)	1.178(3)	
$[(S(CH_2)_2S(CH_2)_2S)Fe(NO)_2]^-$ (**17**)		1694, 1739				61
$[(\{SCH_2\}_2\text{-}o\text{-}C_6H_4)Fe(NO)_2]^-$ (**18**)	2.029[e]	1688, 1734				58
$[(S(CH_2)_3S)Fe(NO)_2]^-$ (**19**)	2.031	1671, 1712	167.4(3)	1.674(2)	1.174(3)	62
	2.048[b], 2.033[b], 2.015[b]		172.8(2)	1.677(2)	1.181(3)	
$[Se_5Fe(NO)_2]^-$ (**20**)	2.064	1697, 1736	167.2	1.670(4)	1.166(4)	63
			170.3	1.680(4)	1.174(5)	
N-Donors						
$[(CN)_2Fe(NO)_2]^-$ (**21**)	2.03	1737, 1810				64
$[(SCN)_2Fe(NO)_2]^-$ (**22**)	2.03	1718, 1786	161.3(2)	1.700(2)	1.167(3)	65
			162.4(2)	1.701(2)	1.171(3)	
$[(OCN)_2Fe(NO)_2]^-$ (**23**)	2.03	1698, 1766	160.9(2)	1.697(2)	1.164(3)	65
			162.3(2)	1.700(2)	1.174(3)	
$[(N_3)_2Fe(NO)_2]^-$ (**24**)	2.027	1698, 1755	160.7[c]	1.694(2)[c]	1.172(3)[c]	39
	2.043[b], 2.023[b], 2.012[b]					
$[(N(Mes)(TMS))_2Fe(NO)_2]^-$ (**25**)[f,g]	2.015	1652, 1707	154.6	1.698(3)	1.180(3)	66
	2.022[b], 2.013[b], 2.005[b]		162.2	1.701(3)	1.191(3)	
$[(OC_7H_4SN)_2Fe(NO)_2]^-$ (**26**)	2.031	1723, 1791	162.8(2)	1.685(2) 1.688(2)	1.172(3)	67
			164.3(2)		1.178(3)	
$[(C_3H_3N_2)_2Fe(NO)_2]^-$ (**27**)	2.027	1712, 1774	166.2(3)	1.683(3)	1.168(4)	68
			166.9(3)	1.688(4)	1.175(4)	
$[(C_{12}H_8N)_2Fe(NO)_2]^-$ (**28**)	2.023	1691, 1748	160.0 (5)	1.663(6)	1.164(7)	69
	2.029[b], 2.018[b], 2.011[b]		165.5 (6)	1.683(6)	1.182(7)	
$[(NO)_2Fe(NO)_2]^-$ (**29**)	2.033	1708, 1776	160.9(8)	1.719(8)	1.154(9)	70
	2.052[b], 2.029[b], 2.014[b]		163.0(7)	1.696(7)	1.174(8)	
		1345	106.8(7)	2.008(11)	1.380(12)	
			113.2(7)	2.045(10)	1.280(12)	
$[(Imid\text{-}iPr)_2Fe(NO)_2]^-$ (**30**)[h]	2.038[e], 2.027[e], 2.008[e]	1699, 1765				71

O-Donors						
$[(\kappa^1\text{-}ONO_2)_2Fe(NO)_2]^-$ (**31**)	2.03[i]	1710, 1790	159.4(3)	1.701(3)	1.172(4)	72
			161.3(3)	1.704(3)	1.178(4)	
$[(ONO)_2Fe(NO)_2]^-$ (**32**)	2.03	1705, 1775	161.1(2)	1.691(2)	1.173(2)	57
	2.052[b], 2.029[b], 2.014[b]		164.3(2)	1.699(2)	1.173(2)	
$[(p\text{-}OPhF)_2]Fe(NO)_2]^-$ (**33**)	2.026	1685, 1751	158.4(2)	1.690(2)	1.173(2)	57
	2.041[b], 2.024[b], 2.013[b]		165.0	1.710(2)	1.185(2)	
$[(OPh)_2Fe(NO)_2]^-$ (**34**)	2.025	1674, 1739	156.4(4)	1.683(3)	1.177(4)	57
	2.041[b], 2.022[b], 2.013[b]		164.1(3)	1.696(3)	1.182(4)	
S–O-Donors						
$[(SPh)(ONO)Fe(NO)_2]^-$ (**35**)	2.028	1697, 1752	162.8(5)	1.641(6)	1.157(6)	57
			165.7(5)	1.686(6)	1.176(6)	
$[(SC_6H_4\text{-}o\text{-}COO)Fe(NO)_2]^-$ (**36**)	2.034	1688, 1742	157.1(2)	1.677(2)	1.166(3)	39
	2.054[b], 2.036[b], 2.012[b]		167.2(2)	1.691(2)	1.172(3)	
$[(OPh)(\text{-}SC_4H_3S)Fe(NO)_2]^-$ (**37**)	2.028	1681, 1740	164.3(6)	1.683(4)	1.144(6)	57
	2.038[b], 2.027[b], 2.013[b]		167.3(4)	1.686(4)	1.185(5)	
S–N-Donors						
$[(SPh)(C_3H_3N_2)Fe(NO)_2]^-$ (**38**)	2.027	1708, 1760	164.4(6)	1.668(7)	1.172(7)	68
			165.1(5)	1.701(6)	1.181(8)	
$[(SEt)(C_3H_3N_2)Fe(NO)_2]^-$ (**39**)	2.027	1693, 1743	166.7(3)	1.675(3)	1.171(3)	68
			168.0(3)	1.681(3)	1.178(3)	
$[(S^tBu)(C_3H_3N_2)Fe(NO)_2]^-$ (**40**)	2.027	1691, 1742	164.5(6)	1.642(7)	1.175(8)	68
			169.7(7)	1.688(6)	1.190(9)	
$[(SC_6H_4\text{-}o\text{-}NCOPh)Fe(NO)_2]^-$ (**41**)	2.034	1690, 1737	162.7(3)	1.679(3)	1.170(4)	39
	2.052[b], 2.033[b], 2.011[b]		168.4(3)	1.688(3)	1.176(4)	
O–N-Donors						
$[(OPh)(C_3H_3N_2)Fe(NO)_2]^-$ (**42**)	2.026	1691, 1755	160.7(3)	1.685(3)	1.173(3)	57
	2.046[b], 2.021[b], 2.013[b]		164.5(2)	1.692(3)	1.181(3)	

(continued)

TABLE I
(*Continued*)

Complex	EPR g-value[a]	ν_{NO} (cm^{-1})	<Fe–N–O	Fe–N(O)	N–O	References
Anionic $\{Fe(NO)_2\}^{10}$ DNICs						
P–N-Donors						
$[(PPh_3)(NO_2)Fe(NO)_2]^-$ (**43**)		1642, 1693	169.81(19)	1.6462	1.202(2)	73
			174.0(4)	1.667(4)	1.210(5)	
S–N-Donors						
$[(SC_6H_4\text{-}o\text{-}NMe_2)Fe(NO)_2]^-$ (**44**)		1610, 1660	168.8(2)	1.645(2)	1.204(4)	74
			172.8(2)	1.649(2)	1.204(3)	
N–N-Donors						
$[(Ar\text{-}nacnac)Fe(NO)_2]^-$ (**45**)[j]		1567, 1627	163.2(5)	1.668(5)	1.191(6)	75, 76
			165.1(5)	1.649(4)	1.218(6)	
Cationic $\{Fe(NO)_2\}^9$ DNICs						
P-Donors						
$[(Ph_3P)_2Fe(NO)_2]^+$ (**46**)	2.037[k]	1766, 1814	166.2(4)[c]	1.661(4)[c]	1.160(6)[c]	77
P–O-Donors						
$[(PPh_3)(OPPh_3)Fe(NO)_2]^+$ (**47**)	2.033[i]	1746, 1809	161.2(13)	1.668	1.171	77
O-Donors						
$[(OPPh_3)_2Fe(NO)_2]^+$ (**48**)	2.037[k]	1734, 1813				77
N-Donors						
$[(sparteine)Fe(NO)_2]^+$ (**49**)	2.032	1746, 1814				62
$[(TMEDA)Fe(NO)_2]^+$ (**50**)[l]		1698, 1775				62, 78
$[(1\text{-}MeIm)_2Fe(NO)_2]^+$ (**51**)[m]	2.015[n]					79
$[(6\text{-}Me_3\text{-}TPA)Fe(NO)_2]^+$ (**52**)[o,p]	2.02[q]	1726, 1801	159.7(4)	1.690(3)	1.165(4)	78, 80
			162.4(4)	1.699(3)	1.168(4)	
$[(iPrPDI)Fe(NO)_2]^+$ (**53**)[p,r]	2.015	1721, 1794	156.0(2)	1.700(2)	1.168(2)	81
			165.6(2)	1.707(2)	1.169(2)	
$[(TPA)Fe(NO)_2]^+$ (**54**)[p,s]	2.018	1619, 1720	138.3(2)	1.728(2)	1.164(3)	81
			153.7(3)	1.781(2)	1.187(3)	

NHC-Donors						
$[(NHC\text{-}Me)_2Fe(NO)_2]^+$ **(55)**[t]	2.057[e]	1733, 1789				71
$[(NHC\text{-}i\text{Pr})_2Fe(NO)_2]^+$ **(56)**[u]	2.028[e]	1723, 1791				71
Neutral $\{Fe(NO)_2\}^9$ DNICs						
Thiolates						
$[(C_9H_{21}N_2S_2)Fe(NO)_2]$ (57)		1695, 1740	169.8	1.678 (3)	1.156 (4)	82
				1.681 (3)	1.165 (4)	
$[(H^+bme\text{-}daco)Fe(NO)_2]$ **(58)**[v]	2.03[e]	1695, 1739	169.8(6)	1.618(6)	1.175(7)	61
			173.1(6)	1.657(6)	1.196(7)	
S–NHC-Donors						
$[(IMes)(SPh)Fe(NO)_2]$ **(59)**[w]	2.032	1715, 1763	167.8(2)	1.667(3)	1.182(3)	83
	2.049[b], 2.029[b], 2.013[b]		169.6(2)	1.677(3)	1.186(3)	
$[(NHC\text{-}i\text{Pr})(SPh)Fe(NO)_2]$ **(60)**[u]	2.026[e]	1712, 1757	166.2(3)	1.667(3)	1.169(3)	71
			167.4(3)	1.669(3)	1.175(3)	
S–N-Donors						
$[(SC_6H_4\text{-}o\text{-}NHCOPh)(Im)Fe(NO)_2]$ **(61)**[x]	2.03	1722, 1786	166.0(5)	1.678(6)	1.162(6)	39, 40
			166.0(5)	1.688(5)	1.171(6)	
$[(PyImiS)Fe(NO)_2]$ **(62)**[p,y]	2.017	1662, 1732	153.8(2)	1.685(2)	1.171(3)	81
			164.7(2)	1.710(2)	1.187(3)	
O–N-Donors						
$[(1\text{-}MeIm)(ONO)Fe(NO)_2]$ **(63)**[m]	2.031[z]	1729, 1800	162.7(3)	1.681(3)	1.171(4)	84
			163.6(3)	1.692(3)	1.173(4)	
$[(1\text{-}MeIm)_2(\eta^2\text{-}ONO)Fe(NO)_2]$ **(64)**[m,p]	2.013[z]	1749, 1820	149.3(3)	1.718(3)	1.155(4)	84
			151.0(3)	1.724(3)	1.172(4)	
O–Cl-Donors						
$[(HMPA)(Cl)Fe(NO)_2]$ **(65)**[aa]		1699, 1761				85
P–Cl-Donors						
$[(pph_2py)(Cl)Fe(NO)_2]$ **(66)**[bb]		1726, 1792				86
N–I-Donors						
$[(TMEDA)Fe(NO)_2I]$ **(67)**[l,p]	2.031	1717, 1775	162.6(5)	1.673(4)	1.169(5)	87

(continued)

TABLE I
(Continued)

Complex	EPR *g*-value[a]	ν_{NO} (cm^{-1})	<Fe–N–O	Fe–N(O)	N–O	References
N-Donors						
[(N(Mes)(TMS))$_2$Fe(NO)$_2$] (**68**)[f,g]		1733, 1786	161.9(3)	1.662(3)	1.163(3)	66
				1.893(2)		
[(Ar-nacnac)Fe(NO)$_2$] (**69**)[j]		1709, 1761	162.7(2)	1.6882	1.174(2)	76
			170.1(2)	1.6964	1.177(2)	
Neutral {Fe(NO)$_2$}^{10} DNICs						
N-Donors						
[(Di-2-pyridyl ketone)Fe(NO)$_2$] (**70**)		1645, 1704				88
[(TMEDA)Fe(NO)$_2$] (**71**)[l]		1644, 1698	166.9(3)	1.637(3)	1.188(4)	62
			169.9(3)	1.639(3)	1.197(4)	
[(dmp)Fe(NO)$_2$] (**72**)[cc]		1628, 1692				89
[(sparteine)Fe(NO)$_2$] (**73**)		1622, 1679	167.1(3)	1.6430	1.206(6)	62
			176.0(3)	1.6501	1.220(6)	
[(terpy)Fe(NO)$_2$] (**74**)[dd]		1621, 1688	172.0(3)	1.643(3)	1.189(3)	90
			172.1(3)	1.652(3)	1.193(3)	
[(bipy)Fe(NO)$_2$] (**75**)[ee]		1619, 1684	166.7(4)	1.647(4)	1.183(5)	90
			169.0(4)	1.652(4)	1.188(5)	
[(1-MeIm)$_2$Fe(NO)$_2$] (**76**)[m]		1616, 1673	167.5(3)	1.648(3)	1.188(4)	79
			170.1(3)	1.650(3)	1.189(3)	
[(phen)Fe(NO)$_2$] (**77**)[ff]		1614, 1686				90
N–C-Donors						
[(Imid-*i*Pr)(CO)Fe(NO)$_2$] (**78**)[h]		1698, 1744				71
N–P-Donors						
[(1-MeIm)(PPh$_3$)Fe(NO)$_2$] (**79**)[m]		1651, 1698	167.5(3)	1.637(3)	1.195(4)	84
			174.4(3)	1.653(3)	1.202(4)	
NHC-Donors						
[(NHC-Me)$_2$Fe(NO)$_2$] (**80**)[t]		1624, 1667	174.0(5)	1.657(5)	1.199(6)	71

		174.0(5)	1.661(5)	1.205(7)	
[(NHC-*i*Pr)$_2$Fe(NO)$_2$] (**81**)[u]	1619, 1664	173.8(2)	1.642(3)	1.204(3)	71
		173.8(2)	1.642(3)	1.204(3)	
NHC–C-Donors					
[(IMes)(CO)Fe(NO)$_2$] (**82**)[w]	1702, 1744	174.4(2)	1.657(2)	1.164(3)	83
		175.9	1.694(3)	1.184(3)	
[(NHC-Me)(CO)Fe(NO)$_2$] (**83**)[t]	1697, 1740	173.7(5)	1.718(6)	1.175(7)	71
		172.8(5)	1.736(6)	1.176(7)	
[(NHC-*i*Pr)(CO)Fe(NO)$_2$] (**84**)[u]	1696, 1738	174.4	1.6525	1.172(3)	71
		177.3(2)	1.698(2)	1.189(2)	
P–C-Donors					
[(CO)(PPh$_3$)Fe(NO)$_2$] (**85**)	1719, 1766	177.5(7)	1.704(6)	1.150(8)	62, 79, 91, 92
		178.9(6)	1.732(8)	1.147	
[(PPh$_3$)(TCNE)Fe(NO)$_2$] (**86**)[p,gg]	1790, 1836	165.8(5)	1.657(6)		92
		178.0(5)	1.665(6)		
P–P-Donors					
[(PPh$_3$)$_2$Fe(NO)$_2$] (**87**)	1678, 1724	178.2(7)[c]	1.650(7)[c]	1.190(0)[c]	62, 79, 91
[(PPh$_2$(C$_6$H$_5$-*m*-SO$_3^-$Na$^+$))$_2$Fe(NO)$_2$] (**88**)	1672, 1715				93
C-Donors					
[(CO)$_2$Fe(NO)$_2$] (**89**)	1767, 1810		1.77(2)	1.12(3)	62, 79, 91, 94

[a] The *g*-value at 298 K unless noted otherwise.
[b] The *g*-value at 77 K.
[c] Average values.
[d] 4-Methylbenzenethiolate = S-*p*-tolyl.
[e] The *g*-value at 10 K.
[f] Mesityl = Mes.
[g] Trimethylsilane = TMS.
[h] 2-Isopropylimidazole = Imid-*i*Pr.
[i] The *g*-value at 220 K.
[j] Anion of [(2,6-diisopropylphenyl)NC(Me)]$_2$CH = Ar-nacnac.

TABLE I
(*Continued*)

[k] The *g*-value at 260 K.
[l] *N,N,N′,N′*-Tertramethylethylenediamine = TMEDA.
[m] 1-Methylimidazole = 1-MeIm.
[n] The *g*-value at 240 K.
[o] Tris[(6-methyl-2-pyridyl)methyl)]amine = 6-Me_3-TPA.
[p] Not a four-coordinate compound.
[q] The *g*-value at 5 K.
[r] 2,6-Bis[1-(2,6-diisopropylphenylimino)ethyl]pyridine = *i*PrPDI.
[s] Tris(2-pyridylmethyl)amine = TPA.
[t] 1,3-Dimethylimidazol-2-ylidene = NHC-Me.
[u] 1,3,-Diisopropylimidazol-2-ylidene = NHC-*i*Pr.
[v] Dianion of *N,N′*-bis(2-mercaptoethyl)-1,5-diazacyclooctane = bme-daco.
[w] 1,3-Bis(2,4,6-trimethylphenyl)imidazol-2-ylidene = IMes.
[x] Imidazole = Im.
[y] 2-((1-(pyridin-2-yl)ethylidene)amino)benzenethiolate = PyImiS.
[z] The *g*-value at 180 K.
[aa] Hexamethylphosphoric triamide = HMPA.
[bb] Pyridine = py.
[cc] 2,9-Dimethyl-1,10-phenanthroline = dmp.
[dd] 2,2′,2″-Terpyridine = terpy.
[ee] 2,2′-Bipyridine = bpy.
[ff] 1,10-Phenanthroline = phen.
[gg] Tetracyanoethylene = TCNE.

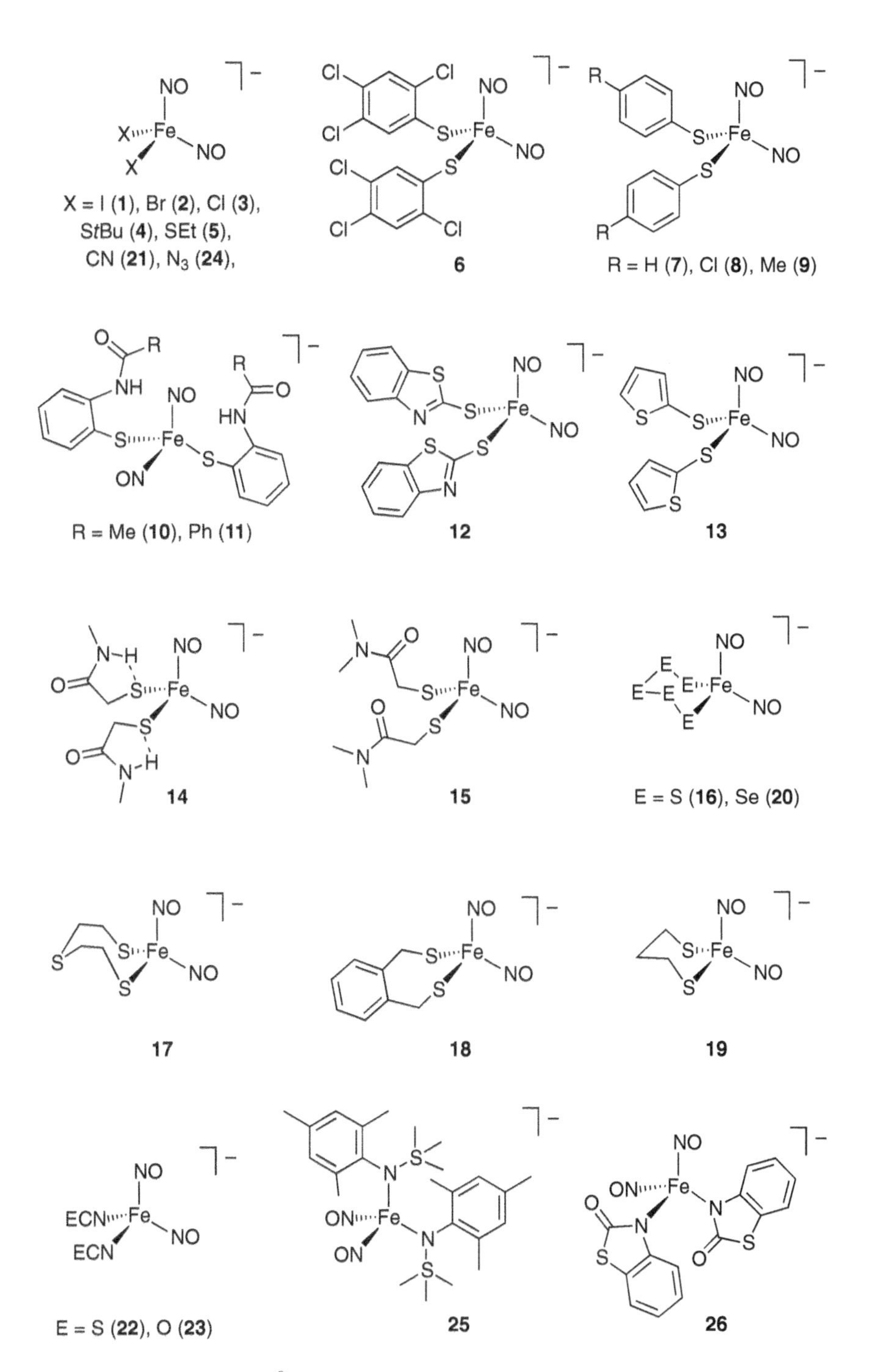

Figure 3. Anionic $\{Fe(NO)_2\}^9$ DNICs. Note that the Fe–N–O angles for all the listed compounds are close to linear except **29** (Table I).

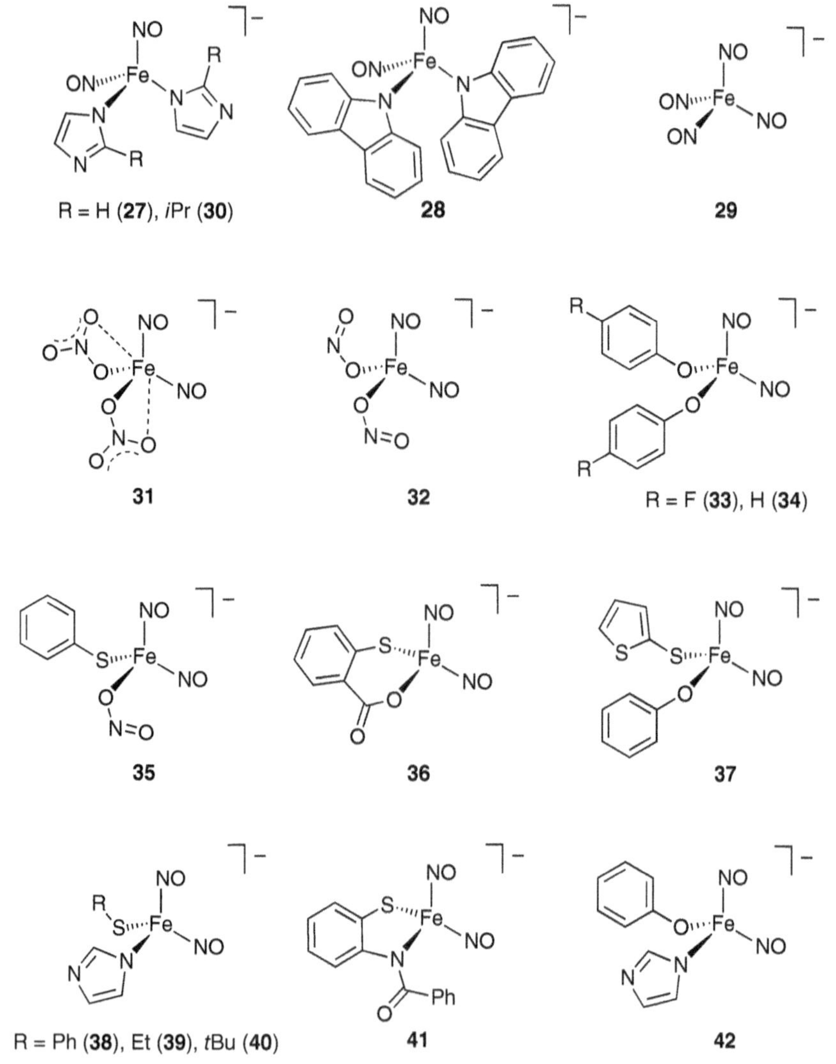

R = H (**27**), *i*Pr (**30**) **28** **29**

31 **32** R = F (**33**), H (**34**)

35 **36** **37**

R = Ph (**38**), Et (**39**), *t*Bu (**40**) **41** **42**

Figure 3. (*Continued*)

the thiolate bound DNICs are the most abundant form. Existence of the anionic $\{Fe(NO_2)\}^9$ DNICs are generally identified by a characteristic isotropic EPR signal of $g = 2.03$ at room temperature and/or a rhombic signal at lower temperatures (Table I). Another helpful tool used to characterize the formation of anionic $\{Fe(NO)_2\}^9$ DNICs is infrared (IR) spectroscopy, where two ν_{NO} stretching

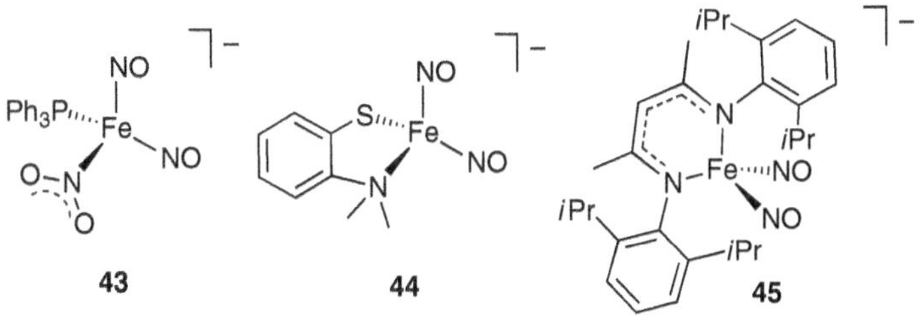

Figure 4. Anionic $\{Fe(NO)_2\}^{10}$ DNICs.

frequencies typically appear between 1670 and 1790 cm^{-1}, with an average separation of 55 cm^{-1}. Liaw and co-workers (95) observed that the separation between the two ν_{NO} stretching frequencies depends on the different types of ligands coordinated to the iron center. Two ν_{NO} stretching frequencies of an anionic $\{Fe(NO)_2\}^9$ DNIC usually have a difference $[\Delta(\nu_{NO})]$ of ~45 cm^{-1} for [S–S] ligation (Table I, compounds **4–19**). The $\Delta(\nu_{NO})$ for [N–N] ligation is ~62 cm^{-1} (Table I, compounds **24, 26–28**) and that for [N–S] ligation is ~50 cm^{-1} (Table I, compounds **38–41**). A significant number of four-coordinate $\{Fe(NO)_2\}^9$ DNICs have been crystallographically characterized. In all cases, iron is in a distorted tetrahedral geometry with two slightly bent NO ligands. The average Fe–N–O bond angle is ~165° where each Fe–N–O bond angle in a DNIC varies slightly (0.1–10°), with the N–O bond distances ranging from 1.144–1.199 Å and the Fe–N(O) bond distances ranging from 1.642–1.721 Å (Table I).

The electronic structures of a group of $\{Fe(NO)_2\}^9$ DNICs were initially investigated by Bryar and Eaton (96), in which the complexes, such as $[I_2Fe(NO)_2]^-$ (**1**) and $[(SPh)_2Fe(NO)_2]^-$ (**7**), were described as 17-electron complexes with d^9 configuration, $\{Fe^{-I}(NO^+)_2\}^9$, based on the EPR *g*-value analysis, X-ray structures, and IR stretching frequencies. More than a decade later, an alternative assignment had been proposed by Liaw and co-workers (60) for a thiolate bound $\{Fe(NO)_2\}^9$ DNIC, $[S_5Fe(NO)_2]^-$ (**16**), where Fe^I ($S_{Fe} = 3/2$) is antiferromagentically coupled to two NO ($S_{NO} = 1/2$) ligands (97). This assignment was based on the oxygen K-edge as well as the iron K-/L-edge X-ray absorption spectroscopy (XAS) of **16** in combinations with DFT calculations, magnetic susceptibility measurement, and EPR spectroscopy. The $\{Fe^I(^{\bullet}NO)_2\}^9$ description for anionic $\{Fe(NO)_2\}^9$ DNICs including $[(SPh)_2Fe(NO)_2]^-$ (**7**), $[S_5Fe(NO)_2]^-$ (**16**), and $[Se_5Fe(NO)_2]^-$ (**20**) was further supported by Dai and Ke (97) from the normal coordinate analysis linking Raman spectroscopy and DFT calculations. More recently, Hsu and co-workers (57) extended their XAS studies and calculations to include a variety of DNICs with S-, N-, and O-donor ligands. They proposed a resonance hybrid of $\{Fe^I(\bullet NO)_2\}^9$, $\{Fe^{II}(^{\bullet}NO)(NO^-)\}^9$, and $\{Fe^{III}(NO^-)_2\}^9$ as a general electronic structure of anionic DNICs, where the population of the

$\{Fe^{III}(NO^-)_2\}^9$ resonance form in O-bound DNICs was greater than that in S-bound DNICs (e.g., $[(OPh)_2Fe(NO)_2]^-$ (**34**) vs $[(SPh)_2Fe(NO)_2]^-$ (**7**)) (57).

The electronic structures of an anionic $\{Fe(NO)_2\}^{10}$ DNIC, [(Ar−nacnac)Fe-$(NO)_2]^-$ (**45**) and its oxidized form, [(Ar−nacnac)Fe$(NO)_2$] (**69**) have been reported by Ye and Neese (98). Complexes **45** and **69** are structurally analogous DNIC redox partners that have been synthesized and fully characterized by Lippard and co-workers (75, 76). Interestingly, complexes **45** and **69** have very similar solid-state structures (Table I) and the Mössbauer parameters ($\delta = 0.22\,mm\ s^{-1}$ and $\Delta E_Q = 1.31\,mm\ s^{-1}$ for **45**; $\delta = 0.19\,mm\ s^{-1}$ and $\Delta E_Q = 0.79\,mm\ s^{-1}$ for **69**) (76). A thorough DFT study (TPSSh Kohn–Sham solutions) by Ye and Neese (98) suggest that **45** is best described as a high-spin ferrous center ($S_{Fe} = 2$) antiferromagnetically coupled to two triplet NO^- ligands ($S_{(NO)_2} = 2$), while **69** is best rationalized by a resonance hybrid consisting of a ferric center ($S_{Fe} = 5/2$) antiferromagnetically coupled to two NO^- ligands ($S_{(NO)_2} = 2$), and a ferrous ion ($S_{Fe} = 2$) coupled to an overall quartet $(NO)_2^-$ ligand ($S_{(NO)_2} = 3/2$) in an antiferromagnetic fashion.

Though we do not describe here dinuclear DNICs (e.g., RREs in Fig. 2) in great detail, advanced spectroscopic techniques have been used to differentiate between mononuclear DNICs and dinuclear DNICs. Nuclear resonance vibration spectroscopy (NRVS) provides diagnostic spectral signatures between 500 and 700 cm^{-1} originated from the symmetric and asymmetric stretching modes for the iron dinitrosyl species (99–100). Sulfur K-edge XAS is another tool to distinguish an S-bound mononuclear DNIC and dinuclear analogues based on the unique pre-edge pattern–energy, where the average pre-edge absorption peak for RREs appears at higher energy than that of DNICs (101).

B. Preparation

There are a variety of synthetic ways to generate anionic $\{Fe(NO)_2\}^9$ DNICs. They can be obtained from the reaction of $[Fe^{II}(SR)_4]^{2-}$ and 2 equiv of NO, which accompanies the loss of thiolate and disulfide (Reaction 1). This reaction was utilized in the synthesis of $[(S\mathit{t}Bu)_2Fe(NO)_2]^-$ (**4**) (52) and $[(SPh)_2Fe(NO)_2]^-$ (**7**) (54). Interestingly, with the strict 1:1 stoichiometry of $[Fe^{II}(SR)_4]^{2-}$ and NO, an $\{Fe(NO)\}^7$ mononitrosyl iron complex (MNIC), $[(SR)_3Fe(NO)]^-$, forms prior to the generation of a DNIC, implying that MNICs are likely the intermediate species in the generation of DNICs. Alternatively, $\{Fe(NO)_2\}^9$ DNICs can also be synthesized from the ferric precursor, $[Fe^{III}(SR)_4]^-$ (Reaction 2), as shown in the synthesis of $[(SEt)_2Fe(NO)_2]^-$ (**5**) and $[(SPh)_2Fe(NO)_2]^-$ (**7**) (54). Similar to Reaction 1, the $\{Fe(NO)\}^7$ MNIC is generated as a reaction intermediate during reaction 2 (54). The difference between Reactions 1 and 2 lies in the first step by which the MNICs are generated. Binding of NO to the electron-deficient ferric $[Fe^{III}(SR)_4]^-$ complexes triggers reductive elimination of RSSR to yield an $\{Fe(NO)\}^7$ MNIC, while that to the ferrous $[Fe^{II}(SR)_4]^{2-}$ species leads to the substitution of an anionic thiolate ligand by NO (54). In the following step, the additional equivalent of NO binds to MNICs to yield DNICs and RSSR.

$$[Fe^{II}(SR)_4]^{2-} \xrightarrow[RS^-]{NO} [(SR)_3Fe(NO)]^- \xrightarrow[1/2\ RSSR]{NO} \underset{(\mathbf{5})}{[(SR)_2Fe(NO)_2]^-} \tag{1}$$

$$[Fe^{III}(SR)_4]^{-} \xrightarrow[1/2\ RSSR]{NO} [(SR)_3Fe(NO)]^- \xrightarrow[1/2\ RSSR]{NO} \underset{(\mathbf{7})}{[(SR)_2Fe(NO)_2]^-} \tag{2}$$

A simple ligand-exchange reaction of $\{Fe(NO_2)\}^9$ DNICs with the appropriate anionic external ligands is a flexible method to synthesize various $\{Fe(NO_2)\}^9$ DNICs, in which the weaker-binder (X) and the stronger-binder (Y) ligands can be halide, nitrite (NO_2^-), phenolate, imidazolate, thiolate, and others (Reactions 3 and 4). The popular precursors for such substitution are nitrosyl iron complex halogenate ligands, such as $[Cl_2Fe(NO)_2]^-$ (**3**) or its dimeric analogue, $[Fe_2(\mu\text{-}Cl)_2(NO)_4]$ (51, 102). Additionally, Liaw and co-workers (55, 57) showed that a $\{Fe(NO)_2\}^9$ DNIC with nitrite, $[(ONO)_2Fe(NO)_2]^-$ (**32**), or the one with 2-benzothiozoyl thiolate, $[(2\text{-}SC_7H_4NS)_2Fe(NO)_2]^-$ (**12**) (Fig. 3), are useful synthons from which a number of $\{Fe(NO)_2\}^9$ DNICs can be synthesized by ligand substitution. The binding affinity of the anionic ligand toward the $\{Fe(NO)_2\}^9$ motif follows the series $[SPh]^- \sim [SC_4H_3S]^- > [Im]^- > [OPh]^- > [NO_2]^-$ (57). Dinitrosyl iron complexes with mixed ligands can be prepared through the route of Reaction 3 upon addition of 1 equiv of an anionic ligand with higher affinity. Numerous $\{Fe(NO)_2\}^9$ DNICs including **10, 12, 13, 18, 19, 24, 27, 32–35, 37,** and **42** (Table I, Fig. 3) have been prepared from Reactions 3 and 4.

$$[(X)_2Fe(NO)_2]^- + Y^- \rightarrow [(X)(Y)Fe(NO)_2]^- + X^- \tag{3}$$

$$[(X)_2Fe(NO)_2]^- + 2Y^- \rightarrow [(Y)_2Fe(NO)_2]^- + 2X^- \tag{4}$$

The ligand substitution can also be achieved by the addition of neutral thiol (103), in which the coordinated anionic weaker binder (X) undergoes protonation followed by substitution (Reaction 5). *Note* that X does not represent halide. For example, $[(SPh)(C_3H_3N_2)Fe(NO)_2]^-$ (**38**) can be synthesized from $[(C_3H_3N_2)_2Fe(NO)_2]^-$ (**27**), during which a weaker binder, imidazolate ($=C_3H_3N_2$), becomes protonated and replaced by thiophenol (HSPh). Applications of this reaction can be found in the synthesis of **38–40** (68, 69).

$$[(X)_2Fe(NO)_2]^- + RSH \rightarrow [(X)(RS)Fe(NO)_2]^- + XH \tag{5}$$

Another general reagent that is commonly used to achieve ligand substitution is disulfide, which is particularly useful to substitute a strong electron-donating ligand bound to DNICs, such as $[(SPh)_2Fe(NO)_2]^-$ (**7**). With an addition of appropriate disulfides, **10**, **12**, and **13** (Table I, Fig. 3) can be synthesized from $[(SPh)_2Fe(NO)_2]^-$ (**7**) following Reaction 6, which is accompanied by the formation of PhSSPh (55). Similarly, an addition of NO to a thiolate bound DNIC, $[(S(CH_2)_3S)Fe(NO)_2]^-$ (**19**), triggers oxidation of thiolate to disulfide and generates a novel homoleptic iron complex, $[(NO^-)_2Fe(NO)_2]^-$ (**29**) (Fig. 3), containing two nitroxyls attached to a $\{Fe(NO)_2\}^9$ motif, Reaction 7 (70).

$$\underset{(7)}{[(SPh)_2Fe(NO)_2]^-} + RS'S'R \rightarrow [(RS')_2Fe(NO)_2]^- + PhSSPh \quad (6)$$

$$\underset{(\mathbf{19})}{[(S(CH_2)_3S)Fe(NO)_2]^-} + 2\,NO \longrightarrow \underset{(\mathbf{29})}{[(NO^-)_2Fe(NO)_2]^-} + \overline{S(CH_2)_3S} \quad (7)$$

In some cases, anionic DNICs may be generated from [Fe–S] clusters or RREs (Fig. 2). Degradation of [2Fe–2S] clusters via NO produces anionic DNICs with elemental sulfur as a byproduct [Reaction 8, where R = *t*Bu, Et, and $(S_5)^{2-}$] as observed in complexes **4, 5** and **16** (52, 54, 60), and an addition of thiolate to RREs (Reaction 9, where R = *t*Bu and Et) yields anionic DNICs (52, 95).

$$[Fe_2S_2(SR)_2]^{2-} + NO \rightarrow 2\,[(RS)_2Fe(NO)_2]^- + 2\,S^0 \quad (8)$$

$$[Fe_2(\mu\text{-}SR)_2(NO)_4] + 2\,[RS]^- \rightarrow 2\,[(SR)_2Fe(NO)_2]^- \quad (9)$$

C. Reactivity

The majority of known reactivity of DNICs deals with ligand substitution of DNICs as part of their synthesis as described above. In this section, chemical reactivity of DNICs outside of synthesis will be reviewed.

In the late 1980s through 1990s, Postel and co-workers (104) made significant efforts to understand reactivity of a series of anionic and neutral DNICs (see also Section IV). One such effort was to study dioxygen reactivity of the $\{Fe(NO)_2\}^9$ DNIC, $[Cl_2(NO)_2Fe]^-$ (**3**), which was cocomplexed with $[Fe(NN)_3]^{2+}$, where NN = bpy, 4,4′-dimethyl-2,2′-bipyridine, or phen. The reaction between $[Fe(NN)_3][Fe(NO)_2Cl_2]_2$ and O_2 resulted in iron nitrates, $[Fe(NO_3)(NN)Cl_2]$ and $[Fe(NO_3)_2(NN)Cl]$. Interestingly, these iron nitrate complexes, as well as $[Cl_2Fe(NO)_2]^-$, showed catalytic activity for the autoxidation of cyclohexene in the presence of O_2 via a radical mechanism. In addition, the nitrato complexes rapidly reacted with PPh_3 and carried out oxygen transfer from nitrate to PPh_3 producing $O{=}PPh_3$ in the absence of O_2 (104).

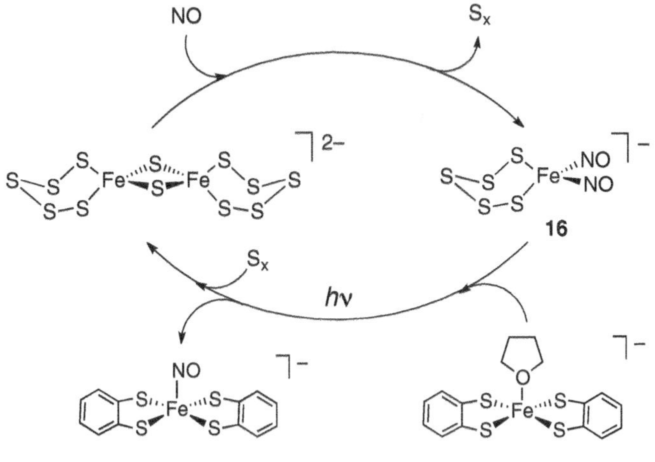

Scheme 1. Nitric oxide donation from $[S_5Fe(NO)_2]^-$ (**16**) to $[(thf)Fe(S,S\text{-}C_6H_4)_2]^-$ triggered by photolysis. (thf = tetrahydrofuran, ligand). [Adapted from (60).]

Major advances of the understanding of the chemistry of DNICs have been made over the past 10–15 years as the importance of such species in biology has become widely recognized. Liaw and co-workers (60), in particular, have uncovered a great deal of information about the generation and reactivity of anionic $\{Fe(NO)_2\}^9$ complexes. They reported the first example of biomimetic degradation and reassembly of an [2Fe–2S] cluster employing the $(S_5)^{2-}$ ligand. As shown in Reaction 8, a DNIC, $[S_5Fe(NO)_2]^-$ (**16**), was generated from nitrosylation of a [2Fe–2S] cluster, $[S_5Fe^{III}(\mu\text{-}S)_2Fe^{III}S_5]^{2-}$ (Scheme 1). Interestingly, **16** released NO when photolyzed in the presence of the NO acceptor reagent, $[(thf)Fe(S,S\text{-}C_6H_4)_2]^-$, resulting in regeneration of $[S_5Fe^{III}(\mu\text{-}S)_2Fe^{III}S_5]^{2-}$ (Scheme 1). This work offers insight into both the actions of DNICs as NO-donors and the repair of DNICs back to Fe–S clusters. In the absence of the NO acceptor, photolysis led only to the formation of an insoluble yellow solid, indicating that its presence is necessary for cluster formation (60).

The studies of NO transfer activity of DNICs were further extended to the carbazolate bound DNIC, $[(C_{12}H_8N)_2Fe(NO)_2]^-$ (**28**), which was shown to donate NO^+, NO, or NO^-, triggered by substitution of the carbazolate with varying chelating ligands, Scheme 2. The electronic and structural properties afforded by the incoming ligands optimized the $\{Fe(NO)_2\}$ core for the donation of NO in its various redox forms (69). This work sheds light onto the action of DNICs *in vivo* because, in that setting, versatility in NO donation could be aided by their ligands and protein environment (69). This idea was further supported by the studies with a series of six- and five-coordinate DNICs. When Liaw and co-workers (81) tried to synthesize a thermally unstable, five-coordinate, anionic $\{Fe(NO)_2\}^9$ complex (not

28

X = H or Na

NR_2 = carbazolate

NO^+ donor

NR_2 = carbazolate

$NO_{(g)}$ donor

NO^- donor

Scheme 2. Effect of incoming substitution ligand on the release of NO^+, NO, or NO^- from $[(C_{12}H_8N)_2Fe(NO)_2]^-$ (**28**) [Adapted from (69).]

shown), it spontaneously released 1 equiv of nitroxyl (NO^-) to yield a dimeric mononitrosyl iron complex. This NO^- releasing activity of a five-coordinate anionic DNIC is drastically different from the reactivity of a four-coordinate anionic DNIC or those of cationic or neutral five- and six-coordinate DNICs, which further supports a notion that the ligand electronics and the geometric structure of DNICs modulate the release of NO from DNICs. The reader is referred to Liaw and co-workers (81) for further detailed discussions.

Darensbourg and co-workers (65) recently reported NO donating activity in a thiocyanate-bound anionic $\{Fe(NO)_2\}^9$ DNIC, $[(SCN)_2Fe(NO)_2]^-$ (**22**) (Fig. 3). Upon exposure of the THF solution of **22** to air, NO was lost and an air-stable complex, *trans*-$[Fe^{III}(thf)_2(NCS)_4]^-$ formed. This facile NO release from **22** may be correlated to its ν_{NO} values (1718 and 1786 cm^{-1}, Table I), which are among the highest of anionic $\{Fe(NO)_2\}^9$ DNICs (65). This work further establishes that NO

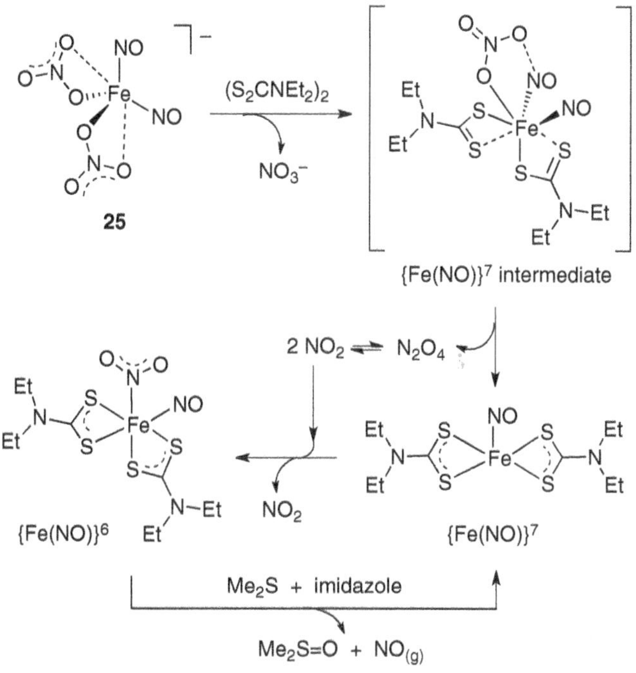

Scheme 3. Conversion of nitrate-to-nitrite-to-nitric oxide modulated by an anionic $\{Fe(NO)_2\}^9$ DNIC. [Adapted from (72).]

transfer reactivity in DNICs can be finely tuned by the electronic structure of the $\{Fe(NO)_2\}$ core.

In addition to the NO transfer from DNICs to the NO acceptors, a very recent report suggests that the DNIC chemistry may be important in the nitrate → nitrite → nitric oxide pathway in cell signaling, in which nitrate bound to the $\{Fe(NO)_2\}^9$ DNIC center can be converted into nitrite and NO (72). In this report by Liaw and co-workers (72), addition of disulfide to a nitrate-containing DNIC, $[(\kappa^1\text{-}ONO_2)Fe(NO)_2]^-$ (**31**), resulted in the formation of the N-bound nitro $\{Fe(NO)\}^6$ complex (Scheme 3), the mechanism of which likely involved the intramolecular association of nitrate and the adjacent NO generating $^{\bullet}NO_2$ (Scheme 3). Addition of dimethyl sulfide (Me_2S) to the N-bound nitro $\{Fe(NO)\}^6$ compound in the presence of imidazole led to the formation of the $\{Fe(NO)\}^7$ species $[(NO)Fe(S_2CNEt_2)_2]$ with release of NO (Scheme 3). These data illuminate the possibility of nitrate-containing $\{Fe(NO)_2\}^9$ DNICs being involved in NO/NO_3^- storage and transfer in biology.

Anionic $\{Fe(NO)_2\}^9$ DNICs also show reactivity other than NO donation, as Postel and co-workers (104) reported earlier. Kim and co-workers (105) studied O_2

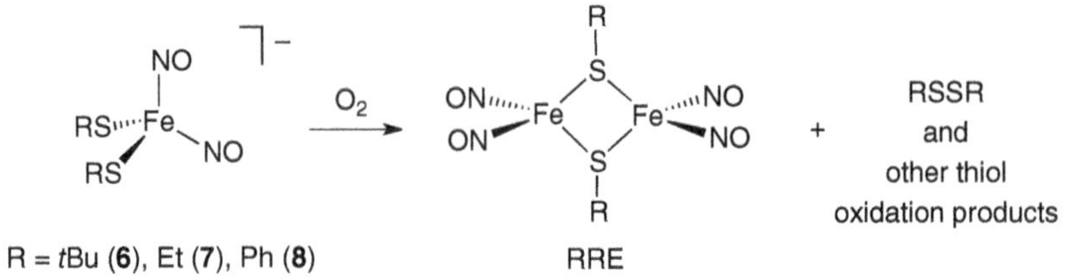

Scheme 4. Dioxygen reactivity of anionic $\{Fe(NO)_2\}^9$ DNICs. [Adapted from (106).]

reactivity of a group of thiolate-containing anionic $\{Fe(NO)_2\}^9$ DNICs, $[(RS)_2Fe(NO)_2]^-$ (**4**, **5**, **7**), where R = *t*Bu, Et, or Ph, and shown that oxidation of these species would occur at the sulfur atoms of the thiolate ligands as opposed to the $\{Fe(NO)_2\}^9$ unit (Scheme 4). This reactivity makes an interesting contrast to the oxygenation chemistry of $[Cl_2Fe(NO)_2]^-$ (**3**) reported by Postel and co-workers (105) (see above) and that of neutral $\{Fe(NO)_2\}^{10}$ complexes (see Section IV). Complexes **4, 5, 7** were converted to the related RREs in the presence of excess dioxygen while thiol oxidation products (e.g., disulfide) were generated at the same time (105). The O_2 reactivity shown with these DNICs suggest that cellular anionic DNICs could lead to the oxidation of cysteine, an important protein posttranslation modifications *in vivo*.

There are few known anionic $\{Fe(NO)_2\}^{10}$ DNICs, but some of their unique reactivity warrants attention. The CO activity of an anionic $\{Fe(NO)_2\}^{10}$ DNIC, $[(SC_6H_4\text{-}o\text{-}NMeFe(NO)_2]^-$ (**44**) was uncovered by the work of Chen et al. (74), Scheme 5. When **44** was reacted with 1 equiv of a neutral $\{Fe(NO)_2\}^{10}$, $[(TMEDA)Fe(NO)_2]$ (**71**), under an atmosphere of CO, a novel CO bridged anionic $\{Fe(NO)_2\}^{10}-\{Fe(NO)_2\}^{10}$ dimer was isolated (Scheme 5). The authors proposed that the dimer was stabilized by the π-accepting nature of CO, which lessened the electron density on the $\{Fe(NO)_2\}^{10}-\{Fe(NO)_2\}^{10}$ centers (74).

The unique acid activity of the anionic $\{Fe(NO)_2\}^{10}$ DNIC possessing a nitro ligand, $[(PPh_3)(NO_2)Fe(NO)_2]^-$ (**43**), has been reported (73). The N-bound nitro

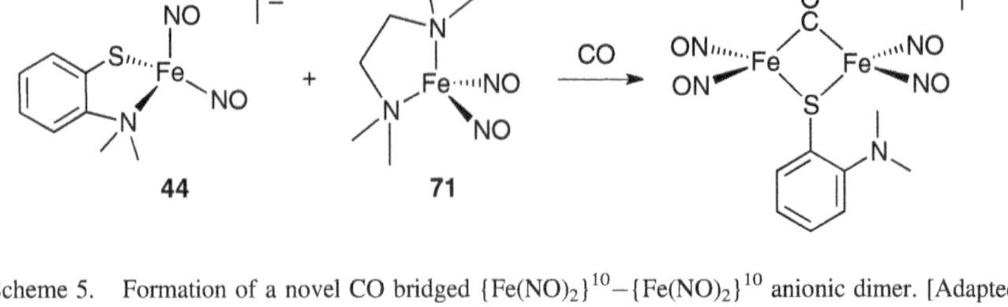

Scheme 5. Formation of a novel CO bridged $\{Fe(NO)_2\}^{10}-\{Fe(NO)_2\}^{10}$ anionic dimer. [Adapted from (74).]

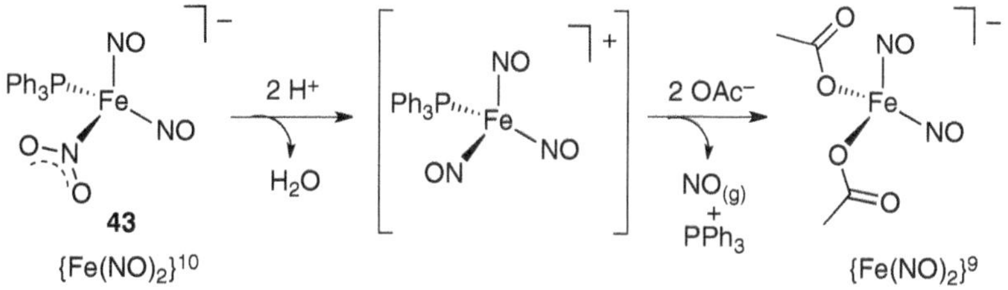

Scheme 6. Conversion of NO_2^- to NO at the anionic $\{Fe(NO)_2\}^{10}$ center. [Adapted from (73).]

ligand acts as a π-acceptor in **43** and can be easily converted to water and NO upon addition of acid with the reducing equivalent provided from the $\{Fe(NO)_2\}^{10}$ center. An addition of 2 equiv of acetic acid to **43** resulted in an anionic $\{Fe(NO)_2\}^9$ DNIC, $[(OAc)_2Fe(NO)_2]^-$, releasing NO, H_2O, and PPh_3 (Scheme 6), in which the formation of nitrosonium-bound intermediate was proposed (73). This work compares very well with the conversion of nitrite to NO catalyzed by the neutral $\{Fe(NO)_2\}^9$ DNIC, which is described in Section IV.

III. CATIONIC DINITROSYL IRON COMPLEXES

A. General Overview

Cationic $\{Fe(NO)_2\}^{10}$ DNICs are not known, while a handful of cationic $\{Fe(NO)_2\}^9$ DNICs have been observed and characterized (Table I, compounds **46–56** and Fig. 5). Due to their inherent instability, four-coordinate cationic $\{Fe(NO)_2\}^9$ DNICs are difficult to isolate and characterize. They typically possess neutral N-, P-, O-, or C-donor ligands that can affect their lifetime, where DNICs with phosphine or a rigid chelate seem to add stability. While these compounds may be difficult to isolate, they can be characterized spectroscopically. Four-coordinate cationic $\{Fe(NO)_2\}^9$ DNICs share the characteristic EPR g-value ($g = 2.03$) with anionic $\{Fe(NO)_2\}^9$ DNICs, which is distinct from the EPR signals observed for six-coordinate iron nitrosyls ($g = 2.018$ for **54** and $g = 2.015$ for **53**), Table I. Similarly, four-coordinate cationic $\{Fe(NO)_2\}^9$ DNICs have comparable IR features to their anionic counterparts, but vary from six-coordinate iron introsyls. Two ν_{NO} peaks from four-coordinate cationic $\{Fe(NO)_2\}^9$ DNICs appear between 1723 and 1814 cm^{-1} with an average separation of ~60 cm^{-1}, which are slightly blue-shifted with a larger ν_{NO} separation compared to those observed for the anionic $\{Fe(NO)_2\}^9$ analogues (Table I). The IR features appear to be considerably affected by the coordination number, in which a significantly greater separation between the two ν_{NO} frequencies (~83 cm^{-1}) is present for five- and six-coordinate cationic DNICs (**52–54**).

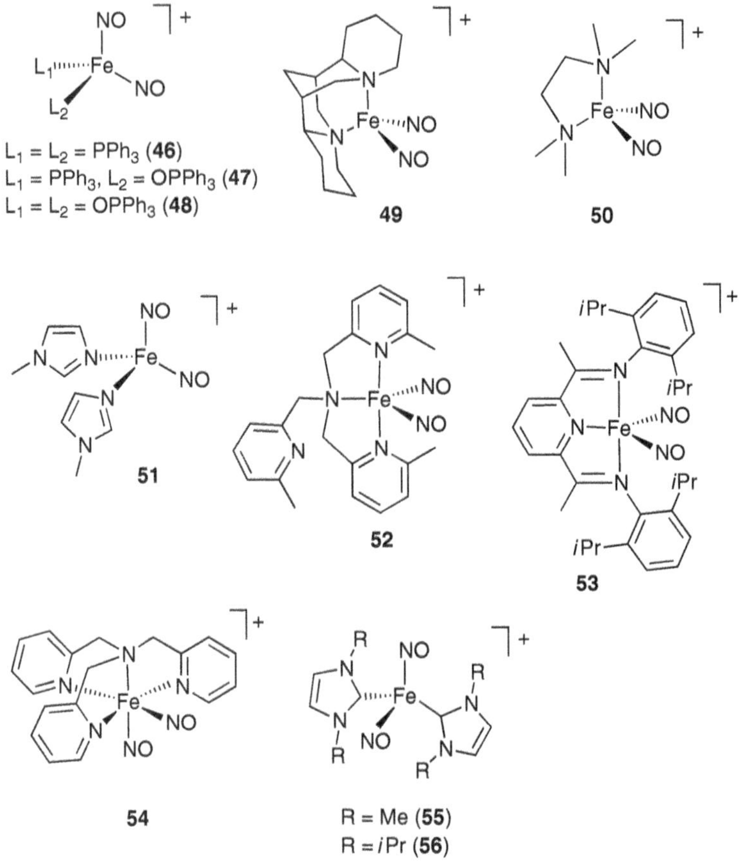

$L_1 = L_2 = PPh_3$ (**46**)
$L_1 = PPh_3$, $L_2 = OPPh_3$ (**47**)
$L_1 = L_2 = OPPh_3$ (**48**)

49 **50** **51** **52** **53** **54**

R = Me (**55**)
R = *i*Pr (**56**)

Figure 5. Cationic $\{Fe(NO)_2\}^9$ DNICs.

There are five X-ray structures known for cationic DNICs (Table I), among which only two are four coordinate. The limited number of reported structures for this group of cationic, four coordinate $\{Fe(NO)_2\}^9$ DNICs, makes it difficult to build a trend about their bond angles and distances compared to their anionic counterparts. Nonetheless, their Fe–N–O bond angles and their Fe–N(O) and N–O bond distances appear to fall in a similar range as those found in anionic $\{Fe(NO)_2\}^{9/10}$ DNICs (Table I).

B. Preparation

Most of the four-coordinate cationic $\{Fe(NO)_2\}^9$ DNICs are synthesized from oxidation of the neutral $\{Fe(NO)_2\}^{10}$ complexes, shown in Reaction 10, where L can be neutral N-, P-, O-, or C-donor ligands. Oxidants (Ox) that have been used include NO^+ (62, 77), ferrocenium (71, 77), and O_2 (71). Similar to the anionic $\{Fe(NO)_2\}^9$ DNICs, ligand subsitution is another practical method to generate various cationic DNICs, Reactions 11 and 12. Five- and six-coordinate cationic $\{Fe(NO)_2\}^9$ DNICs (**52–54**) have been synthesized using a combination of oxidation and ligand substitution, Reaction 12 (81).

$$[L_2Fe(NO)_2] + [Ox] \rightarrow [L_2Fe(NO)_2]^+ \qquad (10)$$

$$\underset{(\mathbf{46})}{[(Ph_3P)_2Fe(NO)_2]^+} + OPPh_3 \rightarrow \underset{(\mathbf{47})}{[(PPh_3)(OPPh_3)Fe(NO)_2]^+} \qquad (11)$$

$$[(CO)_2Fe(NO)_2] + [Ox] + L \rightarrow [(L)Fe(NO)_2]^+ + 2\,CO \qquad (12)$$

C. Reactivity

Chemical-reactivity studies of cationic DNICs are just as rare as their synthesis. Darensbourg and co-workers (71) investigated the NO transfer ability of N-heterocyclic carbene (NHC) containing cationic $\{Fe(NO)_2\}^9$ DNICs, $[(NHC\text{-}Me)_2Fe(NO)_2]^+$ (**55**) and $[(NHC-iPr)_2Fe(NO)_2]^+$ (**56**) as well as their reduced analogues, neutral $\{Fe(NO)_2\}^{10}$ DNICs, $[(NHC-Me)_2Fe(NO)_2]^+$ (**80**) and $[(NHC-iPr)_2Fe(NO)_2]^+$ (**81**). In these experiments, each DNIC was combined with the NO trapping agent, $Co^{II}(TPP)$, where TPP = tetraphenylporphyrinate (2−). In the case of the cationic $\{Fe(NO)_2\}^9$ DNICs, an IR absorption band corresponding to (NO)Co(TPP) quickly appeared at $1683\,cm^{-1}$. However, the neutral $\{Fe(NO)_2\}^{10}$ complexes showed no reaction, indicating that these DNICs were inert to NO transfer in the reduced state (71). Similarly, Kim and co-workers (78) reported that the instability of the cationic $[(TMEDA)Fe(NO)_2]^+$ (**50**) originated from its propensity to lose NO. The oxidation of the neutral $\{Fe(NO)_2\}^{10}$ compound, $[(TMEDA)Fe(NO)_2]$ (**71**), by ferrocenium hexafluorophosphate resulted in the formation of the semistable cationic $\{Fe(NO)_2\}^9$ species (**50**) from which $NO_{(g)}$ evolved. Consistent results were also observed when the iodide ligand was removed from the neutral, five-coordinate $\{Fe(NO)_2\}^9$ complex, $[(TMEDA)Fe(NO)_2I]$ (**67**), Scheme 7.

IV. NEUTRAL DINITROSYL IRON COMPLEXES

A. General Overview

Neutral $\{Fe(NO)_2\}^9$ and $\{Fe(NO)_2\}^{10}$ DNICs are the final classification of the $\{Fe(NO)_2\}$ unit where the latter are more prevalent in the literature (Table I, Figs. 6 and 7). Neutral $\{Fe(NO)_2\}^9$ are typically found with either an anionic bidentate chelate or a combination of anionic and neutral ligands (Fig. 6). Like the

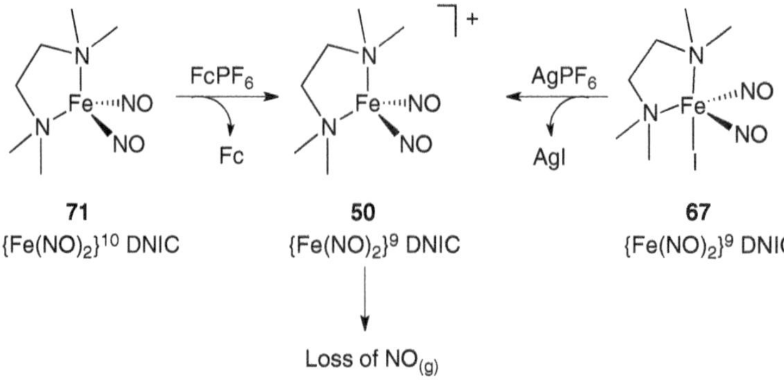

71
$\{Fe(NO)_2\}^{10}$ DNIC

FcPF6
Fc

50
$\{Fe(NO)_2\}^{9}$ DNIC

Loss of $NO_{(g)}$

AgPF6
AgI

67
$\{Fe(NO)_2\}^{9}$ DNIC

Scheme 7. Release of NO from an unstable four-coordinate cationic $\{Fe(NO)_2\}^9$ DNIC. [Adapted from (78).]

anionic and cationic DNICs, neutral $\{Fe(NO)_2\}^9$ DNICs display their characteristic EPR signal of $g = 2.03$ at room temperature (Table I). Two expected strong ν_{NO} bands from neutral $\{Fe(NO)_2\}^9$ DNICs are seen in IR spectroscopy with an average separation of the stretching frequencies to be ~58 cm^{-1}. An interesting observation can be made within the series of $\{Fe(NO)_2\}^9$ DNICs, in which the average separation between the ν_{NO} bands follows the order of cationic > neutral > anionic DNICs (Table I). Neutral $\{Fe(NO)_2\}^9$ DNICs have the same mean Fe–N(O) bond distances of ~1.66 Å as the anionic and cationic $\{Fe(NO)_2\}^9$ DNICs. Similarly, the N–O bond distances for four-coordinate neutral $\{Fe(NO)_2\}^9$ average to ~1.175 Å, as found in both anionic and cationic $\{Fe(NO)_2\}^9$ DNICs.

The $\{Fe(NO)_2\}^{10}$ DNICs are spectroscopically and structurally distinct from $\{Fe(NO)_2\}^9$ DNICs (see also Section II). The neutral $\{Fe(NO)_2\}^{10}$ DNICs are diamagnetic and thus are EPR silent. Similar to the case of a pair of anionic $\{Fe(NO)_2\}^{9/10}$ DNICs, the positions and the separation of the two IR ν_{NO} stretching frequencies of neutral $\{Fe(NO)_2\}^{10}$ DNICs are significantly different from those of neutral $\{Fe(NO)_2\}^9$ DNICs. The ν_{NO} of $\{Fe(NO)_2\}^{10}$ DNICs appear in the range of 1614–1724 cm^{-1} with an average separation of ~51 cm^{-1}, compared to the range of 1695–1800 cm^{-1} with an average separation of ~58 cm^{-1} found in $\{Fe(NO)_2\}^9$ DNICs (Table I). The N–O bond distances of the neutral $\{Fe(NO)_2\}^{10}$ DNICs fall in the range of 1.189–1.214 Å, which are significantly longer (~0.03 Å) compared to the neutral $\{Fe(NO)_2\}^9$ DNICs, whereas the Fe–N(O) distances (1.637–1.698 Å) are ~0.03 Å shorter than those of neutral $\{Fe(NO)_2\}^9$ DNICs. Since only a couple of anionic $\{Fe(NO)_2\}^{10}$ DNICs have been reported, further structural comparisons between neutral and anionic $\{Fe(NO)_2\}^{10}$ DNICs are unwarranted.

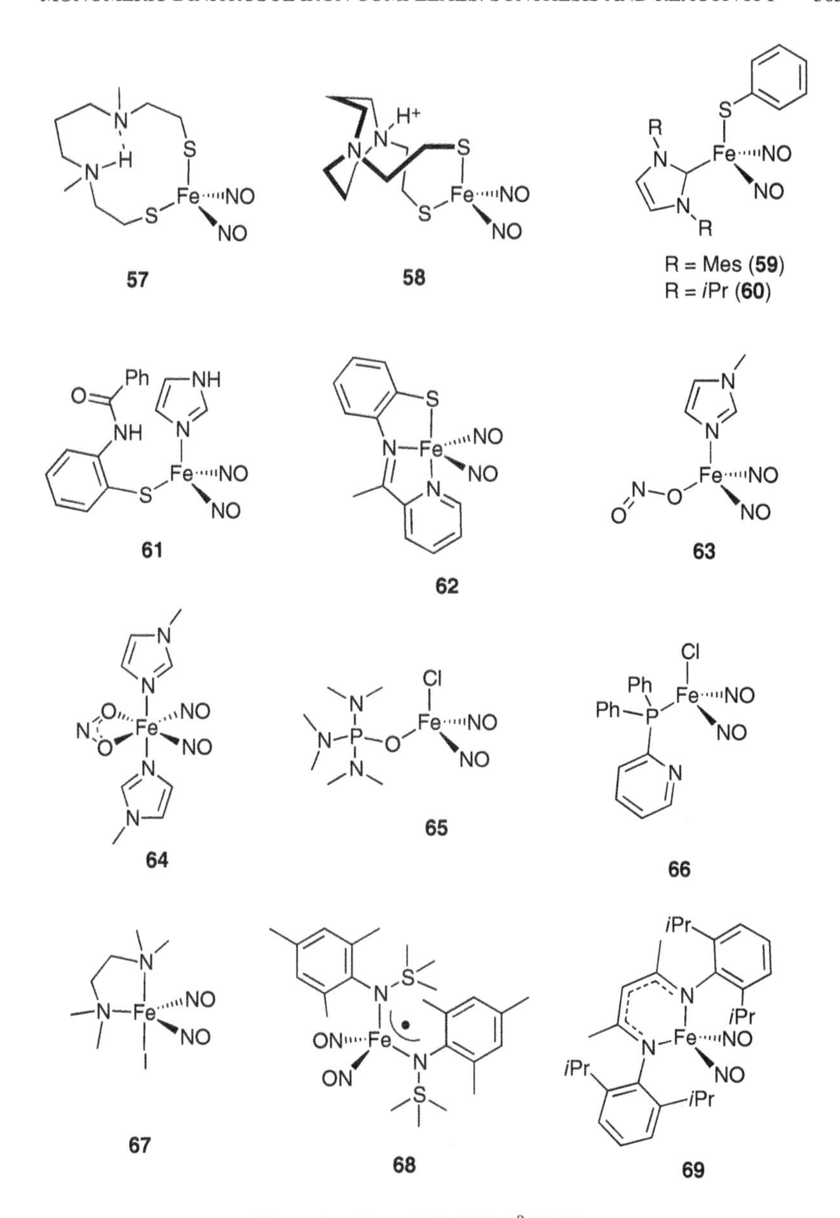

Figure 6. Neutral $\{Fe(NO)_2\}^9$ DNICs.

B. Preparation

The majority of neutral $\{Fe(NO)_2\}^{10}$ DNICs (Table I, Fig. 7) were synthesized from $[(CO)_2Fe(NO)_2]$ (**89**) by ligand substitution using a stoichiometrically controlled addition of neutral ligand, Reaction 13. Alternatively, a DNIC, such as $[(PPh_3)_2Fe(NO)_2]$ (**87**), can be synthesized from RRE, in which bridging thiolate is reductively eliminated upon addition of PPh_3 (39), Reaction 14.

$$[(CO)_2Fe(NO)_2] \xrightarrow[-CO]{L} [(CO)LFe(NO)_2] \xrightarrow[-CO]{L} [L_2Fe(NO)_2] \tag{13}$$

$$[Fe_2(\mu\text{-}SR)_2(NO)_4] + 4\,PPh_3 \rightarrow 2\,\underset{(\mathbf{87})}{[(PPh_3)_2Fe(NO)_2]} + RSSR \tag{14}$$

Likewise, ligand substitution is a common way to synthesize neutral $\{Fe(NO)_2\}^9$ DNICs as well. The most frequently used $\{Fe(NO)_2\}^9$ precursor is $[X_2Fe(NO)_2]^-$ (**1–3**) to which an addition of a neutral ligand or a monoanionic chelate will yield the desired compounds including **57**, **58**, **65**, **68**, and **69** (Table I, Fig. 6). Similar to Reaction 9 (Section II-B), RREs can also be used to synthesize neutral $\{Fe(NO)_2\}^9$ DNICs, as seen for **60** and **61** with an addition of either neutral imidazole or N-heterocyclic carbene, Reaction 15 (39, 40, 62).

$$[Fe_2(\mu\text{-}SR)_2(NO)_4] + 2\,L \rightarrow 2\,[(SR)LFe(NO)_2] \tag{15}$$

Neutral $\{Fe(NO)_2\}^9$ DNICs can also be prepared via oxidation. Formation of $[(N(Mes)(TMS))_2Fe(NO)_2]$ (**68**) (Fig. 6) occurs through a ligand-based one-electron oxidation of an anionic $\{Fe(NO)_2\}^9$ DNIC, $[(N(Mes)(TMS))_2Fe(NO)_2]^-$ (**25**) using $FeCp_2^+$ as an oxidant (66). Another oxidant that can be utilized is $[NO]BF_4$ as seen in the synthesis of $[(IMes)(SPh)Fe(NO)_2]$ (**59**) and $[(NHC\text{-}i\text{Pr})(SPh)Fe(NO)_2]$ (**60**) (Fig. 6). The carbonyl ligand of a neutral $\{Fe(NO)_2\}^{10}$ monocarbonyl compound, $[(CO)LFe(NO)_2]$, can be substituted by an isoelectronic analogue, NO^+, to yield a cationic $\{Fe(NO)_3\}^{10}$ precursor, $[(L)Fe(NO)_3]^+$, Reaction 16. Subsequently, an anionic thiolate ligand replaces one of the NO ligands to yield a neutral $\{Fe(NO)_2\}^9$ DNIC (**59** or **60**) Reaction 17.

$$[(CO)LFe(NO)_2] + NO^+ \rightarrow [LFe(NO)_3]^+ + CO \tag{16}$$

$$[LFe(NO)_3]^+ + RS^- \rightarrow [(RS)LFe(NO)_2] + NO \tag{17}$$

A five-coordinate neutral $\{Fe(NO)_2\}^9$ $[(TMEDA)Fe(NO)_2I]$ (**67**) (Fig. 6) can be synthesized from a neutral $\{Fe(NO)_2\}^{10}$, $[(TMEDA)Fe(NO)_2]$ (**71**) (Fig. 7), via oxidation using I_2 (87). Conversely, reduction of **67** by $CoCp_2$ generates **71** (87).

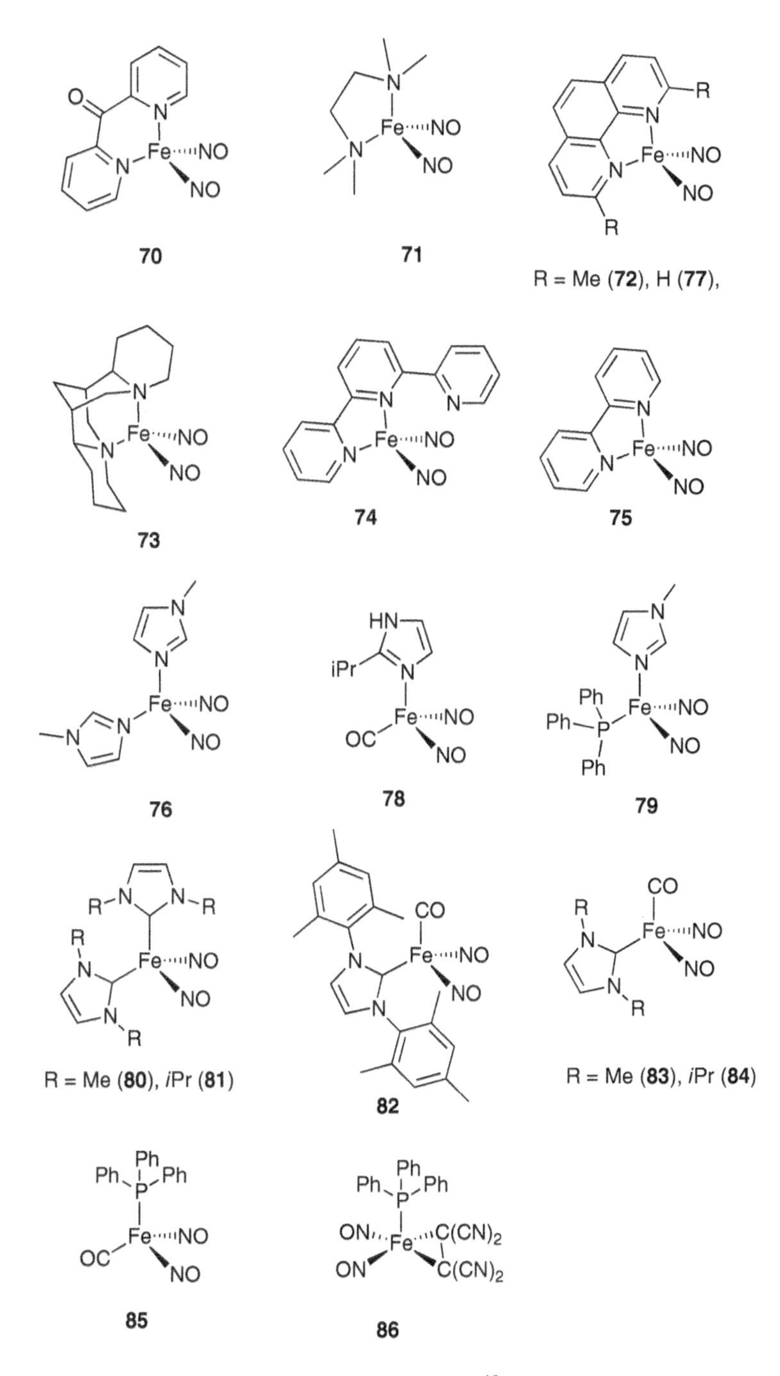

Figure 7. Neutral $\{Fe(NO)_2\}^{10}$ DNICs.

Additionally, **67** can be prepared by ligand substitution from $[Fe_2(\mu\text{-}I)_2(NO)_4]$ with a TMEDA chelate (87).

C. Reactivity

Several distinctive reactivity patterns have been observed for the neutral DNICs over the years, which include dioxygen activity, NO release, and the conversion of NO and nitrite.

As introduced in Section II, one of the projects of Postel's research group in the 1990s was to establish the iron-nitrato–iron-nitrosyl couple as a catalytic system for the activation and transfer of O_2. When the neutral $\{Fe(NO)_2\}^9$ DNIC, $[(Cl)L\ Fe(NO)_2]$ ($L = PPh_3$ or $OPPh_3$) (Scheme 8) was reacted with O_2 in the presence of 1 equiv of PPh_3 or $OPPh_3$, it quantitatively yielded the pentagonal nitrato compound, $[Fe(NO_3)_2Cl(OPPh_3)_2]$ (Scheme 8) (106). The resulting nitrato compound was capable of oxygen transfer to PPh_3 or cyclohexene, demonstrating the first example of oxygen transfer from a nitrato ligand to an olefin (106). An analogous pentagonal nitrato compound, $[Fe(NO_3)_2Cl(OPPh_2py)]$, was reported several years later from oxygenation of a neutral $\{Fe(NO)_2\}^9$ DNICs possessing the 2-(diphenylphosphino)pyridine (PPh_2py) ligand, $[Fe(NO)_2Cl(PPh_2py)]$ (**66**) (Scheme 8). During oxygenation, the denticity of PPh_2Py changed from monodentate to bidentate allowing the formation of a pentagonal nitrato species in the absence of external PPh_3 or $OPPh_3$ (86). The resulting nitrato complex was also

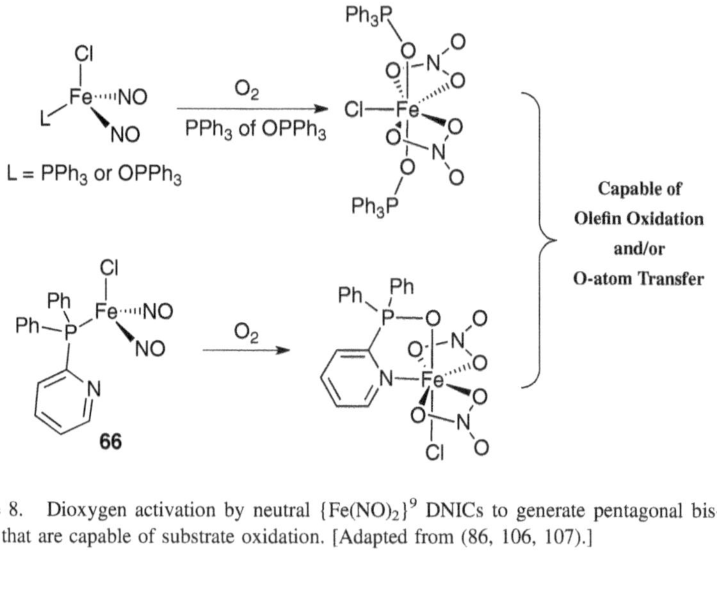

Cl Fe NO NO L
L = PPh3 or OPPh3
O2
PPh3 of OPPh3
Ph3P O N O O Cl Fe O N O Ph3P
Capable of
Olefin Oxidation
and/or
O-atom Transfer
Ph Ph P Fe NO NO Cl N
66
O2
Ph Ph P O N O Fe N O Cl

Scheme 8. Dioxygen activation by neutral $\{Fe(NO)_2\}^9$ DNICs to generate pentagonal bis-nitrato species that are capable of substrate oxidation. [Adapted from (86, 106, 107).]

shown to be capable of phosphine oxidation and to act as a catalyst in cyclohexene autoxidation (107).

Postel and co-workers (85) further noticed that the oxygen-atom transfer ability of the iron nitrato complexes was very much dependent on the number of ligands bound to iron. When they used the neutral $\{Fe(NO)_2\}^9$ with a sterically demanding HMPA ligand, $[(HMPA)(Cl)Fe(NO)_2]$ (**65**), Fig. 6, oxygenation led to several different types of iron nitrato complexes. While the nitrato ligand in the "penta"-coordinated nitrates, such as $[Fe(NO_3)_2Cl(HMPA)_2]$ or $[Fe(NO_3)Cl_2(HMPA)_2]$, were capable of transferring oxygen to phosphines, "tetra"-coordinate nitrates, such as $[Fe(NO_3)_2Cl(HMPA)]$, lost all oxidizing capability. The authors Postel and co-workers (85) suggested that the enhanced electron density on the $Fe(NO_3)_2$ moiety is a key factor for the oxygenation step.

Another interesting, but unexpected, reactivity of neutral DNICs includes reduction of NO to dinitrogen (108). Postel and co-workers (108) prepared dinuclear neutral $\{Fe(NO)_2\}^9$ complexes using bridging diphosphine liagnds, $[Fe(NO)_2Cl](\mu\text{-PP})$ (**90**), where PP = 1,2-bis(diphenylphosphino)ethane or *trans*-1,2-bis(diphenylphosphino)ethene (Scheme 9). The reaction of **90** with excess PP ligands resulted in a mixture of a $\{Fe(NO)_2\}^{10}$ DNIC, $[Fe(NO)_2(PP)]$ (**91**), and an unexpected mononitrosyl iron species, $[Fe(NO)(OPP)]$ (**92**), where OPP = 1,2-bis(diphenylphosphino)ethane monoxide or *trans*-1-2-bis (diphenylphosphino)ethene monoxide (Scheme 9). During the reaction, eveolution of N_2 was detected by gas chromatography (GC). The authors postulated that the

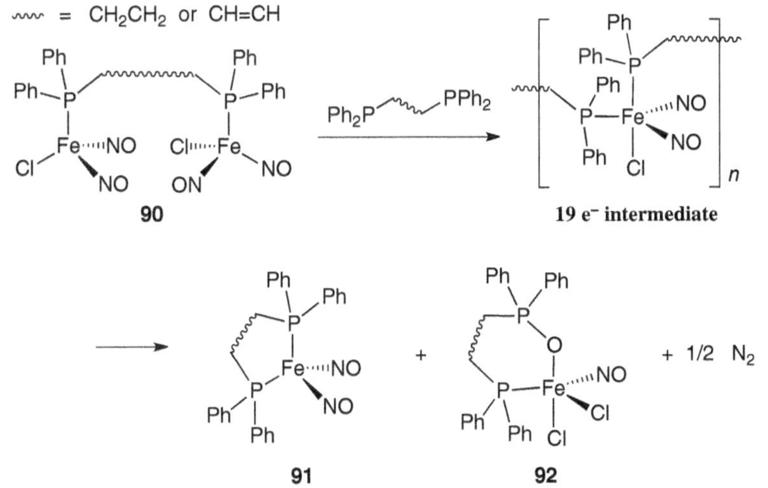

Scheme 9. Monooxygenation of diphonsphine by neutral $\{Fe(NO)_2\}^9$ DNICs. [Adapted from (108).]

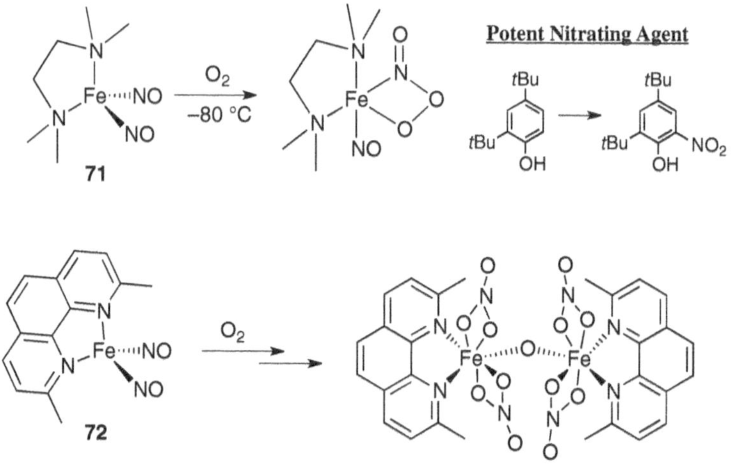

Scheme 10. Dioxygen reactivity of neutral $\{Fe(NO)_2\}^{10}$ DNICs leading to formation of the peroxynitrite intermediate, which isomerizes to nitrate at room temperature. [Adapted from (89, 109).]

formation of **92** along with N_2 proceeded through an intramolecular redox reaction and rearrangement of the five-coordinate, 19-electron intermediate (Scheme 9).

Dioxygen reactivity of neutral DNICs has been recently revisited by the research of Kim and co-workers (110) with a goal of identifying the undefined roles of DNICs in biology. Neutral, N-bound $\{Fe(NO)_2\}^{10}$ DNICs studied by Kim and co-workers (109) show somewhat different activity from the O_2 reactivity discovered by Postel and co-workers (85, 86, 106, 107). In the presence of dioxygen, the N-bound $\{Fe(NO)_2\}^{10}$ DNICs (e.g., **71** and **72**, Scheme 10) are capable of nitrating phenolic substrates. This reaction occurs through an iron-bound peroxynitrite ($ONOO^-$) intermediate that was observed and spectroscopically characterized below −80 °C. Though this peroxynitrite intermediate cannot be spectroscopically observed with all neutral N-bound $\{Fe(NO)_2\}^{10}$ DNICs or at room temperature, its presence is indicated by phenol nitration (89, 109). In the absence of substrate, iron-peroxynitrite isomerizes to a thermally stable nitrato species, which was most fully characterized in the case of $[(dmp)Fe(NO)_2]$ (**72**) (89). These results suggest that DNICs could have multiple physiological or deleterious roles, including that of cellular nitrating agents.

Research groups of Darensbourg and co-workers (61, 110), Lippard and co-workers (75, 76), and Tsou and Liaw (111), have independently studied NO transfer chemistry of neutral DNICs to heme or thiol. Compound **58** (Scheme 11) used by the Darensbourg group has the potential square-planar N_2S_2 ligand that can also serve as a dithiolate bidentate donor for the DNIC (61, 110). The removal of one NO produces a stable mononitrosyl complex (Scheme 11), which can

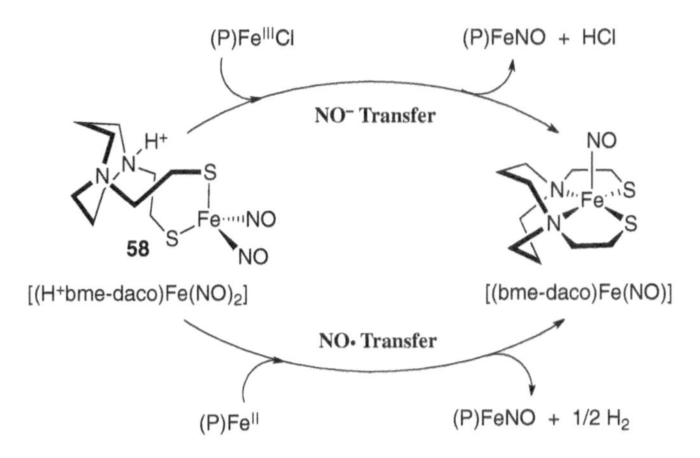

Scheme 11. A neutral $\{Fe(NO)_2\}^9$ DNIC as an NO or NO^- source to Fe^{II} or Fe^{III} prophyrinate (P). [Adapted from (61, 110).]

simplify the studies of NO release mechanisms. They first found that when **58** was mixed with one-half of an equivalent of $[(\text{bme-daco})Fe]_2$ at room temperature (rt), 2 equiv of the mononitrosyl species, [(bme-daco)Fe(NO)] were formed, which was the first clear indication that **58** was capable of donating NO (61). Further studies demonstrated that **58** was capable of donating NO^- (most likely in the form of HNO), as well as NO to, respectively, ferric or ferrous porphyrins (Scheme 11). However, free NO is not evolved from **58** in the absence NO acceptors.

Lippard and co-workers (75, 76) observed comparable NO transfer reactivity with their neutral $\{Fe(NO)_2\}^9$ DNIC, $[(\text{Ar–nacnac})Fe(NO)_2]$ (**69**) (Fig. 6), which reductively nitrosylated $[Fe^{III}(TPP)Cl]$ in either the presence of light or at elevated temperatures (Reaction 20). Interestingly, when an analogous anionic $\{Fe(NO)_2\}^{10}$ DNIC, $[(\text{Ar-nacnac})Fe(NO)_2]^-$ (**45**), was reacted with $[Fe^{III}(TPP)Cl]$, rapid electron transfer from the DNIC to $[Fe^{III}(TPP)Cl]$ produced **69**, which then donated NO to $[Fe^{II}(TPP)Cl]^-$ generating [Fe(TPP)(NO)] (Reactions 21 and 22).

$$\underset{(\mathbf{69})}{[(\text{Ar–nacnac})Fe(NO)_2]} + [Fe^{III}(TPP)Cl] \rightarrow [(\text{Ar–nacnac})Fe(NO)Cl] + [Fe(TPP)(NO)] \quad (20)$$

$$\underset{(\mathbf{45})}{[(\text{Ar–nacnac})Fe(NO)_2]^-} + [Fe^{III}(TPP)Cl] \rightarrow \underset{(\mathbf{69})}{[(\text{Ar–nacnac})Fe(NO)_2]} + [Fe(TPP)Cl]^- \quad (21)$$

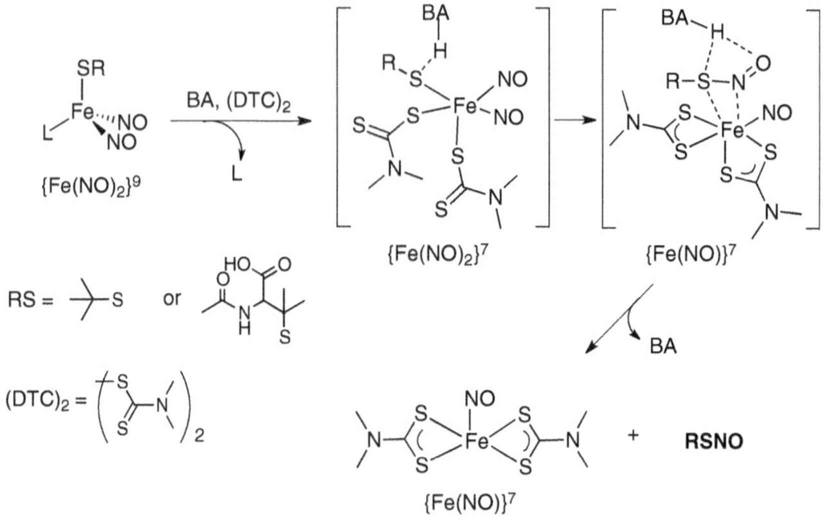

Scheme 12. Transformation of $\{Fe(NO)_2\}^9$ DNICs into RSNOs. $(DTC)_2$ -bis-(dimethylthiocarbamoyl) disulfide [Adapted from (111).]

$$[(\text{Ar}-\text{nacnac})\text{Fe(NO)}_2] + [\text{Fe(TPP)Cl}]^- \rightarrow [(\text{Ar}-\text{nacnac})\text{Fe(NO)Cl}]^- + [\text{Fe(TPP)NO}] \quad (22)$$

(**69**)

Tsou and Liaw (111) studied DNIC mediated S-nitrosation using a group of monothiolate-containing $\{Fe(NO)_2\}^9$ DNICs. Their studies indicate that S-nitrosation of the thiolate ligands was triggered by addition of a disulfide species, (bis-(dimethylthiocarbamoyl) disulfide $(DTC)_2$), along with a Brønsted acid (Scheme 12). The authors proposed that BA was necessary to stabilize the monothiolate-coordinated $\{Fe(NO)_2\}^7$ DNIC (Scheme 12), which was then capable of inducing intramolecular S-nitrosation of the coordinated thiolate to form RSNO. Interestingly, transformation of DNIC into RSNO only occurred on the monothiolate-containing $\{Fe(NO)_2\}^9$ DNICs, not on the dithiolate analogues, such as $[(St\text{Bu})_2Fe(NO)_2]^-$, which dimerized to a RRE upon addition of a BA. This work adds to our understanding of how S-nitrosation of protein-bound thiolates may occur by DNICs, especially in relation to the site specificity for such a modification.

More recently, Liaw and co-workers (73, 84) showed that neutral $\{Fe(NO)_2\}^9$ can serve as the catalytic center for the conversion of nitrite (NO_2^-) to NO chemistry that offers insights into nitrite signaling in cells. Upon addition of 1 equiv of 1-MeIm, the four-coordinate nitrito $\{Fe(NO)_2\}^9$ DNIC [(1-MeIm)(ONO) $Fe(NO)_2$] (**63**) was reversibly converted to the six-coordinate chelating nitrito

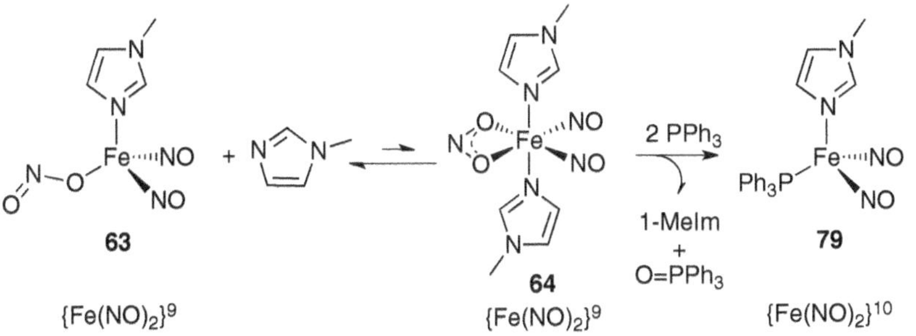

Scheme 13. Transformation of nitrite to nitric oxide at the $\{Fe(NO)_2\}^9$ DNIC center. [Adapted from (73, 84).]

$\{Fe(NO)_2\}^9$ DNIC, $[(1\text{-MeIm})_2(\eta^2\text{-ONO})Fe(NO)_2]$ (**64**), Scheme 13. Addition of PPh_3 to the chelating nitrito species **64** led to conversion of the nitrito ligand into NO, which was released by reductive elimination, forming the $\{Fe(NO)_2\}^{10}$ DNIC, $[(1\text{-MeIm})(PPh_3)Fe(NO)_2]$ (**79**). This study shows how imidazole coordination controls the $\{Fe(NO)_2\}$ center to become suitable for nitrite activation. It implies that the six-coordinate nitrite-containing DNIC species (e.g., **64**) may serve as a transient intermediate in the production of NO in biological system (73, 84).

V. Summary and Perspective

Dinitrosyl iron complexes are a class of molecules found in biological systems that have been synthesized in great numbers in order to gain an understanding of their role in nature and the methods by which they carry out their unique activities. They are considered mostly as NO storage and transfer agents and these effects have been observed *in vivo*. How they accomplish this activity is under investigation and current research indicates that their ability to donate NO in its various redox forms can be controlled by their electronic and structural environment. Dinitrosyl iron complexes may also play roles outside of NO storage and transfer, as is suggested by the ability of neutral $\{Fe(NO)_2\}^{10}$ complexes to promote phenol nitration in the presence of dioxygen. As of today, numerous synthetic DNICs containing a variety of ligands have been characterized (Table I, Figs. 3–7), though the list is still growing and will continue to grow. This brief chapter describes the synthesis, structural characteristics, and reactivity of some of the most well understood examples of these. Examining the trends in electronic and geometric structure in these complexes and how that contributes to their reactivity is vital in forming an understanding of how they behave in a biological setting.

ACKNOWLEDGMENTS

The authors would like to acknowledge Brown University, the National Science Foundation (CAREER CHE-1254733), and the Camille and Henry Dreyfus Foundation for financial support in the preparation of this chapter.

ABBREVIATIONS

Ar-nacnac	Anion of [(2,6-diisopropylphenyl)NC(Me)]$_2$CH
BA	Brønsted acid
bme-daco	Dianion of *N,N′*-bis(2-mercaptoethyl)-1,5-diazacyclooctane
bpy	2,2′-Bipyridine
DFT	Density functional theory
dmp	2,9-Dimethyl-1,10-phenanthroline
DNIC	Dinitrosyl iron complex
(DTC)$_2$	Bis-(dimethylthiocarbanoyl)disulfide
E–F	Enemark–Feltham
EPR	Electron paramagnetic resonance
GC	Gas chromatography
HMPA	Hexamethylphosphoric triamide
Im	Imidazole
IMes	1,3-Bis(2,4,6-trimethylphenyl)imidazol-2-ylidene
Imid-*i*Pr	2-Isopropylimidazole
*i*PrPDI	2,6-Bis[1-(2,6-diisopropylphenylimino)ethyl]pyridine
IR	Infrared
1-MeIm	1-Methylimidazole
Mes	Mesityl
6-Me$_3$-TPA	Tris[(6-methyl-2-pyridyl)methyl)]amine
NHC-*i*Pr	1,3-Diisopropylimidazol-2-ylidene
NHC-Me	1,3-Dimethylimidazol-2-ylidene
MNIC	Mononitrosyl iron complex
NO	Nitric oxide
NRVS	Nuclear resonance vibration spectroscopy
OPP	1,2-Bis(diphenylphosphino)ethane monoxide or *trans*-1-2-bis(diphenylphosphino)ethene monoxide
Ox	Oxidants
phen	1,10-Phenanthroline
PP	1,2-Bis(diphenylphosphinno)ethane or *trans*-1,2-bis(diphenylphosphinno)ethene
PPh$_2$py	2-(Diphenylphosphino)pyridine
py	Pyridine

PyImiS	2-((1-(Pyridin-2-yl)ethylidene)amino)benzenethiolate
RBS	Roussin's black salt
RRE	Roussin's red ester
RRS	Roussin's red salt
RSNO	Nitrosothiol
rt	Room temperature
TCNE	Tetracyanoethylene
terpy	2,2′,2″-Terpyridine
thf	Tetrahydrofuran (ligand)
THF	Tetrahydrofuran (solvent)
S-*p*-tolyl	4-Methylbenzenethiolate
TMEDA	*N*,*N*,*N*′,*N*′-Tetramethylethylenediamine
TMS	Trimethylsilane
TPA	Tris(2-pyridylmethyl)amine
TPP	Tetraphenylporphyrinate(2−)
XAS	X-ray absorption spectroscopy

REFERENCES

1. D. E. Koshland, Jr. and E. Culotta, *Science*, *258*, 1862 (1992).
2. L. J. Ignarro, *Annu. Rev. Pharmacol. Toxicol.*, *30*, 535 (1990).
3. J. E. Brenman and D. S. Bredt, in *Methods in Enzymology*. Vol. 269, P. Lester, Ed. Academic Press, San Diego, 1996, pp. 119–129.
4. J. MacMicking, Q.-w. Xie, and C. Nathan, *Annu. Rev. Immunol.*, *15*, 323 (1997).
5. A. Weichsel, E.M. Maes, J.F. Andersen, J.G. Valenzuela, T.Kh, Shokhireva, F.A. Walker, and W.R. Montfort, *Proc. Natl. Acad. Sci. USA*, *102*, 594 (2005).
6. H. Ding and B. Demple, *Proc. Natl. Acad. Sci. USA*, *97*, 5146 (2000).
7. M. K. Johnson, *Curr. Opin. Chem. Biol.*, *2*, 173 (1998).
8. J. R. Hickok, S. Sahni, H. Shen, A. Arvind, C. Antoniou, L. W. M. Fung, and D. D. Thomas, *Free Radical Biol. Med.*, *51*, 1558 (2011).
9. J. R. Mallard and M. Kent, *Nature (London)*, *204*, 1192 (1964).
10. A. F. Vanin and R. M. Nalbandyan, *Biofizika*, *10*, 167 (1965).
11. A. J. Vithayathil, J. L. Ternberg, and B. Commoner, *Nature (London) 207*, 1246 (1965).
12. A. R. Butler and I. L. Megson, *Chem. Rev.*, *102*, 1155 (2002).
13. X. Duan, J. Yang, B. Ren, G. Tan, and H. Ding, *Biochem. J.*, *417*, 783 (2009).
14. M. C. Kennedy, W. E. Antholine, and H. Beinert, *J. Biol. Chem.*, *272*, 20340 (1997).
15. A. P. Landry, X. Duan, H. Huang, and H. Ding, *Free Radical Biol. Med.*, *50*, 1582 (2011).
16. P. A. Rogers, L. Eide, A. Klungland, and H. Ding, *DNA Repair*, *2*, 809 (2003).

17. V. M. Sellers, M. K. Johnson, and H. A. Dailey, *Biochemistry*, *35*, 2699 (1996).
18. E. T. Yukl, M. A. Elbaz, M. M. Nakano, and P. Moënne-Loccoz, *Biochemistry*, *47*, 13084 (2008).
19. B. D'Autréaux, O. Horner, J.-L. Oddou, C. Jeandey, S. Gambarelli, C. Berthomieu, J.-M. Latour, and I. Michaud-Soret, *J. Am. Chem. Soc.*, *126*, 6005 (2004).
20. E. Cesareo, L. J. Parker, J. Z. Pedersen, M. Nuccetelli, A. P. Mazzetti, A. Pastore, G. Federici, A. M. Caccuri, G. Ricci, J. J. Adams, M. W. Parker, and M. Lo Bello, *J. Biol. Chem.*, *280*, 42172 (2005).
21. J. R. Hickok, D. Vasudevan, G. R. J. Thatcher, and D. D. Thomas, *Antioxid. Redox Signaling*, *17*, 962 (2012).
22. J. C. Toledo, C. A. Bosworth, S. W. Hennon, H. A. Mahtani, H. A. Bergonia, and J. R. Lancaster, Jr., *J. Biol. Chem.*, *283*, 28926 (2008).
23. J. C. Drapier, *Methods*, *11*, 319 (1997).
24. M. A. Keese, M. Böse, A. Mülsch, R. H. Schirmer, and K. Becker, *Biochem. Pharmacol.*, *54*, 1307 (1997).
25. H. C. Lok, Y. S. Rahmanto, C. L. Hawkins, D. S. Kalinowski, C. S. Morrow, A. J. Townsend, P. Ponka, and D. R. Richardson, *J. Biol. Chem.*, *287*, 607 (2012).
26. R. Y. Suryo, D. S. Kalinowski, D. J. Lane, H. C. Lok, V. Richardson, and D. R. Richardson, *J. Biol. Chem.*, *287*, 6960 (2012).
27. D. R. Hyduke, L. R. Jarboe, L. M. Tran, K. J. Y. Chou, and J. C. Liao, *Proc. Natl. Acad. Sci. USA*, *104*, 8484 (2007).
28. J. C. Crack, L. J. Smith, M. R. Stapleton, J. Peck, N. J. Watmough, M. J. Buttner, R. S. Buxton, J. Green, V. S. Oganesyan, A. J. Thomson, and N. E. Le Brun, *J. Am. Chem. Soc.*, *133*, 1112 (2011).
29. H. Cruz-Ramos, J. Crack, G. Wu, M. N. Hughes, C. Scott, A. J. Thomson, J. Green, and R. K. Poole, *EMBO J.*, *21*, 3235 (2002).
30. A. Styś, B. Galy, R.R. Starzynski, E. Smuda, J. C. Drapier, P. Lipinski, and C. Bouton, *J. Biol. Chem.*, *286*, 22846 (2011).
31. J. H. Enemark and R. D. Feltham, *Coord. Chem. Rev.* *13*, 339 (1974).
32. K. H. Hopmann, L. Noodleman, and A. Ghosh, *Chem. Eur. J.*, *16*, 10397 (2010).
33. S. D. Hamilton-Brehm, G. J. Schut, and M. W. W. Adams, *Appl. Environ. Microbiol.*, *75*, 1820 (2009).
34. J. Bourassa, W. DeGraff, S. Kudo, D. A. Wink, J. B. Mitchell, and P. C. Ford, *J. Am. Chem. Soc.*, *119*, 2853 (1997).
35. D. Lloyd, J. C. Harris, S. Maroulis, A. Mitchell, M. N. Hughes, R. B. Wadley, and M. R. Edwards, *J. Appl. Microbiol.*, *95*, 576 (2003).
36. J. Bourassa, B. Lee, S. Bernard, J. Schoonover, and P. C. Ford, *Inorg. Chem.*, *38*, 2947 (1999).
37. J. L. Bourassa and P. C. Ford, *Coord. Chem. Rev.*, *200–202*, 887 (2000).
38. P. C. Ford, *Acc. Chem. Res.*, *41*, 190 (2008).
39. M.-L. Tsai, C.-H. Hsieh, and W.-F. Liaw, *Inorg. Chem.*, *46*, 5110 (2007).

40. M.-L. Tsai and W.-F. Liaw, *Inorg. Chem.*, *45*, 6583 (2006).
41. T. W. Hayton, P. Legzdins, and W. B. Sharp, *Chem. Rev.*, *102*, 935 (2002).
42. D. Seyferth, M. K. Gallager, and M. Cowie, *Organometallics*, *5*, 539 (1986).
43. R. Wang, M. A. Camacho-Fernandez, W. Xi, J. Zhang, and L. Li, *J. Chem. Soc. Dalton Trans.*, *777*, (2009).
44. P. Bladon, M. Dekker, G. R. Knox, D. Willison, G. A. Jaffari, R. J. Doedens, and K. W. Muir, *Organometallics*, *12*, 1725 (1993).
45. H.-H. Chang, H.-J. Huang, Y.-L. Ho, Y.-D. Wen, W.-N. Huang, and S.-J. Chiou, *J. Chem. Soc. Dalton Trans.*, 6396 (2009).
46. C. L. Conrado, J. L. Bourassa, C. Egler, S. Wecksler, and P. C. Ford, *Inorg. Chem.*, *42*, 2288 (2003).
47. A. F. Vanin, *Nitric Oxide*, *21*, 1 (2009).
48. D. R. Richardson and H. C. Lok, *Biochim. Biophys. Acta.*, *1780*, 638 (2008).
49. H. Lewandowska, M. Kalinowska, K. Brzoska, K. Wojciuk, G. Wojciuk, and M. Kruszewski, *J. Chem. Soc. Dalton Trans.*, *40*, 8273 (2011).
50. P. C. Ford, J. Bourassa, B. Lee, I. Lorkovic, K. Miranda, and L. Laverman, *Coord. Chem. Rev.*, *171*, 185 (1998).
51. N. G. Connelly and C. Gardner, *J. Chem. Soc. Dalton Trans.*, 1525 (1976).
52. T. C. Harrop, D. Song, and S. J. Lippard, *J. Am. Chem. Soc.*, *128*, 3528 (2006).
53. C.-C. Tsou, T.-T. Lu, and W.-F. Liaw, *J. Am. Chem. Soc.* *129*, 12626 (2007).
54. T.-T. Lu, S.-J. Chiou, C.-Y. Chen, and W.-F. Liaw, *Inorg. Chem.*, *45*, 8799 (2006).
55. F.-T. Tsai, S.-J. Chiou, M.-C. Tsai, M.-Li. Tsai, H.-W. Huang, M.-H. Chiang, and W.-F. Liaw, *Inorg. Chem.*, *44*, 5872 (2005).
56. H. Strasdeit, B. Krebs, and G. Henkel, *Z. Naturforsch., B: Anorg. Chem. Org. Chem.* *41B*, 1357 (1986).
57. M.-C. Tsai, F.-T. Tsai, T.-T. Lu, M.-L. Tsai, Y.-C. Wei, I.-J. Hsu, J.-F. Lee, and W.-F. Liaw, *Inorg. Chem.*, *48*, 9579 (2009).
58. T. C. Harrop, Z. J. Tonzetich, E. Reisner, and S. J. Lippard, *J. Am. Chem. Soc.*, *130*, 15602 (2008).
59. S.-J. Chiou, C.-C. Wang, and C.-M. Chang, *J. Organomet. Chem.*, *693*, 3582 (2008).
60. M.-L. Tsai, C.-C. Chen, I.-J. Hsu, S.-C. Ke, C.-H. Hsieh, K.-A. Chiang, G.-H. Lee, Y. Wang, J.-M. Chen, J.-F. Lee, and W.-F. Liaw, *Inorg. Chem.*, *43*, 5159 (2004).
61. C.-Y. Chiang, M. L. Miller, J. H. Reibenspies, and M. Y. Darensbourg, *J. Am. Chem. Soc.*, *126*, 10867 (2004).
62. M.-C. Hung, M.-C. Tsai, G.-H. Lee, and W.-F. Liaw, *Inorg. Chem.*, *45*, 6041 (2006).
63. T.-N. Chen, F.-C. Lo, M.-L. Tsai, K.-N. Shih, M.-H. Chiang G.-H. Lee, and W.-F. Liaw, *Inorg. Chim. Acta.*, *359*, 2525 (2006).
64. F. Roncaroli, R. van Eldik, and J. A. Olabe, *Inorg. Chem.*, *44*, 2781 (2005).
65. C. H. Hsieh, S. M. Brothers, J. H. Reibenspies, M. B. Hall, C. V. Popescu, and M. Y. Darensbourg, *Inorg. Chem.*, *52*, 2119 (2013).

66. C.-C. Tsou, F.-T. Tsai, H.-Y. Chen, I. J. Hsu, and W.-F. Liaw, *Inorg. Chem.*, *52*, 1631 (2013).
67. J.-H. Wang and C.-H. Chen, *Inorg. Chem.*, *49*, 7644 (2010).
68. H.-W. Huang, C.-C. Tsou, T.-S. Kuo, and W.-F. Liaw, *Inorg. Chem.*, *47*, 2196 (2008).
69. T.-T. Lu, C.-H. Chen, and W.-F. Liaw, *Chem. Eur. J.*, *16*, 8088 (2010).
70. Z.-S. Lin, T.-W. Chiou, K.-Y. Liu, C.-C. Hsieh, J.-S. K. Yu, and W.-F. Liaw, *Inorg. Chem.*, *51*, 10092 (2012).
71. J. L. Hess, C.-H. Hsieh, J. H. Reibenspies, and M. Y. Darensbourg, *Inorg. Chem.*, *50*, 8541 (2011).
72. F.-T. Tsai, Y.-C. Lee, M.-H. Chiang, and W.-F. Liaw, *Inorg. Chem.*, *52*, 464 (2013).
73. F.-T. Tsai, P.-L. Chen, and W.-F. Liaw, *J. Am. Chem. Soc.*, *132*, 5290 (2010).
74. C.-H. Chen, S.-J. Chiou, and H.-Y. Chen, *Inorg. Chem.*, *49*, 2023 (2010).
75. Z. J. Tonzetich, F. Héroguel, L. H. Do, and S. J. Lippard, *Inorg. Chem.*, *50*, 1570 (2011).
76. Z. J. Tonzetich, L. H. Do, and S. J. Lippard, *J. Am. Chem. Soc.*, *131*, 7964 (2009).
77. F. L. Atkinson, H. E. Blackwell, N. C. Brown, N. G. Connelly, J. G. Crossley, A. G. Orpen, A. L. Rieger, and P. H. Rieger, *J. Chem. Soc. Dalton Trans.*, 3491 (1996).
78. K. M. Skodje, M.-Y. Kwon, S. W. Chung, and E. Kim, *Chem. Sci.*, (2014). DOI: 10.1039/C3SC53319K.
79. N. Reginato, C. T. C. McCrory, D. Pervitsky, and L. Li, *J. Am. Chem. Soc.*, *121*, 10217 (1999).
80. D.-H. Jo, Y.-M. Chiou, and L. Que, *Inorg. Chem.*, *40*, 3181 (2001).
81. W.-C. Shih, T.-T. Lu, L.-B. Yang, F.-T. Tsai, M.-H. Chiang, J.-F. Lee, Y.-W. Chiang, and W.-F. Liaw, *J. Inorg. Biochem.*, *113*, 83 (2012).
82. L. M. Baltusis, K. D. Karlin, H. N. Rabinowitz, J. C. Dewan, and S. J. Lippard, *Inorg. Chem.*, *19*, 2627 (1980).
83. C.-H. Hsieh and M. Y. Darensbourg, *J. Am. Chem. Soc.*, *132*, 14118 (2010).
84. F.-T. Tsai, T.-S. Kuo, and W.-F. Liaw, *J. Am. Chem. Soc.*, *131*, 3426 (2009).
85. H. L. K. Wah, M. Postel, and F. Tomi, *Inorg. Chem.*, *28*, 233 (1989).
86. P. Guillaume and M. Postel, *Inorg. Chim. Acta.*, *233*, 109 (1995).
87. C.-H. Chen, Y.-C. Ho, and G.-H. Lee, *J. Organomet. Chem.*, *694*, 3395 (2009).
88. R. E. Dessy, J. C. Charkoudian, and A. L. Rheingold, *J. Am. Chem. Soc.*, *94*, 738 (1972).
89. K. M. Skodje, P. G. Williard, and E. Kim, *J. Chem. Soc. Dalton Trans.*, *41*, 7849 (2012).
90. R. Wang, X. Wang, E. B. Sundberg, P. Nguyen, G. P. G. Grant, C. Sheth, Q. Zhao, S. Herron, K. A. Kantardjieff, and L. Li, *Inorg. Chem.*, *48*, 9779 (2009).
91. V. G. Albano, A. Araneo, P. L. Bellon, G. Ciani, and M. Manassero, *J. Organomet. Chem.*, *67*, 413 (1974).
92. L. Li, G. D. Enright, and K. F. Preston, *Organometallics*, *13*, 4686 (1994).
93. Y.-J. Chen, W.-C, Ku, L.-T. Feng, M.-L. Tasi, C.-H. Hsieh, W.-H. Hsu, W.-F. Liaw, C.-H. Hung, and Y.-J. Chen, *J. Am. Chem. Soc.*, *130*, 10929 (2008).

94. D. W. McBride, S. L. Stafford, and F. G. A. Stone, *Inorg. Chem.*, *1*, 386 (1962).
95. T.-T. Lu, C.-C. Tsou, H.-W. Huang, I.-J, Hsu, J.-M, Chen, T.-S., Kuo, Y. Wang, and W.-F. Liaw, *Inorg. Chem.*, *47*, 6040 (2008).
96. T. R. Bryar and D. R. Eaton, *Can. J. Chem.*, *70*, 1917 (1992).
97. R. J. Dai and S. C. Ke, *J. Phys. Chem. B*, *111*, 2335 (2007).
98. S. Ye and F. Neese, *J. Am. Chem. Soc.*, *132*, 3646 (2010).
99. C. E. Tinberg, Z. J. Tonzetich, H. Wang, L. H. Do, Y. Yoda, S. P. Cramer, and S. J. Lippard, *J. Am. Chem. Soc.*, *132*, 18168 (2010).
100. Z. J. Tonzetich, H. Wang, D. Mitra, C. E. Tinberg, L. H. Do, F. E. Jenney, Jr., M. W. Adams, S. P. Cramer, and S. J. Lippard, *J. Am. Chem. Soc.*, *132*, 6914 (2010).
101. T. T. Lu, S. H. Lai, Y. W. Li, I. J. Hsu, L. Y. Jang, J. F. Lee, I. C. Chen, and W. F. Liaw, *Inorg. Chem.*, *50*, 5396 (2011).
102. D. Ballivet-Tkatchenko, C. Billard, and A. Revillon, *J. Polym. Sci., Part A: Polym. Chem.* *19*, 1697 (1981).
103. M. A. Marletta, A. R. Hurshman, and K. M. Rusche, *Curr. Opin. Chem. Biol.*, *2*, 656 (1998).
104. H. L. K. Wah, M. Postel, F. Tomi, F. Agbossou, D. Ballivet-Tkatchenko, and F. Urso, *Inorg. Chim. Acta*, *205*, 113 (1993).
105. J. Fitzpatrick, H. Kalyvas, J. Shearer, and E. Kim, *Chem. Commun.*, *49*, 5550 (2013).
106. F. Tomi, H. L. K. Wah, and M. Postel, *New J. Chem.*, *12*, 289 (1988).
107. J.-P. Damiano, V. Munyejabo, and M. Postel, *Polyhedron*, *14*, 1229 (1995).
108. P. Guillaume, H. L. K. Wah, and M. Postel, *Inorg. Chem.*, *30*, 1828 (1991).
109. N. G. Tran, H. Kalyvas, K. M. Skodje, T. Hayashi, P. Moenne-Loccoz, P. E. Callan, J. Shearer, L. J. Kirschenbaum, and E. Kim, *J. Am. Chem. Soc.*, *133*, 1184 (2011).
110. C.-Y. Chiang and M. Y. Darensbourg, *J. Biol. Inorg. Chem.*, *11*, 359 (2006).
111. C.-C. Tsou and W.-F. Liaw, *Chem. Eur. J.*, *17*, 13358 (2011).

Interactions of Nitrosoalkanes/arenes, Nitrosamines, Nitrosothiols, and Alkyl Nitrites with Metals

NAN XU AND GEORGE B. RICHTER-ADDO

University of Oklahoma, Department of Chemistry and Biochemistry, Stephenson Life Sciences Research Center, Norman, OK

CONTENTS

Progress in Inorganic Chemistry, Volume 59, First Edition. Edited by Kenneth D. Karlin.

I. INTRODUCTION

The chemistry of the nitrogen oxides (NO_x) has received increased attention over the last three decades. This interest has, in part, been due to four main areas of interest: (1) understanding the chemical basis of NO_x related environmental pollution and remediation (1–3), (2) enzymatic control of the global N cycle (4–6), (3) the discovery of nitric oxide (NO) as a biological signaling molecule (7, 8), and (4) emerging opportunities for the synthesis of fine chemicals using organonitrogen precursors (1, 9).

The chemistry of NO_x species has a long and rich history (10). Several reviews dealing with their coordination and bioinorganic chemistry (11–16), organometallic chemistry (17–19), structure and bonding (20–24), and reactions and catalysis (25–29) are available.

It is well known that NO_x species react with metals to give metal–NO_x derivatives (30), and that NO_x groups form adducts with organic fragments to give organo–NO_x species (31, 32). However, less attention has been given to the chemistry resulting from the interactions between organo–NO_x species and metal complexes.

In this chapter, we focus our attention on the interactions of four types of organo–NO_x ($x = 1$) compounds with metals. This field was reviewed just over a

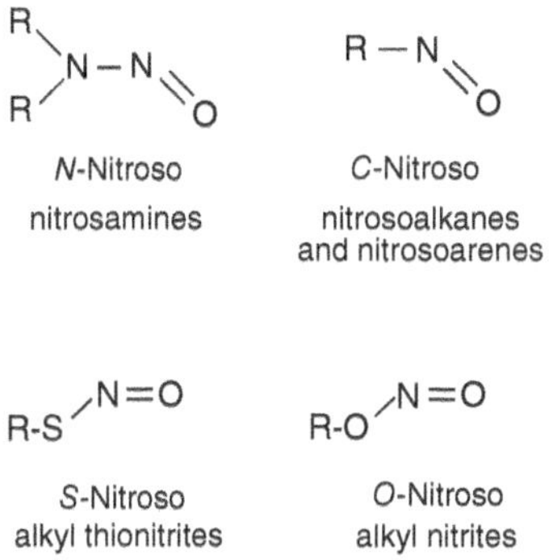

$R_2N{-}N{=}O$ — *N*-Nitroso nitrosamines

$R{-}N{=}O$ — *C*-Nitroso nitrosoalkanes and nitrosoarenes

R-S-N=O — *S*-Nitroso alkyl thionitrites

R-O-N=O — *O*-Nitroso alkyl nitrites

Chart 1.

decade ago (33). This chapter describes the chemistry of *C*-nitroso compounds (RN=O; nitrosoalkanes–arenes), *N*-nitroso compounds ($R_2NN{=}O$; nitrosamines), *S*-nitroso compounds (RSN=O; nitrosothiols, thionitrites), and *O*-nitroso compounds (RON=O; alkyl nitrites). These organo–NO compounds are shown schematically in Chart 1.

Much of the impetus for the study of the interactions of organonitroso compounds with metals derives from the recognition that these interactions have important consequences in biological and synthetic chemistry of NO_x species in general. For example, (1) nitrosoalkanes and nitrosoarenes are known to inhibit critical iron-containing enzymes and other biomolecules by binding to the metal centers (34–38), (2) nitrosamines bind to the heme site of the enzymes cytochrome P450 that metabolically activates them toward carcinogenesis (39–42), (3) nitrosothiol formation is involved in some protein deactivation processes and its chemistry is regulated by metals (43), and (4) alkyl nitrites can serve as oxidants to critical iron-containing biomolecules (44). In this chapter, we describe the fundamental coordination chemistry of these organonitroso compounds with metal centers (33) that form the basis for their observed chemical and biochemical reactivity.

II. NITROSOALKANES AND NITROSOARENES

A. Isolable Metal Complexes

Nitrosoalkanes and nitrosoarenes frequently serve as ligands to metal centers, and in many cases simple adduct formation between the pre-formed *C*-nitroso compounds (45, 46) and metal centers occur with or without the displacement of an existing ligand on the metal. Such *C*-nitroso ligands can also be constructed from precursor components at the metal site. This section describes the synthetic routes to metal-nitrosoalkane and metal-nitrosoarene compounds and the binding modes of the *C*-nitroso ligands that have been established crystallographically (33, 47, 48).

1. Synthesis

a. Addition of Nitrosoalkanes and Nitrosoarenes to Metals. The direct addition of *C*-nitroso compounds to metal complexes is perhaps the most widely used method to prepare metal–RNO complexes. Equation 1 (49), Eq. 2 (TPP = tetraphenylporphyrinato dianion) (50, 51), Eq. 3 (2,4-lut = 2,4-dimethylpyridine or 2,4-lutidine) (52), and Eq. 4 (53) exhibit several recent examples.

$$\mathrm{Pd(CN\{2,6\text{-}(}\mathit{i}\mathrm{\text{-}Pr)_2C_6H_3\})_2 + 2\,PhNO \rightarrow Pd(PhNO)_2(CN\{2,6\text{-}(}\mathit{i}\mathrm{\text{-}Pr)_2C_6H_3\})_2} \quad (1)$$

$$\mathrm{(TPP)Ru(CO) + xs\ PhNO \rightarrow (TPP)Ru(PhNO)_2} \quad (2)$$

$$\mathrm{2\,[Me_2NN]Ni(2,4\text{-}lut) + 3,5\text{-}Me_2C_6H_3NO \rightarrow \{[Me_2NN]Ni\}_2(\mu\text{-}3,5\text{-}Me_2C_6H_3NO)} \quad (3)$$

$$\mathrm{[Cu(MeCN)_4]PF_6 + 3\,}\mathit{p}\mathrm{\text{-}Et_2NC_6H_4NO \rightarrow [Cu(}\mathit{p}\mathrm{\text{-}Et_2NC_6H_4NO)_3]PF_6} \quad (4)$$

The addition of *C*-nitroso compounds to metal complexes does not, however, always result in the formation of the "simple" adduct complexes; the reaction conditions used can alter the reaction pathway to generate more complicated products while retaining the original C–N bonds of the *C*-nitroso moieties. For example, the reactions of the rhenium complexes $(CO)_5ReX$ (X = Cl, Br, I) with 1-nitroso-2-naphthol in refluxing CH_2Cl_2 produced the desired chelated nitroso adduct **1** (υ_{NO} 1380 cm^{-1}) as shown in Scheme 1 (54). In contrast, performing the same reaction in hot toluene resulted in the generation of the N,O-chelating 1,2-naphthaquinone-2-imine complex **2** in which the nitroso functionality had been reduced; the release of CO_2 was confirmed by trapping with $Ba(OH)_2$. In the latter reaction, an initial color change of the reaction mixture to orange, then green, and finally to blue, suggested the formation of relatively stable intermediates along the reaction pathway (54).

Lorenz and co-workers (55) reported the tautomerization of 2-nitroso-*N*-arylanilines upon their coordination to rhenium complexes forming oximine derivatives **3**, as shown in Eq. 5. The authors concluded that the reactions (THF = tetrahydrofuran as solvent) proceeded via the intramolecular proton transfer from the amino group to the nitroso O atom, as judged by proton nuclear magnetic resonance (^1H NMR) spectroscopy and by crystallography; the absence

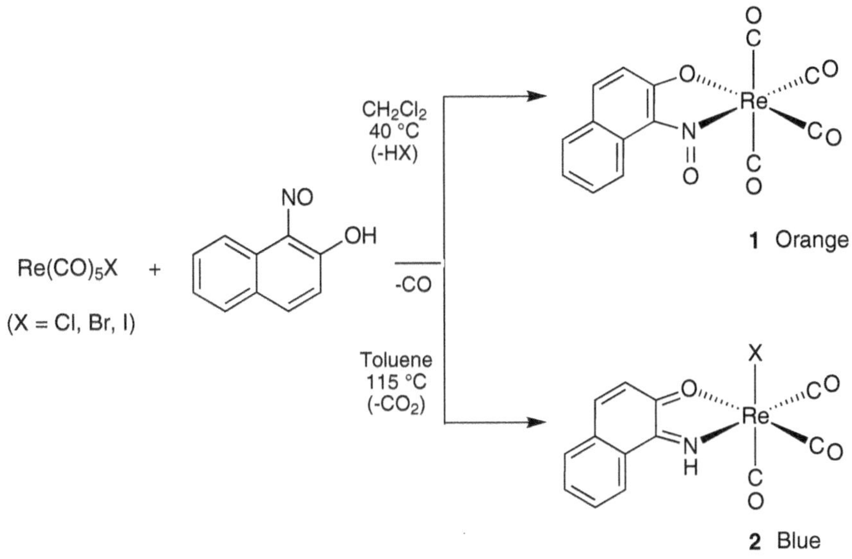

$Re(CO)_5X$ + (X = Cl, Br, I) — CH_2Cl_2, 40 °C (-HX) / Toluene, 115 °C ($-CO_2$); -CO

1 Orange

2 Blue

Scheme 1.

of the nitroso stretching absorption in the infrared (IR) spectra supported the formation of compound **3**.

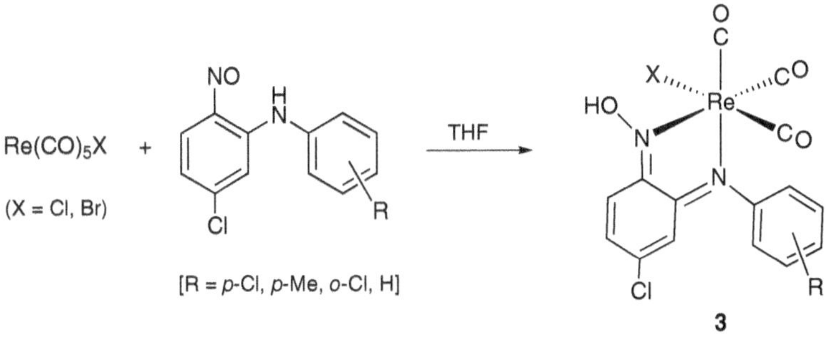

$Re(CO)_5X$ + (X = Cl, Br) [R = *p*-Cl, *p*-Me, *o*-Cl, H] —THF→ **3** (5)

Erker and co-workers (56) demonstrated that nitrosobenzene insertion into a Zr–Me bond occurred when dimethylzirconocene reacted with PhNO; the product was identified by nuclear magnetic resonance (NMR) spectroscopy and crystallography as the hydroxylaminatozirconium complex [η^2-(*O,N*)-ON(Me)Ph]-(Me)$ZrCp_2$ (Cp = η^5-cyclopentadienyl anion). Interestingly, NMR spectroscopic analysis of this complex revealed a rather low activation barrier for the η^2-ON(Me) Ph automerization process.

b. Oxidation of Hydroxylamine Derivatives. Hydroxylamines (RNHOH; R = alkyl, aryl) are formally reduced derivatives of nitroso RN=O compounds. Some metal complexes can participate in the formal oxidations of hydroxylamines. The reaction of the five-coordinate ferric porphyrin (OEP)FeCl (OEP = octaethylporphyrinato dianion) with *i*-PrNHOH in methanol resulted in the formal oxidation of *i*-PrNHOH to generate the six-coordinate mono-alkylnitroso *ferrous* porphyrin derivative with a trans methanol ligand that was hydrogen bonded to a second methanol molecule (Eq. 6) (57).

$$\text{(OEP)FeCl} + i\text{-PrNHOH} + \text{xs MeOH} \rightarrow \text{(OEP)Fe}(i\text{-PrNO})(\text{MeOH}) \cdot \text{MeOH} \quad (6)$$

When the reaction was carried out in the presence of pyridine (py) or 1-methylimidazole (1-MeIm), the (OEP)Fe(*i*-PrNO)(py) and (OEP)Fe(*i*-PrNO)(1-MeIm) derivatives were obtained. The TPP and TTP (TTP = tetratolylporphyrinato dianion) analogues were prepared using this route that followed the procedure reported by Mansuy et al. (58), who proposed that the alkylhydroxylamines served as the reducing agents toward the ferric centers of the metalloporphyrin precursors during these reactions. The υ_{NO}s of the OEP derivatives decrease in the order (OEP)-Fe(*i*-PrNO)(MeOH) (1433 cm^{-1}) > (OEP)Fe(*i*-PrNO)(py) (1429 cm^{-1}) > (OEP)Fe-(*i*-PrNO)(1-MeIm) (1423 cm^{-1}), reflecting the increased relative electron richness in the 1-MeIm derivative. The X-ray crystal structures of several of these complexes were determined (57).

Oxidation of a Pt bis-hydroxylamine complex **4** by Cl_2 in dry chloroform to its bidentate mono-nitrosoalkane derivative **5** has been reported (Scheme 2); in the

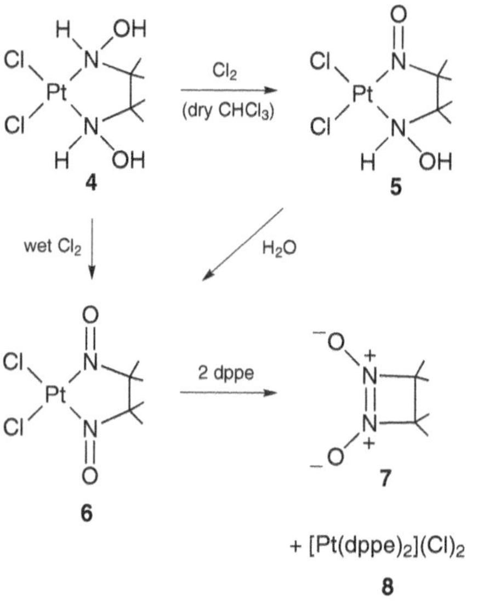

Scheme 2.

presence of water, this reaction proceeded to the dinitrosoalkane derivative **6** (59). The generation of these isolable, and crystallographically characterized, mononitroso and dinitroso derivatives as a function of reaction conditions is remarkable. The nitrosoalkane groups displayed υ_{NO}s in the 1525–1540 cm^{-1} region in their IR spectra. Liberation of the organic dinitrosoalkane ligand as the cyclized 3,3,4,4-tetramethyl-1,2-diazete-1,2-dioxide (**7**) was achieved by ligand displacement using dppe [dppe = 1,2-bis(diphenylphosphino)ethane], forming **8** as the byproduct.

Heating of a hydroxylaminooxime platinum compound in air resulted the generation of a mono-nitrosoalkane complex **9** (R = Me; υ_{NO} 1547 cm^{-1}) whose structure, determined by X-ray crystallography, shows the binding of both the nitroso and oxime N atoms to the metal center (60).

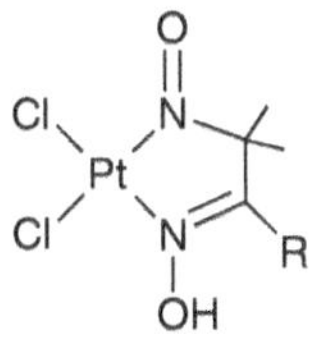

O
N
Cl
Pt
Cl
N
R
OH
9

c. Insertion of the NO Moiety into Metal-Carbon Bonds. Insertion of the nitrosonium cation, NO^+, into the metal–carbon bonds of the $CpCr(NO)_2R$ compounds **10** (R = alkyl, Ph) generated nitrosoalkane and nitrosoarene products (Scheme 3) (61).

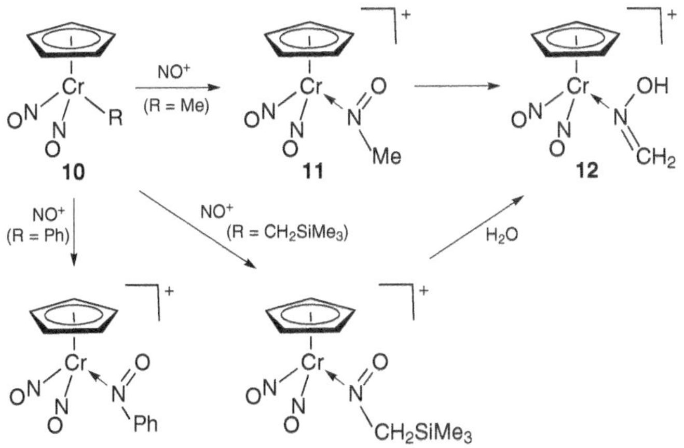

Cr ON N O R **10**
NO^+ (R = Me)
Cr ON N O N O Me **11** +
Cr ON N O N OH CH_2 **12** +
NO^+ (R = Ph)
NO^+ (R = CH_2SiMe_3)
H_2O
Cr ON N O N O Ph **13** +
Cr ON N O N O CH_2SiMe_3 **14** +

Scheme 3.

Specifically, reaction of the dinitrosyl compound $CpCr(NO)_2Me$ (**10**; R = Me) with NO^+ gave the brown complex **11** that isomerized to the green formaldoxime product **12** that was characterized by IR and NMR spectroscopy, and by X-ray crystallography (61, 62). When R = Ph, the red nitrosobenzene complex **13** was generated; the PhNO ligand could then be displaced by chloride to give free PhNO and the $CpCr(NO)_2Cl$ compound that could be reconverted to $CpCr(NO)_2Ph$ to start the process over, constituting a net stoichiometric generation of PhNO. When $R = CH_2SiMe_3$, the C–N bond-formation product **14** was generated as a green viscous oil that was moisture sensitive; it readily converted upon exposure to moisture to the formaldoxime product **12**.

Nitrosonium cation insertion into the metal–carbon bonds of a tricobalt cluster complex was also reported (63). Recently, Wong and co-workers (64) described the insertion of NO^+ into ruthenium–aryl bond of a cyclometalated ruthenium(II) complex (**15**) to give its *C*-nitroso derivative **16** (Eq. 7; υ_{NO} 1374 cm^{-1}). The nitrosonium insertion reaction was quite general and occurred for several S-chelated Ru compounds and py derivatives (64, 65). Kinetics and density functional theory (DFT) studies suggested a direct insertion pathway for NO^+ without the need for the formation of a Ru nitrosyl intermediate. The X-ray crystal structural and IR spectroscopic data (υ_{NO} 1365–1396 cm^{-1}) of the products suggested that the bound nitrosoarenes maintained neutral $(ArNO)^0$ character (see Section II.A.3).

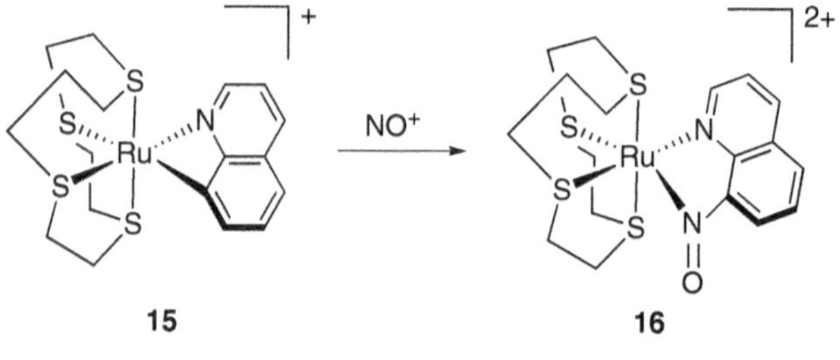

15 $\xrightarrow{NO^+}$ **16** (7)

Metal-mediated C–N bond formation also occurs via *intramolecular* insertion of NO into metal–carbon bonds. Notable examples of such intramolecular NO insertions have been reported previously for several metal complexes including those of W (66), Fe (67, 68), Ru (69), and Co (70–72). Onishi and co-workers (73) recently reported the formation of a dimeric nitrosoethane Ru complex (**18**) [Tp = tris((pyrazolyl)borate] from heating a benzene solution of a Ru nitrosyl ethyl precursor compound (**17**) (Scheme 4); the identity of the dimer **18** was confirmed by X-ray crystallography. When **18** was heated in the presence of PhNO, the products $TpRu(PhNO)_2Cl$ and TpRu(PhNO)(N(OH)=CHMe) were obtained. Heating **17** in the presence of PPh_3 produces the monomeric NO

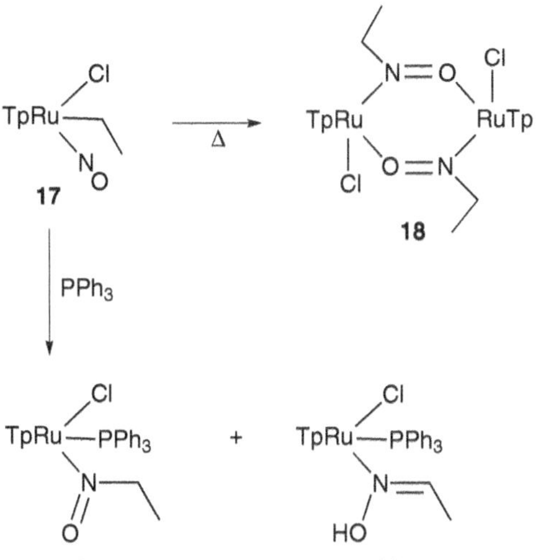

Scheme 4.

insertion product **19** along with a minor product **20** (the latter complex was also obtained when **19** was reacted with ethylamine) (73).

Legzdins and co-workers (74) reported that a hydroperoxide-initiated intramolecular NO insertion reaction occurred in the nitrosyl dialkyl complexes (**21**) of Mo ($R = CMe_3$) and W ($R = SiMe_3$), as shown in Eq. 8. The formation of an oxo metal intermediate was proposed to facilitate the NO insertion process via a resulting reduced back-bonding to the bound NO, making the nitrosyl N atom more electrophilic and more susceptible to the coupling reaction. The resulting η^2-nitrosoalkane complex **22** (M = Mo; $R = CMe_3$) reacted readily with O_2 in benzene to liberate the corresponding nitroalkanes and the dioxo $Cp^*Mo(O)_2CH_2CMe_3$ product ($Cp^* = \eta^5$-pentamethylcyclopentadienyl anion).

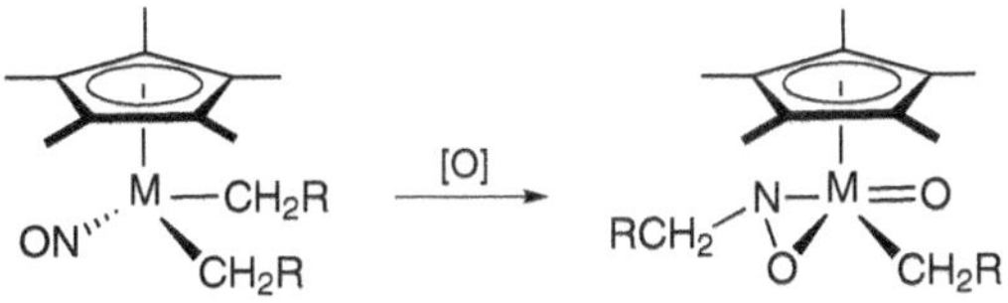

(M = Mo, W) (8)

d. From N–C Bond Coupling Reactions. Brunner and Loskot (75) reported the formation of a dinitrosoalkane complex **23** from the reaction of $[CpCo(NO)]_2$ with norbornene and NO, as shown in Eq. 9 ($Cp = \eta^5$-cyclopentadienyl anion).

Several derivatives containing substituted olefins (cyclic and noncyclic) have been prepared and characterized (75–77). Importantly, the dinitrosyl compound CpCo $(NO)_2$ has been shown to be an intermediate in these alkene–NO coupling reactions (76, 78).

$^1/_2$ $[CpCo(NO)]_2$ —NO→ CpCo(N(=O))2-norbornane **23** (9)

Compounds analogous to **23** have been obtained when $[(TMEDA)Co(NO)_2]BPh_4$ (TMEDA = tetramethylethylenediamine) was reacted with various sodium cyclopentadienyls in the presence of norbornene; extensions to tris-(pyrazolyl)borates were also made (79). Applications of this C–N ligand-coupling reaction in the functionalization of olefinic/vinylic C–H bonds to give elaborated organics have been reported (80–82).

The $Ru(NO)_2Cl_2(thf)$ complex (thf = tetrahydrofuran as ligand) contains both bent [124.0(4)°] and linear [178.5(6)°] RuNO moieties. This feature has been utilized successfully for double C–N bond-coupling reactions to generate dinitrosoalkane derivatives (e.g., **24** Eq. 10) (83). The thf ligands in compound **24** could then be replaced by TMEDA; the product was crystallized and characterized by crystallography. Other alkenes have been employed for this reaction, and the

$Ru(NO)_2Cl_2(thf)$ + norbornadiene —THF→ **24** (10)

authors made an interesting observation that dinitrosoalkane products were obtained when strained and non-enolizable alkenes were used, whereas mixed nitrosoalkane/oxime products were obtained when 1,1-disubstituted and 1,1,2-trisubstituted alkenes were used (83).

The reactions of $TpRu(NO)Cl_2$ with 2-vinylpyridines in the presence of excess Et_3N in refluxing CH_2Cl_2 yielded nitrosovinyl derivatives (e.g., **25**) via N–C coupling of the NO ligand with the 2-vinylpyridine accompanied by a concurrent C–H bond activation (Scheme 5) (84). Complex **25** displays intriguing reactivity with phosphine to generate a π-coordinated nitrile **26** (major) and cationic **27**; all three compounds were characterized crystallographically.

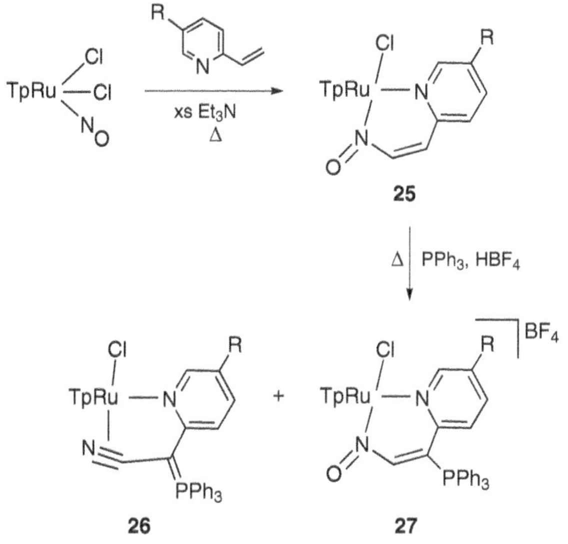

Scheme 5.

The formation of Ru nitrosoethenolato species (**28**) was achieved during the reactions of $TpRu(NO)Cl_2$ with triethylamine in the presence of PPh_3, a process involving C–H bond activation, oxidative dehydrogenation of the amine, and enamine hydrolysis (Scheme 6) (85).

When an alternate tertiary amine is used instead, compound **28** was also isolated but in conjunction with compounds **29** and **30**, with the latter containing the 1-ethyl-3-nitroso-2-pyrroline moiety. Compounds **28–30** were characterized by crystallographically as well.

An intramolecular C–N coupling occurs between a nitrosyl ligand and a chelate during the reactions of the Ru–NO moiety of the *mer* complex **31** with sodium azide in methanol, producing a nitrosoalkane derivative **32** (Eq. 11; υ_{NO} 1408 cm^{-1}); this reaction suggested a C–H bond cleavage with subsequent C–N bond formation (86). The azide ligands were displaceable by chloride to yield the dichloro analogue of **32** (υ_{NO} 1415 cm^{-1}).

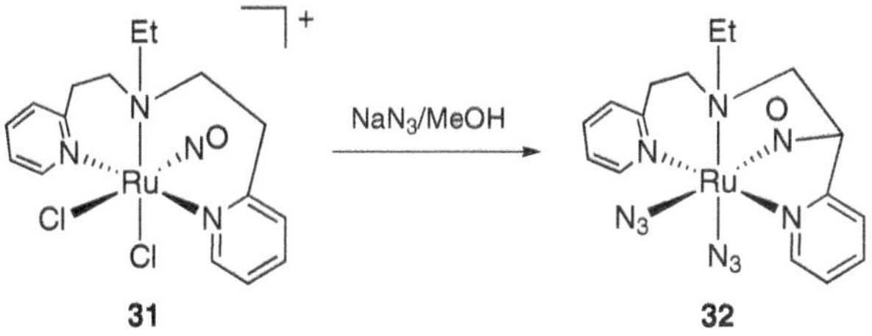

(11)

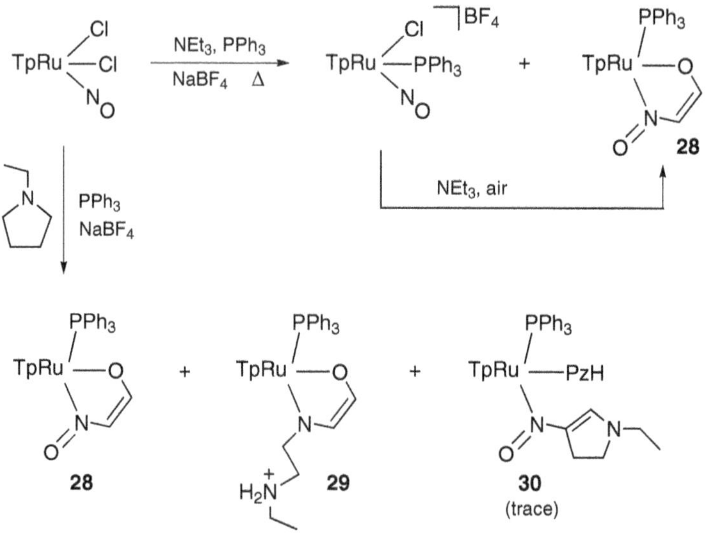

Scheme 6.

Examples of nucleophilic attack of well-defined carbanions on coordinated nitrosyl groups resulting in formation of metal–RNO complexes have been reported. The electrophilic metal nitrosyl complexes, such as nitroprusside $[Fe(CN)_5(NO)]^{2-}$, *trans*-$[RuCl(py)_4(NO)]^{2+}$, and $[Ru(bpy)_2(NO)X]^{2+}$ (bpy = 2,2′,-bipyridine; X = Cl, NO_2) have been reported as targets of carbanion nucleophiles (27, 87–92). The reactions of the NO bridged dinuclear complex $[CpRh(CO)]_2(\mu\text{-}NO)]^+$ with alkynes resulted in the generation of coordinated RNO groups (93).

Ikariya and co-workers (94) reported the synthesis of the nitrosomethane ruthenium complexes **34** (R = H, Me) from nucleophilic attack of the methyl group on the bound nitrosyl in **33** (Scheme 7). Interestingly, they observed a rare reversible (OTf = triflate anion) protonation process to give the hydroxylamido derivatives **35** that could be isolated in two isomeric forms that were subsequently crystallographically characterized (R = Me).

Doctorovich and co-workers (95) reported the reactions of the electron-poor complex $K[Ir(NO)Cl_5]$ **36** (υ_{NO} 2006 cm^{-1}; KBr) with dicyclopentadiene to give the coordinated *C*-nitrosochloroalkane derivative **37** (Eq. 12; υ_{NO} 1520 cm^{-1}). The authors proposed that the nucleophilic attack of the alkene on the highly electrophilic nitrosyl group was accompanied by the *syn* addition of chloride to generate the syn-1-chloro-2-nitroso-1,2-dihydrodicyclopentadiene ligand.

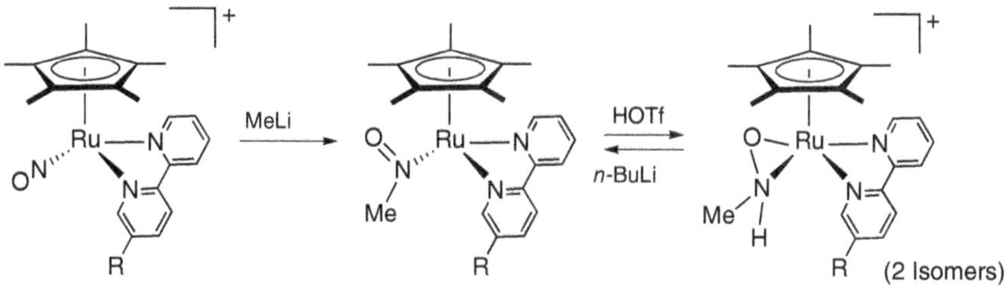

Scheme 7.

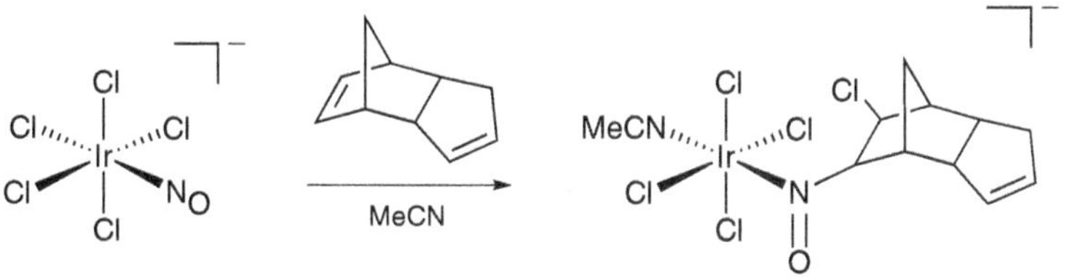

(12)

e. From Organic Nitro Compounds. It is well known that organic nitro compounds can be deoxygenated in the reactions with metal carbonyl or phosphine complexes to afford metal–RNO derivatives. The reactions of $ArNO_2$ with Ru(dppe)-$(CO)_3$ deoxygenated the nitroarenes, generating Ru(dppe)$(CO)_2$(η^2-ONAr) products and CO_2 (96, 97). Skoog et al. proposed that an inner-sphere electron transfer to the nitroarene was involved in these O-atom transfer reactions. Bunker and O'Connor (98) reported the reactions of $CpCo(PPh_3)_2$ with nitroalkyl and nitroarene organics, generating either monomeric nitrosoalkane (η^1-RNO) complexes **38** (R = Me, Et, *i*-Pr; υ_{NO} 1301 cm^{-1} for the MeNO complex) or the dimeric nitrosoarene;η^2-*N*, *O*-RNO derivatives **39** (Scheme 8, TMS = tetramethylsilyl).

It is interesting to note that, in Scheme 8, the reactions employing nitro*alkanes* resulted in the formation of the sole *N*-binding nitroso derivatives, while the use of nitro*arenes* produced μ-η^1-*N*:η^2-*N*,*O*-RNO derivatives.

The deoxygenation of nitrophenols by $Ru(CO)_3(PPh_3)_2$ (99) and $Co(PMe_3)_4$ (100) has been reported, producing chelated nitrosophenolate metal complexes. In the latter case, bimetallic imidonaphtholatodicobalt $Co_2[\eta^3\text{-}O;\mu^2\text{-}N][C_{10}H_6(N)O]$-$(PMe_3)_3$ and amidonaphtholatodicobalt $Co_2[(\eta^3\text{-}O,\mu^2\text{-}NH)(C_{10}H_6(NH)O)]_2(PMe_3)_4$ species were obtained when xs $Co(PMe_3)_4$ was present.

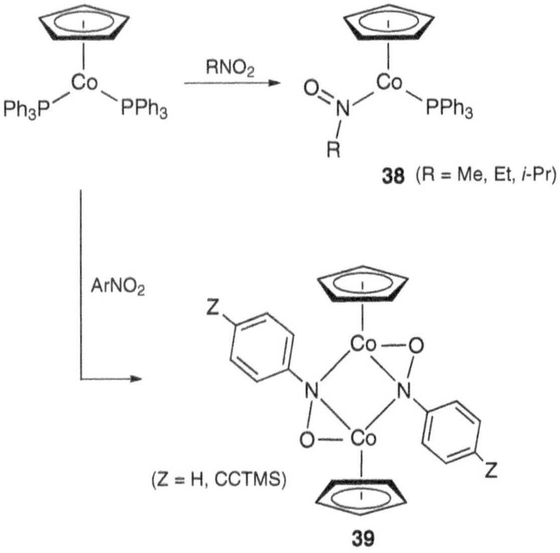

Scheme 8.

An iridium nitroso complex (**41**) (υ_{NO} 1483 cm^{-1}) was obtained from the reaction of the substituted nitroarene $HC{\equiv}C(C_6H_4)NO_2$ containing a terminal alkyne with compound **40** as shown in Eq. 13 (101). The deoxygenation of the nitro group in the terminal alkyne resulted from the intramolecular oxygen transfer from the nitro group to the C≡C bond. In contrast, the use of internal alkynes resulted in the production of iridium hydride anthracil complexes instead.

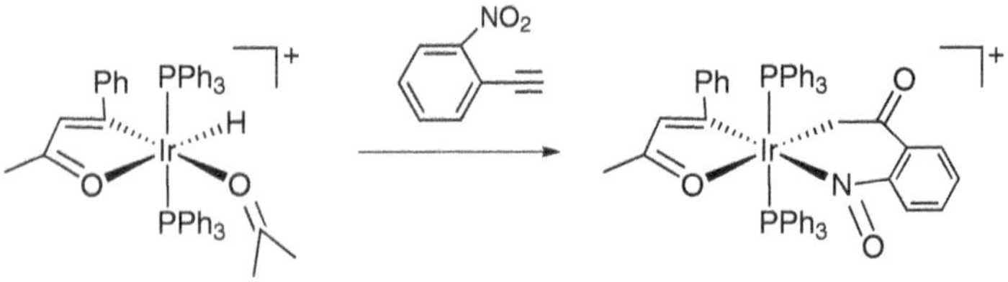

(13)

Note that other routes for the preparation of nitroso–metal complexes involving sulfinilamines (102), nitronyl nitroxides (103), and nitrene precursors (102, 104) have been reported. The radiolysis methodology has been applied for the preparation of metal–RNO complexes from aqueous solutions of $[Fe(CN)_5NO]^{2-}$ and $[Ru(NH_3)_5NO]^{3+}$ in the presence of *t*-BuOH, *t*-BuNH$_2$ and other organic compounds (105–107).

***N*-Binding**

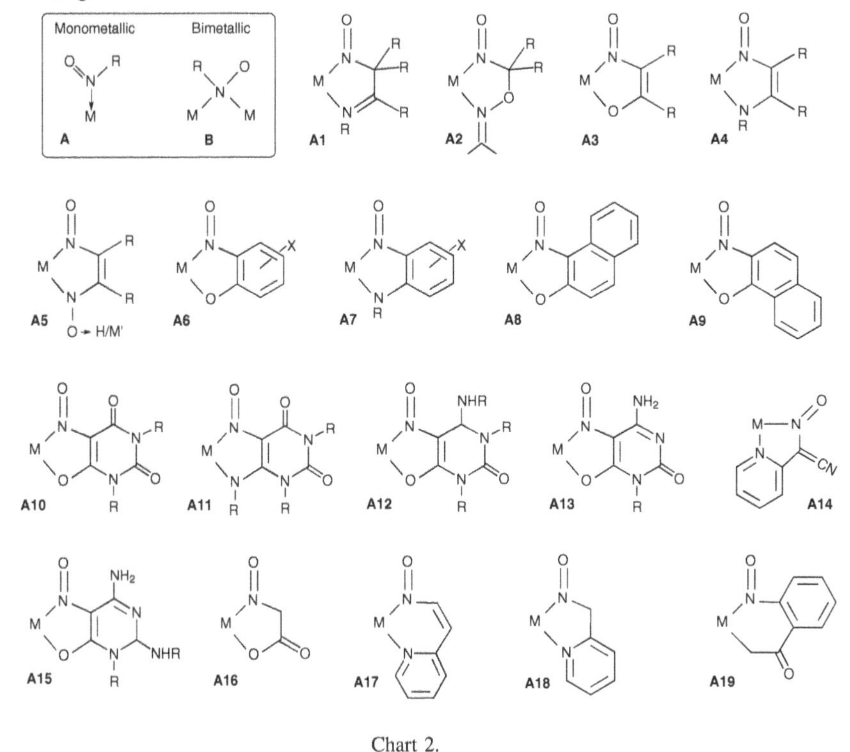

Chart 2.

2. *Coordination Modes of Nitrosoalkane and Nitrosoarene Ligands*

The presence of two heteroatoms (N and O) with electron lone pairs and π-bonding electrons provides a rich coordination chemistry for *C*-nitroso ligands in their metal complexes. There are three main categories for the binding modes of *C*-nitroso compounds: sole *N*-binding, sole *O*-binding, and *N,O*-binding. Chart 2 displays the *N*-binding modes characterized by X-ray crystallography to date, whereas Chart 3 displays the sole *O*-binding and the *N,O*-binding modes characterized to date.

It is impractical to list all the complexes belonging to these groups that have been reported. The reader is referred to an earlier review of this topic for the listings of both the organic *C*-nitroso compounds and the metal–nitroso complexes characterized by crystallography and reported prior to 2002 (33). In this section, only a representative set will be presented and discussed, with a focus on compounds reported after 2002.

***O*-Binding**

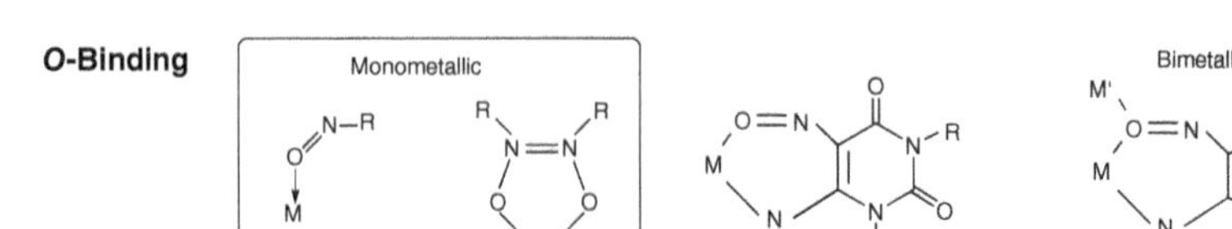

***N,O*-Binding**

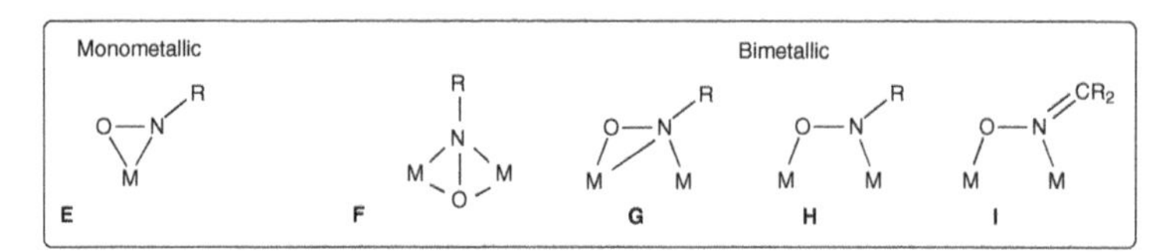

Chart 3.

a. Sole N-binding. The sole *N*-binding mode is by far the most common mode determined in the X-ray crystal structures of metal–RNO complexes. The C–N/N–O bond lengths and C–N–O bond angles of selected metal–RNO complexes in this group are listed in Table I. The data from the crystal structures of this group of *N*-binding complexes reveal N–O bond lengths in the general 1.202–1.300-Å range, which is similar to the range previously reported (33). A longer N–O bond length of 1.344(6) Å was reported, however, for a silver–RNO coordination polymer (108).

Although not specifically covered in this chapter, we note that the sole *N*-binding mode **A** is the only one reported to date for the handful of nitrosoalkane and nitrosoarene derivatives of heme proteins, such as Mb(EtNO) (Mb = myoglobin) (37), Hb(MeNO) (Hb = hemoglobin; PDB accession code 4M4A) (125), Hb(EtNO) (PDB accession code 4M4B) (125), and legHb(PhNO) (leg Hb = Hb found in leguminous plants) (126).

Returning to Chart 2, numerous variations in binding mode **A** have been reported. Indeed, many nitroso-containing ligands exhibit the *N*-binding mode with assistance from chelation by other non-nitroso heteroatoms. Examples for structures **A1–A14** have been presented and discussed previously (33). Representative new examples are shown schematically as **A15–A19**. A few such new nitroso chelates, with NO bond distances shown in square brackets, are now presented. Structure **A1** has been reported for a Pt complex [1.206(15) Å; υ_{NO}

TABLE I
Selected Structural Data for η^1-*N*-Bonded *C*-Nitroso Groups **A** and **B** from Chart 2

Compound	C–N (Å)	N–O (Å)	∠CNO (°)	References
	Monometallic (**A**)			
$W(CO)_5(t\text{-BuNO})$	1.51(3)	1.24(3)	111(2)	109
$[CpFe(CO)(PPh_3)(t\text{-BuNO})]BF_4$	1.554(15)	1.234(10)	111.4(8)	109
$[CpFe(CO)_2(PhNO)]SbF_6$	1.450(3)	1.226(3)	114(2)	110
$[Fe(TIM)(MeCN)(PhNO)](PF_6)_2{}^a$	1.465(2)	1.240(2)	113.0(1)	111
$Cp^*Ru(PPhMe_2)(Ph)(EtNO)$	1.494^b	1.257^b	109.3^b	69
$TpRu(PPh_3)(Cl)(EtNO)$	1.509(19)	1.204(13)	107.9^b	73
$TpRuCl[N(OH){=}CHMe](PhNO)$	1.460^b	1.261(2)	112.4^b	73
$[Cp^*Ru(MeNO)(\eta^2\text{-}C_6H_4py)]OTf$	1.47(1)	1.27(1)	110.7(7)	94
$CoCl_2(ONC_6H_4NMe_2\text{-}p)_2$	1.356(6)	1.261(5)	118.6(4)	112
	1.363(6)	1.272(5)	117.8(4)	
$CpCo(PPh_3)(MeNO)$	1.480(2)	1.285(2)	110.2(2)	98
$CpCo(PPh_3)(EtNO)$	1.484	1.282	109.9^b	71,72
$RhCl(NBD)(ONC_6H_4Br\text{-}p)^c$	1.43(1)	1.23(1)	114.4(7)	113
$(AsPh_4)[IrCl_4(MeCN)(ON\text{-ligand})$ **37**	1.494(9)	1.216(8)	117.3(6)	95
trans-$PdCl_2(PhNO)_2$	1.411(3)	1.209(3)	118.2(2)	114
trans-$PtCl_2(t\text{-BuNO})_2$	1.56(2)	1.21(1)	116(1)	115
$Pd(PhNO)_2(CN\{2,6\text{-}(i\text{-Pr})_2C_6H_3\})_2$	1.382	1.291(2)	118.7	49
$(PMe_3)_2Pt(\mu\text{-PhNO})_2Pt(PMe_3)(\mathit{PhNO})$	1.372^b	1.296(14)	117.3(9)	116
$PtCl\{ON(t\text{-Bu}){=}CHCH_2\}(t\text{-BuNO})$	1.543^b	1.22(2)	114.3^b	117
$[Cu(p\text{-}Et_2NC_6H_4NO)_3]PF_6$	1.358(2)	1.265(2)	118.1(1)	53
	1.378(2)	1.252(2)	117.8(1)	
	1.367(2)	1.258(2)	117.9(1)	
	Bidentate Dinitroso (**A**)			
$(TMEDA)RuCl_2\{(ON)_2C_6H_8\}^d$	1.508(3)	1.231(4)	112.6^b	83
	1.524(3)	1.227(4)	113.2^b	
$(Me_4Cp)Co\{(ON)_2C_{10}H_{14}O_2\}^e$	1.486(3)	1.260(3)	111.9(2)	81
	1.498(3)	1.261(3)	112.2(2)	
$(Me_4Cp)Co\{(ON)_2C_{10}H_{12}O_2\}^e$	1.506(4)	1.266(3)	112.3(3)	81
	1.523(4)	1.257(3)	112.4(2)	
$(Me_4Cp)Co\{(ON)_2C_{10}H_{14}O(NTs)\}^e$	1.507(5)	1.261(4)	112.2(3)	81
	1.511(5)	1.257(4)	111.9(3)	
$Tp^*Co\{(ON)_2C_7H_{10}\}^{e,f}$	1.498(3)	1.256(2)	112.9^b	79
	1.504(3)	1.248(3)	113.0^b	
$CpCo\{(ON)_2C_7H_{10}\}^e$	1.492(5)	1.252(4)	112.4^b	77
	1.487(5)	1.247(4)	112.8^b	
$PtCl_2\{(ON)_2C_2(Me)_4\}$ (**6**)	1.527(9)	1.205(8)	117.3(6)	60
	1.534(9)	1.202(8)	117.1(6)	

(*continued*)

TABLE I
(Continued)

Compound	C–N (Å)	N–O (Å)	∠CNO (°)	References
	Porphyrin Complexes (**A**)			
(TPP)Fe(*i*-PrNO)(*i*-$PrNH_2$)[g]	1.55(2)	1.26(2)	117(1)	58
	1.54(2)	1.26(2)	115(1)	
(OEP)Fe(*i*-PrNO)(MeOH)·MeOH	1.473(3)	1.255(2)	121.2(3)	57
(OEP)Fe(*i*-PrNO)(py)	1.493(4)	1.233(3)	115.6(2)	57
(OEP)Fe(*i*-PrNO)(1-MeIm)	1.372(6)	1.26(2)	112.0(4)	57
(TTP)Fe(*i*-PrNO)(1-MeIm)[g]	1.503(4)	1.246(4)	113.3(3)	57
	1.457(5)	1.284(4)	112.9(3)	
(TPP)Fe(*i*-PrNO)(1-MeIm)	1.497(4)	1.245(4)	113.5(3)	57
(TPP)Fe$(PhNO)_2$	1.468(4)	1.237(3)	112.4(2)	118
	1.467(3)	1.227(3)	112.6(2)	
(TPP)Fe(py)(PhNO)	1.472(4)	1.249(4)	113.2(3)	119
(TPP)Fe(py)($ONC_6H_4NMe_2$-*p*)	1.437(8)	1.252(6)	112.3(6)	119
(TPP)Fe(1-MeIm)(PhNO)[g]	1.444(9)	1.254(8)	109.9(3)	119
	1.453(2)	1.267(3)	111.2(8)	
(OEP)Fe(1-MeIm)(PhNO)[g]	1.448(6)	1.269(5)	111.7(4)	119
	1.454(6)	1.258(4)	110.1(3)	
(TPP)Ru$(PhNO)_2$	1.468(3)	1.243(3)	112.3(2)	50
	1.452(3)	1.248(3)	113.1(2)	
(TTP)Ru(*o*-tolNO)$_2$	1.455(5)	1.247(4)		51
	1.469(5)	1.251(4)		
(TTP)Ru($ONC_6H_2(Me)_2$OMe-*p*)$_2$	1.442(3)	1.252(2)		51
	1.450(3)	1.252(3)		
(OEP)Ru($ONC_6H_4NMe_2$-*p*)$_2$	1.453(7)	1.248(7)	115.9(5)	120
	1.468(6)	1.257(7)	116.7(5)	
(TPP)Ru(*o*-tolNO)(1-MeIm)	1.462(6)	1.254(5)		51
(TPP)Ru(PhNO)(py)	1.475(4)	1.259(3)	112.3(3)	51
(TTP)Ru(PhNO)(py)	1.466(7)	1.271(5)		51
(TPP)Ru(PhNO)(1-MeIm)	1.462(8)	1.267(7)		50
(TTP)Ru($ONC_6H_2(Me)_2$OMe-*p*)-(1-MeIm)	1.453(7)	1.290(6)		51
(TPP)Os$(PhNO)_2$	1.469(5)	1.249(4)	112.4(3)	121
	1.458(5)	1.259(6)	113.1(3)	
(TTP)Os$(PhNO)_2$	1.475(14)	1.298(13)	109.3(11)	121
	1.43(2)	1.278(12)	113.2(10)	
(TMP)Os$(PhNO)_2$[h]	1.40(2)	1.31(2)	110.4(12)	121
	1.45(2)	1.252(12)	112.5(9)	
(TTP)Os(CO)(PhNO)	1.39(3)	1.26(2)	120(2)	121
(OEP)Os(*o*-tolNO)$_2$	1.481(14)	1.273(13)	108.8(9)	121
	1.441(14)	1.219(12)	116.9(10)	
[(TPP)Co($ONC_6H_4NMe_2$-*p*)$_2$]ClO_4	1.35(2)	1.33(2)	109(2)	120
	1.38(2)	1.27(2)	112(2)	
	Nitrosodicyanomethanide Complexes			
Ir(CO)$(PPh_3)_2${ONC$(CN)_2$}	1.342(21) av.	1.264(16) av.	118.2(13) av.	122
Re$(CO)_5${ONC$(CN)_2$}	1.35(1)	1.26(1)	117.0(8)	123

TABLE I
(*Continued*)

Compound	C–N (Å)	N–O (Å)	∠CNO (°)	References
	Bimetallic (**A**)			
$[Cp^*IrX_2]_2(\mu\text{-}\eta^1\text{-}N\text{-BNPP})^i$ (X = Cl, Br, I)	1.38(2)–1.432(9)	1.245(4)–1.270	114.4(7)–116.6(7)	124
$[Cp^*RhCl_2]_2(\mu\text{-}\eta^1\text{-}N\text{-BNPP})^i$	1.389(3)	1.246(3)	117.1(2)	124
	Bimetallic (**B**)			
$Pt_2(PMe_3)_3(PhNO)_2(\mathit{PhNO})$	1.459[e]	1.428(27)	106.6(15)	116

[a] 2,3,9,10-Tetraazacyclotetradeca-1,3,8,10-tetraene = TIM.
[b] Data obtained from the Cambridge Structural Database.
[c] Norbomadiene = NBD.
[d] Analogue of compound **24**, where TMEDA has replaced the THF molecules.
[e] Analogue of compound **23**.
[f] $HB(3,5\text{-Dimethylpyrazolyl})_3$ = Tp*.
[g] Two independent molecules.
[h] Tetramesitylporphyrinato dianion = TMP.
[i] 1,4-Bis(4-nitrosophenyl)piperazine = BNPP.

1545 cm^{-1}] (60). A Co complex [1.300(4) Å] and several Sm complexes [1.278–1.292 Å] containing structure **A6** were reported recently (100). Several **A8** structures, where the metal is Rh [1.242(4) Å; υ_{NO} 1040 cm^{-1}] (127), Ir [1.249–1.256 Å; υ_{NO} ~1157 cm^{-1}] (128), and Re [1.246(11) Å; υ_{NO} 1380 cm^{-1}] (54) have been determined. The related 2-nitroso-1-naphthol based structure **A9** was observed in a Ag complex [1.274(4) Å] (129); a Ag coordination polymer contains the similar **A9** structure, showing the coordination of 2-nitroso-1-naphthol-4-sulfonate to the Ag center via its hydroxyl O atom and nitroso N atom [1.344(6) Å] (108). Structure **A13** has been confirmed for some Zn complexes at ambient and low (120 K) temperatures (130, 131).

Structure **A15** has been determined for a Cd complex [1.277(4) Å] (132), and a related structure (where –NHR is replaced by a thioketone) determined for a Re complex [1.267(6) Å] (133). Structure **A16** was reported in a complex of Re [1.246 Å] (134), and structure **A17** was reported in some Ru complexes [1.235–1.267 Å] (64, 65). An example of structure for **A18** is found in Ru compound **32**. The rare organometallic chelate structure **A19** has been determined in the structure of the Ir complex **41** [1.231(5) Å, υ_{NO} 1483 cm^{-1}] (101).

b. Sole O-binding. The sole *O*-binding mode, structure **C** in Chart 2, is still relatively rare. Structural data for these compounds are shown in Table II. The crystal structure of $Cl_2Me_2Sn(ONC_6H_4NMe_2)_2$ has been redetermined as part of an important study to demonstrate viability of solid-state ^{17}O NMR spectroscopy for nitroso

TABLE II
Selected Structural Data for the *O*-Bonded *C*-Nitroso Groups **C** and **D** (from Chart 3) and Related Groups

Compound	C–N (Å)	N–O (Å)	∠CNO (°)	References
		Nitrosoarene (**C**)		
$Cl_2Me_2Sn(ONC_6H_4NMe_2)_2$	1.337(5)	1.296(4) Major	114.1(4)	135
	1.387(14)	1.283(15) Minor	113.6(11)	
$Cl_2Zn(ONC_6H_4NMe_2)_2$	1.345(4)	1.305(4)	114.9(3)	138
	1.339(4)	1.304(4)	114.6(3)	
$[Ph_3Pb(ONC_6H_3(F)O)]_x$	1.31	1.33	112	139
$[(TPP)Fe(ONC_6H_4NEt_2)_2]SbF_6$	1.323(11)	1.157(6)	118.7(7)	118
	1.413(13)[a]	1.062(8)[a]	121.8(9)[a]	
$[(TPP)Mn(ONC_6H_4NEt_2)_2]SbF_6$	1.36(2)	1.057(10)	121.9(12)	140
	1.356(13)[a]	1.109(8)[a]	122.0(9)[a]	
$[(Me_6tren)Cu(ONPh)]OTf$[b]	1.374(4)	1.337(3)	110.7(2)	141
$[(Me_6tren)Cu(ONPh)]SbF_6$[b]	1.399(5)	1.320(4)	112.2(3)	141
		Nitrosodicyanomethanide Complexes		
$(TTP)Fe(ONC(CN)_2)$	1.300(8)	1.330(5)	113.6(5)	122
$[\{(Me_2N)_3PO\}_4Yb\{ONC(CN)_2\}_2]^+$	1.37(3)	1.20(2)	111(2)	142
$[(NC)_2CNO]^-$	1.55(4)	1.05(3)	94(2)	
$\{(Me_2N)_3PO\}_4Nd\{ONC(CN)_2\}_3$	1.28(4)	1.29(2)	114(2)	143
	1.38(2)	1.28(2)	118(1)	
	1.59(4)	1.13(2)	96(2)	
		Diazenedioxide Form (**D**)		
$[Fe\{[PhNO]_2\}_3][FeCl_4]_2$	1.442(13)–	1.270(10)–	117.3(8)–	144,145
[N=N = 1.276(11)–1.328(12)]	1.490(14)	1.299(11)	120.6(8)	

[a] Data for the disordered component.
[b] Tris(2-dimethylaminoethyl)amine = Me_6tren.

complexes (135). Interactions between the nitroso O atoms and Zn^{2+} (136) and Sr^{2+} (137) ions have been noted in the crystal structures of their nitroso chelate derivatives.

c. N,O-Binding. As shown in the two binding modes above, both the nitroso N and O atoms are capable of binding to metals (Tables III and IV). The structures of two copper complexes, $[Me_2NN]Cu(\eta^2\text{-}3,5\text{-}Me_2C_6H_3NO)$ (1.333(4) Å) and $[Me_2NN]$-$Cu(\eta^2\text{-}PhNO)$ (1.334(5) Å, 1.338(5) Å; υ_{NO} 1113 cm^{-1}; Me_2NN = 2,4-bis(2,6-dimethylphenylimido)pentyl anion) display the *N,O*-binding mode **E** (52); the N–O bond lengths are shorter than the range of 1.386(3)–1.432(6) Å for other mononuclear η^2-ONR metal compounds. The bimetallic structure **F** has been reported for a bimetallic Ni complex $([Me_2NN]Ni)_2(\mu\text{-}\eta^2{:}\eta^2\text{-}ONAr)$ (Ar = 3,5-$Me_2C_6H_3$) with the lengthened N–O bond of 1.440(4) Å (52); a low υ_{NO} at

TABLE III
Selected Structural Data for *N,O*-Bonded *C*-Nitroso Groups **E** and **F** from Chart 3

Compound	C–N (Å)	N–O (Å)	∠CNO (°)	References
	Monometallic (**E**)			
(L)Mo(O)(PhNO)[a]	1.442(4)	1.386(3)	112.0(2)	150
(L′)Mo(HMPA)(O)(PhNO)[a,b]		1.416(7)	112.6(4)	151
$[PPh_4]_2[(CN)_4Mo(O)(o\text{-tolNO})]$	1.471[c]	1.393(10)	112.5(7)	152
Cp*W(NPh)(Ph)(PhNO)	1.435(7)	1.432(6)	111.0(4)	66
$(dppe)Ru(CO)_2\{ONC_6H_3(o\text{-}CF_3)(p\text{-}Cl)\}$	1.42(1)	1.412(9)	112.2(7)	96,97
$Pt(PPh_3)_2(PhNO)$	1.412(9)	1.410(7)	110.9(5)	153
$[Me_2NN]Cu(3,5\text{-}C_6H_3Me_2NO)$	1.441(6)	1.333(4)	111.5(4)	52
$[Me_2NN]Cu(\eta^2\text{-PhNO})$	1.442(6)	1.338(5)	112.3(4)	52
	1.438(6)	1.334(5)	112.3(3)	
	Bimetallic (**F**)			
$[(Cp^*Rh)_2(\mu\text{-Cl})(\mu\text{-PhNO})]BF_4$	1.422(5)	1.422(4)	113.6(3)	154
$[(TC\text{-}3,5)Hf]_2(\mu\text{-O})(\mu\text{-PhNO})$[d]	1.417(9)	1.500(7)	117.0(6)	155
$([Me_2NN]Ni)_2(\mu\text{-}ONC_6H_3Me_2\text{-}3,5)$	1.438(5)	1.440(4)	112.8(3)	52

[a] *S*-Methyl 3-(2-hydroxyphenyl)methylenedithiocarbazate = L; L′ = tridentate pyridine-2,6-dicarboxylate.
[b] Hexamethylphosphoramide = HMPA.
[c] Data obtained from the Cambridge Structural Database.
[d] Tetraazamacrocyclic tropocoronand = TC-3,5.

TABLE IV
Selected Structural Data for the *N,O*-Bonded *C*-Nitroso Groups **G**, **H**, **I**, **J**, and **K** from Chart 3

Compound	C–N (Å)	N–O (Å)	C–N–O (°)	References
	Bimetallic (**G**)			
$[(CO)_3Fe\{ONC_6H_3(o\text{-Me})(m\text{-Cl})\}]_2$	1.43(1)	1.40(1)	112.2(7)	156
$Os_3(CO)_{11}(ONC_6F_5)$	1.506(25)	1.437[a]	107.8[a]	157
$[CpCo(PhNO)]_2$	1.423(5)	1.385(4)	114.4(3)	158
$Cp^*Rh(\mu\text{-Cl})(\mu\text{-PhNO})Rh(Cl)Cp^*$	1.495(9)	1.432(9)	117.4(6)	154
	1.53(1)[b]	1.37(2)[b]	131(1)[b]	
$[\{P(t\text{-Bu})_3\}Pd(PhNO)]_3$	1.43(3)	1.35(1)	113(1)	159
$([Me_2NN]Cu)_2(\mu\text{-}\eta^2\text{:}\eta^1\text{-}ONC_6H_3Me_2\text{-}3,5)$	1.454(7)	1.375(6)	110.3(5)	52
$([Me_2NN]Cu)_2(\mu\text{-}\eta^2\text{:}\eta^1\text{-ONPh})$	1.445(3)	1.368(2)	112.97(16)	52
$[CpCo]_2(\mu\text{-}\eta^1\text{:}\eta^2\text{-ONAr})$[c]	1.411(3)	1.374(3)	115.4(2)	98
	Bimetallic (**H**)			
$[(CO)_3Re(\mu\text{-X})]_2(\eta^2\text{-}ONC_6H_4NR_2)$[d]	1.335(17)–1.377(10)	1.280(8)–1.319(8)	109.5(18)–113.4(7)	147
$[(CO)_3Re(\mu\text{-X})]_2ONPh$	1.440(8)	1.282(7)	110.5(5)	148
$[(CO)_3Re(\mu\text{-X})]_2ONPhYZ_2$[e]	1.440(11)–1.458(19)	1.257(15)–1.299(8)	109.7(5)–110.9(7)	146
$Sr_2(C_5H_5N_4O_3)_4(H_2O)_6$	1.330(3)	1.290(2)	118.02(17)	149

(continued)

TABLE IV
(Continued)

Compound	C–N (Å)	N–O (Å)	C–N–O (°)	References
$[(CpFe(\mu-\eta^2-ONPh)]_2-\mu-NHPh]SbF_6$	1.446(3)	1.318(2)	109.53(17)	110
	1.441(3)	1.319(2)	107.87(19)	
$Ru_3(CO)_7(NPh)_2(PhNO)_2$ *(no Ru–Ru bond)*	1.44(1)	1.334(9)	111.6[a]	160
$Ru_3(CO)_8(ONC_{10}H_6O)_2$[f] *(no Ru–Ru bond)*	1.33(2)	1.30(1)	120.9[a]	161
	1.38(2)	1.31(1)	115.4[a]	
$[\{Cp^*Ru(S(i-Pr))\}_2(PhNO)]OTf$ *(Ru–Ru bond)*	1.391(6)	1.353(5)	112.8(4)	162
$[\{Cp^*Rh(S(i-Pr))\}_2(PhNO)]BPh_4$ *(no Rh–Rh bond)*	1.391(8)	1.333(6)	109.3(6)	162
$[\{Cp^*Ir(S(i-Pr))\}_2(PhNO)]BPh_4$ *(no Ir–Ir bond)*	1.40(1)	1.339(8)	109.5(7)	162
$Pt_2(PMe_3)_3(PhNO)_2(PhNO)$ *(no Pt–Pt bond)*	1.357[a]	1.433(15)	116.1(9)	116
$(PEt_3)_2Pt(PhNO)Ge\{N(SiMe_3)_2\}_2$ *(Pt–Ge bond)*	1.381(7)	1.498(6)	109.8(4)	163
$ArGe(\eta^1:\eta^1-\mu-PhNO)_2GeAr$				164
$Ar = C_6H_3-2,6-(C_6H_3-2,6-iPr_2)_2$		1.449(5)		
		1.459(4)		
$Ar = C_6H-2,6-(C_6H_3-2,6-iPr_2)_2-3,5-i-Pr_2$		1.4366(19)		
		1.4372(18)		
$ArGe(\eta^1:\eta^1-\mu-PhNO)_3GeAr$		1.4531(18)		164
$Ar = C_6H_3-2,6-(C_6H_3-2,6-i-Pr_2)_2$		1.4519(19)		
		1.4640(19)		
Related Oximato $M_2(\mu-O-N{=}CR_2)$ (**I**)				
$Fe_2(CO)_6(NHi-Pr)(ONCMe_2)$ *(Fe–Fe bond)*	1.278(7)	1.371(5)	115.5(4)	165
$Fe_2(CO)_6(NCMe_2)(ONCMe_2)$ *(Fe–Fe bond)*	1.291(7)	1.351(6)	115.1(5)	166
$Cp_2Co_2(NCPh_2)(ONCPh_2)$ *(Co–Co bond)*	1.299(8)	1.343(6)	118.4(5)	167
Tetrametallic (**J**)				
$Fe_4(CO)_{11}(NEt)(ONEt)$	1.479[a]	1.446[a]	105.9[a]	168
Related Oximato (**K**)				
$(NBu_4)_2[Mo_4O_{12}\{MeC(NH_2)NO\}_2]$	1.298(5)	1.420(4)		169
	1.328(5)	1.418(4)		
$\{MeC(NH_2)NHOH\}_2[Mo_4O_{12}\{MeC(NH_2)NO\}_2]$	1.328(7)	1.428(5)	109.8[a]	169
$(NBu_4)_2[W_4O_{12}\{MeC(NH_2)NO\}_2]$	1.33(1)	1.42(1)		169

[a] Data obtained from the Cambridge Structural Database.
[b] Data for the disordered component.
[c] $Ar = p-C_6H_4CC{\equiv}TMS$.
[d] X = Cl, Br, I; R = Me, Et.
[e] X = Cl, Br, I; Y = H, Cl; Z = Cl, Br.
[f] The NO group is part of a bidentate 1,2-naphthoquinone-2-oximato ligand (binding mode **A8**).

915 cm^{-1} of the related PhNO derivative ($[Me_2NN]Ni)_2(\mu$-η^2:η^2-ONPh) was reported. Structure **G** was determined in the ($[Me_2NN]Cu)_2(\mu$-η^2:η^1-ONAr), ($[Me_2NN]Cu)_2(\mu$-η^2:η^1-ONPh) and $[(\eta^5$-$C_5H_5)Co]_2[\mu$-η^2:η^1-ONAr] complexes (52). Lengthening of the N−O bonds in ($[Me_2NN]Cu)_2(\mu$-η^2:η^1-ONAr) [1.375(6) Å] (52), in ($[Me_2NN]Cu)_2(\mu$-η^2:η^1-ONPh) [1.368(2) Å; υ_{NO} 1040 cm^{-1}] (52), and in $[CpCo]_2[\mu$- η^2:η^1-ONAr] (1.374(3) Å) (98) were observed. In the bimetallic structure **H**, the nitroso N and O atoms bind to two separate metal centers; examples for **H** have the N−O bond lengths in the 1.257–1.319 Å range (110, 146–149). The υ_{NO} range from 1392–1421 cm^{-1} for the $[(CO)_3Re(\mu$-$X)]_2(\eta^2$-$ONC_6H_4NR_2)$ (X = Cl, Br, I; R = Me, Et) complexes (147) and 1329–1350 cm^{-1} for the $[(CO)_3Re$-$(\mu$-$X)]_2ONPhYZ_2$ complexes (X = Cl, Br, I; Y = H, Cl; Z = Cl, Br) (146).

3. Nature of the Interaction Between the C-Nitroso Moieties and Metals

The π-acid character of RN=O ligands was demonstrated previously using IR spectroscopy; the most intense υ_{CO} in the η^1-N complex $W(CO)_5$(*t*-BuNO) (1966 cm^{-1}) is *higher* than that for $W(CO)_5$(piperidine) (1929 cm^{-1}), reflecting less electron density at the metal center in the *C*-nitroso derivative (109). This π-acid character is also evident when comparing the precursor (TTP)Os(CO) $(solv)_x$ (υ_{CO} = 1916 cm^{-1}) with its η^1-*N*-nitrosobenzene adduct (TTP)Os(CO) (PhNO) (υ_{CO} 1972 cm^{-1}); the π-acid PhNO *reduces* overall electron density at the metal center and back-donation to the CO ligand, thus increasing υ_{CO} (121).

In general, RNO ligands are, however, less π acidic than the valence isoelectronic HNO ligand as computed by Zhang and co-workers (170) using the DFT method mPWVWM that accurately reproduced the IR and NMR spectral properties of various HNO, RNO, and NO metalloporphyrins.

Weighardt and co-workers (111) proposed a useful tool to characterize the nature of PhNO binding to metal centers, and this is shown schematically in Table V. Essentially, the bound PhNO ligand can be classified as inherently noninnocent by virtue of the extent of its π-acid character as a function of the complex.

Given this classification, the complex *trans*-$PdCl_2(\eta^1$-*N*-$PhNO)_2$ (υ_{NO} 1496 cm^{-1}; N−O = 1.209 Å) could then best be described as containing essentially

TABLE V
Character of the Bound PhNO in Metal Complexes and Correlation with N−O Bond Distance and IR υ_{NO}

	$(PhNO)^0$ Little or no Backbonding	PhNO*[a] π Acid	$(PhNO)^{\bullet -}$ π Radical	$(PhNO)^{2-}$ Dianion
N–O Distance (Å)	~1.20–1.23	~1.23–1.26	~1.26–1.31	>1.32
υ_{NO} (cm^{-1})	~1500–1400	~1400–1300	~1300–1150	~1150–900

[a] An asterisk indicates considerable metal-to-ligand π back-bonding.

neutral PhNO ligands (111, 114). However, the related $Pd(L)_2(\eta^1\text{-}N\text{-}PhNO)_2$ ($L = CN\text{-}2,6\text{-}\{2,6\text{-}(i\text{-}Pr)_2C_6H_3\}_2C_6H_3$; υ_{NO} 1316 cm^{-1}; N–O = 1.291 Å) can best be described as having a $(PhNO)^{\bullet -}$ radical anion character; this electronic description of the complex was confirmed by using a superconducting quantum interference device (SQUID) (111). Using a similar reasoning, the authors argue that the side-on PhNO groups in the related compounds $Pd(L)_2(\eta^2\text{-}PhNO)$ [N–O = 1.364(4) Å] may thus best be described as possessing substantial dianionic character. Clearly, the structural and spectroscopic data for metal–RNO compounds suggest varying degrees of "electron donation" from the metal centers to these potentially π-acid ligands, a feature that is expressed in the wide range of coordination modes and reactivity patterns for bound RNO groups.

B. Reactivity Resulting from the Metal–RNO Interaction

The C–N–O moiety in *C*-nitroso compounds can undergo a variety of reactions at each site of this C–N–O unit. In this section, we will briefly examine the major classes of such reactions. We will not discuss the simplest reaction, namely, that of simple displacement of the *C*-nitroso ligands from the metal centers.

1. *Isomerization*

Organic nitrosoalkanes of the form $R_2CHN{=}O$ are prone to isomerize to their oxime derivatives $R_2C{=}NOH$, and this also occurs in the presence of some metal centers (33, 171–173). An example was shown earlier in Scheme 3 in the conversion of the nitrosomethane complex **11** to its formaldoxime product **12** (61). The conversion of $Cp^*Ru(PMe_3)(Ph)(ONEt)$ in the presence of added PMe_3 to its oximate derivative $Cp^*Ru(PMe_3)_2(N(O){=}CHMe)$ occurs with a formal loss of PhH (69). The somewhat related conversion of compound **17–20** (Scheme 4) has been reported (73).

2. *Carbon–Nitrogen Bond Cleavage*

Osborn and co-workers (174) reported, in 1968, the reactions of CF_3NO with $Pt(PPh_3)_4$ or Cp_2Ni to form the corresponding $Pt(PPh_3)_2(NO)(CF_3)$ and $CpNi(NO)$ complexes, respectively. Pizzotti et al. (153) subsequently isolated the nitrosoalkane $Pt(PPh_3)_2(\eta^2\text{-}ONCF_3)$ compound from this reaction, and were able to isolate the related Pd analogue.

Other C–N bond cleavage reactions have been observed in a cobalt cluster complex to form its nitrosyl product (63), and during the formal ligand-exchange reactions in analogues of compound **23,** as shown schematically in Eq. 14 that involve C–N bond cleavage and formation (75, 78).

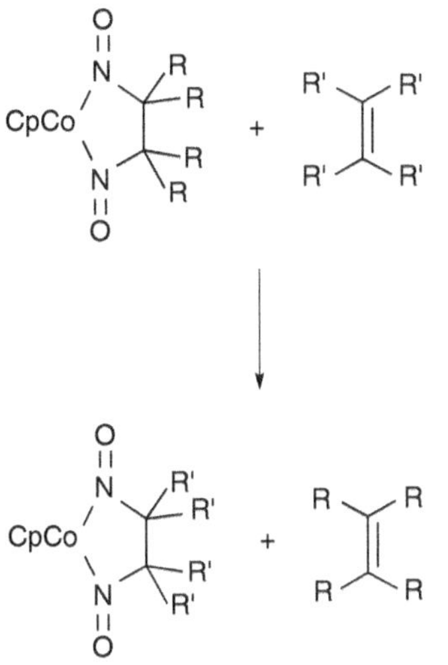

(14)

3. *Nitrogen–Oxygen Bond Cleavage*

Metal-mediated activation of RNO compounds can also result in the breakage of the nitroso RN=O double bond. For instance, a μ-oxo/imido bridged bimetallic complex $\{[Me_2NN]Co\}_2(\mu\text{-O})(\mu\text{-NAr})$ ($Ar = 3{,}5\text{-}Me_2C_6H_3$) was formed by the reaction of the precursor β-diketiminate complex with the O=NAr reagent as shown in Eq. 15 (175).

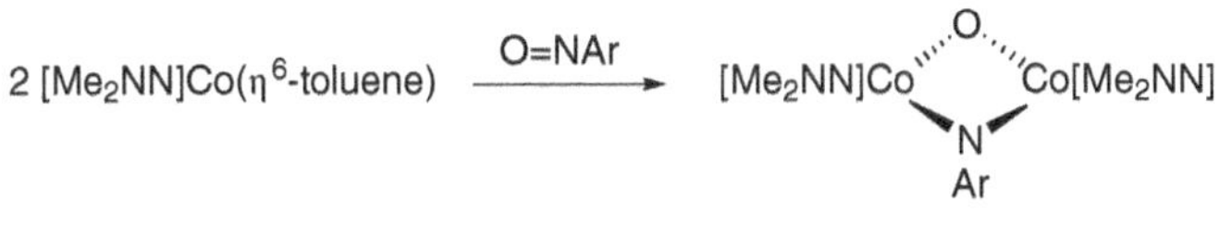

$$2\,[Me_2NN]Co(\eta^6\text{-toluene}) \xrightarrow{O=NAr} [Me_2NN]Co(\mu\text{-O})(\mu\text{-NAr})Co[Me_2NN] \quad (15)$$

Activation of the RN=O bond has been reported in the reaction of nitrosotoluene with a dinuclear germanium complex (**42**) ($Ar' = C_6H_3\text{-}2{,}6(C_6H_3\text{-}2{,}6\text{-}i\text{-}Pr_2)_2$) to give a stable diradical **43** that was crystallographically characterized (Scheme 9) (176). A reaction pathway was proposed that involved initial η^1-N coordination of the nitrosoarene (intermediate **44**) followed by a bidentate μ-η^1-N;η^1-O coordination (intermediate **45**) prior to the N–O bond cleavage.

Interestingly, different products were obtained when PhNO was used in the reaction (Scheme 10). The nitroso-bridged species **46**, with a weak Ge–Ge interaction [Ge–Ge = 2.4731(7) Å], was generated when 2 equiv of PhNO were used (164). The tris-PhNO bridged species **47** formed when 3 equiv of PhNO were used.

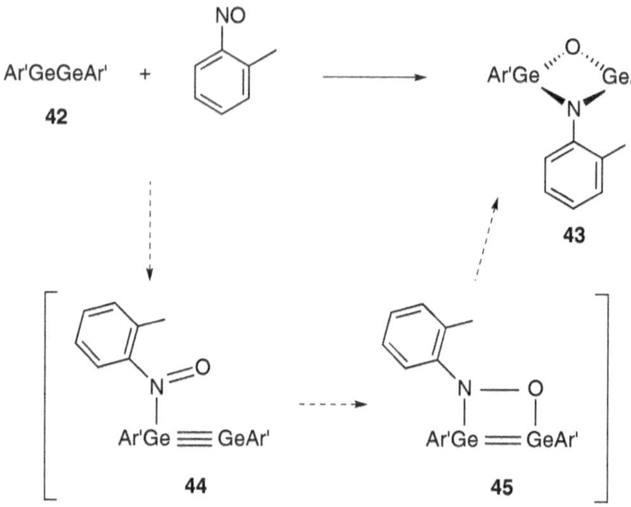
NO
Ar'GeGeAr'
+
42
Ar'Ge
O
GeAr'
N
43
N
O
Ar'Ge
GeAr'
44
N
O
Ar'Ge
GeAr'
45

Scheme 9.

The use of *C*-nitroso compounds as precursor –NR transfer agents for the formation of new carbon–nitrogen bonds mediated by metal complexes have been studied. For example, Nicholas and Srivastava and their co-workers (53, 177) reported the amination of α-methyl styrene by reaction with $[Cu(PhNO)_3]PF_6$. The $[(alkene)Cu(RNO)_3]^+$ and $[(allyl)Cu(RNO)_2(RNHOH)]^+$ species were proposed as intermediates during these amination reactions based on the experimental and computational results.

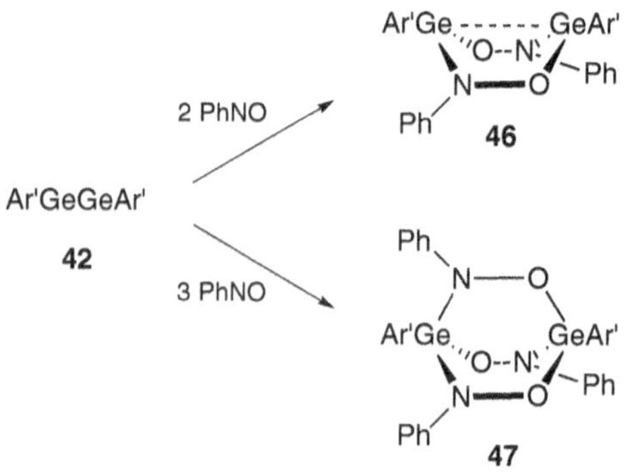
Ar'Ge
GeAr'
O--N
N
O
Ph
Ph
46
2 PhNO
Ar'GeGeAr'
42
3 PhNO
Ph
N
O
Ar'Ge
GeAr'
O--N
N
O
Ph
Ph
47

Scheme 10.

4. *Reactions with NO*

The monometallic complex $[Me_2NN]Cu(\eta^2\text{-}ONAr)$ ($Ar = 3,5\text{-}C_6H_3Me_2$) reacted with NO to yield a diazeniumdiolate product **48** (Eq. 16) (52).

$$[Me_2NN]Cu(\eta^2\text{-}ONAr) \xrightarrow{NO} [Me_2NN]Cu(O_2N_2Ar)\ \mathbf{48} \tag{16}$$

The bimetallic nickel complex $\{[Me_2NN]Ni\}_2(\mu\text{-}\eta^2{:}\eta^2\text{-}ONAr)$ also reacted with NO gas to give the related Ni diazeniumdiolate derivative $[Me_2NN]Ni(\eta^2\text{-}O_2N_2Ar)$ and the nitrosyl $[Me_2NN]Ni(NO)$ (52). The double insertion of NO into a metal–carbon bond of a Ni complex **49** to form the diazeniumdiolate derivative **50** also has been reported (Eq. 17; L = 2,4-lut)] (178). Surprisingly, only a few such double insertions of NO into metal–carbon bonds have been reported previously (28, 30, 179).

$$[Me_2NN]Ni(L)(Et)\ \mathbf{49} \xrightarrow{2\ NO} [Me_2NN]Ni(O_2N_2Et)\ \mathbf{50} \tag{17}$$

5. *Protonation of Coordinated RNO*

The coordinated nitrosomethane group in the Ru complex (**34**) (R = H, Me) undergoes a rare reversible *N*-protonation to produce the hydroxylamido ruthenium complexes **35**, as shown in Scheme 7 (94).

6. *Coupling Reactions of Bound RNO*

The side-on bound RNO ligands in the complex $Pt(PPh_3)(RNO)$ (R = Ph, *t*-Bu, CF_3) can be attacked by various reagents to yield RNO coupled products, as shown in Scheme 11 (153, 180–185). The diversity of such reactions at monometallic and bimetallic centers is quite impressive (96, 163).

7. *Carbon–Hydrogen Functionalization of Coordinated Dinitrosoalkanes*

Elaboration of coordinated dinitrosoalkane ligands in Co derivatives analogous to compound **23** has been accomplished elegantly by Bergman and co-workers (80, 81). In particular, such cobalt dinitrosoalkane complexes have been employed

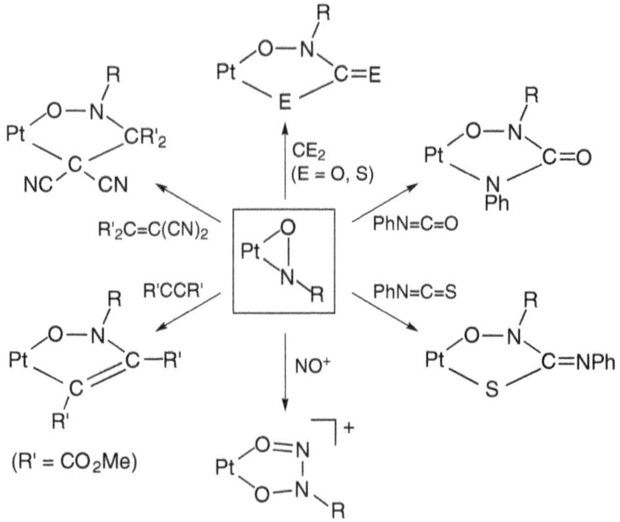

Scheme 11.

successfully for the overall C−H functionalization of alkenes (80) and the [3 + 2]-annulation of alkenes with α,β-unsaturated ketones and imines (81).

III. NITROSAMINES

The most recognized *N*-nitroso compounds are the nitrosamines with the general formula R_2NNO, where the "R_2N" fragment is derived from an amine or amine-like species. Nitrosamines are generally regarded as toxic and/or carcinogenic, and their biological effects have been reviewed extensively (39, 41, 42, 186, 187). The metabolic activation of nitrosamines by the cytochrome P450 class of enzymes is necessary for the expression of the carcinogenic properties of these molecules (188–193).

Nitrosamines interact with the P450 enzymes via the normal distal pocket binding and/or via the direct interaction with the heme-Fe center (188). Importantly, DeVore and Scott (194) recently determined the X-ray crystal structures of tobacco-specific nitrosamines in the active site of human cytochrome P450s, and showed how the distal pocket binding resulted in nitrosamine metabolism. The unexplored direct interaction of *N*-nitroso compounds with the metal centers in heme proteins is partly responsible for the interest in the metal-mediated bioinorganic chemistry of the nitrosamine class of compounds (33).

Historically, nitrosamines have been employed as NO donor molecules to metal centers in inorganic coordination compounds (30), a process that necessarily involves the cleavage of the nitrosamine N–N bonds. Indeed, nitrosamines, such as Et_2NNO, Ph_2NNO, $(PhCH_2)_2NNO$, and Ph(Me)NNO have been employed as nitrosylating agents for metal–NO derivatives for various metals (30, 195), as have the diazeniumdiolates (NONOates) such as $(Et_2N)(O^-)NNO$ (i.e., DEANO, 2-(*N*, *N*-diethylamino)diazenolate-2-oxide) (196, 197). A synthetically useful dinitrosamine $R(ON)NC_6H_4N(NO)R$ (R = Me, CH_2COOH) serves as a convenient precursor for photoinduced NO release to metal centers in heme models and heme proteins (198–200).

Zhu et al. (201, 202) determined, using titration calorimetry and thermodynamic cycles, that the homolytic cleavage of the N–N bonds in Ph_2NNO and in *p*-substituted benzenesulfonamides $(p\text{-}X)C_6H_4S(O)_2N(NO)Me$ (X = Me, OMe, H, Cl, Br, etc.) is favored over the related heterolytic cleavage.

A. Interactions Between Metal Complexes and Nitrosamines

1. *Synthesis and Reactivity*

There are two general routes used for the preparation of metal nitrosamine complexes. The first involves the direct addition of a nitrosamine to metal complexes to form metal–(R_2NNO) adducts, and the second involves nucleophilic attack of N-containing species (e.g., amines) on the nitrosyl N-atom of a metal–NO compound.

a. Addition of Nitrosamines to Metals. Addition of *N*-nitroso compounds to metal complexes can result in either simple adduct formation or fragmentation of the incoming *N*-nitroso reagent.

It is well known that nitrosamines can form simple adducts with Lewis acids (e.g., BF_3 and BCl_3) (203–205), and with metal complexes (e.g., $SbCl_5$ and $CuCl_2$) (206). Isolable adducts between nitrosamines and ferric porphyrins of the form $[(por)Fe(ONNR_2)_2]^+$ (por = porphyrinato dianion) have been obtained when the cations $[(por)Fe(thf)_2]^+$ (por = TPP, TTP) were reacted with the dialkylnitrosamines R_2NNO [R_2 = Me_2, Et_2, (*cyclo*-$CH_2)_4$, *cyclo*-$CH_2)_5$, $(PhCH_2)_2$] (207–209). In the case of the octaethylporphyrin analogue $[(OEP)Fe(ONNMe_2)]^+$, the observed five coordination at Fe was stabilized by a "dimeric" structure in the crystal in which the porphyrin planes on the non-ligated side were stacked ~3.3 Å away from each other, essentially preventing sixth ligand coordination (209). In contrast, the ferrous porphyrins $(TPP)Fe^{II}$ and the ferric porphyrin $[(TPP)Fe^{III}(thf)_2]^+$ were *nitrosylated* by *aryl*nitrosamines such as Ph(Me)NNO, Ph(Et)NNO, and Ph_2NNO to form the known five-coordinate (TPP)FeNO complex (209).

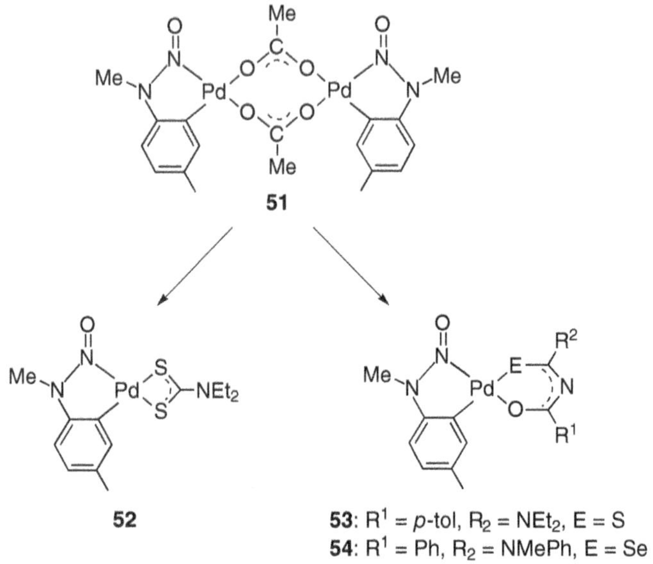

51

52

53: R^1 = *p*-tol, R_2 = NEt_2, E = S
54: R^1 = Ph, R_2 = NMePh, E = Se

Scheme 12.

A cyclo-(ortho-)-palladation of Ph(Me)NNO occurs when the nitrosamine was reacted with Na_2PdCl_4 to give a chloro-bridged dimer $[Pd(\eta^2\text{-}N,C\text{-}N(O)N(Me)\text{-}C_6H_4)]_2(\mu\text{-}Cl)_2$ (210). Reactions of this complex with ligands (L; e.g., phosphines) generated the $L(Cl)Pd(\eta^2\text{-}N,C\text{-}N(O)N(Me)C_6H_4)$ derivatives. The related acetate-bridged complex **51** has been prepared, and it underwent ligand-exchange reactions to produce derivatives (e.g., **52–54**) that have been structurally characterized (Scheme 12) (211, 212).

Although many reactions between nitrosamines and metal complexes have been reported, it remains somewhat surprising that only a few metal–nitrosamine adducts have been isolated and structurally characterized. The tendency of the N−N bond to cleave is certainly a driving factor for the vast range of products that can result from these reactions, some of which retain the "R_2N" and "NO" components of the nitrosamines substrates used. For example, the reaction of the iron carbonyl $Fe_2(CO)_9$ with the nitrosamines R_2NNO [$R_2 = Me_2$, Et_2, and Ph(Me)] resulted in the formal amination of complexed CO, generating a mixture of products including the dicarbonyldinitrosyl $Fe(NO)_2(CO)_2$ and derivatives with new C−N bonds (compounds **55–58**, Chart 4) in different amounts depending on the nitrosamine used (213).

Warren and co-workers (214) reported that Ph_2NNO was activated by $[Me_3NN]Ni(2,4\text{-lutidine})$ to give the isolable compounds $[Me_3NN]Ni(NPh_2)$

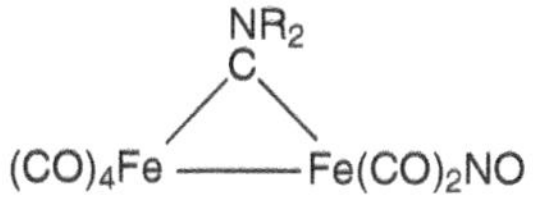

$(CO)_4Fe$ — $Fe(CO)_2NO$ bridged by $C(NR_2)$

55

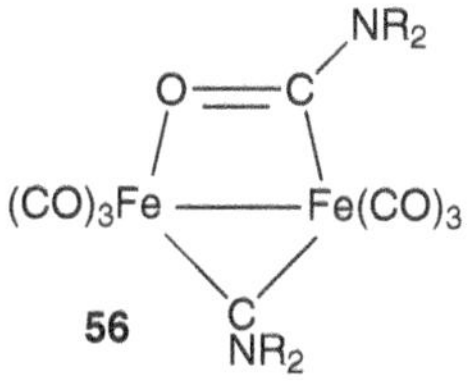

$(CO)_3Fe$ — $Fe(CO)_3$ bridged by $O{=}C(NR_2)$ and $C(NR_2)$

56

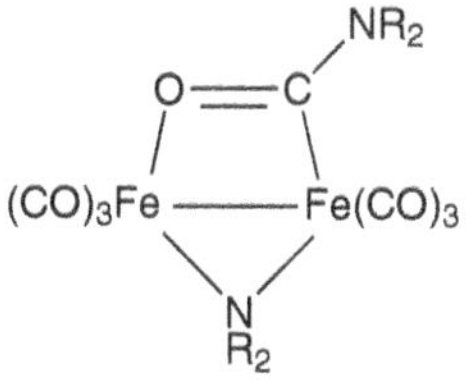

$(CO)_3Fe$ — $Fe(CO)_3$ bridged by $O{=}C(NR_2)$ and NR_2

57

Me N, C=O, Fe $(CO)_4$

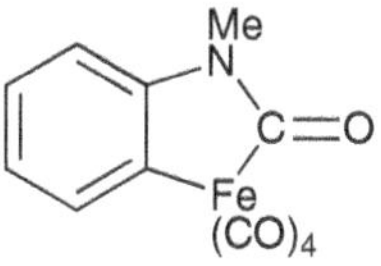

58

Chart 4.

and $[Me_3NN]Ni(NO)$. The related reaction of Ph_2NNO with $[Me_2NN]Cu$ at low temperature resulted in an unusual nitrosation of the methine carbon of the β-diketiminate ligand to generate a bimetallic complex $[Me_2NN]Cu^{II}(ON-Me_2NN)_2$-($\mu$-$Cu^{I}$) with the O-bonded modified ligand coordinating to one of the Cu centers; the three-coordinate $[Me_2NN]Cu(NPh_2)$ was also produced (215).

b. Formation of Nitrosamine Compounds by Attack of N-Based Nucleophiles on Coordinated NO. The attack of amines on coordinated NO ligands has a long and rich history (27, 28), and surprisingly, has only been exploited recently for the generation of isolable metal–nitrosamine products that have been structurally characterized by crystallography (216–218).

In 1971, Maltz et al. (219) analyzed the organic products generated when nitroprusside $[Fe(CN)_5NO]^{2-}$ reacted with amines, and characterized the products (e.g., Et_2NNO, when Et_2NH was used) as deriving from nitrosamines; a determination that was consistent with nitroprusside functioning as a nitrosating agent toward amines. In fact, nitroprusside is now well established as containing a reactive NO ligand that is susceptible to attack by several nucleophiles (92, 220). Doctorovich and co-workers (221, 222) reinvestigated the reaction of nitroprusside with amines, and probed the reaction of the nitroprusside with the *n*-butylamide anion. Their isolation of the *n*-butyldiazote anion provided strong evidence for initial attack of the anion on the coordinated NO (compound **59**) and the occasional involvement of diazoates (compound **60**) as intermediates during the formation of the final products isolated from such reactions (Scheme 13) (221).

Doctorovich et al. (222, 223) further reexamined Meyer's report of the reaction of $[Ru(bpy)_2(NO)Cl]^{2+}$ with amines. They proposed, using *n*-butylamine as the

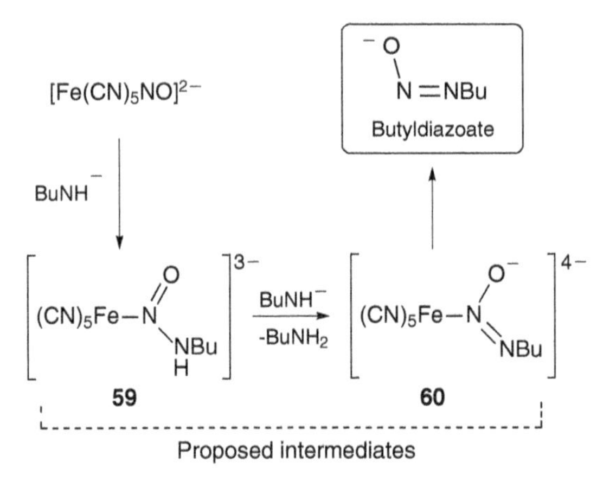

Scheme 13.

nucleophile and based on product isolation and DFT calculations, that the reaction occurred via the initial attack of the amine on the coordinated NO; this was then followed by isomerization of the modified ligand to the diazoic acid "Ru−N(OH)=NBu", and subsequent loss of hydroxide to give the coordinated diazenido "Ru−NNBu" prior to its final conversion to the dinitrogen derivative "Ru−N_2" (222). Leung and co-workers (224) also proposed a similar attack of primary amines on a coordinated NO to yield a bimetallic dinitrogen-bridged "Ru−N_2−Ru" derivative.

Doctorovich and co-workers (216, 217) provided the first crystallographically characterized metal–nitrosamine complexes derived from amine attack on a coordinated nitrosyl ligand. They reported that primary aliphatic and aromatic amines reacted with the highly electrophilic nitrosyl in K[Ir(NO)Cl_5] (υ_{NO} 2006 cm^{-1}; KBr) to give the stable derivatives **61–63** that were characterized by X-ray crystallography as their tetraphenylphosphonium salts (Scheme 14).

Similar complexes resulted when *n*-butylamine, cyclopropylamine, and 9-octyladenine were used. The data from the crystal structures were most consistent with the bound nitrosamine formulation; however, DFT calculations revealed a preferred stability of the diazoic acid HO−N=N−R forms of the ligands (216). Electrospray ionization mass spectrometry (ESIMS) and tandem (ESIMS/MS) have proven useful for the identification of these product species (225).

2. *Coordination Chemistry of Nitrosamine Ligands*

The multiple-binding sites in the R_2N–*N=O* molecule ensures its rich coordination chemistry that is expanded even further if the "R" group possesses

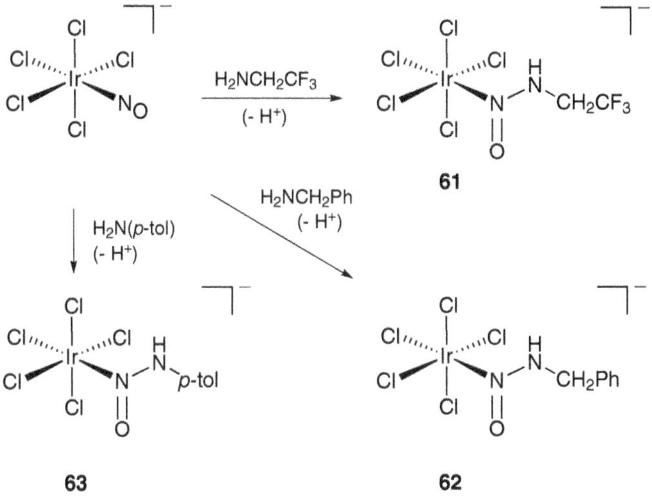

Scheme 14.

coordination sites as well (33). So far, four binding modes in metal–nitrosamine complexes have been established by single-crystal X-ray crystallography, as shown in Chart 5, and selected bond lengths and angles of the structurally characterized metal nitrosamine complexes are listed in Table VI.

The sole *N*-binding mode **A** was not established by crystallography until 2005 (216), although this sole *N*-binding mode is notably the most common in the related structurally characterized metal adducts of *C*-nitroso compounds (Chart 1). Scheme 14 shows the reaction pathway to the sole *N*-bonded nitrosamine compounds reported to date (216–218, 225). The authors recognized and considered the inherent ambiguity in the assignment of the nitrosamine vs diazoic acid (HO)N=NR formulation using X-ray crystallography. However, their assignment was only made after careful consideration of the metrical parameters,

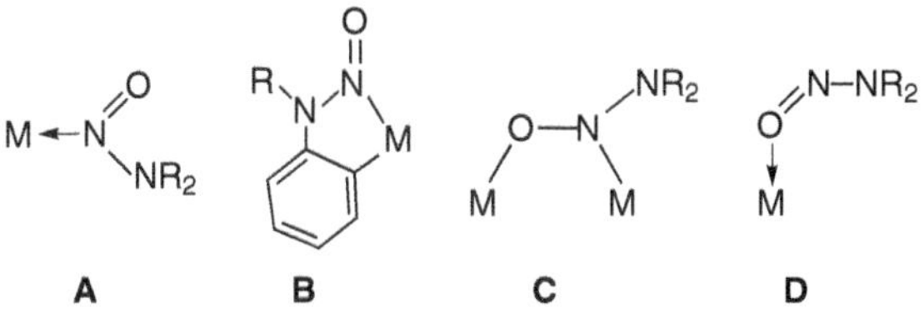

Chart 5.

TABLE VI
Selected Structural Data for Metal–Nitrosamines from Chart 5

Compound	N–N (Å)	N–O (Å)	N–N–O (°)	References
	Structure A			
$[PPh_4][IrCl_5(ONN(H)Ph(Me))]$ (**63**)	1.309(10)	1.225(10)	118.4(7)	216
$[PPh_4][IrCl_5(ONN(H)CH_2CF_3)]$ (**61**)	1.32(3)	1.14(2)	120(2)	216
$[PPh_4][IrCl_5(ONN(H)CH_2Ph)]$ (**62**)	1.315 (9)	1.232 (8)	116.8 (6)	217
	Structure B			
$Pd\{N(O)N(Me)C_6H_4\}Cl(PPh_3)$	1.276(19)	1.243(18)	117.8(11)	210
$Pd\{N(O)N(Me)C_6H_4\}_2$	1.301(7)	1.246(6)	117.4(4)	227
	1.331(6)	1.226(5)	116.6(4)	
$[Pd\{N(O)N(Me)C_6H_2(2\text{-}Cl)\text{-}(5\text{-}Me)\}]_2(\mu\text{-}O_2CCF_3)_2$	1.328(11)	1.242(9)	118.1[a]	228
$Pd(\eta^2\text{-}C,N\text{-}4\text{-}MeC_6H_3N(Me)NO)\text{-}(\eta^2\text{-}S,O\text{-}4\text{-}MeC_6H_4C(O)NC(S)NEt_2)$	1.3306(12)	1.2334(12)		211
$Pd(\eta^2\text{-}C,N\text{-}4\text{-}MeC_6H_3N(Me)NO)(\eta^2\text{-}Se,O\text{-}C_6H_5C(O)NC(Se)NMePh)$	1.328(2)	1.2288(19)		211
$Pd(\eta^2\text{-}C,N\text{-}4\text{-}MeC_6H_3N(Me)NO)_2$	1.3209(19)	1.2422(18)		211
$Pd(\eta^2\text{-}C,N\text{-}C_6H_4N(Me)NO)(\eta^2\text{-}Se,N\text{-}N(tol)C(Ph)NC(Se)NEt_2)$	1.333	1.225	118.41	212
	Structure C			
$(Me_2NNO)CuCl_2$	1.29(2)	1.22(2)	115.89(5)	229, 230
$\{(CH_2)_5NNO\}CuCl_2$	1.288(3)	1.279(3)	114.9(2)	206
	Structure D			
$(Me_2NNO)SbCl_5$	1.262(5)	1.310(5)	112.0(3)	206
$[(TPP)Fe(ONNEt_2)_2]ClO_4$	1.276(10)	1.260(9)	113.9(7)	207, 208
	1.288(9)[b]	1.285(8)[b]	113.6(7)[b]	
$[(OEP)Fe(ONNMe_2)]ClO_4$	1.275(3)	1.251(3)	115.9(2)	209
$[(TPP)Fe(ONNMe_2)_2]ClO_4$	1.284(1)	1.271(1)	114.2(1)	209
$[(TPP)Fe(ONN(CH_2Ph)_2)_2]ClO_4]\text{-}[(TPP)Fe(OClO_3)_2)\cdot(CH_2Cl_2)_2$	1.288(2)	1.271(2)	114.5(1)	209
$[(TPP)Fe(ONN(c\text{-}CH_2)_5)_2]ClO_4\cdot(CH_2Cl_2)_2$[c]				209
cation 1	1.271(2)	1.275(2)	114.2(2)	
cation 2	1.282(2)	1.271(2)	114.6(2)	
$(OEP)Ru(CO)(ONNEt_2)$	1.287(9)	1.243(8)	113.9(7)	208
$(TTP)Os(CO)(ONNEt_2)$	1.294(14)	1.241(13)	113.3(12)	208

[a] Data obtained from the Cambridge Structural Database.
[b] Data for the second molecule.
[c] Two independent cations.

spectroscopic, and DFT computational data, and reactivity patterns for these coordinated ligands (217).

Interestingly, the metrical data in Table VI show that in the *N*-bonded compounds with structure **A** (1.31–1.32 Å) and **B** (1.28–1.33 Å), the N–N(O) bonds are longer than those for the other coordination modes listed that are in the 1.26–1.29 Å range, perhaps suggesting an easier reactivity toward N–N bond cleavage in the former complexes.

The binding mode **B** has been determined for a Mn compound [only a preliminary structure was reported (226)] and a series of Pd complexes, and these result from the cyclometalation of aromatic nitrosamines (e.g., Ph_2NNO). The binding mode **C** was the first to be established for a metal–nitrosamine complex by X-ray crystallography, and this was for an oligomeric copper complex, $[(Me_2NNO)CuCl_2]_n$. The sole η-O binding mode **D** was demonstrated in the six-coordinate heme model complexes, $[(TPP)Fe(ONNEt_2)_2]ClO_4$, $(OEP)Ru(CO)(ONNEt_2)$, and $(TTP)Os(CO)(ONNEt_2)$ (207–209). This was then extended to noncyclic and cyclic dialkylnitrosamine derivatives of ferric porphyrins. These complexes were obtained from adduct formation between the preformed R_2NNO compounds and the metalloporphyrin precursors. The lengthening of the N–O bonds (1.251–1.275 Å) and shortening of the N–N bonds (1.271–1.288 Å) in the coordination nitrosamines compared to those of the corresponding free nitrosamines suggested a resonance hybrid contribution for the coordinated nitrosamines.

$$R_2N{-}N{=}O \longleftrightarrow R_2N^{+}{=}N{-}O^{-}$$

The five-coordinate complex $[(OEP)Fe(ONNMe_2)]ClO_4$ displays a pseudo π–π interaction between two adjacent porphyrin rings, which may block the sixth coordination site and prevent the trans addition of another nitrosamine ligand (209).

B. Metal Mediated Formation of Nitrosamines from Metal–Amine Precursors

In Section III.A.1.b, we outlined the reactions of incoming amine nucleophiles on coordinated NO ligands in metal complexes to form metal–(R_2NNO) moieties. An alternate route for the formation of nitrosamines is presented by the reaction of various NO_x species, including NO itself, on coordinated amines and amine-type ligands. For these reactions, the pathways for the formation of the

N−N bonded products can be varied and diverse. For example, Pell and Armor (231) studied the kinetics of the formation of the Ru–dinitrogen complex resulting from the reaction of NO on an amine ligand in $[Ru(NH_3)_6]^{2+}$ as a function of pH, and propose that in alkaline solution the NO attacks a coordinated amine ligand to form a new N−N bond and ultimately the $[Ru(NH_3)_5N_2]^{2+}$ complex. Alternatively, amine ligands can react with inorganic nitrite to result in the nitrosation of the coordinated amine group to produce metal nitrosamine complexes (232). In a study linked to the possible relationship between red meat and colon cancer, it was reported that heme can assist in the nitrosation of the genotoxic heterocyclic amine, 2-amino-3-methylimidazo[4,5-*f*]quinoline (IQ), in the presence of H_2O_2 to yield a carcinogenic intestinal nitrosamine product, although the exact mechanism of nitrosation remains to be fully elucidated (233). Brackman and Smit (234) reported a Cu(II)-catalyzed reaction between diethylamine and NO to generate Et_2NNO and N_2O as products, and proposed a "Cu^{II}−NO" intermediate that acts as a nitrosating agent toward the amine.

Ford and co-workers (235) reported the first example for the intramolecular nitrosation of the $(DAC)Cu^{II}$ complex [DAC = 1,8-bis(9-anthracylmethyl) derivative of the macrocyclic cyclam (1,4,8,11-tetraazacyclotetradecane)] to form the corresponding organic mononitrosamine compound **64** identified on the basis of its IR (υ_{NO} 1430 cm^{-1}), NMR, and ESIMS data (Eq. 18). A concurrent reduction of Cu(II) to Cu(I) and subsequent dissociation of the Cu(I) ion from the macrocycle, presumably due to its weaker interaction with this nitrosamine derivative, was demonstrated. The general mechanisms of reductive nitrosylations have been reviewed by Ford et al. (236).

(18)

They proposed, based on detailed kinetics and DFT studies, an attack of NO at a reversibly deprotonated amine site in aqueous methanol or acetonitrile (237). Importantly, this reaction set the stage for applications in NO detection methodology; the luminescence of the pendant anthracenyl group in the initial Cu(II) complex is intramolecularly quenched by the precursor paramagnetic Cu(II) center, and the release of Cu in the presence of NO (to form the nitrosated product) restores the luminescence of the organic moiety (235).

Lippard and co-workers (238, 239) employed similar methodology for NO detection using Cu(II)–fluorescein derivatives. Analysis of the data from their detailed kinetics studies also favors nitrosamine formation through the direct attack of NO on a deprotonated amine, rather than through a Cu–NO intermediate (238). Warren and co-workers (215) demonstrated that the amide nitrogen of the three-coordinate $[Me_2NN]Cu^{II}{-}NPh_2$ complex is attacked by NO with resulting N–N bond formation and subsequent generation of Ph_2NNO.

In a series of related papers, Mondal and co-workers (240–246) demonstrated related mono-, di-, and trinitrosation of coordinated N-based ligands to generate isolable nitrosamines. Using the chelating N-based ligand in $[(H_2NCH_2CH_2)_3NCu(CNMe)]^{2+}$, they showed single *N*-nitrosation at one of the amine moieties, and proposed that the reaction likely proceeded via a "Cu(NO)" intermediate (υ_{NO} 1650 cm^{-1}) (240). They extended this work to a related substituted tripodal ligand and showed that a *tri*nitrosation of the chelate was possible (**65**; Eq. 19) (241).

(R = Et, *i*-Bu) — NO, MeCN → **65** (X-ray; R = Me) + Other products (19)

They reported, using these ligands on other *N*-based chelate ligands, that observable "Cu(NO)" complexes were active participants in these *N*-nitrosation reactions, based on fourier transform infrared (FTIR) and ultraviolet–visible (UV–vis) spectroscopy (242). The flexibility of choice of chelate ligands allowed them to isolate various organic products (e.g., **66**, **67**) that differed in the extent of nitrosation (e.g., eq 20) (245).

$$\text{[Cu}^{II}\text{(RNH(CH}_2)_2\text{NHR)}_2]^{2+} \xrightarrow[\text{MeCN}]{\text{NO}} \text{RNH(CH}_2)_2\text{N(R)NO } (\mathbf{66}) + \text{ON(R)N(CH}_2)_2\text{N(R)NO } (\mathbf{67}) + \text{Other products} \tag{20}$$

(R = Me, Et, *i*-Bu) (X-ray; R = Me)

Mondal and co-workers (243) suggested that the nature of the ligand framework in these Cu(II) complexes plays a key role in determining the pathway and extent of nitrosation of the coordinated N atoms. They proposed that the nitrosation of a macrocyclic tetradentate *N*-based liganded Cu(II) complex would likely proceed via attack of the incoming NO on a deprotonated N atom of the ring, whereas nitrosation with a Cu(II) complex containing a non-macrocyclic but chelate ligand would likely proceed via initial $Cu^{(II)}$–NO bond formation prior to the nitrosation event (243).

IV. NITROSOTHIOLS

A. Interactions Between Metal Complexes and RSNO

1. Decomposition of RSNO

The decomposition of RSNO compounds by various metal ions has been studied extensively (247). For example, it is known that the active species for Cu induced catalytic decomposition of RSNOs is the Cu(I) ion (Eqs. 21, 22) (248, 249).

$$Cu^{2+} + RS^- \rightarrow Cu^+ + 1/2\,RSSR \tag{21}$$

$$Cu^+ + RSNO \rightarrow Cu^{2+} + RS^- + NO \tag{22}$$

Studies of the Cu catalyzed decomposition of RSNOs in aqueous or biological media have been extended to bound Cu complexes as well (250–252). The Hg(II) and Ag(I) salts are known to decompose RSNO compounds, most probably via initial S-coordination of the RSNOs (253). Interestingly, a series of Ru complexes with nitrogen-containing macrocycles have been found to mediate *S*-nitroso-*N*-acetylpenicillamine (SNAP) decomposition without scavenging the NO via Ru−NO bond formation (254).

There are some reports that Cu^{2+} can also mediate the formation of RSNO from NO under appropriate conditions. For example, Inoue et al. (255) reported that ceruloplasmin (a major multi-copper-containing plasma protein) catalyzes the formation of *S*-nitrosoglutathione (GSNO) from NO under physiological conditions. Another report suggests that although Cu^{+} destabilizes RSNO, the Cu^{2+} ion, but not Fe^{3+}, induces RSNO formation from the thiol groups (of bovine serum albumin and human Hb) and NO.

2. *Formation of Metal Nitrosyls*

These RSNO compounds are often useful as nitrosylating agents for metal complexes to form metal–NO groups, that is, where the metal effectively captures the released NO (256–258). Certainly, RSNO compounds (e.g., SNAP, nitrosocysteine, and GSNO) have been used effectively in some instances for the delivery of NO to biologically relevant metal centers.

Louro and co-workers (259) determined rate constants for NO transfer from SNAP to water-soluble anionic and cationic iron porphyrins. They proposed a mechanism involving hydrolysis of SNAP by an Fe coordinated H_2O molecule to eventually yield the nitrosyl and sulfenic acid (Eq. 23).

$$[(\text{por})\text{Fe}^{\text{III}}(\text{H}_2\text{O})]^+ + \text{RSNO} \rightarrow (\text{por})\text{Fe}^{\text{II}}(\text{NO}) + \text{RSOH} + \text{H}^+ \qquad (23)$$

Their conclusion regarding an assisted RSNO decomposition pathway was made stronger by their observation that the rate of reduction to give the ferrous nitrosyl was faster that the rate for simple RSNO decomposition, and suggested a direct NO^- transfer from RSNO to Fe (260).

Indeed, such a "direct NO transfer" from RSNO to metal centers was established earlier by Butler et al. (261) who examined several reactions of RSNOs derived from *N*-acetylcysteine, cysteine, homocysteine, and captopril, with the metal centers in $[(DMPS)_2Fe]^{4-}$ and [(MGD)Fe] (DMPS = 2,3-dimercaptopropane-1-sulfonate; MGD = *N*-methyl-D-glucamine dithiocarbamate). They concluded, based on their results, that the reactions proceeded via an Fe initiated decomposition of the RSNOs, and that the "NO" was formally transferred as NO^+ in the case of the DMPS compound, whereas it was transferred as neutral NO in the case of the MGD compound.

Zhu et al. (262) reported that NO transfer from tritylthionitrite Ph_3CSNO to structurally related Co tetraarylporphyrins occurred most likely via a concerted homolytic S–NO cleavage with Co–NO bond formation; presumably via a "Co–N(=O)SCPh$_3$" intermediate containing a weak S–N bond. In a separate study on NO transfer from RSNO (e.g., SNAP) to Fe porphyrins in methanol, they concluded that the NO transfer reactions likely proceeded via heterolytic S–N bond cleavage and subsequent NO$^+$ transfer to the metal centers (263).

3. *Formation of Metal Thiolates and Nitrosyl Thiolates*

On occasion, both or either of the "RS" and "NO" fragments of the incoming RSNO reagents can get captured by metal centers during the reactions of metal complexes with RSNOs. For example, ethyl thionitrite reacts with the iron carbonyl $Fe_2(CO)_9$ to give a thiolate-bridged diiron complex $[Fe(NO)_2]_2$-(μ-SEt)$_2$ as one of the products together with $Fe(CO)_2(NO)_2$ (213).

In 1996, Richter-Addo and co-workers (264) reported the first well-defined reactions between an RSNO and a metalloporphyrin. They demonstrated a then-unprecedented formal trans addition of the RSNO, namely, *N*-acetyl-*S*-nitroso-L-cysteine methyl ester, across the metal center of the precursor (OEP)Ru(CO) complex to give the (OEP)Ru(NO)(*S*-NACysMe) (NACysMe = *N*-acetyl-L-cysteinate methyl ester) product that was characterized by crystallography (264). They proposed a metal-assisted decomposition of the RSNO that involved initial *S*-coordination of the RSNO to the metal center followed by homolytic S–N bond cleavage and subsequent NO capture by the metal center, as sketched at the top of Scheme 15. An alternate pathway, shown at the bottom of Scheme 15, involving N-binding of RSNO with subsequent SR radical release was discounted, due to the lack of observation of any disulfide RSSR products in the reaction mixture.

These RSNO formal trans additions to metalloporphyrins of both Ru and Os to give reasonable-to-good yields of (por)M(NO)(SR) products are now well established (265–272). Additional evidence for the proposed reaction pathway shown at the top of Scheme 15 came from a series of reactions involving Os porphyrins. In one example, the (OEP)Os(CO) compound was reacted with *in situ* generated PhSNO; IR monitoring of the reaction mixture revealed the appearance, in addition to those of the starting (OEP)Os(CO) at 1883 cm^{-1} and product (OEP)Os(NO)(SPh) at 1766 cm^{-1}, of a new band at 1957 cm^{-1} that disappeared with time as the product formed (266). This band at 1957 cm^{-1} was assigned to that of the intermediate carbonyl thiolate (OEP)Os(CO)(SPh), as shown in Scheme 16. The $\Delta\upsilon_{NO}$ of +74 cm^{-1} was consistent with a formal change in oxidation state of the Os center as the thiolate binds the metal.

Indeed, the use of the valence isoelectronic arylazo sulfide reagent PhSN=NC$_6$H$_4$(*p*-NO$_2$) in place of PhSNO revealed a similar formation of the

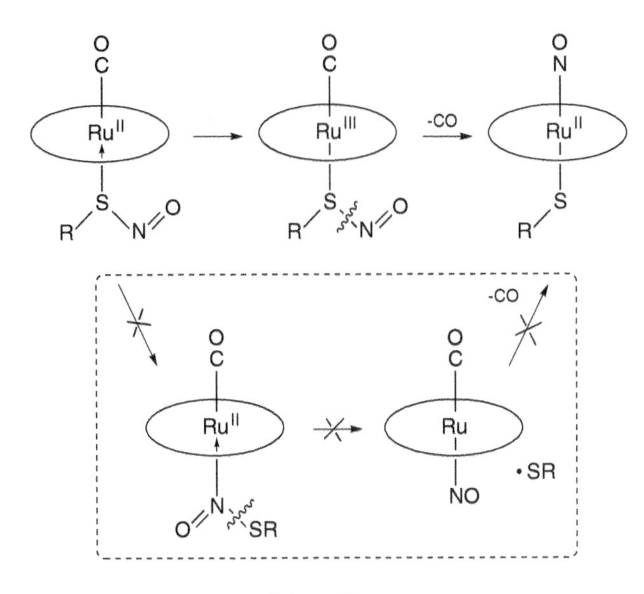

Scheme 15.

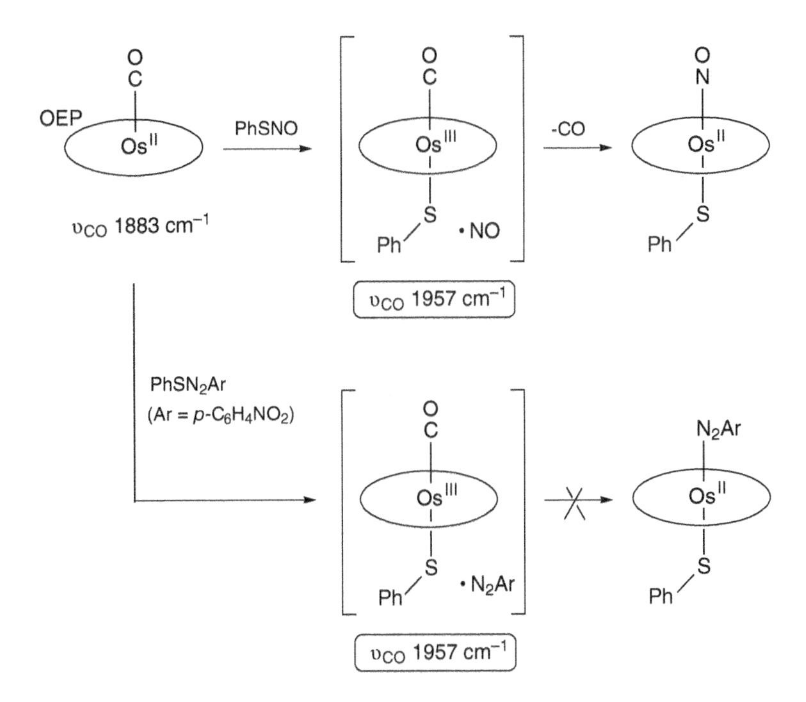

Scheme 16.

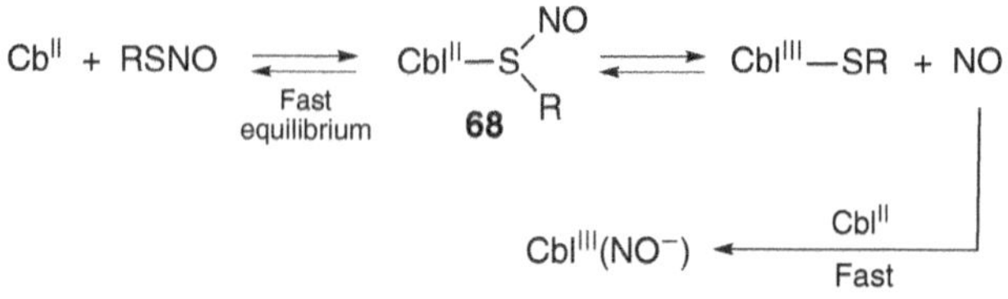

$$Cb^{II} + RSNO \underset{\text{equilibrium}}{\overset{\text{Fast}}{\rightleftharpoons}} Cbl^{II}{-}S(NO)(R)\ (\mathbf{68}) \rightleftharpoons Cbl^{III}{-}SR + NO$$

$$NO \xrightarrow[\text{Fast}]{Cbl^{II}} Cbl^{III}(NO^{-})$$

Scheme 17.

intermediate complex with υ_{CO} 1957 cm^{-1}, confirming the proposed *S*-binding of the reagents to the metal center (266).

The formal trans addition of RSNO to Ru porphyrins was explored mechanistically by Ford and Richter-Addo and co-workers (273) for the reaction of (OEP)-Ru(CO) with *N*-acetyl-1-amino-2-methylpropyl-2-thionitrite, and the pathway defined by the sequence at the top of Scheme 15 was confirmed.

The reaction of cobalamin (Cbl) with SNAP and GSNO has been found to parallel that of initial *S*-binding of the RSNOs to the metal centers of the Ru and Os porphyrins described above. Thus, the net decomposition of RSNO by reduced cobalamin(II) [Cbl(II)] proceeded via initial adduct formation involving a proposed *S*-binding of the RSNO [**68**; due to the soft nature of reduced Cbl(II)] followed by S–N bond cleavage and subsequent fast trapping of released NO by Cbl(III) (Scheme 17) (274).

In the case of aquacobalamin(III), two decomposition pathways were considered for the observed RSNO decomposition (Scheme 18). The first involved *S*-binding of the RSNO compound (**69**) eventually to yield the thiolato cobalamin, and the second involved *N*-binding of the RSNO (i.e., **70**) eventually to yield the nitrosyl product. Given that the nitrosyl cobalamin was the major product isolated from the reaction, the authors proposed that the major pathway for the reaction involved the *N*-binding of the RSNO to the Cbl(III) center (i.e., bottom of Scheme 17) (274).

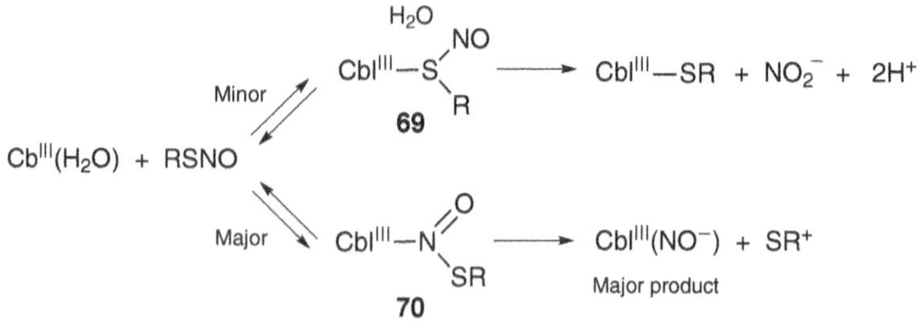

$$Cb^{III}(H_2O) + RSNO \overset{\text{Minor}}{\rightleftharpoons} H_2O\ Cbl^{III}{-}S(NO)(R)\ (\mathbf{69}) \longrightarrow Cbl^{III}{-}SR + NO_2^{-} + 2H^{+}$$

$$Cb^{III}(H_2O) + RSNO \overset{\text{Major}}{\rightleftharpoons} Cbl^{III}{-}N(=O)(SR)\ (\mathbf{70}) \longrightarrow Cbl^{III}(NO^{-}) + SR^{+}\ \text{(Major product)}$$

Scheme 18.

Warren and co-workers (214) demonstrated that the reaction of $[Me_3NN]Ni$-(2,4-lut) with AdSNO (Ad = adamantyl) in toluene generates $\{[Me_3NN]Ni\}$-$(\mu\text{-}SAd)_2$, whereas the same reaction in benzene-d_6 provided $[Me_3NN]Ni(NO)$. They also reported that the $\{[Me_2NN]Cu\}_2$ complex reacted with 2 equiv of Ph_3CSNO in cold toluene to generate the thiolate $[Me_2NN]Cu{-}SCPh_3$ that was unstable in the presence of NO, and also the $[Me_2NN]Cu\{ON[Me_2NN]\}$ complex in which a β-diketiminate ligand had been C-nitrosated at the backbone methine position (275).

B. Metal-Mediated Formation of RSNO

1. *Transnitrosation*

A "thiolate-exchange" reaction occurs between thiolates and RSNO compounds. Houk et al. (276) determined, using DFT calculations, that a nitroxyl disulfide adduct **71** was a viable intermediate in this process (Scheme 19); they also presented mass spectral evidence for such an intermediate in the gas phase when MeSNO and its thiolate were used for the reaction.

Doctorovich and co-workers (277) later obtained the first direct NMR (^{1}H, ^{13}C, and ^{15}N) spectroscopic evidence for such a nitroxyl disulfide intermediate during this transnitrosation process (Scheme 19) when *S*-nitroso-L-cysteine ethyl ester and its thiol were used for the reaction.

Warren and co-workers (278–280) presented perhaps the most clarity regarding such thiolate–RSNO exchange reactions involving metal thiolate complexes. For example, they showed that although the $[Me_2NN]ZnSR$ (R = *t*-Bu, CPh_3) complexes did not react with free NO to form RSNO, they underwent transnitrosation reactions as sketched in Scheme 20 (cy = cyclohexyl) (278).

They also showed that some (L)Cu(SR) (L = neutral *N*-heterocyclic carbene; R = Bn (Bn = benzyl), *t*-Bu, $CH_2C_6H_4(t\text{-}Bu\text{-}p)$) complexes undergo transnitrosations with BnSNO to generate the corresponding (L)Cu(SBn) and the free RSNO compounds, a process that necessarily also involved the breaking and formation of S–N bonds (279). Similar transnitrosations occurred with various (tris(pyrazolyl) borate)Zn–thiolates when reacted with RSNOs (280).

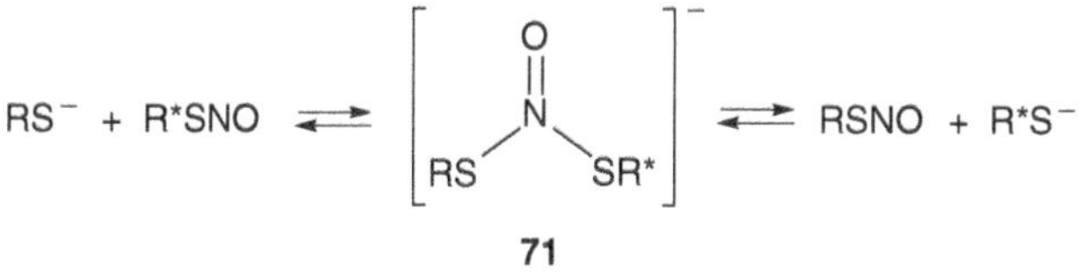

$$RS^- + R^*SNO \rightleftharpoons \left[RS{-}N(=O){-}SR^* \right]^- \rightleftharpoons RSNO + R^*S^-$$

71

Scheme 19.

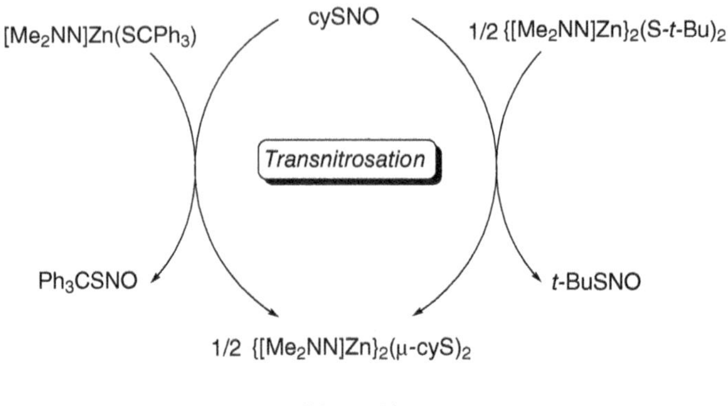

Scheme 20.

2. *Attack of NO on Metal Thiolates*

Hoff and co-workers (281, 282) reported the reactions of group 6 metal carbonyl thiolates with NO that resulted in the generation of free RSNO species, as shown in Eqs. 24 (281) and 25 (282).

$$\mathrm{W(phen)(CO)_2(SPh)_2 + 2\,NO \rightarrow W(phen)(CO)_2(NO)(SPh) + PhSNO} \quad (24)$$

$$\mathrm{Cp^*Cr(CO)_3SPh + 2\,NO \rightarrow Cp^*Cr(CO)_2NO + PhSNO} \quad (25)$$

The reaction of the formally W(II) bis(-thiolate) complex $W(phen)(CO)_2(SPh)_2$, where phen = 1,10-phenanthroline, with NO in CH_2Cl_2 resulted in the simultaneous generation of the formally W(0) product $W(phen)(CO)_2(NO)(SPh)$ and free PhSNO that was detected spectroscopically (υ_{NO} 1560 cm^{-1}) prior to its decomposition (281). A similar reactivity was observed for the analogous Mo(phen) $(CO)_2(S\text{-tolyl})_2$ complex.

The generation of free PhSNO was also observed when the organometallic compound $Cp^*Cr(CO)_3SPh$ was reacted with 1–2 atm NO in toluene (282). In this latter case, the intermediacy of a Cr–$\{\eta^2\text{-C(=O)SPh}\}$ complex was proposed based on the kinetics data prior to its elimination of the SPh radical that was trapped by a second NO molecule.

Zheng and Birke (283) reported that the reaction of glutathionylcobalamin (proposed to be a major component of vitamin B_{12} in mammals) with NO resulted in the formation of the nitrosylcobalamin product, via a caged structure involving an GSNO intermediate (Scheme 21; SG = glutathionyl).

Nitric oxide is known to inhibit the heme enzyme cytochrome P450 and to deactivate it (284–288). Van Eldik and co-workers (289) used a synthetic iron

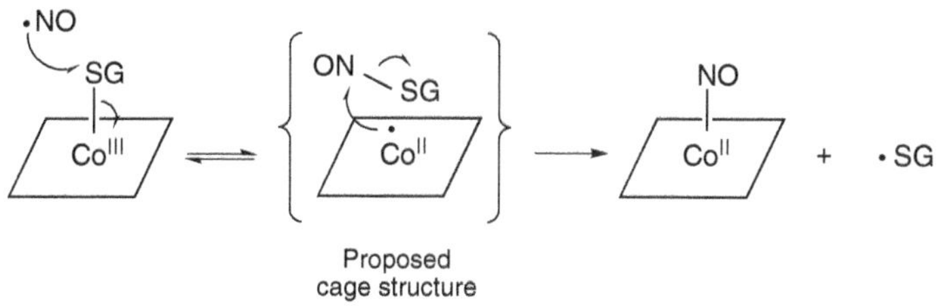

Proposed cage structure

Scheme 21.

porphyrin with a tethered thiolate ligand to model the P450 inactivation process by NO, and concluded, based on kinetics and spectroscopy, that the first incoming NO molecule coordinated the ferric center of the enzyme (**72** in Scheme 22) to generate the ferric nitrosyl thiolate (**73**). They determined that the attack of a second NO molecule then occurred at the sterically protected proximal thiolate ligand to result in RSNO formation (complex **75**) via the intermediacy of a caged complex (**74**).

Such an attack of NO on a coordinated thiolate has been demonstrated unequivocally by X-ray crystallography during NO binding to the heme thiolate active site of the *Cimex* nitrophorins (290).

Interestingly, Warren and co-workers (279) showed that the nitrosonium ion NO^+ can attack the coordinated thiolates in some (L)Cu–SR (L = neutral *N*-heterocyclic carbene; SR = SBn, S-*t*-Bu) complexes to liberate the free RSNO compounds; they reported similar reactivity of NO^+ (but not NO) with a

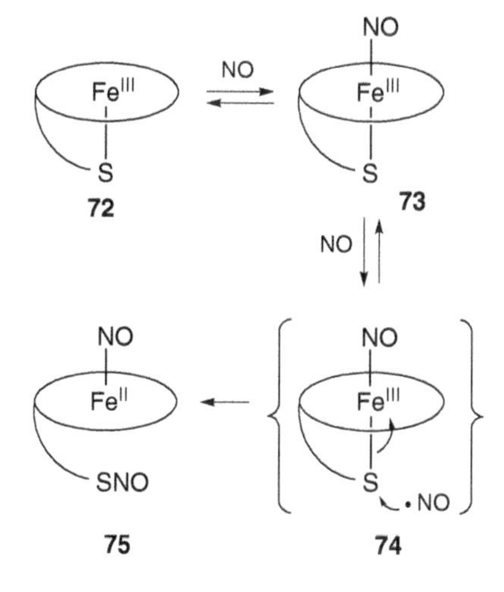

Scheme 22.

[tris(pyrazolyl)borate)Zn(S-*t*-Bu] complex to release the free tertiary alkyl thionitrite *t*-BuSNO (280). Lippard and co-workers (291, 292) demonstrated related reactivity of NO^+ with some Zn–thiolates. An intramolecular nitrosation of a coordinated thiolate by bound NO was demonstrated by Tsou and Liaw (293) when the dinitrosyl iron compounds $Fe(NO)_2(SR)(L)$ (SR = S-*t*-Bu, thiolate from *N*-acetylpenicillamine; L = 2-MeIm, dmso (dimethyl sulfoxide), 2-OAc) were reacted with Brønsted acid–Lewis base pairs to release free RSNO.

3. *Attack of Thiols and Thiolates on Coordinated NO*

The attack of thiols on the coordinated NO ligand of nitroprusside $[Fe(CN)_5NO]^{2-}$ has been studied extensively. It is well established that the initial product results from the attack of thiolate anion on the N atom of the coordinated NO to give a metal–N(O)SR species **76** (Eq. 26) (92, 294), and that the stability of the product is dependent on a number of factors that include the nature of the R group.

$$[Fe(CN)_5NO]^{2-} + RS^- \longrightarrow \left[(NC)_5Fe \leftarrow N{\overset{\displaystyle /\!\!/ O}{\underset{\displaystyle \backslash SR}{}}}\right]^{3-} \tag{26}$$

76

For example, Stasicka and co-workers (295) determined that electron-withdrawing R groups stabilized the metal–N(O)SR adduct, whereas electron-donating groups destabilized it. The nucleophilic attack with biologically relevant thiols (e.g., cysteine, *N*-acetylcysteine, and glutathione) resulted in the generation of such adducts that subsequently decomposed mainly via homolytic N–S splitting due to the weak interaction of the incoming –SR groups with the NO ligand (294).

Well-defined metal–N(O)SR adducts were obtained when the anion $[IrCl_5NO]^-$ is reacted with thiols (296, 297). For example, the reaction of the anion $[IrCl_5NO]^-$ with benzylthiol, thiophenol, mercaptosuccinic acid, *N*-acetyl-L-cysteine, and L-cysteine ethyl ester in acetonitrile resulted in the generation of the Ir–N(O)-SR species (**77** Scheme 23) (297).

A trans labilization of a chloride ligand occurred when the R group of the thiol contained an aromatic ring. For example, the reaction of $[IrCl_5NO]^-$ with benzylthiol resulted initially in the production of a green product that turned dark red with time, to ultimately generate the *trans*-$[(MeCN)IrCl_4\{N(=O)SCH_2Ph\}]^-$ product **78** (υ_{NO} 1431 cm^{-1}) characterized by X-ray crystallography as its K^+ and $PPh_4{}^+$ salts (Scheme 23) (297). The DFT studies on this complex reveal an increased stability of the coordinated RSNO compared with its unliganded form.

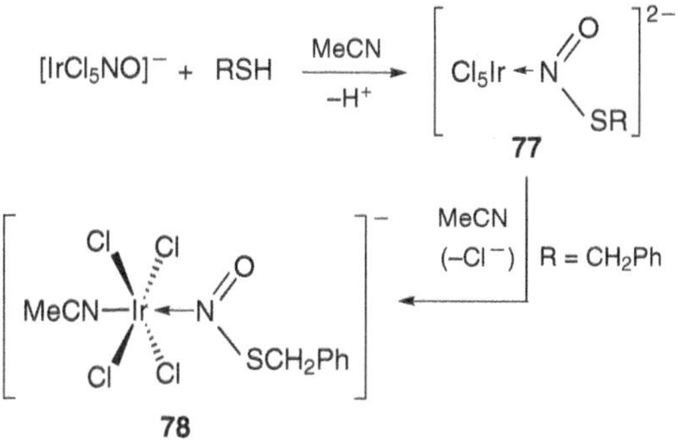

Scheme 23.

C. Coordination Modes of Metal–RSNO Compounds

To date, the only crystallographically characterized metal–RSNO complex is Doctovorich's $[(MeCN)IrCl_4\{N(=O)SCH_2Ph\}]^-$ anion **78**, shown in Scheme 23, as both its K^+ and PPh_4^+ salts (296, 297). The crystal structure clearly reveals the sole *N*-binding mode of the alkyl thionitrite to the metal center, and reveals a syn conformation of the bound RSNO.

V. ALKYL NITRITES

A. Interactions Between Metals and RONO

1. *Decomposition of RONO on Metal Surfaces*

Adsorption of alkyl nitrites on metal surfaces frequently results in the generation of alkoxy radicals and NO, and can lead to follow-up reactions with other substrates. Examples include studies on alkyl nitrite adsorption on Pt(111) (298–300) and Ag(111) (301–305). A recent quantum chemical molecular dynamics investigation of MeONO and *t*-BuONO photodissociation on copper surfaces concludes that the internal O atom in Me*O*NO preferentially contacts the copper surface during the photodissociation process, whereas it is the terminal O atom *t*-BuON*O* that contacts the surface during the reaction presumably due to steric hindrance of the *t*-butyl group (306).

In a somewhat related study, Li et al. (307) demonstrated a $[AuCl_2(phen)]^+$ catalyzed carbonylation of alkyl nitrites to give ROC(=O)OR products and NO, a process proposed to involve initial coordination of the internal O atom (e.g., of Et*O*NO) to the Au^{3+} center.

2. *Formation of Metal Nitrosyls*

The relatively weak bonds in RO–NO compounds provide a convenient entry into the use of these compounds as NO donors to metal centers. Indeed, alkyl nitrites have been used as nitrosylating agents for the preparation of metal–NO derivatives (308, 309).

3. *Formation of Metal Alkoxides and Nitrosyl Alkoxides*

Alkyl nitrites react with group 8 metalloporphyrins of the form (por)M(CO) (M = Ru, Os) to give the formal trans addition products (por)M(NO)(OR), where both the NO and alkoxide ligands are attached to the metal center trans to each other (33, 264–266, 269, 270, 310). The pathway for this reaction is analogous to that described previously for RSNO addition (see Scheme 24) to these metalloporphyrins. Infrared spectroscopic evidence for an intermediate resulting from the initial *O*-binding of the RONO group followed by alkoxide formation and subsequent NO binding was obtained for the reaction of (TTP)Ru(CO) with isoamyl nitrite in dichloromethane (270). The higher υ_{CO} for the intermediate, compared with the (TTP)Ru(CO) precursor, is consistent with a formal change in oxidation state of the Ru center as the alkoxide binds.

The reaction of the cationic $[(TPP)Fe(thf)_2]^+$ compound with isoamyl nitrite yielded the cationic nitrosyl alcohol complex $[(TPP)Fe(NO)(HO\text{-}i\text{-}C_5H_{11})]^+$ derivative (270).

Another example of alkyl nitrites serving to provide both the alkoxide and nitrosyl to metals is evident from Warren's (214) report of the preparation of the bimetallic complex $\{[Me_3NN]Ni\}_2(\mu\text{–}OCy)_2$ (CyOH = cyclohexyl alcohol) and

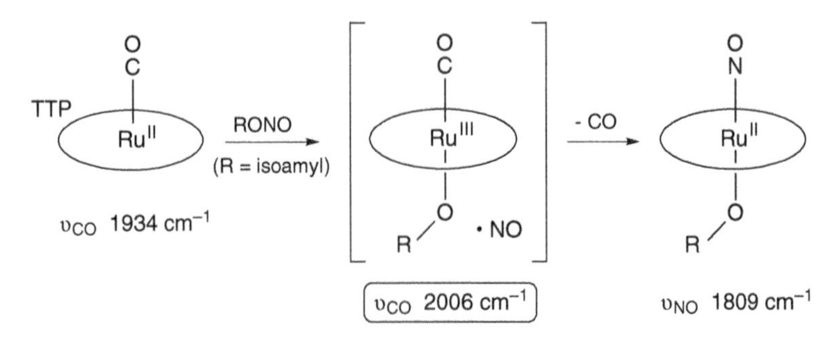

Scheme 24.

the nitrosyl $[Me_3NN]Ni{-}NO$ from the reaction of $[Me_3NN]Ni$(2,4-lutidine) with CyONO in toluene at room temperature.

B. Metal-Mediated Formation of RONO

1. Coupling of Nitro and Carbon-Based Ligands

An unusual reversible coupling of a nitro group and alkenes has been reported for a Pd compound **79**, as shown in Scheme 25 (311). The internal cyclization results in the formation of heterocyclopentane compounds **80–81**; this formation occurred in seconds when isobutylene or norbornene was used as the incoming alkenes, and in minutes if ethylene or propylene was used. The crystal structure of a product deriving from a substituted norbornene dicyclopentadiene confirmed the metallacycle formation.

2. Nucleophilic Attack of Alcohols and Alkoxides on Metal–NO Groups

Reed and Roper (312) demonstrated that the coordinated ligands in a number of Ir complexes of the form $IrCl_3(NO)L_2$ ($L = PPh_3$, $AsPh_3$) were attacked by alcohols ROH (R = Me, Et, *i*-Pr) to give neutral $IrCl_3L_2$(RONO) that were

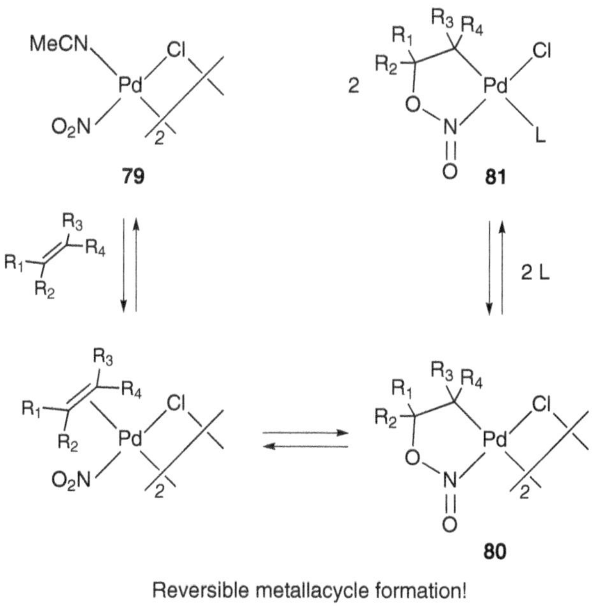

Reversible metallacycle formation!

Scheme 25.

characterized by IR and ^{1}H NMR spectroscopy. Walsh et al. (313) reported similar nucleophilic attack of alkoxides (ROH–collidine; R = Me, Et, *n*-Bu, *i*-Pr) on the coordinated NO ligands of the $[Ru(bpy)_2(py)NO]^{3+}$ and $[Ru(bpy)_2(Cl)NO]^{2+}$ compounds to give Ru–(RONO) products that were characterized by IR and ^{1}H NMR spectroscopy and by electrochemistry.

The reaction of the ferric porphyrin (TPP)FeCl with NO in the presence of methanol and toluene that resulted in the formation of MeONO and the reduced iron porphyrin was reported earlier by Wayland and Olson (314).

Ford and co-workers (315–317) showed that the $[(DMP)_2Cu(H_2O)]^{2+}$ (DMP = 2,9-dimethyl-1,10-phenanthroline) complex reacted with NO in methanol to release MeONO and a reduced Cu^+ product, a process that involved an inner-sphere attack of methanol on a Cu bound NO group. Such chemistry has been exploited by Ford and others (315–320) for the fluorescence-based detection of NO (see also Section III.B).

C. Coordination Modes of Metal–RONO Compounds

There is only one published metal–RONO crystal structure, namely, that of **82** shown below (311), which is a derivative of compound **80** in Scheme 25. Compound **82** displays N-binding of the chelate RONO moiety, with a υ_{NO} of 1612 cm^{-1} in its IR spectrum, and with N=O distances of 1.19(2) and 1.18(2) Å, O–N distances of 1.36(2) and 1.40(2) Å, and an O–N=O bond angle of 113(1)°.

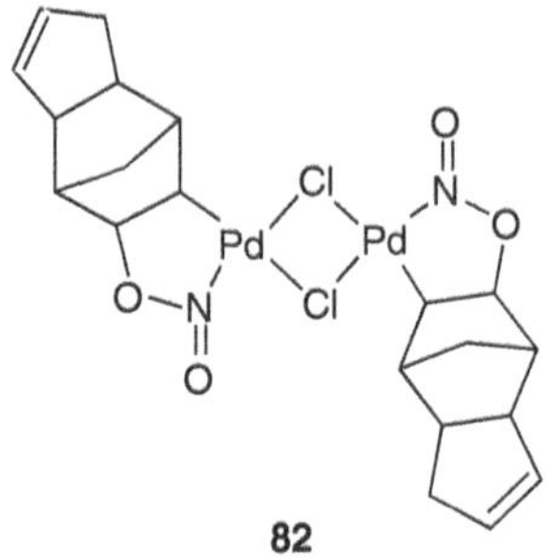

82

VI. OUTLOOK

Although much is known about metal–NO_x compounds and organo–NO_x chemistry, the ability of metals to change the course of reactions of organo–NO_x compounds is a relatively unexplored area of research and still in its infancy.

The varied π acidity of the organo–NO fragment (e.g., in nitrosoarenes) may find utility in the potential control of subsequent metal-based reactivity for the production of fine chemicals. In addition, and given the increased attention that nitrosative stress is having in modulating disease states, there are now increased opportunities to uncover the complex chemistry that biological metals (e.g., Fe, Cu, and Zn) exhibit in the metabolic chemistry of organo–NO_x species. Several factors, including the dichotomy between homolytic and heterolytic cleavage of the organo–NO bonds in the presence of metals, need to be explored further to better assess the stabilities and/or reactivities of these compounds under reaction stress conditions. We fully expect that this area of research will increase in intensity over the next few years.

ACKNOWLEDGMENTS

We are grateful to the U.S. National Science Foundation (CHE-1213674) for funding for this work.

ABBREVIATIONS

Ad	Adamantyl
Bn	Benzyl
BNPP	1,4-Bis(4-nitrosophenyl)piperazine
bpy	2,2-Bipyridine
Cbl	Cobalamin
Cp	η^5-Cyclopentadienyl anion
Cp*	η^5-Pentamethylcyclopentadienyl anion
cy	Cyclohexyl
CyOH	Cyclohexyl alcohol
DAC	1,8-Bis(9-anthracylmethyl) derivative of the macrocyclic cyclam (1,4,8,11-tetraazacyclotetradecane)
DEANO	2-(*N*,*N*-Diethylamino)diazenolate-2-oxide
DFT	Density functional theory
DMP	2,9-Dimethyl-1,10-phenanthroline
DMPS	2,3-Dimercaptopropane-1-sulfonate
dmso	Dimethyl sulfoxide (ligand)
dppe	1,2-Bis(diphenylphosphino)ethane
ESIMS	Electrospray ionization mass spectrometry

FTIR	Fourier transform infrared
GSNO	*S*-Nitrosoglutathione
^{1}H NMR	Proton nuclear magnetic resonance
Hb	Hemoglobin
HMPA	Hexamethylphosphoramide
IQ	2-amino-3-methylimidazole[4,5-*f*]quinoline
IR	Infrared
leg Hb	Hb found in leguminous plants
2,4-lut	2,4-dimethylpyridine or 2,4-lutidine
1-MeIm	1-methylimidazole
Mb	Myoglobin
Me_2NN	2,4-Bis(2,6-dimethylphenylimido)pentyl anion
Me_6tren	Tris(2-dimethylaminoethyl)amine
MGD	*N*-Methyl-D-Glucamine dithiocarbamate
NACysMe	*N*-Acetyl-L-cysteinate methyl ester
NBD	Norbornadiene
NMR	Nuclear magnetic resonance
NO	Nitric oxide
OEP	Octaethylporphyrinato dianion
OTf	Triflate anion
phen	1,10-Phenanthroline
por	Porphyrinato dianion
py	Pyridine
SG	Glutathionyl
SNAP	*S*-Nitroso-*N*-acetylpenicillamine
SQUID	Superconducting quantum interference device
TC-3,5	Tetraazamacrocylic tropocoronand
thf	Tetrahydrofuran (ligand)
THF	Tetrahydrofuran (solvent)
TIM	2,3,9,10-Tetramethyl-1,4,8,11-tetraazacyclotetradeca-1,3,8,10-tetraene
TMEDA	Tetramethylethylenediamine
TMP	Tetramesitylporphyrinato dianion
TMS	Trimethylsilyl
Tp	Tris(pyrazolyl)borate
Tp*	$HB(3,5\text{-Dimethylpyrazolyl})_3$
TPP	Tetraphenylporphyrinato dianion
TTP	Tetratolylporphyrinato dianion
Ts	Tosyl
UV–vis	Ultraviolet–visible

REFERENCES

1. R. L. Klimisch and J. G. Larson, Eds., *The Catalytic Chemistry of Nitrogen Oxides*, Plenum Press, New York, 1975.
2. D. G. Capone, R. Popa, B. Flood, and K. H. Nealson, *Science 312*, 708 (2006).
3. V. Smil, *Sci. Amer. 277*, 76 (1997).
4. I. M. Wasser, S. de Vries, P. Moenne-Loccoz, I. Schroder, and K. D. Karlin, *Chem. Rev. 102*, 1201 (2002).
5. P. M. Vitousek, J. D. Aber, R. W. Howarth, G. E. Likens, P. A. Matson, D. W. Schindler, W. H. Schlesinger, and D. G. Tilman, *Ecol. Appl. 7*, 737 (1997).
6. B. A. Averill, *Chem. Rev. 96*, 2951 (1996).
7. A. R. Butler and D. L. H. Williams, *Chem. Soc. Rev. 22*, 233 (1993).
8. A. R. Butler, *Chem. Bri.*, 419 (1990).
9. T. W. J. Taylor and W. Baker *The Organic Chemistry of Nitrogen*, Revised and Rewritten from N. V. Sidgwick's First ed., Clarendon Press, Oxford, UK, 1937.
10. E. W. Ainscough and A. M. Brodie, *J. Chem. Educ. 72*, 686 (1995).
11. K. G. Caulton, *Coord. Chem. Rev. 14*, 317 (1975).
12. N. G. Connelly, *Inorg. Chim. Acta 6*, 47 (1972).
13. B. F. G. Johnson and J. A. McCleverty, *Prog. Inorg. Chem. 7*, 277 (1966).
14. A. R. Butler, C. Glidewell, and M.-H. Li, *Adv. Inorg. Chem. 32*, 335 (1988).
15. L. Cheng and G. B. Richter-Addo, in *The Porphyrin Handbook*, Guilard, R., Smith, K., and Kadish, K. M. Eds., Vol. *4*, Chap. 33, pp. 219–291. Academic Press, New York. 2000.
16. A. R. Butler and I. L. Megson, *Chem. Rev. 102*, 1155 (2002).
17. G. B. Richter-Addo and P. Legzdins, *Chem. Rev. 88*, 991 (1988).
18. W. P. Griffith, *Adv. Organomet. Chem. 7*, 211 (1968).
19. T. W. Hayton, P. Legzdins, and W. B. Sharp, *Chem. Rev. 102*, 935 (2002).
20. J. H. Enemark and R. D. Feltham, *Coord. Chem. Rev. 13*, 339 (1974).
21. J. H. Enemark, R. D. Feltham, J. Riker-Nappier, and K. F. Bizot, *Inorg. Chem. 14*, 624 (1975).
22. R. D. Feltham and J. H. Enemark, *Top. Stereochem. 12*, 155 (1981).
23. B. L. Westcott and J. H. Enemark, in *Inorganic Electronic Structure and Spectroscopy*, A. B. P. Lever and E. I. Solomon, Eds., Chap. 7, John Wiley and Sons Inc., New York, 1999.
24. D. M. P. Mingos and D. J. Sherman, *Adv. Inorg. Chem. 34*, 293 (1989).
25. R. Eisenberg and D. E. Hendricksen, *Adv. Catal. 28*, 79 (1979).
26. J. A. McCleverty, *J. Mol. Catal. 13*, 309 (1981).
27. F. Bottomley, *Acc. Chem. Res. 11*, 158 (1978).
28. F. Bottomley, in *Reactions of Coordinated Ligands*, P.S. Braterman, Ed., pp. 115–222, Plenum Press, New York, 1989.

29. M. J. Clarke and J. B. Gaul, *Struct. Bonding 81*, 147 (1993).
30. G. B. Richter-Addo and P. Legzdins, *Metal Nitrosyls*, Oxford University Press, New York. 1992.
31. D. L. H. Williams *Nitrosation*, Cambridge University Press, Cambridge, UK, 1988.
32. D. L. H. Williams, *Nitrosation Reactions and the Chemistry of Nitric Oxide*, Elsevier, Amsterdam, The Netherlands, 2004.
33. J. Lee, L. Chen, A. H. West, and G. B. Richter-Addo, *Chem. Rev. 102*, 1019 (2002).
34. D. Mansuy, J. C. Chottard, and G. Chottard, *Eur. J. Biochem. 76*, 617 (1977).
35. A. Renodon, J.-L. Boucher, C. Wu, R. Gachhui, M.-A. Sari, D. Mansuy, and D. Stuehr, *Biochemistry 37*, 6367 (1998).
36. R. Ricoux, J.-L. Boucher, D. Mansuy, and J.-P. Mahy, *Biochem. Biophys. Res. Commun. 278*, 217 (2000).
37. D. M. Copeland, A. H. West, and G. B. Richter-Addo, *Proteins: Struct. Func. Genet. 53*, 182 (2003).
38. D. Mansuy, P. Beaune, J. C. Chottard, J. F. Bartoli, and P. Gans, *Biochem. Pharmacol. 25*, 609 (1976).
39. C. S. Yang and T. J. Smith, *Adv. Exp. Med. Biol. 387*, 385 (1996).
40. R. N. Loeppky, A. Fuchs, C. Janzowski, C. Humberd, P. Goelzer, H. Schneider, and G. Eisenbrand, *Chem. Res. Toxicol. 11*, 1556 (1998).
41. W. Lijinsky *Chemistry and Biology of N-Nitroso Compounds*, Cambridge University Press, Cambridge, UK, 1992.
42. M. J. Hill, Ed., *Nitrosamines: Toxicology and Microbiology*, VCH Ellis Horwood Ltd., Chichester, UK, 1988.
43. J. S. Stamler, in *Current Topics in Microbiology and Immunology. The Role of Nitric Oxide in Physiology and Pathophysiology*, H. Koprowski and H., Maeda, Eds., pp. 19–35, Springer-Verlag, Berlin, 1995.
44. M. P. Doyle, R. A. Pickering, and J. da Conceição, *J. Biol. Chem. 259*, 80 (1984).
45. B. G. Gowenlock and G. B. Richter-Addo, *Chem. Rev. 104*, 3315 (2004).
46. B. G. Gowenlock and G. B. Richter-Addo, *Chem. Soc. Rev. 34*, 797 (2005).
47. M. Cameron, B. G. Gowenlock, and G. Vasapollo, *Chem. Soc. Rev. 19*, 355 (1990).
48. B. G. Gowenlock and W. Luttke, *Quart. Rev. Chem. Soc. 12*, 321 (1958).
49. L. A. Labios, M. D. Millard, A. L. Rheingold, and J. S. Figueroa, *J. Am. Chem. Soc. 131*, 11318 (2009).
50. J. Lee, B. Twamley, and G. B. Richter-Addo, *Can. J. Chem. 80*, 1252 (2002).
51. J. Lee, B. Twamley, and G. B. Richter-Addo, *Dalton Trans.* 189 (2004).
52. S. Wiese, P. Kapoor, K. D. Williams, and T. H. Warren, *J. Am. Chem. Soc. 131*, 18105 (2009).
53. R. S. Srivastava, M. A. Khan, and K. M. Nicholas, *J. Am. Chem. Soc. 127*, 7278 (2005).
54. C. Krinninger, S. Wirth, J. C. G. Ruiz, P. Klufers, H. Noth, and I. P. Lorenz, *Eur. J. Inorg. Chem.*, 4094 (2005).

55. S. Wirth, A. U. Wallek, A. Zernickel, F. Feil, M. Sztiller-Sikorska, K. Lesiak-Mieczkowska, C. Brauchle, I. P. Lorenz, and M. Czyz, *J. Inorg. Biochem. 104*, 774 (2010).
56. S. A. Cummings, R. Radford, G. Erker, G. Kehr, and R. Frohlich, *Organometallics 25*, 839 (2006).
57. C. D. Sohl, J. Lee, S. S. Alguindigue, M. A. Khan, and G. B. Richter-Addo, *J. Inorg. Biochem. 98*, 1238 (2004).
58. D. Mansuy, P. Battioni, J.-C. Chottard, C. Riche, and A. Chiaroni, *J. Am. Chem. Soc. 105*, 455 (1983).
59. K. V. Luzyanin, P. V. Gushchin, A. J. L. Pombeiro, M. Haukka, V. I. Ovcharenko, and V. Y. Kukushkin, *Inorg. Chem. 47*, 6919 (2008).
60. K. V. Luzyanin, V. Y. Kukushkin, M. L. Kuznetsov, A. D. Ryabov, M. Galanski, M. Haukka, E. V. Tretyakov, V. I. Ovcharenko, M. N. Kopylovich, and A. J. L. Pombeiro, *Inorg. Chem. 45*, 2296 (2006).
61. P. Legzdins, G. B. Richter-Addo, B. Wassink, F. W. B. Einstein, R. H. Jones, and A. C. Willis, *J. Am. Chem. Soc. 111*, 2097 (1989).
62. P. Legzdins, G. B. Richter-Addo, F. W. B. Einstein, and R. H. Jones, *Organometallics 6*, 1807 (1987).
63. A. Goldhaber, K. P. C. Vollhardt, E. C. Walborsky, and M. Wolfgruber, *J. Am. Chem. Soc. 108*, 516 (1986).
64. S.-C. Chan, H.-Y. Cheung, and C.-Y. Wong, *Inorg. Chem. 50*, 11636 (2011).
65. S. C. Chan, P. K. Pat, T. C. Lau, and C. Y. Wong, *Organometallics 30*, 1311 (2011).
66. E. B. Brouwer, P. Legzdins, S. J. Rettig, and K. J. Ross, *Organometallics 13*, 2088 (1994).
67. M. D. Seidler and R. G. Bergman, *Organometallics 2*, 1897 (1983).
68. B. N. Diel, *J. Organomet. Chem. 284*, 257 (1985).
69. J. Chang, M. D. Seidler, and R. G. Bergman, *J. Am. Chem. Soc. 111*, 3258 (1989).
70. S. Niu and M. B. Hall, *J. Phys. Chem. A 101*, 1360 (1997).
71. W. P. Weiner, M. A. White, and R. G. Bergman, *J. Am. Chem. Soc. 103*, 3612 (1981).
72. W. P. Weiner and R. G. Bergman, *J. Am. Chem. Soc. 105*, 3922 (1983).
73. Y. Arikawa, S.-Y. Tashita, S. Yamaguchi, H. Miyahara, T. Asayama, K. Umakoshi, and M. Onishi, *Chem. Lett. 39*, 940 (2010).
74. P. M. Graham, M. S. A. Buschhaus, R. A. Baillie, S. P. Semproni, and P. Legzdins, *Organometallics 29*, 5068 (2010).
75. H. Brunner and S. Loskot, *Angew. Chem. Int. Ed. Engl. 10*, 515 (1971).
76. P. N. Becker and R. G. Bergman, *Organometallics 2*, 787 (1983).
77. G. Evrad, R. Thomas, B. R. Davis, and I. Bernal, *J. Organomet. Chem. 124*, 59 (1977).
78. P. N. Becker and R. G. Bergman, *J. Am. Chem. Soc. 105*, 2985 (1983).
79. M. R. Crimmin, L. E. Rosebrugh, N. C. Tomson, T. Weyhermuller, R. G. Bergman, F. D. Toste, and K. Wieghardt, *J. Organomet. Chem. 696*, 3974 (2011).

80. J. M. Schomaker, W. C. Boyd, I. C. Stewart, F. D. Toste, and R. G. Bergman, *J. Am. Chem. Soc. 130*, 3777 (2008).
81. J. M. Schomaker, F. D. Toste, and R. G. Bergman, *Org. Lett. 11*, 3698 (2009).
82. C. Zhao, F. D. Toste, and R. G. Bergman, *J. Am. Chem. Soc. 133*, 10787 (2011).
83. M. R. Crimmin, R. G. Bergman, and F. D. Toste, *Angew. Chem., Int. Ed. 50*, 4484 (2011).
84. Y. Arikawa, T. Asayama, K. Itatani, and M. Onishi, *J. Am. Chem. Soc. 130*, 10508 (2008).
85. T. Asayama, Y. Arikawa, T. Murabe, S. Agari, K. Umakoshi, and M. Onishi, *Chem. Asian. J. 7*, 664 (2012).
86. S. Fukui, A. Kajihara, T. Hirano, F. Sato, N. Suzuki, and H. Nagao, *Inorg. Chem. 50*, 4713 (2011).
87. W. L. Bowden, W. F. Little, and T. J. Meyer, *J. Am. Chem. Soc. 96*, 5605 (1974).
88. W. L. Bowden, W. F. Little, and T. J. Meyer, *J. Am. Chem. Soc. 98*, 444 (1976).
89. W. L. Bowden, W. F. Little, T. J. Meyer, and D. Salmon, *J. Am. Chem. Soc. 97*, 6897 (1975).
90. F. Bottomley, P. S. White, M. Mukaida, K. Shimura, and H. Kakihana, *J. Chem. Soc., Dalton Trans.*, 2965 (1988).
91. A. Ishigaki, M. Oue, Y. Matsushita, I. Masuda, and T. Shono, *Bull. Chem. Soc. Jpn. 50*, 726 (1977).
92. A. R. Butler and C. Glidewell, *Chem. Soc. Rev. 16*, 361 (1987).
93. S. Clamp, N. G. Connelly, J. A. K. Howard, I. Manners, J. D. Payne, and W. E. Geiger, *J. Chem. Soc., Dalton Trans.*, 1659 (1984).
94. S. Kura, S. Kuwata, and T. Ikariya, *Angew. Chem., Int. Ed. 44*, 6406 (2005).
95. N. Escola, A. Llebaria, G. Leitus, and F. Doctorovich, *Organometallics 25*, 3799 (2006).
96. S. J. Skoog, J. P. Campbell, and W. L. Gladfelter, *Organometallics 13*, 4137 (1994).
97. S. J. Skoog and W. L. Gladfelter, *J. Am. Chem. Soc. 119*, 11049 (1997).
98. J. M. O'Connor and K. D. Bunker, *Organometallics 22*, 5268 (2003).
99. M. Pizzotti, C. Crotti, and F. Demartin, *J. Chem. Soc., Dalton Trans.*, 735 (1984).
100. H. F. Klein, A. Schmidt, U. Florke, and H. J. Haupt, *Inorg. Chim. Acta 342*, 171 (2003).
101. X. Li, C. D. Incarvito, T. Vogel, and R. H. Crabtree, *Organometallics 24*, 3066 (2005).
102. G. La Monica and S. Cenini, *J. Chem. Soc., Dalton Trans.*, 1145 (1980).
103. F. Hintermaier, S. Helding, L. B. Volodarsky, K. Sunkel, K. Polborn, and W. Beck, *Z. Naturforsch. B. 53B*, 101 (1998).
104. G. La Monica and S. Cenini, *Inorg. Chim. Acta 29*, 183 (1978).
105. J. N. Armor, R. Furman, and M. Z. Hoffman, *J. Am. Chem. Soc. 97*, 1737 (1975).
106. R. P. Cheney, S. D. Pell, and M. Z. Hoffman, *J. Inorg. Nucl. Chem. 41*, 489 (1979).
107. R. P. Cheney, M. G. Simic, M. Z. Hoffman, I. A. Taub, and K.-D. Asmus, *Inorg. Chem. 16*, 2187 (1977).

108. H. Wu, X. W. Dong, J. F. Ma, H. Y. Liu, J. Yang, and H. Y. Bai, *Dalton Trans.* 3162 (2009).
109. R. S. Pilato, C. McGettigan, G. L. Geoffroy, A. L. Rheingold, and S. J. Geib, *Oranometallics 9*, 312 (1990).
110. J. C. Stephens, M. A. Khan, and K. M. Nicholas, *J. Organomet. Chem. 690*, 4727 (2005).
111. N. C. Tomson, L. A. Labios, T. Weyhermuller, J. S. Figueroa, and K. Wieghardt, *Inorg. Chem. 50*, 5763 (2011).
112. D. B. Sams and R. J. Doedens, *Inorg. Chem. 18*, 153 (1979).
113. G. Vasapollo, A. Sacco, C. F. Nobile, M. A. Pellinghelli, and M. Lanfranchi, *J. Organomet. Chem. 353*, 119 (1988).
114. R. G. Little and R. J. Doedens, *Inorg. Chem. 12*, 537 (1973).
115. D. Mansuy, M. Dreme, J. C. Chottard, and J. Guilhem, *J. Organomet. Chem. 161*, 207 (1978).
116. D. L. Packett, W. C. Trogler, and A. L. Rheingold, *Inorg. Chem. 26*, 4308 (1987).
117. D. Mansuy, M. Dreme, J.-C. Chottard, J.-P. Girault, and J. Guilhem, *J. Am. Chem. Soc. 102*, 844 (1980).
118. L.-S. Wang, L. Chen, M. A. Khan, and G. B. Richter-Addo, *Chem. Commun.*, 323 (1996).
119. N. Godbout, L. K. Sanders, R. Salzmann, R. H. Havlin, M. Wojdelski, and E. Oldfield, *J. Am. Chem. Soc. 121*, 3829 (1999).
120. L. Chen, J. B. Fox, Jr., G.-B. Yi, M. A. Khan, and G. B. Richter-Addo, *J. Porph. Phthalocyanines 5*, 702 (2001).
121. L. Chen, M. A. Khan, G. B. Richter-Addo, V. G. Young, Jr., and D. R. Powell, *Inorg. Chem. 37*, 4689 (1998).
122. D. S. Bohle, B. J. Conklin, and C.-H. Hung, *Inorg. Chem. 34*, 2569 (1995).
123. E. Fritsch, K. Polborn, K. Sunkel, W. Beck, H. Kohler, and L. Jager, *Z. Anorg. Allg. Chem. 617*, 110 (1992).
124. S. Wirth, F. Barth, and I. P. Lorenz, *Dalton Trans. 41*, 2176 (2012).
125. J. Yi, G. Ye, L. M. Thomas, and G. B. Richter-Addo, *Chem. Commun.*, *49*, 11179 (2013).
126. I. P. Kuranova, A. V. Teplyakov, G. V. Obmolova, A. A. Voronova, A. N. Popov, D. M. Kheiker, and E. G. Arutyunyan, *Bioorg. Khim. 8*, 1625 (in Russian). *Chem. Abstr.* CA1698:48994. PDB accession codes 1LH4 and 2LH4 (1982).
127. Y. N. Liu, W. Z. Liang, X. G. Sang, Y. Q. Huo, L. Sze-To, K. F. Yung, and X. X. Liu, *Inorg. Chim. Acta 363*, 949 (2010).
128. S. Wirth, C. J. Rohbogner, M. Cieslak, J. Kazmierczak-Baranska, S. Donevski, B. Nawrot, and I. P. Lorenz, *J. Biol. Inorg. Chem. 15*, 429 (2010).
129. H. Wu, X. W. Dong, H. Y. Liu, J. F. Ma, S. L. Li, J. Yang, Y. Y. Liu, and Z. M. Su, *Dalton Trans.* 5331 (2008).

130. R. L. Garzon, M. L. G. Salido, J. N. Low, and C. Glidewell, *Acta Crystallogr. Sect. C59*, M291 (2003).
131. M. N. Moreno, J. M. Salas, E. Colacio, M. P. Sanchez, and F. Nieto, *Acta Crystallogr. Sect. C42*, 407 (1986).
132. P. Arranz-Mascaros, M. D. Gutierrez-Valero, R. Lopez-Garzon, M. D. Lopez-Leon, M. L. Godino-Salido, A. Santiago-Medina, and H. Stoekckli-Evans, *Polyhedron 27*, 623 (2008).
133. N. A. Illan-Cabeza, A. R. Garcia-Garcia, and M. N. Moreno-Carretero, *Inorg. Chim. Acta 366*, 262 (2011).
134. A. M. Kirillov, M. Haukka, M. R. C. G. da Silva, and A. J. Pombeiro, *Eur. J. Inorg. Chem.*, 2071 (2005).
135. G. Wu, J. F. Zhu, R. Y. Wang, and V. Terskikh, *J. Am. Chem. Soc. 132*, 5143 (2010).
136. R. Lopez-Garzon, M. L. Godino-Salido, P. Arranz-Mascaros, M. A. Fontecha-Camara, M. D. Gutierrez-Valero, R. Cuesta, J. M. Moreno, and H. Stoeckli-Evans, *Inorg. Chim. Acta 357*, 2007 (2004).
137. C. Glidewell, J. N. Low, P. A. Mascaros, R. C. Martos, and M. D. G. Valero, *Acta Crystallogr. Sect. C58*, M61 (2002).
138. S. Hu, D. M. Thompson, P. O. Ikekwere, R. J. Barton, K. E. Johnson, and B. E. Robertson, *Inorg. Chem. 28*, 4552 (1989).
139. N. G. Bokii, A. I. Udel'nov, Y. T. Struchkov, D. N. Kravtsov, and V. M. Pachevskaya, *J. Struct. Chem. 18*, 814 (1977).
140. S. J. Fox, L. Chen, M. A. Khan, and G. B. Richter-Addo, *Inorg. Chem. 36*, 6465 (1997).
141. M. S. Askari, B. Girard, M. Murugesu, and X. Ottenwaelder, *Chem. Commun. 47*, 8055 (2011).
142. Y. L. Zub, G. G. Sadikov, V. V. Skopenko, M. A. Porai-Koshits, and V. P. Nikolaev, *Russ. J. Coord. Chem. 11*, 304 (1985).
143. G. G. Sadikov, Y. L. Zub, V. V. Skopenko, V. P. Nikolaev, and M. A. Porai-Koshits, *Russ. J. Coord. Chem. 10*, 690 (1984).
144. R. S. Srivastava, M. A. Khan, and K. M. Nicholas, *J. Am. Chem. Soc. 118*, 3311 (1996).
145. R. S. Srivastava and K. M. Nicholas, *J. Am. Chem. Soc. 119*, 3302 (1997).
146. C. Krinninger, S. Wirth, P. Klufers, P. Mayer, and I. P. Lorenz, *Eur. J. Inorg. Chem.*, 1060 (2006).
147. C. Krinninger, C. Hogg, H. Noth, J. C. G. Ruiz, P. Mayer, O. Burkacky, A. Zumbusch, and I. P. Lorenz, *Chem. Eur. J. 11*, 7228 (2005).
148. R. Wilberger, C. Krinninger, H. Piotrowski, P. Mayer, and I. P. Lorenz, *Eur. J. Inorg. Chem.*, 2488 (2004).
149. J. N. Low, P. Arranz, R. Cuesta, M. D. Gutierrez, and C. Glidewell, *Acta Crystallogr. Sect. C59*, M21 (2003).
150. S. K. Dutta, D. B. McConville, W. J. Youngs, and M. Chaudhury, *Inorg. Chem. 36*, 2517 (1997).

151. L. S. Liebeskind, K. B. Sharpless, R. D. Wilson, and J. A. Ibers, *J. Am. Chem. Soc. 100*, 7061 (1978).

152. F. Ridouane, J. Sanchez, H. Arzoumanian, and M. Pierrot, *Acta Crystallogr. Sect. C46*, 1407 (1990).

153. M. Pizzotti, F. Porta, S. Cenini, F. Demartin, and N. Masciocchi, *J. Organomet. Chem. 330*, 265 (1987).

154. D. W. Hoard and P. R. Sharp, *Inorg. Chem. 32*, 612 (1993).

155. M. J. Scott and S. J. Lippard, *Organometallics 17*, 466 (1998).

156. M. J. Barrow and O. S. Mills, *J. Chem. Soc. (A)*, 864 (1971).

157. H. G. Ang, W. L. Kwik, and K. K. Ong, *J. Fluorine Chem. 60*, 43 (1993).

158. S. Stella, C. Floriani, A. Chiesi-Villa, and C. Guastini, *J. Chem. Soc., Dalton Trans.*, 545 (1988).

159. M. Calligaris, T. Yoshida, and S. Otsuka, *Inorg. Chim. Acta 11*, L15 (1974).

160. K. K. H. Lee and W. T. Wong, *J. Chem. Soc., Dalton Trans.*, 3911 (1996).

161. K. K.-H. Lee and W.-T. Wong, *J. Chem. Soc., Dalton Trans.*, 2987 (1997).

162. T. Iwasa, H. Shimada, A. Takami, H. Matsuzaka, Y. Ishii, and M. Hidai, *Inorg. Chem. 38*, 2851 (1999).

163. K. E. Litz, J. W. Kampf, and M. M. B. Holl, *J. Am. Chem. Soc. 120*, 7484 (1998).

164. X. P. Wang, Y. Peng, Z. L. Zhu, J. C. Fettinger, P. P. Power, J. D. Guo, and S. Nagase, *Angew. Chem. Int. Ed. 49*, 4593 (2010).

165. S. Aime, G. Gervasio, L. Milone, R. Rossetti, and P. L. Stanghellini, *J. Chem. Soc., Dalton Trans.*, 534 (1978).

166. G. P. Khare and R. J. Doedens, *Inorg. Chem. 15*, 86 (1976).

167. T. Carofiglio, S. Stella, C. Floriani, A. Chiesi-Villa, and C. Guastini, *J. Chem. Soc., Dalton Trans.*, 1957 (1989).

168. G. Gervasio, R. Rossetti, and P. L. Stanghellini, *J. Chem. Soc., Chem. Commun.*, 387 (1977).

169. V. Chilou, P. Gouzerh, Y. Jeannin, and F. Robert, *J. Chem. Soc., Chem. Commun.*, 1469 (1987).

170. Y. Ling, C. Mills, R. Weber, L. Yang, and Y. Zhang, *J. Am. Chem. Soc. 132*, 1583 (2010).

171. M. Alcamí, O. Mó, M. Yáñez, A. Luna, J.-P. Morizur, and J. Tortajada, *J. Phys. Chem. A 102*, 10120 (1998).

172. B. G. Gowenlock and L. Batt, *J. Mol. Struct. (THEOCHEM) 454*, 103 (1998).

173. M. W. Schoonover, E. C. Baker, and R. Eisenberg, *J. Am. Chem. Soc. 101*, 1880 (1979).

174. M. Green, R. B. L. Osborn, A. J. Rest, and F. G. A. Stone, *J. Chem. Soc. (A)*, 2525 (1968).

175. X. L. Dai, P. Kapoor, and T. H. Warren, *J. Am. Chem. Soc. 126*, 4798 (2004).

176. X. P. Wang, Y. Peng, M. M. Olmstead, J. C. Fettinger, and P. P. Power, *J. Am. Chem. Soc. 131*, 14164 (2009).

177. R. S. Srivastava, N. R. Tarver, and K. M. Nicholas, *J. Am. Chem. Soc. 129*, 15250 (2007).
178. S. C. Puiu and T. H. Warren, *Organometallics 22*, 3974 (2003).
179. H. Lei, B. D. Ellis, C. B. Ni, F. Grandjean, G. J. Long, and P. P. Power, *Inorg. Chem. 47*, 10205 (2008).
180. F. Demartin, M. Pizzotti, F. Porta, and S. Cenini, *J. Chem. Soc., Dalton Trans.*, 605 (1987).
181. S. Cenini, F. Pots, M. Pizzotti, and G. La Monica, *J. Chem. Soc., Dalton Trans.*, 355 (1984).
182. S. Cenini, F. Porta, M. Pizzotti, and C. Crotti, *J. Chem. Soc., Dalton Trans.*, 163 (1985).
183. C. J. Jones, J. A. McCleverty, and A. S. Rothin, *J. Chem. Soc., Dalton Trans.*, 401 (1985).
184. P. L. Bellon, S. Cenini, F. Demartin, M. Pizzotti, and F. Porta, *J. Chem. Soc., Chem. Commun.*, 265 (1982).
185. P. Fantucci, M. Pizzotti, and F. Porta, *Inorg. Chem. 30*, 2277 (1991).
186. R. Preussman, and B. W. Stewart, in *Chemical Carcinogens*, C. E. Searle, Ed., 2nd ed., pp. 643–828, American Chemical Society, Washington, DC, 1984.
187. S. S. Hecht, *Chem. Res. Toxicol. 11*, 559 (1998).
188. K. E. Appel, H. H. Ruf, B. Mahr, M. Schwarz, R. Rickart, and W. Kunz, *Chem.-Biol. Interact. 28*, 17 (1979).
189. C. Ioannides and G. G. Gibson, in *Safety Evaluation of Nitrosatable Drugs and Chemicals*, G. G. Gibson, and C. Ioannides, Eds., pp. 257–278, Taylor and Francis, London, 1981.
190. C. S. Yang, T. J. Smith, J.-Y. Hong, and S. Zhou, in *Nitrosamines and Related N-Nitroso Compounds. Chemistry and Biochemistry* R. N. Loeppky, and C. J. Michejda, Eds., pp. 169–178 (Chap. 114), American Chemical Society, Washington, DC, 1994.
191. C. J. Patten, T. J. Smith, M. J. Friesen, R. E. Tynes, C. S. Yang, and S. E. Murphy, *Carcinogenesis 18*, 1623 (1997).
192. M. Mochizuki, E. Okochi, K. Shimoda, and K. Ito, in *Nitrosamines and Related N-Nitroso Compounds*, R. N. Loeppky, and C. J. Michejda, Eds., pp. 158–168 (Chap. 113), American Chemical Society, Washington, DC, 1994.
193. F. P. Guengerich. *Human Cytochrome P450 Enzymes*, 3rd ed., Kluwer Academic/Plenum, New York, 2005.
194. N. M. DeVore and E. E. Scott, *J. Biol. Chem. 287*, 26576 (2012).
195. M. I. Khan and U. C. Agarwala, *Bull. Chem. Soc. Jpn. 59*, 1285 (1986).
196. L. K. Keefer, R. W. Nims, K. M. Davies, and D. A. Wink, *Methods Enzymol. 268*, 281 (1996).
197. J. A. Hrabie and L. K. Keefer, *Chem. Rev. 102*, 1135 (2002).
198. S. Namiki, T. Arai, and K. Fujimori, *J. Am. Chem. Soc. 119*, 3840 (1997).

199. M. Z. Cabail and A. A. Pacheco, *Inorg. Chem. 42*, 270 (2003).
200. N. Purwar, J. M. McGarry, J. Kostera, A. A. Pacheco, and M. Schmidt, *Biochemistry 50*, 4491 (2011).
201. X.-Q. Zhu, W.-F. Hao, H. Tang, C.-H. Wang, and J.-P. Cheng, *J. Am. Chem. Soc. 127*, 2696 (2005).
202. X.-Q. Zhu, J.-Q. He, Q. Li, M. Xian, J. Lu, and J.-P. Cheng, *J. Org. Chem. 65*, 6729 (2000).
203. A. Schmidpeter, *Chem. Ber. 96*, 3275 (1963).
204. D. Klamann and W. Koser, *Angew. Chem., Int. Ed. (Engl.) 2*, 741 (1963).
205. A. Meller, W. Maringgele, and H.-G. Kohn, *Monatsh. Chem. 107*, 89 (1976).
206. A. Schmidpeter and H. Nöth, *Inorg. Chim. Acta 269*, 7 (1998).
207. G.-B. Yi, M. A. Khan, and G. B. Richter-Addo, *J. Am. Chem. Soc. 117*, 7850 (1995).
208. L. Chen, G.-B. Yi, L.-S. Wang, U. R. Dharmawardana, A. C. Dart, M. A. Khan, and G. B. Richter-Addo, *Inorg. Chem. 37*, 4677 (1998).
209. N. Xu, L. Goodrich, N. Lehnert, D. R. Powell, and G. B. Richter-Addo, *Inorg. Chem. 49*, 4405 (2010).
210. A. G. Constable, W. S. McDonald, and B. L. Shaw, *J. Chem. Soc., Dalton Trans.*, 2282 (1980).
211. F. Fuge, C. W. Lehmann, J. Rust, and J. Mohr, *J. Organomet. Chem. 694*, 2395 (2009).
212. A. Bredenkamp, X. Zeng, and F. Mohr, *Polyhedron 33*, 107 (2012).
213. P. Bladon, M. Dekker, G. R. Know, D. Willison, G. A. Jaffari, R. J. Doedens, and K. W. Muir, *Organometallics 12* 1725 (1993).
214. M. M. Melzer, S. Jarchow-Choy, E. Kogut, and T. H. Warren, *Inorg. Chem. 47*, 10187 (2008).
215. M. M. Melzer, S. Mossin, X. Dai, A. M. Bartell, P. Kapoor, K. Meyer, and T. H. Warren, *Angew. Chem., Int. Ed. 49*, 904 (2010).
216. F. Doctorovich, F. Di Salvo, N. Escola, C. Trapani, and L. Shimon, *Organometallics 24*, 4707 (2005).
217. F. Di Salvo, D. A. Estrin, G. Leitus, and F. Doctorovich, *Organometallics 27*, 1985 (2008).
218. F. Doctorovich and F. Di Salvo, *Acc. Chem. Res. 40*, 985 (2007).
219. H. Maltz, M. A. Grant, and M. C. Navaroli, *J. Org. Chem. 36*, 363 (1971).
220. R. Nast and J. Schmidt, *Z. Naturforsch. B. 32*, 469 (1977).
221. C. Trapani, N. Escola, and F. Doctorovich, *Organometallics 21* 2021 (2002).
222. F. Doctorovich, N. Escola, C. Trapani, D. A. Estrin, N. C. G. Lebrero, and A. G. Turjanski, *Organometallics 19*, 3810 (2000).
223. F. Di Salvo, A. Crespo, D. A. Estrin, and F. Doctorovich, *Tetrahedron 58*, 4237 (2002).
224. H.-F. Ip, X.-Y. Yi, W.-Y. Wong, I. D. Williams, and W.-H. Leung, *Dalton Trans. 40*, 11043 (2011).

225. N. Escola, F. Di Salvo, R. Haddad, L. Perissinotti, M. N. Eberlin, and F. Doctorovich, *Inorg. Chem. 46*, 4827 (2007).
226. M. B. Dinger, L. Main, and B. K. Nicholson, *J. Organomet. Chem. 565*, 125 (1998).
227. A. Albinati, S. Affolter, and P. S. Pregosin, *J. Organomet. Chem. 395*, 231 (1990).
228. W. Mossi, A. J. Klaus, P. Rys, A. Currao, and R. Nesper, *Acta Crystallogr. Sect. C51*, 2549 (1995).
229. U. Klement, *Acta Cryst. B25*, 2460 (1969).
230. U. Klement and A. Schmidpeter, *Angew. Chem., Int. Ed. (Engl.) 7*, 470 (1968).
231. S. D. Pell and J. N. Armor, *J. Am. Chem. Soc. 95*, 7625 (1973).
232. O. N. Adrianova, N. S. Gladkaya, and V. N. Vorotnikova, *Zh. Neorg. Khim. (Russ.) 15*, 2770. CA2774:9183x (1970).
233. V. M. Lakshmi, M. L. Clapper, W.-C. Chang, and T. V. Zenser, *Chem. Res. Toxicol. 18*, 528 (2005).
234. W. Brackman and P. J. Smit, *Rec. Trav. Chim. Pays-B. 84*, 357 (1965).
235. K. Tsuge, F. DeRosa, M. D. Lim, and P. C. Ford, *J. Am. Chem. Soc. 126*, 6564 (2004).
236. P. C. Ford, B. O. Fernandez, and M. D. Lim, *Chem. Rev. 105*, 2439 (2005).
237. C. Khin, M. D. Lim, K. Tsuge, A. Iretskii, G. Wu, and P. C. Ford, *Inorg. Chem. 46*, 9323 (2007).
238. L. E. McQuade, M. D. Pluth, and S. J. Lippard, *Inorg. Chem. 49*, 8025 (2010).
239. M. H. Lim, B. A. Wong, W. H. Pitcock, D. Mokshagundam, M.-H. Baik, and S. J. Lippard, *J. Am. Chem. Soc. 128*, 14364 (2006).
240. M. Sarma, A. Singh, S. G. G., G. Das, and B. Mondal, *Inorg. Chim. Acta 363*, 63 (2010).
241. M. Sarma, A. Kalita, P. Kumar, A. Singh, and B. Mondal, *J. Am. Chem. Soc. 132*, 7846 (2010).
242. M. Sarma and B. Mondal, *Inorg. Chem. 50*, 3206 (2011).
243. A. Kalita, P. Kumar, R. C. Deka, and B. Mondal, *Inorg. Chem. 50*, 11868 (2011).
244. P. Kumar, A. Kalita, and B. Mondal, *Dalton Trans. 40*, 8656 (2011).
245. M. Sarma and B. Mondal, *Dalton Trans. 41*, 2927 (2012).
246. M. Sarma, V. Kumar, A. Kalita, R. C. Deka, and B. Mondal, *Dalton Trans. 41*, 9543 (2012).
247. D. L. H. Williams, *Acc. Chem. Res. 32*, 869 (1999).
248. S. C. Askew, D. J. Barnett, J. McAninly, and D. L. H. Williams, *J. Chem. Soc., Perkin Trans. 2*, 741 (1995).
249. A. P. Dicks, H. R. Swift, D. L. H. Williams, A. R. Butler, H. H. Al-Sa'doni, and B. G. Cox, *J. Chem. Soc., Perkin Trans. 2*, 481 (1996).
250. S. Hwang, W. Cha, and M. E. Meyerhoff, *Angew. Chem. Int. Ed. 45*, 2745 (2006).
251. B. K. Oh and M. E. Meyerhoff, *J. Am. Chem. Soc. 125*, 9552 (2003).
252. D. Jourd'heuil, F. S. Laroux, A. M. Miles, D. A. Wink, and M. B. Grisham, *Arch. Biochem. Biophys. 361*, 323 (1999).

253. H. R. Swift and D. L. H. Williams, *J. Chem. Soc., Perkin Trans. 2*, 1933 (1997).
254. C.-M. Ho, H.-C. Leung, C.-N. Lok, and C.-M. Che, *Chem. Asian. J. 5*, 1965 (2010).
255. K. Inoue, T. Akaike, Y. Miyamoto, T. Okamoto, T. Sawa, M. Otagiri, S. Suzuki, T. Yoshimura, and H. Maeda, *J. Biol. Chem. 274*, 27069 (1999).
256. D. S. Pandey and U. C. Agarwala, *Inorg. Chim. Acta 159*, 197 (1989).
257. D. S. Pandey and U. C. Agarwala, *Synth. React. Inorg. Met.-Org. Chem. 21*, 361 (1991).
258. D. S. Pandey, S. K. Saini, and U. C. Agarwala, *Bull. Chem. Soc. Jpn. 60*, 3031 (1987).
259. F. S. D. S. Vilhena, A. W. S. da Silva, and S. R. W. Louro, *J. Inorg. Biochem. 100*, 1722 (2006).
260. F. S. D. S. Vilhena, and S. R. W. Louro, *J. Inorg. Biochem. 98*, 459 (2004).
261. A. R. Butler, S. Elkins-Daukes, D. Parkin, and D. L. H. Williams, *Chem. Commun.*, 1732 (2001).
262. X.-Q. Zhu, J.-Y. Zhang, and J.-P. Cheng, *Inorg. Chem. 46*, 592 (2007).
263. X.-Q. Zhu, J.-Y. Zhang, L.-R. Mei, and J.-P. Cheng, *Org. Lett. 8*, 3065 (2006).
264. G.-B. Yi, M. A. Khan, and G. B. Richter-Addo, *Chem. Commun.*, 2045 (1996).
265. S. M. Carter, J. Lee, C. A. Hixson, D. R. Powell, R. A. Wheeler, M. J. Shaw, and G. B. Richter-Addo, *Dalton Trans.* 1338 (2006).
266. L. Chen, M. A. Khan, and G. B. Richter-Addo, *Inorg. Chem. 37*, 533 (1998).
267. J. Lee, G.-B. Yi, M. A. Khan, and G. B. Richter-Addo, *Inorg. Chem. 38*, 4578 (1999).
268. J. Lee, G.-B. Yi, D. R. Powell, M. A. Khan, and G. B. Richter-Addo, *Can. J. Chem. 79*, 830 (2001).
269. G. B. Richter-Addo, *Acc. Chem. Res. 32*, 529 (1999).
270. G.-B. Yi, L. Chen, M. A. Khan, and G. B. Richter-Addo, *Inorg. Chem. 36*, 3876 (1997).
271. G.-B. Yi, M. A. Khan, D. R. Powell, and G. B. Richter-Addo, *Inorg. Chem. 37*, 208 (1998).
272. J. Yi, M. A. Khan, J. Lee, and G. B. Richter-Addo, *Nitric Oxide 12*, 261 (2005).
273. L. V. Andreasen, I. M. Lorkovic, G. B. Richter-Addo, and P. C. Ford, *Nitric Oxide 6*, 228 (2002).
274. M. Wolak, G. Stochel, and R. van Eldik, *Inorg. Chem. 45*, 1367 (2006).
275. M. M. Melzer, S. Mossin, A. J. P. Cardenas, K. D. Williams, S. Zhang, K. Meyer, and T. H. Warren, *Inorg. Chem. 51*, 8658 (2012).
276. K. N. Houk, B. N. Hietbrink, M. D. Bartberger, P. R. McCarren, B. Y. Choi, R. D. Voyksner, J. S. Stamler, and E. J. Toone, *J. Am. Chem. Soc. 125*, 6972 (2003).
277. L. L. Perissinotti, A. G. Turjanski, D. A. Estrin, and F. Doctorovich, *J. Am. Chem. Soc. 127*, 486 (2005).
278. M. S. Varonka and T. H. Warren, *Inorg. Chim. Acta 360*, 317 (2007).
279. M. M. Melzer, E. Li, and T. H. Warren, *Chem. Commun.*, 5847 (2009).
280. M. S. Varonka and T. H. Warren, *Inorg. Chem. 48*, 5605 (2009).

281. K. B. Capps, A. Bauer, K. A. Abboud, and C. D. Hoff, *Inorg. Chem. 38*, 6212 (1999).
282. K. B. Capps, A. Bauer, K. Sukcharoenphon, and C. D. Hoff, *Inorg. Chem. 38*, 6206 (1999).
283. D. Zheng, and R. L. Birke, *J. Am. Chem. Soc. 124*, 9066 (2002).
284. Y. Minamiyama, S. Takemura, S. Imaoka, Y. Funae, Y. Tanimoto, and M. Inoue, *J. Pharmacol. Exp. Ther. 283*, 1479 (1997).
285. J. Stadler, W. A. Schmalix, and J. Doehmer, *Adv. Exptl. Med. Biol. 387*, 187 (1996).
286. J. Stadler, J. Trockfeld, W. A. Schmalix, T. Brill, J. R. Siewert, H. Greim, and J. Doehmer, *Proc. Natl. Acad. Sci. USA 91*, 3559 (1994).
287. R. E. Ebel, D. H. O'Keeffe, and J. A. Peterson, *FEBS Lett. 55*, 198 (1975).
288. D. A. Wink, Y. Osawa, J. F. Darbyshire, C. R. Jones, S. C. Eshenaur, and R. W. Nims, *Arch. Biochem. Biophys. 300*, 115 (1993).
289. A. Franke, G. Stochel, N. Suzuki, T. Higuchi, K. Okuzono, and R. van Eldik, *J. Am. Chem. Soc. 127*, 5360 (2005).
290. A. Weichsel, E. M. Maes, J. F. Andersen, J. G. Valenzuela, T. K. Shokhireva, F. A. Walker, and W. R. Montfort, *Proc. Natl. Acad. Sci., USA 102*, 594 (2005).
291. J. Kozhukh and S. J. Lippard, *Inorg. Chem. 51*, 7346 (2012).
292. J. Kozhukh, J. F. Lopes, H. F. D. Santos, and S. J. Lippard, *Organometallics 31*, 8063 (2012).
293. C.-C. Tsou and W.-F. Liaw, *Chem. Eur. J. 17*, 13358 (2011).
294. K. Szacilowski, A. Wanat, A. Barbieri, E. Wasielewska, M. Witko, G. Stochel, and Z. Stasicka, *New. J. Chem. 26*, 1495 (2002).
295. K. Szacilowski, G. Stochel, and Z. Stasicka, *New. J. Chem. 21*, 893 (1997).
296. L. L. Perissinotti, D. A. Estrin, G. Leitus, and F. Doctorovich, *J. Am. Chem. Soc. 128*, 2512 (2006).
297. L. L. Perissinotti, G. Leitus, L. Shimon, D. Estrin, and F. Doctorovich, *Inorg. Chem. 47*, 4723 (2008).
298. J. W. Peck, D. I. Mahon, D. E. Beck, B. Bansenaur, and B. E. Koel, *Surface Sci. 410*, 214 (1998).
299. J. W. Peck, D. I. Mahon, D. E. Beck, and B. E. Koel, *Surface Sci. 410*, 170 (1998).
300. H. Ihm, J. W. Medlin, M. A. Barteau, and J. M. White, *Langmuir 17*, 798 (2001).
301. H. G. Jenniskens, W. van Essenberg, M. Kadodwala, and A. W. Kleyn, *Surface Sci. 402–404*, 140 (1998).
302. H. G. Jenniskens, L. Philippe, M. Kadodwala, and A. W. Kleyn, *J. Phys. Chem. B 102*, 8736 (1998).
303. J. E. Fieberg and J. M. White, *Chem. Phys. Lett. 306*, 103 (1999).
304. J. E. Fieberg and J. M. White, *J. Chem. Phys. 113*, 3839 (2000).
305. C. Kim, W. Zhao, and J. M. White, *Surface Sci. 464*, 240 (2000).
306. X. Wang, W. Wang, P. Han, M. Kubo, and A. Miyamoto, *Appl. Surface. Sci. 254*, 6991 (2008).
307. J. Li, J. Hu, and G. Li, *Cat. Commun. 12*, 1401 (2011).

308. D. S. Pandey, M. I. Khan, and U. C. Agarwala, *Indian. J. Chem. 26A*, 570 (1987).

309. S. D. Robinson and M. F. Uttley, *J. Chem. Soc., Dalton Trans.*, 1 (1972).

310. L. Cheng, D. R. Powell, M. A. Khan, and G. B. Richter-Addo, *Inorg. Chem. 40*, 125 (2001).

311. M. A. Andrews, T. C.-T. Chang, C.-W. F. Cheng, T. J. Emge, K. P. Kelly, and T. F. Koetzle, *J. Am. Chem. Soc. 106*, 5913 (1984).

312. C. A. Reed and W. R. Roper, *J. Chem. Soc., Dalton Trans.*, 1243 (1972).

313. J. L. Walsh, R. M. Bullock, and T. J. Meyer, *Inorg. Chem. 19*, 865 (1980).

314. B. B. Wayland and L. W. Olson, *J. Chem. Soc., Chem. Commun.*, 897 (1973).

315. D. Tran and P. C. Ford, *Inorg. Chem. 35*, 2411 (1996).

316. D. Tran, B. W. Skelton, A. H. White, L. E. Laverman, and P. C. Ford, *Inorg. Chem. 37*, 2505 (1998).

317. M. D. Lim, K. B. Capps, T. B. Karpishin, and P. C. Ford, *Nitric Oxide 12*, 244 (2005).

318. P. Kumar, A. Kalita, and B. Mondal, *Dalton Trans. 41*, 10543 (2012).

319. M. H. Lim and S. J. Lippard, *J. Am. Chem. Soc. 17*, 12170 (2005).

320. R. C. Smith, A. G. Tennyson, A. C. Won, and S. J. Lippard, *Inorg. Chem. 45*, 9367 (2006).

Aminopyridine Iron and Manganese Complexes as Molecular Catalysts for Challenging Oxidative Transformations

ZOEL CODOLA, JULIO LLORET-FILLOL, AND MIQUEL COSTAS

Grup de Química Bioinorgànica i Supramolecular (QBIS), Institut de Química Computacional i Catàlisi (IQCC) and Departament de Química, Universitat de Girona, Campus Montilivi, Girona, Catalonia, Spain

CONTENTS

Progress in Inorganic Chemistry, Volume 59, First Edition. Edited by Kenneth D. Karlin.

I. AMINOPYRIDINE IRON COMPLEXES AS MOLECULAR CATALYSTS FOR HYDROCARBON OXIDATIONS

A. Introduction

Hydrocarbon oxidation reactions mediated by iron centers are important in biology, bulk and fine chemistry, and consequently of great research interest (1–4). Iron holds a prevalent place at the active site of enzymes involved in oxidation reactions (5–10), and is a very attractive metal for developing synthetic methods because it is highly available and biocompatible (11–15). Peroxides can be seen as two-electron reduced forms of dioxygen, and compare favorably in terms of atom economy against most common organic oxidants. This consideration is especially obvious for hydrogen peroxide. Therefore, the combination of peroxides with iron coordination compounds has been seen as a rather simple and convenient strategy to mimic and create simple models for oxidation reactions taking place at iron-dependent oxygenases (16), and also to develop more sustainable oxidation technologies for oxidizing hydrocarbons in chemical synthesis.

While the approach may look conceptually simple, the combination of iron species and peroxides has proven to raise a very rich and subtle chemistry. Its interpretation often remains difficult (17). Complexities in the mechanistic landscape of these reactions are rooted in the rich reduction–oxidation (red-ox) chemistry of iron, peroxides, and the possibility of multiple oxidation and spin states. Differences in energy between electronic states and reaction paths are small and susceptible in a dramatic manner to the coordination structure of the iron site, and specific experimental conditions. Consequently, this complexity calls for great care with extrapolation of mechanistic analyses among apparently similar systems.

Fenton-type free diffusing radical processes dominate the chemistry in most of the cases where iron species are reacted with peroxides, although sometimes this happens in a complex and rather disguised manner that has led to misinterpretation. Concerning this complex scenario, the first question that the field needed to address in the mid-1990s was if the chemistry of non-heme iron-dependent oxygenases could be reproduced by using simple coordination compounds, without the aid of the elaborate machinery provided by enzyme active sites. Specifically, chemists in the field needed to clarify if challenging reactions (e.g., the stereospecific hydroxylation of alkanes) respond to chemical paths distinct from free diffusing radicals, or if these were shaped by enzymatic cavities, controlling the trajectory of otherwise free diffusing radicals. A second question that remained unsettled was if non-heme iron centers, without the assistance of strongly donating dianionic porphyrin ligands, which also could act as a red–ox noninnocent pool for delocalizing positive charge, could support higher oxidation states than the common ferric state. That also appeared as a very important question because in the absence of these highly electrophilic high-valent states, iron was confined to cycle between Fe(II) and Fe(III), in a Fenton process.

In this scenario, well-defined coordination complexes containing tetradentate aminopyridine ligands have become a very useful platform to investigate reaction mechanisms at mononuclear iron sites. They have provided positive proof of the principle that both aspects can indeed be accomplished with synthetic iron compounds. By doing so, they paved the way toward the development of bioinspired oxidation catalysts as tools for C−H and C=C oxidation in organic synthesis. Initial reports in this endeavor made use of a large excess of substrate to minimize substrate overoxidation reactions, but some iron complexes with tetradentate aminopyridine ligands have already proven competent for catalyzing reactions of selective oxidation of alkanes and alkenes providing yields amenable for preparative purposes. In comparison with organometallic catalysis (18–20), and oxidations with metalloporphyrins (21–24), the use of iron-coordination complexes as oxidation catalysts for organic synthesis is still very minor, but the potential structural versatility of these compounds is promising with regard to catalysts design. Iron complexes with aminopyridine ligands are emerging as privileged structures for mediating this challenging reactivity. Relevant results focused in the stereospecific hydroxylation of alkanes, olefin epoxidation, and cis-dihydroxylation are discussed in this chapter.

B. Class A Complexes: Iron Coordination Complexes Containing Aminopyridine Ligands That Enforce Strong Ligand Fields

A number of iron-coordination complexes have been studied as potential C−H and C=C oxidation catalysts (25), but among them, mono-iron complexes containing aminopyridine ligands deserve special consideration because of their

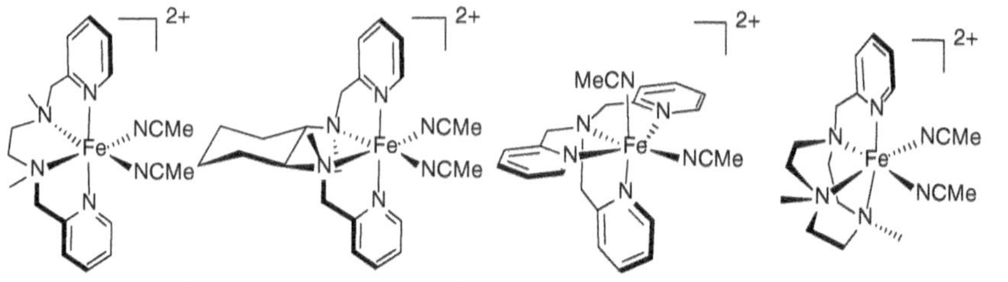

$[Fe(men)(MeCN)_2]^{2+}$ α-$[Fe(mcp)(MeCN)_2]^{2+}$ $[Fe(tpa)(MeCN)_2]^{2+}$ $[Fe(^{Me,H}Pytacn)(MeCN)_2]^{2+}$

Scheme 1. Structure of prototypical iron complexes with strong field tetradentate N-based ligands. *Note*: These complexes are usually isolated with triflate anions instead of acetonitrile, but in acetonitrile solution, where they are employed as catalysts, acetonitrile molecules are bound [men = *N,N′*-dimethyl-*N,N′*-bis(2-pyridylmethyl)ethane-1,2-diamine; mcp = *N,N′*-dimethyl-*N,N′*-bis(2-pyridylmethyl)cyclohexane-*trans*-1,2-diamine; tpa = tris(2-pyridylmethyl)amine; ^{Me,H}Pytacn = 1,4-dimethyl-7-(2-pyridylmethyl)-1,4,7-triazacyclononane].

ability to reproduce basic reactivity aspects of oxygenases, and because some of them recently have found success in organic synthesis. Among them, compounds $[Fe^{II}(tpa)(MeCN)_2]^{2+}$ (26–28), $[Fe^{II}(men)(MeCN)_2]^{2+}$ (26, 29), α-$[Fe^{II}(mcp)(MeCN)_2]^{2+}$ (30), $[Fe^{II}(^{Me,H}Pytacn)(MeCN)_2]^{2+}$ (31–34), (Schemes 1 and 2), serve as prototypical examples of catalysts of C–H and C=C oxidation reactions. The mechanistic studies solidly establish that these catalysts do not proceed through free diffusing radical transformations, but instead operate as metal-centered based systems in C–H and C=C oxidation reactions.

R = H: men
R = Me: $6Me_2$-men

mcp

pdp

bispidine, $BpdL_1$

R=R′=H : tpa
R=H, R′=Me : $5Me_3$-tpa
R=Me, R′=H : $6Me_3$-tpa

bqen

R = H: ^{Me,H}Pytacn
R = Me: ^{Me,Me}Pytacn

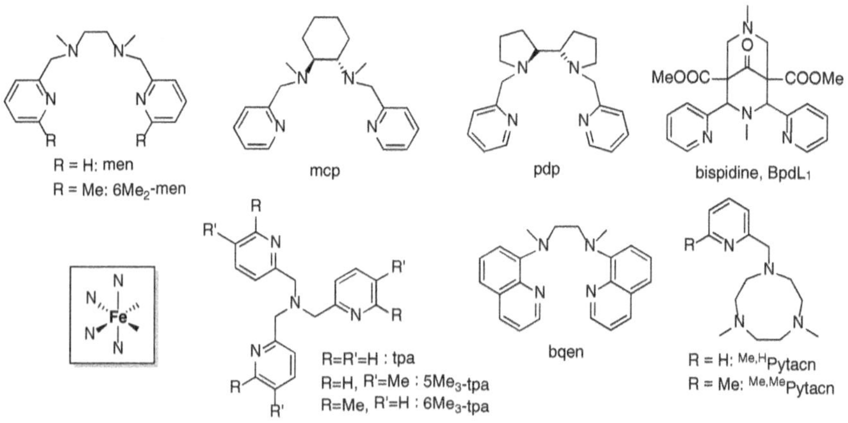

Scheme 2. Representation of most commonly employed tetradentate ligands used to prepare mononuclear iron(II) complexes that could perform the oxidation of alkanes. For chiral C_2-symmetric ligands, only one of the two enantiomers is shown [bqen = *N,N′*-dimethyl,*N,N′*-di(quinolin-8-yl)ethane-1,2-diamine; pdp = *N,N′*-bis(2-pyridylmethyl)methanamine; BpdL = dimethyl-3,7-dimethyl-9-oxo-2,4-di(pyridin-2-yl)-3,7-diazabicyclo[3.3.1]nonane-1,5-dicarboxylate].

Despite of their obvious differences, these catalysts share important structural similarities. They contain octahedral iron sites ligated to aminopyridine ligands that are robust under oxidative conditions. These ligands form very stable iron-coordination complexes via multiple chelation to the metal center. Binding of the iron to the ligand takes place through formation of kinetically very stable five-membered chelate rings. Furthermore, the ligands provide strong ligand fields, facilitating access to low-spin states. Finally, the tetradentate nature of the ligands leaves two-coordination positions at the iron centers that are enforced in a relative cis disposition. These positions are occupied by easily exchangeable ligands. They are the locus of peroxide-binding and activation. In acetonitrile solution, where catalytic oxidations are commonly studied, these positions are initially fulfilled with two molecules of this solvent. The presence of labile acetonitrile or triflate ligands is required for metal-based oxidation pathways. Instead, more strongly binding ligands (e.g., chloride) lead to Fenton-type reactions (35). The presence of two labile sites in cis-relative position also appears to be a crucial element in dictating the catalytic ability of this type of non-heme iron complexes in hydrocarbon oxidation reactions. Complexes that contain trans-labile sites appear to be much less active (36).

Table I collects results from the catalytic oxidation of alkanes with a series of this type of iron complexes. Scheme 2 shows a line drawing of the corresponding

TABLE I
Oxidation of Alkanes by Tetradentate Iron(II) Complexes, Ligands Represented in Scheme 2

Catalyst[b]	Cyclohexane A + K (A/K)[c]	Cyclohexane KIE[d]	*cis*-1,2-DMCH[a] RC (%)[e]	Adamantane 3°/2°[f]	Reference
$[Fe^{II}(tpa)(MeCN)_2]^{2+}$	3.2 (6)	3.5	100	17	26
$[Fe^{II}(5Me_3\text{-tpa})(MeCN)_2]^{2+}$ [g]	4.0 (5)	3.8	100	21	26
$[Fe^{II}(6Me_3\text{-tpa})(MeCN)_2]^{2+}$ [h]	2.9 (2)	3.3	54	15	26
$[Fe^{II}(men)(MeCN)_2]^{2+}$	6.3 (5)	3.2	96	15	26
$[Fe^{II}(bqen)(MeCN)_2]^{2+}$	5.1 (5)				37
α-$[Fe^{II}(mcp)(MeCN)_2]^{2+}$	5.9 (9)	3.2	>99	15	30
β-$[Fe^{II}(mcp)(MeCN)_2]^{2+}$	1.9 (0.9)	4.0	68	17	30
$[Fe^{II}(BpdL_1)(MeCN)_2]^{2+}$	24 (1.5)[i]	5.2		17	38
$[Fe^{II}(CF_3SO_3)_2(^{Me,H}Pytacn)]$	6.5 (12)	4.3	93	30	39
$[Fe^{II}(CF_3SO_3)_2(^{Me,Me}Pytacn)]$	7.6 (10)	3.4	94	20	39

[a] 1,2-*cis*-Dimethylcyclohexane = DMCH.
[b] Conditions, catalyst:H_2O:alkane = 1:10:1000.
[c] Turnover number (TON, mol of product/mol catalyst), A = cyclohexanol, K = cyclohexanone, A/K = (mol A/mol K).
[d] Kinetic isotope effect = KIE.
[e] Retention of configuration (RC) in the oxidation of *cis*-1,2-dimethylcyclohexane.
[f] 3°/2° = 3 × (1-adamantanol)/(2-adamantanol+2-adamantanone).
[g] Tris(5-methyl-2-pyridylmethyl)amine = $5Me_3$-tpa.
[h] Tris(6-methyl-2-pyridylmethyl)amine = $6Me_3$-tpa.
[i] 100 equiv H_2O_2.

ligands. Upon reaction with H_2O_2, usually delivered by syringe pump to an acetonitrile solution of iron complex and substrate under air at room temperature, these catalysts oxidize hydrocarbons with a relatively high efficiency in terms of oxidant converted into products. Reactions are fast and complete once the addition of the oxidant is finished, suggesting that the oxidant reacts almost instantaneously with the catalyst. Under an excess of substrate, alcohols are the primary oxidation product in these reactions, although overoxidation of the more reactive alcohols to the corresponding ketones also takes place. This observation is important, as alcohols are the genuine oxidation products from a $2e^-$ hydroxylation of a C–H bond, which occur in monooxygenases. Instead, reaction patterns characterized by equimolar amounts of alcohol and ketones are indicative of Russell–Saunders terminations of free diffusing radicals. Therefore, the monooxygenase-like character of the alkane oxidation reactions mediated by these catalysts constitutes a first evidence of the fidelity of their chemistry to a monooxygenase enzymatic hydroxylation (40).

Oxidation reactions catalyzed by these complexes are characterized by relatively high KIE measured in the competitive catalytic oxidation of cyclohexane/cyclohexane-d_{12}, and a large normalized 3°/2° C–H selectivity ratios in the oxidation of adamantane (Table I and Scheme 3). These reactivity patterns are indicative of the implication of a selective oxidant. More interesting is the observation that the hydroxylation of the tertiary C–H bond in *cis*-1,2-dimethylcyclohexane is stereospecific, and yields *trans*-1,2-dimethylcyclohexanol, with minimum formation of its cis epimer. In conclusion, this data constitutes strong indication that the oxidation of the C–H bond is a metal-centered reaction.

Isotopic-labeling experiments indicate minimum incorporation of O_2 in the oxidation products, demonstrating that free diffusing carbon centered radicals are

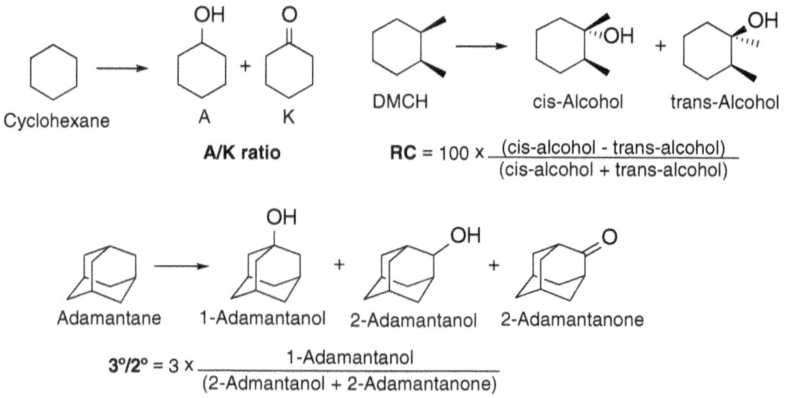

Cyclohexane → A + K

A/K ratio

DMCH → cis-Alcohol + trans-Alcohol

$$\textbf{RC} = 100 \times \frac{(\text{cis-alcohol} - \text{trans-alcohol})}{(\text{cis-alcohol} + \text{trans-alcohol})}$$

Adamantane → 1-Adamantanol + 2-Adamantanol + 2-Adamantanone

$$\textbf{3°/2°} = 3 \times \frac{\text{1-Adamantanol}}{(\text{2-Admantanol} + \text{2-Adamantanone})}$$

Scheme 3. Mechanistic probes employed in mechanistic studies in C–H oxidation reactions.

not significantly formed. Instead, the origin of the oxygen atoms incorporated into products is the oxidant (H_2O_2) and water. Since peroxide type of species cannot exchange their oxygen atoms with water, the incorporation of oxygen from water into the alcohol product derived from stereospecific alkane oxidation is taken as an indication for the involvement of a high-valent iron-oxo species, because metal-oxo species are known to engage in water-exchange reactions.

a. A P450-Like Reaction Mechanism. The current understanding of the mechanism of stereospecific C–H hydroxylation catalyzed with non-heme iron complexes with strong-field tetradentate ligands was initially proposed by Chen and Que (26) and is shown in Scheme 4.

The mechanism has strong resemblance to that of P450, and involves the initial formation of a ferric hydroperoxide species (**Ia**) that evolves via heterolytic O–O cleavage to form a high-valent oxoiron intermediate species (**Ib**) that is responsible for the oxidation of the substrate. Besides C–H hydroxylation, the same mechanistic scenario can be used to understand olefin epoxidation and cis-dihydroxylation by the same catalysts (28).

Several experimental and spectroscopic evidences give support to this mechanism;

1. Ferric species have been identified as the resting state of the catalysts (26, 28, 33).
2. For some specific complexes, metastable ferric hydroperoxide species have been spectroscopically trapped at low temperatures after reacting the iron catalysts with H_2O_2 (27, 41, 42). These type of species have proven to be kinetically not competent for reacting with alkane and alkene

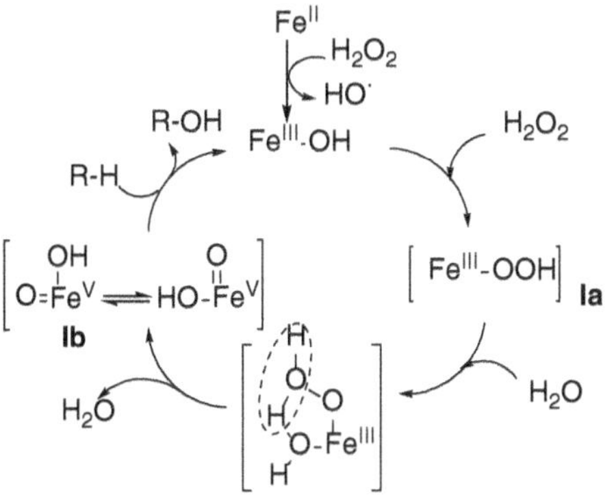

Scheme 4. Reaction mechanism proposed for the generation of $Fe^V(O)(OH)$ species involved in C–H and C=C oxidation reactions.

substrates (43, 44), discarding their implication as final oxidants. Instead, they appear to be precursors of the true oxidizing species (41, 45, 46). Nevertheless, recent studies showed that $[Fe^{III}(tmc)(OOH)]^{2+}$ (tmc = tetramethylcyclam) has a similar reactivity to the iron(IV)–oxo complex in C–H bond activation of alkylaromatics (47). This observation suggests that the factors that rule the reactivity of these species are not completely understood and require further exploration.

3. Evidence for a water-assisted cleavage of the O–O bond in hydroperoxide species has been provided by kinetic analysis of the decay of the $[Fe^{III}(OOH)(tpa)]$ intermediate (41). Hydroperoxide decay was found dependent on water concentration. The reaction exhibits a KIE = 2.5 when D_2O is employed instead of H_2O, highlighting the important role of a proton in the rate-determining cleavage of the O–O bond. Decay rate matches product formation rate. Activation parameters associated with this reaction are $\Delta H^{\neq} = 45(2)\ kJ\ mol^{-1}$ and $\Delta S^{\neq} = -95(10)\ J\ K^{-1}\ mol^{-1}$. The activation enthalpy appears to be distinctive for the type of O–O lysis mechanism. Heterolytic O–O cleavage in a previously described ferric hydroperoxide species exhibits $\Delta H^{\neq} = 44(2)\ kJ\ mol^{-1}$ (48), while homolytic O–O cleavage in mononuclear ferric peroxide systems are characterized consistently by larger activation enthalpies ranging from 52(1) to 56(2) $kJ\ mol^{-1}$ (49–52).
4. Indirect evidence for water assistance in O–O cleavage is deduced by the observation that related $[Fe(OOH)(L^{N_5})]$ complexes with neutral pentadentate N-rich ligands, that cannot bind a water molecule because the coordination sphere is complete, are not capable of stereospecific C–H hydroxylation, presumably because heterolytic O–O cleavage cannot occur (26, 53).
5. Evidence for the implication of a $Fe^{V}(O)(OH)$ intermediate is derived from the oxygen inventory of the cis-dihydroxylation reaction. Corresponding syn diols contain one oxygen atom that originates from water and one oxygen atom that originates from the H_2O_2 (28, 32).
6. Labeling experiments also show that hydroxylation and epoxidation reactions result in products that contain oxygen atoms that originate from water (26, 28, 33, 54). As reactions are stereospecific, these products cannot be formed by water trapping of a cationic carbon-centered intermediate. Instead, these products constitute evidence of the implication of an oxidant that can exchange with water, presumably a metal-oxo species.
7. Evidence in favor of the $Fe^{V}(O)(OH)$ intermediate formed by water assisted O–O lysis, as competent in cis-dihydroxylation reactions, has been provided by cryospray-ionization mass spectrometry (CSI-MS), in combination with isotopic labeling (55).

8. Computational analysis has shown that the water-assisted mechanism has a small barrier, and leads to the formation of $Fe^{V}(O)(OH)$ species. This analysis has been performed for at least three different $[Fe^{III}(OOH)(H_2O)(L^{N_4})]$ systems, L^{N_4} = tpa, men, and ^{Me,H}Pytacn (33, 56, 57).

The C–H hydroxylation and olefin epoxidation by the $Fe^{V}(O)(OH)$ species bears strong resemblance to that occurring in hemes (5). In the case of C–H hydroxylation, the reaction involves initial hydrogen-atom abstraction from the C–H of the substrate by the Fe^{V}=O, forming a carbon-centered radical and Fe^{IV}-OH, and then ligand transfer (rebound), to form the alcohol and the Fe(III). Computational analysis indicates that the reaction in non-heme systems is best described as a concerted asynchronous process (33, 58). That means that cleavage of the C–H and formation of the Fe–O–H bond precede the lysis of the Fe–OH unit and formation of the C–OH moiety, but Fe^{IV}–OH and the alkyl radical are not stable reaction intermediates in the process. Since the substrate radical has virtually no lifetime, it cannot epimerize, and consequently the overall reaction occurs with retention of stereochemistry.

On the other hand, computational analyses indicate that reaction of the $Fe^{V}(O)(OH)$ with olefins can occur via two different paths (58); if the attack is initiated by the oxo ligand, reaction leads to the formation of epoxides. Alternatively, if the hydroxyl moiety attacks the olefin, a syn-dihydroxylation results, via a concerted asynchronous reaction where the two C–O bonds are formed sequentially (55).

Further subtleties have been evidenced in mechanistic studies on hydroxylation and epoxidation reactions catalyzed by $[Fe^{II}(CF_3SO_3)_2(^{Me,H}Pytacn)]$. Isotopic-labeling experiments showed that remarkably large percentages of water are incorporated into alcohol and epoxide products (up to 79%) (33, 54). The level of water incorporation proved to be independent of substrate concentration, but instead it depends on the particular structure of the substrate. The observation that oxygen atoms from water end up in the epoxide products should be understood as indirect evidence that iron-oxo species are responsible for the oxygen-atom transfer event. However, unlike in synthetic heme systems, these observations also discard a competition between the reactions of water-exchange and oxygen-atom transfer. The latter is the classical behavior observed in heme systems, where water exchange is explained via the so-called oxo-hydroxo tautomerism (59, 60). To account for these results, it has been proposed that two tautomeric $Fe^{V}(O)(OH)$ species $\mathbf{O_A}/\mathbf{O_B}$, that are in rapid equilibrium, mediate the oxidation reaction (Scheme 5) (33, 54). One of the two tautomers contains an oxo ligand that originates from a water molecule, but in the other isomer, the oxo ligand comes from the peroxide oxidant. In this scenario, the level of water incorporation reflects the relative amount of each of the two tautomeric species, and their relative reactivity against a given substrate (33, 61).

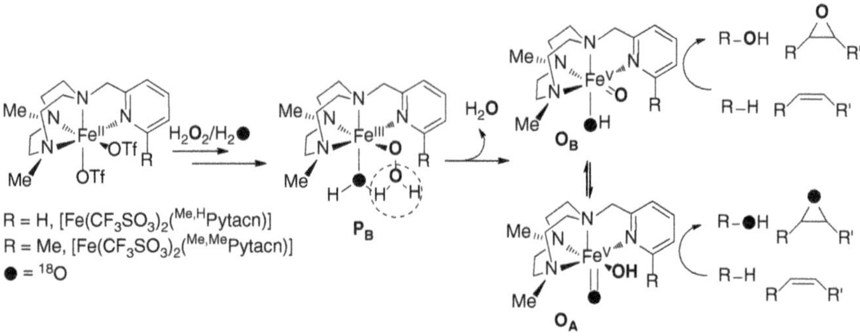

Scheme 5. Mechanistic scheme of the formation of the two tautomeric forms of the $Fe^{V}(O)(OH)$ oxidant (**O_A** and **O_B**), and corresponding epoxidation and hydroxylation reactions (OTf= trifluoromethanesulfonate).

Substitutions at position α of the pyridine ring of the Pytacn backbone have been shown to exert important effects in the relative reactivity of the two tautomeric $Fe^{V}(O)(OH)$ species, $\mathbf{O_A}/\mathbf{O_B}$ (33, 54). Stereospecific C−H hydroxylation of tertiary alkyl C−H bonds mediated by $[Fe^{II}(CF_3SO_3)_2(^{Me,H}Pytacn)]$ is predominantly performed by $\mathbf{O_A}$ as shown by the large extent of oxygen atoms originating from water in the alcohol product (up to 79%). Instead, hydroxylation of secondary C−H bonds occurs with incorporation of ~40% of the oxygen atoms from water, suggesting comparable reactivity of both tautomers. In sharp contrast, hydroxylations catalyzed by $[Fe^{II}(CF_3SO_3)_2(^{Me,Me}Pytacn)]$ exhibit small extent (~10%) of water incorporation in hydroxylation of secondary C−H bonds, and negligible (<3%) incorporation in the hydroxylation of tertiary C−H sites, indicating that hydroxylation is predominantly performed by $\mathbf{O_B}$. Analogous observations are made in epoxidation reactions. For example, incorporation of water into cyclooctene epoxide product is very high for $[Fe^{II}(CF_3SO_3)_2(^{Me,H}Pytacn)]$ (~70%), while for $[Fe^{II}(CF_3SO_3)_2(^{Me,Me}Pytacn)]$ the amount of ^{18}O-labeled epoxide dramatically decreases to $\sim 7 \pm 3\%$.The origin of this difference may reflect steric protection of the oxo ligand in $\mathbf{O_B}$ by the 6-methyl group of the pyridine, which prevent substrate approach at this position.

C. Class B Complexes: Iron-Coordination Complexes Containing Aminopyridine Ligands that Enforce Weak Ligand Fields

The subtlety of the chemistry is best exemplified by the activity of a second group or by iron complexes $[Fe^{II}(6Me_3{-}tpa)(MeCN)_2]^{2+}$, β-$[Fe^{II}(mcp)(MeCN)_2]^{2+}$, and $[Fe^{II}(BpdL_1)(MeCN)_2]^{2+}$ (Scheme 6) as catalysts in analogous C−H oxidation reactions. Despite the obvious structural similarities between the complexes shown

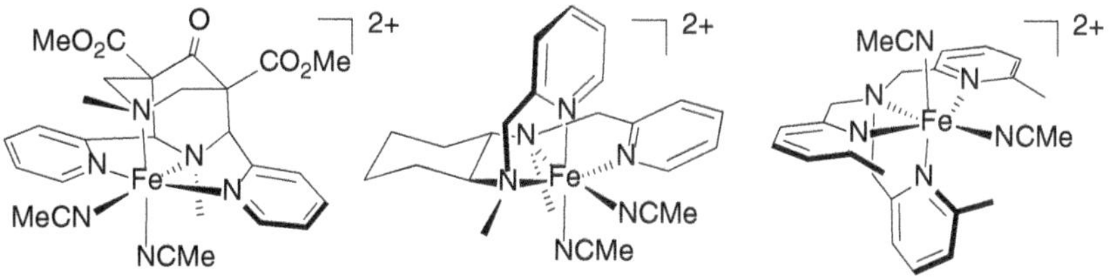

$[Fe^{II}(BpdL_1)(MeCN)_2]^{2+}$ $\beta-[Fe^{II}(mcp)(MeCN)_2]^{2+}$ $[Fe^{II}(6Me_3\text{-}tpa)(MeCN)_2]^{2+}$

Scheme 6. Structure of prototypical iron complexes with weak field tetradentate *N*-based ligands. *Note*: These complexes are usually isolated with triflate anions instead of acetonitrile, but in acetonitrile solution where they are employed as catalysts, acetonitrile molecules are bound.

in Scheme 6 and those previously shown in Scheme 1, in terms of iron-coordination sphere, the patterns exhibited by the later in C−H oxidation reactions are different. Thus, exogenous O_2 is incorporated into oxidation products, but not H_2O, as ascertained by isotopic-labeling analysis, A/K ratios are close to 1, and there is important stereoscrambling in the oxidation of *cis*-1,2-dimethylcyclohexane (26, 30). Therefore, these reactions exhibit patterns characteristic of the implication of free diffusing carbon-centered radicals. However, C−H regioselectivity (KIE and C3/C2 in adamantane oxidation) remain very similar in all complexes described in Table I. This finding suggests that differences may not be in the nature of the C−H oxidizing species, but instead are found in the lifetime of the carbon-centered radical that forms after H-atom abstraction. However, this interpretation is in conflict with the one derived from mechanistic studies in olefin oxidation reactions catalyzed by this class of complexes (see below).

The origins of the differences between the two types of catalysts therefore remain unclear. A common aspect of this second group of complexes is that they all contain structural elements that disfavor short Fe−N distances. This common aspect is translated into coordination environments that exert weaker ligand fields than in **1–4**, and higher red-ox potentials, presumably because the atomic radius of low-spin ions are smaller than their high-spin counterparts, another reason is because the built up of positive charge in higher oxidation states require shortening of the Fe–ligand distances.

Comba and co-workers (61, 62) proposed an alternative mechanism to explain the experimental results obtained in the oxidation of olefins catalyzed by $[Fe^{II}(BpdL_1)(MeCN)_2]^{2+}$ (Scheme 7), and that could serve as a paradigm for Class B complexes. This alternative pathway suggests that these catalysts operate through a Fe(II)/Fe(IV) cycle. The active species is an Fe(IV) compound formed by O−O homolysis of a $Fe^{II}(H_2O_2)$ intermediate. In particular, it was found by density functional theory (DFT) calculations that two high-valent isomers are energetically favored: $Fe^{IV}(OH)_2$ and $H_2O{-}Fe^{IV}{=}O$. The former is responsible for

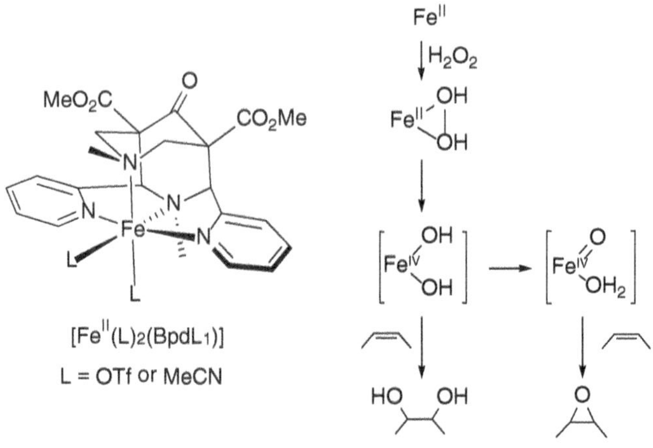

Scheme 7. Schematic diagram of the Fe(II)/Fe(IV) catalytic cycle proposed by Comba and coworkers (61, 62).

the cis-dihydroxylation and the latter affords the epoxide product (Fe(II)/Fe(IV) mechanism; Scheme 7). Consistent with this scenario, isotopic-labeling experiments observed for the cis-dihydroxylation reaction catalyzed by Class B complexes show that both oxygen atoms originate from the same H_2O_2 molecule.

The different nature of the active species in Class A and B catalysts is also evidenced by their marked different electrophilicity, as shown in a comparative study of complexes $[Fe^{II}(tpa)(MeCN)_2]^{2+}$ and $[Fe^{II}(6Me_3-tpa)(MeCN)_2]^{2+}$ (63). These two complexes were selected as prototypes for Class A and B, respectively. Competition experiments between electron-rich and electron-deficient olefins revealed that the Class A catalyst preferentially oxidizes the more electron-rich olefins, while the Class B catalyst shows the opposite behavior. For example, in a competition experiment between cyclooctene (electron-rich) and *tert*-butyl acrylate (electron-deficient), by $[Fe^{II}(tpa)(MeCN)_2]^{2+}$ favors cyclooctene oxidation (epoxidation and cis-dihydroxylation) by a factor of 4, while $[Fe^{II}(6Me_3-tpa)(MeCN)_2]^{2+}$ favors *tert*-butyl acrylate oxidation by a factor of 4. These opposite preferences exhibited by both catalysts most likely imply distinct oxidants, although their identity could not be further clarified in these studies.

The details of the mechanism of C–H oxidation by Class B catalysts remains less explored than for **1–4**, presumably because these reactions give low-product yields and isotopic labeling and mechanistic probes do not provide a simple interpretation. It remains unclear if an Fe(II)/Fe(IV) cycle can be operating for these catalysts in C–H oxidation reactions and if some other unconsidered reasons justify their poor reactivity in these reactions.

D. The Role of Acetic Acid in Catalytic Oxidation Reactions

Acetic acid has been shown to play a prominent positive role in iron-catalyzed C–H and olefin epoxidation reactions with H_2O_2. In general, acetic acid increases overall product yields (64, 65), and in the particular case of the oxidation of olefins, addition of this acid inhibits olefin cis-dihydroxylation (see below), enhancing the chemoselectivity toward epoxidation (66). Acetic acid can be replaced by other carboxylic acids in iron-catalyzed asymmetric epoxidation reactions, improving the enantioselectivity of the reactions (67).

Mas-Balleste and Que (66) proposed that acetic acid plays a key role in promoting fast O–O bond cleavage in a Fe^{III}(OOH) species (Scheme 8). This reaction leads to the formation of formal Fe^{V}(O)(OAc) species (**IIb** in Scheme 8) finally responsible for the oxidation reaction. Experimental spectroscopic evidence in favor of this mechanistic scenario also has been recently provided by Talsi, Bryliakov, and co-workers (68, 69), and further computational support for the formation and oxidative competence of these species has been built by Rajaraman and co-workers (70) in a study of the ortho hydroxylation of aromatic compounds by non-heme Fe complexes.

Recently, this mechanism has been reconsidered on the basis of computational analysis performed by Shaik, Que and co-workers (71) on the reaction mechanism of complex [Fe^{II}(*S*,*S*)-pdp)$(MeCN)_2$]$(SbF_6)_2$ (see Scheme 2 for the structure of the pdp ligand). The DFT calculations suggest that the barrier for the formation of the putative Fe^{V}(O)(OAc) oxidant is too high for it to be feasible. Instead, a much lower barrier is found for the formation of a [Fe^{III}(κ^2-peracetate)((*S*,*S*)-pdp)] compound.

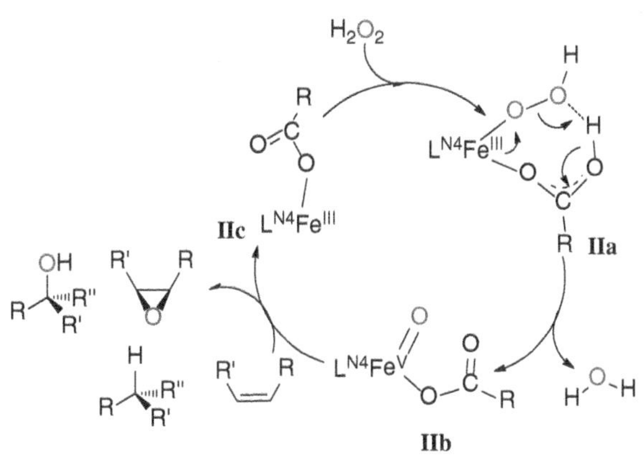

Scheme 8. Schematic diagram of the mechanism proposed for carboxylic acid assisted Fe catalyzed oxidations. Here L^{N4} stands for a tetradentate N-based ligand.

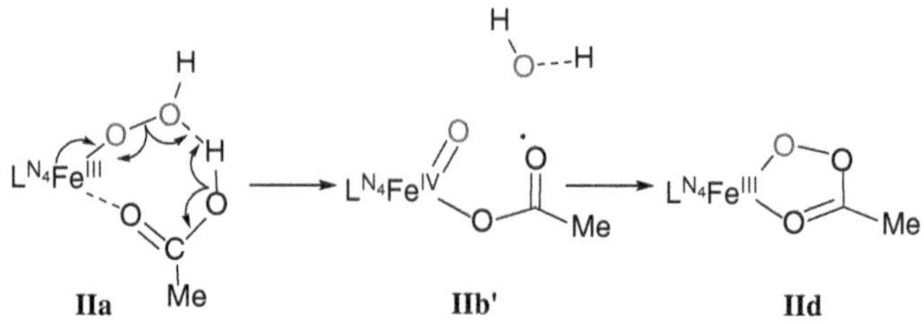

Scheme 9. Schematic diagram of the mechanism proposed by Shaik and Que (71) for the formation of a peracetate intermediate.

In the course of C–H oxidation reactions, this complex undergoes O–O bond homolysis to become a transient [((*S*,*S*)-pdp)Fe^{IV}(O)(AcO•)] species (**IIb**′, Schemes 9 and 10). This complex performs the hydroxylation of alkanes, via initial hydrogen-atom transfer by the oxo group, and rapid hydroxyl rebound to the incipient carbon-centered radical. Thus, C–H bond breakage is initiated by interaction of Fe^{IV}(O)(AcO•) with the C–H bond, biasing the formation of a highly reactive Fe^{V} species.

E. Application in Chemical Synthesis

The discovery that C–H oxidation reactions are stereospecific with some selected iron catalysts constitutes a conceptual breakthrough because it represents a clear cut with the chemistry of free diffusing radicals (including Fenton reactions), and demonstrates that reactions are metal based (27). In this scenario, one can envision that selectivity in C–H oxidation reactions could be introduced by means of catalyst design, drawing a very interesting analogy to the tailored reactivity in C–H oxidation mediated by enzymes. Despite this magnificent potential, until 2007 selective C–H hydroxylation with non-heme iron catalysts had little synthetic relevance because reactions employed a large excess of substrate relative to the oxidant. However, in 2007, a keystone work by Che and White (65) described the Fe catalyzed hydroxylation of unactivated alkyl C–H

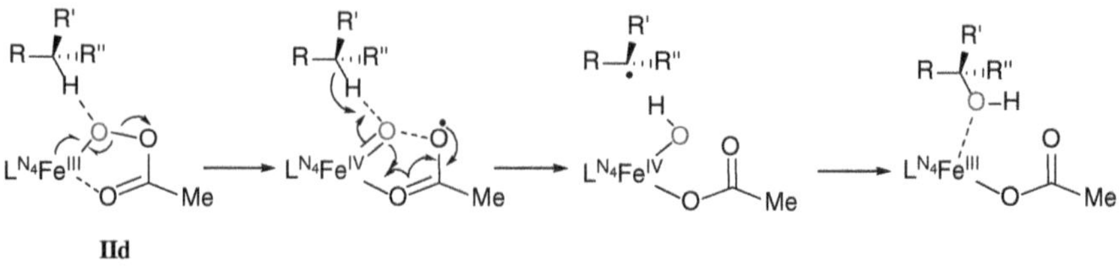

Scheme 10. Schematic diagram of the mechanism proposed by Shaik and Que (71) for the hydroxylation of alkane C–H bonds by the [Fe(pdp)] catalyst.

bonds achieving both synthetically useful yields and predictable selectivities. The catalyst employed in these transformations is the chiral bipyrrolidine-based complex $[Fe^{II}(pdp)(MeCN)_2](SbF_6)_2$. This complex is structurally related to $[Fe^{II}(men)(MeCN)_2]^{2+}$ (Scheme 2), but appeared to be more active. Hydrogen peroxide was employed as oxidant, and acetic acid was identified as a key additive to provide synthetically valuable product yields. The catalyst appears to be tolerant with diverse functionalities that include esters, amides, halides, and epoxides. The most important factors regulating site selectivity in substrates with multiple C–H bonds were clarified; electronic and sterics are recognized as major deciding factors (Scheme 11). The former responds to the electrophilic nature of the oxidizing species, being e-rich C–H bonds preferentially oxidized. The later

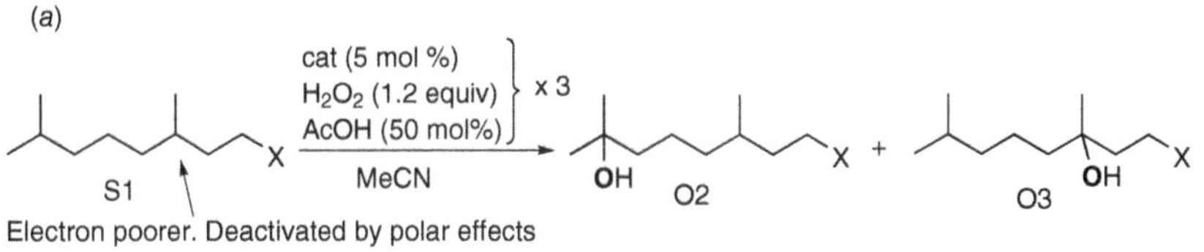

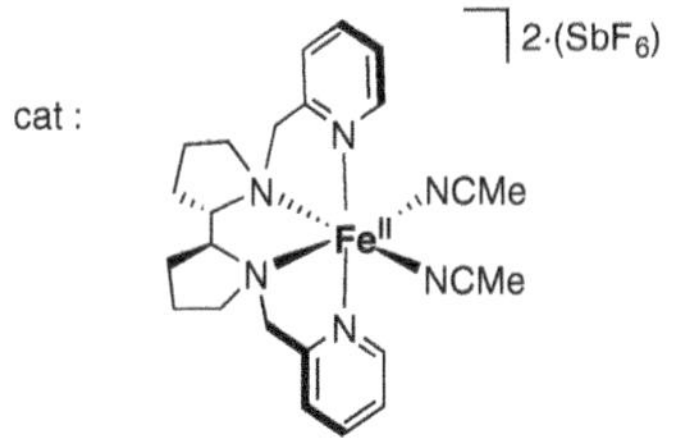

S1	Yield(%)	O2/O3
X = H	48	1/1
OAc	43	5/1
Br	39	9/1
F	43	6/1

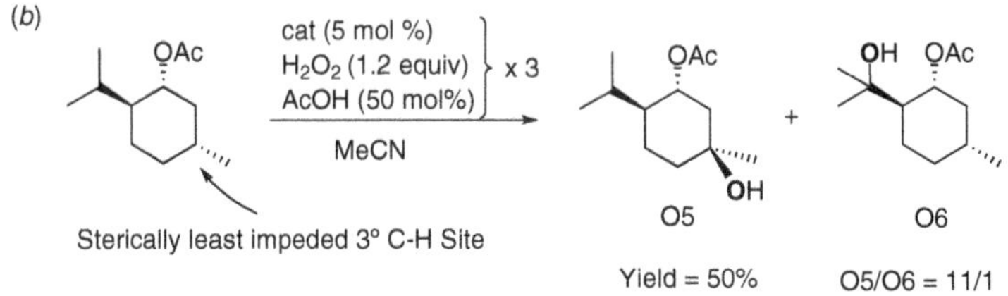

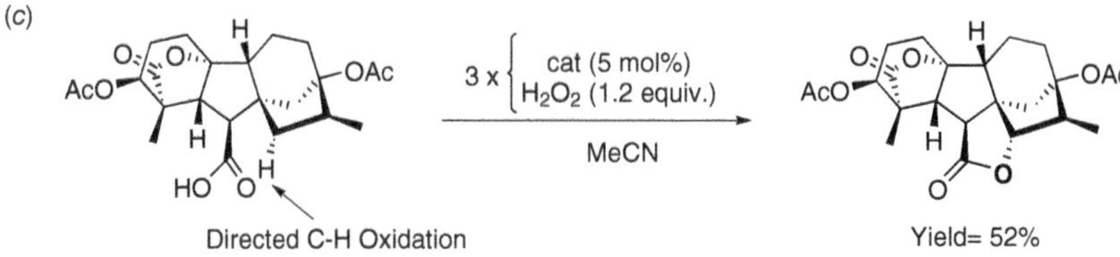

Scheme 11. Alkane hydroxylation using $[Fe^{II}(pdp)(MeCN)_2](SbF_6)_2$. (*a*) Substrate electronic effects on site selectivity in hydroxylation of multiple 3°C–H bonds. (*b*) Selective hydroxylation based on steric effects. (*c*) Directing role of carboxylic acids.

accounts for the spatial demands imposed by the relatively bulky iron catalyst, which causes preferential oxidation of the spatially least crowded C–H sites. Molecules bearing carboxylic acid moieties are a special class of substrates because this functional group can bind to the iron site, acting as the C–H oxidation site directing groups. Following site-directed hydroxylation, rapid lactonization takes place (72). This methodology could be applied in the selective oxidation of complex organic molecules. For example tetrahydrogibberellic acid is hydroxylated with high site selectivity and diastereoselectivity (Scheme 11*c*). The corresponding lactone product was obtained in 52% yield.

The authors (65, 72) showed that analysis of these factors allows us to predict the regioselectivity in the oxidation of complex organic molecules (e.g., natural products) even in the absence of the carboxylic acid directing group. This finding is best illustrated in the hydroxylation of (+)-artemisin (Scheme 12), which occurs regio- and stereospecifically at the most e-rich tertiary C–H bond. By employing an experimental protocol involving three iterative additions of 5 mol% of catalyst and 1.2 equiv of H_2O_2, hydroxide-artemisin was obtained in 35% isolated yield. By recycling the starting material twice, the diasteriometrically pure product was afforded in 54% yield.

The directing role of the carboxylic acid allows the catalyst to bias the innate C–H reactivity based on electronic, steric, and stereoelectronic factors. This finding was nicely illustrated in the site-selective oxidation of the paclitaxel framework at position C2, reproducing the naturally occurring stereoconfiguration, despite the fact that position C1 is electronically favored (72) (Scheme 13).

Non-heme iron catalysts generate highly electrophilic species that are also capable of oxidizing stronger C–H bonds (e.g., the nonactivated methylene sites in Scheme 14) (73). Chen et al. (73) described that site selectivity in methylene oxidation also responds in a predictive manner to steric, electronic, and inductive factors. Modest selectivities were observed in the oxidation of substrates where only one of the directive factors is involved. However, in complex organic molecules remarkable site selectivities could be achieved by synergistic interplay

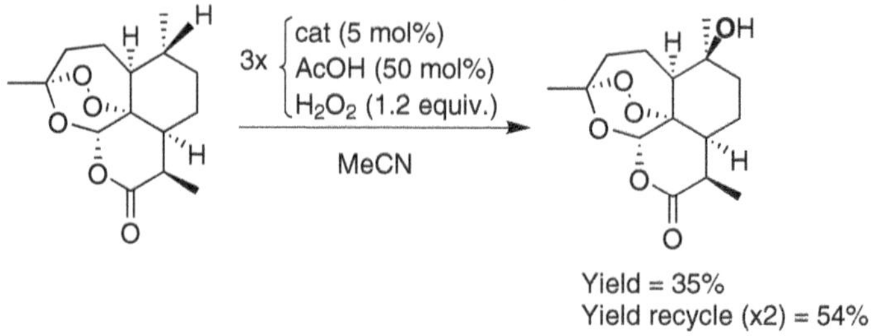

3x { cat (5 mol%), AcOH (50 mol%), H_2O_2 (1.2 equiv.) }
MeCN

Yield = 35%
Yield recycle (x2) = 54%

Scheme 12. Selective hydroxylation of (+)-artemisin by $[Fe^{II}(pdp)(MeCN)_2](SbF_6)_2$, where cat = catalyst.

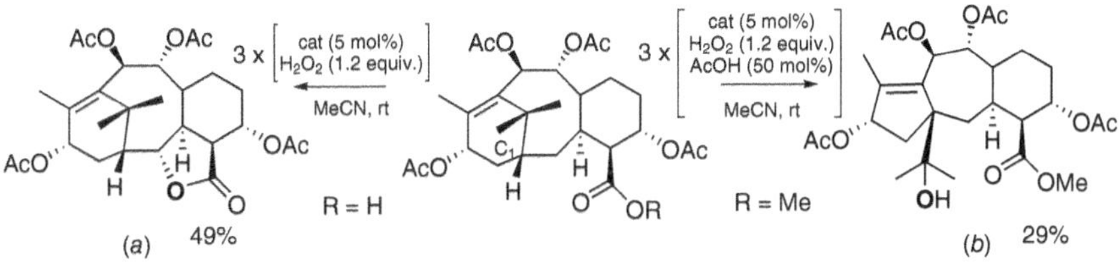

Scheme 13. Carboxylic acid directed hydroxylation of C–H bonds (R = H, *a*), overriding innate C1–H preferential reactivity occurring in the absence of the directing group (R = Me, *b*) [cat = [Fe^{II}(pdp)$(MeCN)_2$]$(SbF_6)_2$.]

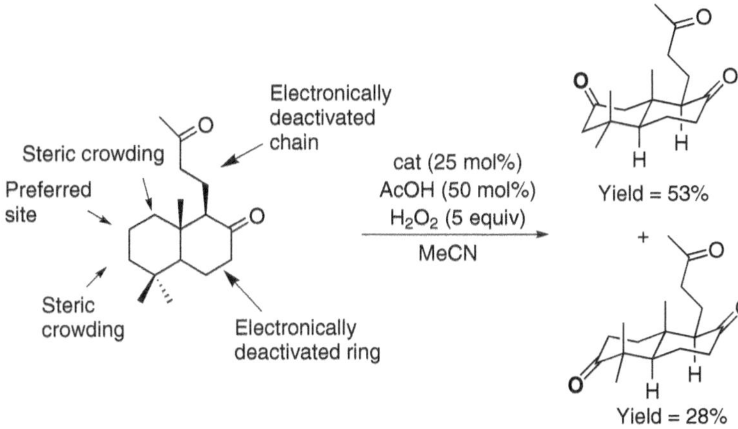

Scheme 14. The combination of steric and electronic effects allows the selective oxidation of the methylene group by the [Fe^{II}(pdp)$(MeCN)_2$]$(SbF_6)_2$.

of multiple factors. Combination of these factors could also promote preferential oxidation of a secondary C–H bond in front of the weaker tertiary C–H bonds.

White and co-workers (74) developed a system that uses 15% of the catalyst loadings, an elaborated synthetic protocol via iterative additions of reagents, and in some cases substrate recycling. With the aim of developing iron catalysts more resistant to bimolecular decomposition pathways, Gómez et al. (75) designed the structurally related [$Fe^{II}(CF_3SO_3)_2$((*S*,*S*,*R*)-mcpp), (Scheme 15), where mcpp = *N*-((7,7-dimethyl-5,6,7,8-tetrahydro-6,8-methanoisoquinolin-3-yl)methyl)-*N*′-(7,7-dimethyl-5,6,7,8-tetrahydro-6,8-methanoisoquinolin-3-yl)-*N*,*N*′-dimethyl-cyclohexane-1,2-diamine], which contains bulky pinene groups attached at positions 4 and 5 of the pyridine ligands. Alteration of the most obvious sixth position was discarded owing to the well-established deactivation effect that

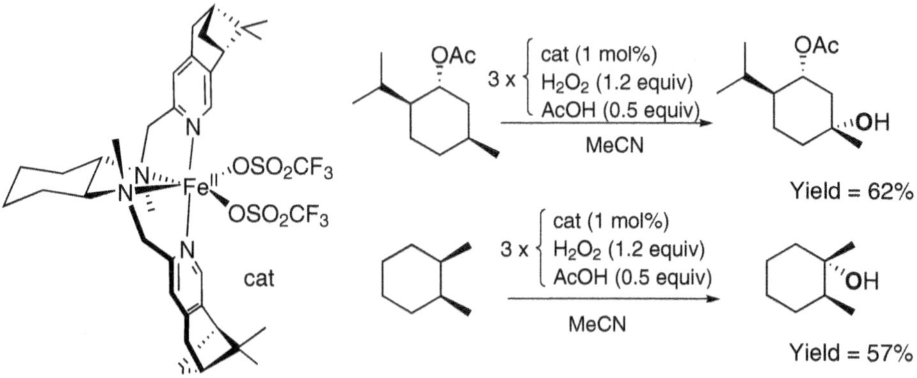

Scheme 15. Complex [$Fe^{II}(CF_3SO_3)_2$(*S*,*S*,*R*)-mcpp)] described by Gómez et al. (75), and results in catalytic C–H oxidation reactions.

occurs when steric bulk is increased at this position. Instead, bulky groups at positions 4 and 5 create a robust cavity that isolate the metal center, and prevent self-decomposition pathways, without compromising catalyst stability. This catalyst indeed shows a high efficiency in C–H hydroxylation, giving yields comparable to [Fe^{II}(pdp)$(MeCN)_2$]$(SbF_6)_2$, although using much lower catalyst loadings (3 mol%).

Furthermore, under identical experimental conditions, employing low-catalyst loadings, these highly structured iron catalysts provide substantially improved yields with regard to simpler iron catalysts in the oxidation of complex organic molecules. For example, the rigid bridged tricyclic sesquiterpene (+)-cedryl acetate is selectively hydroxylated at the least sterically hindered tertiary C–H site C2-H (Scheme 16) (76), which is also the most distant site from the

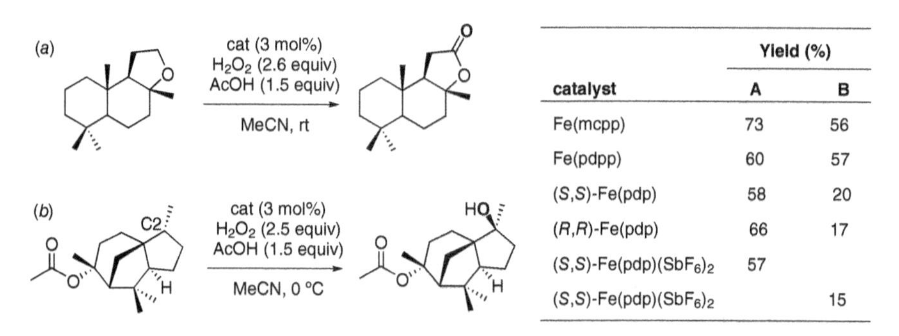

catalyst	Yield (%) A	Yield (%) B
Fe(mcpp)	73	56
Fe(pdpp)	60	57
(*S*,*S*)-Fe(pdp)	58	20
(*R*,*R*)-Fe(pdp)	66	17
(*S*,*S*)-Fe(pdp)$(SbF_6)_2$	57	
(*S*,*S*)-Fe(pdp)$(SbF_6)_2$		15

Scheme 16. Comparative performance of non-heme iron catalysts in the C–H oxidation of complex organic molecules, Fe(pdpp) = [Fe(CF3SO3)2(pdpp)], pdpp=N-((7,7-dimethyl-5,6,7,8-tetrahydro-6,8-methanoisoquinolin-3-yl)methyl)-N′-(7,7-dimethyl-5,6,7,8-tetrahydro-6,8-methanoisoquinolin-3-yl)2,2-bipyrrolidine.

deactivating acetate group, in 57% yield employing [$Fe^{II}(CF_3SO_3)_2$(*S,S,R*)-mcpp)] [Fe(mcpp) in Scheme 2]. Instead [$Fe^{II}(CF_3SO_3)_2$(pdp)] [Fe(pdp) in Scheme 2] (both enantiomers) and [Fe^{II}(pdp)$(MeCN)_2$]$(SbF_6)_2$ ([Fe(pdp) $(SbF_6)_2$] in Scheme 16) afford more modest 15–20% product yields, and Gif systems provide only 1% yield (77). Best yields are also obtained with [$Fe^{II}(CF_3SO_3)_2$(*S,S,R*)-mcpp)] in the oxidation of (+)-ambroxide, when a series of iron catalysts are compared under analogous low-catalyst loading conditions are employed, although the initial report describes a 80% yield with the use of 15 mol% of the Fe(pdp) system.

The versatile structure of non-heme complexes allows for relatively straightforward access to a number of structurally diverse catalysts. In principle, such structure diversity can open the door to the design of catalysts exhibiting novel selectivities in C–H oxidation reactions, responding to catalyst–substrate interactions. This effect will be analogous to those operating in enzymes, where site selectivity may respond to subtle interactions between the structural elements of the enzyme cavity and the substrate. The ability to manipulate the regioselectivity in C–H oxidations constitutes a very attractive target for chemical synthesis because C–H oxidation is most commonly dominated by the innate properties of C–H bonds, and the nature of the oxidizing agent has little influence (2). Non-heme iron catalysts showing novel selectivities in their catalytic activity in C–H oxidations have started to appear.

1. Modulation between Oxidation of Tertiary and Secondary C–H Bonds

Tertiary C–H bonds are generally preferentially oxidized in front of non-activated methylenic sites on the basis of their relative bond strength. Activation of secondary sites by means of hyperconjugation or ring strain may lower the energy of these bonds making them more readily oxidizable in the presence of tertiary sites, but in the absence of these activation mechanisms, C–H oxidation by common oxidizing reagents occurs preferentially at tertiary sites (2). On the other hand, tertiary C–H bonds are sterically more impeded than secondary sites. This result may provide a handle on developing reagents that could preferentially oxidize secondary in front of tertiary C–H bonds. Traditional oxidizing agents have little structure and therefore are affected by steric factors to a very limited extend. However, iron-coordination complexes can in principle be elaborated for creating sterically more sensitive oxidizing agents. Creating steric bulk at the iron catalyst can therefore be seen as a simple strategy to obtain these catalysts. The main challenge of this approach is that the stability of iron-coordination compounds under oxidative conditions has been shown to be very sensitive to modifications of the structure of the complexes (25, 40). Paths for overcoming this lack of stability are not obvious, although some pioneering examples, where steric bulk is used to modify innate C–H relative reactivity, have started to appear.

In a recent work, Goldsmith and co-workers (78) designed the tetradentate bbpc ligand, which is related to the mcp ligand by replacing the N–Me group by the *N*-benzyl arm. This ligand gives rise to two iron complexes, α-[Fe(bbpc)$(MeCN)_2]^{2+}$ where bbpc = *N,N'*-dibenzyl-*N,N'*-bis(pyridylmethyl)-cyclohexane-*trans*-1,2-diamine, which adopts a cis-α topology, with the acetonitrile ligands bound cis- to each other, and β-[Fe$(CF_3SO_3)_2$(bbpc)], where the ligand adopts a trans conformation, with the four nitrogen atoms of the ligand in the same coordination plane, and with the triflate ligands bound trans- to each other (Scheme 17). The use of the two complexes in catalytic C–H oxidation reactions shows that the former is a better catalyst and was further explored. It was found that the presence of the sterically demanding benzyl groups in close proximity to the iron center nicely translates into an enhanced selectivity toward oxidation of methylene sites in *trans*-1,2-dimethylcyclohexane, when compared with [Fe(L)$(MeCN)_2]^{2+}$, L = men, mcp, and pdp under identical experimental conditions.

Enhanced selectivity toward methylene sites also has been documented for the triazacyclononane-based catalyst [Fe$(CF_3SO_3)_2$(^{Me,Me}Pytacn)], **[Fe(Pytacn)**, Scheme 18] (79). This catalyst mediates the oxidation of alkyl C–H bonds in

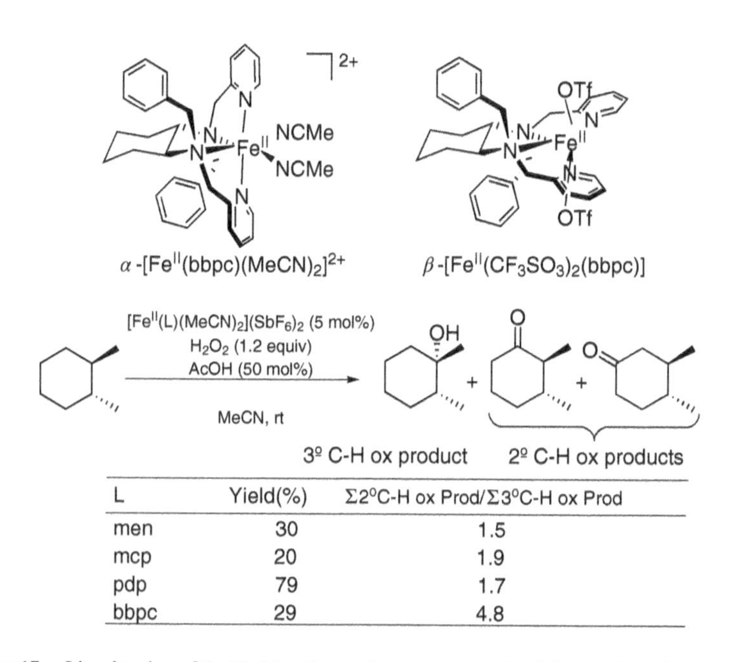

L	Yield(%)	Σ2ºC-H ox Prod/Σ3ºC-H ox Prod
men	30	1.5
mcp	20	1.9
pdp	79	1.7
bbpc	29	4.8

Scheme 17. Line drawing of the [Fe(bbpc)] complexes and regioselectivity results in the oxidation of *cis*-1,2-dimethylcyclohexane.

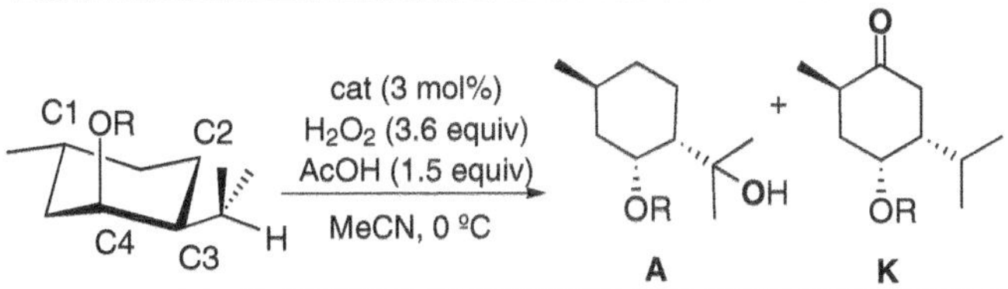

cat (3 mol%), H_2O_2 (3.6 equiv), AcOH (1.5 equiv), MeCN, 0 ºC

Catalyst	Conversion(%)	A(%)	K(%)	[A]:[K]
R = AcO				
Fe(pdp)	66	22	22	**1 : 1**
Fe(mcpp)	78	36	7	**5 : 1**
Fe(Pytacn)	72 (15)	15(14)	33(32)	**1 : 2**
R = Piv				
Fe(pdp)	50	15	23	**1 : 1.5**
Fe(mcpp)	71	43	11	**4 : 1**
Fe(Pytacn)	62 (31)	11 (11)	30 (30)	**1 : 3**
R = $COCF_3$				
Fe(pdp)	61	13	20	**1 : 1.5**
Fe(mcpp)	63	29	6	**5 : 1**
Fe(Pytacn)	64	7	31	**1 : 4.5**

Scheme 18. Schematic diagram of the structure of (+)-neomenthyl esters, and comparisons between different iron catalysts in its oxidation.

hydrocarbons, and also in more complex natural products reaching product yields that could be amenable for synthesis.

An illustrative example of the ability of iron catalysts to modulate selectivity between the oxidation of secondary and tertiary C–H bonds has been shown by employing ester derivatives of (+)-neomentol. (+)-Neomenthyl esters contain four tertiary and six secondary different C–H bonds (Scheme 18). Of these, secondary C2 is the most remote site from the electron-withdrawing ester group, while tertiary sites C3 and C1 are two C–C bonds away from this deactivating group. In the case of C1, although the C–H is located in axial position, it is deactivated by the sterically hindering of the ester. The conclusion of this analysis is that C2 and C3 are envisioned as the C–H sites most prone to oxidation. Oxidation of neomenthyl esters by a series of iron catalysts does confirm this analysis, but most interestingly relative oxidation at C2 or C3 depends on the structure of the catalyst. Oxidation with the $[Fe^{II}(CF_3SO_3)_2(pdp)]$ **[Fe(pdp),**

Scheme 18] system yields a roughly equimolar mixture of the ketone from oxidation at C2, and a tertiary alcohol from oxidation at C3. But, by employing catalyst [$Fe^{II}(CF_3SO_3)_2$(*S*,*S*,*R*)-mcpp)] [**Fe(mcpp)**, Scheme 18] oxidation occurs preferentially at 3° C—H (3:1 < A:K < 5:1). Remarkably, when [$Fe(CF_3SO_3)_2$ (^{Me,Me}Pytacn)] is employed, oxidation preferentially occurs at the methylene site (1:4.5 < A/K < 1:2).

The chiral nature of catalysts offers an additional structural aspect that exerts modulation in C—H regioselectivity when chiral substrates are used. This can be illustrated as well in the oxidation of (+)-neomenthyl esters. While catalysts exhibiting a Λ chirality at the metal, [$Fe^{II}(CF_3SO_3)_2$(*S*,*S*,*R*)-mcpp)] and [$Fe^{II}(CF_3SO_3)_2$(*S*,*S*,*R*)-mcpp)] do effectively hydroxylate selectively at a tertiary site (91% conversion, 48% oxidation at C3, 10% oxidation at C2), use of the Δ-esteroisomers results in preferential oxidation at the secondary site (89% conversion, 29% oxidation at C2, and 20% oxidation at C3).

2. *Modulation of Selectivity among Multiple Methylene Sites*

Modulation of regioselectivity in the oxidation of nonactivated methylene moieties in organic molecules containing multiple, nonequivalent sites constitutes a major challenge. Regioselective oxidation of linear alkanes is arguably the most difficult problem because of the strength of these C—H bonds and because there are only subtle differences among multiple sites. Selective oxidation of this type of substrates has been accomplished by using very sterically hindered porphyrins (80, 81), but analogous levels of selectivity have not been reached yet when non-heme iron complexes are employed. However, changes in site selectivity in the functionalization of structurally richer substrates have been documented (82). For example, (+)-sclareolide constitutes a convenient platform. In this substrate, the carbonyl group deactivates the surrounding C—H bonds, and because of that oxidation reactions should concentrate at the most remote cyclohexane ring, which contains three chemically distinct methylene sites (C2, C3, and C4 in Scheme 19). Selectivity among these three sites in C—H functionalization reactions is believed to be governed by steric factors. Preferential functionalization at C2 was initially described with catalyst [Fe^{II}((*R*,*R*)-pdp)$(MeCN)_2](SbF_6)_2$ (73). More recently, site selectivity in the oxidation of (+)-sclareolide was described by using highly structured non-heme iron catalysts (76). (+)-2-Oxo-sclareolide, (+)-3-oxo-sclareolide, and (+)-4-oxo-sclareolide can be obtained as the main product in 36% (61% selectivity), 40% (55% selectivity), and 34% (47% selectivity) yield, respectively, by choosing the appropriate experimental conditions and catalyst (Scheme 19). While those yields may seem modest, they are competitive with enzymatic oxidations. For example (+)-2-oxo-sclareolide is obtained enzymatically with reduced yields and longer reaction times (83).

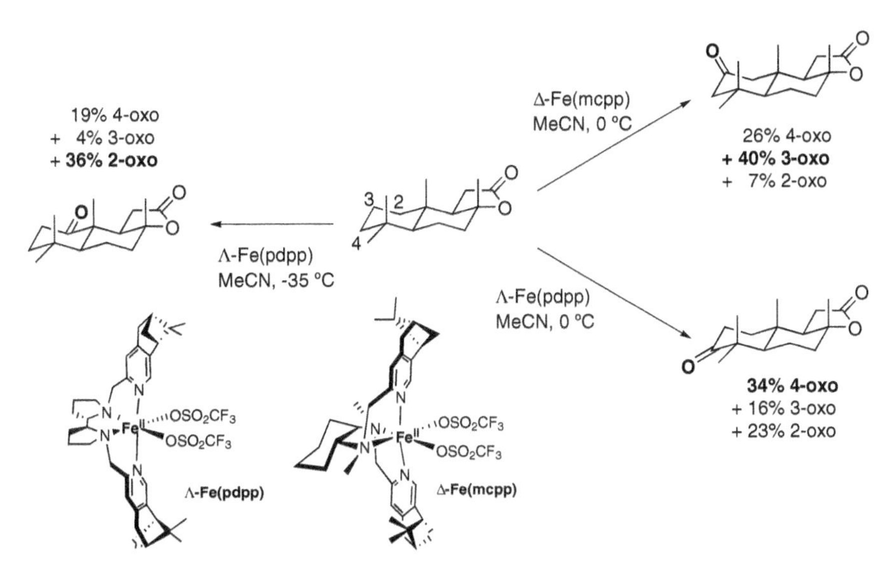

Scheme 19. Selective oxidation of (+)-sclareolide with non-heme iron catalysts.

3. *Desaturation of C−H*

Formation of C−O bonds is the most C−H common oxidation reaction. More recently, $[Fe^{II}(pdp)(MeCN)_2](SbF_6)_2$ has been shown to mediate the catalytic desaturation of alkane C−H bonds of carboxylic acids to form the corresponding alkenes (84). The presence of a carboxylic acid moiety appeared crucial for the reaction, as analogous ester substrates yield corresponding hydroxylated products. The reaction is proposed to entail an initial hydrogen-atom abstraction executed by a high-valent iron-oxo species, generating a carbon-centered radical intermediate. This intermediate can undergo either hydroxyl ligand rebound to give hydroxylation products or a second one-electron oxidation to provide the final olefin product. Under the reaction conditions, epoxidation of the olefinic site follows, which is then opened by intermolecular attack of the carboxylate moiety. Desaturation and hydroxylation appear to be competitive reactions, and the former is so far always the minor path. Since a carboxylic acid functionality on the substrate is required for this reactivity, the interaction between the iron center and the carboxylate ligand was deemed critical to avoid the −OH rebound. Moreover, both desaturation and hydroxylation reactions exhibit site selectivity, because binding of the carboxylic acid to the iron center site directs C−H oxidation toward the γ C−H site, resulting in the formation of γ-butyrolactones (Scheme 20) (84).

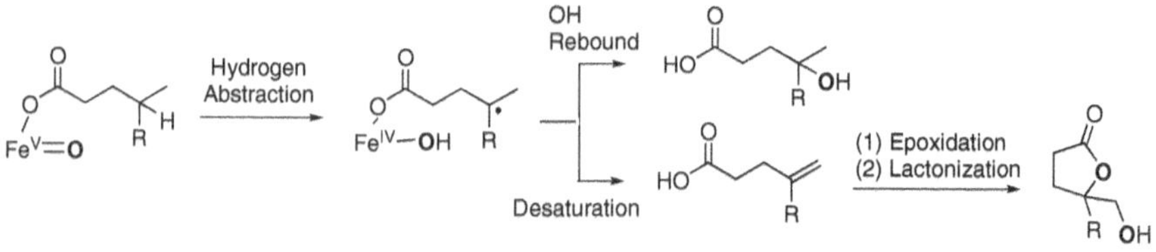

Scheme 20. Proposed mechanism for both hydroxylation and dehydrogenation catalyst by complex $[Fe^{II}(pdp)(MeCN)_2](SbF_6)_2$.

F. Stereospecific C–H Hydroxylation With Other Catalysts

Until recently, stereospecific C–H hydroxylation with non-heme iron systems was limited to coordination complexes containing aminopyridine tetradentate ligands. However, two recent works have added structural diversity to iron catalysts suitable for this reaction.

A recent report by Kodera and co-workers (85) described an N_5-pentadentate ligand, dpaq = 2-[bis(pyridin-2-ylmethyl)]amino-*N*-quinolin-8-yl-acetamide, with the peculiarity that a nitrogen atom is part of an amidate moiety (Scheme 21). The complex $[Fe^{III}(dpaq)(H_2O)]^{2+}$ reacts with H_2O_2 oxidizing alkyl C–H bonds with selectivities similar of the ones described for tetradentate iron(II) complexes, and slightly inferior efficiencies. For example, by using a 1% catalyst, 1 equiv of H_2O_2 and 1.2 equiv of *cis*-4-methylcyclohexyl-1-pivalate in acetonitrile afforded the tertiary alcohol product with a 38% yield and 94% stereoretention.

In the oxidation of cyclohexane, this catalyst gives high ratios of A/K (~12), and a KIE of 3.9 was determined for the oxidation of cyclohexane vs cyclohexane-d_{12}, indicative of an oxidant that responds sensitively to the strength of the C–H bond (85).

The pentadentate nature of the dpaq ligand constitutes an important difference with regard to iron complexes with neutral tetradentate ligands. Without the presence of two labile cis-sites, internal water and carboxylic acid assisted O–O lysis in $Fe^{III}(OOH)(L)$, $L = OH_2$ or $MeCO_2H$ intermediates is not possible. Alternatively, Hitomi et al. (85) propose that the amide ligand is essential to

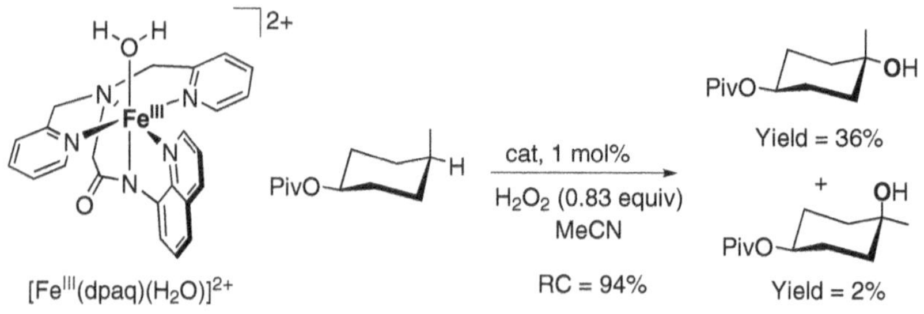

Scheme 21. Selective oxidation of *cis*-4-methylcyclohexyl-1-pivalate by a $[Fe^{III}(dpaq)(H_2O)]^{2+}$ complex.

facilitate the heterolysis of the O–O bond via a trans effect, affording an iron(V)-oxo species responsible for the C–H bond oxidation reaction. Coldspray-ionization mass spectroscopy was used to trap catalytic intermediates. The mixture of $[Fe^{III}(dpaq)(H_2O)]^{2+}$ and 50 equiv of H_2O_2 shows two important peaks that correspond to $[Fe^{III}(dpaq)(OOH)]^+$ and $[Fe^{V}(dpaq)(O)](ClO_4)^+$, both of which are recognized as important intermediates in the oxidative catalytic cycle.

The iron complex $[Fe(qpy)](ClO_4)_2$ (Scheme 22, where qpy = 2,2′:6′,2″:6″,2‴:6‴,2⁗-quinquepyridine) has been described as a catalyst for the

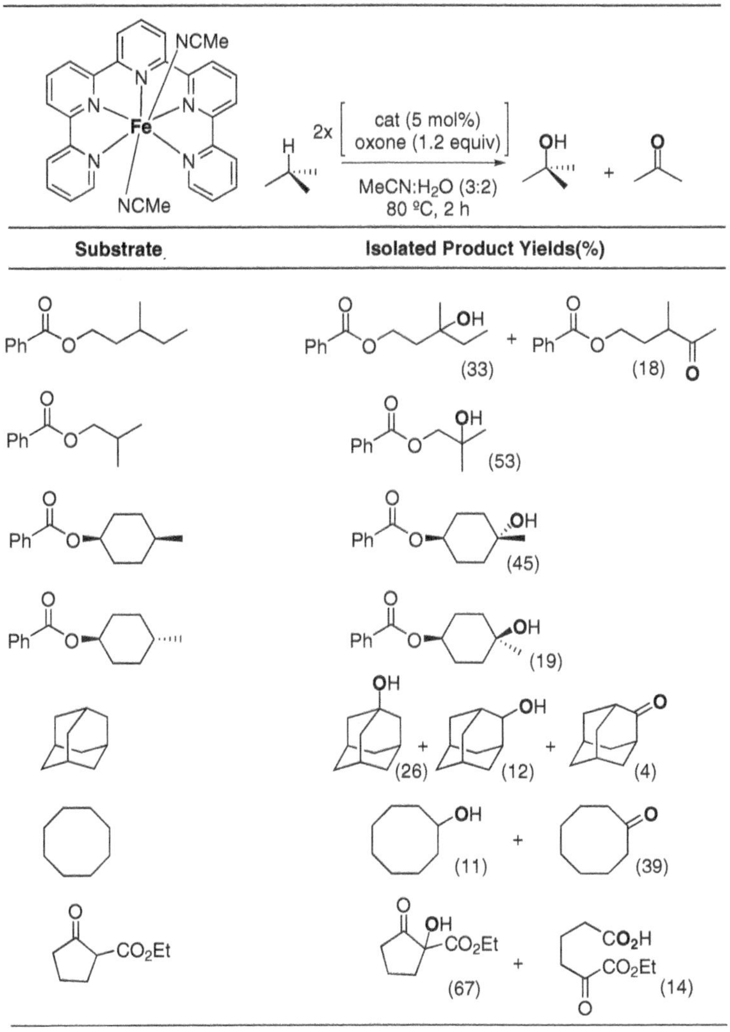

Scheme 22. Alkane oxidation with $[Fe(qpy)](ClO_4)_2$, where the volumes in parentheses are isolated product yields.

oxidation of alkanes to the corresponding alcohols and ketones with oxone©. Oxidation occurred with stereoretention (86).

G. Alkene Epoxidation

Epoxides are very interesting products because they are versatile synthons and suitable precursors for a number of more elaborated products with interest in organic synthesis, polymer science and pharmaceutical industry, among other fields. While effective epoxidation methods are well established in the literature, iron-catalyzed epoxidation remains as a particularly interesting reaction because of the lack of toxicity of this metal, and its availability. Olefin epoxidation is indeed a very common reaction resulting from the combination of iron complexes and peroxide oxidants. A number of oxidizing species can epoxidize olefins, and this includes metal-based reagents, but also free-radical chains. However, a limited number of iron catalysts have shown potential synthetic utility. The following lines collect representative examples.

In situ generation of catalysts by the combination of commercially available reagents constitutes a common practice in a number of fields of catalysis, but the subtleties of iron chemistry in redox processes makes this approach uncommon in selective oxidation reactions. Successful application of this approach to olefin epoxidation constitutes a remarkable exception. In this regard, particularly relevant developments have been accomplished by Beller and co-workers (87, 88). Excellent epoxidation catalysts that employ H_2O_2 as oxidant were prepared by the combination *in situ* of $FeCl_3 \cdot 6H_2O$, pyridine-2,6-dicarboxylic acid, and distinct nitrogen ligands. When pyrrolidine was employed, the resulting catalyst epoxidizes aromatic olefins with high efficiency and selectivity. Instead, use of benzylamine derivatives allowed epoxidation of a high variety of aliphatic and aromatic olefins (Scheme 23).

The use of formamide ligands forms a convenient system for epoxidizing styrenes and conjugated dienes (89). Finally, Imidazoles are also excellent ligands

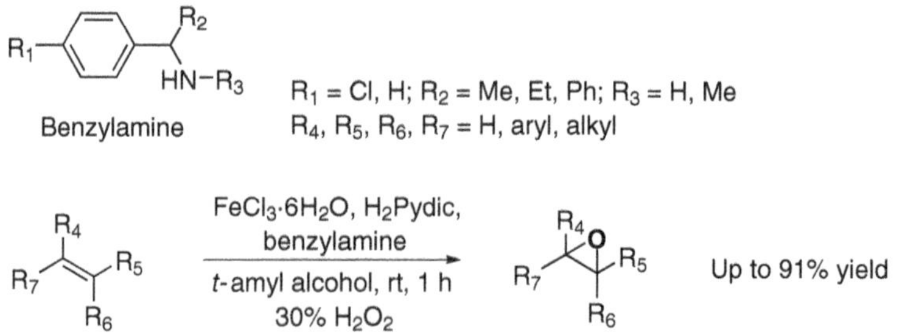

Scheme 23. *In situ* generated catalyst that performs epoxidation of a high range of olefins (H_2Pydic = dipicolinic acid.).

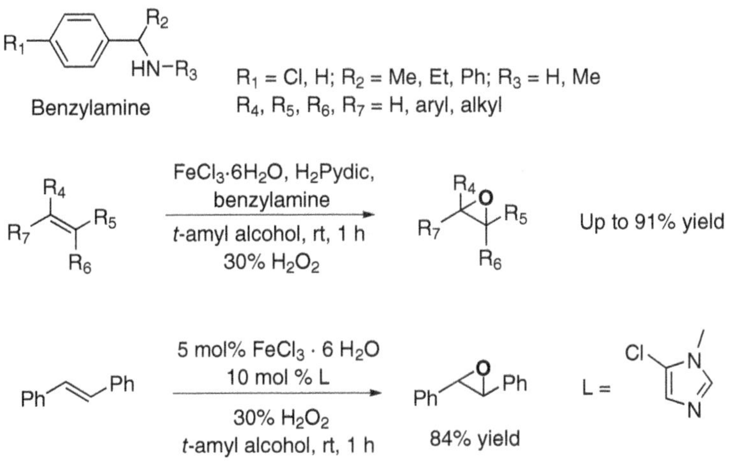

Scheme 24. Best results obtained in the epoxidation of *trans*-stilbene using 5-cloro-1-methylimidazole (5-Cl-1-MeIm) ligand.

and the resulting catalysts epoxidize aliphatic, aromatic olefins, and conjugated dienes with moderate-to-excellent yields and high chemoselectivity (89–91). The nature of the catalyst formed when imidazoles are used as N-based ligands was explored in detail. By testing a number of different imidazoles, it was discovered that a free 2-position of the imidazole ligand is essential for high catalytically activity. Moreover, it was found that best yields were obtained with 5-chloro-1-methylimidazole (Scheme 24). Isolated complex *trans*-$[FeCl_2(5\text{-Cl-1-MeIm})_4]Cl$ was also prepared and it was shown that it exhibits a catalytic activity similar to that obtained from the *in situ* catalysis (91).

Pioneering use of iron complexes with aminopyridine ligands as epoxidation catalysts useful in preparative scale was described by Jacobsen and co-workers (64). The mononuclear complex $[Fe^{II}(men)(MeCN)_2](SbF_6)_2$, was reported as a very active catalyst for the epoxidation of aliphatic olefins. High-epoxide yields (60–90%) were obtained in short reaction times by using 3 mol% catalyst loading, 1.5 equiv of H_2O_2 and acetic acid as a key additive to ensure high-product yields (Scheme 25*a*). While the system proved particularly effective with aliphatic olefins, aromatic alkenes were not tolerated.

The original report suggested that an *in situ* generated oxo-carboxylate bridged diiron(III) complex was the active catalyst. This structure constitutes a very common motif in O_2 activating diiron enzymes, and in particular is reminiscent to the active site of soluble oxidized methane monooxygenase (sMMO) enzyme (92). However, mechanistic work by Que and co-workers (28) demonstrated the importance of the presence of exchangeable sites at the catalysts for activating the H_2O_2 oxidant. Since the oxo-carboxylate dimer contains

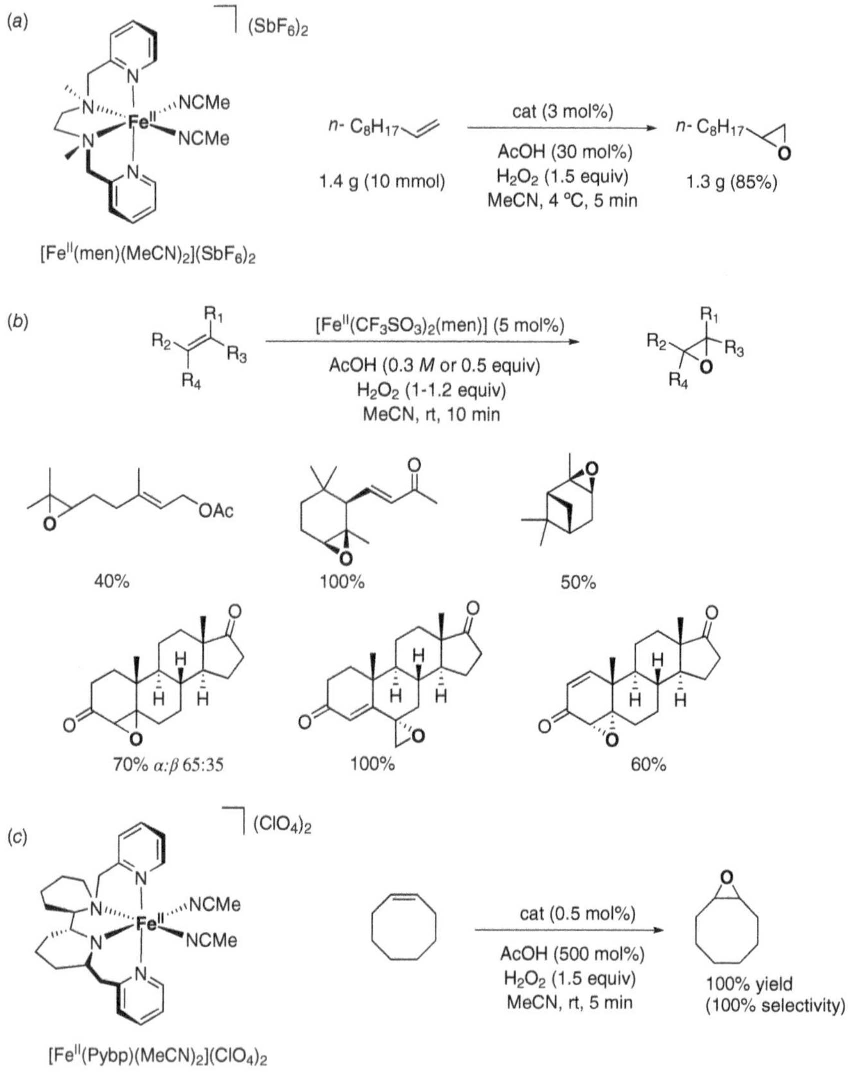

(a) $[Fe^{II}(men)(MeCN)_2](SbF_6)_2$

n-C_8H_{17} (1.4 g (10 mmol)) → cat (3 mol%), AcOH (30 mol%), H_2O_2 (1.5 equiv), MeCN, 4 °C, 5 min → epoxide 1.3 g (85%)

(b) $[Fe^{II}(CF_3SO_3)_2(men)]$ (5 mol%), AcOH (0.3 *M* or 0.5 equiv), H_2O_2 (1-1.2 equiv), MeCN, rt, 10 min

40% 100% 50%

70% α:β 65:35 100% 60%

(c) $[Fe^{II}(Pybp)(MeCN)_2](ClO_4)_2$

cat (0.5 mol%), AcOH (500 mol%), H_2O_2 (1.5 equiv), MeCN, rt, 5 min → 100% yield (100% selectivity)

Scheme 25. Epoxidation by $[Fe^{II}(men)(MeCN)_2](SbF_6)_2$ and $[Fe^{II}(Pybp)(MeCN)_2](ClO_4)_2$ (Pybp = *N,N'*-bis(pyridylmethyl)-2,2-bipyperidine).

coordinatively saturated ferric sites, its competence as an oxidation catalyst was disputed. Indeed, the isolated pure differric complex proved incompetent for mediating the reaction. Instead, Fujita and Que (93) proposed that a mononuclear iron(III) complex is the real catalyst. Acetic acid was proposed to play a key role by reacting with H_2O_2 and forming peracetic acid *in situ*, which in turn is the oxidant that reacts with the mononuclear iron center forming the olefin epoxidation species.

More recently, the triflate bound form of this catalyst $[Fe^{II}(CF_3SO_3)_2(men)]$ has been applied to the epoxidation of terpenoid (94), and steroidal substrates (Scheme 25*b*) (95), highlighting the potential utility of these catalysts in organic synthesis of natural products.

Excellent epoxidation activity has also been described by Rybak-Akimova and co-workers (42) for the iron catalyst $[Fe^{II}(Pybp)(MeCN)_2](ClO_4)_2$ (Scheme 25*c*). The catalyst is structurally related to $[Fe^{II}(men)(MeCN)_2](SbF_6)_2$, but the ethylenediamine moiety has been replaced by a more rigid bis-piperidine, which confers higher catalytic activity to the complex. At low-catalyst loading (0.5 mol%) with respect to substrate, and in the presence of 500 mol% of acetic acid (HOAc), nearly quantitative conversion of aliphatic olefins occurred rapidly (5–10 min) at room temperature. Under these conditions, TON reach 200, and can be further improved up to 715 in the epoxidation of cyclooctene, by lowering the catalyst loading down to 0.1 mol%. A particularly interesting aspect of this catalysts is that it performs the selective monoepoxidation of cyclic dienes (1,3-cyclooctadiene and 1,4-cyclohexadiene), and that epoxidation of cyclohexene takes place with minimum formation of allylic oxidation products (1–3%). A particularly interesting aspect of this complex is that a metastable reaction intermediate with spectroscopic features characteristic of low-spin ferric hydroperoxide species ($\lambda_{max} = 560$ nm, $g = 2.185$, 1.98, and 1.955) could be spectroscopically characterized when the reaction was performed at low temperature. Rybak-Akimova and co-workers (42) proposed that the reaction mechanism involved formation of $Fe^{V}(O)(OAc)$ species as the final epoxidizing agent, in line with an earlier proposal by Mas-Balleste and Que (66).

A set of iron complexes of general formula $[Fe^{II}(CF_3SO_3)_2(Mac\text{-}N5)$, where Mac-N5 stands for a series of pentadentate ligands based on a triazapyridine macrocycle, with an appended aminopropyl arm has also been described as efficient epoxidation catalysts (Scheme 26, Mac-N5 = 3-(3,7,11-triaza-1(2,6)-pyridinacyclododecaphane-7-yl)propan-1-amine) (96). X-ray analysis of one of the complexes reveals a square-pyramidal geometry, with the aminopropyl arm binding at the axial position. Strong, poorly coordinating acids like triflic acid, cause reversible dissociation of the pendant arm. Under these conditions, the complexes efficiently catalyze olefin epoxidation using hydrogen peroxide as the oxidant, under mild conditions and short reaction times. Epoxide yields of 66–89% with 90–98% selectivity were obtained with cyclooctene as substrate. However, in the deprotonated form the ligand arm binds to the iron center stopping the catalytic activity. Coordinating acids like HCl also stop the epoxidation activity, presumably via chloride binding to the iron center. Taktak et al. (96) propose that epoxidation is assisted by a “pull effect” mechanism that involves a hydrogen-

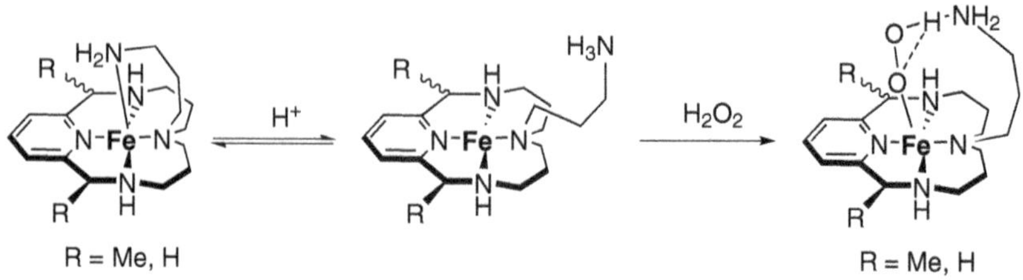

Scheme 26. Acid–base and epoxidation activity of $[Fe(CF_3SO_3)_2(Mac\text{-}N_5)]$ complexes [Adapted from [(96).]

bonding interaction between the metal-bound hydroperoxide and the amine pendant arm. This interaction enhances the electrophilicity of the peroxide.

Oxo-bridged diiron systems are very common structural motifs in oxidative enzymes, and synthetic models have been also described as robust and efficient epoxidation catalysts. A prototypical example is the μ-oxoiron(III) dimer with phenanthroline ligands described by Stack and co-workers (97). This complex can be prepared *in situ* by simply mixing common ferric salts and phenanthroline as a ligand, and epoxidizes a wide range of alkenes, including terminal and electron-deficient ones, using peracetic acid as the oxidant. Reactions are fast, require small catalyst loadings, and operate at relatively large substrate concentration, converting this methodology in a synthetically valuable protocol for olefin epoxidation (Scheme 27*a*).

Che and co-workers (98) described a mononuclear octahedral iron complex with two *p*-chloro terpyridine ligands as an efficient epoxidation catalyst with oxone (Scheme 27*b*). The catalyst has a broad substrate scope and epoxidizes

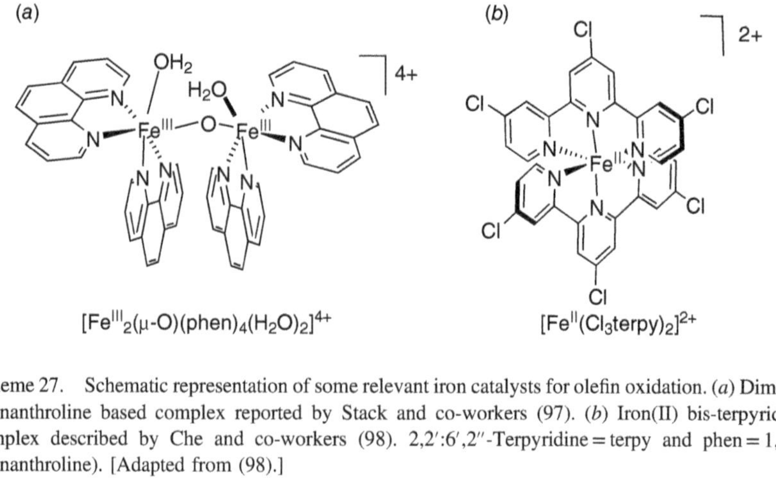

Scheme 27. Schematic representation of some relevant iron catalysts for olefin oxidation. (*a*) Dimeric phenanthroline based complex reported by Stack and co-workers (97). (*b*) Iron(II) bis-terpyridine complex described by Che and co-workers (98). 2,2′:6′,2″-Terpyridine = terpy and phen = 1,10-phenanthroline). [Adapted from (98).]

aliphatic, aromatic, terminal, electron-rich and electron-deficient olefins with good-to-excellent yields, in short reaction times and at room temperature. The potential applicability of the system was demonstrated by performing oxidations in gram scale, and by reuse of the ligand. Specifically, following catalyst deactivation, catalytic activity can be rescued by the addition of a new batch of iron(II) salts. Evidence for the implication of high-valent oxo-iron species was obtained by means of MS experiments (98).

H. Asymmetric Epoxidation

Asymmetric epoxidation with non-heme iron complexes remains as a very challenging reaction, extensively explored with porphyrin catalysts (13, 99). However, examples of non-heme iron-dependent catalysts that provide good product yields and enantiomeric excess (ee) remain rare (13, 99).

Pioneering work in the field was performed by Beller and co-workers (100, 101) who developed an operationally simple olefin asymmetric epoxidation methodology using catalysts generated *in situ* by combining $FeCl_3 \cdot 6H_2O$, pyridine-2,6-dicarboxylic acid and chiral benzylamines as ligands (Scheme 28*a*). The catalytic system reached good yields and high ee, up to 97% for *trans*-stilbene substrates using H_2O_2 as oxidant (Scheme 28*a*).

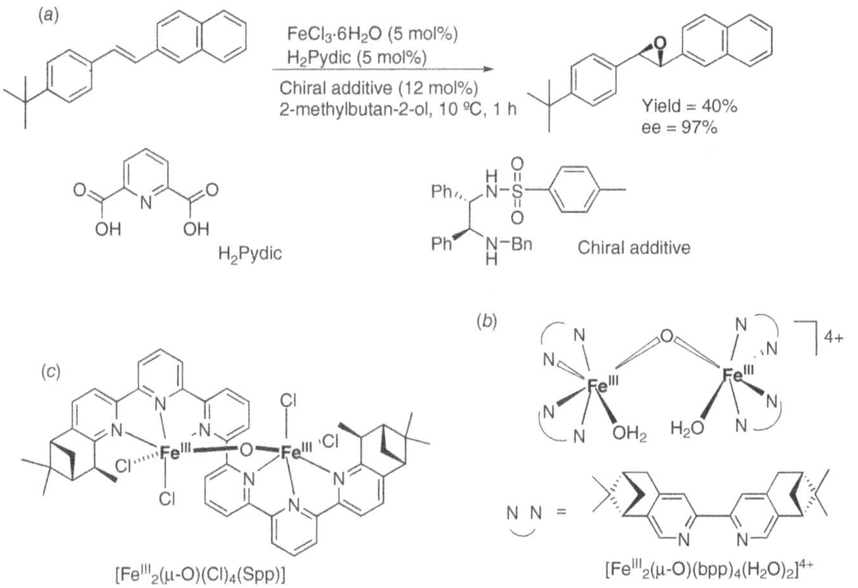

Scheme 28. Iron-based asymmetric epoxidation catalysts [Spp = 6,6‴-bis((5*R*,7*R*,8*R*)-6,6,8-trimethyl-5,6,7,8-tetrahydro-5,7-methanoquinolin-2-yl-2,2′:6′,2″:6″,2″-quaterpyridine.]

Menage and co-workers (102) described a non-heme oxo-bridged diiron complex using chiral 2,2′-bipyridine (bpy) ligands, that enantioselectivity catalyzes epoxidation of alkenes (Scheme 28*b*). The complex oxidizes olefins with high efficiencies, up to 850 TON, and moderate ees ranging from 9 to 63%. Best stereoselectivity results were obtained when electron-deficient substrates (e.g., chalcone) were used (Scheme 28*b*).

Ménage and co-workers (103) also introduced carboxylate ligand arms in the chiral mcp ligand. Corresponding iron complexes have been studied in the epoxidation of alkenes. Epoxidation of *trans*-2-heptene with H_2O_2 in a MeCN:AcOH 1:2 solvent mixture provides the corresponding epoxide in 40% yield and 34% ee (103).

In a recent work, Kwong and co-workers (104) reported a chiral six pyridine iron complex as an efficient catalyst for the epoxidation of alkenes with H_2O_2 in the presence of acetic acid, providing moderate ee in the oxidation of aromatic olefins. Up to 43% ee was obtained in the oxidation of styrene (Scheme 28*c*).

More recently, Sun and co-workers (105) reported a family of chiral iron(II) complexes based on N_4-ligands with a chiral cyclohexane diamine backbone that catalyze the enantioselective epoxidation of α,β-enones. By using 2 mol% of catalyst, 5 equiv of acetic acid (with respect to the substrate), and the slow injection of 2 equiv of H_2O_2, afford the epoxidation of a high number of α,β-enone substrates with moderated-to-good yields (33–90%) and with an ee up to 87% (Scheme 29, complexes **1** and **2**). The same group reported a second catalyst based on a more rigid chiral diamine derived from L-proline and two benzimidazole donor groups. Using similar conditions as before and α,β-enones as substrates, Sun and co-workers (105) could improve yields (up to 99%) and enantioselectivities (up to 97%) (Scheme 29, complexes **3**–**5**) (106).

Finally, Talsi and co-workers (67) reported the enantioselective epoxidation of alkenes by using 1 mol% of catalyst, $[Fe^{II}(CF_3SO_3)_2(pdp)]$, and H_2O_2 as oxidant. Using these simple conditions, without carboxylic acid addition, chalcone was epoxidized in 13% yield and 61% ee. The addition of carboxylic acid (1.1 equiv) improves the efficiency and the enantioselectivity of this system giving yields up to 100%. The nature of the carboxylic acid appears also to affect the

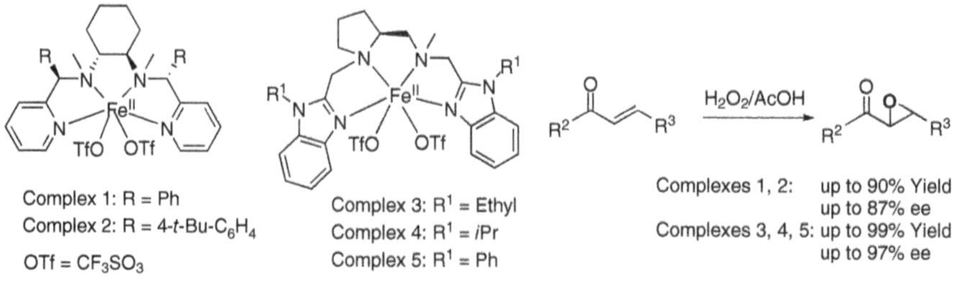

Complex 1: R = Ph
Complex 2: R = 4-*t*-Bu-C_6H_4
OTf = CF_3SO_3

Complex 3: R^1 = Ethyl
Complex 4: R^1 = *i*Pr
Complex 5: R^1 = Ph

H_2O_2/AcOH

Complexes 1, 2: up to 90% Yield, up to 87% ee
Complexes 3, 4, 5: up to 99% Yield, up to 97% ee

Scheme 29. Iron(II) complexes employed for the enantioselective epoxidation of α,β-enones (105).

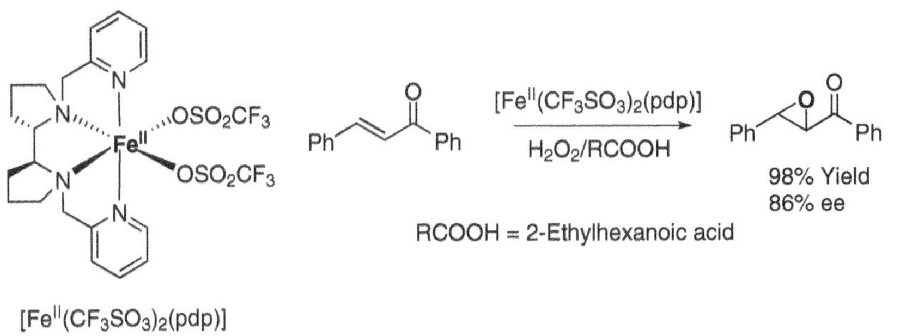

Scheme 30. Bipyrrolidine-based iron(II) complex catalyze the asymmetric epoxidation of various olefins in the presence of a carboxylic acid [*N*,*N*′-bis(2-pyridylmethyl)-2,2′-bipyrrolidine = pdp].

stereodiscrimination and the best results were obtained by using 2-ethylhexanoic acid, epoxidizing chalcone in 98% yield and 86% ee (Scheme 30).

An interesting catalyst for the asymmetric epoxidation of *β*,*β*-substituted enones based in carefully elaborated chiral phenanthrolines (phen*) was recently described by Nishikawa and Yamamoto (107) (Scheme 31). The phenanthroline heterocylic contain a binaphthyl group attached at position 2, and a systematic exploration of a series of ligand analogues showed that introduction of arene groups at the 3- and 8-positions in the phenanthroline moiety was a key element to ensure good ee. These arene groups were expected to restrict the rotation of the bond between the binaphthyl group and the phenanthroline moiety. Crystallographic analysis could be performed for one of the catalysts of the series. The mixture of $Fe(CF_3SO_3)_2$ and 2 equiv of a racemic mixture of a phen* ligand in acetonitrile provided crystals of a mononuclear $[Fe(CF_3SO_3)(phen^*)_2(MeCN)](CF_3SO_3)$ complex, where two homochiral phenanthroline ligands coordinate the iron center in a cis topology, affording a pseudo-C_2-symmetric iron complex.

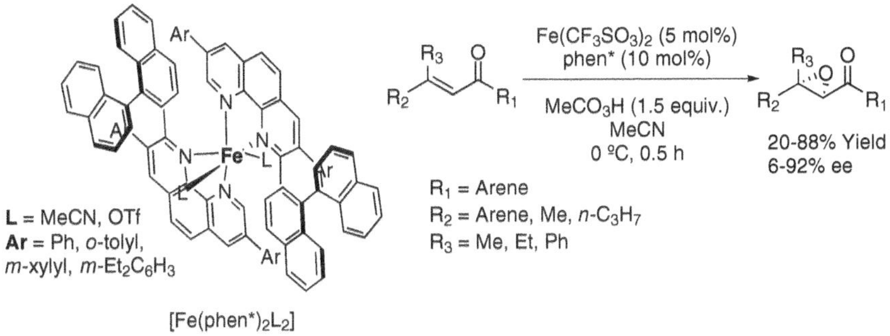

Scheme 31. Chiral phenanthroline iron complex as an asymmetric epoxidation catalyst.

Since the two N-heterocyles of the phen* moieties are not equivalent, and a racemic mixture of the ligand was used in this analysis, octahedral complexes of general formula [Fe(X_2)(phen*)$_2$] (X = MeCN or CF_3SO_3) in principle could exist as multiple diastereoisomers, resulting from the combination of hetero- or homo-combination of the ligand, but only one diastereomer was selectively crystallized. This selective complexation was explained by the $\pi\pi$ interactions between the naphthyl groups and diphenylphenanthroline.

Applying the optimal catalyst several β,β-disubstituted enones were epoxidized with peracetic acid in good yields with excellent enantioselectivity (90–92% ee). An electron-rich olefin (e.g., *trans*-α-methylstilbene) was also oxidized with good enantioselectivity. Peracetic acid proved to be crucial for activity as other oxidants [e.g., H_2O_2 or TBHP (*tert*-butylhydroperoxide)] were ineffective. Reactions were stereospecific and only one diastereomer of the epoxide was obtained in each reaction. Competitive oxidation of a mixture of an electron-rich and an electron-deficient alkene showed preferential oxidation of the more electron-rich olefin, while retaining good stereoselectivity, demonstrating the electrophilic nature of the oxidizing species.

An interesting non-heme iron asymmetric epoxidation catalysts that relies in N-based ligands, but showing remarkable analogies in its oxidation chemistry with iron porphyrins, is a mononuclear iron(III) complex (Scheme 32) containing a 1,8-(bisoxazolyl)-carbazole tridentate ligand (108). This complex catalyzes highly enantioselective asymmetric epoxidation of trans-alkenes at room temperature employing PhIO as oxidant. Unlike the most common octahedrally coordinated iron complexes, this iron(III) complex has a five-coordinated trigonal-bipyramidal structure. Three coordination sites are occupied by the two oxazoline heterocycles, and the carbazole moieties of the tridentate ligand binding in a meridional fashion.

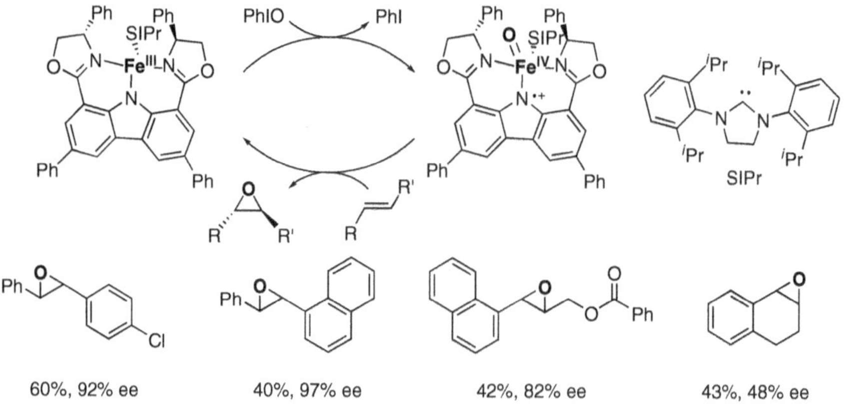

Scheme 32. Iron-oxazolyl carbazole complex in asymmetric epoxidation [SIPr = *N,N'*-bis(2,6-diiso-propylphenyl)-4,5-dihydro-1*H*-imidazol-2-ylidene].

A heterocyclic carbene ligand binds at the fourth site and the fifth coordination site contains a chloride ligand. This later ligand is removed by addition of NaBArF [BArF = tetrakis(3,5-trifluoromethylphenyl)borate] before oxidation catalysis starts, opening a binding site where a terminal oxo ligand is presumably formed upon reaction of the iron center with the oxidant. A particularly interesting aspect of this complex is that the ligands are red–ox noninnocent, and the two-electron oxidized state of the catalyst has the similar electronic structure as that of iron porphyrins, with one of the oxidizing equivalents residing in the carbazole ligand.

I. Alkene cis-Dihydroxylation

Arene cis-dihydroxylation is performed by bacterial enzymes of the family of Rieske oxygenases, as a first step in the degradation of these persistent pollutants (8, 109). Inspired by these family of non-heme iron-dependent enzymes, the ability of iron complexes to catalyze olefin cis-dihydroxylation was first described by Que and co-workers (28, 110) for the $[Fe^{II}(tpa)]^{2+}$ family of complexes, employing H_2O_2 as oxidant. This catalytic activity has been found quite common for iron complexes containing tetradentate N-based ligands, which have been widely explored (28, 32, 34, 36, 61–63, 111–117). The accumulated data shows that the ligand structure modulates the syn-diol/epoxide ratio, but the origins of these changes in chemoselectivity are not understood. As a general trend, introduction of methyl groups in the α position of the pyridine ring of polypyridine complexes promote higher syn-diol/epoxide ratios than those attained with the corresponding complexes where the pyridine contains a hydrogen atom at this position. On the other hand, cis-dihydroxylation requires that the complexes contain two binding sites at the iron center in the cis-relative position, and are occupied by easily exchangeable ligands. Ligands that leave only one coordination site (110), or where the two exchangeable sites are in a relative trans disposition, elicit epoxidation reactions. Based on these observations it seems that two cis-labile sites are required to perform cis-dihydroxylation (113).

While iron-catalyzed olefin cis-dihydroxylation is not uncommon, so far there is only one example of arene cis-dihydroxylation, which indeed constitutes the reaction catalyzed by naphthalene dioxygenase (NDO) (118). $[Fe^{II}(tpa)(NCMe)_2]^{2+}$ (Scheme 33) catalyzes the oxidation of naphthalene with H_2O_2

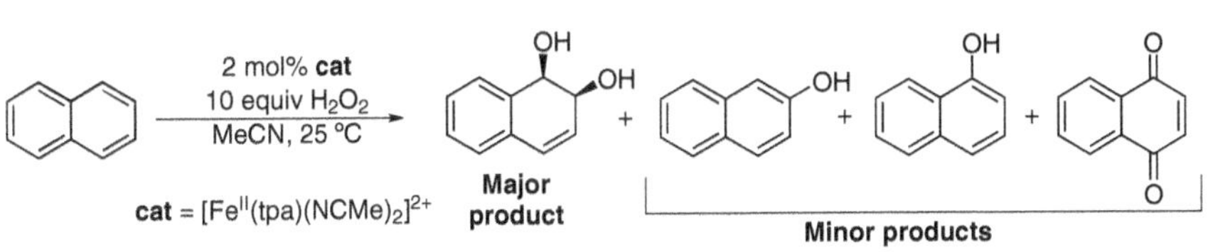

2 mol% **cat**
10 equiv H_2O_2
MeCN, 25 °C

cat = $[Fe^{II}(tpa)(NCMe)_2]^{2+}$

OH OH + OH + OH + O O

Major product

Minor products

Scheme 33. Arene cis-dihydroxylation catalyzed by $[Fe^{II}(tpa)(NCMe)_2]^{2+}$.

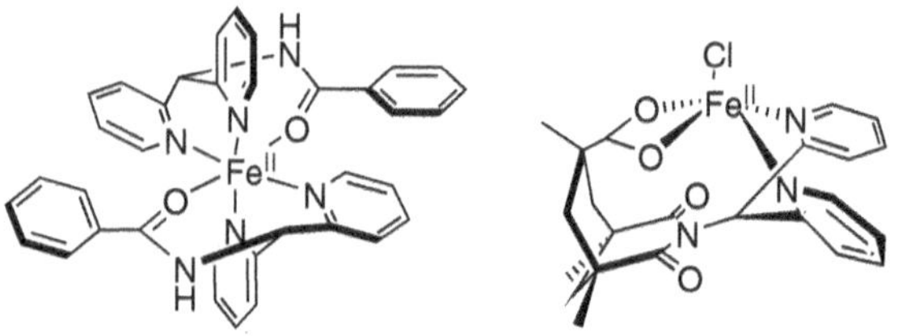

$[Fe^{II}(Ph\text{-}DPAH)_2(CF_3SO_3)_2]$ $[Fe^{II}(Bpka)(Cl)]$

Scheme 34. Complexes described as functional and structural models of Rieske oxygenases [Bpka = 3-(di(pyridin-2-yl)methyl)-1,5,7-trimethyl-2,4-dioxo-3-azabicyclo[3.3.1]nonane-7-carboxylic acid; Ph-DPAH = di(pyridin-2-yl)methyl benzamide].

affording *cis*-1,2-dihydro-1,2-naphthalenediol, 1-naphthol, 2-naphthol, and 1,4-naphthoquinone. The first is the major product, as in the NDO catalyzed reaction.

By means of ligand design, oxygen-binding groups have been placed in the first coordination structure of iron complexes with the aim of reproducing the *N,N,O*-ligand (2His-1-carboxylate) environment found at the enzyme active site. Prototypical examples are collected in Scheme 34 (119–124). These complexes catalyze the cis-dihydroxylation of olefins using H_2O_2 as oxidant and, in selected cases, excellent chemoselectivity toward cis-dihydroxylation was attained, although large excess of substrate was required (119, 121).

Iron-catalyzed cis-dihydroxylation is a particularly attractive reaction because it has the potential to replace osmium-based methodologies. However, two major drawbacks limit the synthetic utility of the bioinspired iron catalysts; (1) most of them require a large excess of substrate and provide very modest substrate conversion and product yields. (2) A second important aspect of concerns is the limited chemoselectivity of this reaction, because epoxides are also commonly formed in parallel, so mixtures of epoxides and diols are obtained.

Few examples of iron catalysts have already shown the potential to become synthetic tools. Moderate-to-good product yields and chemoselectivity toward syn-diols in the oxidation of aliphatic olefins with H_2O_2 could be obtained with $[Fe^{II}(5Me_3\text{-tpa})(MeCN)_2](ClO_4)_2$ (Scheme 35, $5Me_3$-tpa = tris(5-methyl-2-pyridylmethyl)amine (115), and chemoselectivity toward cis-dihydroxylation appears to be comparable to that obtained under conditions of large excess of substrate. Under analogous conditions, other tpa derivatives included oxo-bridged dimers, exerted reduced yields, but the origin of the improved performance of this catalyst is not understood.

More recently, after screening a broad family of Fe-tacn based complexes as potential olefin cis-dihydroxylation catalysts with H_2O_2, $[Fe^{II}(CF_3SO_3)_2(^{Me,Me}Py\text{-}tacn)]$ was identified as the optimum catalyst among the series in terms of yields and chemoselectivity, and it was subjected to further study (111). It was found that

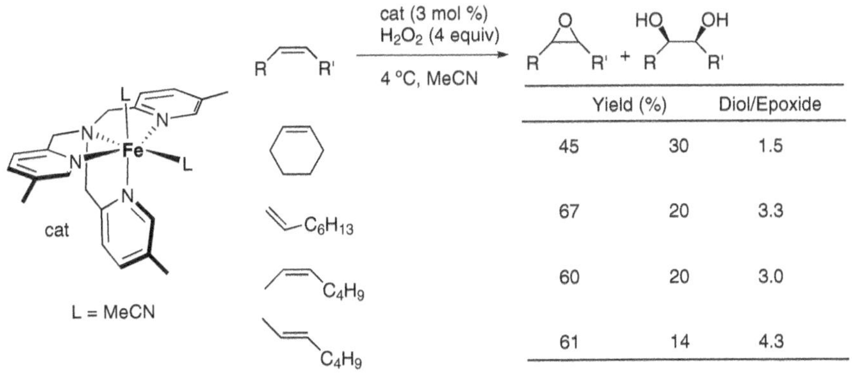

cat (3 mol %)
H_2O_2 (4 equiv)
4 °C, MeCN

cat

L = MeCN

Substrate	Yield (%)		Diol/Epoxide
	45	30	1.5
C_6H_{13}	67	20	3.3
C_4H_9	60	20	3.0
C_4H_9	61	14	4.3

Scheme 35. Olefin oxidation with $[Fe^{II}(5Me_3{-}tpa)(MeCN)_2](ClO_4)_2$.

addition of water (1000 equiv with regard to catalyst) has a beneficial effect in the chemoselectivity of the reaction toward cis-dihydroxylation, presumably because it facilitates product release. Syn-diols also were found competitive substrates for the oxidizing species and because of that a two-step protocol was employed, involving diol removal from the reaction mixture. Best yields were obtained when terminal or cis aliphatic olefins are oxidized. Instead, in the oxidation of trans olefins, tri-substituted olefins, and 1,1-disubstituted olefins the epoxide product was obtained preferentially (Table II). The oxidizing species generated with this catalyst have electrophilic character, and the oxidation of dienes occurs at the more electron-rich site. In addition, cis-dihydroxylation occurs preferentially in cis over trans substrates.

Most interesting is the system described by Che and co-workers (125) that employs a chemically robust iron complex $[Fe^{III}(Cl)_2(c\text{-}PyNMe_2)]^+$, where c-PyNMe$_2$ = N,N'-dimethyl-2,11-diaza[3.3](2,6)pyridinophane as olefin cis-dihydroxylation catalyst with oxone (Scheme 36). By using 0.7–3.5 mol% of catalyst loading, and limiting amounts of substrate, with 2 equiv of oxone a large range of alkenes could be oxidized. Best yields were obtained in the oxidation of electron-deficient olefins, from 56 to 99%, yielding high syn-diol/epoxide ratios (8.5–20.3). On the other hand, electron-rich olefins are oxidized in moderate-to-good yields (16–64%) and lower syn-diol/epoxide selectivities.

This method could be applied in large scale, for example, the cis-dihydroxylation of methyl cinnamate (9.7 g) with oxone (2 equiv) afforded the syn-diol product (8.4 g), in 84% yield (Scheme 36). This reaction was performed by adding each of the substrates and the catalyst in two equal portions, and each portion was reacted with oxone. The second addition of the catalyst could be replaced by the iron salt, $Fe(ClO_4)_2{\cdot}4H_2O$, which allows the *in situ* regeneration of the catalyst obtaining the same results. Epoxides, hydroxyketones, and ketones (or aldehydes) resulting from oxidative C–C cleavage are side products of the reaction.

TABLE II
Olefin Oxidation Catalysis With the $[Fe^{II}(CF_3SO_3)_2(^{Me,Me}Pytacn)]$ Catalyst

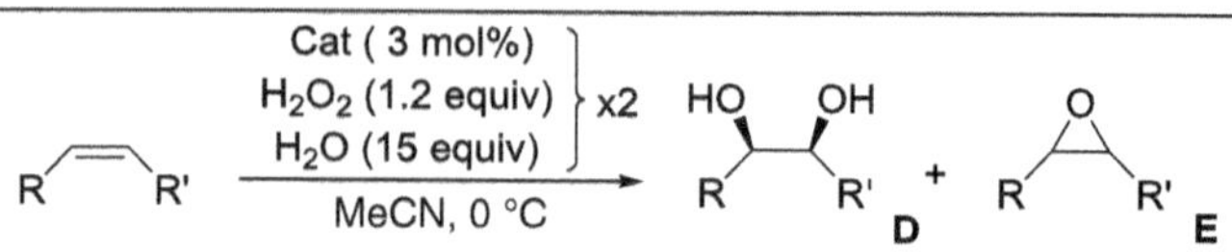

Cat (3 mol%), H_2O_2 (1.2 equiv), H_2O (15 equiv) } x2; MeCN, 0 °C; R–CH=CH–R' → diol **D** + epoxide **E**

Substrate	Conversion	Yield (%) D	E	D/E	Yield (%)
 cyclooctene [a]	96	60	11	5.7	71
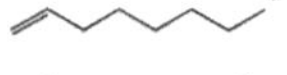 1-octene [a]	73	53	20	2.7	73
 *cis*-2-octene [a]	98	56	26	2.2	82
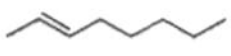 *trans*-2-octene [a]	69	19	19	1	38
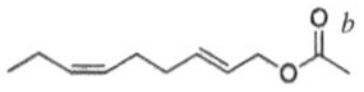 diene acetate [b]	88	56	17	3.3	73
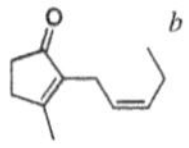 *cis*-jasmone [b]	90	45	7[c]	6.7	52

[a] Gas chromatography (GC) yield.
[b] Isolated yields.
[c] *Trans*-Diol product originated from the ring opening of the epoxide product.

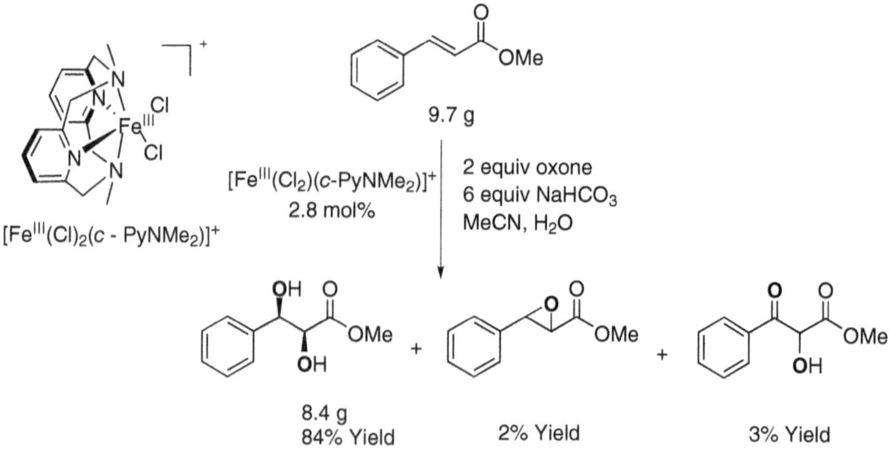

$[Fe^{III}(Cl)_2(c\text{-}PyNMe_2)]^+$

Methyl cinnamate 9.7 g; $[Fe^{III}(Cl_2)(c\text{-}PyNMe_2)]^+$ 2.8 mol%; 2 equiv oxone, 6 equiv $NaHCO_3$, MeCN, H_2O → diol (8.4 g, 84% Yield) + epoxide (2% Yield) + α-hydroxy ketone (3% Yield)

Scheme 36. Large-scale catalysis performed by $[Fe^{III}(Cl)_2(c\text{-}PyNMe_2)]^+$.

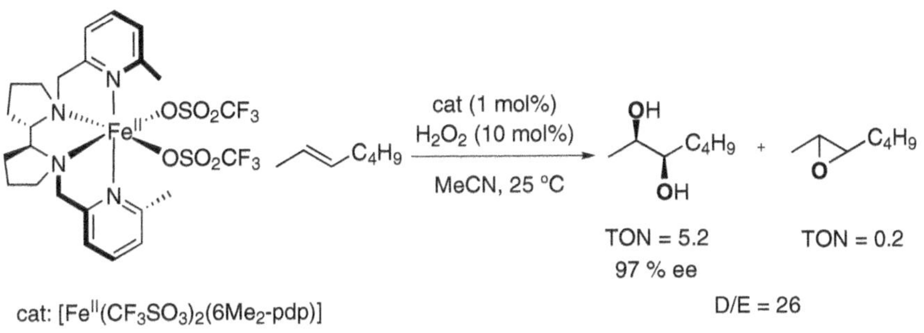

OSO_2CF_3 Fe^{II} OSO_2CF_3

cat (1 mol%)
H_2O_2 (10 mol%)
MeCN, 25 °C

C_4H_9 OH OH C_4H_9 + C_4H_9

TON = 5.2
97 % ee

TON = 0.2

D/E = 26

cat: $[Fe^{II}(CF_3SO_3)_2(6Me_2\text{-pdp})]$

Scheme 37. Asymmetric cis-dihydroxylation using $[Fe^{II}(CF_3SO_3)_2(6Me_2\text{-pdp})]$ complex, obtaining syn-diol product in 97% ee.

An interesting mechanistic aspect of this work is that Che and co-workers (125) identify by electrospray ionization mass spectrometry (ESI–MS) analyses on catalytic reactions a cluster ion that could be simulated as $[Fe^{V}(O)_2(c\text{-PyNMe}_2)]^+$, and that was proposed to be the species responsible for the cis-dihydroxylation of olefins on the basis of DFT analysis.

J. Asymmetric cis-Dihydroxylation

While Os catalyzed asymmetric cis-dihydroxylation is recognized as a well-established, very reliable methodology, the prospect of developing iron-catalyzed methodologies has great interest. Good ee in olefin cis-dihydroxylation reactions have been obtained with the two structurally related iron catalysts $[Fe^{II}(CF_3SO_3)_2(6Me_2\text{-pdp})]$, where *N,N′*-bis(6-methyl-2-pyridymethyl)-2,2′-bipyrrolidine, and $[Fe^{II}(CF_3SO_3)_2(6Me_2\text{-mcp})]$ where $6Me_2$-mcp = *N,N′*-diamethyl-*N,N′*-bis-(6-methyl-2-pyridylmethyl)-cyclohexane-*trans*-1,2-diamine (Scheme 37) (116, 126). Good chemoselectivity toward the syn-diol and excellent enantioselectivity can be obtained with the former in the cis-dihydroxylation of linear trans-aliphatic alkenes, but the substrate conversion of these systems is very poor (<5%) precluding their synthetic value.

II. AMINOPYRIDINE MANGANESE COMPLEXES AS MOLECULAR CATALYSTS FOR HYDROCARBON OXIDATIONS

Manganese presents a rich chemistry in biochemical oxidation processes. An example of one of the most remarkable of its extraordinary catalytic activity and capacity to control oxidation reactions is the role of the tetranuclear manganese cluster in the oxidation of water in the photosystem II (PSII). Among the pioneer examples of the application of a manganese complex in oxidation catalysis was the seminal report by Hage et al. (127) on the use of $[Mn_2(\mu\text{-}O)_3(tmtacn)_2](PF_6)_2$ (tmtacn = 1,4,7-trimethyl-1,4,7-triazacyclononane, Scheme 42) (128), as bleaching agent and alkene epoxidation catalyst using H_2O_2 as oxidant (127). Since then, manganese has emerged as a very attractive transition metal ion for oxidation catalysis in terms of reactivity, chemo-, and stereoselectivities when using H_2O_2 as oxidant (129, 130). Furthermore, its oxidation chemistry finds important similarities with that of iron, a matter that is commonly invoked to explain mechanistic aspects of manganese catalyzed oxidations, but that is not completely understood, presumably because of the challenge that manganese represents in terms of spectroscopic analyses.

A. Catalytic Oxidation of Alkanes by Mn Complexes

1. *Oxygenation of C–H Bonds*

Manganese complexes have been used as catalysts for C–H oxidation reactions in a handful of examples. Pioneer work in the field was done by Shul'pin et al. (131) by studying the use of $[Mn_2(\mu\text{-}O)_3(tmtacn)_2](PF_6)_2$ as catalyst. The oxidation of alkanes using $MeCN/H_2O_2/AcOH$ mixtures was accomplished with high TON (up to 2000 in the oxidation of *n*-butane) and alcohol/ketone rations ~1.3–1.4. In the oxidation of *cis*-1,2-dimethylcyclohexane partial loss the stereochemistry was observed, and cyclohexyl-hydroperoxide was obtained in the oxidation of cyclohexane. Both observations are suggestive of a free-diffusing radical process (131, 132). In regard to the selective C–H oxidation reactions, the seminal work by Crabtree, Brudgvig, and co-workers (133) deserves special attention. These authors designed a very selective system that recognized the substrate by hydrogen-bond interactions, thus directing an specific C–H bond to the putative metal-oxo center (Scheme 38). Using this strategy, the oxidation of ibuprofen as a model substrate was achieved in good conversions (50–70%) and excellent selectivity (96–99%) using oxone as the terminal oxidant.

Nam and co-workers (134) described a mononuclear manganese(II) complex that catalyzed oxidations of olefins, alcohols, and alkanes using peracetic acid as

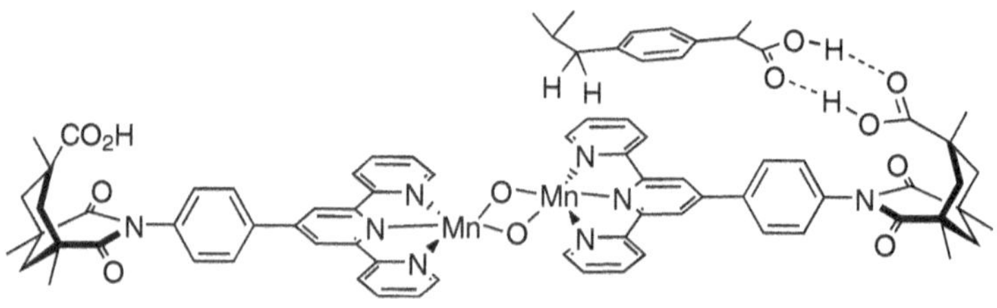

Scheme 38. Catalyst–substrate interaction that facilitate the selective C–H oxidation of ibuprofene.

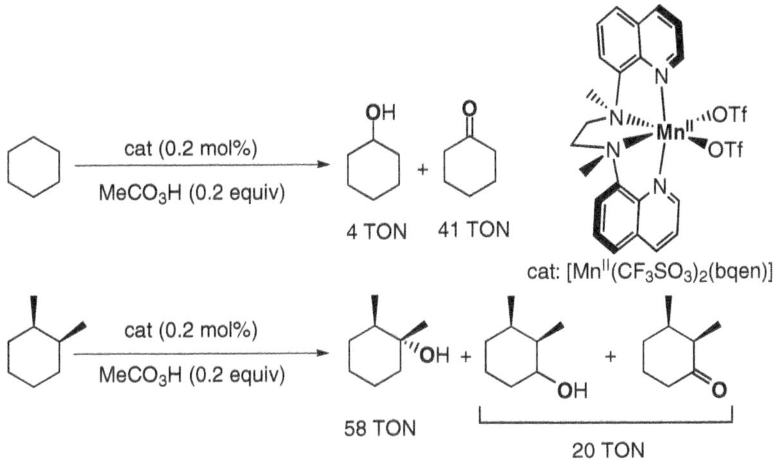

cat (0.2 mol%)
$MeCO_3H$ (0.2 equiv)
4 TON 41 TON
cat: $[Mn^{II}(CF_3SO_3)_2(bqen)]$
58 TON 20 TON

Scheme 39. The C–H oxidation reported by Nam and co-workers (134) when using a manganese complex, a tetradentate N-based ligand.

oxidant. The reactions were carried out using 1 mol% catalyst, 1 equiv of oxidant, and 5 equiv of substrate (Scheme 39). Under conditions of excess of substrate, cyclohexane was oxidized to a mixture of cyclohexanol (4% based on oxidant) and cyclohexanone (82%). These values represent an excellent efficiency in oxidant transformation into oxidized products (86% conversion of peracetic acid). However, the use of an excess of substrate and the rather low-substrate conversion hamper the use of this system for preparative purposes.

Very recently Bryliakov and co-workers (135) reported that aminopyridine manganese complexes $[Mn^{II}(CF_3SO_3)_2L]$ [L = mcp, pdp, and bpeb, where bpeb = *N,N′*-dimethyl-*N,N′*-bis(1-*R*-1*H*-benzo[*d*]imidazol-2-yl)methyl)cyclohexane-*trans*-1,2-diamine], demonstrate excellent efficiency (up to 970 turnovers) and stereospecificity in oxidation of aliphatic C–H groups with H_2O_2 in the presence of acetic acid (Fig. 40). Oxidations appear to be metal based as evidenced by a cyclohexanol/cyclohexanone ratio of ≈5 in the oxidation of cyclohexane and by the retention of configuration when using *cis*-1,2-dimethylcyclohexane as substrate. Steric influences are an important contribution in deciding regioselectivity, as shown in the selective oxidation of (+)-methyl acetate. Likewise electronic effects do also regulate C–H selectivity and demonstrate that the active species have electrophilic character. While the oxidation of 2,6-dimethyloctane yields an equimolar mixture of tertiary C–H oxidation products, introduction of electron acceptors deactivates the proximal tertiary C–H group.

Recent progress in non-heme manganese catalyzed enantioselective epoxidations, cis-dihydroxylation, and C–H hydroxylation shows that catalytic systems of

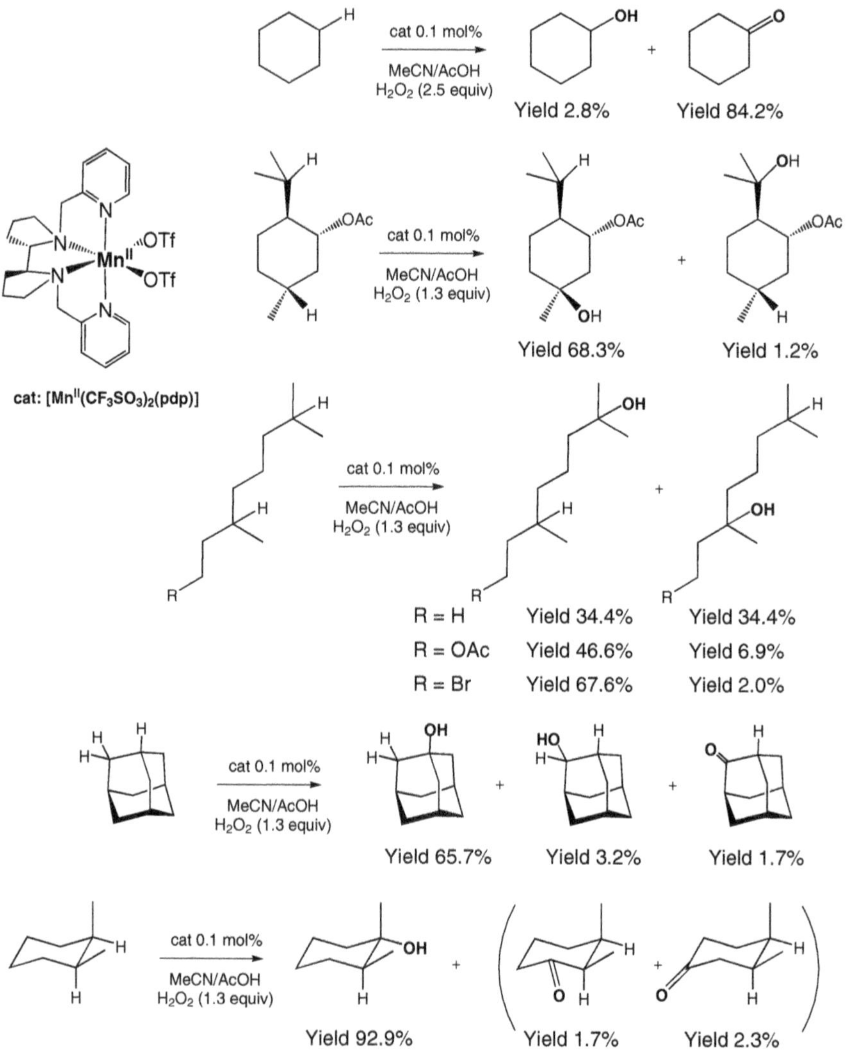

cat 0.1 mol%
MeCN/AcOH
H_2O_2 (2.5 equiv)
Yield 2.8% + Yield 84.2%

cat: $[Mn^{II}(CF_3SO_3)_2(pdp)]$

cat 0.1 mol%
MeCN/AcOH
H_2O_2 (1.3 equiv)
Yield 68.3% + Yield 1.2%

cat 0.1 mol%
MeCN/AcOH
H_2O_2 (1.3 equiv)

R = H	Yield 34.4%	Yield 34.4%
R = OAc	Yield 46.6%	Yield 6.9%
R = Br	Yield 67.6%	Yield 2.0%

cat 0.1 mol%
MeCN/AcOH
H_2O_2 (1.3 equiv)
Yield 65.7% + Yield 3.2% + Yield 1.7%

cat 0.1 mol%
MeCN/AcOH
H_2O_2 (1.3 equiv)
Yield 92.9% + (Yield 1.7% + Yield 2.3%)

Scheme 40. Catalytic oxidation of various aliphatic alkanes with $[Mn^{II}(CF_3SO_3)_2(pdp)]$.

this type can approach sustainable and efficient chemo- and stereoselective transformations by using the convenient H_2O_2 as oxidant. Note that the scope of "good" substrates for non-heme manganese systems is rather narrow at present. However, although the area is still in its infancy, it is developing quickly, and the potential of non-heme manganese-based systems will hopefully be clearer within the next few years.

2. *Desaturation of C–H Bonds*

Likewise, iron complexes, the oxidation of C–H bonds of specific substrates catalyzed by manganese complexes can also result in desaturation transformations. Compared with other oxidation reactions this transformation has been seldom studied with manganese complexes that contain aminopyridine ligands (136, 137). In this line, Crabtree and co-workers (137) studied the C–H desaturation of different substrates via manganese complexes in a high oxidation state. Complexes tested included [Mn(tpp)(Cl)], [Mn(salen)(Cl)], and $[Mn_2(\mu\text{-}O)(terpy)_2(H_2O)_2]$ (Scheme 41 where salen = *N,N′*-ethylenebis(salicylimine), tpp = tetraphenylporphyrin), all of which, upon reaction with an oxidant, showed good selectivity for the desaturation process. In the oxidation of the model substrate

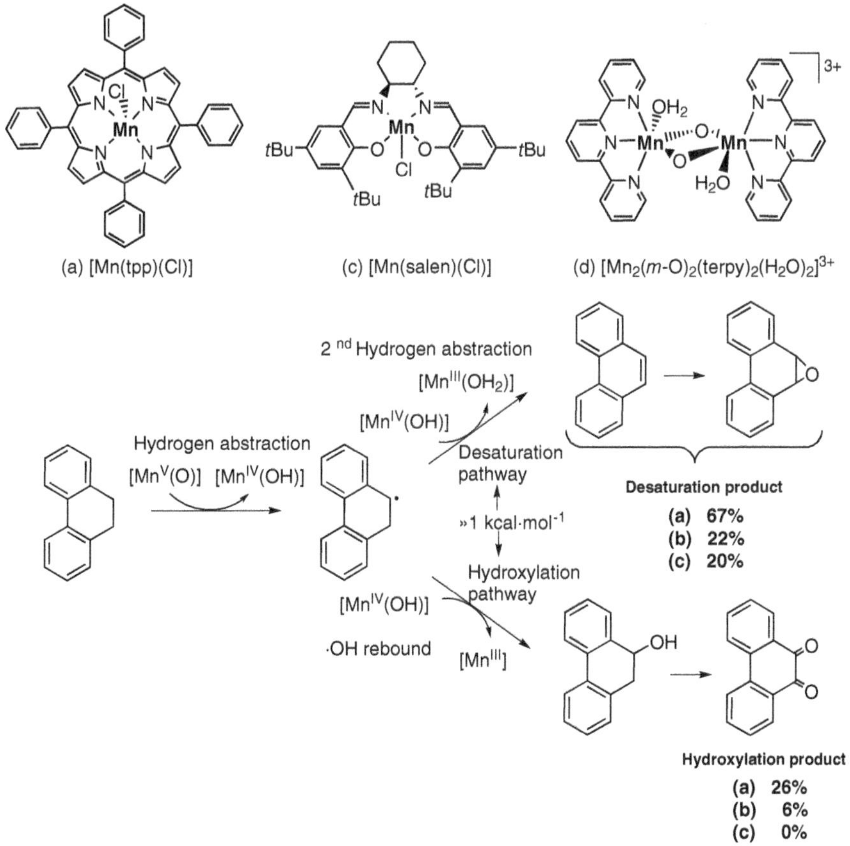

Scheme 41. Diagram of the desaturation and hydroxylation pathways and the oxidation of 9,10-dihydrophenanthrene by selected Mn complexes.

9,10-dihydrophenanthrene, three different products were observed: the product derived from the desaturation, the epoxide, and the diketone products. Crabtree and co-workers (137) demonstrated that the epoxide was formed from epoxidation of the desaturated product, and that the diketone was formed via overoxidation of the hydroxylation product. In general, the best yields were obtained when the porphyrinic complex [Mn(tpp)(Cl)] was employed as catalyst.

The DFT modeling of the reaction showed that in the case of the porphyrinic–Mn system the preferred reaction is the desaturation (Scheme 41). Initially, the substrate suffers a hydrogen-atom abstraction by the [Mn^{V}(O)(tpp)(Cl)] complex, forming a carbon-centered radical. Then the [Mn^{IV}(OH)(tpp)(Cl)] is still active to perform a second hydrogen-abstraction reaction, which leads to the formation of the desaturated product. Alternatively, the carbon-centered radical can react with the hydroxyl ligand of the manganese ion, forming the two-electron hydroxylated product, like a rebound mechanism. The DFT modeling indicates that both pathways are possible because corresponding energy barriers differ in the 1-kcal·mol^{-1} range.

Marked C–H desaturation selectivity and activity was also found by Prokop et al. (138) for the manganese(V)–oxo corrolazine complex [(TBP_8Cz)Mn^{V}(O)(X)] (TBP_8Cz = octakis(*p-tert*-butylphenyl)corrolazinato^{3-} and X = none, F^- or CN^-) when exploring their hydrogen atom transfer (HAT) reactivity toward 9,10-dihydroanthracene (DHA). Gas-phase DFT calculations indicate that the transformation consists of two-stepwise HAT reactions. First, is the formation of [(TBP_8Cz)Mn^{IV}(OH)(X)] and the organic radical and then a second HAT reaction forming [(TBP_8Cz)Mn^{III}(OH_2)(X)] and anthracene.

B. Catalytic Epoxidation of Alkenes by Mn Complexes

A remarkable progress in the epoxidation of alkenes with manganese catalysts using convenient oxidants (e.g., H_2O_2) has been accomplished over the last decade. Simple manganese salts combined with cocatalysts (139), but more interestingly manganese complexes containing mainly nitrogen donor-based ligands (140–143) (Schemes 42, 43) have been found to be useful epoxidation catalysts.

In 1996, the groups of Bein and co-workers and Kerschner and co-workers (145–148) reported that both *in situ* generated or the synthesized dimeric manganese complexes based on substituted 1,4,7-triazacyclononane [Mn_2O_3-(Me_3tacn)$_2$] were catalytically active for the epoxidation of olefins performing up to 1000 TON when using H_2O_2 as oxidant. This complex and modifications thereof have been largely studied. For instance, structural modifications of the basic 1,4,7-triazacyclononane (tacn) motive produced active Mn catalysts with similar performances to Me_3tacn in the epoxidation of alkenes. More interestingly, the catalytic activity was improved by introducing carboxylic acid as an

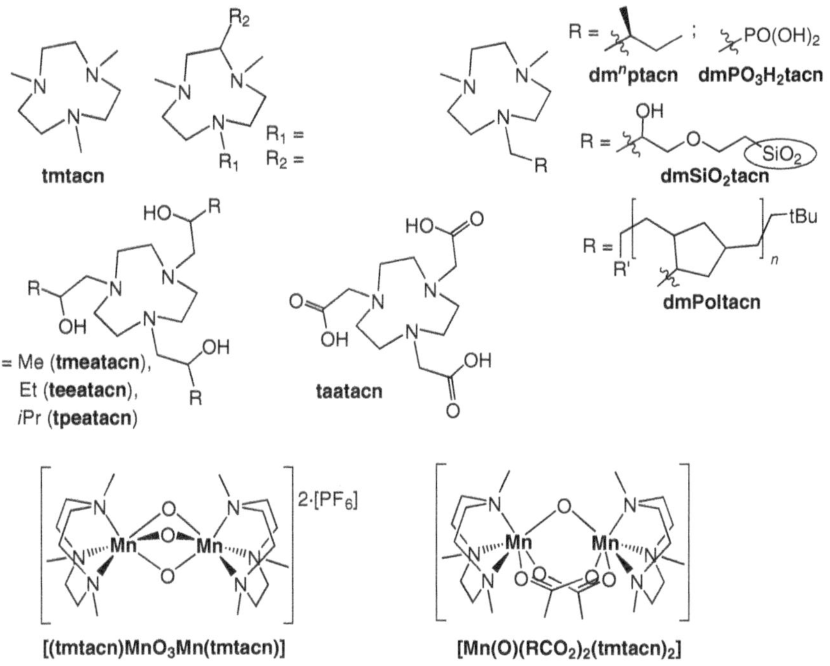

Scheme 42. Selected manganese complexes based on tacn were studied as catalysts for the epoxidation of alkenes (taatacn = 1,4,7-triazacyclononane-1,4,7-triyl-triacetic acid; tmeatacn = 1,4,7-triazacyclononane-1,4,7-triyl-tris(propan-2-ol), tpeatacn = 1,4,7-triazacyclononane-1,4,7-triyl-tris-2-methylbutan-2-ol, teeatacn = 1,4,7-triazacyclononane-1,4,7-triyl-tris(butan-2-ol).

additive (82, 149–154). Furthermore, carboxylic acid *additives* allowed for both the suppression of catalase activity and an increase in selectivity toward epoxidation, but also toward cis-dihydroxylation (see below). This change in the activity and selectivity has been attributed to the *in situ* formation of a new catalytic species identified as a carboxylate-bridged dinuclear Mn complexes of the type $[Mn^{III}{}_2O(RCO_2)_2(tmtacn)_2]^{2+}$ (Scheme 42 and Table III) (152). Later, formation of carboxylate-bridged dinuclear complexes were used and applied to immobilize $[Mn_2O_3(Me_3tacn)_2]$ on various SiO_2 and AlO_3 supports (155, 156). The immobilization through the carboxylic acids is a clear improvement since former heterogenization processes involved more laborious tacn ligand modification (157, 158). Nevertheless, in most cases the heterogeneous catalysts produced similar catalytic performances as the homogeneous counterparts, but improves catalyst recyclability.

A seminal contribution to this field was made by Stack and co-workers (160) in 2003, by describing the use of a mononuclear manganese complex

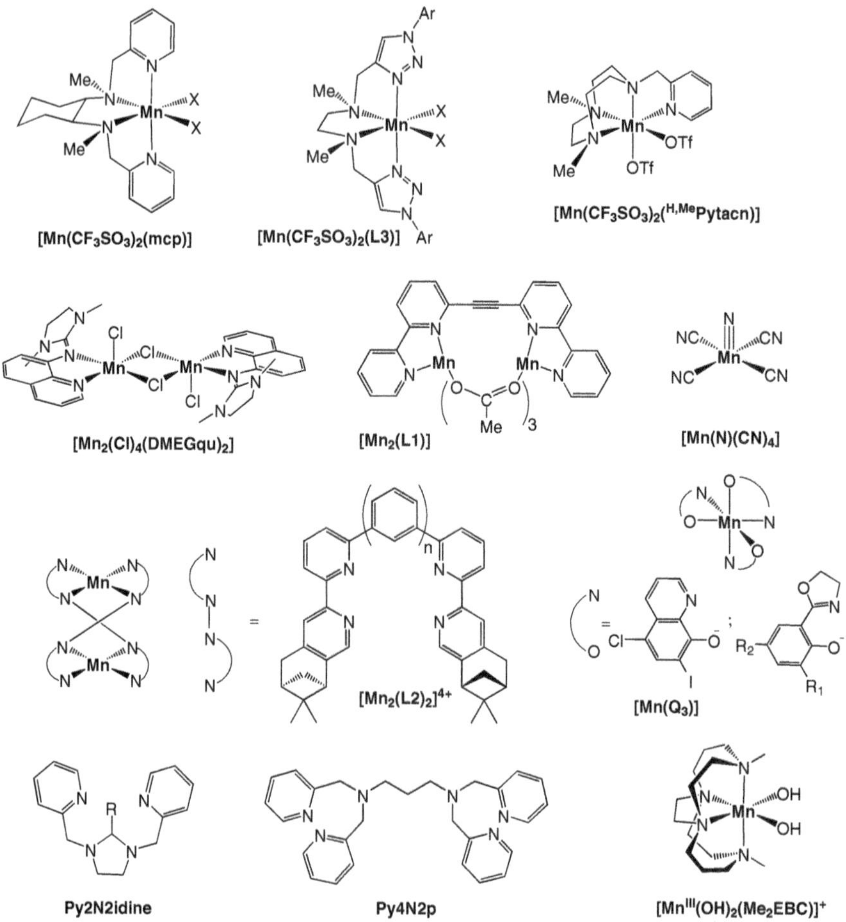

Scheme 43. Selected manganese complexes and ligands studied in the Mn catalyzed epoxidation of alkenes [DMEGqu = 1,3-dimethyl-*N*-(1-(3-methylpyridin-2-yl)vinyl)imidazolidin-2-imine; Me_2EBC = 4,11-dimethyl-1,4,8,11-tetraazabicyclo[6.6.2]hexadecane; Py2N2idine = 1,3-bis(pyridin-2-ylmethyl) imidazolidine; Py4N2p = 1,3-bis(bis(2-pyridylmethyl)amino)propane].

$[Mn(CF_3SO_3)_2((R,R)\text{-mcp})]$, as a very active epoxidation catalyst. This complex, in combination with peracetic acid efficiently catalyzes the epoxidation of olefins. The use of 0.1 mol% of catalyst, 1.2 equiv of $MeCO_3H$ and MeCN as solvent lead to the epoxidation of a wide range of alkenes in excellent yields (90–100%): cyclooctene (99%), *cis*-2-heptene (99%), 1-heptene (95%), 2-cyclohexen-1-one (97%), or *cis*-β-methylstyrene (90%). The system showed excellent chemoselectivity toward electron-rich olefins in dienes [e.g., (*R*)-carvone]. Later on, the same

TABLE III
Summary of an Olefin Epoxidation Catalyzed by Manganese Systems Based on the tacn Moiety and Using H_2O_2 as Oxidant

System	Additive	Substrate	TON[a]	Reference
tmtacn + $MnSO_4 \cdot H_2O$[b]		Styrene	1000	147
tmtacn + $Mn(OAc)_2 \cdot 4H_2O$[c]	3 equiv Oxalic acid Sodium oxalate (1:1)	1-Hexene	660	149
tmtacn + $Mn(OAc)_2 \cdot 4H_2O$[c]	8 equiv Sodium ascorbate	Methyl acrylate	3230	150
$[Mn_2(\mu\text{-}O)_3(tmtacn)_2](PF_6)_2$[c]	250 equiv GMHA[d]	Styrene	860	151
$[Mn_2(\mu\text{-}O)_3(tmtacn)_2](PF_6)_2$[c]	1 equiv Salicyclic acid	Cyclooctene	695	152
teeatacn + $MnSO_4 \cdot H_2O$[e,f]		Cyclohexene	82	146
$[Mn(CF_3SO_3)_2(^{Me,H}Pytacn)]$[c]	14 equiv $MeCO_2H$	Styrene	4500	159

[a] Number of mol of formed epoxide/mol of Mn used = TON.
[b] In acetone.
[c] In MeCN.
[d] In GMHA = Glyoxylic acid methyl ester hemiacetal.
[e] 1,4,7-Triazacyclononane-1,4,7-triyl-tris(butan-2-ol) = teeatacn.
[f] In Me_3OH.

authors tested a wide variety of *in situ* formed manganese catalysts with N-based ligands for the oxidation of 1-octene (161). On the other hand, it is known that commercially available peracetic acid solutions are highly acidic, and this factor prevented the use of complexes with more labile ligands than the mcp chelate. This problem was solved by preparation of peracetic acid solutions using an acidic resin as catalyst. The resulting peracetic acid solutions are much less acidic, permitting simple systems, such as $[Mn(CF_3SO_3)_2(bpy)_2]$ to catalyze the epoxidation of 1-octene under low-catalyst loadings (0.1 mol%), providing excellent conversion and yield.

This precedent constituted the starting point for the study of manganese complexes containing nitrogen-based ligands as epoxidation catalysts. Following in this path, the triazacyclononane-derived manganese complex $[Mn(CF_3SO_3)_2(^{Me,H}Pytacn)]$ (Scheme 43) was introduced as an epoxidation catalyst. This complex performed fast and efficient catalytic alkene epoxidations of a wide array of olefins by using low-catalyst loadings (0.1 mol%) (Scheme 43) (162). The authors initially reported the use of peracetic acid as oxidant. Under these conditions, the system showed a good selectivity for cis-aliphatic olefins over trans-aliphatic ones (cis/trans = 9.0). Afterward, the same authors explored the use of H_2O_2 as oxidant, because it is more attractive from the point of view of atom economy and because it is relatively easy and safe to handle. The disproportionation reaction of H_2O_2 could be avoided by the use of acetic acid (14 equiv), which permitted epoxidation to proceed efficiently with nearly stoichiometric amounts (1.2 equiv) of the oxidant. Excellent conversions and yields (80–100%) were obtained, comparable to those obtained with peracetic acid, while maintaining a low catalyst loading (0.1%). Interestingly, the use of the H_2O_2/AcOH combination expanded the substrate scope to acid-sensitive alkenes (e.g., stilbenes in Scheme 44) (159). Under these H_2O_2/$MeCO_2H$ conditions, the system maintained the preferential chemoselectivity for cis- over trans- aliphatic olefins (cis/trans = 5.0), but this selectivity was inverted for aromatic stilbene olefins (cis/trans = 0.17). That observation constitutes a rare example of reversal of cis/trans selectivity depending on the aliphatic or aromatic nature of the olefin.

Besides manganese complexes with aminopyridine ligands, some Mn based systems that employ much simpler ligands have been described as efficient epoxidation catalysts (Table IV and Scheme 43). For example, Lau and coworkers (163) used a Mn(V) nitride complex $[Mn(N)(CN)_4](PPh_4)_2$ (1 mol%) to catalyze the oxidation of several alkenes using H_2O_2. The use of acetic acid accelerated the reaction while retaining the efficiency. The $[Mn(N)(CN)_4](PPh_4)_2$ complex was proposed to act as a Lewis acid facilitating oxygen-atom transfer from H_2O_2 to the alkene. This work was based on the prior work of Goldberg's group on the corrolazine complex $[(TBP_8Cz)Mn^V(NMes)]$, where it was found that the Mn(V) imido complex is an active catalyst for the epoxidation of olefins (164).

Particularly remarkable is the extremely efficient epoxidation activity (up to 300,000 TON) and with a broad substrate scope of a manganese-based catalyst based on the *in situ* combination of a Mn(II) salt, pyridine-2-carboxylic acid and substoichiometric amounts of 1,2-diketones were added as cocatalysts (165). This system follows previous observations by the same group that several Mn complexes with N-based ligands, upon reacting with the oxidant, undergo initial

TABLE IV
Comparison Among Selected Manganese Catalysts in the Epoxidation of 1-Octene

Complex	Catalyst Loading (%)	Oxidant (equiv)	Conversion (%)	Yield (%)	TON	Reference
$[Mn_2(\mu\text{-}O)_3(tmtacn)_2](PF_6)_2$	1	H_2O_2 (100)	99[a]	99[a]	99[a]	127
$[Mn(CF_3SO_3)_2(R,R)\text{-}mcp]$	0.1	$MeCO_3H$ (1.2)	95[b]	89[b]	890[b]	160
$[Mn(CF_3SO_3)_2(^{Me,H}Pytacn)]$	0.15	$MeCO_3H$ (1.4)	100	96	640	162
$[Mn(CF_3SO_3)_2(^{Me,H}Pytacn)]$	0.1	H_2O_2 (1.2)	90	90	900	159
$[Mn(Q_3)]$	2	H_2O_2 (1.5)	95	93	47	140
[Mn(L1)]	3.3	$MeCO_3H$ (2.0)	91	91	28	141
$[Mn_2(DMEGqu)_2]^{2+}$	0.1	$MeCO_3H$ (2.0)	85	62	620	142
$[Mn_2(L2)_2]^{4+}$	1	$MeCO_3H$ (2.0)	100[c]	99[c]	99[c]	143
$[Mn(CF_3SO_3)_2(L3)]$	0.5	$MeCO_3H$ (2.0)		93[d]	465[d]	144
$[Mn(N)(CN)_4]$	1	H_2O_2 (2.5)	98	74	74	163

[a] Oxidation of styrene.
[b] Oxidation of 1-heptene.
[c] Oxidation of 1-pentene.
[d] Oxidation of 1-decene.

oxidative degradation to form picolinic acid, which is the actual ligand in the true Mn epoxidizing species.

C. Catalytic Asymmetric Epoxidation of Alkenes With Manganese Complexes

Catalytic asymmetric epoxidation of alkenes was pioneered by Jacobsen and co-workers (166) and Katsuki (167) in the 1990s by employing Mn–salen complexes. Although excellent enantioselectivities were obtained, other aspects (e.g., catalyst loadings, substrate scope, and oxidants) employed are still susceptible to important improvement. In particular, more robust catalysts that could rely on the use of H_2O_2 continue to be a very attractive target. A selection of ligands developed since then for asymmetric epoxidation of olefins with manganese complexes are depicted in Scheme 44.

The initial report by Stack and co-workers (160) in the catalytic activity of [Mn $(CF_3SO_3)_2$(*R*,*R*)-mcp] did not describe asymmetric epoxidation reactions but the chiral nature of this catalyst prompted further development this aspect. With this interest in mind, Gomez et al. (168) designed two novel chiral ligands based on the (*R*,*R*)-mcp system by replacing the pyridine (py) by a [4,5]-pinene pyridine ring

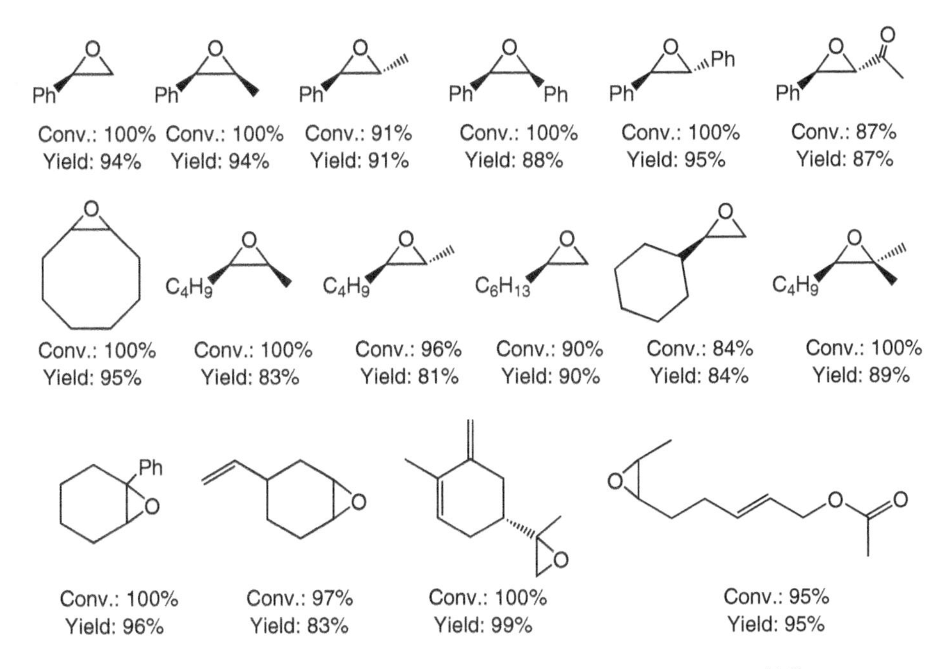

Scheme 44. Selected substrate scope of alkene epoxidation mediated by $[Mn(CF_3SO_3)_2(^{Me,H}Pytacn)]$ using 14 equiv of $MeCO_2H$ with respect to the substrate and H_2O_2 as oxidant.

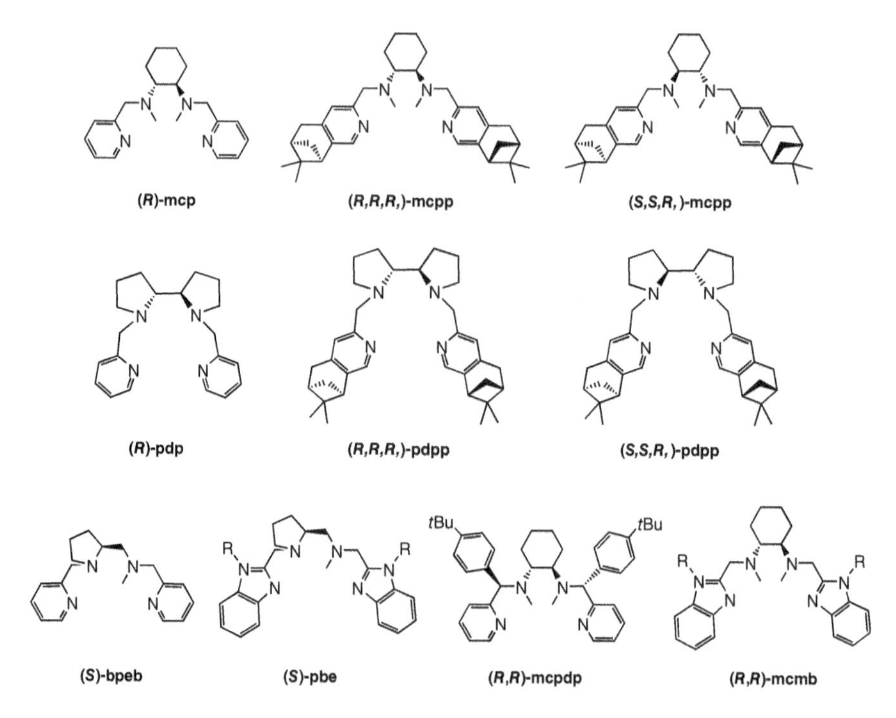

Scheme 45. Chiral ligands developed for epoxidation of alkenes with manganese complexes.

(Scheme 45). The resulting diastereoisomerically related ligands (*S*,*S*,*R*)-mcpp and (*R*,*R*,*R*)-mcpp formed the two corresponding manganese complexes Λ-[Mn $(CF_3SO_3)_2$(*S*,*S*,*R*)-mcpp)] and Δ-[Mn$(CF_3SO_3)_2$((*R*,*R*,*R*)-mcpp)], where Λ and Δ refer to the chirality at the metal. Both complexes were studied in the asymmetric epoxidation of styrene, and in this reaction Λ-[Mn$(CF_3SO_3)_2$((*S*,*S*,*R*)-mcpp)] provided the highest stereoselectivity (46% ee). Complex Λ-[Mn$(CF_3SO_3)_2$((*S*, *S*,*R*)-mcpp)] (0.5 mol%) was subsequently used in the epoxidation of different alkenes using peracetic acid (2 equiv) as oxidant. Despite achievement of good conversions (80–100%) and yields (60–100%), the stereoselectivities were moderate at best (ee 40–50%). Later on, the same authors reported other two novel chiral N4-tetradentate ligands combining the (*R*)-[4,5]-pinenepyridine moiety with the (*R*,*R*)- and (*S*,*S*)-bipyrroline backbone, and corresponding complexes Λ-[Mn $(CF_3SO_3)_2$((*S*,*S*,*R*)-pdpp)] and Δ-[Mn$(CF_3SO_3)_2$((*R*,*R*,*R*)-pdpp)] improved ees up to (50–70%) (169).

Very important advances in the field were contributed by Xia and co-workers (170–172) in a series of papers where the tetradentate ligand structure of mcp was retained, but different modifications where introduced, mcpdp = *N*,

N′-bis(*R*)-(4-(*tert*-butyl)phenyl)(pyridin-2-yl)methyl)-*N*,*N′*-dimethylcyclohexane-1,2-diamine; pbe = *N*,*N′*-bis-(1-R-benzymidazol-2-yl)methyl-2-aminomethylpyrrolidine leading to novel generations of catalysts with improved performance. A first report described catalysts created by introducing aryl groups at the pseudobenzylic methylenic groups of the mcp ligand, which in turn represented an extra element of chirality, closer to the chiral center. With catalyst [Mn(CF_3SO_3)$_2$((*R*,*R*)-mcpdp)] Sun and co-workers (170) obtained good-to-excellent enantioselectivities in the epoxidation of α,β-enones (70–90%), but moderate stereoselectivities (30–50%) were accomplished for other substrates. More recently, Sun and co-workers (171) introduced the chiral C_1-symmetric (*S*)-bpeb and (*S*)-bpe ligands, and the corresponding [Mn(CF_3SO_3)(L)], L = (*S*)-bpeb and (*S*)-bpe complexes. Manganese complex [Mn(CF_3SO_3)((*S*)-bpeb)] produced moderate enantioselective epoxidation of *trans*-chalcone (68% ee), but [Mn(CF_3SO_3)((*S*)-bpe)], where the pyridine rings have been replaced by benzylimidazoles, achieved excellent stereoinduction (up to 92% ee). Following the same strategy, the same authors obtained excellent enantioselectivities in a wide variety of substituted chalcones (75–96% ee) and derivatives with a novel complex of the series, with a new ligand (*R*,*R*)-mcmb, containing N-alkyl substituted benzylimidazol arms and the cyclohexane-1,2-diamine backbone. Unfortunately, this catalyst was ineffective in the case of styrene (39% ee) or aliphatic olefins (e.g., 1-octene) (171,172).

Parallel efforts by Talsi and co-workers (173) involved the preparation and study of the chiral complex [Mn(CF_3SO_3)$_2$((*S*,*S*)-pdp)] (173) based on the chiral bipyrrolidine diamine (Scheme 45), which catalyzed alkene epoxidation using the H_2O_2/AcOH conditions early described by Garcia-Bosch et al. (162). Remarkable stereoselectivites were obtained for electron-deficient substrates, such as *trans*-chalcone (78% ee) or 2,2-dimethyl-2*H*-1-benzopyran-6-carbonitrile (76% ee). Modest selectivity was achieved for substrates, such as styrene (39% ee), 1,2-dihydronaphthalene (27% ee), and *trans*-stilbene (8% ee). Importantly, the same authors revealed that the use of different carboxylic acids can affect enantioselectivity. The enantioselectivities increased with the steric demand of the carboxylic acids (e.g., the following order was generally found: formic acid < acetic acid < *n*-butyric acid = *n*-valeric acid = *β*-caproic acid < *iso*-butyric < pivalic acid < 2-ethylhexanoic acid). Significant improvement of the enantioselection was found in the case of the 2,2-dimethyl-2*H*-1-benzopyran-6-carbonitrile from a value of 73–93% ee (174,175).

Regarding the reaction mechanism, three possible mechanisms have been proposed to rule the manganese-based epoxidation catalysis: (1) direct oxo transfer from a high-valent manganese species, (2) a pathway where peroxyl radicals are involved (3) a high-valent manganese species acting as a Lewis acid, which binds and activates the peroxide, and delivers one oxygen atom to the olefin in a concerted mechanism (129, 174, 175).

D. Catalytic cis-Dihydroxylation. of Alkenes by Mn Complexes

Early work by De Voss et. al. (176) involved support of the manganese tmtacn-based catalyst in silica, showing that these materials have the capacity of oxidizing alkenes to obtain epoxide/cis-diol mixtures with TONs up to 500. These results combine with the observation that certain additives promoted the capability of Mn complexes for catalyzing cis-dihydroxylation of alkenes (151, 176).

A major breakthrough was reported by Feringa and co-workers (152) showing that $[Mn_2(\mu\text{-}O)_3(tmtacn)_2](PF_6)_2$ (Scheme 42) is a highly active cis-dihydroxylation catalyst. In a first approach, the authors used H_2O_2 as oxidant, and different carboxylic acids as cocatalysts. The use of carboxylic acids suppressed the inherent catalase activity of the complex turning the consumption of H_2O_2 toward alkene oxidation. More importantly the selectivity of the catalytic system to cis-hydroxylation and epoxidation can be tuned by the judicious choice of the carboxylic acid. In the case of the cyclooctene by using CCl_3CO_2H as cocatalyst, a ratio of 1.8 cis-diol/epoxide was observed, and with 2,6-dichlorobenzoic acid the ratio was 7. Interestingly, by using salicylic acid as an additive this ratio was reversed, and epoxide was the major product (diol/epoxide ratio = 0.09). This methodology was applied to a wide variety of alkenes (Table V). Despite good yields and selectivity toward cis-dihydroxylation was observed for electron-rich aliphatic substrates

TABLE V

Selected Results from the Reported cis-Diol and Olefin Epoxidation Catalyzed by $[Mn_2(\mu\text{-}O)_3(tmtacn)_2](PF_6)_2$ and Using H_2O_2 as Oxidant

Olefin —[$[Mn_2(\mu\text{-}O)_3(tmtacn)_2](PF_6)_2$ (0.1 mol%); carboxylic acid / H_2O_2]→ cis-Diol + Epoxide

Substrate	Cocatalyst	cis-Diol (TON)	Epoxide (TON)	cis-Diol/epoxide
Cyclooctene	Salicylic acid	60	695	0.1
Cyclooctene	CCl_3CO_2H	440	225	2
Cyclooctene	2,6-Dichlorobenzoic acid	525	75	7
Cis-2-heptene	2,6-Dichlorobenzoic acid	450	130	3.5
Trans-2-heptene	CCl_3CO_2H	240	110	2.2
Cyclohexene	CCl_3CO_2H	110	400	0.28
Cyclopentene	2,6-Dichlorobenzoic acid	305	360	0.85
1-Octene	2,6-Dichlorobenzoic acid	125	295	0.42
Styrene	CCl_3CO_2H	35	615	0.06
Dimethylmaleate	Salicylic acid	0	20	
Dimethylfumarate	Salicylic acid	105	0	

(e.g., cyclooctene, *cis*-2-heptene, or *trans*-2-heptene), the system failed to produce selective cis-dihydroxylation of aromatic substrates (e.g., styrene). In addition, electron-poor substrates (e.g., dimethylmaleate and dimethylfumarate) were also unsuitable for the system.

The first hint on the nature of the active intermediates in the cis-dihydroxylation is related to the intrinsic nature of the reaction. The retention of the stereochemistry forming exclusively cis-diols can be regarded as evidence for the involvement of high-valent manganese oxo species as responsible for the cis-dihydroxylation. Insights into the mechanism were obtained by extensive mechanistic and spectroscopic studies; The complex $[Mn_2(\mu\text{-}O)_3(tmtacn)_2](PF_6)_2$ was shown to be reduced during catalysis, presumably by H_2O_2, and reacted with carboxylic acids, in the case of CCl_3CO_2H, to form a new dimeric species $[Mn_2(OH)_2(\mu\text{-}CCl_3CO_2)_2(tmtacn)_2]^{2+}$. This complex acted as a catalyst resting state (Scheme 46) (177). Subsequently, this dimeric Mn(III) species is proposed to be responsible for the H_2O_2 activation reaction by forming a

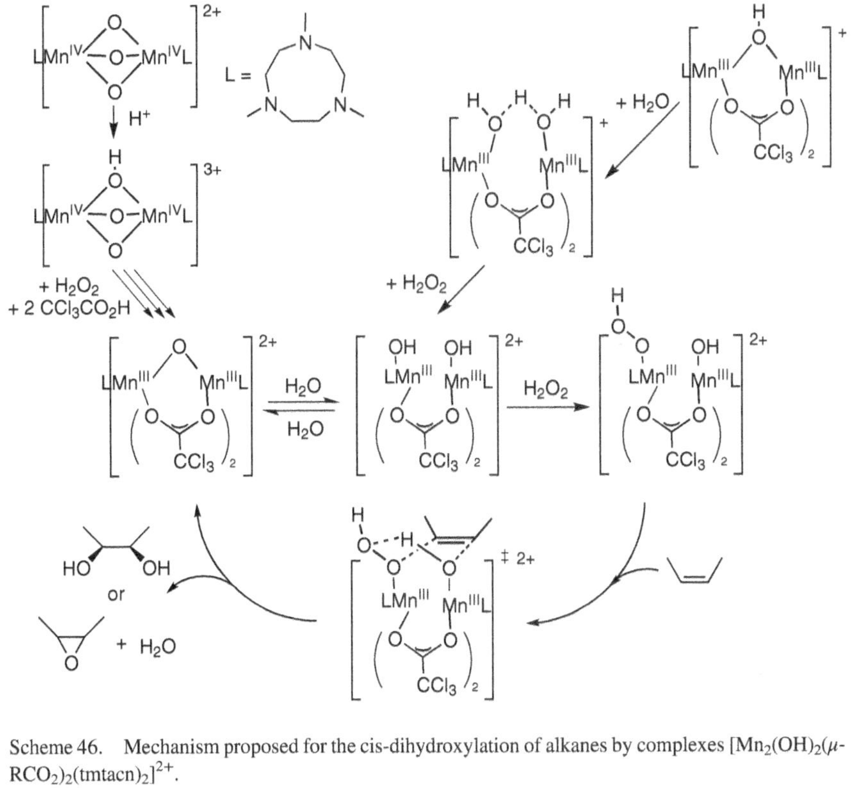

Scheme 46. Mechanism proposed for the cis-dihydroxylation of alkanes by complexes $[Mn_2(OH)_2(\mu\text{-}RCO_2)_2(tmtacn)_2]^{2+}$.

Mn(III)–hydroperoxo species that oxidizes the alkene moiety. This mechanism could account for the role of the carboxylic acid to turn the selectivity toward cis-hydroxylation or epoxidation depending on their influence in the evolution of the transition state.

Recently, Browne and co-workers (178) found that manganese-based catalysts employing N-rich ligands [e.g., Py2N2idine = 1,3-bis(2-pyridylmethyl) imidazlidine] and derivatives in Scheme 43), and whose structures incorporate pyridyl groups decompose *in situ* under oxidative conditions to produce pyridine-2-carboxylic acid. The decomposition occurs prior to the catalyzed oxidation of organic substrates. This observation lead the authors to develop a convenient and simple method for cis-dihydroxylate electron-deficient olefins (179). Hydrogen peroxide (2.0 equiv) was combined with $Mn(ClO_4)_2$ (0.3 mol%), pyridine-2-carboxylic acid (1.8 mol%) and NaOAc (3.0 mol%) in acetone. The substrate scope of this methodology was restricted to electron-poor alkenes, because electron-rich olefins were mostly oxidized to the corresponding epoxide (Scheme 47).

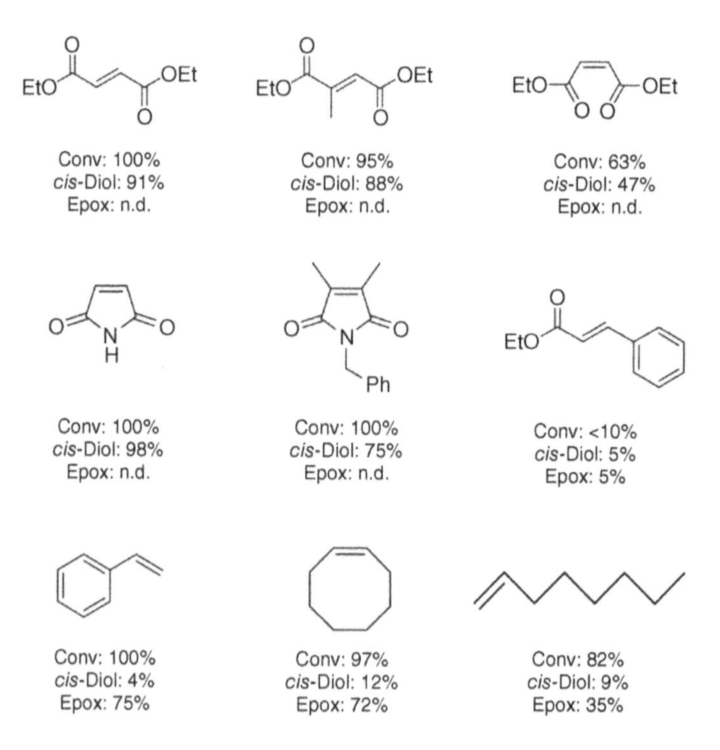

Scheme 47. Selected examples of the cis-dihydroxylation alkane scope of $Mn^{II}(ClO_4)_2$/pyridine-2-carboxylic acid/NaOAc/H_2O_2 in acetone, where n.d. = not determined.

E. Asymmetric cis-Dihydroxylation

A stereoselective version of the cis-hydroxylation system $[Mn_2(\mu\text{-}O)_3(tmtacn)_2](PF_6)_2$ was reported by Feringa et al. (180). Chirality in the catalytic system was introduced by the use of chiral carboxylic acids (R*COOH) as cocatalysts (e.g., Boc-phenylalanine or Boc-phenylglycine, where Boc = *tert*-butyloxycarbonyl, that *in situ* generate the chiral $[Mn_2(\mu\text{-}O)(\mu\text{-}R^*CO_2)_2(tmtacn)_2]^{2+}$ complex. Even when the oxidation of 2,2-dimethychromene was achieved with remarkable enantioselectivity (54% ee, Scheme 48), the methodology was not used in other substrate oxidations. Nevertheless, the high TONs achieved, the ready tunable of the system and the use of H_2O_2 as oxidant suggest that useful methods based on manganese complexes could be achieved.

A very interesting Mn based cis-dihydroxylating system was recently described by Che and co-workers (181) employing a manganese complex with a tetradentate ligand and oxone as oxidant. The complex $[Mn(Cl)_2((S,S)\text{-bqcn})]$, where bqcn = *N,N′*-dimethyl-*N,N′*-di(quinolin-8-yl)cyclohexane-1,2-*cis*-diamine, adopts a C_2-symmetric structure with a cis-α topology, analogous to that described for $[Mn(CF_3SO_3)_2(mcp)]$. By employing 2–5 mol% catalyst loading and 2 equiv of oxone, a series of trans-substituted electron-deficient olefins were oxidized to the corresponding syn-diols in 65–95% yields and 87–96% ee (Scheme 49). Cis-substituted analogues and styrene were oxidized with reduced yields and enantioselectivities. The system could be applied to gram-scale asymmetric cis-dihydroxylations, proving the synthetic value of the catalyst.

From a mechanistic perspective, evidence of the active species was provided by ESI–MS analysis of the reaction of the catalyst with oxone, which showed the formation of a cluster ion interpreted as a $cis\text{-}[Mn^V(O)_2(S,S)\text{-bqcn})]^+$ reacting intermediate.

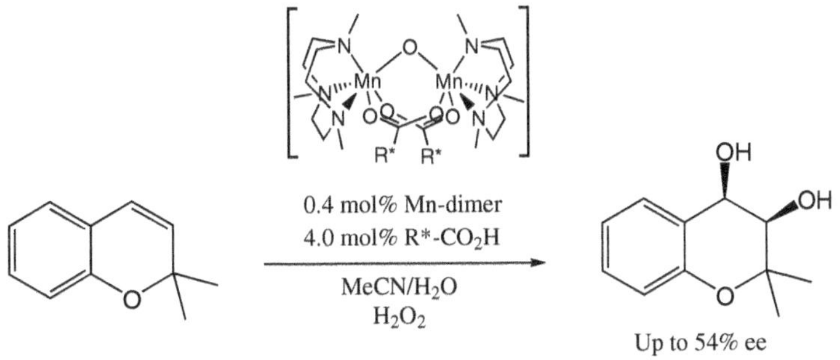

0.4 mol% Mn-dimer
4.0 mol% $R^*\text{-}CO_2H$
$MeCN/H_2O$
H_2O_2
Up to 54% ee

Scheme 48. Stereoselective cis-hydroxylation of 2,2-dimethychromene using $[Mn_2(\mu\text{-}O)_3(tmtacn)_2](PF_6)_2$ as catalyst.

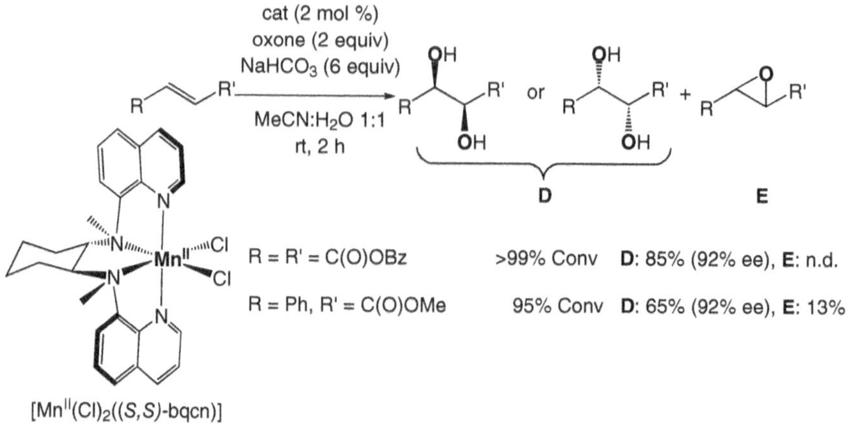

cat (2 mol %)
oxone (2 equiv)
$NaHCO_3$ (6 equiv)
$MeCN:H_2O$ 1:1
rt, 2 h

$[Mn^{II}(Cl)_2((S,S)\text{-bqcn})]$

R = R' = C(O)OBz — >99% Conv — **D**: 85% (92% ee), **E**: n.d.

R = Ph, R' = C(O)OMe — 95% Conv — **D**: 65% (92% ee), **E**: 13%

Scheme 49. Asymmetric cis-dihydroxylation with catalyst $[Mn(Cl)_2((S,S)\text{-bqcn})]$

III. AMINOPYRIDINE IRON AND MANGANESE COMPLEXES AS MOLECULAR CATALYSTS FOR WATER OXIDATION

During the past 30 years a significant progress has been achieved in the understanding and development of catalysts for water oxidation based on first-row transition metals. However, the development of well-defined systems remains challenging, as many examples of early considered "homogeneous catalysts" are taken into reconsideration because of the possible implication of nanoparticules operating in heterogeneous regimes. Certainly, some remarkable examples of homogeneous catalysis with molecular manganese and iron complexes have been recently reported. This key point is necessary in order to understand, control, and develop new and more efficient water oxidation catalysts with the ultimate aim of applying them to a technologically elaborated artificial photosynthesis system. On the other hand, these studies are envisioned to help us to understand how Nature uses the water oxidation reaction to extract electrons from water (182–191).

A. Aminopyridine Manganese Complexes as Molecular Catalysts for Water Oxidation

Nature, through the so-called PSII in cyanobacteria, algae, and higher green plants, has been able to solve the water oxidation reaction in an elegant manner. Water oxidation takes place at a tetramanganese calcium cluster, the oxygen-evolving center (OEC), localized in a large transmembrane enzyme of the thylakoid membrane. The OEC structure is constituted by three manganese atoms and a calcium atom bridged by four oxygen atoms in a cubane-distorted cluster

($CaMn_3O_4$). An external Mn atom is bound through an oxo bridge to this cluster, completing the overall Mn_4O_5Ca structure. The spectroscopic and chemical characterization of this polymetallic system is extremely complicated, and therefore the use of coordination complexes is a useful tool to unravel the basic principles that operate in the $CaMn_3O_4$ cluster.

Synthetic construction of such a complex chemical architecture was not accomplished until 2011, when a first accurate model of $[Mn_3CaO_4]^{6+}$ cluster was successfully synthetized and characterized. Agapie and co-workers (192) used a 1,3,5-triarylbenzene subunit that contained two pyridines and one alcohol per arm to form a rigid trinucleating ligand framework (Scheme 50). The study of the different oxidation states shows rich cluster redox chemistry. The utility of this synthetic model has also been demonstrated by the synthesis of the homologous clusters obtained by replacement of the calcium, capping metal, by several Lewis acid ions, Na^+, Sr^{2+}, Zr^{2+}, and Y^{3+}. Electrochemical studies of this series of clusters showed that the reduction potentials are highly and directly dependent on the Lewis acidity of the redox-inactive metal. This correlation offers a rationale for the role of the Ca^{2+} ion in the OEC, which can serve as a modulator of the redox potential of the $[Mn_3CaO_4]^{+6}$ cluster (193, 194).

Model systems based on di- and tri-nuclear complexes with pyridine ligands are very appealing to simplify structural motifs and have been explored as potential catalysts for the oxidation of water. The first example was reported in 1986, when Kaneko and co-workers (195) observed water oxidation by the

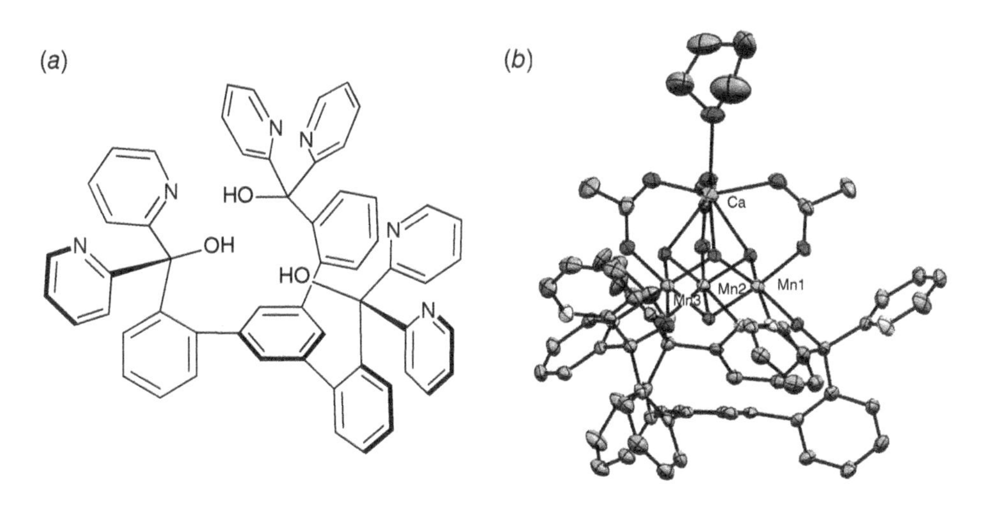

Scheme 50. (*a*) Ligand employed by Agapie and co-workers (192) and (*b*) the synthetized complex, a biomimetic $[Mn_3CaO_4]^{6+}$ cluster.

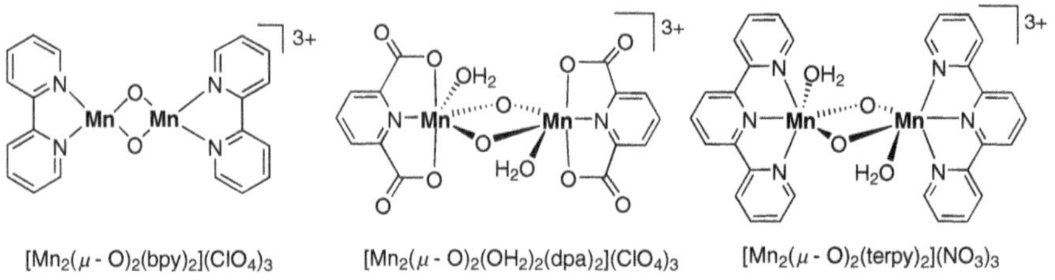

$[Mn_2(\mu\text{-}O)_2(bpy)_2](ClO_4)_3$ $[Mn_2(\mu\text{-}O)_2(OH_2)_2(dpa)_2](ClO_4)_3$ $[Mn_2(\mu\text{-}O)_2(terpy)_2](NO_3)_3$

Scheme 51. Selected examples of pyridine-based manganese complexes studied in oxygen evolution.

reaction of $[Mn_2(\mu\text{-}O)_2(bpy)_2]$ with CAN [cerium(IV) ammonium nitrate in Scheme 51]. Interestingly, oxygen evolution was only observed at the surface of the unsolved complex, strongly suggesting that water oxidation was a heterogeneous reaction. However, the process was not catalytic.

The replacement of CAN by alternative milder oxidants [e.g., OCl^- or HSO_5^- (oxone)] resulted in an improvement of the stability of the complexes, which then is reflected in the catalytic O_2 evolution activity. Indeed, the use of CAN only produced small amounts of O_2, but in any case is catalytic (196). This very low reactivity was related to the severe oxidizing conditions and the low pH imposed by CAN, which was thought to cause the instability of the complexes (197). In this regard, Crabtree and co-workers (198) reported that the $[Mn_2^{III,IV}(\mu\text{-}O)_2(terpy)_2](NO_3)_3$ complex (Scheme 51) was catalytically active for the formation of O_2 by using OCl^- or oxone ($[HSO_5^-]$) as sacrificial oxidants (4 TON and >50 TON, respectively).

Mechanistic studies performed by the same authors showed a first-order dependence of reaction rates in the concentration of $[Mn_2^{III,IV}(\mu\text{-}O)_2(terpy)_2]^{3+}$ and zero order in oxone. The resting state was determined by ultraviolet–visible (UV–vis) and electron paramagnetic resonance (EPR) spectroscopy to be the $[Mn_2^{III,IV}(\mu\text{-}O)_2(terpy)_2]^{3+}$ complex. The thermodynamically favorable coordination of Oxone leads to the formation of two possible isomers, produced by the two coordination vacancies available at the $[Mn_2^{III,IV}(\mu\text{-}O)_2(terpy)_2]^{3+}$ complex, one located at the Mn (IV) and the other at the Mn(III) metal center. As expected, the oxidant ($[HSO_5]^-$) is preferably coordinated to the more electrophilic metal center, the Mn(IV), but the high energetic barrier for the O–O bond rupture and the concomitant two-electron oxidation of Mn(IV) to Mn(VI) inhibit the reaction. This situation produces an accumulation of the catalytically inactive $[(terpy)Mn^{III}(\mu\text{-}O)_2Mn^{IV}\text{–}(HSO_5)(terpy)]^{3+}$, but in equilibrium with the less stable and catalytically competent $[(terpy)(HSO_5)\text{-}Mn^{III}(\mu\text{-}O)_2Mn^{IV}(terpy)]^{3+}$ intermediate. The heterolytic rupture of the O–O bond in $[(terpy)(HSO_5)\text{–}Mn^{III}(\mu\text{-}O)_2Mn^{IV}(terpy)]^{3+}$ to form $[(terpy)Mn^{V}(=O)(\mu\text{-}O)_2Mn^{IV}(terpy)]^{3+}$ (or best described as $Mn_2^{IV,IV}$ oxyl radical) was determined to be the rate-determining step of the reaction under an oxone excess. Finally, the highly electrophilic $[(terpy)Mn^{V}(=O)(\mu\text{-}O)_2Mn^{IV}(terpy)]^{3+}$ reacts with H_2O or another molecule of oxone, liberating oxygen (Scheme 52) (199, 200).

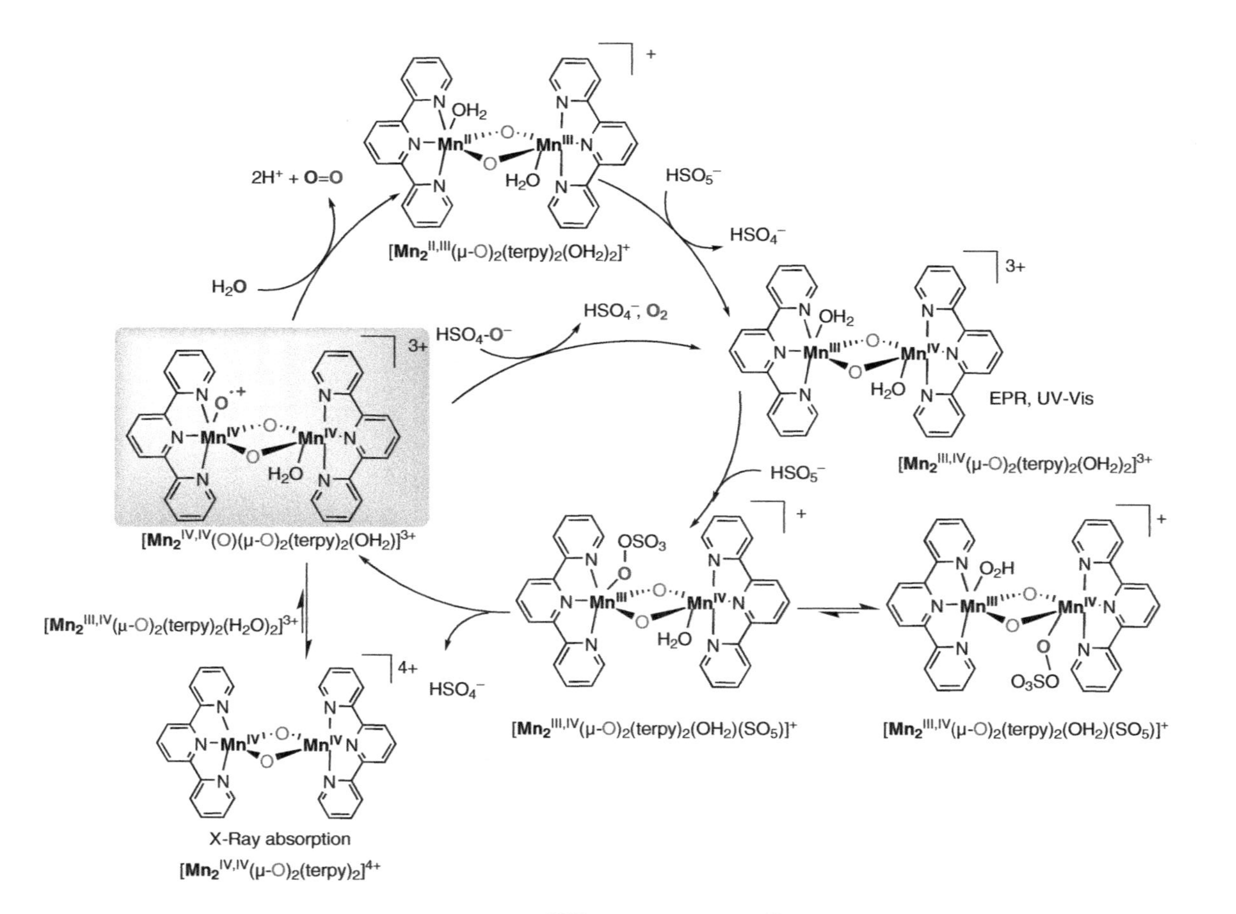

Scheme 52. Mechanism proposed for the oxidation of water with $[Mn_2^{III,IV}(\mu\text{-}O)_2(terpy)_2(H_2O)_2]^{3+}$ when using oxone as oxidant. The proposed intermediate responsible for the oxygen release has been highlighted.

Competing with this pathway, $[(terpy)Mn^{V}(=O)(\mu\text{-}O)_2Mn^{IV}(terpy)]^{3+}$ can react with the resting state $[Mn_2^{III,IV}(\mu\text{-}O)_2(terpy)_2]^{3+}$ to form the catalytically inactive $[Mn_2^{IV,IV}(\mu\text{-}O)_2(terpy)_2]^{4+}$ species. Independently synthesized $[Mn_2^{IV,IV}(\mu\text{-}O)_2(terpy)_2]^{4+}$ was found catalytically inert. Furthermore, speciation studies under catalytic conditions, performed by X-ray absorption spectroscopy, found an inverse correlation between the catalytic activity and the concentration of $[Mn_2^{IV,IV}(\mu\text{-}O)_2(terpy)_2]^{4+}$ in solution. Therefore, the high concentrations of these species measured during the oxygen evolution reaction was rationalized by the faster comproportionation between the active species $[(terpy)Mn^{V}(=O)(\mu\text{-}O)_2Mn^{IV}(terpy)]^{3+}$ and the intermediate $[Mn_2^{III,IV}(\mu\text{-}O)_2(terpy)_2]^{3+}$ than the water oxidation reaction that results in oxygen evolution. Hence, the real catalytic activity of $[(terpy)Mn^{V}(=O)(\mu\text{-}O)_2Mn^{IV}(terpy)]^{3+}$ must be higher than that inferred by considering that all the manganese species are active. The unexpected lack of catalytic activity toward oxygen evolution of $[Mn_2^{IV,IV}(\mu\text{-}O)_2(terpy)_2]^{4+}$ in the presence of Oxone arise from the impossibility of two-electron oxidation of Mn(IV) to Mn(VI) by Oxone. Besides, mono-electronic oxidation of $[Mn_2^{IV,IV}(\mu\text{-}O)_2(terpy)_2]^{4+}$ to $[Mn_2^{V,IV}(\mu\text{-}O)_2(terpy)_2]^{5+}$ was discarded based on the two-electron oxidizing behavior of Oxone (Scheme 52) (200).

Lundberg et al. (201) examined the O–O bond formation capacity of $[Mn^{V}(=O)(\mu\text{-}O)_2\text{–}Mn^{IV}]^{3+}$ by theoretical studies at the DFT level, and concluded that the attack of a water molecule to the highly electrophilic oxo ligand is viable under catalytic conditions ($23\ kcal\ mol^{-1}$). The studies also manifested radical character of the Mn(V)–oxo moiety better described as Mn(IV)–oxyl that have a large positive impact on the formation of the O–O bond. Later on, *ab inito* molecular dynamic simulations suggest that the thermodynamic accessibility of the hydroperoxo species is considered a critical and rate-limiting intermediate. Moreover, the rupture of one μ-oxo bridge to generate a $[Mn^{IV}O(\mu\text{-}O)Mn^{IV}(O)]^{4+}$ dimer, may be an alternative pathway for the oxygen production (Scheme 53) (202).

Labeling experiments showed that incorporation of ^{18}O from $H_2{}^{18}O$ in O_2 is dependent on the oxone concentration; the lower the concentration of oxone the higher the amounts of ^{18}O-labeled O_2 observed. Therefore, the attack of the water molecule to $[Mn^{V}(=O)(\mu\text{-}O)_2\text{-}Mn^{IV}]^{3+}$ to give $[Mn^{III}(OOH)(\mu\text{-}O)_2\text{–}Mn^{IV}]^{2+}$ is favored at low oxone concentration, concluding that water oxidation take place to a certain extend (203, 204). Additional evidences for water oxidation were provided by Yagi and co-workers (205, 206). They studied the catalytic activity of $[Mn_2^{III,IV}(\mu\text{-}O)_2(terpy)_2]^{3+}$ as an intercalation adsorbed complex in different layered compounds (e.g., kaolin or mica). This strategy was implemented to allow the monoelectronic oxidation of CAN to form the Mn(V) avoiding the ligand decoordination due to the low pH imposed by aqueous solutions of CAN. The reaction of the $[Mn_2^{III,IV}(\mu\text{-}O)_2(terpy)_2]^{3+}$/clay material with CAN generates O_2 with a maximum TON of 15. K-edge extended X-ray absorption fine structure

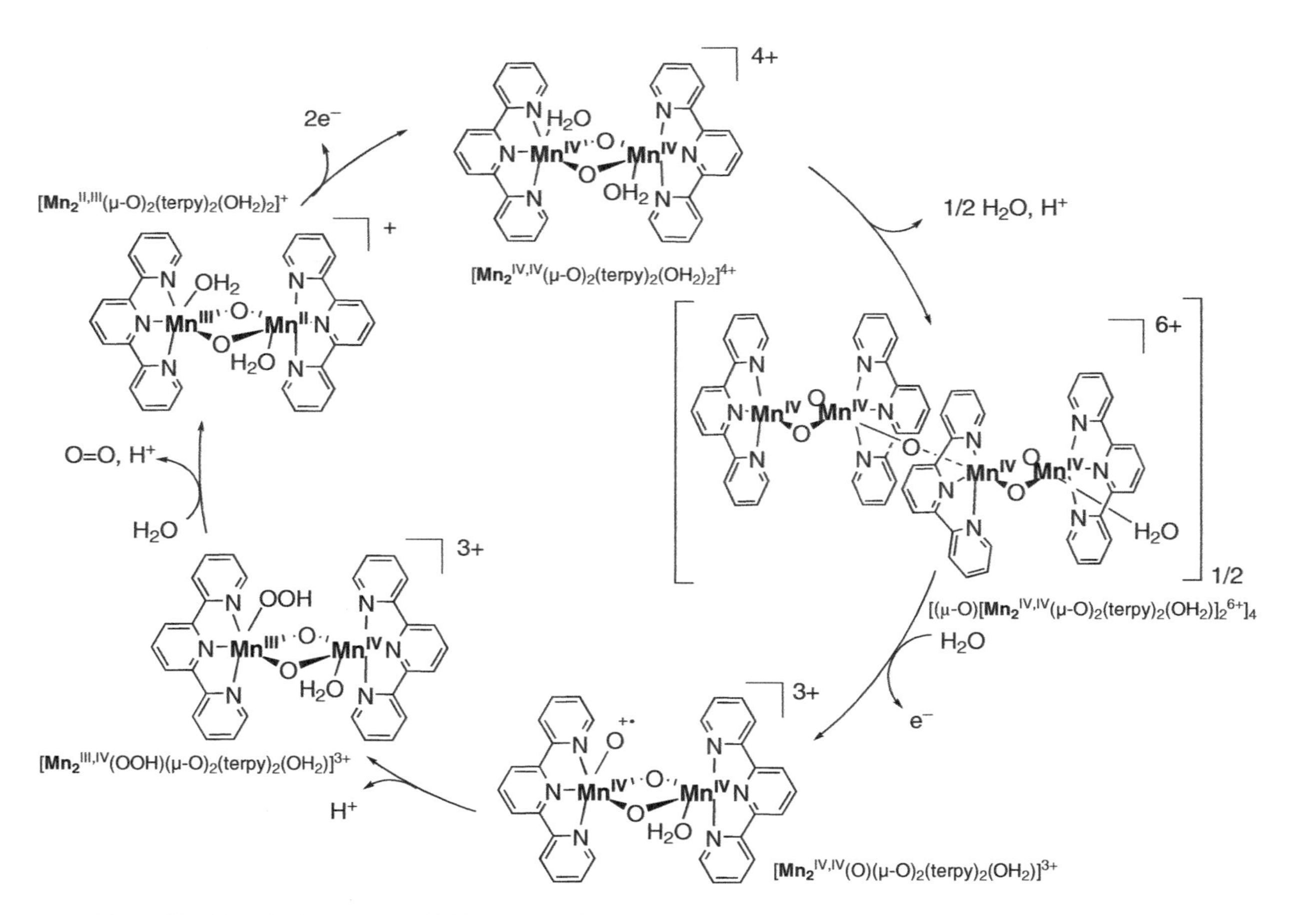

Scheme 53. Dimerization process and DFT proposed cleavage to the monomeric species active in the water oxidation reaction.

(EXAFS) spectrum of the $[Mn_2^{III,IV}(\mu\text{-}O)_2(terpy)_2]^{3+}$/clay material confirmed the presence of the entire compound before and after 30 min of the treatment with CAN (196, 205). This was a remarkable result since it indicates a real catalytic water oxidation activity. However, the real nature of the active catalytic species is difficult to be identified. Nevertheless, clay adsorbed complexes can be used to elucidate fundamental details in the water oxidation ability of Mn complexes. For instance, electronic effects, introduced by ligand modification, could be evaluated by studying the catalytic activity in water oxidation (WO) of a series of $[Mn_2^{III,IV}(\mu\text{-}O)_2(R\text{-}terpy)_2]^{3+}$ (where R-terpy stands for 4′-substituted 2,2′:6′,2″-terpyridine) in the presence of CAN. Strong correlation between the $E_{1/2}$ of $Mn_2^{III,IV}/Mn_2^{IV,IV}$ and a second-rate constant (k_2) for the catalytic activity was observed. The higher the reduction potential, the higher the k_2 observed (206).

A very promising, simple yet efficient methodology reported by Nepal and Das (207) solved the degradative side reactions while sustaining TON in the case of the pH sensitive $[Mn_2^{III,IV}(\mu\text{-}O)_2(R\text{-}terpy)_2]^{3+}$ complex. The damaging intramolecular reactions between manganese water oxidation catalyst (WOCs), were avoided by assembling the catalyst into a well-defined pore, of a highly stable MOF ($(\mu\text{-}O_3)Cr_3(COO)_6$) to acidic and oxidant conditions, but small enough to block the catalyst diffusion between pores or out of the structure. Analytical data [Fourier transform infrared (FTIR), EPR, UV–vis, and elemental analysis] supported the intact presence of $[Mn_2^{III,IV}(\mu\text{-}O)_2(R\text{-}terpy)_2]^{3+}$ after the catalytic water oxidation test when using $KHSO_5$ as sacrificial oxidant. This methodology resulted in a 20-fold improvement of the catalytic activity with respect to the example previously reported by Crabtree and co-workers (198).

Complementarily, an electrochemical study revealed the quantitative dimerization of the electrochemically formed $[Mn_2^{IV,IV}(\mu\text{-}O)_2(terpy)_2]^{4+}$ dimer to a thermodynamically stable linear tetranuclear Mn cluster $[Mn_4O_5(terpy)_4(H_2O)_2]^{6+}$, where two Mn dimers are linked by a mono-μ-oxo bridge. Furthermore, this compound was shown to be spontaneously formed from $[Mn_2^{IV,IV}(\mu\text{-}O)_2(terpy)_2]^{4+}$ when dissolved in water. The resulting tetramer cannot be further oxidized either under electrochemical conditions or by oxone. Moreover, attempts to perform light-driven water oxidation using $[Ru(bpy)_3]^{2+}$ and $Na_2S_2O_8$ produced a similar amount of O_2 as in the blank reaction (208). On the other hand, Brudvig and co-workers (209) showed that the tetranuclear Mn complex adsorbed on kaolin is indeed active for water oxidation when using CAN as an sacrificial oxidant.

In 2005, McKenzie and co-workers (210) reported O_2 evolution when reacting $[Mn^{II}_2(mcbpen)_2(H_2O)_2](ClO_4)_2$ (mcbpen = *N*-methyl-*N*′-carboxymethyl-*N,N*′-bis(2-pyridylmethyl)ethane-1,2-diamine) with a large excess of TBHP as sacrificial oxidant. Based on ESI–MS spectrometric and UV spectroscopic studies, the authors proposed the mechanism shown in Scheme 54: the initial Mn(II) dimer, under catalytic conditions, is oxidized and quickly transformed to a monomeric $[Mn^{III}(mcbpen)(OH)]^+$ species. Then, a dimerization takes place with extrusion of

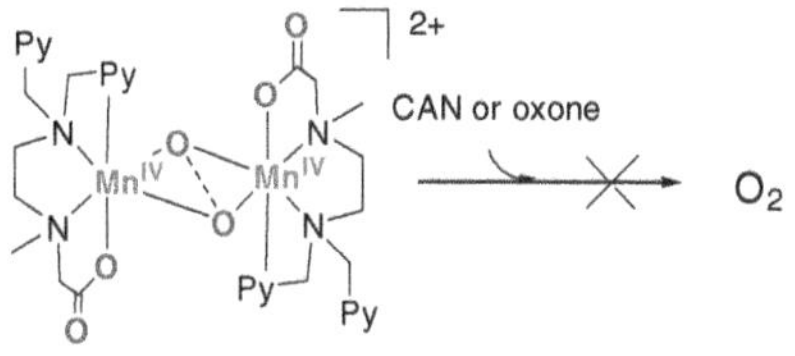

Scheme 54. (*a*) Collapse of the diamond core proposed by McKenzie and co-workers [212] and (*b*) the intermediate isolated by Anderlund and co-workers [211] and their reactivity with CAN or Oxone.

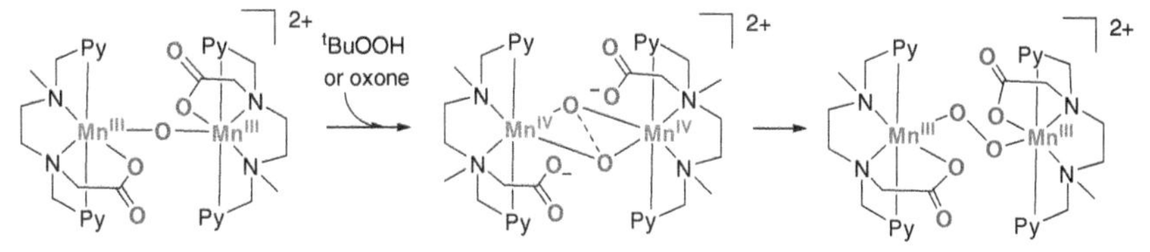

a water molecule, to give a short-lived $[Mn^{III}_2(mcbpen)_2(\mu\text{-}O)]^{2+}$ intermediate, which was detected by ESI–MS and UV–vis. McKenzie and co-workers (210) postulated that further oxidation of $[Mn^{III}_2(mcbpen)_2(\mu\text{-}O)]^{2+}$ with TBHP produced a new intermediate with a diamond core-like structure $[Mn^{IV}_2(mcbpen)_2(\mu\text{-}O)_2]^{2+}$ that after collapsing to form a linear $Mn^{IV}_2(\mu\text{-}O_2)$ released O_2, regenerating the initial $[Mn^{II}_2(mcbpen)_2(H_2O)_2]^{2+}$ species, thus closing the catalytic cycle. Surprisingly, labeling studies in 95% $H_2{}^{18}O$ showed mainly formation of the mixed labeled $^{34}O_2$ along with some $^{32}O_2$, but no reported trace of $^{36}O_2$ (211, 212). Further optimization of the catalytic conditions with TBHP produces up to 15000 TONs without apparent oxidative damage of the ligand. However, TBHP is known to evolve O_2 via one-electron radical-type pathways, not necessarily involving the Mn complex in the key O–O bond-forming step. Therefore, the exact contribution of the water oxidation reaction to the amount of O_2 evolved could not be established. The use of oxone also produced O_2 in catalytic amounts, but the use of one-electron oxidants, such as CAN or $[Ru(bpy)_3]^{3+}$, results in substoichiometric O_2 evolution activity (210).

In contrast, an independent study carried out by Anderlund and co-workers (213) showed that the stable dimeric $[Mn^{IV}_2(mcbpen)_2(\mu\text{-}O)_2](ClO_4)_2$ and $[Mn^{IV}_2(bpmg)_2(\mu\text{-}O)_2](ClO_4)_2$ (bpmg = 2-[[2-[bis(pyridin-2-pyridylmethyl) amino]-ethyl](methyl)amino]acetic acid) complexes containing a $(\mu\text{–}O)_2$ diamond-like core did not produce any detectably O_2 after treatment with CAN or oxone. They concluded that in this particular system the collapse of the core hypothesized by McKenzie and co-workers (212) does not operate. Unfortunately, the sum of observations did not shed light into the nature of the O_2 formed.

An attempt to compare the water oxidation activity of selected dimeric manganese complexes under homogeneous conditions was carried out by Styring and co-workers (214). The results were summarized in Table VI and show that the

TABLE VI
Reporter Comparison for Selected Manganese Complexes in Catalyzed O_2 Evolution Under the Same Catalytic Conditions

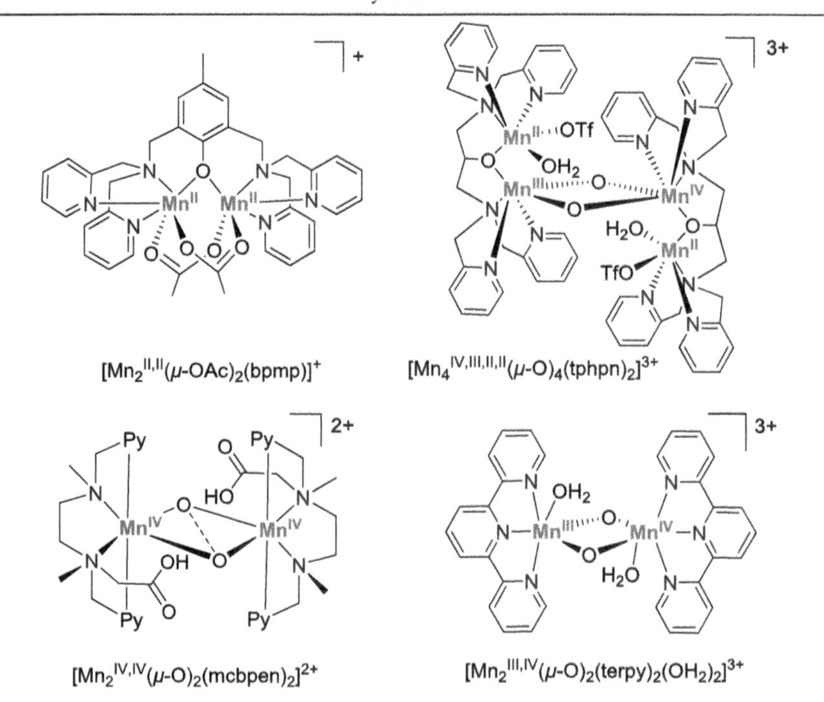

Compound	H_2O_2 [a]	TBHP	HSO_5^-	ClO^- [b]	Ce^{4+} [b,c]
$[Mn_2^{II,II}(\mu\text{-}OAc)_2(bpmp)_2]^{+}$ [d]	33	1	16	n.d.	n.d.
$[Mn_4^{III,IV,II,II}(\mu\text{-}O)_4(tphpn)_2]^{3+}$ [e]	>500[b]	34	105	n.d.	n.d.
$[Mn_2^{IV,IV}(\mu\text{-}O)_2(mcbpen)_2]^{2+}$	7.5	Traces	38	Traces	n.d.
$[Mn_2^{III,IV}(\mu\text{-}O)_2(terpy)_2(OH_2)_2]^{3+}$	1.8	36	>500[b]	7	n.d.

[a] Added 2 equiv of H_2O_2.
[b] Non-oxygen detected (n.d.) above the lower detection limit of 1 mM_{O_2} min^{-1} M_{metal}^{-1}.
[c] Dioxygen evolution was faster than the upper detection limit of the Clark electrode.
[d] 2,6-Bis((bis(2-pyridylmethyl)amino)methyl)-4-methylphenol = bpmp.
[e] 1,3-Bis(bis(2-pyridylmethyl)amino)propan-2-ol = tphpn.

complex reported by Crabtree and co-workers (198) can be considered superior in terms of O_2 evolution rates. Again, oxo-transfer sacrificial oxidants like H_2O_2, oxone, OCl^-, or *t*BuOOH produce catalytic amounts of O_2, while the use of one-electron oxidants that are not oxygen-transfer agents (CAN or $[Ru(bpy)_3]^{3+}$) do not produce significant amounts of O_2 under homogeneous conditions.

While the main goal is to finally achieve the light driven water oxidation, there are only a few successful examples. Since 1985 only a few examples of various publications that reported the ability of Mn(III) salen complexes to liberate oxygen in the presence of an excess of *p*-benzoquinone and irradiation with visible light. The O_2 evolution proceeded with a concomitant reduction of the *p*-benzoquinone to semiquinone. The activity is highly sensitive to the structure since $[Mn(salpd)(H_2O)]^{2+}$ [salpd = propane-1,3-diylbis(salicylideneiminate)] is one of the most active complexes reported. Kinetic studies showed that the rate of dioxygen evolution is first-order dependent on the Mn(III) complex and half-order in quinone concentrations. Rates are dependent of the pH of the reaction medium, but are independent of solvent (215–217).

An effort to mimic the electron transfer between chlorophyll, tyrosine, and the Mn OEC that drives the water oxidation reaction at PSII, where done by Styring and co-workers (218–220). They assembled a photosensitizer (e.g., a ruthenium-based complex), through a tyrosine to a monomeric and dimeric manganese-based complexe. The authors demonstrated, on the basis of EPR spectroscopy and optical flash photolysis techniques, photon driven stepwise electron transfer from the Mn complex to the tyrosine and finally to the Ru complexes mimicking the electron-acceptor site of PSII.

In 2011, Åkewmark and co-workers (221) presented the first synthetic manganese complexe $[Mn_2(dClP)(OMe)(MeCO_2)]$ (Fig. 55, where dClP = 2-(3-(7-carboxy-1*H*-3λ^4-benzo[*d*]imidazol-2-yl)-2-hydroxyphenyl)-1*H*-benzo[*d*]imidazole-4-carboxylic acid), which by refluxing in MeOH afforded a tetra-manganese X-ray structure reminiscent of the OEC cluster. Compound $[Mn_2(dClP)(OMe)(MeCO_2)]$ was found capable of performing the water oxidation under homogeneous conditions using a single-electron oxidant, in particular when using $[Ru^{III}(bpy)_3]^{3+}$ 0.027 s^{-1} TOF (turnover frequency) and 25 TON values were found. The ^{18}O labeling experiments (^{18}O-water 5.8% enriched) demonstrate that the isotopic distribution of the O_2 evolved ($^{16,18}O_2/^{16,16}O_2$ ratio observed = 0.10, expected = 0.12) was consistent with a water oxidation process. More interestingly, this complex also catalyzed the photochemical oxidation of water using $[Ru^{II}(deep)_3](PF_6)_2$ [deep = 4,4′-bis(ethoxy-carbonyl)-2,2′-bipyridine] as photosensitizer and $Na_2S_2O_8$ as the ultimate electron acceptor (TON = 4).

Recently, Brudvig and co-workers (222) studied $[Mn(PaPy_3)(NO_3)](ClO_4)$ ($PaPy_3H$ = *N*,*N*-bis(2-pyridylmethyl)-amine-*N*-ethyl-2-pyridine-2-carboxamide), which shows activity in the oxygen-evolution reaction when using Oxone in water. Up to 5 TON and 50 TON h^{-1} were obtained. In contrast, $[Mn(N4Py)CF_3SO_3](CF_3SO_3)]$,

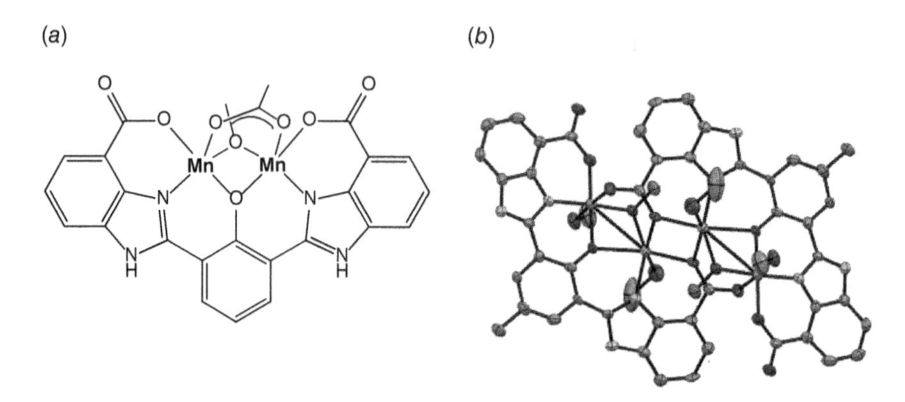

Scheme 55. Line drawing of the synthesized monomeric Mn complex (*b*) and X-ray crystal structure of dimer Mn complex at a 50% probability level (*a*) [221].

where N4Py = 1,1-di(pyridin-2-yl)-*N*,*N*-bis(pyridin-2-ylmethyl)methanamine, produces very low oxygen evolution activity under the same conditions and $[Mn(PY5(OH_2)](ClO_4)_2$ was inactive (Scheme 56). The authors suggested that $[Mn(PaPy_3)(NO_3)](ClO_4)]$ is robust enough not to be protonated during catalysis and that the anionic N-donor moiety increases the stability of the high-valent intermediate competent for oxygen formation.

B. Aminopyridine Iron Complexes as Molecular Catalysts for Water Oxidation

Iron can be seen as a particularly attractive metal for designing oxidation catalysts because it is abundant, environmentally benign, and inexpensive. Early reports exploring iron-based coordination complexes for a water oxidation reaction were authored by Elizarova et al. (223, 224). They reported water oxidation activity for a large family of iron complexes along with cobalt, copper, nickel, and manganese complexes based on chloride, bipyridyl, ethylendiamine, and amino

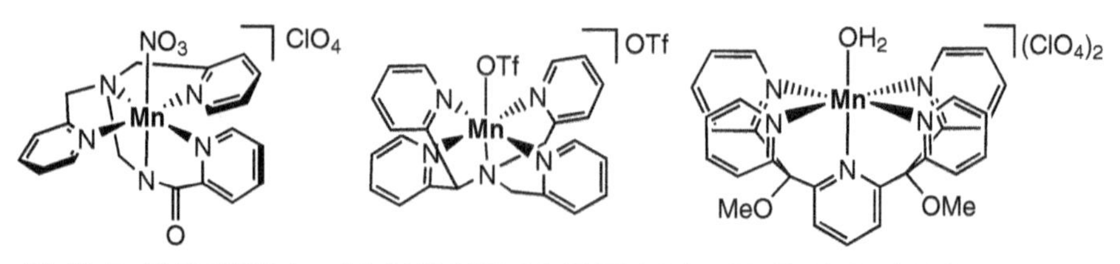

Scheme 56. Line drawing of Mn complexes tested in oxygen evolution using oxone as oxidant [PY5 = 2,6-bis(methoxydi(pyridin-2-yl)pyridine].

ligands, upon reaction with polypyridine ruthenium and iron tricationic oxidants with the general formula $[ML_3]^{3+}$ (where L stands for bpy or phen and M = Ru or Fe) at basic pH (Eq. 1). These studies showed formation of substoichiometric amounts of O_2 with respect to the metal complexes, but unfortunately the nature of the formed O_2 was not studied (224–227). Elizarova et al. (223, 224) reported that $[Fe_2O(phen)_4]^{4+}$ dissolved in a basic $[Ru(bpy)_3]^{3+}$ solution yielded 55% of O_2 relative to the stoichiometric expected amount (TON ≈ 1).

$$[4ML_3]^{3+} + 2H_2O \xrightarrow{\text{coord.complex}} [4ML_3]^{2+} + O_2 + 4H^+ \quad (1)$$

It was 2010 that the first example of truly catalyzed oxygen production by an iron complex was reported. Collins and co-workers (228) reported that coordination complexes of Fe(III) with the tetra anionic amido macrocyclic ligands (**$^{a\text{-}d}$TAML**, tetra anionic amido macrocyclic ligands, Scheme 57) react rapidly with CAN to produce O_2 at very fast rates (up to 4680 h^{-1}) (229). Unfortunately, the high active catalyst vanished quickly to give a maximum of 16 TON, in the best case (Scheme 57, *d*). The authors also detected catalytic oxygen evolution using $NaIO_4$ as an oxidant. Due to the insignificant UV–vis adsorption of $NaIO_4$, it was found a very useful oxidant for UV–vis spectroscopic studies. Indeed, a characteristic signature of a $(TAML)Fe^{IV}{-}O{-}Fe^{IV}(TAML)$ dimer was observed in the UV–vis spectrum of catalytic reactions when using this oxidant at pH 5.5 (229). Later on, a theoretical study performed by Cramer and co-workers (230) suggested that two proton coupled electron-transfer steps result in the generation of $Fe^{V}(O)(^{d}TAML)$ species from $Fe^{III}(OH_2)(^{d}TAML)$. The DFT calculations reported by Cramer and co-workers (230) showed that the attack of a water molecule to $Fe^{V}(O)(^{d}TAML)$ is too highly energetic ($\Delta G = 41.5\ \text{kcal mol}^{-1}$) to occur but $[Fe^{V}(O)(^{d}TAML^{\bullet+})]$ (accessible by CAN oxidation) presented a considerably lower activation barrier ($\Delta G = 30.0\ \text{kcal mol}^{-1}$). The author suggested that it may be the rate-determining step, resulting in the formation of the $Fe^{IV}(OOH)(^{d}TAML)$ intermediate.

A significant increase in catalytic efficiency and lifetime has been reported by a family of iron complexes based on neutral tetradentate aminopyridine ligands.

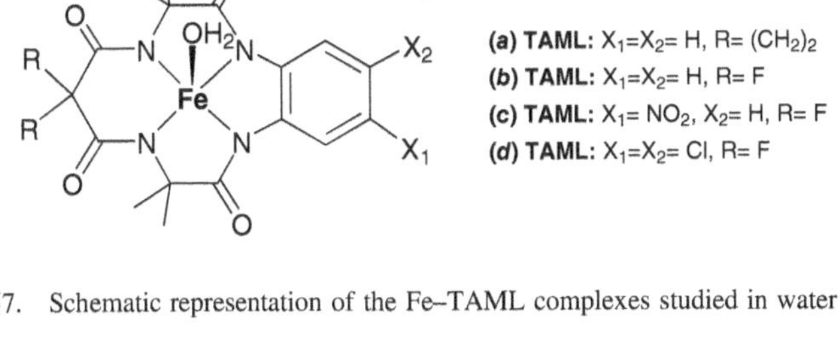

(*a*) TAML: X_1=X_2= H, R= $(CH_2)_2$
(*b*) TAML: X_1=X_2= H, R= F
(*c*) TAML: X_1= NO_2, X_2= H, R= F
(*d*) TAML: X_1=X_2= Cl, R= F

Scheme 57. Schematic representation of the Fe–TAML complexes studied in water oxidation.

Turnover numbers >350 and >1000 were obtained using CAN at pH 1 ($E^\circ = 1.70$ V) and sodium periodate at pH 2 ($E^\circ = IO_4^-/IO_3^- = 1.60$ V), respectively. These are among the highest TONs per metal center described so far for any homogeneous water oxidation reaction based on a first-row transition metal (231) These catalysts contain readily available and modular tetradentate nitrogen ligands (Scheme 58), which leads to a broad accessibility of nitrogen-based ligands allowed to extract preliminary data about the scope, efficiency of the reaction, as well as specific mechanistic information for the iron complexes in catalytic water oxidation reactions. Along this line, it was found that iron complexes with tetradentate nitrogen-based ligands that leave two exchangeable cis-positions

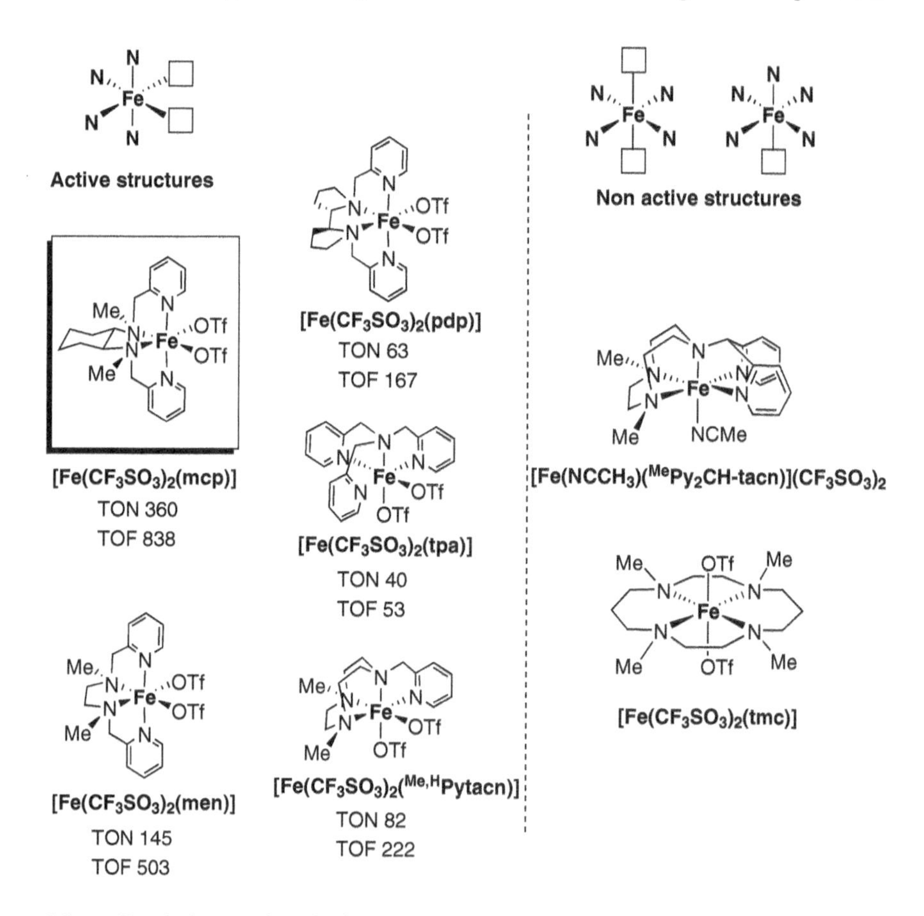

Scheme 58. Active complexes having neutral tetradentate nitrogen ligands leading to cis exchangeable coordination sites. No activity was found for complexes with tetradentate ligands leaving trans free coordination sites and pentadentate neutral nitrogen ligands. The highest active structure has been highlighted. Reaction conditions: Catalyst : CAN = 12.5 μ*M* : 125 m*M* at 25 °C, pH 0.8.

were effective water oxidation catalysts (e.g., $[Fe(CF_3SO_3)_2(^{Me,H}Pytacn)]$ and $[Fe(CF_3SO_3)_2(mcp)]$) when using CAN as the oxidant. On the contrary, iron complexes with tetradentate nitrogen ligands that leave two available trans-positions, $[Fe(CF_3SO_3)_2(tmc)]$ or with pentadentate nitrogen ligands $[Fe(NCMe)(^{Me}Py_2CH{-}tacn)](CF_3SO_3)_2$, do not form significant amounts of O_2 (Scheme 58).

Isotopic-labeling experiments with $H_2{}^{18}O$, under catalytic conditions when using CAN, showed good agreement between the experimental and theoretical O_2 isotopic distribution. In addition no N_2 formation was detected, discarding the possible oxidation of ammonium ions. This isotopic-labeling experiment showed evidence that all of the O_2 originated exclusively from the water oxidation reaction. Under catalytic conditions (CAN, pH 1), the absence of significant amounts of CO_2, together with the absence of nanoparticles is an indication that no major decomposition of the system takes place (231). At low pH values, $[Fe(OH_2)_6]^{3+}$ is the most stable Fe(III) speciation in water rather than Fe_2O_3 nanoparticles, but $FeCl_2$, $Fe(CF_3SO_3)_2$, $FeCl_3$, or $Fe(ClO_4)_3$ do not catalyzed the oxidation of water with CAN or $NaIO_4$. This result indicates that at low pH values, even if some iron–ligand dissociation occurs from the iron complexes, the resulting $[Fe(OH_2)_6]^{3+}$ complex will not be able to catalyze the water oxidation, and because of this finding the catalytic activity must be exclusively ascribed to the molecular complexes.

Lau and co-workers (232) reported that the reaction of $[FeCl_2(mcp)]$ with $[Ru(bpy)_3](ClO_4)_3$ ($E° = 1.21$ V) at pH 8.5 in borate buffer, did not produce significant amounts of O_2. This result probably is due to fact that $[Ru(bpy)_3](ClO_4)_3$ may not be able to oxidize $[Fe(OH_2)_2(mcp)]^{2+}$ to the high-valent iron-active species required for the O–O bond-formation reaction. However, the use of $[Ru(bpy)_3]Cl_2$ as the photosensitizer and visible light and $Na_2S_2O_8$ as the sacrificial oxidant at pH 7.5–9 in borate buffer generates up to 194 TON of O_2. Lau and co-workers (232) proposed that the strongly oxidizing $SO_4{}^{\bullet -}$ radical ($E° = 2.4$ V) generated under photochemical conditions is involved in the oxidation of $[Fe(mcp)(OH)_2]^+$ to the high-valent iron oxo species, which at high pH conditions evolved to the formation of Fe_2O_3 nanoparticles.

Lloret-Fillol et al. (231) presented some preliminary information about the possible resting state and proposed a catalytic cycle. Titration experiments of $[Fe(CF_3SO_3)_2(^{Me,H}Pytacn)]$ with CAN (6 equiv) showed that $[Fe^{IV}(O)(OH_2)(^{Me,H}Pytacn)]^{2+}$ is formed, but no increase of pressure was detected under these conditions. Moreover, the analysis of the reaction headspace does not show any O_2. This finding together with the observed stability (by UV–vis and ESI–MS) of $[Fe^{IV}(O)(OH_2)(^{Me,H}Pytacn)]^{2+}$ under these conditions suggested that this intermediate was not directly responsible for the O–O bond formation. However, further addition of 75 equiv of CAN produced a concomitant Ce(IV) consumption and O_2 evolution while keeping the UV–vis signature of a low-spin ($S = 1$) oxoiron (IV) species, as the major product in solution. Stoichiometric O_2 formation versus

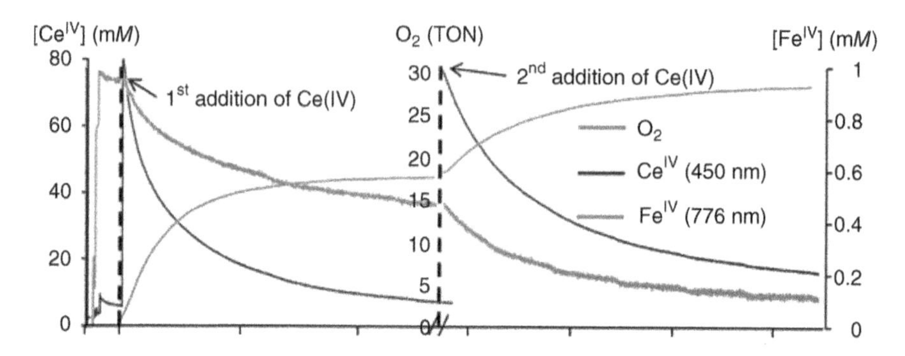

Scheme 59. Water oxidation reaction monitored by UV–vis spectroscopy and a pressure sensor for the observation of the Ce(IV) consumption (blue), the complex degradation (green), and the oxygen evolution (red). Fe^{IV}=O was fully formed by addition of CAN (6 equiv) over $[Fe(CF_3SO_3)_2(^{Me,H}Pytacn)]$ (1 mM in MiliQ water). Oxygen was not detected and the Fe^{IV}=O species were stable for >2 h. A second addition of oxidant (75 equiv) allowed the oxygen production. (See the color version of this scheme in color plates section.)

the consumed Ce(IV) indicates that the oxidizing equivalents are essentially used in the water oxidation reaction. Kinetic studies (Scheme 59) showed that the rate of Ce(IV) consumption was first order in $[Fe^{IV}(O)(OH_2)(^{Me,H}Pytacn)]^{2+}$ concentration and first order in low concentrations of Ce(IV). Therefore, the authors concluded that $[Fe^{IV}(O)(OH_2)(^{Me,H}Pytacn)]^{2+}$ is an intermediate and most probably the resting state. Then further oxidation produced the $[Fe^{V}(O)(OH)(^{Me,H}Pytacn)]^{2+}$ species responsible for the O–O bond formation. Attack of the water molecule in the $[Fe^{V}(O)(OH)(^{Me,H}Pytacn)]^{2+}$ may produce the $[Fe^{III}(OOH)(OH_2)(^{Me,H}Pytacn)]^{2+}$ intermediate, which by reaction with excess CAN under catalytic conditions rapidly evolves to form O_2 and recovers the $[Fe^{IV}(O)(OH)(^{Me,H}Pytacn)]^{2+}$, closing the catalytic cycle (Scheme 60).

The modular nature of the ^{Me,H}Pytacn ligand allows the authors to tune the electronic properties of the ligands by modifying the substituents located at the para position of the pyridine ring. Indeed, complexes derived from the substitution at the pyridine, $[Fe(CF_3SO_3)_2(^{Me,X}Pytacn)]$ (OTf = CF_3SO_3, X = –H, –Cl, –CO_2Et, –NO_2), were highly active molecular catalysts for cerium(IV)-driven WO and more interestingly an improvement of the efficiency of this family of catalysts was achieved by introducing electron-withdrawing groups. Clearly, electronic effects have an important influence on the water oxidation activity (Scheme 61) Furthermore, the rates of O_2 evolution and Ce(IV) consumption correlate with the electron-donating nature of the ligand. The systematic electronic effects observed on the catalytic activity and kinetics in $[Fe(CF_3SO_3)_2(^{Me,X}Pytacn)]$ (X = H, Cl, CO_2Et, NO_2) complexes strongly supports that their WO activity originates from molecular complexes operating in a homogeneous phase (233).

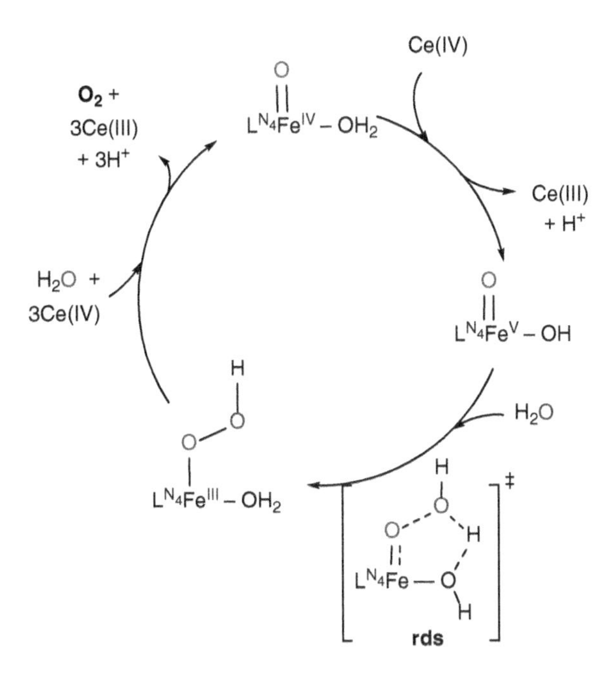

Scheme 60. Mechanism postulated for the iron water oxidation. The only observed intermediate was Fe^{IV}=O. All the other intermediates were postulated (rds = rate-determining step).

Mechanistic studies point out that a (Fe^{IV}=O–Ce^{IV}) adduct is formed in the presence of a large excess of CAN. Therefore the resting state may change as a function of the exact catalytic conditions. In addition, experimental and computational methods suggest that the O–O bond-formation event is the rate-determining step, via nucleophilic attack of a H_2O molecule toward a Fe^V(O)(OH) species. The cis-labile sites of the oxo/hydroxo groups at the iron center seem crucial for O–O

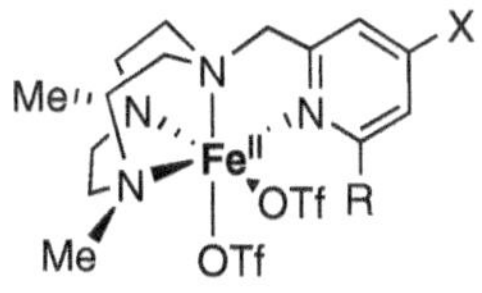

[$Fe^{II}(CF_3SO_3)_2$(^{Me,X}Pytacn)]

X: H, Cl, CO_2Et, NO_2

X	TON	TOF(h^{-1})
H	80	230
Cl	100	380
CO_2Et	145	560
NO_2	180	820

Scheme 61. Iron complexes with different four-subtituted pyridines show an increasing activity in water oxidation related to the electron-withdrawing property. Reaction conditions for water oxidation: Catalyst : CAN = 12.5 μ*M* : 125 m*M* at 25 °C, pH 0.8.

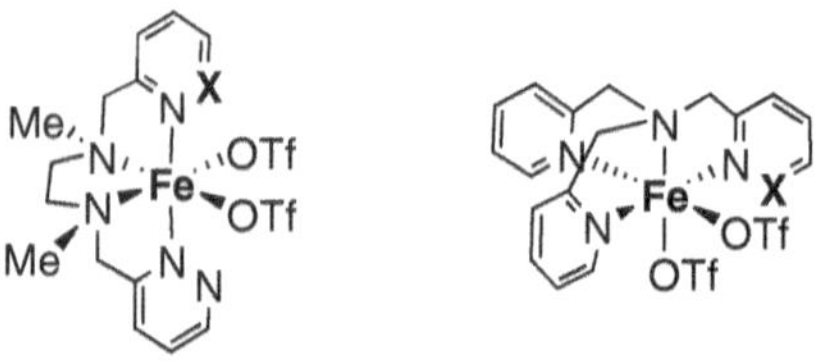

X	TON	TOF
N	67	141
CH	72	228

X	TON	TOF
N	20	42
CH	14	36

Scheme 62. Representation of the iron complexes for base-assisted substrate deprotonation and comparison of the catalytic water oxidation activity with the previous reported iron complexes. Reaction conditions for water oxidation: Catalyst : CAN = 100 μM : 125 mM at 25 °C, pH 0.8.

bond formation. Codola et al. (233) suggested that the water molecule that attacks the oxo group to form the O–O bond is activated by the OH group that then acts as an internal base. Along this line, Yang and co-workers (234) reported that by incorporation of hydrogen-bonding functionalities into the second-coordination sphere of the iron complex can slightly affect the TON (Scheme 62). Nevertheless, the observed differences in reactivity can also be attributed to electronic effects. Therefore more studies are needed to clarify if the external base can aid in the water oxidation activity in this family of iron-coordination complexes.

Furthermore, Garcia-Bosch et al. (112) have recently observed that complexes [Fe(CF_3SO_3)$_2$(Pytacn)] and [Fe(CF_3SO_3)$_2$(mcp)], which contain cis-labile binding sites were catalytically active in stereospecific hydroxylation of alkanes and cis-dihydroxylation of olefins using water as a source of oxygen and CAN as oxidant. However, under the same catalytic conditions [Fe(CF_3SO_3)$_2$(tmc)] and [Fe(NCMe)(MePy$_2$CH–tacn)](CF_3SO_3)$_2$ do not produce any oxidation of *cis*-decaline and *cis*-2-octene. Indeed, C–H and C=C and water oxidation appear to require iron complexes containing a tetradentate ligand that leaves two available cis sites (112). Therefore, this correlation strongly suggests the involvement of common iron intermediates mediating substrate oxidation in both catalytic pathways. Nevertheless, the mechanism is far from being completely understood.

IV. CONCLUSION

Iron and manganese coordination complexes containing aminopyridine ligands have emerged as privileged platforms to sustain highly challenging oxidation reactions. Evidence has accumulated supporting the idea that these processes imply very reactive high-valent metal-oxo intermediates. The overall robust nature of this type of ligands, as well as its highly basic character can be regarded as key issues to sustain the formation of these highly reactive species. Alternatively, the highly modular nature of this type of ligand has allowed the preparation of structurally diverse examples, which in turn have provided an excellent starting point for interrogating fundamental aspects of their reactivity. Some of these complexes are turning into valuable tools in organic synthesis that serve as an alternative to less sustainable classical oxidation methods. Iron coordination complexes with this type of ligand are also promising catalysts for water splitting. Still, full exploration of the capacities of this type of compounds in catalytic transformations is at its infancy, and there is a long road ahead before these compounds are discovered.

ACKNOWLEDGMENTS

We would like to thank the European Research Foundation for project FP7-PEOPLE-2010-ERG-268445 (J.Ll.), MINECO for projects CTQ2012-37420-C02-01/BQU and CSD2010-00065 (M.C.) and for a Ramon y Cajal contract (J.Ll.), Generalitat de Catalunya for an ICREA Academia Award and the European Research Foundation for Project ERC-2009-StG-239910 (M.C.).

ABBREVIATIONS

5-Cl-1-MeIm	5-Cloro-1-methylimidazole
$6Me_2$-pdp	*N,N'*-Bis(6-methyl-2-pyridylmethyl)-2,2'-bipyrrolidine
$6Me_2$-mcp	*N,N'*-Dimethyl-*N,N'*-bis-(6-methyl-2-pyridylmethyl)-cyclohexane-*trans*-1,2-diamine
$6Me_2$-men	*N,N'*-Dimethyl-*N,N'*-bis(6-methyl-2-pyridylmethyl)ethane-1,2-diamine
$5Me_3$-tpa	Tris(5-methyl-2-pyridylmethyl)amine
$6Me_3$-tpa	Tris(6-methyl-2-pyridylmethyl)amine
BArF	tetrakis(3,5-trifluoromethylphenyl)borate
bbpc	*N,N'*-dibenzyl-*N,N'*-bis(2-pyridylmethyl)-cyclohexane-*trans*-1,2-diamine
Bn	Benzyl
BOC	*tert*-Butyloxycarbonyl
BpdL1	Dimethyl 3,7-dimethyl-9-oxo-2,4-di(pyridin-2-yl)-3,7-diazabicyclo[3.3.1]nonane-1,5-dicarboxylate
bpeb	*N,N'*-Dimethyl-*N,N'*-bis(1-*R*-1*H*-benzo[*d*]imidazol-2-yl)methyl) cyclohexane-*trans*-1,2-diamine

Bpka	3-(Di(pyridin-2-yl)methyl)-1,5,7-trimethyl-2,4-dioxo-3-azabicyclo[3.3.1]nonane-7-carboxylic acid
bpmg	2-[[2-[Bis(2-pyridylmethyl)amino]-ethyl](methyl)amino]acetic acid
bpmp	2,6-Bis((bis(2-pyridylmethyl)amino)methyl)-4-methylphenol
bpp	2,2′-Bis(7,7-dimethyl-5,6,7,8-tetrahydro-6,8-methanoisoquinoline)
bpy	2,2′-Bipyridine
bqcn	*N,N′*-Dimethyl-*N,N′*-di(quinolin-8-yl)cyclohexane-1,2-*trans*-diamine
bqen	*N,N′*-Dimethyl-*N,N′*-di(quinolin-8-yl)ethane-1,2-diamine
Bz	Benzoyl
CAN	Cerium(IV) ammonium nitrate
cat	Catalyst
Cl_3terpy	4,4′,4″-Trichloro-2,2′:6,2″,-terpyridine
CSI-MS	Cryospray-Ionization mass spectrometry
dCIP	2-(3-(7-Carboxy-1*H*-3λ^4-benzo[*d*]imidazol-2-yl)-2-hydroxyphenyl)-1*H*-benzo[*d*]imidazole-4-carboxylic acid
deep	4,4′-Bis(ethoxy-carbonyl)-2,2′-bipyridine
DFT	Density functional theory
DHA	Dihydroanthracene
DMCH	1,2-*cis*-Dimethylcyclohexane
DMEGqu	1,3-Dimethyl-*N*-(l-(3-methylpyridin-2-yl)vinyl)imidazolidin-2-imine
dpaq	2-(Bis(pyridin-2-ylmethyl)amino)-*N*-(quinolin-8-yl)acetamide
ee	Enantiomeric excess
EPR	Electronic paramagnetic resonance
ESI–MS	Electrospray ionization mass spectrometry
EXAFS	Extended X-ray absorption fine structure
FTIR	Fourier transform infrared spectroscopy
GC	Gas chromatography
GMHA	Glyoxylic acid methyl ester hemiacetal
HAT	Hydrogen atom transfer
H_2Pydic	Dipicolinic acid
KIE	Kinetic isotopic effect
Mac-N5	3-(3,7,11-Triaza-1(2,6)-pyridinacyclododecaphane-7-yl)propan-1-amine
mcbpen	*N*-Methyl-*N′*-carboxymethyl-*N,N′*-bis(2-pyridylmethyl)ethane-1,2-diamine
mcmb	*N,N′*-Dimethyl-*N,N′*-bis((1-*R*-benzymidazol-2-yl)methyl)cyclohexane-*trans*-1,2-diamine
mcp	*N,N′*-Dimethyl-*N,N′*-bis(2-pyridylmethyl)cyclohexane-*trans*-1,2-diamine

mcpp	*N*-((7,7-Dimethyl-5,6,7,8-tetrahydro-6,8-methanoisoquinolin-3-yl)methyl)-*N′*-(7,7-dimethyl-5,6,7,8-tetrahydro-6,8-methanoiso-quinolin-3-yl)-methyl)-*N*,*N′*-dimethylcyclohexane-1,2-diamme
mcpdp	*N*,*N′*-Dimethyl-*N*,*N′*-bis(((4-(*tert*-butyl)phenyl)(2-pyridylmethyl) cyclohexane-*trans*-1,2-diamine
Me_2EBC	4,11-Dimethyl-1,4,8,11-tetraazabicyclo[6.6.2]hexadecane
men	*N*,*N′*-Dimethyl-*N*,*N′*-bis(2-pyridylmethyl)ethane-1,2-diamine
mepp	*N*-((7,7-Dimethyl-5,6,7,8-tetrahydro-6,8-methanoisoquinolin-3-yl)methyl)-*N′*-(-7,7-dimethyl-5,6,7,8-tetrahydro-6,8-methanoiso-quinolin-3-yl)methyl)-*N*,*N′*-dimethylethylen-1,2-diamine
MMO	Methane monooxygenase
MOF	Metal organic framework
n.d.	Not determined
NDO	Naphthalene dioxygenase
N4Py	1,1-Di(pyridin-2-yl)-*N*,*N*-bis(2-pyridylmethyl)methanamine
OEC	Oxygen-evolving complex
OTf	Trifluoromethanesulfonate
ox	Oxidation
$PyNMe_2$	*N*,*N′*-Dimethyl-2,11-diaza[3.3](2,6)pyridinophane
PaPy3H	*N*,*N*-Bis(2-pyridylmethyl)-amine-*N*-ethyl-2-pyridine-2-carboxamide
pbe	*N*-Methyl-1-(1-*R*-1*H*-benzo[*d*]imidazol-2-yl)-*N*-((1-((1-*R*-1*H*-benzo[*d*]imidazol-2-yl)methyl)pyrrolidin-2-yl)methyl) methanamine
pdp	*N*,*N′*-Bis(2-pyridylmethyl)-2,2′-bipyrrolidine
pdpp	*N*-((7,7-Dimethyl-5,6,7,8-tetrahydro-6,8-methanolisoquinolin-3-yl)*N′*-(7,7-dimethyl-5,6,7,8-tetrahydro-6,8-methanolisoquinolin-3-yl)methyl)-2,2′-biplyrrolidine
Ph-DPAH	Di(pyridin-2-yl)methyl benzamide
phen	1,10-Phenantroline
piv	Pivaloyl
PSII	Photosystem II
py	Pyridine
Py2N2idine	1,3-Bis(2-pyridylmethyl)imidazolidine
PY5	2,6-Bis(bis(2-pyridyl)methoximethane)pyridine
Pybp	*N*,*N′*-Bis(2-pyridylmethyl)-2,2′-bipyperidine
Py4N2p	1,3-Bis(bis(2-pyridylmethyl)amino)propane
Pytacn	1,4-Dimethyl-7-(2-pyridylmethyl)-1,4,7-triazacyclononane
qpy	2,2′:6′,2″:6″,2‴:6‴,2⁗-Quinquepyridine
rds	Rate determining step
red	Reduction
rt	Room temperature

RC	Retention of configuration
salen	*N*,*N*′-Ethylenebis(salicylimine)
salpd	Propane-1,3-diylbis(salicylideneiminate)
SIPr	1,3-Bis(2,6-diisopropylphenyl)-4,5-dihydro-1*H*-imidazol-2-ylidene
Spp	6,6‴-Bis((5*R*,7*R*,8*S*)-6,6,8-trimethyl-5,6,7,8-tetrahydro-5,7-methanoquinolin-2-yl)-2,2′:6′,2″:6″,2‴-quaterpyridine
taatacn	1,4,7-Triazacyclononane-1,4,7-triyl-triacetic acid
tacn	1,4,7-Triazacyclononane
TAML	Tetra anionic amido macrocyclic ligands
TBHP	*tert*-Butylhydroperoxide
TBP_8Cz	[Octakis(*p*-*tert*-butylphenyl)corrolazinato]$^{3-}$
teeatacn	1,4,7-Triazacyclononane-1,4,7-triyl-tris(butan-2-ol)
terpy	2,2′:6′,2′-Terpyridine
tmc	Tetramethylcyclam
tmeatacn	1,4,7-Triazacyclononane-1,4,7-triyl-tris(propan-2-ol)
tmtacn	1,4,7-Trimethyl-1,4,7-triazacyclononane
TOF	Turnover frequency
TON	Turnover number
tpa	Tris(2-pyridylmethyl)amine
tpeatacn	1,4,7-Triazacyclononane-1,4,7-triyl-tris-2-methylbutan-2-ol
tphpn	1,3-Bis(bis(2-pyridylmethyl)amino)propan-2-ol
tpp	Tetraphenylporphyrin
UV–Vis	Ultraviolet–visible
WO	Water oxidation
WOC	Water oxidation catalysis

REFERENCES

1. A. Company, J. Lloret-Fillol, L. Gómez, and M. Costas, Alkane C–H Oxygenation Catalyzed by Transition Metal Complexes, in *Alkane C–H Activation by Single-Site Metal Catalysis*, P. J. Perez, Ed., Springer Netherlands, Dordrecht, The Netherlands, Vol. 38, 2013.
2. T. Newhouse and P. S. Baran, *Angew Chem. Int. Ed.*, *50*, 3362 (2011).
3. G. B. Shul'pin, *Org. Biomol. Chem.*, *8*, 4217 (2010).
4. G. Dyker, *Handbook of C–H Transformations*, Wiley–VCH, Weinheim, Germany, Vols. 1–2, 2005.
5. P. R. Ortiz de Montellano, *Chem. Rev.*, *110*, 932 (2010).
6. P. R. Ortiz de Montellano, *Cytochrome P450: Structure, Mechanism and Biochemistry*. 3rd ed.; Kluwer Academic/Plenum Publishers, New York, 2005.
7. S. Shaik, S. Cohen, Y. Wang, H. Chen, D. Kumar, and W. Thiel, *Chem. Rev.*, *110*, 949 (2010).

8. E. G. Kovaleva and J. D. Lipscomb, *Nat. Chem. Biol.*, *4*, 186 (2008).
9. P. C. A. Bruijnincx, G. v. Koten, and R. J. M. K. Gebbink, *Chem. Soc. Rev.*, *12*, 2716 (2008).
10. M. H. Sazinsky and S. J. Lippard, *Acc. Chem. Res.*, *39*, 558 (2006).
11. S. Enthaler, K. Junge, and M. Beller, *Angew. Chem. Int. Ed.*, *47*, 3317 (2008).
12. C.-L. Sun, B.-J. Li, and Z.-J. Shi, *Chem. Rev.*, *111*, 1293 (2011).
13. K. Gopalaiah, *Chem. Rev.*, *113*, 3248 (2013).
14. A. Correa, O. G. Mancheno, and C. Bolm, *Chem. Soc. Rev.*, *37*, 1108 (2008).
15. E. B. Bauer, *Curr. Org. Chem.*, *12*, 1341 (2008).
16. L. Que Jr., and W. B. Tolman, *Nature (London)*, *455*, 333 (2008).
17. K. U. Ingold and P. A. MacFaul, Distinguishing Biomimetic Oxidations from Oxidations Mediated by Freely Diffusing Radicals, in *Biomimetic Oxidations Catalyzed by Transition Metal Complexes*, B. Meunier, Ed., Imperial College Press: London, UK, pp. 45, 2000.
18. A. Boudier, L. O. Bromm, M. Lotz, and P. Knochel, *Angew. Chem. Int. Ed.*, *39*, 4414 (2000).
19. *Applied Homogeneous Catalysis with Organometallic Compounds*, John Wiley & Sons, Inc., Weinheim, Germany, 2002.
20. R. H. Crabtree, *The Organometallic Chemistry of the Transition Metals*, John Wiley & Sons, Inc, Hoboken, NJ, 2009.
21. B. Meunier, S. P. de Visser, and S. Shaik, *Chem. Rev.*, *104*, 3947 (2004).
22. H. Lu and P. Zhang, *Chem. Soc. Rev.*, *40*, 1899 (2011).
23. M. Costas, *Coord. Chem. Rev.*, *255*, 2912 (2011).
24. C.-M. Che, V. K.-Y. Lo, C.-Y. Zhou, and J.-S. Huang, *Chem. Soc. Rev.*, *40*, 1950 (2011).
25. A. Company, L. Gomez, and M. Costas, Bioinspired Non-heme Iron Catalysts, in C–H and C=C Oxidation Reactions, in *Iron-Containing Enzymes, Versatile Catalysts of Hydroxylation Reactions in Nature*, S. P. De Visser and D. Kumar, Eds., RSC: Cambridge, UK, 2011.
26. K. Chen and L. Que, Jr., *J. Am. Chem. Soc.*, *123*, 6327 (2001).
27. C. Kim, K. Chen, J. Kim, and L. Que, Jr., *J. Am. Chem. Soc.*, *119*, 5964 (1997).
28. K. Chen, M. Costas, J. Kim, A. K. Tipton, and L. Que, Jr., *J. Am. Chem. Soc.*, *124*, 3026 (2002).
29. K. Chen and L. Que, Jr., *Chem. Commun. 1375* (1999).
30. M. Costas and L. Que, Jr., *Angew. Chem. Int. Ed.*, *12*, 2179 (2002).
31. A. Company, L. Gómez, M. Güell, X. Ribas, J. M. Luis, L. Que, Jr., and M. Costas, *J. Am. Chem. Soc.*, *129*, 15766 (2007).
32. A. Company, L. Gómez, X. Fontrodona, X. Ribas, and M. Costas, *Chem. Eur. J.*, *14*, 5727 (2008).
33. I. Prat, A. Company, V. Postils, X. Ribas, L. Que, Jr., J. M. Luis, and M. Costas, *Chem. Eur. J.*, *19*, 6724 (2013).

34. A. Company, Y. Feng, M. Güell, X. Ribas, J. M. Luis, L. Que, Jr., and M. Costas, *Chem. Eur. J.*, *15*, 3359 (2009).
35. Y. Mekmouche, S. Ménage, J. Pécaut, C. Lebrun, L. Reilly, V. Schuenemann, A. Trautwein, and M. Fontecave, *Eur. J. Inorg. Chem.*, *2004*, 3163 (2004).
36. R. Mas-Ballesté, M. Costas, T. v. d. Berg, and L. Que, Jr., *Chem. Eur. J.*, *12*, 7489 (2006).
37. J. England, G. J. P. Britovsek, N. Rabadia, and A. J. P. White, *Inorg. Chem.*, *46*, 3752 (2007).
38. P. Comba, M. Maurer, and P. Vadivelu, *Inorg. Chem.*, *48*, 10389 (2009).
39. A. Company, PhD Thesis, Universitat de Girona, Girona, January 2008.
40. M. Costas, K. Chen, and L. Que, Jr., *Coord. Chem. Rev.*, *200–202*, 517 (2000).
41. W. N. Oloo, A. J. Fielding, and L. Que, Jr., *J. Am. Chem. Soc.*, *135*, 6438 (2013).
42. E. A. Mikhalyova, O. V. Makhlynets, T. D. Palluccio, A. S. Filatov, and E. V. Rybak-Akimova, *Chem. Commun.*, *48*, 687 (2012).
43. M. S. Seo, T. Kamachi, T. Kouno, K. Murata, M. J. Park, K. Yoshizawa, and W. Nam, *Angew. Chem. Int. Ed.*, *46*, 2291 (2007).
44. M. J. Park, J. Lee, Y. Suh, J. Kim, and W. Nam, *J. Am. Chem. Soc.*, *128*, 2630 (2006).
45. O. V. Makhlynets and E. V. Rybak-Akimova, *Chem. Eur. J.*, *16*, 13995 (2010).
46. A. Thibon, V. Jollet, C. Ribal, K. Senechal-David, L. Billon, A. B. Sorokin, and F. Banse, *Chem. Eur. J.*, *18*, 2715 (2012).
47. J. Cho, S. Jeon, S. A. Wilson, L. V. Liu, E. A. Kang, J. J. Braymer, M. H. Lim, B. Hedman, K. O. Hodgson, J. S. Valentine, E. I. Solomon, and W. Nam, *Nature (London)*, *478*, 502 (2011).
48. F. F. Li, K. K. Meier, M. A. Cranswick, M. Chakrabarti, K. M. Van Heuvelen, E. Munck, and L. Que, Jr., *J. Am. Chem. Soc.*, *133*, 7256 (2011).
49. J. Kaizer, M. Costas, and L. Que, Jr., *Angew. Chem. Int. Ed.*, *42*, 3671 (2003).
50. M. P. Jensen, A. Mairata i. Payeras, A. T. Fiedler, M. Costas, J. Kaizer, A. Stubna, E. Münck, and L. Que, Jr., *Inorg. Chem.*, *46*, 2398 (2007).
51. A. Mairata i Payeras, R. Y. N. Ho, M. Fujita, and L. Que, Jr., *Chem. Eur. J.*, *10*, 4944 (2004).
52. L. V. Liu, S. Hong, J. Cho, W. Nam, and E. I. Solomon, *J. Am. Chem. Soc.*, *135*, 3286 (2013).
53. S. Sahu, L. R. Widger, M. G. Quesne, S. P. de Visser, H. Matsumura, P. Moënne-Loccoz, M. A. Siegler, and D. P. Goldberg, *J. Am. Chem. Soc.*, *135*, 10590 (2013).
54. I. Prat, Anna Company, T. Corona, X. Ribas, and M. Costas *Inorg. Chem.*, *52*, 9229 (2013).
55. I. Prat, J. S. Mathieson, M. Güell, X. Ribas, J. M. Luis, L. Cronin, M. Costas, *Nat. Chem.*, *3*, 788 (2011).
56. A. Bassan, M. R. A. Blomberg, P. E. M. Siegbahn, and L. Que, Jr., *J. Am. Chem. Soc.*, *124*, 11056 (2002).
57. D. Quinonero, K. Morokuma, D. G. Musaev, R. Mas-Balleste, and L. Que, Jr., *J. Am. Chem. Soc.*, *127*, 6548 (2005).

58. A. Bassan, M. R. A. Blomberg, P. E. M. Siegbahn, and L. Que, Jr., *Angew. Chem. Int. Ed.*, *44*, 2939 (2005).
59. B. Meunier and J. Bernadou, *Top. Catal.*, *21*, 47 (2002).
60. J. Bernadou and B. Meunier, *Chem. Commun.*, 2167 (1998).
61. P. Comba and G. Rajaraman, *Inorg. Chem.*, *47*, 78 (2008).
62. M. R. Bukowski, P. Comba, A. Lienke, C. Limberg, C. L. de Laorden, R. Mas-Balleste, M. Merz, and L. Que, Jr., *Angew. Chem. Int. Ed.*, *45*, 3446 (2006).
63. M. Fujita, M. Costas, and L. Que, Jr., *J. Am. Chem. Soc.*, *125*, 9912 (2003).
64. M. C. White, A. G. Doyle, and E. N. Jacobsen, *J. Am. Chem. Soc.*, *123*, 7194 (2001).
65. M. S. Chen and M. C. White, *Science*, *318*, 783 (2007).
66. R. Mas-Balleste and L. Que, Jr., *J. Am. Chem. Soc.*, *129*, 15964 (2007).
67. O. Y. Lyakin, R. V. Ottenbacher, K. P. Bryliakov, and E. P. Talsi, *Acs Catalysis*, *2*, 1196 (2012).
68. O. Y. Lyakin, K. P. Bryliakov, G. J. P. Britovsek, and E. P. Talsi, *J. Am. Chem. Soc.*, *131*, 10798 (2009).
69. O. Y. Lyakin, I. Prat, K. P. Bryliakov, M. Costas, and E. P. Talsi, *Catal. Commun.*, *29*, 105 (2012).
70. A. Ansari, A. Kaushik, and G. Rajaraman, *J. Am. Chem. Soc.*, *135*, 4235 (2013).
71. Y. Wang, D. Janardanan, D. Usharani, K. Han, L. Que, Jr., and S. Shaik, *Acs Catalysis*, *3*, 1334 (2013).
72. M. A. Bigi, S. A. Reed, and M. C. White, *J. Am. Chem. Soc.*, *134*, 9721 (2012).
73. M. S. Chen and M. C. White, *Science*, *327*, 566 (2010).
74. N. A. Vermeulen, M. S. Chen, and M. C. White, *Tetrahedron*, *65*, 3078 (2009).
75. L. Gomez, I. Garcia-Bosch, A. Company, J. Benet-Buchholz, A. Polo, X. Sala, X. Ribas, and M. Costas, *Angew. Chem. Int. Ed.*, *48*, 5720 (2009).
76. L. Gomez, M. Canta, D. Font, I. Prat, X. Ribas, and M. Costas, *J. Org. Chem.*, *78*, 1421 (2013).
77. D. H. R. Barton, J.-C. Beloeil, A. Billion, J. Boivin, J.-Y. Lallemand, P. Lelandais, and S. Mergui, *Helv. Chim. Acta*, *70*, 2187 (1987).
78. Y. He, J. D. Gorden, and C. R. Goldsmith, *Inorg. Chem.*, *50*, 12651 (2011).
79. I. Prat, L. Gómez, M. Canta, X. Ribas, and M. Costas, *Chem. Eur. J.*, *19*, 1908 (2013).
80. J. F. Bartoli, O. Brigaud, P. Battioni, and D. Mansuy, *J. Chem. Soc. Chem. Commun.*, 440 (1991).
81. B. R. Cook, T. J. Reinert, and K. S. Suslick, *J. Am. Chem. Soc.*, *108*, 7281 (1986).
82. L. Bin, C. Yi, Y. Cheng-Zhi, and S. Zheng-Wu, *Chin. J. Chem.*, *21*, 833 (2010).
83. A. Cano, M. T. Ramírez-Apan, and G. Delgado, *J. Braz. Chem. Soc.*, *22*, 1177 (2011).
84. M. A. Bigi, S. A. Reed, and M. C. White, *Nat. Chem.*, *3*, 216 (2011).
85. Y. Hitomi, K. Arakawa, T. Funabiki, and M. Kodera, *Angew. Chem. Int. Ed.*, *51*, 3448 (2012).
86. P. Liu, Y. Liu, E. L.-M. Wong, S. Xiang, and C.-M. Che, *Chem. Sci.*, *2*, 2187 (2011).

87. B. Bitterlich, G. Anilkumar, F. G. Gelalcha, B. Spilker, A. Grotevendt, R. Jackstell, M. K. Tse, and M. Beller, *Chem. As. J.*, *2*, 521 (2007).
88. B. Bitterlich, K. Schröder, M. K. Tse, and M. Beller, *Eur. J. Org. Chem.*, 4867 (2008).
89. K. Schröder, S. Enthaler, B. Join, K. Junge, and M. Beller, *Adv. Synth. Catal.*, 1771 (2010).
90. K. Schröder, X. Tong, B. Bitterlich, M. K. Tse, F. G. Gelalcha, A. Brücknerc, and M. Beller, *Tetrahedron Lett.*, 6339 (2007).
91. K. Schröder, S. Enthaler, B. Bitterlich, T. Schulz, A. Spannenberg, M. K. Tse, K. Junge, and M. Beller, *Chem. Eur. J.*, *15*, 5471 (2009).
92. S. Friedle, E. Reisner, and S. J. Lippard, *Chem. Soc. Rev.*, *39*, 2768 (2010).
93. M. Fujita and L. Que, Jr., *Adv. Synth. Catal.*, *346*, 190 (2004).
94. D. Clemente-Tejeda, A. López-Moreno, and F. A. Bermejo, *Tetrahedron*, *69*, 2977 (2013).
95. D. Clemente-Tejeda, A. López-Moreno, and F. A. Bermejo, *Tetrahedron*, *68*, 9249 (2012).
96. S. Taktak, Y. Wanhua, A. M. Herrera, and E. V. Rybak-Akimova, *Inorg. Chem.*, *46*, 2929 (2007).
97. G. Dubois, A. Murphy, and T. D. P. Stack, *Org. Lett.*, *5*, 2469 (2003).
98. P. Liu, E. L. Wong, A. W. Yuen, and C. Che, *Org. Lett.*, *10*, 3275 (2008).
99. M. Darwish and M. Wills, *Catal. Sci. Technol.*, *2*, 243 (2012).
100. F. G. Gelalcha, B. Bitterlich, G. Anilkumar, M. K. Tse, and M. Beller, *Angew. Chem. Int. Ed.*, *46*, 7293 (2007).
101. F. G. Gelalcha, G. Anilkumar, M. K. Tse, A. Brückner, and M. Beller, *Chem. Eur. J.*, *14*, 7687 (2008).
102. C. Marchi-Delapierre, A. Jorge-Robin, A. Thibon, and S. Ménage, *Chem. Commun.*, 1166 (2007).
103. F. Oddon, E. Girgenti, C. Lebrun, C. Marchi-Delapierre, J. Pecaut, and S. Menage, *Eur. J. Inorg. Chem.*, 85 (2012).
104. H.-L. Yeung, K.-C. Sham, C.-S. Tsang, T.-C. Lau, and H.-L. Kwong, *Chem. Commun.*, 3801 (2008).
105. M. Wu, C.-X. Miao, S. Wang, X. Hu, C. Xia, F. E. Kühn, and W. Sun, *Adv. Synth. Catal.*, *353*, 3014 (2011).
106. B. Wang, S. Wang, C. Xia, and W. Sun, *Chem. Eur. J.*, *18*, 7332 (2012).
107. Y. Nishikawa and H. Yamamoto, *J. Am. Chem. Soc.*, *133*, 8432 (2011).
108. T. Niwa and M. Nakada, *J. Am. Chem. Soc.*, *134*, 13538 (2012).
109. D. T. Gibson, *Microbial Degradation of Organic Compounds*. Marcel Dekker, New York, 1984.
110. K. Chen and L. Que, Jr., *Angew. Chem. Int. Ed.*, *38*, 2227 (1999).
111. I. Prat, D. Font, A. Company, K. Junge, X. Ribas, M. Beller, and M. Costas, *Adv. Synth. Catal.*, *355*, 947 (2012).

112. I. Garcia-Bosch, Z. Codolà, I. Prat, X. Ribas, J. Lloret-Fillol, and M. Costas, *Chem. Eur. J.*, *18*, 13269 (2012).
113. Y. Feng, J. England, and L. Que, Jr., *Acs Catalysis*, *1*, 1035 (2011).
114. K. Suzuki, P. D. Oldenburg, and L. Que, Jr., *Angew. Chem. Int. Ed.*, *47*, 1887 (2008).
115. J. Y. Ryu, J. Kim, M. Costas, K. Chen, W. Nam, and L. Que, Jr., *Chem. Commun.*, 1288 (2002).
116. M. Costas, A. K. Tipton, K. Chen, D.-H. Jo, and L. Que, Jr., *J. Am. Chem. Soc.*, *123*, 6722 (2001).
117. J. Bautz, P. Comba, C. L. D. Laorden, M. Menzel, and G. Rajaraman, *Angew. Chem. Int. Ed.*, *46*, 8067 (2007).
118. Y. Feng, C.-Y. Ke, G. Xue, and L. Que, Jr., *Chem. Commun.*, 50 (2009).
119. P. D. Oldenburg, A. A. Shteinman, and L. Que, Jr., *J. Am. Chem. Soc.*, *127*, 15672 (2005).
120. P. C. A. Bruijnincx, I. L. C. Buurmans, S. Gosiewska, M. A. H. Moelands, M. Lutz, A. L. Spek, G. van Koten, and R. J. M. K. Gebbink, *Chem. Eur. J.*, *14*, 1228 (2008).
121. P. D. Oldenburg, Y. Feng, I. Pryjomska-Ray, D. Ness, and L. Que, Jr., *J. Am. Chem. Soc.*, *132*, 17713 (2010).
122. V. J. Dungan, Y. Ortin, H. Mueller-Bunz, and P. J. Rutledge, *Org. Biomol. Chem.*, *8*, 1666 (2010).
123. V. J. Dungan, S. M. Wong, S. M. Barry, and P. J. Rutledge, *Tetrahedron*, *68*, 3231 (2012).
124. P. D. Oldenburg, C.-Y. Ke, A. A. Tipton, A. A. Shteinman, and L. Que, Jr., *Angew Chem. Int. Ed.*, *45*, 7975 (2006).
125. T. W. S. Chow, E. L.-M. Wong, Z. Guo, Y. Liu, J.-S. Huang, C. M. Che, *J. Am. Chem. Soc.*, *132*, 13229 (2010).
126. K. Suzuki, P. D. Oldenburg, and L. Que, Jr., *Angew. Chem. Int. Ed.*, *47*, 1887 (2008).
127. K. Wieghardt, U. Bossek, B. Nuber, J. Weiss, J. Bonvoisin, M. Corbella, S. E. Vitols, and J. J. Girerd, *J. Am. Chem. Soc.*, *110*, 7398 (1988).
128. R. Hage, J. E. Iburg, J. Kerschner, J. H. Koek, E. L. M. Lempers, R. J. Martens, U. S. Racherla, S. W. Russell, T. Swarthoff, M. R. P. van Vliet, J. B. Warnaar, L. van der Wolf, and B. Krijnen, *Nature (London)*, *369*, 637 (1994).
129. P. Saisaha, J. W. de Boer, and W. R. Browne, *Chem. Soc. Rev.*, *42*, 2059 (2013).
130. E. T. Talsi and K. P. Bryliakov, *Coord. Chem. Rev.*, *256*, 1418 (2012).
131. G. B. Shul'pin, G. Süss-Fink, and J. R. L. Smith, *Tetrahedron*, *55*, 5345 (1999).
132. G. B. Shul'pin and J. R. Lindsay-Smith, *Russ. Chem. Bull.*, *47*, 2379 (1998).
133. S. Das, C. D. Incarvito, R. H. Crabtree, and G. W. Brudvig, *Science*, *312*, 1941 (2006).
134. K. Nehru, S. J. Kim, I. Y. Kim, M. S. Seo, Y. Kim, S.-J. Kim, J. Kim, and W. Nam, *Chem. Commun.*, 4623 (2007).
135. R. V. Ottenbacher, D. G. Samsonenko, E. P. Talsi, and K. P. Bryliakov, *Org. Lett.*, *14*, 4310 (2012).
136. S. Shi, Y. Wang, A. Xu, H. Wang, D. Zhu, S. B. Roy, T. A. Jackson, D. H. Busch, and G. Yin, *Angew. Chem. Int. Ed.*, *50*, 7321 (2011).

137. J. F. Hull, D. Balcells, E. L. O. Sauer, C. Raynaud, G. W. Brudvig, R. H. Crabtree, and O. Eisenstein, *J. Am. Chem. Soc.*, *132*, 7605 (2010).
138. K. A. Prokop, S. P. de Visser, and D. P. Goldberg, *Angew. Chem. Int. Ed.*, *49*, 5091 (2010).
139. B. S. Lane and K. Burgess, *J. Am. Chem. Soc.*, *123*, 2933 (2001).
140. S. Zhong, Z. Fu, Y. Tan, Q. Xie, F. Xie, X. Zhou, Z. Ye, G. Peng, and D. Yin, *Adv. Synth. Catal.*, *350*, 802 (2008).
141. E. Hao, Z. Wang, L. Jiao, and S. Wang, *Dalton Trans.*, *39*, 2660 (2010).
142. R. Wortmann, U. Flörke, B. Sarkar, V. Umamaheshwari, G. Gescheidt, S. Herres-Pawlis, and G. Henkel, *Eur. J. Inorg. Chem.*, 121 (2011).
143. V. Madhu, B. Ekambaram, L. J. W. Shimon, Y. Diskin, G. Leitus, and R. Neumann, *Dalton Trans.*, *39*, 7266 (2010).
144. K.-C. Sham, H.-L. Weung, S.-M. Yiu, T.-C. Lau, and H.-L. Kwong, *Dalton Trans.*, *39*, 9469 (2010).
145. D. E. De Vos, J. L. Meinershagen, and T. Bein, *Angew. Chem. Int. Ed.*, *35*, 2211 (1996).
146. D. E. De Vos and T. Bein, *J. Organomet. Chem.*, *520*, 195 (1996).
147. D. E. De Vos and T. Bein, *Chem. Commun.*, 917 (1996).
148. V. Chin Quee-Smith, L. DelPizzo, S. Jureller, J. Kerschner, and R. Hage, *Inorg. Chem.*, *35*, 6461 (1996).
149. D. E. De Vos, B. F. Sels, M. Reynaers, Y. V. Subba Rao, and P. A. Jacobs, *Tetrahedron Lett.*, *39*, 3221 (1998).
150. A. Berkessel, C. A. Sklorz, *Tetrahedron Lett.*, *40*, 7965 (1999).
151. J. Brinksma, L. Schmieder, G. v. Vliet, R. Boaron, R. Hage, D. E. De Vos, P. L. Alsters, and B. L. Feringa, *Tetrahedron Lett.*, *43* (2002).
152. J. W. de. Boer, J. Brinksma, W. R. Browne, A. Meetsma, P. L. Alsters, R. Hage, and B. L. Feringa, *J. Am. Chem. Soc.*, *127*, 7990 (2005).
153. C. B. Woitiski, Y. N. Kozlov, D. Mandelli, G. V. Nizova, U. Schuchardt, and G. B. Shul'pin, *J. Mol. Catal. A: Chem.*, *222*, 103 (2004).
154. H. Kilic, W. Adam, and P. L. Alsters, *J. Org. Chem.*, *74*, 1135 (2009).
155. N. Schoenfeldt, Z. Ni, A. Korinda, R. Meyer, and J. Notestein, *J. Am. Chem. Soc.*, *133*, 18684 (2011).
156. N. Schoenfeldt, A. Korinda, and J. Notestein, *Chem. Commun.*, *46*, 1640 (2010).
157. D. E. De Vos, S. de Wildeman, B. F. Sels, P. J. Grobet, and P. A. Jacobs, *Angew. Chem. Int. Ed.*, *38*, 980 (1999).
158. A. Grenz, S. Ceccarelli, and C. Bolm, *Chem. Commun.*, 1726 (2001).
159. I. Garcia-Bosch, X. Ribas, and M. Costas, *Adv. Synth. Catal.*, *351*, 348 (2009).
160. A. Murphy, G. Dubois, and T. D. P. Stack, *J. Am. Chem. Soc.*, *125*, 5250 (2003).
161. A. Murphy, A. Pace, and T. D. P. Stack, *Org. Lett.*, *6*, 3119 (2004).
162. I. Garcia-Bosch, A. Company, X. Fontrodona, X. Ribas, and M. Costas, *Org. Lett.*, *10*, 2095 (2008).
163. H.-K. Kwong, P.-K. Lo, K.-C. Lau, and T.-C. Lau, *Chem. Commun.*, *47*, 4273 (2011).

164. P. Leeladee and D. P. Goldberg, *Inorg. Chem.*, *49*, 3083 (2010).
165. J. J. Dong, P. Saisaha, T. G. Meinds, P. L. Alsters, E. G. Ijpeij, R. P. van Summeren, B. Mao, M. Fañanás-Mastral, J. W. de Boer, R. Hage, B. L. Feringa, and W. R. Browne, *ACS Catal.*, *2*, 1087 (2012).
166. W. Zhang, J. L. Loebach, S. R. Wilson, and E. N. Jacobsen, *J. Am. Chem. Soc.*, *112*, 2801 (1990).
167. T. Katsuki, *Coord. Chem. Rev.*, *140*, 189 (1995).
168. L. Gómez, I. Garcia-Bosch, A. Company, X. Sala, X. Fontrodona, X. Ribas, and M. Costas, *Dalton Trans.*, 5539 (2007).
169. I. Garcia-Bosch, L. Gómez, A. Polo, X. Ribas, and M. Costas, *Adv. Synth. Catal.*, *354*, 65 (2012).
170. M. Wu, B. Wang, S. Wang, C. Xia, and W. Sun, *Org. Lett.*, *11*, 3622 (2009).
171. B. Wang, C. Miao, S. Wang, C. Xia, and W. Sun, *Chem. Eur. J.*, *18*, 6750 (2012).
172. X. Wang, C. Miao, S. Wang, C. Xia, and W. Sun, *ChemCatChem*, *5*, 2489 (2013).
173. R. V. Ottenbacher, K. P. Bryliakov, and E. P. Talsi, *Adv. Synth. Catal.* 353, 885 (2011).
174. R. V. Ottenbacher, K. P. Bryliakov, and E. P. Talsi, *Inorg. Chem.*, *49*, 8620 (2010).
175. O. Y. Lyakin, R. Ottenbacher, K. P. Bryliakov, and E. P. Talsi, *Top. Catal.*, *56*, 939 (2013).
176. D. E. De Vos, S. de Wildeman, B. F. Sels, P. J. Grobet, and P. A. Jacobs, *Angew. Chem. Int. Ed.*, *38*, 980 (1999).
177. J. W. de Boer, W. R. Browne, J. Brinksma, P. L. Alsters, R. Hage, and B. L. Feringa, *Inorg. Chem.*, *46*, 6353 (2007).
178. D. Pijper, P. Saisaha, J. W. de Boer, R. Hoen, C. Smit, A. Meetsma, R. Hage, R. P. van Summeren, P. L. Alsters, B. L. Feringa, and W. R. Browne, *Dalton Trans.*, *39*, 10375 (2010).
179. P. Saisaha, D. Pijper, R. P. v. Summerren, R. Hoen, C. Smit, J. W. de. Boer, R. Hage, P. L. Alsters, B. L. Feringa, and W. R. Browne, *Org. Biomol. Chem.*, *8*, 4444 (2010).
180. J. W. de Boer, W. R. Browne, S. R. Harutyunyan, L. Bini, T. D. Tiemersma-Wegman, P. L. Alsters, R. Hage, and B. L. Feringa, *Chem. Commun.*, 3747 (2008).
181. T. W.-S. Chow, Y. Liu, and C.-M. Che, *Chem. Commun.*, *47*, 11204 (2011).
182. M. Yagi and M. Kaneko, *Chem. Rev.*, *101*, 21 (2000).
183. A. Singh and L. Spiccia, *Coord. Chem. Rev.*, *257*, 2607 (2013).
184. D. G. Nocera, *Acc. Chem. Res.*, *45*, 767 (2012).
185. H. Yamazaki, A. Shouji, M. Kajita, and M. Yagi, *Coord. Chem. Rev.*, *254*, 2483 (2010).
186. T. A. Betley, Q. Wu, T. Van Voorhis, and D. G. Nocera, *Inorg. Chem.*, *47*, 1849 (2008).
187. X. Sala, I. Romero, M. Rodríguez, L. Escriche, and A. Llobet, *Angew. Chem. Int. Ed.*, *48*, 2842 (2009).
188. T. R. Cook, D. K. Dogutan, S. Y. Reece, Y. Surendranath, T. S. Teets, and D. G. Nocera, *Chem. Rev.*, *110*, 6474 (2010).
189. T. A. Betley, Y. Surendranath, M. V. Childress, G. E. Alliger, R. Fu, C. C. Cummins, and D. G. Nocera, *Philos. Trans. R. Soc. B: Biol. Sci.*, *363*, 1293 (2008).

190. H. Dau and I. Zaharieva, *Acc. Chem. Res.*, *42*, 1861 (2009).
191. M. M. Najafpour and Govindjee, *Dalton Trans.*, *40*, 9076 (2011).
192. J. S. Kanady, E. Y. Tsui, M. W. Day, and T. Agapie, *Science*, *333*, 733 (2011).
193. E. Y. Tsui, R. Tran, J. Yano, and T. Agapie, *Nat. Chem.*, *5*, 293 (2013).
194. J. S. Kanady, J. L. Mendoza-Cortes, E. Y. Tsui, R. J. Nielsen, W. A. Goddard, and T. Agapie, *J. Am. Chem. Soc.*, *135*, 1073 (2012).
195. R. Ramaraj, A. Kira, and M. Kaneko, *Angew. Chem. Int. Ed.*, *25*, 825 (1986).
196. M. Yagi and K. Narita, *J. Am. Chem. Soc.*, *126*, 8084 (2004).
197. R. Tagore, H. Chen, H. Zhang, R. H. Crabtree, and G. W. Brudvig, *Inorg. Chim. Acta*, *360*, 2983 (2007).
198. J. Limburg, J. S. Vrettos, L. M. Liable-Sands, A. L. Rheinglod, R. H. Crabtree, and G. H. Brudvig, *Science*, *283*, 1524 (1999).
199. J. Limburg, J. S. Vrettos, H. Chen, J. C. de Paula, R. H. Crabtree, and G. W. Brudvig, *J. Am. Chem. Soc.*, *123*, 423 (2000).
200. R. Tagore, R. H. Crabtree, and G. W. Brudvig, *Inorg. Chem.*, *47*, 1815 (2008).
201. M. Lundberg, M. R. A. Blomberg, and P. E. M. Siegbahn, *Inorg. Chem.*, *43*, 264 (2003).
202. J. L. Valles-Pardo, H. J. M. de Groot, and F. Buda, *Phys. Chem. Chem. Phys.*, *14*, 15502 (2012).
203. H. Chen, R. Tagore, G. Olack, J. S. Vrettos, T. C. Weng, J. Penner-Hahn, R. H. Crabtree, and G. W. Brudvig, *Inorg. Chem.*, *46*, 34 (2007).
204. Chen, R. Tagore, G. Olack, J. S. Vrettos, T.-C. Weng, J. Penner-Hahn, R. H. Crabtree, and G. W. Brudvig, *Inorg. Chem.*, *46*, 34 (2006).
205. K. Narita, T. Kuwabara, K. Sone, K.-i. Shimizu, and M. Yagi, *Russ. J. Phys. Chem. B*, *110*, 23107 (2006).
206. H. Yamazaki, S. Igarashi, T. Nagata, and M. Yagi, *Inorg. Chem.*, *51*, 1530 (2012).
207. B. Nepal and S. Das, *Angew. Chem. Int. Ed.*, *52*, 7224 (2013).
208. C. Baffert, S. Romain, A. Richardot, J.-C. Leprêtre, B. Lefebvre, A. Deronzier, and M.-N. Collomb, *J. Am. Chem. Soc.*, *127*, 13694 (2005).
209. H. Chen, R. Tagore, S. Das, C. Incarvito, J. W. Faller, R. H. Crabtree, and G. W. Brudvig, *Inorg. Chem.*, *44*, 7661 (2005).
210. A. K. Poulsen, A. Rompel, and C. J. McKenzie, *Angew. Chem. Int. Ed.*, *44*, 6916 (2005).
211. R. K. Seidler-Egdal, A. Nielsen, A. D. Bond, M. J. Bjerrum, and C. J. McKenzie, *Dalton Trans.*, *40*, 3849 (2011).
212. W. M. C. Sameera, C. J. McKenzie, and J. E. McGrady, *Dalton Trans.*, *40*, 3859 (2011).
213. G. Berggren, A. Thapper, P. Huang, L. Eriksson, S. Styring, and M. F. Anderlund, *Inorg. Chem.*, *50*, 3425 (2011).
214. P. Kurz, G. Berggren, M. F. Anderlund, and S. Styring, *Dalton Trans.*, 4258 (2007).
215. F. M. Ashmawy, C. A. McAuliffe, R. V. Parish, and J. Tames, *J. Chem. Soc., Dalton Trans.*, 1391 (1985).

216. M. Watkinson, A. Whiting, and C. A. McAuliffe, *J. Chem. Soc., Chem. Commun.*, 2141 (1994).
217. G. Gonzalez-Riopedre, M. I. Fernandez-Garcia, A. M. Gonzalez-Noya, M. A. Vazquez-Fernandez, M. R. Bermejo, and M. Maneiro, *Phys. Chem. Chem. Phys.*, *13*, 18069 (2011).
218. L. Hammarström, L. Sun, B. Åkermark, and S. Styring, *Catal. Today*, *58*, 57 (2000).
219. C. Jegerschöld and S. Styring, *Biochemistry*, *35*, 7794 (1996).
220. R. S. F. Mamodov and S. Styring *Biochemistry*, *37*, 14245 (1988).
221. E. A. Karlsson, B.-L. Lee, T. Åkermark, E. V. Johnston, M. D. Kärkäs, J. Sun, Ö. Hansson, J.-E. Bäckvall, and B. Åkermark, *Angew. Chem. Int. Ed.*, *50*, 11715 (2011).
222. K. J. Young, M. K. Takase, and G. W. Brudvig, *Inorg. Chem.*, *52*, 7615 (2013).
223. G. L. Elizarova, L. G. Matvienko, V. N. Parmon, and K. I. Zamaraev, *Dokl. Akad. Nauk SSSR*, *249*, 863 (1979).
224. G. L. Elizarova, L. G. Matvienko, N. V. Lozhkina, V. N. Parmon, and K. I. Zamaraev, *React. Kinet. Catal. Lett.*, *16*, 191 (1981).
225. G. L. Elizarova, L. G. Matvienko, N. V. Lozhkina, V. E. Maizlish, and V. N. Parmon, *React. Kinet. Catal. Lett.*, *16*, 285 (1981).
226. V. N. Parmon, G. L. Elizarova, and T. V. Kim, *React. Kinet. Catal. Lett.*, *21*, 1985 (1982).
227. T. V. Kim, G. L. Elizarova, and V. N. Parmon, *React. Kinet. Catal. Lett.*, *26*, 57 (1984).
228. W. C. Ellis, N. D. McDaniel, S. Berhard, and T. J. Collins, *J. Am. Chem. Soc.*, *132*, 10990 (2010).
229. F. T. de Oliveira, A. Chanda, D. Banerjee, X. Shan, S. Mondal, L. Que, Jr., E. L. Bominaar, E. Münck, and T. J. Collins, *Science*, *315*, 835 (2007).
230. M. Z. Ertem, L. Gagliardi, and C. J. Cramer, *Chem. Sci.*, *3*, 1293 (2012).
231. J. Lloret-Fillol, Z. Codolà, I. Garcia-Bosch, L. Gómez, J. J. Pla, and M. Costas, *Nat. Chem.*, *3*, 807 (2011).
232. G. Chen, L. Chen, S.-M. Ng, W.-L. Man, and T.-C. Lau, *Angew. Chem. Int. Ed.*, *52*, 1789 (2013).
233. Z. Codolà, I. Garcia-Bosch, F. Acuña-Parés, J. M. Lluis, M. Costas, and J. Lloret-Fillol, *Chem. Eur. J.*, *19*, 8042 (2013).
234. W. A. Hoffert, M. T. Mock, A. M. Appel, and J. Y. Yang, *Eur. J. Inorg. Chem.*, *2013*, 3846 (2013).

Subject Index

Progress in Inorganic Chemistry, Volume 59, First Edition. Edited by Kenneth D. Karlin.

Cumulative Index, Volumes 1–59

Progress in Inorganic Chemistry, Volume 59, First Edition. Edited by Kenneth D. Karlin.